AutoCAD® 2008

2008

COMPANION

Essentials of AutoCAD
Plus Solid Modeling

The McGraw-Hill Graphics Series

Providing you with the highest quality textbooks that meet your changing needs requires feedback, improvement, and revision. The team of authors and McGraw-Hill are committed to this effort. We invite you to become part of our team by offering your wishes, suggestions, and comments for future editions and new products and texts.

Please mail or fax your comments to: James A. Leach
c/o McGraw-Hill Higher Education
Engineering Division
2460 Kerper Blvd
Dubuque, IA 52001
FAX: 563-589-2825

Titles in the McGraw-Hill Graphics Series

Bertoline, *Introduction to Graphics Communications for Engineers*

Bertoline/Wiebe, *Fundamentals of Graphic Communication*

Bertoline/Wiebe, *Technical Graphics Communication*

French/Vierck/Foster, *Engineering Drawing and Graphic Technology*

Howard/Musto, *Introduction to Solid Modeling Using SolidWorks*

Jensen/Helsel/Short, *Engineering Drawing and Design Student Edition*

Kelley, *Pro/Engineering Wildfire*

Leach, *AutoCAD 2008 Instructor*

Leake, *Autodesk Inventor*

AutoCAD® 2008
COMPANION

Essentials of AutoCAD
Plus Solid Modeling

James A. Leach
University of Louisvile

Edited and Abridged by
James H. Dyer

Mc Graw Hill **Higher Education**

Burr Ridge, IL   Dubuque, IA   New York   San Francisco   St. Louis
Bangkok   Bogotá   Caracas   Kuala Lumpur   Lisbon   London   Madrid   Mexico City
Milan   Montreal   New Delhi   Santiago   Seoul   Singapore   Sydney   Taipei   Toronto

The McGraw-Hill Companies

McGraw-Hill
Higher Education

AUTOCAD 2008 COMPANION: ESSENTIALS OF AUTOCAD PLUS SOLID MODELING

 This book is printed on recycled, acid-free paper containing 10% postconsumer waste.

1 2 3 4 5 6 7 8 9 0 QPD/QPD 0 9 8 7

ISBN 978–0–07–340246–8
MHID 0–07–340246–X

Global Publisher: *Raghothaman Srinivasan*
Executive Editor: *Michael Hackett*
Senior Sponsoring Editor: *Bill Stenquist*
Director of Development: *Kristine Tibbetts*
Developmental Editor: *Lora Kalb*
Executive Marketing Manager: *Michael Weitz*
Project Manager: *Joyce Watters*
Lead Production Supervisor: *Sandy Ludovissy*
Associate Media Producer: *Christina Nelson*
Designer: *John Joran/Brenda Rowles*
Cover Designer: *Brenda A. Rowles*
(USE) Cover Image: © *Nicholas Pitt/Digital Vision/Getty Images*
Compositor: *Visual Q*
Typeface: 10.5/12 *Palatino*
Printer: *Quebecor World Dubuque, IA*

Library of Congress Control Number: 2007930550

TABLE OF CONTENTS

TABLE OF CONTENTS

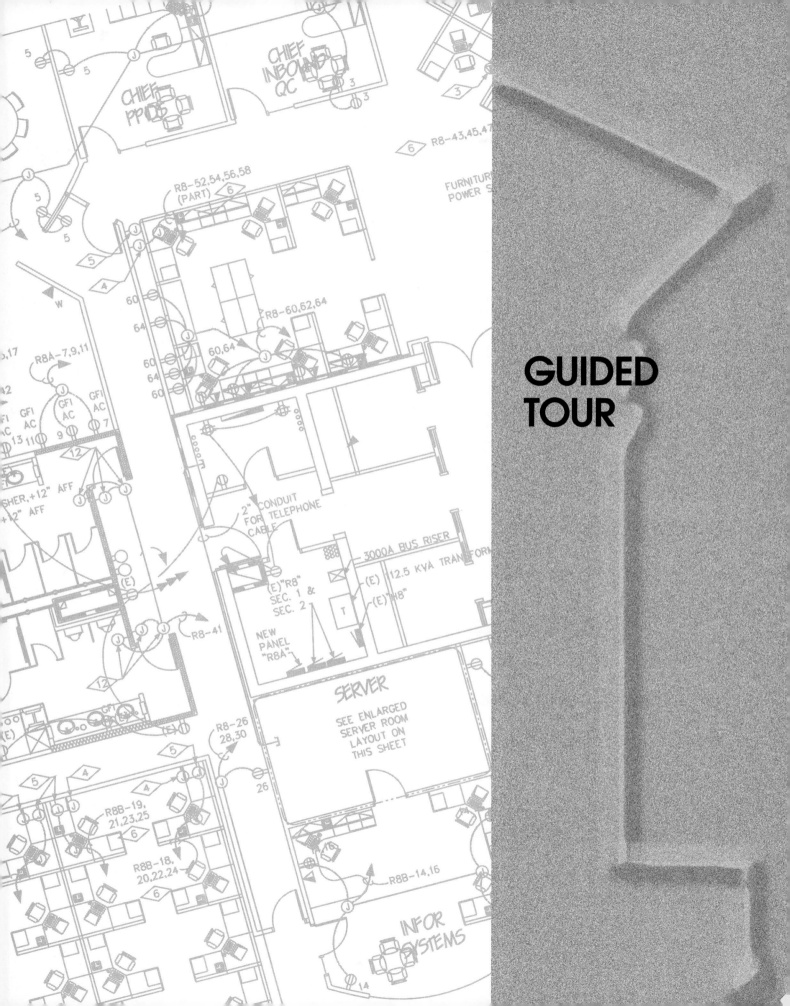

GUIDED
TOUR

GUIDED TOUR

Welcome to *AutoCAD 2008 Companion*. Here are some features you will find in this book to help you learn AutoCAD 2008.

Chapter Objectives

Each chapter opens with a list of new concepts you can expect to learn. Having objectives in mind helps you focus on the important ideas as you move through each chapter.

CHAPTER 9

MODIFY COMMANDS I

CHAPTER OBJECTIVES

After completing this chapter you should:

1. be able to *Erase* objects;
2. be able to *Move* objects from a base point to a second point;
3. know how to *Rotate* objects about a base point;
4. be able to enlarge or reduce objects with *Scale*;
5. be able to *Stretch* objects and change the length of *Lines* and *Arcs* with *Lengthen*;
6. be able to *Trim* away parts of objects at cutting edges and *Extend* objects to boundary edges;
7. know how to use the four *Break* options;
8. be able to *Copy* objects and make *Mirror* images of selected objects;
9. know how to create parallel copies of objects with *Offset*;
10. be able to make *rectangular* and *Polar Arrays* of objects;
11. be able to create a *Fillet* and a *Chamfer* between two objects.

Selection Sets   63

Using *Move*

The *Move* command specifically prompts you to (1) select objects, (2) specify a "base point," a point to move from, and (3) specify a "second point of displacement," a point to move to.

Move

Pull-down Menu	Command (Type)	Alias (Type)	Short-cut	Screen (side) Menu	Tablet Menu
Modify Move	Move	M	(Edit Mode) Move	MODIFY2 Move	V,19

1. Draw a *Circle* and two *Lines*. Use the *Move* command to practice selecting objects and to move one *Line* and the *Circle* as indicated in the following table.

STEPS	COMMAND PROMPT	PERFORM ACTION	COMMENTS
1.	Command:	type **M** and press **Spacebar**	M is the command alias for *Move*
2.	MOVE Select objects:	use pickbox to select one *Line* and the *Circle*	objects are highlighted
3.	Select objects:	press **Spacebar** or **Enter**	
4.	Specify base point or displacement:	**PICK** near the *Circle* center	base point is the handle, or where to move *from*
5.	Specify second point:	**PICK** near the other *Line*	second point is where to move *to*

2. Use *Move* again to move the *Circle* back to its original position. Select the *Circle* with the *Window* option.

STEPS	COMMAND PROMPT	PERFORM ACTION	COMMENTS
1.	Command:	select the *Modify* pull-down, then *Move*	
2.	MOVE Select objects:	type **W** and press **Spacebar**	select only the circle; object is highlighted
3.	Select objects:	press **Spacebar** or **Enter**	
4.	Specify base point or displacement:	**PICK** near the *Circle* center	base point is the handle, or where to move *from*
5.	Specify second point:	**PICK** near the original location	second point is where to move *to*

Practice Tables

Step-by-step practice exercises in the early chapters help you systematically tackle fundamental skills.

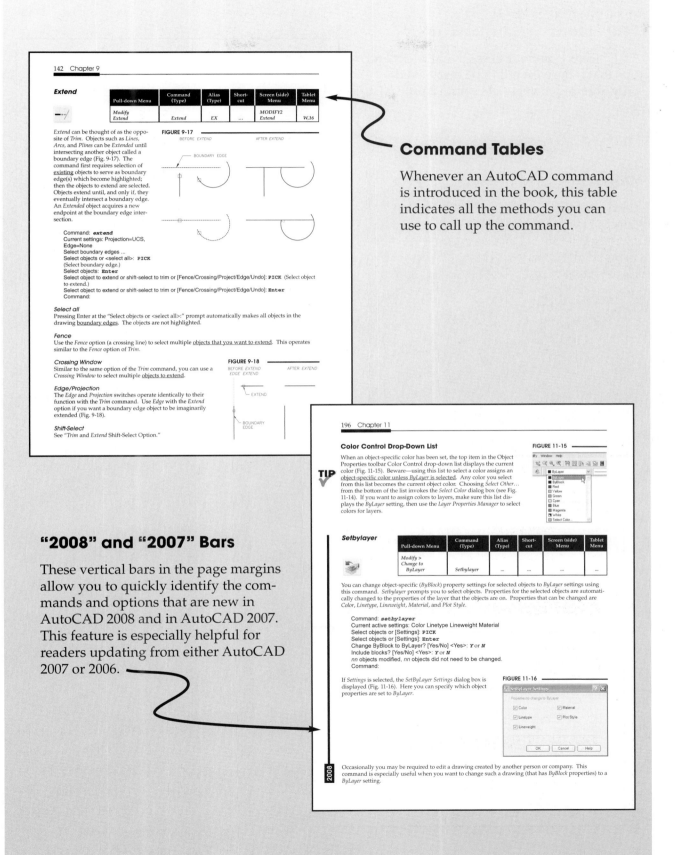

142   Chapter 9

Extend

Pull-down Menu	Command (Type)	Alias (Type)	Short-cut	Screen (side) Menu	Tablet Menu
Modify Extend	Extend	EX	...	MODIFY2 Extend	W,16

Extend can be thought of as the opposite of *Trim*. Objects such as *Lines*, *Arcs*, and *Plines* can be *Extended* until intersecting another object called a boundary edge (Fig. 9-17). The command first requires selection of existing objects to serve as boundary edge(s) which become highlighted; then the objects to extend are selected. Objects extend until, and only if, they eventually intersect a boundary edge. An *Extended* object acquires a new endpoint at the boundary edge intersection.

```
Command: extend
Current settings: Projection=UCS,
Edge=None
Select boundary edges ...
Select objects or <select all>: PICK
(Select boundary edge.)
Select objects: Enter
Select object to extend or shift-select to trim or [Fence/Crossing/Project/Edge/Undo]: PICK (Select object
to extend.)
Select object to extend or shift-select to trim or [Fence/Crossing/Project/Edge/Undo]: Enter
Command:
```

Select all
Pressing Enter at the "Select objects or <select all>:" prompt automatically makes all objects in the drawing boundary edges. The objects are not highlighted.

Fence
Use the *Fence* option (a crossing line) to select multiple objects that you want to extend. This operates similar to the *Fence* option of *Trim*.

Crossing Window
Similar to the same option of the *Trim* command, you can use a *Crossing Window* to select multiple objects to extend.

Edge/Projection
The *Edge* and *Projection* switches operate identically to their function with the *Trim* command. Use *Edge* with the *Extend* option if you want a boundary edge object to be imaginarily extended (Fig. 9-18).

Shift-Select
See "*Trim* and *Extend* Shift-Select Option."

FIGURE 9-17

BEFORE *EXTEND*   AFTER *EXTEND*

BOUNDARY EDGE

FIGURE 9-18

BEFORE *EXTEND*   AFTER *EXTEND*
EDGE EXTEND

EXTEND

BOUNDARY EDGE

Command Tables

Whenever an AutoCAD command is introduced in the book, this table indicates all the methods you can use to call up the command.

---

196   Chapter 11

Color Control Drop-Down List

When an object-specific color has been set, the top item in the Object Properties toolbar Color Control drop-down list displays the current color (Fig. 11-15). Beware—using this list to select a color assigns an object-specific color unless *ByLayer* is selected. Any color you select from this list becomes the current object color. Choosing *Select Other...* from the bottom of the list invokes the *Select Color* dialog box (see Fig. 11-14). If you want to assign colors to layers, make sure this list displays the *ByLayer* setting, then use the *Layer Properties Manager* to select colors for layers.

FIGURE 11-15

Setbylayer

Pull-down Menu	Command (Type)	Alias (Type)	Short-cut	Screen (side) Menu	Tablet Menu
Modify > Change to ByLayer	Setbylayer	...	...	...	...

You can change object-specific (*ByBlock*) property settings for selected objects to *ByLayer* settings using this command. *Setbylayer* prompts you to select objects. Properties for the selected objects are automatically changed to the properties of the layer that the objects are on. Properties that can be changed are *Color*, *Linetype*, *Lineweight*, *Material*, and *Plot Style*.

```
Command: setbylayer
Current active settings: Color Linetype Lineweight Material
Select objects or [Settings]: PICK
Select objects or [Settings]: Enter
Change ByBlock to ByLayer? [Yes/No] <Yes>: Y or N
Include blocks? [Yes/No] <Yes>: Y or N
nn objects modified, nn objects did not need to be changed.
Command:
```

If *Settings* is selected, the *SetByLayer Settings* dialog box is displayed (Fig. 11-16). Here you can specify which object properties are set to *ByLayer*.

FIGURE 11-16

2008

Occasionally you may be required to edit a drawing created by another person or company. This command is especially useful when you want to change such a drawing (that has *ByBlock* properties) to a *ByLayer* setting.

"2008" and "2007" Bars

These vertical bars in the page margins allow you to quickly identify the commands and options that are new in AutoCAD 2008 and in AutoCAD 2007. This feature is especially helpful for readers updating from either AutoCAD 2007 or 2006.

"TIP" Icons

"TIP" icons point out time-saving tricks and tips that otherwise might be discovered only after much experience with AutoCAD.

TIP The *Find* command searches a drawing globally and operates with all types of text, including dimensions, tables, and *Block* attributes. In contrast, the *Find/Replace* option in the *Text Formatting* Editor right-click menu operates only for *Mtext* and for only one *Mtext* object at a time.

FIGURE 18-42

Find Text String, Find/Find Next
To find text only (not replace), enter the desired string in the *Find Text String* edit box and press *Find* (Fig. 18-42). When matching text is found, it appears in the context of the sentence or paragraph in the *Search Results* area. When one instance of the text string is located, the *Find* button changes to *Find Next*.

Replace With, Replace
You can find any text string in a drawing, or you can search for text and replace it with other text. If you want to search for a text string and replace it with another, enter the desired text strings in the *Find Text String* and *Replace With* edit boxes, then press the *Find/Find Next* button. You can verify and replace each instance of the text string found by alternately using *Find Next* and *Replace*, or you can select *Replace All* to globally replace all without verification. The status area confirms the replacements and indicates the number of replacements that were made.

Select Objects
The small button in the upper-right corner of the dialog box allows you to select objects with a pickbox or window to determine your selection set for the search. Press Enter when objects have been selected to return to the dialog box. Once a selection set has been specified, you can search the *Entire drawing* or limit the search to the *Current selection* by choosing these options in the *Search in:* drop-down list.

Select All
This options finds and selects (highlights) all objects in the current selection set containing instances of the text that you enter in *Find Text String*. This option is available only when you use *Select Objects* and set *Search In* to *Current Selection*. When you choose *Select All*, the dialog box closes and AutoCAD displays a message on the Command line indicating the number of objects that it found and selected. Note that *Select All* does not replace text; AutoCAD ignores any text in *Replace With*.

TIP *Zoom To*
This feature is extremely helpful for finding text in a large drawing. Enter the desired text string to search for in the *Find Text String* edit box, select *Find/Find Next* button, then use *Zoom to*. AutoCAD automatically locates the desired text string and zooms in to display the text (Fig. 18-43). Although AutoCAD searches model space and all layouts defined for the drawing, you can zoom only to text in the current *Model* or *Layout* tab.

FIGURE 18-43

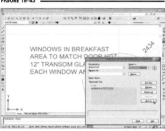

---

6. Select the *Preview* button to temporarily close the dialog box and return to the drawing where you can view the array and chose to *Accept* or *Modify* the array. Selecting *Modify* returns you to the *Array* dialog box.

Figure 9-39 illustrates a *Polar Array* created using 8 as the *Total number of items*, 360 as the *Angle to fill*, and selecting *Rotate items as copied*.

FIGURE 9-39

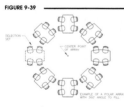

Figure 9-40 illustrates a *Polar Array* created using 5 as the *Total number of items*, 180 as the *Angle to fill*, and selecting *Rotate items as copied*.

FIGURE 9-40

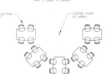

Occasionally an unexpected pattern may result if you remove the check from *Rotate items as copied*. In such a case, the objects may not seem to rotate about the specified center, as shown in Figure 9-41. The reason is that AutoCAD must select a single base point from the set of objects to use for the array such as the end of a line. This base point is not necessarily at the center of the set of objects.

The solution to this problem is to use the *More* button near the bottom of the *Array* dialog box to open the lower section of the dialog box. In this area, remove the check for *Set to object's default*, then specify a *Base point* at the center of the selected set of objects by either entering values or PICKing a point interactively.

FIGURE 9-41

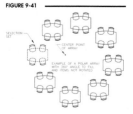

Graphically Driven Page Layout

Approximately 1200 illustrations are given in the book. Explanatory text is located directly next to the related figure.

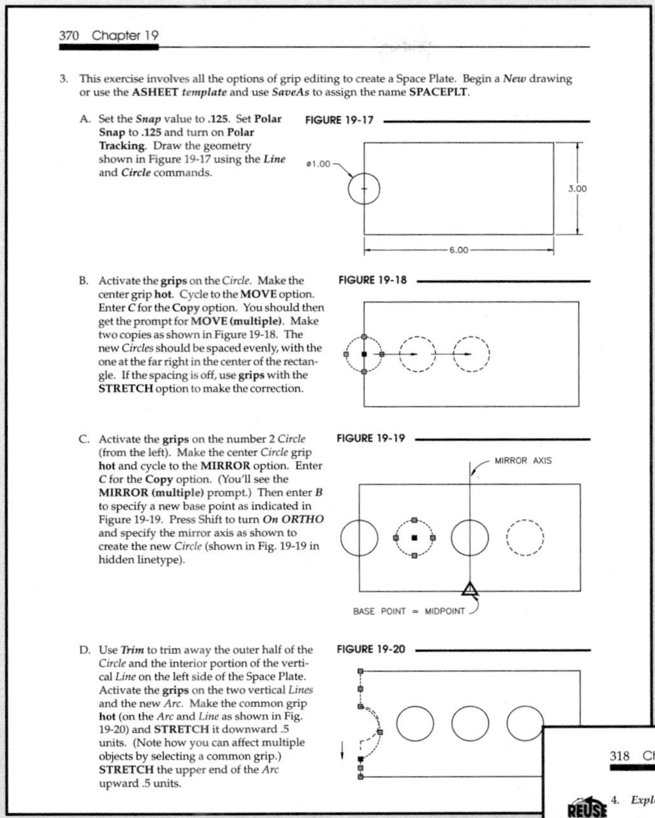

370   Chapter 19

3. This exercise involves all the options of grip editing to create a Space Plate. Begin a *New* drawing or use the **ASHEET** *template* and use *SaveAs* to assign the name **SPACEPLT**.

A. Set the *Snap* value to **.125**. Set **Polar Snap** to **.125** and turn on **Polar Tracking**. Draw the geometry shown in Figure 19-17 using the *Line* and *Circle* commands.

FIGURE 19-17

B. Activate the **grips** on the *Circle*. Make the center grip **hot**. Cycle to the **MOVE** option. Enter *C* for the **Copy** option. You should then get the prompt for **MOVE (multiple)**. Make two copies as shown in Figure 19-18. The new *Circles* should be spaced evenly, with the one at the far right in the center of the rectangle. If the spacing is off, use **grips** with the **STRETCH** option to make the correction.

FIGURE 19-18

C. Activate the **grips** on the number 2 *Circle* (from the left). Make the center *Circle* grip **hot** and cycle to the **MIRROR** option. Enter *C* for the **Copy** option. (You'll see the **MIRROR (multiple)** prompt.) Then enter *B* to specify a new base point as indicated in Figure 19-19. Press Shift to turn *On ORTHO* and specify the mirror axis as shown to create the new *Circle* (shown in Fig. 19-19 in hidden linetype).

FIGURE 19-19

D. Use *Trim* to trim away the outer half of the *Circle* and the interior portion of the vertical *Line* on the left side of the Space Plate. Activate the **grips** on the two vertical *Lines* and the new *Arc*. Make the common grip **hot** (on the *Arc* and *Line* as shown in Fig. 19-20) and **STRETCH** it downward .5 units. (Note how you can affect multiple objects by selecting a common grip.) **STRETCH** the upper end of the *Arc* upward .5 units.

FIGURE 19-20

Chapter Exercises

Exercises help you try out the commands discussed in each chapter. Exercises begin with simple, step-by-step use of commands and progress to more advanced drawings based on synthesis of earlier ideas.

Multi-chapter "Reuse" Exercises

Exercise drawings that are used again in other chapters are designated with this "Reuse" icon. The associated drawing file names are printed in a reversed font. Using these multi-chapter exercises maximizes the student's efforts, creates connections between concepts, and makes the most of the natural pedagogical progression of the material.

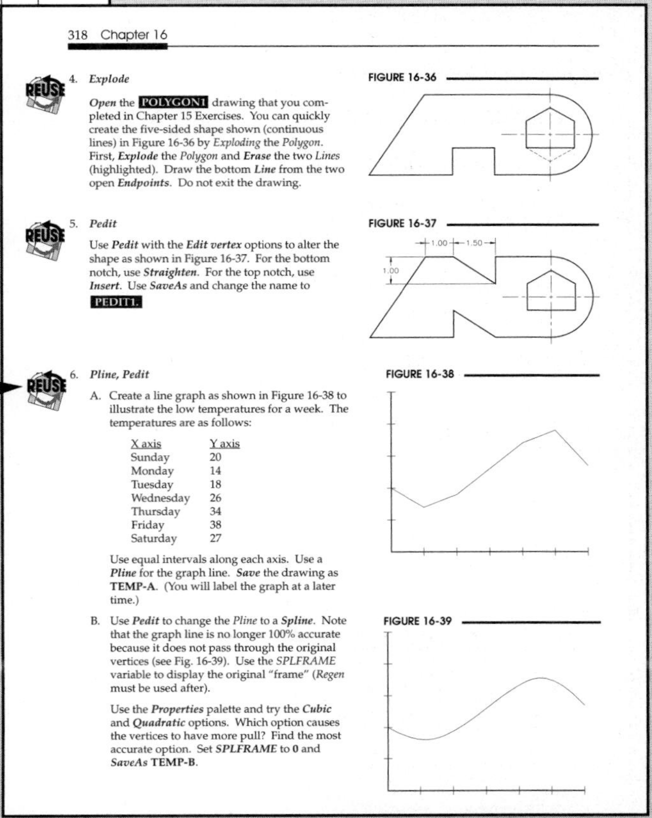

318   Chapter 16

4. *Explode*

Open the **POLYGON1** drawing that you completed in Chapter 15 Exercises. You can quickly create the five-sided shape shown (continuous lines) in Figure 16-36 by *Exploding* the *Polygon*. First, *Explode* the *Polygon* and *Erase* the two *Lines* (highlighted). Draw the bottom *Line* from the two open *Endpoints*. Do not exit the drawing.

FIGURE 16-36

5. *Pedit*

Use *Pedit* with the *Edit vertex* options to alter the shape as shown in Figure 16-37. For the bottom notch, use *Straighten*. For the top notch, use *Insert*. Use *SaveAs* and change the name to **PEDIT1**.

FIGURE 16-37

6. *Pline, Pedit*

A. Create a line graph as shown in Figure 16-38 to illustrate the low temperatures for a week. The temperatures are as follows:

X axis	Y axis
Sunday	20
Monday	14
Tuesday	18
Wednesday	26
Thursday	34
Friday	38
Saturday	27

Use equal intervals along each axis. Use a *Pline* for the graph line. *Save* the drawing as **TEMP-A**. (You will label the graph at a later time.)

FIGURE 16-38

B. Use *Pedit* to change the *Pline* to a *Spline*. Note that the graph line is no longer 100% accurate because it does not pass through the original vertices (see Fig. 16-39). Use the *SPLFRAME* variable to display the original "frame" (*Regen* must be used after).

Use the *Properties* palette and try the *Cubic* and *Quadratic* options. Which option causes the vertices to have more pull? Find the most accurate option. Set *SPLFRAME* to 0 and *SaveAs* **TEMP-B**.

FIGURE 16-39

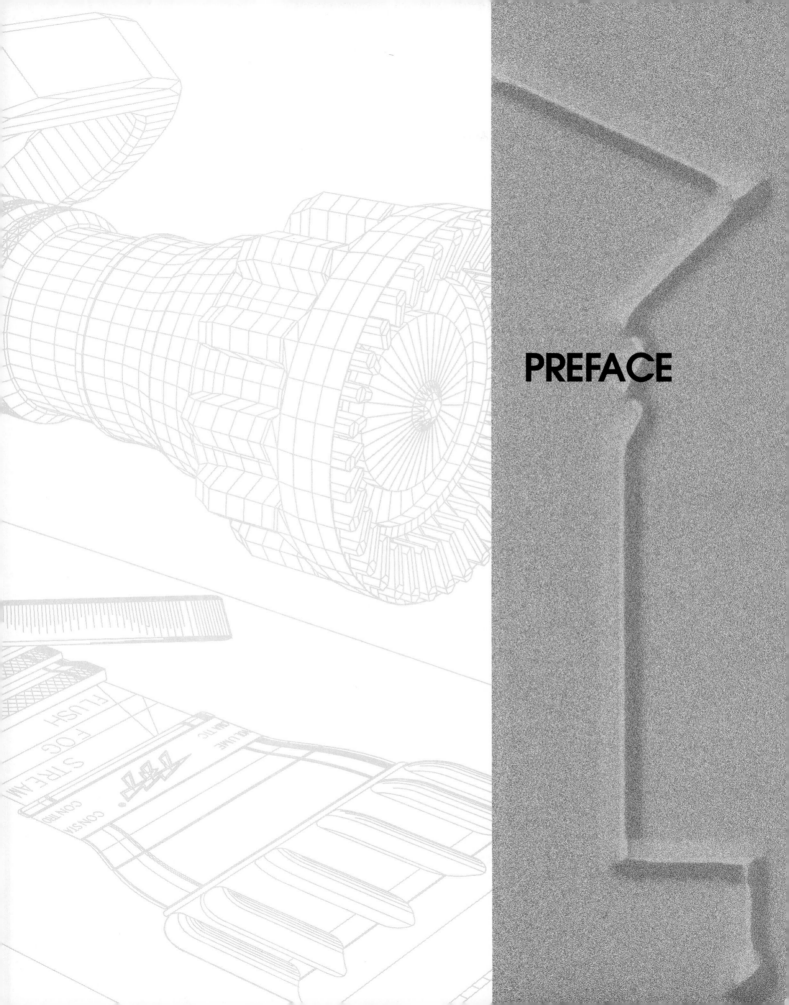

PREFACE

ABOUT THIS BOOK

This Book Is Your AutoCAD 2008 Instructor

The objective of *AutoCAD 2008 Companion* is to provide you the material typically covered in a one-semester AutoCAD course. *AutoCAD 2008 Companion* covers the essentials of 2D design and drafting as well as solid modeling.

Graphically Oriented

Because *AutoCAD 2008 Companion* discusses concepts that are graphical by nature, many illustrations (approximately 1200) are used to communicate the concepts, commands, and applications.

Easy Update from AutoCAD 2006 and AutoCAD 2007

AutoCAD 2008 Companion is helpful if you are already an AutoCAD user but are updating from AutoCAD 2006 or 2007. All new commands, concepts, features, and variables are denoted on the edges of the pages by a vertical "2007" bar (denoting an update since 2006) or a "2008" bar (denoting an update since 2007).

Pedagogical Progression

AutoCAD 2008 Companion is presented in a pedagogical format by delivering the fundamental concepts first, then moving toward the more advanced and specialized features of AutoCAD. The book begins with small pieces of information explained in a simple form and then builds on that experience to deliver more complex ideas, requiring a synthesis of earlier concepts. The chapter exercises follow the same progression, beginning with a simple tutorial approach and ending with more challenging problems requiring a synthesis of earlier exercises.

Multi-chapter "Reuse" Exercises

About half of the Chapter Exercises are used again later in subsequent chapters. This concept emphasizes the natural pedagogical progression of the text, creates connections between concepts, and maximizes the student's efforts. Reuse Exercises are denoted with a "Reuse" (disk) icon and the related drawing file names are printed in a reversed font.

Important "Tips"

Tips, reminders, notes, and cautions are given in the book and denoted by a "TIP" marker in the margin. This feature helps you identify and remember important concepts, commands, procedures, and tricks used by professionals that would otherwise be discovered only after much experience.

Web-Based Materials

Please visit our web page at www.mhhe.com/leach. For students, the AutoCAD 2008 site offers online chapters quizzes (multiple choice, true-false, and essay) with online grading and electronic routing to instructor's email. In addition, important terms, commands, and options for each chapter are given for self-instruction and review. Instructors can download solution drawings to the chapter exercises (in .DWG format), answers to the review questions, and have access to an image library containing all screen captures from the text. A variety of information and instructional content is available for other Leach books including *AutoCAD 2008 Instructor*.

Have Fun

I predict you will have a positive experience learning AutoCAD. Although learning AutoCAD is not a trivial endeavor, you will have fun learning this exciting technology. In fact, I predict that more than once in your learning experience you will say to yourself, "Sweet!" (or something to that effect).

James A. Leach

ABOUT THE AUTHOR

James A. Leach (B.I.D., M.Ed.) is Professor of Engineering Graphics at the University of Louisville. He began teaching AutoCAD at Auburn University early in 1984 using Version 1.4, the first version of AutoCAD to operate on IBM personal computers. Jim is currently Director of the AutoCAD Training Center established at the University of Louisville in 1985, one of the first fifteen centers to be authorized by Autodesk.

In his 31 years of teaching Engineering Graphics and AutoCAD courses, Jim has published numerous journal and magazine articles, drawing workbooks, and textbooks about Autodesk and engineering graphics instruction. He has designed CAD facilities and written AutoCAD-related course materials for Auburn University, University of Louisville, the AutoCAD Training Center at the University of Louisville, and several two-year and community colleges. Jim is the author of 16 AutoCAD textbooks published by Richard D. Irwin and McGraw-Hill.

EDITING AUTHOR

James Dyer is an Autodesk Author and Developer having held senior positions at Autodesk Incorporated, Revit Technologies, and Bentley Systems Incorporated. He is the founding author of the "Productivity Corner" columns in CAE Mechanical Solutions, Cadence, and Cadalyst magazines. His background includes over twenty years of managerial experience in the design, engineering, analysis, and systems integration within the design, manufacturing, construction, and data management industries.

During the past twenty years James has published numerous journals and courses, hundreds of magazine articles, and edited several textbooks related to computer-aided design throughout the global design community. He also has over ten years of experience as a professional educator.

He is an internationally recognized public speaker in the field of design, business, management, and information technology.

James' credentials include and Electro-Mechanical Design degree from the America Institute of Design, in Philadelphia, Pennsylvania.

CONTIBUTORS

Tom Heffner has been presently the Coordinator and AutoCAD Instructor at Winnipeg Technical College, Continuing Education since 1989. Tom prepares the curriculum and teaches the Continuing Education AutoCAD Program to Level I, II, and III (3D) students. Over the past 33 years Tom has worked for several engineering companies in Canada, spending the last 19 years with The City of Winnipeg, Public Works Department, Engineering Division as Supervisor of Drafting and Graphic Services. Tom was instrumental in introducing and training City of Winnipeg Engineering employees AutoCAD starting with Version 2.6 up to the current Autodesk release. Tom has been a contributing author for McGraw-Hill for *AutoCAD 2008 Instructor, AutoCAD 2007 Instructor,* and *AutoCAD 2006 Companion* chapter exercises and chapter test/review questions.

Steven H. Baldock is an engineer at a consulting firm in Louisville and offers CAD consulting as well. Steve is an Autodesk Certified Instructor and teaches several courses at the University of Louisville AutoCAD Training Center. He has 21 years experience using AutoCAD in architectural, civil, and structural design applications. Steve has degrees in engineering, computer science, and mathematics from the University of Louisville and the University of Kentucky. Steve Baldock prepared material for several sections of *AutoCAD 2008 Companion* and *AutoCAD 2008 Instructor,* such as groups, text, multiline drawing and editing, and dimensioning. Steve also created several hundred figures used in *AutoCAD 2008 Companion* and *AutoCAD 2008 Instructor.*

ACKNOWLEDGMENTS

I want to thank all of the contributing authors for their assistance in writing *AutoCAD 2008 Companion.* Without their help, this text could not have been as application-specific nor could it have been completed in the short time frame.

I would like to give thanks to the excellent editorial and production group at McGraw-Hill who gave their talents and support during this project, especially Bill Stenquist and Lora Kalb.

A special thanks goes to Laura Hunter for the page layout work of *AutoCAD 2008 Companion.* She was instrumental in fulfilling my objective of providing the most direct and readable format for conveying concepts. I also want to thank Matt Hunter for his precise and timely proofing and for preparing the Table of Contents and Index for *AutoCAD 2008 Companion.*

It is wonderful to have an additional pair of eyes when proofing hundreds of pages of technical material. Rose Kernan, of RPK Editorial Services, Inc., not only served that function, but was responsible for achieving the optimum visual balance of text and figures on every page. Thanks, Rose.

I also want to thank all of the readers that have contacted me with comments and suggestions on specific sections of this text. Your comments help me improve this book, assist me in developing new ideas, and keep me abreast of ways AutoCAD and this text are used in industrial and educational settings.

I also acknowledge: my colleague and friend Robert A. Matthews for his support of this project and for doing his job well; Charles Grantham of Contemporary Publishing Company of Raleigh, Inc., for generosity and consultation; and Speed School of Engineering Dean's Office for support and encouragement.

James A. Leach

I would like to thank Theresa Curry, Harry McNeal, and Teresa McNeal for providing a positive, stable, and encouraging environment. Additionally, I want to thank Kimberly Fabio for her support and encouragement. Finally, I extend my sincere appreciation to James Leach for providing the opportunity to collaborate in the effort to evolve the reach of the AutoCAD Companion series.

James H. Dyer

TRADEMARK AND COPYRIGHT ACKNOWLEDGMENTS

The following drawings used for Chapter Exercises appear courtesy of James A. Leach, *Problems in Engineering Graphics Fundamentals, Series A* and *Series B*, ©1984 and 1985 by Contemporary Publishing Company of Raleigh, Inc.: Gasket A, Gasket B, Pulley, Holder, Angle Brace, Saddle, V-Block, Bar Guide, Cam Shaft, Bearing, Cylinder, Support Bracket, Corner Brace, and Adjustable Mount. Reprinted or redrawn by permission of the publisher.

The following are trademarks of Autodesk, Inc. in the United States. For additional trademark information, please contact the Autodesk, Inc., Legal Department: 415-507-6744. The following are trademarks of Autodesk, Inc. in the United States. For additional trademark information, please contact the Autodesk, Inc., Legal Department: 415-507-6744. 3DEC (design/logo)®, 3December®, 3December.com®, 3ds Max®, ActiveShapes®, ActiveShapes (logo)®, Actrix®, ADI®, AEC Authority (logo)®, Alias®, AliasStudio™, [Alias] "Swirl Design" (design/logo)®, Alias | Wavefront (design/logo)®, ATC®, AUGI® [Note: AUGI is a registered trademark of Autodesk, Inc., licensed exclusively to the Autodesk User Group International.], AutoCAD®, AutoCAD Learning Assistance™, AutoCAD LT®, AutoCAD Simulator™, AutoCAD SQL Extension™, AutoCAD SQL Interface™, Autodesk®, Autodesk Envision®, Autodesk Insight™, Autodesk Intent™, Autodesk Inventor®, Autodesk Map®, Autodesk MapGuide®, Autodesk Streamline®, AutoLISP®, AutoSketch®, AutoSnap™, AutoTrack™, Backburner™, Backdraft®, Built with ObjectARX (+product) (logos)™, Burn™, Buzzsaw®, CAiCE™, Can You Imagine®, Character Studio®, Cinestream™, Civil 3D®, Cleaner®, Cleaner Central™, ClearScale™, Colour Warper™, Combustion®, Communication Specification™, Constructware®, Content Explorer™, Create>what's>Next> (design/logo)®, Dancing Baby, The (image)™, DesignCenter™, Design Doctor™, Designer's Toolkit™, DesignKids™, DesignProf™, Design Server™, DesignStudio®, Design | Studio (design/logo)®, Design Your World®, Design Your World (design/logo)®, Discreet®, DWF®, DWG™, DWG (design/logo)™, DWG TrueConvert™, DWG TrueView™, DXF™, EditDV®, Education by Design®, Extending the Design Team™, FBX®, Field to Field®, Filmbox®, FMDesktop™, GDX Driver™, Gmax®, Heads-up Design™, Heidi®, HOOPS®, HumanIK®, i-drop®, iMOUT™, Incinerator®, IntroDV®, Inventor™, Kaydara®, Kaydara (design/logo)®, LocationLogic™, Lustre®, Maya®, Mechanical Desktop®, MotionBuilder™, ObjectARX®, ObjectDBX™, Open Reality®, PolarSnap™, PortfolioWall®, Powered with Autodesk Technology™, Productstream®, ProjectPoint®, Reactor®, RealDWG™, Real-time Roto™, Render Queue™, Revit®, Showcase™, SketchBook®, Topobase™, Toxik™, Visual®, Visual Bridge™, Visual Construction®, Visual Drainage®, Visual Hydro®, Visual Landscape®, Visual LISP®, Visual Roads®, Visual Survey®, Visual Syllabus™, Visual Toolbox®, Visual Tugboat®, Voice Reality®, Volo®, Wiretap™

All other brand names, product names, or trademarks belong to their respective holders.

LEGEND

The following special treatment of characters and fonts in the textual content is intended to assist you in translating the meaning of words or sentences in *AutoCAD 2008 Companion*.

<u>Underline</u>	Emphasis of a word or an idea.
Helvetica font	An AutoCAD prompt appearing on the <u>screen</u> at the command line or in a text window.
Italic (Upper and Lower)	An AutoCAD command, option, menu, toolbar, or dialog box name.
UPPER CASE	A file name.
UPPER CASE ITALIC	An AutoCAD system variable or a drawing aid (*OSNAP, SNAP, GRID, ORTHO*).

Anything in **Bold** represents user input:

Bold	What you should <u>type</u> or press on the keyboard.
Bold Italic	An AutoCAD <u>command</u> that you should type or <u>menu item</u> that you should select.
BOLD UPPER CASE	A <u>file name</u> that you should type.
BOLD UPPER CASE ITALIC	A <u>system variable</u> that you should type.
PICK	Move the cursor to the indicated position on the screen and press the <u>select</u> button (button #1 or left mouse button).

INTRODUCTION

WHAT IS CAD?

CAD is an acronym for Computer-Aided Design. CAD allows you to accomplish design and drafting activities using a computer. A CAD software package, such as AutoCAD, enables you to create designs and generate drawings to document those designs.

Design is a broad field involving the process of making an idea into a real product or system. The design process requires repeated refinement of an idea or ideas until a solution results—a manufactured product or constructed system. Traditionally, design involves the use of sketches, drawings, renderings, two-dimensional (2D) and three-dimensional (3D) models, prototypes, testing, analysis, and documentation. Drafting is generally known as the production of drawings that are used to document a design for manufacturing or construction or to archive the design.

CAD is a tool that can be used for design and drafting activities. CAD can be used to make "rough" idea drawings, although it is more suited to creating accurate finished drawings and renderings. CAD can be used to create a 2D or 3D computer model of the product or system for further analysis and testing by other computer programs. In addition, CAD can be used to supply manufacturing equipment such as lathes, mills, laser cutters, or rapid prototyping equipment with numerical data to manufacture a product. CAD is also used to create the 2D documentation drawings for communicating and archiving the design.

The tangible result of CAD activity is usually a drawing generated by a plotter or printer but can be a rendering of a model or numerical data for use with another software package or manufacturing device. Regardless of the purpose for using CAD, the resulting drawing or model is stored in a CAD file. The file consists of numeric data in binary form usually saved to a magnetic or optical device such as a diskette, hard drive, tape, or CD.

WHY SHOULD YOU USE CAD?

Although there are other methods used for design and drafting activities, CAD offers the following advantages over other methods in many cases:

1. Accuracy
2. Productivity for repetitive operations
3. Sharing the CAD file with other software programs

Accuracy

Since CAD technology is based on computer technology, it offers great accuracy. When you draw with a CAD system, the graphical elements, such as lines, arcs, and circles, are stored in the CAD file as numeric data. CAD systems store that numeric data with great precision. For example, AutoCAD stores values with fourteen significant digits. The value 1, for example, is stored in scientific notation as the equivalent of 1.0000000000000. This precision provides you with the ability to create designs and drawings that are 100% accurate for almost every case.

Productivity for Repetitive Operations

It may be faster to create a simple "rough" drawing, such as a sketch by hand (pencil and paper), than it would by using a CAD system. However, for larger and more complex drawings, particularly those involving similar shapes or repetitive operations, CAD methods are very efficient. Any kind of shape or operation accomplished with the CAD system can be easily duplicated since it is stored in a CAD file. In short, it may take some time to set up the first drawing and create some of the initial geometry, but any of the existing geometry or drawing setups can be easily duplicated in the current drawing or for new drawings.

Likewise, making changes to a CAD file (known as editing) is generally much faster than creating the original geometry. Since all the graphical elements in a CAD drawing are stored, only the affected components of the design or drawing need to be altered, and the drawing can be plotted or printed again or converted to other formats.

As CAD and the associated technology advance and software becomes more interconnected, more productive developments are available. For example, it is possible to make a change to a 3D model that automatically causes a related change in the linked 2D engineering drawing. One of the main advantages of these technological advances is productivity.

Sharing the CAD File with Other Software Programs

Of course, CAD is not the only form of industrial activity that is making technological advances. Most industries use computer software to increase capability and productivity. Since software is written using digital information and may be written for the same or similar computer operating systems, it is possible and desirable to make software programs with the ability to share data or even interconnect, possibly appearing simultaneously on one screen.

For example, word processing programs can generate text that can be imported into a drawing file, or a drawing can be created and imported into a text file as an illustration. (This book is a result of that capability.) A drawing created with a CAD system such as AutoCAD can be exported to a finite element analysis program that can read the computer model and compute and analyze stresses. CAD files can be dynamically "linked" to spreadsheets or databases in such a way that changing a value in a spreadsheet or text in a database can automatically make the related change in the drawing, or vice versa.

Another advance in CAD technology is the automatic creation and interconnectivity of a 2D drawing and a 3D model in one CAD file. With this tool, you can design a 3D model and have the 2D drawings automatically generated. The resulting set has bi-directional associativity; that is, a change in either the 2D drawings or the 3D model is automatically updated in the other.

With the introduction of the new Web technologies, designers and related professionals can more easily collaborate by viewing and transferring drawings over the Internet. CAD drawings can contain Internet links to other drawings, text information, or other related Web sites. Multiple CAD users can even share a single CAD session from remote locations over the Internet.

CAD, however, may not be the best tool for every design related activity. For example, CAD may help develop ideas but probably won't replace the idea sketch, at least not with present technology. A 3D CAD model can save much time and expense for some analysis and testing but cannot replace the "feel" of an actual model, at least not until virtual reality technology is developed and refined. With everything considered, CAD offers many opportunities for increased accuracy, productivity, and interconnectivity. Considering the speed at which this technology is advancing, many more opportunities are rapidly obtainable. However, we need to start with the basics. Beginning by learning to create an AutoCAD drawing is a good start.

WHY USE AutoCAD?

CAD systems are available for a number of computer platforms: laptops, personal computers (PCs), workstations, and mainframes. AutoCAD, offered to the public in late 1982, was one of the first PC-based CAD software products. Since that time, it has grown to be the world leader in market share for all CAD products. Autodesk, the manufacturer of AutoCAD, is the world's leading supplier of PC design software and multimedia tools. At the time of this writing, Autodesk is one of the largest software producers in the world and has several million customers in more than 150 countries.

Learning AutoCAD offers a number of advantages to you. Since AutoCAD is the most widely used CAD software, using it gives you the highest probability of being able to share CAD files and related data and information with others.

As a student, learning AutoCAD, as opposed to learning another CAD software product, gives you a higher probability of using your skills in industry. Likewise, there are more employers who use AutoCAD than any other single CAD system. In addition, learning AutoCAD as a first CAD system gives you a good foundation for learning other CAD packages because many concepts and commands introduced by AutoCAD are utilized by other systems. In some cases, AutoCAD features become industry standards. The .DXF file format, for example, was introduced by Autodesk and has become an industry standard for CAD file conversion between systems.

As a professional, using AutoCAD gives you the highest possibility that you can share CAD files and related data with your colleagues, vendors, and clients. Compatibility of hardware and software is an important issue in industry. Maintaining compatible hardware and software allows you the highest probability for sharing data and information with others as well as offering you flexibility in experimenting with and utilizing the latest technological advancements. AutoCAD provides you with great compatibility in the CAD domain.

This introduction is not intended as a selling point but to remind you of the importance and potential of the task you are about to undertake. If you are a professional or a student, you have most likely already made up your mind that you want to learn to use AutoCAD as a design or drafting tool. If you have made up your mind, then you can accomplish anything. Let's begin.

1

GETTING STARTED

CHAPTER OBJECTIVES

After completing this chapter you should:

1. understand how the X, Y, Z coordinate system is used to define the location of drawing elements in digital format in a CAD drawing file;

2. understand why you should create drawings full size in the actual units with CAD;

3. be able to start AutoCAD to begin drawing;

4. recognize the areas of the AutoCAD Drawing Editor and know the function of each;

5. be able to use the many methods of entering commands;

6. be able to turn on and off the *SNAP*, *GRID*, *ORTHO*, *POLAR*, and *DYN* drawing aids.

CONCEPTS

Coordinate Systems

Any location in a drawing, such as the endpoint of a line, can be described in X, Y, and Z coordinate values (Cartesian coordinates). If a line is drawn on a sheet of paper, for example, its endpoints can be charted by giving the distance over and up from the lower-left corner of the sheet (Fig. 1-1).

These distances, or values, can be expressed as X and Y coordinates; X is the horizontal distance from the lower-left corner (origin) and Y is the vertical distance from that origin. In a three-dimensional coordinate system, the third dimension, Z, is measured from the origin in a direction perpendicular to the plane defined by X and Y.

Two-dimensional (2D) and three-dimensional (3D) CAD systems use coordinate values to define the location of drawing elements such as lines and circles (called <u>objects</u> in AutoCAD).

In a 2D drawing, a line is defined by the X and Y coordinate values for its two endpoints (Fig. 1-2).

In a 3D drawing, a line can be created and defined by specifying X, Y, and Z coordinate values (Fig. 1-3). Coordinate values are always expressed by the X value first separated by a comma, then Y, then Z.

FIGURE 1-1

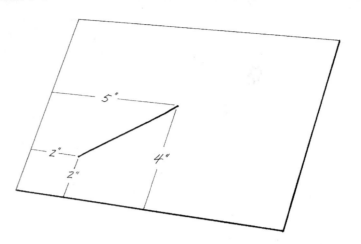

FIGURE 1-2

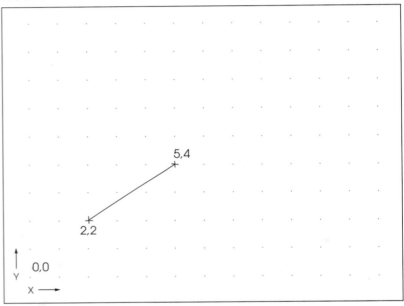

FIGURE 1-3

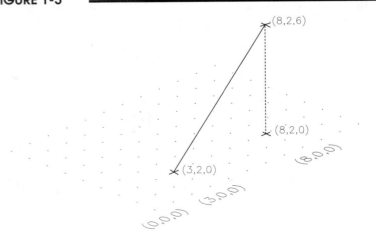

The CAD Database

A CAD (Computer-Aided Design) file, which is the electronically stored version of the drawing, keeps data in binary digital form. These digits describe coordinate values for all of the endpoints, center points, radii, vertices, etc. for all the objects composing the drawing, along with another code that describes the kinds of objects (line, circle, arc, ellipse, etc.). Figure 1-4 shows part of an AutoCAD DXF (Drawing Interchange Format) file giving numeric data defining lines and other objects. Knowing that a CAD system stores drawings by keeping coordinate data helps you understand the input that is required to create objects and how to translate the meaning of prompts on the screen.

FIGURE 1-4

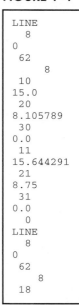

```
LINE
   8
 0
   62
          8
   10
   15.0
   20
   8.105789
   30
   0.0
   11
   15.644291
   21
   8.75
   31
   0.0
    0
 LINE
   8
 0
   62
          8
   18
```

Angles in AutoCAD

Angles in AutoCAD by default are measured in a <u>counterclockwise direction</u>. Angle 0 is positioned in a positive X direction, that is, horizontally from left to right. Therefore, 90 degrees is in a positive Y direction, or straight up; 180 degrees is in a negative X direction, or to the left; and 270 degrees is in a negative Y direction, or straight down (Fig. 1-5).

The position and direction of measuring angles in AutoCAD can be changed; however, the defaults listed here are used in most cases.

FIGURE 1-5

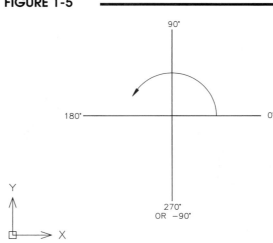

Draw True Size

When creating a drawing with pencil and paper tools, you must first determine a scale to use so the drawing will be proportional to the actual object and will fit on the sheet (Fig. 1-6). However, when creating a drawing on a CAD system, there is no fixed size drawing area. The number of drawing units that appear on the screen is variable and is assigned to fit the application.

The CAD drawing is not scaled until it is physically transferred to a fixed size sheet of paper by plotter or printer.

FIGURE 1-6

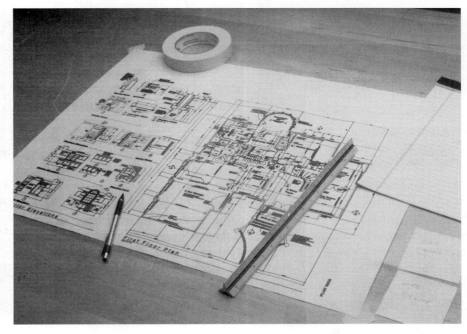

TIP

The rule for creating CAD drawings is that the drawing should be created <u>true size</u> using real-world units. The user specifies what units are to be used (architectural, engineering, etc.) and then specifies what size drawing area is needed (in X and Y values) to draw the necessary geometry (Fig. 1-7).

FIGURE 1-7

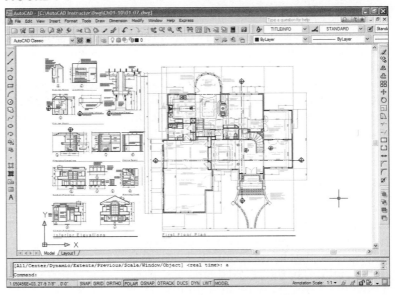

Whatever the specified size of the drawing area, it can be displayed on the screen in its entirety (Fig. 1-7) or as only a portion of the drawing area (Fig. 1-8).

FIGURE 1-8

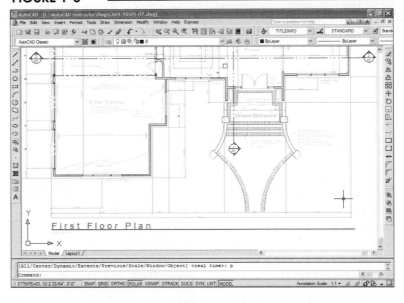

Plot to Scale

As long as a drawing exists as a CAD file or is visible on the screen, it is considered a virtual, full-sized object. Only when the CAD drawing is transferred to paper by a plotter or printer is it converted (usually reduced) to a size that will fit on a sheet. A CAD drawing can be automatically scaled to fit on the sheet regardless of sheet size; however, this action results in a plotted drawing that is not to an accepted scale (not to a regular proportion of the real object). Usually it is desirable to plot a drawing so that the resulting drawing is a proportion of the actual object size. The scale to enter as the plot scale (Fig. 1-9) is simply the proportion of the <u>plotted drawing</u> size to the <u>actual object</u>.

FIGURE 1-9

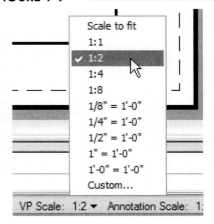

STARTING AutoCAD

Assuming that AutoCAD has been installed and configured properly for your system, you are ready to begin using AutoCAD.

To start AutoCAD 2008, locate the "AutoCAD 2008" shortcut icon on the desktop (Fig. 1-10). Double-clicking on the icon launches AutoCAD 2008.

FIGURE 1-10

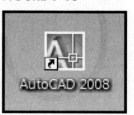

If you cannot locate the AutoCAD 2008 shortcut icon on the desktop, select the "Start" button, highlight "Programs" and search for "Autodesk." From the list that appears select "AutoCAD 2008" (Fig. 1-11).

FIGURE 1-11

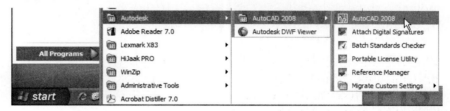

THE AutoCAD DRAWING EDITOR

When you start AutoCAD, you see the Drawing Editor. "Drawing Editor" is simply the name of the AutoCAD screen that allows you to create and edit drawings. The Drawing Editor is composed of a central drawing area, called the "graphics" area, and an array of toolbars, menus, a command line, and other elements, depending on the workspace and other settings that you can change.

Workspaces

AutoCAD 2008 offers three workspaces:

> *AutoCAD Classic*
> *2D Drafting & Annotation*
> *3D Modeling*

You can set the workspace by selecting the desired option from the *Workspaces* drop-down list in the upper-left corner above the graphics area (Fig. 1-12). For creating 2-dimensional (2D) drawings, use either the *AutoCAD Classic* or the *2D Drawing & Annotation* workspace. The features and capabilities of these two workspaces are essentially the same except for the configuration and location of the tools (icon buttons). These two workspaces provide alternate methods for you to accomplish the same goal—create and edit 2D drawings.

FIGURE 1-12

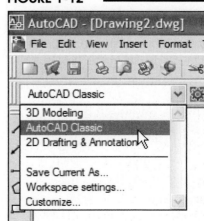

2008

AutoCAD Classic Workspace

The *AutoCAD Classic* workspace is the traditional layout used by most AutoCAD users since it was the only 2D drawing workspace in previous releases of AutoCAD (Fig. 1-13). The tools (icon buttons) are located in toolbars such as the *Draw* and *Modify* toolbars located along the top and extreme left and right sides of the graphics (drawing) area, respectively. Additional toolbars can be activated. By default, the *Tool Palettes* are activated (right side) but are normally closed to provide an unobstructed drawing area. The advantage of this workspace is the large drawing area.

FIGURE 1-13

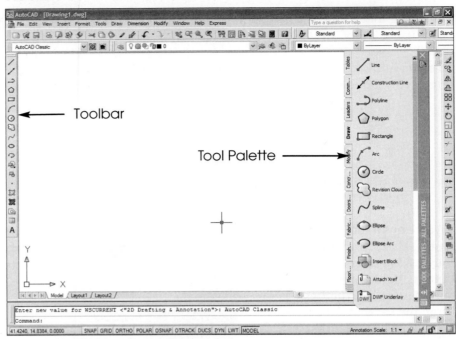

NOTE: In order to provide the largest possible drawing area, most of the illustrations in this book depicting the AutoCAD Drawing Editor display the *AutoCAD Classic* workspace without the *Tool Palettes*.

2D Drawing & Annotation Workspace

This workspace is new for AutoCAD 2008. Rather than individual toolbars along the top, left, and right of the drawing area, the Dashboard (right side) is active (Fig. 1-14). The Dashboard contains several "control panels," such as the *Layers*, *2D Draw*, and *Dimensions* control panels. The control panels contain tools (similar to toolbars) as well as drop-down lists and sliders. Although this workspace offers a couple of small advantages, valuable drawing area is occupied by the Dashboard.

FIGURE 1-14

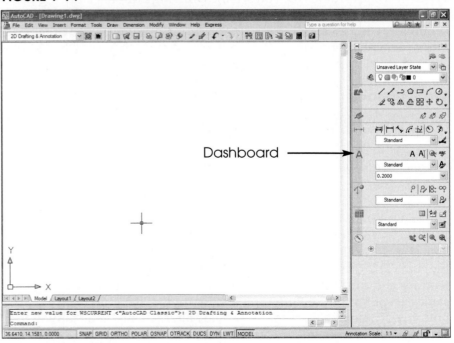

Beginning a Drawing

When you open AutoCAD for the first time, normally a "blank" drawing appears in the graphics area ready for you to begin drawing lines, circles, and so on. When you start AutoCAD or begin a new drawing, AutoCAD uses a "template" drawing as a starting point. A template drawing contains no geometry (lines, arcs, circles, etc.) but has specific settings based on what and how you intend to draw. Template drawing files have a .DWT file extension. For example, if you intend to use inch units, use the *acad.dwt* template drawing; whereas, if you intend to use metric units, use the *acadiso.dwt*.

The *Select Template* Dialog Box

If you use the *New* command from the *File* pull-down menu, the *Select Template* dialog box appears (Fig. 1-15). Here you can select the desired template to use for beginning a drawing. For creating 2D drawings with no other specific settings, use the *acad.dwt* for inch units and the *acadiso.dwt* for metric units. If you want to create a 3D model, select either the *acad3D.dwt* or *acadiso3D.dwt* template, then change to the *3D Modeling* workspace.

FIGURE 1-15

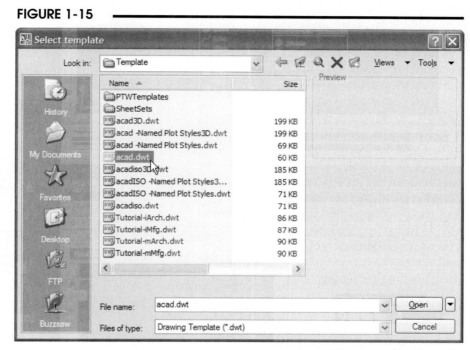

The *Startup* Dialog Box

Depending on how your system is configured, it is possible that the *Startup* dialog box may appear when you begin AutoCAD or when you use the *New* command (Fig. 1-16). The *Startup* dialog box allows you to begin a new drawing or open an existing drawing. You can begin a new drawing by selecting the *Start from Scratch* option, the *Use a Template* option, or the *Use a Wizard* option. If you want to begin a 2D drawing with the default inch or metric settings, select one of these options:

FIGURE 1-16

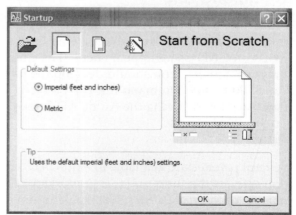

1.   Select the *Start from Scratch* button, select either *Imperial (feet and inches)* or *Metric*, then press *OK*.
2.   Select the *Use a Template* button, select either *acad.dwt* or *acadiso.dwt*, then press *OK*.

Changing the *STARTUP* system variable to 1 causes AutoCAD to display the *Startup* dialog box when AutoCAD is launched and the *Create New Drawing* dialog box when the *New* command is used. See Chapters 2, 6, and 7 for more information on the *Startup* and *Create New Drawing* dialog boxes.

NOTE: If your session automatically starts with a template other than *acad.dwt* or *acadiso.dwt*, a different template was specified by the *Qnew* setting in the *Options* dialog box (see "*Qnew*" in Chapter 2). For the examples and exercises in Chapters 1–5, use one of the options explained above to begin a drawing.

Graphics Area

The large central area of the screen is the <u>Graphics area</u>. It displays the lines, circles, and other objects you draw that will make up the drawing. The <u>cursor</u> is the intersection of the crosshairs (vertical and horizontal lines that follow the mouse or puck movements). The default size of the graphics area for English settings is 12 units (X or horizontal) by 9 units (Y or vertical). This usable drawing area (12 x 9) is called the drawing *Limits* and can be changed to any size to fit the application. As you move the cursor, you will notice the numbers in the <u>Coordinate display</u> change (Fig. 1-17, bottom left).

FIGURE 1-17

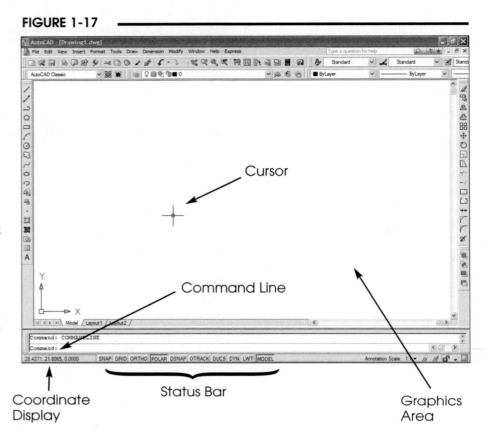

Command Line

The <u>Command line</u> consists of the two text lines at the bottom of the screen (by default) and is the most important area other than the drawing itself (see Fig. 1-17). Any command that is entered or any prompt that AutoCAD issues appears here. The Command line is always visible and gives the current state of drawing activity. You should develop the habit of glancing at the Command line while you work in AutoCAD. The Command line can be set to display any number of lines and/or moved to another location (see "Customizing the AutoCAD Screen" later in this chapter).

Palettes

Several palettes are available in AutoCAD, and one or more palettes may appear on your screen when you activate the *AutoCAD Classic* workspace, such as the *Sheet Set Manager* (Fig. 1-18), the *Properties* palette, or the *Tool Palettes*. Each palette serves a particular function as explained in later chapters. You can close the palettes by clicking on the "X" in the upper corner of the palette.

Toolbars

In the *AutoCAD Classic* workspace, AutoCAD provides a variety of <u>toolbars</u> (Fig. 1-18). Each toolbar contains a number of icon buttons (tools) that can be PICKed to invoke commands for drawing or editing objects (lines, arcs, circles, etc.) or for managing files and other functions.

The <u>Standard toolbar</u> is the row of icons nearest the top of the screen. The Standard toolbar contains many standard icons used in other Windows applications (like *New, Open, Save, Print, Cut, Paste,* etc.) and other icons for AutoCAD-specific functions (like *Zoom* and *Pan*). The <u>Layers and Properties toolbars</u>, located beneath the standard toolbar, are used for managing properties of objects, such as *Layers* and *Linetypes*. The <u>Draw and Modify toolbars</u> also appear in the *AutoCAD Classic* workspace. As shown in Figure 1-18, the Draw and Modify toolbars (extreme left and right side) and the Standard, Layer, and Properties toolbars (top) are <u>docked</u>. Many other toolbars are available and can be made to float, resize, or dock (see "Customizing the AutoCAD Screen" later in this chapter).

FIGURE 1-18

Standard Toolbar   Properties Toolbar

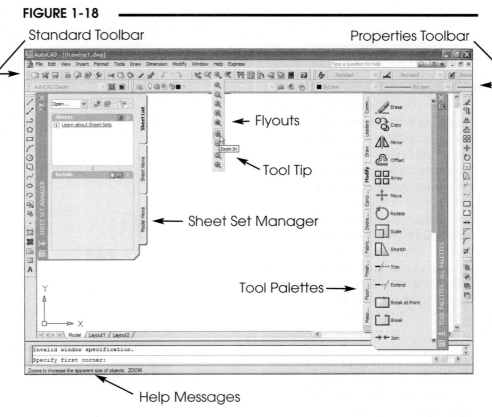

← Flyouts

Tool Tip

Sheet Set Manager

Tool Palettes →

Help Messages

If you place the pointer on any icon and wait momentarily, a <u>Tool Tip</u> and a <u>Help message</u> appear. Tool Tips pop out by the pointer and give the command name (see Fig. 1-18). The Help message appears at the bottom of the screen, giving a short description of the function. <u>Flyouts</u> are groups of related icons that pop out in a row or column when one of the group is selected. PICKing any icon that has a small black triangle in its lower-right corner causes the related icons to fly out.

Dashboard

An alternate method for activating a command or option is to use the dashboard. The dashboard appears when you use the *2D Drafting & Annotation* workspace or use the *Dashboard* command (see Fig. 1-14, on page 6). The dashboard is composed of several "control panels," such as the *Layers, 2D Draw,* and *Dimensions* control panels. Each control panel contains icon buttons (tools) and may include a drop-down list. Selecting a tool from the dashboard accomplishes the same action as selecting the same tool from a toolbar. For example, you can invoke the *Line* command by selecting the *Line* tool from the *Draw* toolbar or from the *2D Make* control panel of the dashboard.

Pull-down Menus

The pull-down menu bar is at the top of the screen just under the title bar (Fig. 1-19). Selecting any of the words in the menu bar activates, or pulls down, the respective menu. Selecting a word appearing with an arrow activates a cascading menu with other options. Selecting a word with ellipsis (. . .) activates a dialog box (see "Dialog Boxes"). Words in the pull-down menus are not necessarily the same as the formal command names used when typing commands. Menus can be canceled by pressing Escape or PICKing in the graphics area. The pull-down menus do not contain all of the AutoCAD commands and variables but contain the most commonly used ones.

FIGURE 1-19

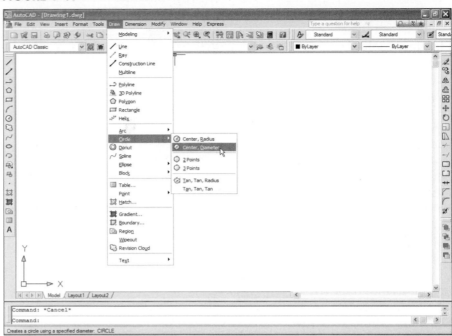

NOTE: Because the words that appear in the pull-down menus are not always the same as the formal commands you would type to invoke a command, AutoCAD can be confusing to learn. The Help message that appears just below the Command line (when a menu is pulled down or the pointer rests on an icon button) can be instrumental in avoiding this confusion. The Help message gives a description of the command followed by colon (:), then the formal command name (see Fig. 1-18 and Fig. 1-19).

Dialog Boxes

Dialog boxes provide an interface for controlling complex commands or a group of related commands. Depending on the command, the dialog boxes allow you to select among multiple options and sometimes give a preview of the effect of selections. The *Layer Properties Manager* dialog box (Fig. 1-20) gives complete control of layer colors, linetypes, and visibility.

FIGURE 1-20

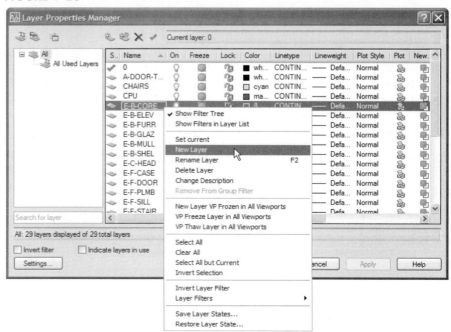

Dialog boxes can be invoked by typing a command, selecting an icon button, or PICKing from the menus. For example, typing *Layer,* PICKing the *Layer Properties Manager* button, or selecting *Layer* from the *Format* pull-down menu causes the *Layer Properties Manager* dialog box to appear. In the pull-down menu, all commands that invoke a dialog box end with ellipsis points (…).

The basic elements of a dialog box are listed below.

Button	Resembles a push button and triggers some type of action.
Edit box	Allows typing or editing of a single line of text.
Image tile	A button that displays a graphical image.
List box	A list of text strings from which one or more can be selected.
Drop-down list	A text string that drops down to display a list of selections.
Radio button	A group of buttons, only one of which can be turned on at a time.
Checkbox	A checkbox for turning a feature on or off (displays a check mark when on).

Shortcut Menus

AutoCAD makes use of shortcut menus that are activated by pressing the right mouse button (sometimes called right-click menus). Shortcut menus give quick access to command options. There are many shortcut menus to list since they are based on the active command or dialog box. The menus fall into five basic categories listed here.

Default Menu

The default menu appears when you right-click in the drawing area and no command is in progress. Using this menu, you can repeat the last command, select from recent input (like using the up and down arrows), use the Windows Cut, Copy, Paste functions, and select from other viewing and utility commands (Fig. 1-21).

FIGURE 1-21

Edit-Mode Menu

This menu appears when you right-click when <u>objects have been selected</u> but no command is in progress. Note that several of AutoCAD's *Modify* commands are available on the menu such as *Erase, Move, Copy, Scale,* and *Rotate* (Fig. 1-22).

NOTE: Edit mode shortcut menus do not appear if the *PICKFIRST* system variable is set to 0.

FIGURE 1-22

Command-Mode Menu

These menus appear when you right-click <u>when a command is in progress</u>. This menu changes since the options are specific to the command (Fig. 1-23).

FIGURE 1-23

Dialog-Mode Menu

When the pointer is in a dialog box or tab and you right-click, this menu appears. The options on this menu can change based on the current dialog box (Fig. 1-24 and 1-20).

FIGURE 1-24

Other Menus

There are other menus that can be invoked. For example, this menu appears if you right-click in the Command line area (Fig. 1-25).

Because there are so many shortcut menus, don't be too concerned about learning these until you have had some experience. The best advice at this time is just remember to experiment by right-clicking often to display the possible options.

FIGURE 1-25

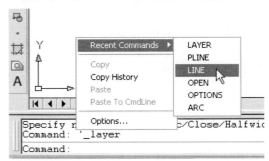

Screen (Side) Menu

The screen menu, available since the first version of AutoCAD, allows you to select commands and options from the side of the Drawing Editor. Although the screen menu is rarely used today, it can be made to appear by selecting this option in the *Display* tab of the *Options* dialog box. Keep in mind that this menu is not needed if you intend to use any other method of invoking commands.

Digitizing Tablet Menu

If you have a digitizing tablet, the AutoCAD commands are available by making the desired selection from the AutoCAD digitizing tablet menu. Digitizing tablets were popular in years past as a means of command input but are not used much today. For this reason, the digitizing tablet menu supplied with AutoCAD has not been updated to include most of the newer commands.

Command Entry Using the Keyboard

Although commands are accessible using toolbars, the dashboard, pull-down menus, and shortcut menus, you can enter the command using the keyboard. As you type the letters, they appear on the Command line (near the bottom of the screen) or in the Dynamic Input box near the cursor (see "Drawing Aids," "*DYN*"). Using the keyboard to type in commands offers several features:

Command name	Type in the full command name.
Command alias	Type the one- or two-letter shortcut for the command (see App. B).
Accelerator keys	Type a Ctrl key plus another key to invoke the command (see App. C).
Up and down arrows	Use the up and down arrows to cycle through the most recent input.
Tab key	Type in a few letters and press the Tab key to "AutoComplete" the command.

Accelerator Keys (Control Key Sequences)

Several control key sequences (holding down the Ctrl key or Alt key and pressing another key simultaneously) invoke regular AutoCAD commands or produce special functions (see App. D).

Special Key Functions

Esc	The Esc (escape) key cancels a command, menu, or dialog box or interrupts processing of plotting or hatching.
Space bar	In AutoCAD, the space bar performs the same action as the Enter key. Only when you are entering text into a drawing does the space bar create a space.
Enter	If Enter or the space bar is pressed when no command is in use (the open Command: prompt is visible), the last command used is invoked again.

Mouse Buttons

Depending on the type of mouse used for cursor control, a different number of buttons is available. In any case, the buttons normally perform the following tasks:

left button	PICK	Used to select commands or pick locations on the screen.
right button	Enter or shortcut menu	Depending on the status of the drawing or command, this button either performs the same function as the enter key or produces a shortcut menu.
wheel	*Pan*	If you press and drag, you can pan the drawing about on the screen.
	Zoom	If you turn the wheel, you can zoom in and out centered on the location of the cursor.

COMMAND ENTRY

Methods for Entering Commands

There are many possible methods for entering commands in AutoCAD depending on your system configuration. Generally, most of the methods can be used to invoke a particular commonly used command or dialog box.

1. Keyboard		Type the command name, command alias, or accelerator keys at the keyboard.

2. Pull-down menus   Select the command or dialog box from a pull-down menu.

3. Tools (icon buttons)   Select the command or dialog box by PICKing a tool (icon button) from a toolbar, palette, or control panel on the dashboard.

4. Shortcut menus   Select the command from the right-click shortcut menu. Right-clicking produces a shortcut menu depending on whether a command is active, objects are selected, or the pointer is in a dialog box.

5. Screen (side) menu   Select the command or dialog box from the screen menu (if activated).

6. Digitizing tablet menu   Select the command or dialog box from the tablet menu (if available).

All these methods generally accomplish the same goals; however, one method may offer a slightly different option or advantage over another depending on the command. Keep in mind that the screen menu and the digitizing tablet menu are older input methods and have not been updated to include newer commands.

Using the "Command Tables" in this Book to Locate a Particular Command

Command tables, like the one below, are used throughout this book to show the possible methods for entering a particular command. The table shows the icon used in the toolbars, the palettes, and the dashboard, which gives the selections to make for the pull-down menu, gives the correct spelling for entering commands and command aliases at the keyboard, and gives the shortcut menu and option. This example uses the *Copy* command.

Copy

Pull-down Menu	Command (Type)	Alias (Type)	Short-cut	Screen (side) Menu	Tablet Menu
Modify *Copy*	*Copy*	*CO* or *CP*	**(Edit Mode)** *Copy Selection*	*MODIFY1* *Copy*	*V,15*

AutoCAD DRAWING AIDS

This section explains several features that appear near the bottom of the Drawing Editor including the Coordinate display, Status bar (Drawing Aids), Model and Layout tabs, and the Text window.

Coordinate Display (Coords)

The Coordinate display is located in the lower-left corner of the AutoCAD screen (see Fig. 1-26). The Coordinate display (*Coords*) displays the current position of the cursor in one of two possible formats explained below. This display can be very helpful when you draw because it can give the X, Y, and Z coordinate position of the cursor or give the cursor's distance and angle from the last point established. The format of *Coords* is controlled by clicking on the *Coords* display (numbers) at the bottom left of the screen. *Coords* can also be toggled off.

Cursor tracking

When *Coords* is in this position, the values display the current location of the cursor in absolute Cartesian (X, Y, and Z) coordinates (see Fig. 1-26).

FIGURE 1-26

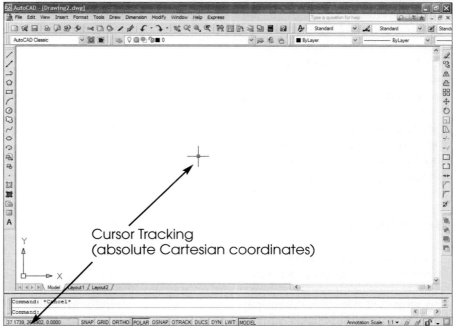

Cursor Tracking
(absolute Cartesian coordinates)

Relative polar display

This display is possible only if a draw or edit command is in use. The values give the distance and angle of the "rubberband" line from the last point established (not shown).

Status Bar

The Status bar is a set of toggles that allows you to turn drawing aids on and off. The Status bar appears at the very bottom of the screen (see Fig. 1-26). The following drawing aids can be toggled on or off by single-clicking (pressing the left mouse button once) on the desired word or by using Function keys or Ctrl key sequences. The following drawing aids are explained in this or in following chapters:

SNAP, GRID, ORTHO, POLAR, OSNAP, OTRACK, DUCS, DYN, LWT, MODEL

Drawing Aids

This section gives a brief introduction to AutoCAD's Drawing Aids. For a full explanation of the related commands and options, see Chapters 3 and 6.

SNAP (F9)

SNAP has two modes in AutoCAD: Grid Snap and Polar Snap. Only one of the two modes can be active at one time. Grid Snap is a function that forces the cursor to "snap" to regular intervals (.5 units is the default English setting), which aids in creating geometry accurate to interval lengths. You can use the *Snap* command or the *Drafting Settings* dialog box to specify any value for the Grid Snap increment. Figure 1-27 displays a Snap setting of .25 (note the values in the coordinate display in the lower-left corner).

FIGURE 1-27

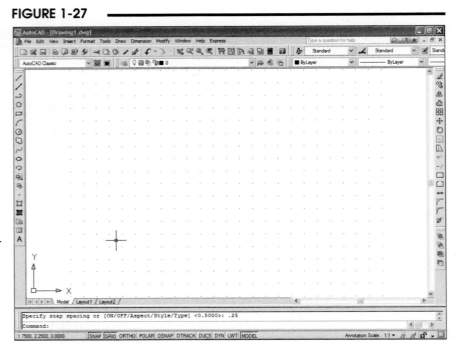

The other mode of Snap is Polar Snap. Polar Snap forces the cursor to snap to regular intervals along angular lines. Polar Snap is functional only when *POLAR* is also toggled on since it works in conjunction with *POLAR*. The Polar Snap interval uses the Grid Snap setting by default but can be changed to any value using the *Snap* command or the *Drafting Settings* dialog box. Polar Snap is discussed in detail in Chapter 3.

Since you can have only one *SNAP* mode on at a time (Grid Snap or Polar Snap), you can select which of the two is on by right-clicking on the word "*SNAP*" on the Status bar (Fig. 1-28). You can also access the *Drafting Settings* dialog box by selecting *Settings...* from the menu.

FIGURE 1-28

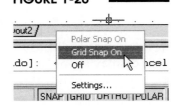

GRID (F7)

A drawing aid called *GRID* can be used to give a visual reference of units of length. The *GRID* default value for English settings is .5 units. The *Grid* command or *Drawing Aids* dialog box allows you to change the interval to any value. The *GRID* is not part of the geometry and is not plotted. Figure 1-29 displays a *GRID* of .5. *SNAP* and *GRID* are independent functions—they can be turned on or off independently.

FIGURE 1-29

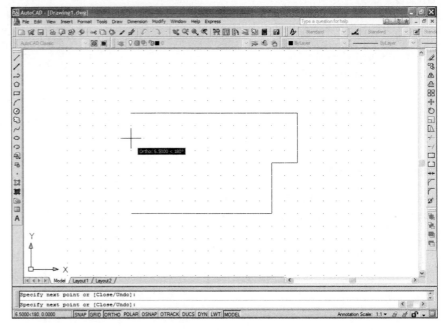

ORTHO (F8)

If *ORTHO* is on, lines are forced to an orthogonal alignment (horizontal or vertical) when drawing (Fig. 1-29). *ORTHO* is often helpful since so many drawings are composed mainly of horizontal and vertical lines. *ORTHO* can be turned only on or off.

POLAR (F10)

POLAR makes it easy to draw lines at regular angular increments, such as 30, 45, or 90 degrees. Using the F10 key or *POLAR* button toggles Polar Tracking on or off. When Polar Tracking is on, a polar tracking vector (a faint dotted line) appears when the rubber band line approaches the desired angular increment as shown in Figure 1-30. By default, *POLAR* is set to 90 degrees, but can be set to any angular increment by right-clicking on the *POLAR* button and selecting *Settings*.

FIGURE 1-30

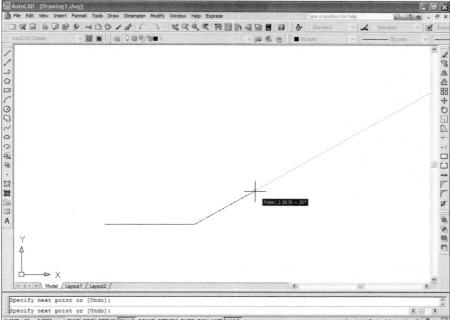

OSNAP (F3)

The *OSNAP* toggle is discussed in detail in Chapter 7.

OTRACK (F11)

The *OTRACK* toggle is also discussed in Chapter 7.

DYN (F12)

DYN, or Dynamic Input, is a feature that helps you visualize and specify coordinate values and angular values when drawing lines, arcs, circles, etc. *DYN* may display absolute Cartesian coordinates (X and Y values) or relative polar coordinates (distance and angle) depending on the current command prompt and the settings you prefer. Dynamic Input is explained further in Chapter 3 (Fig. 1-31).

FIGURE 1-31

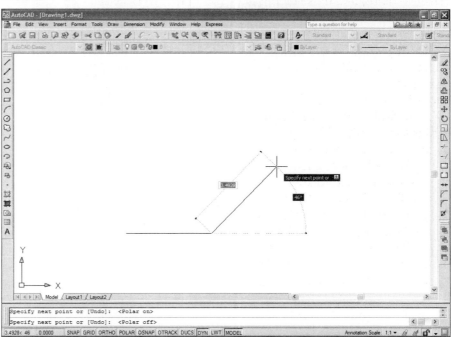

You can toggle Dynamic Input on or off by using the *DYN* button or the F12 key. You can also change the format of the display during input by right-clicking on the *DYN* button and selecting *Settings* from the menu. For example, you can set Dynamic Input to display a "tool tip" that gives the current coordinate location of the cursor (cursor tracking mode) when no commands are in use.

LWT
This feature is discussed in Chapter 11.

MODEL
See Chapter 13 for an explanation of the *MODEL/PAPER* toggle.

Status Bar Control
Right-clicking on the down arrow in the lower-right corner of the screeen produces the shortcut menu shown in Figure 1-32. This menu gives you control of which of these drawing aids appear in the Status bar. Clearing a check mark removes the associated drawing aid from the Status bar. It is recommended for most new AutoCAD users to keep all the drawing aids visible on the Status bar.

FIGURE 1-32

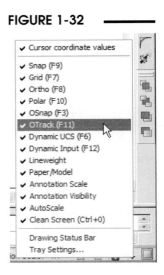

Function Keys

Several function keys are usable with AutoCAD. They offer a quick method of turning on or off (toggling) drawing aids and other features.

F1	*Help*	Opens a help window providing written explanations on commands and variables.
F2	*Text window*	Activates the text window showing the previous command line activity (command history).
F3	*OSNAP*	Turns *OSNAP* (Running Object Snaps) on or off. If no Running Object Snaps are set, F3 produces the *Osnap Settings* dialog box (discussed in Chapter 7).
F4	*Tablet*	Turns the *TABMODE* variable on or off. If *TABMODE* is on, the digitizing tablet can be used to digitize an existing paper drawing into AutoCAD.
F5	*Isoplane*	When using an *Isometric* style *SNAP* and *GRID* setting, toggles the cursor (with *ORTHO* on) to draw on one of three isometric planes.
F6	*DUCS*	Turns *Dynamic UCS* on or off for 3D modeling (see Chapters 28 and 29).
F7	*GRID*	Turns the *GRID* on or off.
F8	*ORTHO*	Turns *ORTHO* on or off.
F9	*SNAP*	Turns *SNAP* on or off.
F10	*POLAR*	Turns *POLAR* (Polar Tracking) on or off.
F11	*OTRACK*	Turns *OTRACK* (Object Snap Tracking) on or off.
F12	*DYN*	Turns *DYN* (Dynamic Input) on or off.

AutoCAD Text Window

Pressing the F2 key activates the *AutoCAD Text Window,* sometimes called the Command History. Here you can see the text activity that occurred at the Command line—kind of an "expanded" Command line. Press F2 again to close the text window. The *Edit* pull-down menu in this text window provides several options. If you highlight text in the window (Fig. 1-33), you can then *Paste to Cmdline* (Command line), *Copy* it to another program such as a word processor, *Copy History* (entire command history) to another program, or *Paste* text into the window. The *Options* choice invokes the *Options* dialog box.

FIGURE 1-33

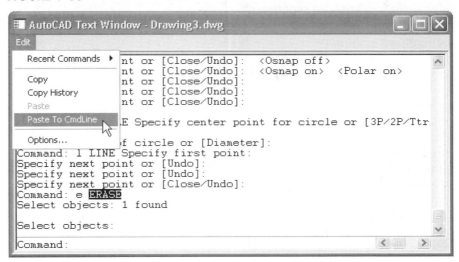

Model Tab and *Layout* Tabs

Model Tab
When you create a new drawing, the *Model* tab is the current tab (see Fig. 1-34, bottom left). This area is also known as <u>model space</u>. In this area you <u>should create the geometry representing the subject of your drawing</u>, such as a floor plan, a mechanical part, or an electrical schematic. Dimensions are usually created and attached to your objects in model space.

FIGURE 1-34

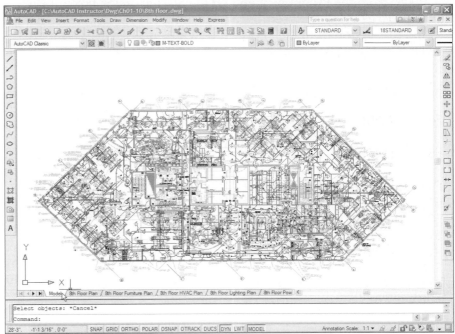

Layout Tabs

When you are finished with your drawing, you can plot it directly from the *Model* tab or switch to a *Layout* tab (Fig. 1-35). Layout tabs, sometimes known as paper space, represent sheets of paper that you plot on. You must use several commands to set up the layout to display the geometry and set all the plotting options such as scale, paper size, plot device, and so on. Plotting and layouts are discussed in detail in Chapters 13 and 14.

FIGURE 1-35

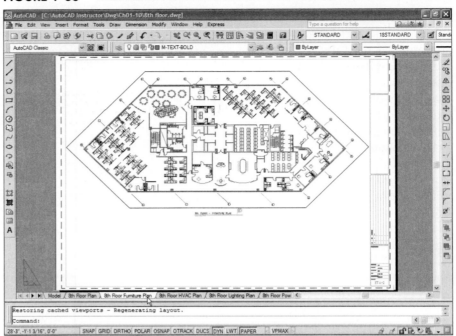

CHAPTER EXERCISES

1. **Starting and Exiting AutoCAD**

    Start AutoCAD by double-clicking the "AutoCAD 2008" shortcut icon or selecting "AutoCAD 2008" from the Programs menu. If the *Startup* dialog box appears, select *Start from Scratch*, choose *Imperial* as the default setting, and click the *OK* button. Draw a *Line*. Exit AutoCAD by selecting the *Exit* option from the *Files* pull-down menu. Answer *No* to the "Save changes to Drawing1.dwg?" prompt. Repeat these steps until you are confident with the procedure.

2. **Using Drawing Aids**

    Start AutoCAD. Turn on and off each of the following modes:

    SNAP, GRID, ORTHO, POLAR, DYN

3. **Understanding Coordinates**

    Begin drawing a **Line** by PICKing a "Specify first point:". Toggle *DYN* on and then toggle *Coords* to display each of the <u>three</u> formats. PICK several other points at the "Specify next point or [Undo]:" prompt. Pay particular attention to the coordinate values displayed for each point and visualize the relationship between that point and coordinate 0,0 (absolute Cartesian value) or the last point established and the distance and angle (relative polar value). Finish the command by pressing Enter.

4. **Using the Text Window**

   Use the Text Window (**F2**) to toggle between the text window and the graphics screen.

5. **Drawing with Drawing Aids**

   Draw four *Lines* using <u>each</u> Drawing Aid: ***GRID, SNAP, ORTHO, POLAR***. Toggle on and off each of the drawing aids one at a time for each set of four *Lines*. Next, draw *Lines* using combinations of the Drawing Aids, particularly ***GRID + SNAP*** and ***GRID + SNAP + POLAR***.

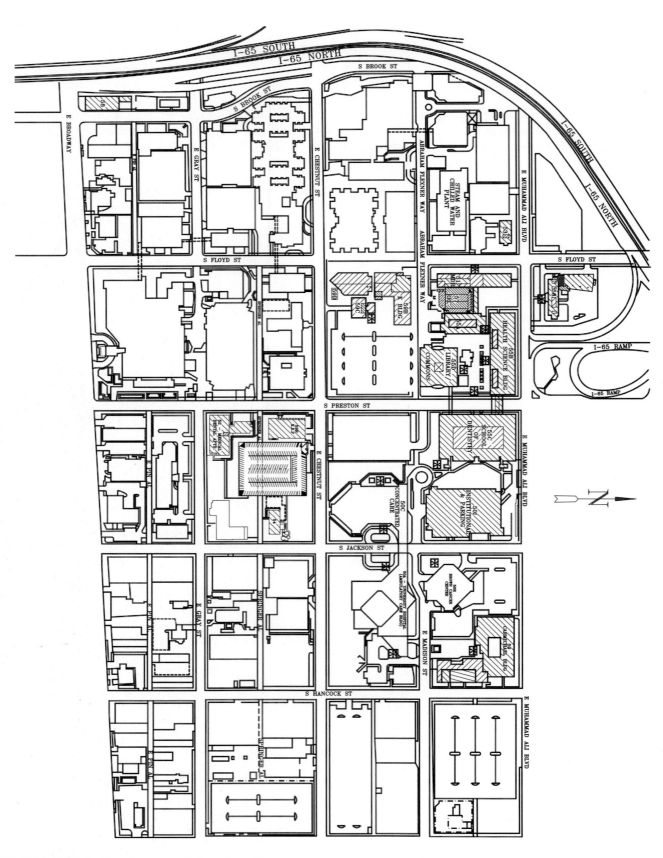

HSCMAP.DWG, Courtesy of Michael Anderson

2

WORKING
WITH FILES

CHAPTER OBJECTIVES

After completing this chapter you should:

1. be able to name drawing files;

2. be able use file-related dialog boxes;

3. be able to use the Windows right-click shortcut menus in file dialog boxes;

4. be able to create *New* drawings;

5. be able to *Open* and *Close* existing drawings;

6. be able to *Save* drawings;

7. be able to use *SaveAs* to save a drawing under a different name, path, and/or format.

AutoCAD DRAWING FILES

Naming Drawing Files

What is a drawing file? A CAD drawing file is the electronically stored data form of a drawing. The computer's hard disk is the principal magnetic storage device used for saving and restoring CAD drawing files. Disks, compact disks (CDs), networks, and the Internet are used to transport files from one computer to another, as in the case of transferring CAD files among clients, consultants, or vendors in industry. The AutoCAD commands used for saving drawings to and restoring drawings from files are explained in this chapter.

An AutoCAD drawing file has a name that you assign and a file extension of ".DWG." An example of an AutoCAD drawing file is:

PART-024.DWG
file name   extension

The file name you assign must be compliant with the Windows file name conventions; that is, it can have a <u>maximum</u> of 256 alphanumeric characters. File names and directory names (folders) can be in UPPER-CASE, Title Case, or lowercase letters. Characters such as _ - $ # ( ) ^ and spaces can be used in names, but other characters such as \ / : * ? < > | are not allowed. AutoCAD automatically appends the extension of .DWG to all AutoCAD-created drawing files.

NOTE: The chapter exercises and other examples in this book generally list file names in UPPERCASE letters for easy recognition. The file names and directory (folder) names on your system may appear as UPPERCASE, Title Case, or lowercase.

Beginning and Saving an AutoCAD Drawing

TIP When you start AutoCAD, the drawing editor appears and allows you to begin drawing even before using any file commands. As you draw, you should develop the habit of saving the drawing periodically (about every 15 or 20 minutes) using *Save*. *Save* stores the drawing in its most current state to disk.

The typical drawing session would involve using *New* to begin a new drawing or using *Open* to open an existing drawing. Alternately, the *Startup* dialog box that may appear when starting AutoCAD would be used to begin a new drawing or open an existing one. *Save* would be used periodically, and *Exit* would be used for the final save and to end the session.

FIGURE 2-1

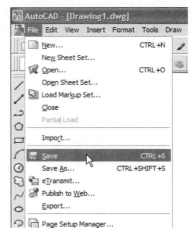

Accessing File Commands

Proper use of the file-related commands covered in this chapter allows you to manage your AutoCAD drawing files in a safe and efficient manner. Although the file-related commands can be invoked by several methods, they are easily accessible via the first pull-down menu option, *File* (Fig. 2-1). Most of the selections from this pull-down menu invoke dialog boxes for selection or specification of file names.

The Standard toolbar at the top of the AutoCAD screen has tools (icon buttons) for *New, Open,* and *Save*. File commands can also be entered at the keyboard by typing the formal command name, the command alias, or using the Ctrl keys sequences. File commands and related dialog boxes are also available from the *File* screen menu, or by selection from the digitizing tablet menu.

File Navigation Dialog Box Functions

There are many dialog boxes appearing in AutoCAD that help you manage files. All of these dialog boxes operate in a similar manner. A few guidelines will help you use them. The *Save Drawing As* dialog box (Fig. 2-2) is used as an example.

- The top of the box gives the title describing the action to be performed. It is <u>very important</u> to glance at the title before acting, especially when saving or deleting files.
- The desired file can be selected by PICKing it, then PICKing *OK*. Double-clicking on the selection accomplishes the same action. File names can also be typed in the *File name:* edit box.
- Every file name has an extension (three letters following the period) called the <u>type.</u> File types can be selected from the *Files of Type:* section of the dialog boxes, or the desired file extension can be entered in the *File name:* edit box.
- The current folder (directory) is listed in the drop-down box near the top of the dialog box. You can select another folder (directory) or drive by using the drop-down list displaying the current path, by selecting the *Back* arrow, or by selecting the *Up One Level* icon to the right of the list. (Rest your pointer on an icon momentarily to make the tool tip appear.)

FIGURE 2-2

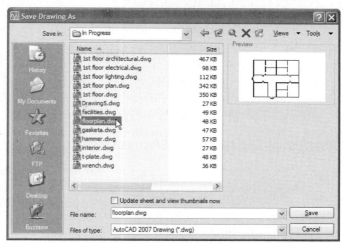

FIGURE 2-3

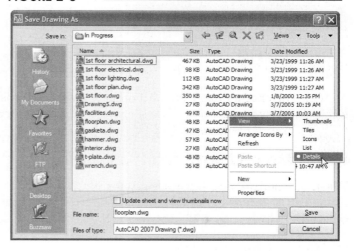

- Selecting one of the folders displayed on the left side of the file dialog boxes allows you to navigate to the following locations. The files or locations found appear in the list in the central area.
  History: Lists a history of files you have opened from most to least recent.
  My Documents: Lists all files and folders saved in the My Documents folder.
  Favorites: Lists all files and folders saved in the Favorites folder.
  Desktop: Shows files, folders, and shortcuts on your desktop.
  FTP: Allows you to browse through the FTP sites you have saved.
  Buzzsaw: Connects you to Buzzsaw—an Autodesk Web site for collaboration of files and information.
- You can use the *Search the Web* icon to display the *Browse the Web* dialog box (not shown). The default site is http://www.autodesk.com/.
- Any highlighted file(s) can be deleted using the "X" (*Delete*) button.
- A new folder (subdirectory) can be created within the current folder (directory) by selecting the *New Folder* icon.
- The *View* drop-down list allows you to toggle the listing of files to a *List* or to show *Details*. The *List* option displays only file names (see Fig. 2-2), whereas the *Details* option gives file-related information such as file size, file type, time, and date last modified (see Fig. 2-3). The detailed list can also be sorted alphabetically by name, by file size, alphabetically by file type, or chronologically by time and date. Do this by clicking the *Name, Type, Size,* or *Date Modified* tiles immediately above the list. Double-clicking on one of the tiles reverses the order of the list.

You can resize the width of a column (*Name, Size, Type,* etc.) by moving the pointer to the "crack" between two columns, then sliding the double arrows that appear in either direction. You can also use the *Preview* option to show a preview of the highlighted file (see Fig. 2-2) or to disable the preview (see Fig. 2-3).

- Use the *Tools* drop-down list to add files or folders to the *Favorites* or *FTP* folder. *Places* adds files or folders to the folder listing on the left side of the dialog box. *Options* allows you to specify formats for saving DWG and DXF files, and *Security Options* allows you to apply a password or Digital Signature to the file (see "*Saveas*").

Windows Right-Click Shortcut Menus

AutoCAD utilizes the right-click shortcut menus that operate with the Windows operating systems. Activate the shortcut menus by pressing the right mouse button (right-clicking) <u>inside the file list area</u> of a dialog box. There are two menus that give you additional file management capabilities.

Right-Click, No Files Highlighted

When no files are highlighted, right-clicking produces a menu for file list display and file management options (Fig. 2-3). The menu choices are as follows:

View	Shows *Thumbnails* (see Fig. 2-4), *Tiles* (large icons), *Icons* (small icons), *List* (see Fig. 2-3), or *Details* (see Fig. 2-3).
Arrange Icons By	Sorts by file *Name*, file *Size*, extension *Type*, or date *Modified*.
Paste	If the *Copy* or *Cut* function was previously used (see "Right-Click, File Highlighted," Fig. 2-4), *Paste* can be used to place the copied file into the displayed folder.
Paste Shortcut	Use this option to place a *Shortcut* (to open a file) into the current folder.
New	Creates a new *Folder, Shortcut,* or document.
Properties	Displays a dialog box listing properties of the current folder.

Right-Click, File Highlighted

When a file is highlighted, right-clicking produces a menu with options for the selected file (Fig. 2-4). The menu choices are as follows:

FIGURE 2-4

Select	Processes the file (like selecting the *OK* button) according to the dialog box function. For example, if the *SaveAs* dialog box is open, *Select* saves the highlighted file; or if the *Select File* dialog box is active, *Select* opens the drawing.
Open	Opens the application (AutoCAD or other program) and loads the selected file. Since AutoCAD can have multiple drawings open, you can select several drawings to open with this feature.
Enable/Disable Digital...	This option allows you to toggle the display of a small signature designator on the .DWG icon for drawings that have been signed with a digital signature.
Print	Sends the selected file to the configured system printer.
Convert to Adobe PDF	Drawings can be automatically converted to a .PDF file using this option. The *Adobe PDF Status* dialog box appears.
Publish DWF	Select this option to open the drawing and produce the *Plot* dialog box.

Send To	Copies the selected file to the selected device. <u>This is an easy way to copy a file from your hard drive to a disk in A: drive.</u>
Cut	In conjunction with *Paste,* allows you to move a file from one location to another.
Copy	In conjunction with *Paste,* allows you to copy the selected file to another location. You can copy the file to the same folder, but Windows renames the file to "Copy of . . .".
Create Shortcut	Use this option to create a *Shortcut* (to open the highlighted file) in the current folder. The shortcut can be moved to another location using drag and drop.
Delete	Sends the selected file to the Recycle Bin.
Rename	Allows you to rename the selected file. Move your cursor to the highlighted file name, click near the letters you want to change, then type or use the backspace, delete, space, or arrow keys.
Properties	Displays a dialog box listing properties of the selected file.

Specific features of other dialog boxes are explained in the following sections.

AutoCAD FILE COMMANDS

When you start AutoCAD, one of three situations occur based on how the *Options* are set on your system:

1. The *Startup* dialog box appears.
2. No dialog box appears and the session starts with the ACAD.DWT or the ACADISO.DWT template (determined by the *MEASUREINIT* system variable setting, 0 or 1, respectively).
3. No dialog box appears and the session starts with a template specified by the *Qnew* setting in the *Options* dialog box (see "QNEW").

To control the appearance of the *Startup* dialog box (and the *Create New Drawing* dialog box), use the *STARTUP* system variable. A setting of 1 (on) enables the dialog boxes, whereas a setting 0 (off) disables the dialog boxes.

The *Startup* dialog box allows you to create new drawings and open existing drawings. These options are the same as using the *New* and *Open* commands. Therefore, a description of the *New* and *Open* commands explains these functions in the *Startup* dialog box.

New

Pull-down Menu	Command (Type)	Alias (Type)	Short-cut	Screen (side) Menu	Tablet Menu
File *New...*	*New*	*...*	*Ctrl+N*	FILE *New*	*T,24*

The *New* command begins a new drawing. The new drawing can be a completely "blank" drawing or it can be based on a template that may already have a title block and some additional desired settings. Based on the settings for your system, the *New* command produces one of two methods to start a new drawing:

1. The *Create New Drawing* dialog box appears (identical to the *Startup* dialog box).
2. The *Select Template* dialog box appears.

Control these two actions by setting the *STARTUP* system variable. Setting the variable to 1 (on) causes the *Create New Drawing* dialog box to appear when you use the *New* command. Setting the variable to 0 (off) causes the *Select Template* dialog box to appear when you use the *New* command.

The *Create New Drawing* Dialog Box

In the *Create New Drawing* dialog box (Fig. 2-5), the options for creating a new drawing are the same as in the *Startup* dialog box (shown previously in Fig. 1-16): *Start from Scratch, Use a Template,* and *Use a Wizard*.

Start from Scratch

In the *Create New Drawing* dialog box select the *Start from Scratch* option (Fig. 2-5). Next, choose from either the *Imperial (feet and inches)* or *Metric* default settings. Selecting *Imperial* uses the default ACAD.DWT template drawing with a drawing area (called *Limits*) of 12 x 9 units. Choosing *Metric* uses the ACADISO.DWT template drawing which has *Limits* settings of 420 x 297.

FIGURE 2-5

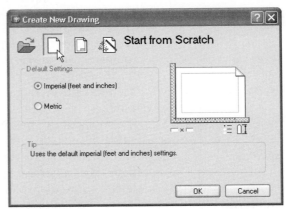

Use a Wizard

In the *Create New Drawing* dialog box select the *Use a Wizard* option (Fig. 2-6). Next, choose between a *Quick Setup Wizard* and an *Advanced Setup Wizard*. The *Quick Setup Wizard* prompts you to select the type of drawing *Units* you want to use and to specify the drawing *Area* (*Limits*). The *Advanced Setup* offers options for units, angular direction and measurement, and area. The *Quick Setup Wizard* and the *Advanced Setup Wizard* are discussed in Chapter 6, Basic Drawing Setup.

FIGURE 2-6

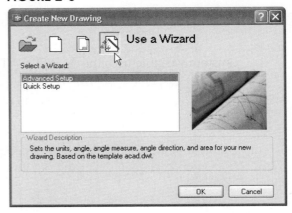

Use a Template

Use this option if you want to create a new drawing based on an existing template drawing (Fig. 2-7). A template drawing is one that may have some of the setup steps performed but contains no geometry (graphical objects).

Selecting the ACAD.DWT template begins a new drawing using the Imperial default settings with a drawing area of 12 x 9 units (identical to using the *Start from Scratch, Imperial* option). Selecting the ACADISO.DWT template begins a new drawing using the metric settings having a drawing area of 420 x 279 units (identical to using the *Start from Scratch, Metric* option).

FIGURE 2-7

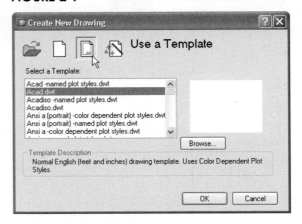

The *Select Template* Dialog Box

The *Select Template* dialog box (Fig. 2-8) appears when you use the *New* command and your setting for the *STARTUP* system variable is 0 (off). The *Select Template* dialog box has the identical outcome as using the *Use a Template* option in the *Create New Drawing* dialog box (see Fig. 2-7); that is, you can select from all AutoCAD-supplied templates (see previous discussion).

NOTE: For the purposes of learning AutoCAD starting with the basic principles and commands discussed in this text, it is helpful to use the <u>Imperial (inch) settings</u> when beginning a new drawing. Until you read Chapters 6 and 12, you can begin new drawings for completing the exercises and practicing the

FIGURE 2-8

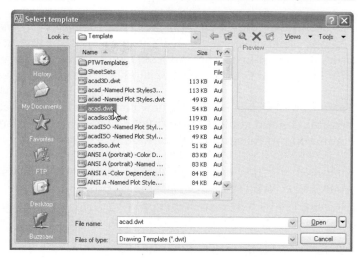

examples in Chapters 1 through 5 by any of the following methods.

Select *Imperial (feet and inches)* in the *Start from Scratch* option.
Select the ACAD.DWT template drawing using any method <u>then perform a *Zoom All*</u>.
Cancel the *Startup* dialog when AutoCAD starts (the default setting of 0 for the *MEASURINIT* system variable uses the ACAD.DWT).

Qnew

Pull-down Menu	Command (Type)	Alias (Type)	Short-cut	Screen (side) Menu	Tablet Menu
...	*Qnew*	...	...	...	...

Qnew (Quick New) is intended to immediately begin a new drawing using the template file specified in the *Files* tab of the *Options* dialog box. <u>If the proper settings are made</u>, *Qnew* is faster to use than the *New* command since it does not force the *Create New Drawing* or the *Select Template* dialog box to open, as the *New* command does.

To set *Qnew* to operate quickly when opening a drawing, you must first make two settings (if you are using AutoCAD in a school or office setting, these settings may have already been made):

1. Specify a default template to use when *Qnew* is invoked. Open the *Options* dialog box by selecting *Options* from the *Tools* pull-down menu. Select the

FIGURE 2-9

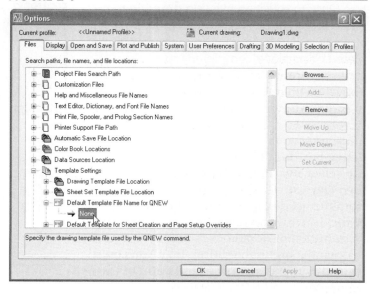

Files tab and expand the *Default Template File Name for QNEW* line (Fig. 2-9). The default setting is *None*. Highlight *None*, then pick the *Browse* button to locate the template file you want to use.

2. Set the *STARTUP* system variable to 0.

NOTE: The default setting for the *Default Template File Name for QNEW* line (in the *Options* dialog box) is "None." That means if you use *Qnew* without specifying a template drawing, as will be the case for all users upon first installing AutoCAD, it operates similarly to the *New* command by opening the *Create New Drawing* or *Select Template* dialog box, depending on your setting for the *STARTUP* system variable.

Since Autodesk has made this new command relatively complex to understand and control, the three possibilities and controls are explained simply here. Based on the settings for your system, the *Qnew* command produces one of three methods to start a new drawing:

Method 1.   The *Create New Drawing* dialog box appears.
Method 2.   The *Select Template* dialog box appears.
Method 3.   A new drawing starts with the *Qnew* template specified in the *Options* dialog box.

Control the previous actions by the following settings:

Method 1.   Set the *STARTUP* system variable to 1.
Method 2.   Set the *STARTUP* system variable to 0 and ensure the *Default Template File Name for QNEW* is set to *None*.
Method 3.   Set the *STARTUP* system variable to 0 and set the *Default Template File Name for QNEW* to the desired template file.

Open

Pull-down Menu	Command (Type)	Alias (Type)	Short-cut	Screen (side) Menu	Tablet Menu
File *Open...*	*Open*	*...*	*Ctrl+O*	*FILE* *Open*	*T,25*

Use *Open* to select an existing drawing to be loaded into AutoCAD. Normally you would open an existing drawing (one that is completed or partially completed) so you can continue drawing or to make a print or plot. You can *Open* multiple drawings at one time in AutoCAD.

The *Open* command produces the *Select File* dialog box (Fig. 2-10). In this dialog box you can select any drawing from the current directory list. PICKing a drawing name from the list displays a small bitmap image of the drawing in the *Preview* tile. Select the *Open* button or double-click on the file name to open the highlighted drawing. You could instead type the file name (and path) of the desired drawing in the *File name:* edit box and press Enter, but a preview of the typed entry will not appear.

FIGURE 2-10

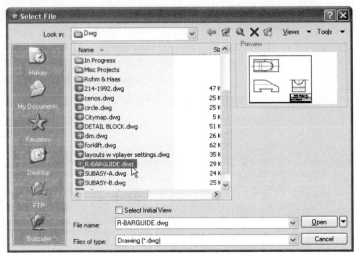

You can open multiple drawings at one time by holding down the Shift key to select a range (all files between and including the two selected) or holding down the Ctrl key to select multiple drawings not in a range.

Using the *Open* drop-down list (lower right) you can select from these options:

Open	Opens a drawing file and allows you to edit the file.
Open Read-Only	Opens a drawing file for viewing, but you cannot edit the file.
Partial Open	Allows you to open only a part of a drawing for editing.
Partial Open Read-Only	Allows you to open only a part of a drawing for viewing, but you cannot edit the drawing.

Other options in this dialog box are similar to the *Save Drawing As* dialog box described earlier in "File Navigation Dialog Box Functions." For example, to locate a drawing in another folder (directory) or on another drive on your computer or network, select the drop-down list on top of the dialog box next to *Look in:*, or use the *Up one level* button. Remember that you can also display the file details (size, type, modified) by toggling the *Details* option from the *Views* drop-down list. You can open a .DWG (drawing), .DWS (drawing standards), .DXF (drawing interchange format), or .DWT (drawing template) file by selecting from the *Files of type:* drop-down list.

Selecting the *Find...* option from the *Tools* drop-down list in the *Select File* dialog box (upper right) invokes the *Find* dialog box (Fig. 2-11). The dialog box offers two criteria to locate files: *Name & Location* and *Date Modified*.

FIGURE 2-11

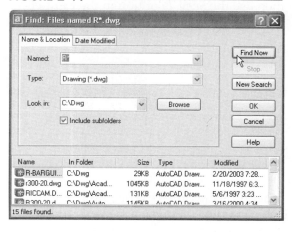

Using the *Name & Location* tab, you can search for .DWG, .DWS, .DWT, or .DXF files in any directory, including subdirectories if desired. In the *Named:* edit box, enter a drawing name to find. Wildcards can be used.

The *Date Modified* tab of the *Find* dialog box enables you to search for files meeting specific date criteria that you specify (Fig. 2-12). This feature helps you search for files *between* two dates, *during the previous months,* or *during the previous days,* based on when the files were last saved.

After specifying the search criteria in either tab, select the *Find Now* button. All file names that are found matching the criteria are displayed in the list at the bottom of the dialog box. Double-clicking on the name or using the *OK* button passes the name to the *File Name* section of the *Select File* dialog box.

FIGURE 2-12

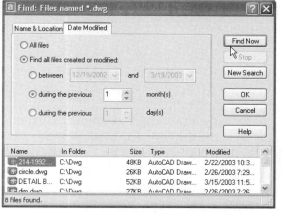

NOTE: It is considered <u>poor practice to *Open* a drawing from a disk in A: drive</u>. Normally, drawings should be copied to a directory on a fixed (hard) drive, then opened from that directory. Opening a drawing from a disk and performing saves to a disk are much slower and less reliable than operating from a hard drive. In addition, some temporary files may be written to the drive or directory from which the file is opened. See other "NOTE"s at the end of the discussion on *Save* and *Saveas.*

Save

Pull-down Menu	Command (Type)	Alias (Type)	Short-cut	Screen (side) Menu	Tablet Menu
File Save	Save	...	Ctrl+S	...	U,24-U,25

TIP The *Save* command is intended to be used periodically during a drawing session (every 15 to 20 minutes is recommended). When *Save* is selected from a menu, the current version of the drawing is saved to disk without interruption of the drawing session. The first time an unnamed drawing is saved, the *Save Drawing As* dialog box (see Fig. 2-2 and Fig. 2-3) appears, which prompts you for a drawing name. Typically, however, the drawing already has an assigned name, in which case *Save* actually performs a *Qsave* (quick save). A *Qsave* gives no prompts or options, nor does it display a dialog box.

When the file has an assigned name, using *Save* by selecting the command from a menu or icon button automatically performs a quick save (*Qsave*). Therefore, *Qsave* automatically saves the drawing in the same drive and directory from which it was opened or where it was first saved. In contrast, <u>typing *Save* always <u>produces the *Save Drawing As* dialog box,</u> where you can enter a new name and/or path to save the drawing or press Enter to keep the same name and path (see *SaveAs*).

TIP NOTE: If you want to save a drawing directly to a disk in A:, first *Save* the drawing to the hard drive, then use the *Send To* option in the right-click menu in the *Save, Save Drawing As*, or *Select File* dialog box (see "Windows Right-Click Shortcut Menus" and Fig. 2-4).

Qsave

Pull-down Menu	Command (Type)	Alias (Type)	Short-cut	Screen (side) Menu	Tablet Menu
...	Qsave	...	Ctrl+S	FILE Qsave	...

Qsave (quick save) is normally invoked automatically when *Save* is used (see *Save*) but can also be typed. *Qsave* saves the drawing under the previously assigned file name. No dialog boxes appear nor are any other inputs required. This is the same as using *Save* (from a menu), assuming the drawing name has been assigned. However, if the drawing has not been named when *Qsave* is invoked, the *Save Drawing As* dialog box appears.

Saveas

Pull-down Menu	Command (Type)	Alias (Type)	Short-cut	Screen (side) Menu	Tablet Menu
File Save As...	Saveas	...	...	FILE Saveas	V,24

The *SaveAs* command can fulfill four functions:
1. Save the drawing file under a new name if desired.
2. Save the drawing file to a new path (drive and directory location) if desired.
3. In the case of either 1 or 2, assign the new file name and/or path to the current drawing (change the name and path of the current drawing).
4. Save the drawing in a format other than the default AutoCAD 2007 format or as a different file type (.DWT, .DWS, or .DXF).

Therefore, assuming a name has previously been assigned, *SaveAs* allows you to save the current drawing under a different name and/or path; but, beware, <u>*SaveAs* sets the current drawing name and/or path to the last one entered.</u> This dialog box is shown in Figures 2-2 and 2-3.

SaveAs can be a benefit when creating two similar drawings. A typical scenario follows. A design engineer wants to make two similar but slightly different design drawings. During construction of the first drawing, the engineer periodically saves under the name DESIGN1 using *Save*. The first drawing is then completed and *Saved*. Instead of starting a *New* drawing, *SaveAs* is used to save the current drawing under the name DESIGN2. *SaveAs* also resets the current drawing name to DESIGN2. The designer then has two separate but identical drawing files on disk which can be further edited to complete the specialized differences. The engineer continues to work on the current drawing DESIGN2.

NOTE: If you want to save the drawing to a disk in A: drive, do not use *SaveAs*. Invoking *SaveAs* by any method resets the drawing name and path to whatever is entered in the *Save Drawing As* dialog box, so entering A:NAME would set A: as the current drive. This could cause problems because of the speed and reliability of a disk as opposed to a hard drive. Instead, you should save the drawing to the hard drive (usually C:), then close the drawing (by using *Close*). Next, use a right-click shortcut menu to copy the drawing file to A:. (See "Windows Right-Click Shortcut Menus" and Fig. 2-4.)

You can save an AutoCAD 2008 drawing in several formats other than the default format (*AutoCAD 2007 Drawing *.dwg*). Use the drop-down list at the bottom of the *Save* (or *Save Drawing As*) dialog box (Fig. 2-13) to save the current drawing as an earlier version drawing (.DWG) of AutoCAD or LT, a template file (.DWT), or a .DWS or a .DXF file. In AutoCAD 2008, a drawing can be saved to an earlier release and later *Opened* in AutoCAD 2008 and all new features are retained during the "round trip."

FIGURE 2-13

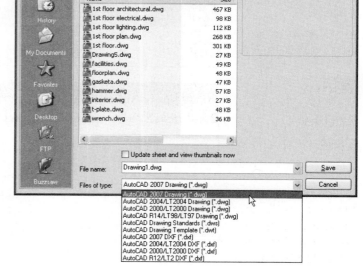

Passwords

An added security feature allows you to save your drawings with password protection. If a password is assigned, the drawing can be opened again in AutoCAD <u>only</u> if you enter the correct password when using the *Open* command. Beware: <u>If you lose or forget the password, the drawing cannot be opened or recovered in any way</u>!

To assign a password during the *Save* or *SaveAs* command, select the *Tools* drop-down menu in the *Save Drawing As* dialog box, then select *Security Options…* (Fig. 2-14) to produce the *Security Options* dialog box. Passwords are not case sensitive.

FIGURE 2-14

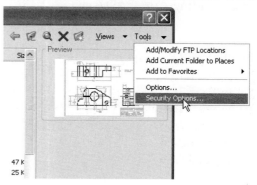

Digital Signatures

You can also attach digital signatures to drawings using the *Security Options* dialog box. If a drawing has a digital signature assigned, it is designated as an original signed and dated drawing. Users are assured that the drawing is original and has not been changed in any way. If such a drawing is later modified, the digital signature is then invalidated, and anyone opening the drawing is notified that the drawing has been changed and the digital signature is invalid.

Close

Pull-down Menu	Command (Type)	Alias (Type)	Short-cut	Screen (side) Menu	Tablet Menu
File Close	Close	...	...	...	...

Use the *Close* command to <u>close the current drawing</u>. Because AutoCAD allows you to have several drawings *Open*, *Close* gives you control to close one drawing while leaving others open. If the drawing has been changed but not saved, AutoCAD prompts you to save or discard the changes. In this case, a warning box appears (Fig. 2-15). *Yes* causes AutoCAD to *Save* then close the drawing; *No* closes the drawing without saving; and *Cancel* aborts the close operation so the drawing stays open.

FIGURE 2-15

Closeall

Pull-down Menu	Command (Type)	Alias (Type)	Short-cut	Screen (side) Menu	Tablet Menu
Window Closeall	Closeall	...	...	...	...

The *Closeall* command closes all drawings currently open in your AutoCAD session. If any of the drawings have been changed but not saved, you are prompted to save or discard the changes for each drawing.

Exit

Pull-down Menu	Command (Type	Alias (Type)	Short-cut	Screen (side) Menu	Tablet Menu
File Exit	Exit	...	Ctrl+Q	...	Y,25

This is the simplest method to use when you want to exit AutoCAD. If any changes have been made to the drawings since the last *Save, Exit* invokes a warning box asking if you want to *Save changes to . . .?* before ending AutoCAD (see Fig. 2-15).

An alternative to using the *Exit* option is to use the standard Windows methods for exiting an application. The two options are (1) select the "**X**" in the extreme upper-right corner of the AutoCAD window, or (2) select the AutoCAD logo in the extreme upper-left corner of the AutoCAD window. Selecting the logo in the upper-left corner produces a pull-down menu allowing you to *Minimize, Maximize,* etc., or to *Close* the window. Using this option is the same as using *Exit*.

AutoCAD Backup Files

When a drawing is saved, AutoCAD creates a file with a .DWG extension. For example, if you name the drawing PART1, using *Save* creates a file named PART1.DWG. The next time you save, AutoCAD makes a new PART1.DWG and renames the old version to PART1.BAK. One .BAK (backup) file is kept automatically by AutoCAD by default. You can disable the automatic backup function by changing the *ISAVEBAK* system variable to 0 or by accessing the *Open and Save* tab in the *Options* dialog box (Fig. 2-16).

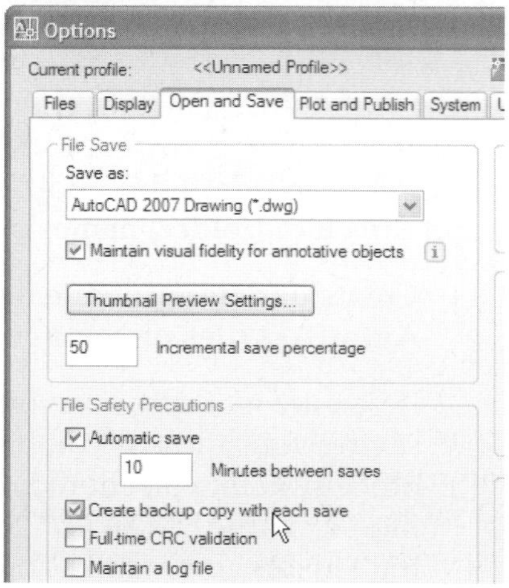

You cannot *Open* a .BAK file. It must be renamed to a .DWG file. Remember that you already have a .DWG file by the same name, so rename the extension <u>and</u> the filename. For example, PART1.BAK could be renamed to PART1OLD.DWG. Use the Windows Explorer, My Computer, or the *Select File* dialog box (use *Open*) with the right-click options to rename the file. Also see "Drawing Recovery."

FIGURE 2-16

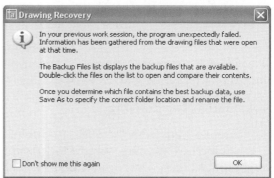

The .BAK files can also be deleted without affecting the .DWG files. The .BAK files accumulate after time, so you may want to periodically delete the unneeded ones to conserve disk space.

Drawing Recovery

In the case of an unexpected or accidental shutdown, such as a power outage, drawings that were open when the shutdown occurred can be recovered. If a shutdown does occur, restart your computer system and begin AutoCAD. A *Drawing Recovery* message automatically appears giving instructions on recovering the drawings that were open before the system shutdown (Fig. 2-17). Click *OK* to bypass the message and begin recovery.

The *Drawing Recovery Manager* also automatically appears and allows you to recover any or all drawings that were open when the shutdown occurred (Fig. 2-18). Highlight each of the .DWG (drawing), .BAK (backup), and SV\$ (temporary backup) files that appear in the *Backup Files* list to determine which has the most current *Last Saved* data. Then simply double-click on the most current file to *Open* the drawing. If you select a .BAK or .SV\$ file, AutoCAD automatically locates the file from its previously stored location, renames it with a .DWG file extension, and relocates the file in the folder where the last .DWG was saved. In this case, you should then use *SaveAs* to rename the file.

Although the *Drawing Recovery Manager* is available at any time by using the *DRAWINGRECOVERY* command or by selecting it from the *File*, *Drawing Utilities* pull-down menu, the manager has no use until an accidental shutdown occurs. Anytime after a shutdown you can invoke the command and recover the drawings.

FIGURE 2-17

FIGURE 2-18

CHAPTER EXERCISES

Start AutoCAD. Use the ACAD.DWT template, then select *Zoom All* from the *View* pull-down menu. If the *Startup* dialog box appears, select *Start from Scratch,* then use the *Imperial* default settings. NOTE: The chapter exercises in this book list file names in UPPERCASE letters for easy recognition. The file names and directory (folder) names on your system may appear as UPPERCASE, Title Case, or lower case.

1. **Create or determine the name of the folder ("working" directory) for opening and saving AutoCAD files on your computer system**

   If you are working in an office or laboratory, a folder (directory) has most likely been created for saving your files. If you have installed AutoCAD yourself and are learning on your home or office system, you should create a folder for saving AutoCAD drawings. It should have a name like "C:\ACAD\DWG," "C:\My Documents\Dwgs," or "C:\Acad\Drawing Files." (HINT: Use the *SaveAs* command. The name of the folder last used for saving files appears at the top of the *Save Drawing As* dialog box.)

2. *Save* **and name a drawing file**

   Draw 2 vertical *Lines*. Select *Save* from the *File* pull-down, from the Standard toolbar, or by another method. (The *Save Drawing As* dialog box appears since a name has not yet been assigned.) Name the drawing "**CH2 VERTICAL**."

3. **Using** *Qsave*

   Draw 2 more vertical *Lines*. Select *Save* again from the menu or by any other method except typing. (Notice that the *Qsave* command appears at the Command line since the drawing has already been named.) Draw 2 more vertical *Lines* (a total of six lines now). Select *Save* again. Do not *Close*.

4. **Start a** *New* **drawing**

   Invoke *New* from the *File* pull-down, the Standard toolbar, or by any other method. If the *Startup* dialog box appears, select **Start from Scratch,** then use the *Imperial* default settings. Draw 2 horizontal *Lines*. Use *Save*. Enter "**CH2 HORIZONTAL**" as the name for the drawing. Draw 2 more horizontal *Lines*, but <u>do not</u> *Save*. Continue to exercise 5.

5. *Close* **the current drawing**

   Use *Close* to close CH2 HORIZONTAL. Notice that AutoCAD first forces you to answer *Yes* or *No* to *Save changes to CH2 HORIZONTAL.DWG?* PICK *Yes* to save the changes.

6. **Using** *SaveAs*

   The CH2 VERTICAL drawing should now be the current drawing. Draw 2 inclined (angled) *Lines* in the CH2 VERTICAL DRAWING. Invoke *SaveAs* to save the drawing under a new name. Enter "**CH2 INCLINED**" as the new name. Notice the current drawing name displayed in the AutoCAD title bar (at the top of the screen) is reset to the new name. Draw 2 more inclined *Lines* and *Save*.

7. **Find and rename a backup file**

   Close all open drawings. Use the *Open* command. When the *Select File* dialog box appears, locate the folder where your drawing files are saved, then enter "*.BAK" in the *File name:* edit box (* is a wildcard that means "all files"). All of the .BAK files should appear. Search for a backup file that was created the last time you saved **CH2 VERTICAL.DWG**, named **CH2 VERTICAL.BAK**. **Right-click** on the file name. Select *Rename* from the shortcut menu and rename the file "**CH2 VERT 2.DWG**." Next, enter *.DWG in the *File name:* edit box to make all the .DWG files reappear. Highlight **CH2 VERT 2.DWG** and the bitmap image should appear in the preview display. You can now *Open* the file if you wish.

C H A P T E R

3

DRAW
COMMAND
CONCEPTS

CHAPTER OBJECTIVES

After completing this chapter you should:

1. be able to understand the formats and types of coordinates;

2. be able to draw lines and circles using mouse input with drawing aids such as *SNAP* and *GRID*;

3. be able to draw lines and circles using Dynamic Input;

4. be able to draw lines and circles using Command Line Input;

5. be able to use Polar Tracking.

AutoCAD OBJECTS

The smallest component of a drawing in AutoCAD is called an <u>object</u> (sometimes referred to as an entity). An example of an object is a *Line,* an *Arc,* or a *Circle* (Fig. 3-1). A rectangle created with the *Line* command would contain four objects.

Draw commands <u>create</u> objects. The draw command names are the same as the object names.

Simple objects are *Point, Line, Arc,* and *Circle.*

Complex objects are shapes such as *Polygon, Rectangle, Ellipse, Polyline,* and *Spline* which are created with one command (Fig. 3-2). Even though they appear to have several segments, they are <u>treated</u> by AutoCAD as one object.

FIGURE 3-1

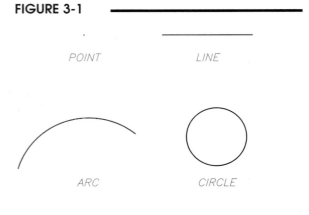

POINT   LINE

ARC   CIRCLE

FIGURE 3-2

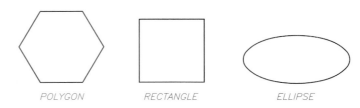

POLYGON   RECTANGLE   ELLIPSE

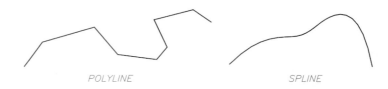

POLYLINE   SPLINE

It is not always apparent whether a shape is composed of one or more objects. However, if you "hover over" an object with the cursor, an object is "highlighted," or shown in a broken line pattern (Fig. 3-3). This highlighting reveals whether the shape is composed of one or several objects.

FIGURE 3-3

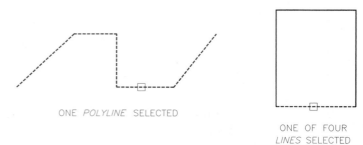

ONE *POLYLINE* SELECTED

ONE OF FOUR
LINES SELECTED

LOCATING THE DRAW COMMANDS

To invoke *Draw* commands, any of these five command entry methods can be used depending on your computer setup.

1. Keyboard   Type the command name, command alias, or accelerator keys at the keyboard.
2. Pull-down menus   Select the command or dialog box from a pull-down menu.
3. Tools (icon buttons)   Select the command or dialog box by PICKing a tool (icon button) from a toolbar, palette, or control panel on the dashboard.
4. Screen (side) menu   Select the command or dialog box from the screen menu (if activated).
5. Digitizing tablet menu   Select the command or dialog box from the tablet menu (if available).

To start a draw command, the *Line* command for example, you can type **LINE** or the command alias, **L**, at the keyboard.

If you want to select a drawing command from a menu, choose from the *Draw* pull-down menu. Options for drawing *Arcs* and *Circles* are found on cascading menus (Fig. 3-4).

FIGURE 3-4

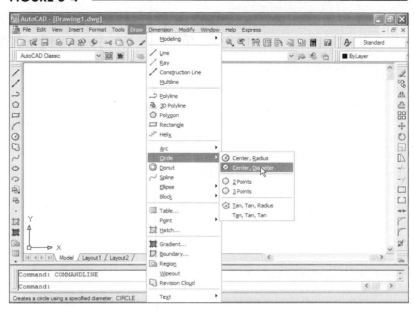

Alternately, if you prefer selecting a tool to activate a draw command, PICK its icon button. If you are using the *AutoCAD Classic* workspace, select the command from the *Draw* toolbar (Fig. 3-5). If you are using the *2D Drafting & Annotation* workspace, select the command from the *2D Draw* control panel on the dashboard (Fig. 3-6). You can also select draw commands from the *Draw Tool Palette* (not shown).

FIGURE 3-5

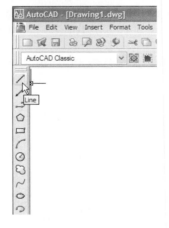

FIGURE 3-6

COORDINATE ENTRY

All drawing commands prompt you to specify points, or locations, in the drawing. For example, the *Line* command prompts you to give the "Specify first point:" and "Specify next point:," expecting you to specify locations for the first and second endpoints of the line. After you specify those points, AutoCAD stores the coordinate values to define the line. A two-dimensional line in AutoCAD is defined and stored in the database as two sets of X, Y, and Z values (with Z values of 0), one for each endpoint.

Coordinate Formats and Types

There are many ways you can specify coordinates, or tell AutoCAD the location of points, when you draw or edit objects. You can also use different types and formats for the coordinates that you enter.

Coordinate Formats

1. Cartesian Format   3,7   Specifies an X and Y value.
2. Polar Format   6<45   Specifies a distance and angle, where 6 is the distance value, the "less than" symbol indicates "angle of," and 45 is the angular value in degrees.

The Cartesian format is useful when AutoCAD prompts you for a point and you know the exact X,Y coordinates of the desired location. The Polar format is useful when you know the specific length and angle of a line or other object you are drawing.

Coordinate Types

1. Absolute Coordinates   3,7 or 6<45   Values are relative to 0,0 (the origin).
2. Relative Coordinates   @3,7 or @6<45   Values are relative to the "last point," where the "@" (at) symbol is interpreted by AutoCAD as the last point specified.

Absolute coordinates are typically used to specify the "first point" of a line or other object since you are concerned with the location of the point with respect to the origin (0,0) in the drawing. Relative coordinates are typically used for the "next point" of a line or other object since the location of the next point is generally given <u>relative</u> to the first point specified. For example, the "next point" might be given as "@6<45", meaning "relative to the last point, 6 units at an angle of 45 degrees."

Typical Combinations

Although you can use either coordinate type in either format, the typical combinations are:

> Absolute Cartesian coordinates (such as 3,7 for the "first point").
> Relative Polar coordinates (such as @6<45 for the "next point").
> Relative Cartesian coordinates (such as @3,7 for the "next point").

Even though you can specify coordinates in any type or format using any of the following three methods, AutoCAD stores the data in the drawing file in absolute Cartesian coordinate format.

Three Methods of Coordinate Input

While drawing or editing lines, circles, and other objects, you have the choice to use the following three methods to tell AutoCAD the location of points. Even though you may be simply picking points on the screen, technically you are entering coordinates.

1. Mouse Input   PICK (press the left <u>mouse</u> button to specify a point on the screen).
2. Dynamic Input   When *DYN* is on, you can <u>key values into the edit boxes</u> near the cursor and press Tab to change boxes.
3. Command Line Input   You can <u>key values</u> in any format <u>at the Command prompt</u>.

Mouse Input

Use mouse input to specify coordinates by simply moving the cursor and pressing the left mouse button to pick points. You can use the mouse to input points at any time regardless of the Status Bar settings; however, you should always use a drawing aid with mouse input, such as *SNAP* and *GRID* or *OSNAP*. With *SNAP* and *GRID* on, you can watch the coordinate display in the lower-left corner of the screen to locate points (see Fig. 3-7).

FIGURE 3-7

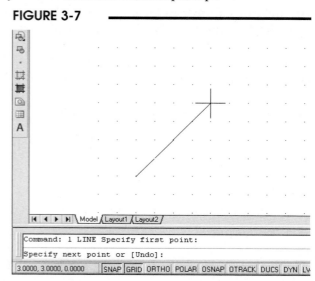

1. Draw a horizontal *Line* of 2 units length starting at 1,1.

Steps	Command Prompt	Perform Action	Comments
1.	Command:	select or type *Line*	use any method
2.	LINE Specify first point:	**PICK** location 1,1	watch coordinate display, lower left
3.	Specify next point or [Undo]:	**PICK** location 3,1	watch coordinate display
4.	Specify next point or [Undo]:	press **ENTER**	completes command

Check your work with the first *Line* as shown in Figure 3-10. Then proceed to construct the other two *Lines* and the *Circle*.

2. Draw a vertical *Line* of 2 units length starting at 4,1.

Steps	Command Prompt	Perform Action	Comments
1.	Command:	select or type *Line*	use any method
2.	LINE Specify first point:	**PICK** location 4,1	watch coordinate display, lower left
3.	Specify next point or [Undo]:	**PICK** location 4,3	watch coordinate display
4.	Specify next point or [Undo]:	press **ENTER**	completes command

3. Draw a diagonal *Line* starting at 5,1.

Steps	Command Prompt	Perform Action	Comments
1.	Command:	select or type *Line*	use any method
2.	LINE Specify first point:	**PICK** location 5,1	watch coordinate display, lower left
3.	Specify next point or [Undo]:	**PICK** location 7,3	watch coordinate display
4.	Specify next point or [Undo]:	press **ENTER**	completes command

4. Draw a _Circle_ of 1 unit radius with the center at 9,2.

Steps	Command Prompt	Perform Action	Comments
1.	Command:	select or type _Circle_	use any method
2.	CIRCLE Specify center point...	**PICK** location 9,2	watch coordinate display
3.	Specify radius of circle...	**PICK** location 10,2	press F6 to change coordinate display to Cartesian coordinates if necessary; completes command

Check your work with the objects shown in Figure 3-10. Remember that when you use mouse input you should use _SNAP_ and _GRID_, _OSNAP_, or other appropriate drawing aids to ensure you pick exact points.

Using Dynamic Input to Draw Lines and Circles

Turn on Dynamic Input by depressing the _DYN_ button on the Status Bar (bottom of AutoCAD screen). For this exercise, _DYN_ and _MODEL_ should be on (depressed) and all other settings (_SNAP_, _GRID_, _ORTHO_, _POLAR_, _OSNAP_, _OTRACK_, _DUCS_, and _LWT_) should be off (buttons appear protruding).

Using the default setting for Dynamic Input, the "first point" specified is absolute Cartesian format and the "next point:" is relative polar format.

In the following 4 exercises you will draw another set of three _Lines_ and one _Circle_ located directly above the previous set as shown in Figure 3-11.

FIGURE 3-11

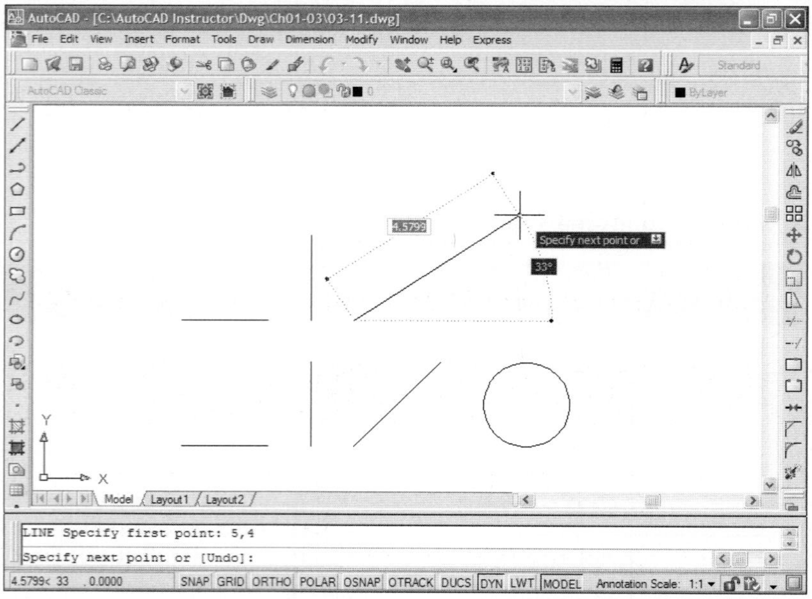

5. Draw a horizontal *Line* of 2 units length starting at 1,4.

Steps	Command Prompt	Perform Action	Comments
1.	Command:	select or type *Line*	use any method
2.	LINE Specify first point:	Enter **1** into first edit box, then press **Tab**	this is the X value for Cartesian format
3.		Enter **4** into second edit box, the press **Enter**	this is the Y value; establishes first endpoint
4.	Specify next point or [Undo]:	Enter **2** into first edit box, then press **Tab**	this is distance value for relative polar format
5.		Enter **0** into second edit box, then press **Enter**	this is angular value; establishes second endpoint
6.	Specify next point or [Undo]:	press **Enter**	completes command

6. Draw a vertical *Line* of 2 units length starting at 4,4.

Steps	Command Prompt	Perform Action	Comments
1.	Command:	select or type *Line*	use any method
2.	LINE Specify first point:	Enter **4** into first edit box, then press **Tab**	this is the X value for Cartesian format
3.		Enter **4** into second edit box, the press **Enter**	this is the Y value; establishes first endpoint
4.	Specify next point or [Undo]:	Enter **2** into first edit box, then press **Tab**	this is distance value for relative polar format
5.		Enter **90** into second edit box, then press **Enter**	this is angular value; establishes second endpoint
6.	Specify next point or [Undo]:	press **Enter**	completes command

7. Draw a diagonal *Line* of 2.8284 units length starting at 5,4.

Steps	Command Prompt	Perform Action	Comments
1.	Command:	select or type *Line*	use any method
2.	LINE Specify first point:	Enter **5** into first edit box, then press **Tab**	this is the X value for Cartesian format
3.		Enter **4** into second edit box, the press **Enter**	this is the Y value; establishes first endpoint
4.	Specify next point or [Undo]:	Enter **2.8284** into first edit box, then press **Tab**	this is distance value for relative polar format

(continued)

Steps	Command Prompt	Perform Action	Comments
5.		Enter **45** into second edit box, then press **Enter**	this is angular value; establishes second endpoint
6.	Specify next point or [Undo]:	press **Enter**	completes command

See Figure 3-11 for an illustration of this exercise.

8. Draw a *Circle* of 1 unit radius with the center at 9,5.

Steps	Command Prompt	Perform Action	Comments
1.	Command:	select or type *Circle*	use any method
2.	CIRCLE Specify first point...	Enter **9** into first edit box, then press **Tab**	this is the X value for Cartesian format
3.		Enter **5** into second edit box, then press **Enter**	this is the Y value; establishes circle center
4.	Specify radius of circle...	Enter **1** into edit box, then press **Enter**	this is distance value; completes command

Using Command Line Input to Draw Lines and Circles

You can use Command line input when *DYN* is on or off; however, for clarity in this exercise it is best to toggle *DYN* off. Although the settings for the other drawing aids do not affect Command line input, in this exercise it is best to set only *MODEL* on (the button appears depressed) and all other settings should be off (buttons appear protruding) (Fig. 3-12).

With Command line input you can key in coordinates at the Command line in any type or format. When using relative format, you must key in the "@" (at) symbol.

FIGURE 3-12

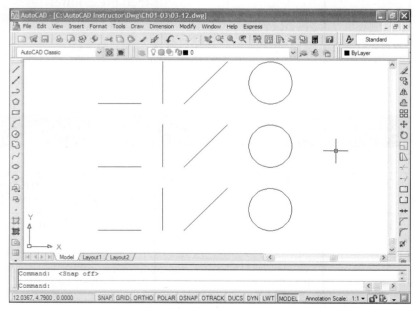

9. Draw a horizontal *Line* of 2 units length starting at 1,7. Use absolute Cartesian coordinates.

Steps	Command Prompt	Perform Action	Comments
1.	Command:	select or type *Line*	use any method
2.	LINE Specify first point:	type **1,7** and press **Enter**	establishes first endpoint
3.	Specify next point or [Undo]:	type **3,7** and press **Enter**	establishes second endpoint
4.	Specify next point or [Undo]:	press **Enter**	completes command

10. Draw a vertical *Line* of 2 units length. Use relative Cartesian coordinates.

Steps	Command Prompt	Perform Action	Comments
1.	Command:	select or type *Line*	use any method
2.	LINE Specify first point:	type **@1,0** and press **Enter**	@ means "last point"
3.	Specify next point or [Undo]:	type **@0,2** and press **Enter**	establishes second endpoint
4.	Specify next point or [Undo]:	press **Enter**	completes command

11. Draw a diagonal *Line* of 2.8284 units length. Use absolute Cartesian and relative polar coordinates.

Steps	Command Prompt	Perform Action	Comments
1.	Command:	select or type *Line*	use any method
2.	LINE Specify first point:	type **5,7** and press **Enter**	establishes first endpoint
3.	Specify next point or [Undo]:	type **@2.8284<45** and press **Enter**	< means "angle of"; establishes second endpoint
4.	Specify next point or [Undo]:	press **Enter**	completes command

12. Draw a *Circle* of 1 unit radius with the center at 9,8. Use absolute Cartesian coordinates.

Steps	Command Prompt	Perform Action	Comments
1.	Command:	select or type *Circle*	use any method
2.	CIRCLE Specify center point...	type **9,8**	establishes center
3.	Specify radius of circle...	type **1**	

Direct Distance Entry and Angle Override

When prompted for a "next point," instead of using relative polar values preceded by the "@" (at) symbol to specify a distance and an angle, you can key in a distance-only value (called Direct Distance Entry) or key in an angle-only value (Angle Override). With these features, you can key in one value (either distance or angle) and specify the other value using mouse input. This capability represents an older, less graphic version of Dynamic Input. Both Direct Distance Entry and Angle Override can be used with Command Line Input and with Dynamic Input.

Direct Distance Entry
Direct Distance Entry allows you to directly key in a distance value. Use Direct Distance Entry in response to the "next point:" prompt. With Direct Distance Entry, you point the mouse so the "rubber-band" line indicates the angle of the desired line, then simply key in the desired distance value (length) at the Command line in response to the "next point:" prompt. For example, if you wanted to draw a horizontal line of 7.5 units, you would first turn on *POLAR*, point the cursor in a horizontal direction, key in "7.5," and press Enter. Direct Distance Entry can be used with *DYN* on or off.

Direct Distance Entry is best used in conjunction with Polar Tracking (turn on *POLAR* on the Status Bar). Polar Tracking assists you in creating a "rubberband" line that is horizontal, vertical, or at a regular angular increment you specify such as 30 or 45 degrees. Indicating the angle of the line with the mouse without using *POLAR* or *ORTHO* will not likely result in a precise angle. See "Polar Tracking and Polar Snap."

Angle Override
If you want to specify that a line is drawn at a specific angle, but specify the line length using mouse input, use Angle Override. At the "next point" prompt, key in the "<" (less than) symbol and an angular value, such as "<30", then press Enter. AutoCAD responds with "Angle Override: 30." The rubberband line is then constrained to 30 degrees but allows you to pick a point to specify the length. Angle Override can be used with Command Line Input and with Dynamic Input.

Dynamic Input Settings

The default setting for Dynamic Input is absolute Cartesian coordinate input for the "first point" and relative polar coordinates for the "second point." To change these settings, right-click on the word *DYN* on the Status Bar and select *Settings....* This action produces the *Dynamic Input* tab of the *Drafting Settings* dialog box (Fig. 3-13). You can also select *Drafting Settings* from the *Tools* pull-down menu or type the command alias *DS* to produce this dialog box.

The *Pointer Input* section allows you to change settings for the type and format of coordinates that are used during Dynamic Input. The *Dimension Input* section controls the appearance of the dimension lines that are shown around the edit boxes during the "next point" specification.

FIGURE 3-13

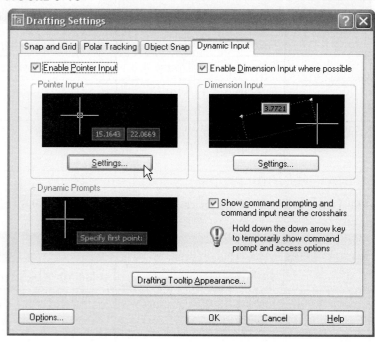

Select *Settings...* in the *Pointer Input* section to produce the *Polar Input Settings* dialog box (Fig. 3-14). Note that *Polar format* and *Relative coordinates* are the default settings for the "next point" specification. You cannot change the format for the "first point" specification, which is always absolute Cartesian format.

FIGURE 3-14

POLAR TRACKING AND POLAR SNAP

Remember that all draw commands prompt you to specify locations, or coordinates, in the drawing. You can indicate these coordinates interactively (using the cursor), entering values (at the keyboard), or using a combination of both. Features in AutoCAD that provide an easy method for specifying coordinate locations interactively are Polar Tracking and Polar Snap. Another feature, Object Snap Tracking, is discussed in Chapter 7. These features help you draw objects at specific angles and in specific relationships to other objects.

Polar Tracking (*POLAR*) helps the rubberband line snap to angular increments such as 45, 30, or 15 degrees. Polar Snap can be used in conjunction to make the rubberband line snap to incremental lengths such as .5 or 1.0. You can toggle these features on and off with the *SNAP* and *POLAR* buttons on the Status Bar or by toggling F9 and F10, respectively. First, you should specify the settings in the *Drafting Settings* dialog box.

Technically, <u>Polar Tracking and Polar Snap would be considered a variation of the Mouse Input method</u> since the settings are determined and specified beforehand, then points on the screen are PICKed using the cursor.

To practice with Polar Tracking and Polar Snap, you should turn *DYN*, *OSNAP*, and *OTRACK* off. Since some of these features generate alignment vectors, it can be difficult to determine which of the features is operating when they are all on.

Polar Tracking

FIGURE 3-15

Polar Tracking (*POLAR*) simplifies drawing *Lines* or performing other operations such as *Move* or *Copy* at specific angle increments. For example, you can specify an increment angle of 30 degrees. Then when Polar Tracking is on, the rubberband line "snaps" to 30-degree increments when the cursor is in close proximity (within the *Aperture* box) to a specified angle. A dotted "tracking" line is displayed and a "tracking tip" appears at the cursor giving the current distance and angle (Fig. 3-15). In this case, it is simple to draw lines at 0, 30, 60, 90, or 120-degree angles and so on.

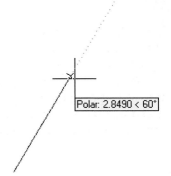

Figure 3-16 displays possible positions for Polar Tracking <u>when a 30-degree angle is specified</u>. Available angle options are 90, 45, 30, 22.5, 18, 15, 10, and 5 degrees, or you can specify any user-defined angle. This is a tremendous aid for drawing lines at typical angles.

FIGURE 3-16

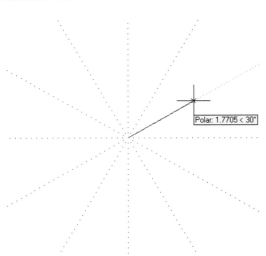

TIP
You can access the settings using the *Polar Tracking* tab of the *Drafting Settings* dialog box (Fig. 3-17). Invoke the dialog box by the following methods:

FIGURE 3-17

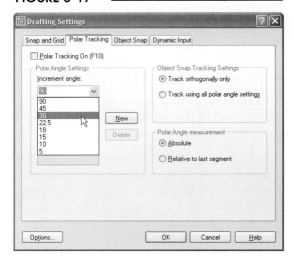

1. Enter *Dsettings* or *DS* at the Command line, then select the *Polar Tracking* tab.
2. Right-click on the word *POLAR* on the Status Bar, then choose *Settings…* from the menu.
3. Select *Drafting Settings* from the *Tools* pull-down menu, then select the *Polar Tracking* tab.

In this dialog box, you can select the *Increment Angle* from the drop-down list (as shown) or specify *Additional Angles*. Use the *New* button to create user-defined angles. Highlighting a user-defined angle and selecting the *Delete* button removes the angle from the list.

The *Object Snap Tracking Settings* are used only with Object Snap (discussed in Chapter 7). When the *Polar Angle Measurement* is set to *Absolute*, the angle reported on the "tracking tip" is an absolute angle (relative to angle 0 of the current coordinate system).

TIP
Polar Tracking with Direct Distance Entry

FIGURE 3-18

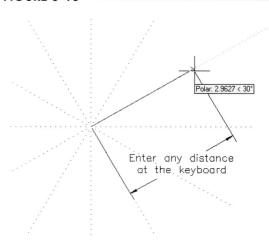

With normal Polar Tracking (Polar Snap is off), the line "snaps" to the set angles but can be drawn to any length (distance). This option is particularly useful in conjunction with Direct Distance Entry since you specify the angular increment in the *Polar Tracking* tab of the *Drafting Settings* dialog box, but indicate the distance by <u>entering values at the Command line</u>. In other words, when drawing a *Line*, move the cursor so it "snaps" to the desired angle, then enter the desired distance at the keyboard (Fig. 3-18). See previous discussion, "Direct Distance Entry and Angle Override."

Polar Tracking Override

When Polar Tracking is on, you can override the previously specified angle increment and draw to another specific angle. The new angle is valid only for one point specification. To enter a polar override angle, enter the left angle bracket (<) and an angle value whenever a command asks you to specify a point. The following command prompt sequence shows a 12-degree override entered during a *Line* command.

```
Command: line
Specify first point: PICK
Specify next point or [Undo]: <12
Angle Override:  12
Specify next point or [Undo]: PICK
```

Polar Tracking Override can be used with Command Line Input or Dynamic Input.

Polar Snap

<u>Polar Tracking</u> makes the rubberband line "snap" to <u>angular increments</u>, whereas <u>Polar Snap</u> makes the rubberband line "snap" to <u>distance increments</u> that you specify. For example, setting the distance increment to 2 allows you to draw lines at intervals of 2, 4, 6, 8, and so on. Therefore, Polar Tracking with Polar Snap allows you to draw at specific angular <u>and</u> distance increments (Fig. 3-19). Using Polar Tracking with Polar Snap off, as described in the previous section, allows you to draw at specified angles but not at any specific distance intervals.

FIGURE 3-19

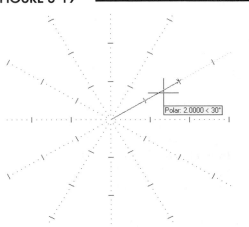

<u>Only one type of Snap, Polar Snap, or Grid Snap, can be used at any one time</u>. Toggle either *Polar Snap* or *Grid Snap* on by setting the radio button in the *Snap and Grid* tab of the *Drafting Settings* dialog box (Fig. 3-20, lower left). Alternately, right-click on the word *SNAP* at the Status Bar and select either *Polar Snap On* or *Grid Snap On* from the shortcut menu (see Fig. 3-21). Set the Polar Snap distance increment in the *Polar Distance* edit box (Fig. 3-20, left-center).

FIGURE 3-20

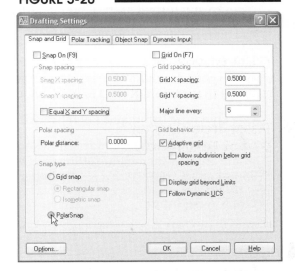

To utilize Polar Snap, both *SNAP* (F9) and *POLAR* (F10) must be turned on, Polar Snap must be on, and a Polar Distance value must be specified in the *Drafting Settings* dialog box. As a check, settings should be as shown in Figure 3-21 (*SNAP* and *POLAR* are recessed), with the addition of some Polar Distance setting in the dialog box. *SNAP* turns on or off whichever of the two snap types is active, Grid Snap or Polar Snap. For example, if Polar Snap is the active snap type, toggling F9 turns only Polar Snap off and on.

FIGURE 3-21

CHAPTER EXERCISES

1. **Start a *New* Drawing**

   Start a *New* drawing. Use the **ACAD.DWT** template, then select *Zoom All* from the *View* pull-down menu. If the *Startup* dialog box appears, select **Start from Scratch** and the **Imperial** default settings. Next, use the *Save* command and save the drawing as **CH3EX1.** Remember to *Save* often as you complete the follow-ing exercises. The completed exer-cise should look like Figure 3-22.

   FIGURE 3-22

   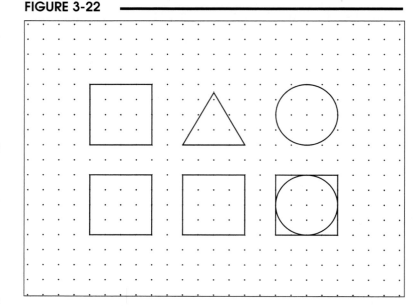

2. **Use Mouse Input**

   Draw a square with sides of 2 unit length. Locate the lower-left corner of the square at **2,2**. Use the *Line* command with mouse input. (HINT: Turn on *SNAP*, *GRID*, and *POLAR*.)

3. **Use Dynamic Input**

   Draw another square with 2 unit sides using the *Line* command. Turn on *DYN*. Enter the correct values into the edit boxes.

4. **Use Command Line Input**

   Draw a third square (with 2 unit sides) using the *Line* command. Turn off *DYN*. Enter coordi-nates in any format at the Command line.

5. **Use Direct Distance Entry**

   Draw a fourth square (with 2 unit sides) beginning at the lower-left corner of **2,5**. Ensure that *DYN* is off. Complete the square drawing *Lines* using direct distance coordinate entry. Don't forget to turn on *ORTHO* or *POLAR*.

6. **Use Dynamic Input and Polar Tracking**

   Draw an equilateral triangle with sides of 2 units. Turn on *POLAR* and *DYN*. Make the appro-priate angular increment settings in *Drafting Settings* dialog box. Begin by locating the lower-left corner at **5,5**. (HINT: An equilateral triangle has interior angles of 60 degrees.)

7. **Use Mouse Input**

   Turn on *SNAP* and *GRID*. Use mouse input to draw a *Circle* with a 1 unit <u>radius</u>. Locate the center at **9,6**.

8. **Use any method**

   Draw another *Circle* with a 2 unit <u>diameter</u>. Using any method, locate the center 3 units below the previous *Circle*.

9. ***Save* your drawing**

   Use *Save*. Compare your results with Figure 3-22. When you are finished, *Close* the drawing.

10. In the next series of steps, you will create the Stamped Plate shown in Figure 3-23. Begin a *New* drawing. Use the **ACAD.DWT** template, then select *Zoom All* from the *View* pull-down menu. If the *Startup* dialog box appears, select *Start from Scratch* and the *Imperial* default settings. Next, use the *Save* command and assign the name **CH3EX2.** Remember to *Save* often as you work.

FIGURE 3-23

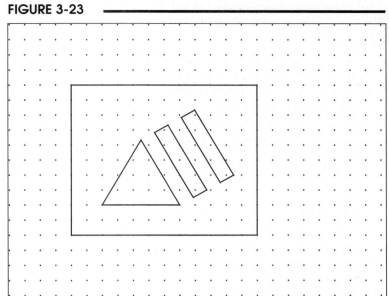

11. First, you will create the equilateral triangle as shown in Figure 3-24. All sides are equal and all angles are equal. To create the shape easily, you should use Polar Tracking with Dynamic Input. Access the *Polar Tracking* tab of the *Drafting Settings* dialog box and set the *Increment Angle* to **30**. (HINT: Right-click on the word *POLAR* and select *Settings* from the shortcut menu.) On the Status Bar, make sure *POLAR* and *DYN* are on (appear recessed), but not *SNAP*, *ORTHO*, *OSNAP*, *OTRACK* or *DUCS*. *GRID* is optional.

FIGURE 3-24

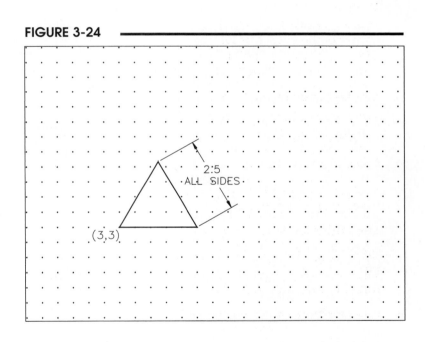

12. Use the *Line* command to create the equilateral triangle. Start at position **3,3** by entering absolute coordinates. All sides should be **2.5** units, so enter the distance values at the keyboard as you position the mouse in the desired direction for each line. The drawing at this point should look like that in Figure 3-24 (not including the notation). *Save* the drawing but do not *Close* it.

13.   Next, you should create the outside shape (Fig. 3-25). Since the dimensions are all at even unit intervals, use Grid Snap to create the rectangle. (HINT: Right-click on the word *SNAP* and select *Grid Snap On*.) At the Status Bar, make sure that *SNAP* and *POLAR* are on. Use the *Line* command to create the rectangular shape starting at point **2,2** (enter absolute coordinates).

FIGURE 3-25

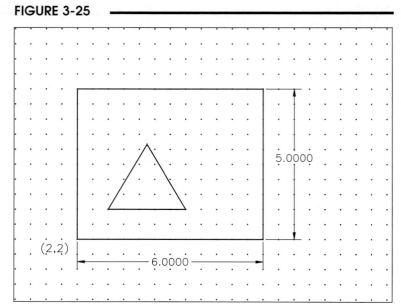

14.   The two inside shapes are most easily created using Polar Tracking in combination with Polar Snap. Access the *Grid and Snap* tab of the *Drafting Settings* dialog box. (HINT: Right-click on the word *SNAP* and select *Settings*.) Set the *Snap Type* to *Polar Snap* and set the *Polar Distance* to **.5**. At the Status Bar, make sure that *SNAP, POLAR,* and *DYN* are on. Next, use the *Line* command and create the two shapes as shown in Figure 3-26. Specify the starting positions for each shape (**5.93,3.25** and **6.80,3.75**) by entering absolute coordinates. Each rectangular shape is .5 units wide.

FIGURE 3-26

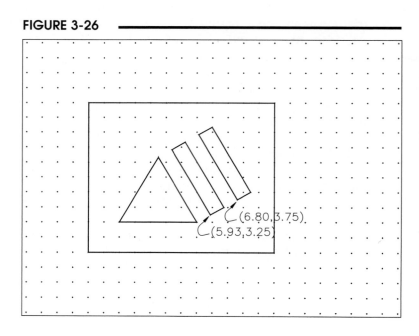

15.   When you are finished, *Save* the drawing and *Exit* AutoCAD.

4

SELECTION SETS

CHAPTER OBJECTIVES

After completing this chapter you should:

1. know that *Modify* commands and many other commands require you to select objects;

2. be able to create a selection set using each of the specification methods;

3. be able to *Erase* objects from the drawing;

4. be able to *Move* and *Copy* objects from one location to another.

MODIFY COMMAND CONCEPTS

Draw commands create objects. Modify commands <u>change existing objects</u> or <u>use existing objects to create new ones</u>. Examples are *Copy* an existing *Circle*, *Move* a *Line*, or *Erase* a *Circle*.

Since all of the Modify commands use or modify <u>existing</u> objects, you must first select the objects that you want to act on. The process of selecting the objects you want to use is called building a <u>selection set</u>. For example, if you want to *Copy, Erase,* or *Move* several objects in the drawing, you must first select the set of objects that you want to act on.

Remember that any of these five command entry methods (depending on your setup) can be used to invoke Modify commands.

1. Keyboard   Type the command name or command alias at the keyboard.
2. Pull-down menus   Select from the *Modify* pull-down menu.
3. Tools (icon buttons) Select from the *Modify* or *Modify II* toolbar, *Modify* tool palette, or *2D Draw* control panel on the dashboard.
4. Screen (side) menu   Select from the *MODIFY1* or *MODIFY2* screen menu (if active=ated).
5. Digitizing tablet menu Select the tablet menu (if available.

FIGURE 4-1

All of the Modify commands will be discussed in detail later, but for now we will focus on how to build selection sets.

Two *Circles* and five *Lines*, as shown in Figure 4-1, will be used to illustrate the selection process in this chapter. In Figure 4-1, no objects are selected.

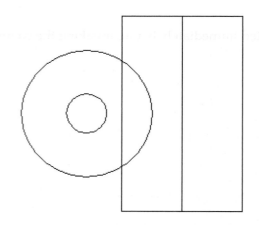

SELECTION SETS

Selection Preview
When you pass the crosshairs or pickbox over an object such as a *Line* or *Circle*, the object becomes thick and dashed (broken) as shown by the vertical *Line* in Figure 4-2. This feature, called selection preview, or sometimes "rollover highlighting," assists you by indicating the objects you are about to select before you actually PICK them. Selection preview may occur both during a modify command at the "Select objects:" prompt (when only the pickbox appears) and also when no commands are in use (when the "crosshairs" appear). This feature, turned on with the default settings, can be turned off or changed if desired.

FIGURE 4-2

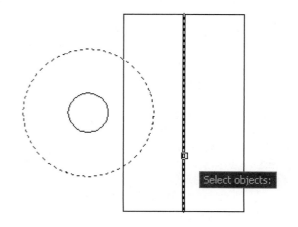

Highlighting
<u>After</u> you PICK an object using the pickbox or any of the methods explained next, the objects become "highlighted," or displayed in a dashed linetype, as shown by the large *Circle* in Figure 4-2. Highlighted (selected) objects become part of the <u>selection set</u>.

Verb/Noun Selection

There are two times you can select objects. You can select objects before using a command, called Noun/Verb (object/command) order or you can select objects during a Modify command when the "Select objects:" prompt appears, called Verb/Noun (command/object) order. It is recommended that you begin learning the selection options by using the Modify command first (command/object), then select objects during the "Select objects:" prompt. (See "Noun/Verb Syntax" for more information.)

Grips

When you select objects before using a command (Noun/Verb selection), blue squares called Grips may appear on the selected objects. Grips offer advanced editing features that are discussed in Chapter 19. When you begin AutoCAD, Grips are most likely turned on by default. It is recommended to <u>turn off Grips</u> until you learn the basic selection options and Modify commands. Turn Grips off by typing in *Grips* at the Command line and changing the setting to *0* (zero).

Selection

No matter which of the five methods you use to invoke a Modify command, you must specify a selection set during the command operation. For example, examine the command syntax that would appear at the Command line when the *Erase* command is used.

    Command: **erase**
    Select objects:

The "Select objects:" prompt is your cue to use any of several methods to PICK the objects to erase. As a matter of fact, most Modify commands begin with the same "Select objects:" prompt (unless you selected immediately before invoking the command). When the "Select objects:" prompt appears, the "crosshairs" disappear and only a small, square pickbox appears at the cursor.

Use the pickbox to select one object at a time. Once an object is selected, it becomes highlighted. The Command line also verifies the selection by responding with "1 found." You can select as many additional objects as you want, and you can use many selection methods as described on the next few pages. Therefore, object selection is a <u>cumulative process</u> in that selected objects remain highlighted while you pick additional objects. AutoCAD continually responds with the "Select objects:" prompt, so press Enter to indicate when you are finished selecting objects and are ready to proceed with the command.

Selection Set Options

The pickbox is the default option, which can automatically be changed to a window or crossing window by PICKing in an open area (PICKing no objects). Since the pickbox, window, and crossing window (sometimes known as the *AUto* option) are the default selection methods, no action is needed on your part to activate these methods. The other methods can be invoked by typing the capitalized letters shown in the option names following.

All the possible selection set options are listed if you enter a "?" (question mark symbol) at the "Select objects:" prompt.

    Command: **move**
    Select objects: **?**
    Invalid selection
    Expects a point or
    Window/Last/Crossing/BOX/ALL/Fence/WPolygon/CPolygon/Add/Remove/Multiple/Previous/
      Undo/AUto/Single/SUboject/Object
    Select objects:

Use any option by typing the indicated uppercase letters and pressing Enter.

pickbox

This default option is used for selecting <u>one object</u> at a time. Locate the pickbox so that an object crosses through it and PICK (Fig. 4-3). You do not have to type or select anything to use this option.

FIGURE 4-3

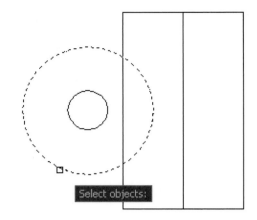

window (*AUto* option)

Only objects <u>completely within</u> the window are selected, not objects that cross through or are outside the window (Fig. 4-4). The window border is a <u>solid linetype</u> rectangular box filled with a transparent blue color.

To use this option, you do not have to type or select any-thing from a menu. Instead, when the pickbox appears, position it in an open area (so that no objects cross through it); then PICK to start a window. If you <u>drag to the right</u>, a window is created. Next, PICK the other diagonal corner.

Window

You can also type the letter *W* at the "Select objects:" prompt to create a selection window as shown in Figure 4-4. In this case, no pickbox appears, and you can PICK the diagonal corners of the window by dragging in either direction.

FIGURE 4-4

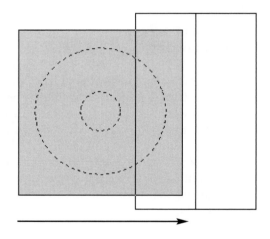

crossing window (*AUto* option)

Note that all objects <u>within and crossing through</u> a crossing window are selected (compare Fig. 4-5 with Fig. 4-4). The crossing window is displayed as a <u>broken linetype</u> rectangu-lar box filled with a transparent green color.

This option is similar to using a window in that when the pickbox appears you can position it in an open area and PICK to start the window. However, <u>drag to the left</u> to form a crossing window. Dragging to the right creates a window instead.

Crossing Window

You can also type the letters *CW* at the "Select objects:" prompt to create a crossing window as shown in Figure 4-5. With this option, no pickbox appears, and you can PICK the diagonal corners by dragging in either direction to form the crossing window.

FIGURE 4-5

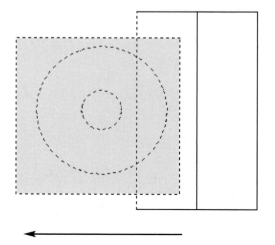

Window Polygon

Type the letters **WP** at the "Select objects:" prompt to use this option. Note that the *Window Polygon* operates like a *Window* in that only objects <u>completely within</u> the window are selected, but the box can be *any* irregular polygonal shape (Fig. 4-6). You can pick <u>any number of corners to form any shape</u> rather than picking just two diagonal corners to form a regular rectangular *Window*.

FIGURE 4-6

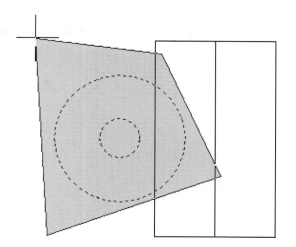

Crossing Polygon

Type the letters **CP** at the "Select objects:" prompt to use this option. The *Crossing Polygon* operates like a *Crossing Window* in that all objects <u>within and crossing through</u> the *Crossing Polygon* are selected; however, the polygon can have any number of corners (Fig. 4-7).

FIGURE 4-7

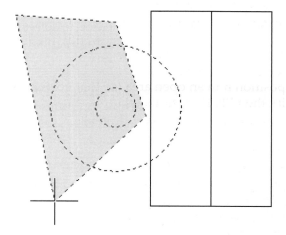

Fence

To invoke this option, type the letter **F** at the "Select objects:" prompt. A *Fence* operates like a <u>crossing line</u>. Any <u>objects crossing</u> the *Fence* are selected. The *Fence* can have any number of segments (Fig. 4-8).

FIGURE 4-8

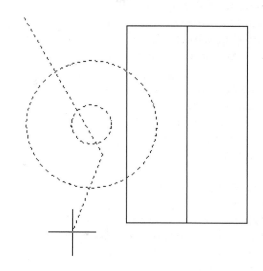

Last

This option automatically finds and selects <u>only</u> the last object <u>created</u>. *Last* does not find the last object modified (with *Move, Stretch,* etc.).

Previous

Previous finds and selects the <u>previous selection set</u>, that is, whatever was selected during the previous command (except after *Erase*). This option allows you to use several editing commands on the same set of objects without having to respecify the set.

ALL

This option selects <u>all objects</u> in the drawing except those on *Frozen* or *Locked* layers (*Layers* are covered in Chapter 11).

BOX

This option is equivalent to the *AUto* window/crossing window option <u>without</u> the pickbox. PICKing diagonal corners from left to right produces a window and PICKing diagonal corners from right to left produces a crossing window (see *AUto*).

Multiple

The *Multiple* option allows selection of objects with the pickbox only; however, the selected objects are not highlighted. Use this method to select very complex objects to save computing time required to change the objects' display to highlighted.

SIngle

This option allows only a <u>single selection</u> using one of the *AUto* methods (pickbox, window, or crossing window), then automatically continues with the command. Therefore, you can select multiple objects (if the window or crossing window is used), but only one selection is allowed. You do not have to press Enter after the selection is made.

Undo

Use *Undo* to cancel the selection of the object(s) most recently added to the selection set.

Remove

Selecting this option causes AutoCAD to <u>switch</u> to the "Remove objects:" mode. Any selection options used from this time on remove objects from the highlighted set.

Add

The *Add* option switches back to the default "Select objects:" mode so additional objects can be added to the selection set.

Shift + Left Button

Holding down the **Shift** key and pressing the left mouse button simultaneously <u>removes</u> objects selected from the highlighted set. This method is generally quicker than, but performs the same action as, *Remove*. The advantage here is that the *Add* mode is in effect unless Shift is held down.

Group

The *Group* option selects groups of objects that were previously specified using the *Group* command. Groups are selection sets to which you can assign a name.

SUobject/Object

3D solids can be composed of several faces, edges, and primitive solids. These options are used for selecting components of 3D solids.

Select

Pull-down Menu	Command (Type)	Alias (Type)	Short-cut	Screen (side) Menu	Tablet Menu
...	*Select*	...	...	...	...

The *Select* command can be used to PICK objects to be saved in the selection set buffer for subsequent use with the *Previous* option. Any of the selection methods can be used to PICK the objects.

Command: **select**
Select objects: **PICK** (use any selection option)
Select objects: **Enter** (completes the selection process)
Command:

The selected objects become unhighlighted when you complete the command by pressing Enter. The objects become highlighted again and are used as the selection set if you use the *Previous* selection option in the <u>next</u> editing command.

NOUN/VERB SYNTAX

An object is the <u>noun</u> and a command is the <u>verb</u>. Noun/Verb syntax order means to pick objects (nouns) first, then use an editing command (verb) second. If you select objects first (at the open Command: prompt) and then immediately choose a Modify command, AutoCAD recognizes the selection set and passes through the "Select objects:" prompt to the next step in the command.

Verb/Noun means to invoke a command and then select objects within the command. For example, if the *Erase* command (verb) is invoked first, AutoCAD then issues the prompt to "Select objects:"; therefore, objects (nouns) are PICKed second. In the previous examples, and with much older versions of AutoCAD, only Verb/Noun syntax order was used.

You can use <u>either</u> order you want (Noun/Verb or Verb/Noun) and AutoCAD automatically understands. If objects are selected first, the selection set is passed to the next editing command used, but if no objects are selected first, the editing command automatically prompts you to "Select objects:".

If you use Noun/Verb order, you are limited to using only the *AUto* options for object selection (pickbox, *Window*, and *Crossing Window*). You can only use the other options (e.g., *Crossing Polygon, Fence, Previous*) if you invoke the desired Modify command first, then select objects when the "Select objects:" prompt appears.

The *PICKFIRST* variable (a very descriptive name) enables Noun/Verb syntax. The default setting is 1 (*On*). If *PICKFIRST* is set to 0 (*Off*), Noun/Verb syntax is disabled and the selection set can be specified only <u>within</u> the editing commands (Verb/Noun).

Setting *PICKFIRST* to 1 provides two options: Noun/Verb and Verb/Noun. You can use <u>either</u> order you want. If objects are selected first, the selection set is passed to the next editing command, but if no objects are selected first, the editing command prompts you to select objects.

SELECTION SETS PRACTICE

NOTE: While learning and practicing with the editing commands, it is suggested that *GRIPS* be turned off. This can be accomplished by typing in **GRIPS** and setting the *GRIPS* variable to a value of **0**. The AutoCAD default has *GRIPS* on (set to 1). *GRIPS* are covered in Chapter 19.

TIP

Begin a *New* drawing to complete the selection set practice exercises. Use the ACAD.DWT template, then select *Zoom All* from the *View* pull-down menu. If the *Startup* dialog box appears, select *Start from Scratch* and the *Imperial* default settings. The *Erase, Move,* and *Copy* commands can be activated by any one of the methods shown in the command tables that follow.

Using *Erase*

Erase is the simplest editing command. *Erase* removes objects from the drawing. The only action required is the selection of objects to be erased.

Erase

Pull-down Menu	Command (Type)	Alias (Type)	Short-cut	Screen (side) Menu	Tablet Menu
Modify *Erase*	*Erase*	*E*	(Edit Mode) *Erase*	*MODIFY1* *Erase*	*V,14*

1. Draw several *Lines* and *Circles*. Practice using the object selection options with the *Erase* command. The following sequence uses the pickbox, *Window*, and *Crossing Window*.

STEPS	COMMAND PROMPT	PERFORM ACTION	COMMENTS
1.	Command:	type *E* and press **Spacebar**	*E* is the alias for *Erase*, Spacebar can be used like Enter
2.	ERASE Select objects:	use pickbox to select one or two objects	objects are highlighted
3.	Select objects:	type *W* to use a ***Window***, then select more objects	objects are highlighted
4.	Select objects:	type *C* to use a ***Crossing Window***, then select objects	objects are highlighted
5.	Select objects:	press **Enter**	objects are erased

2. Draw several more *Lines* and *Circles*. Practice using the *Erase* command with the *AUto Window* and *AUto Crossing Window* options as indicated below.

STEPS	COMMAND PROMPT	PERFORM ACTION	COMMENTS
1.	Command:	select the *Modify* pull-down, then *Erase*	
2.	ERASE Select objects:	use pickbox to **PICK** an open area, drag *Window* to the <u>right</u> to select objects	objects inside *Window* are highlighted
3.	Select objects:	**PICK** an open area, drag *Crossing Window* to the <u>left</u> to select objects	objects inside and crossing through *Window* are highlighted
4.	Select objects:	press **Enter**	objects are erased

Using *Move*

The *Move* command specifically prompts you to (1) select objects, (2) specify a "base point," a point to move <u>from</u>, and (3) specify a "second point of displacement," a point to move <u>to</u>.

Move

Pull-down Menu	Command (Type)	Alias (Type)	Short-cut	Screen (side) Menu	Tablet Menu
Modify *Move*	*Move*	*M*	(Edit Mode) *Move*	*MODIFY2* *Move*	*V,19*

1. Draw a *Circle* and two *Lines*. Use the *Move* command to practice selecting objects and to move one *Line* and the *Circle* as indicated in the following table.

STEPS	COMMAND PROMPT	PERFORM ACTION	COMMENTS
1.	Command:	type *M* and press **Spacebar**	*M* is the command alias for *Move*
2.	MOVE Select objects:	use pickbox to select one *Line* and the *Circle*	objects are highlighted
3.	Select objects:	press **Spacebar** or **Enter**	
4.	Specify base point or displacement:	**PICK** near the *Circle* center	base point is the handle, or where to move <u>from</u>
5.	Specify second point:	**PICK** near the other *Line*	second point is where to move <u>to</u>

2. Use *Move* again to move the *Circle* back to its original position. Select the *Circle* with the *Window* option.

STEPS	COMMAND PROMPT	PERFORM ACTION	COMMENTS
1.	Command:	select the *Modify* pull-down, then *Move*	
2.	MOVE Select objects:	type *W* and press **Spacebar**	select only the circle; object is highlighted
3.	Select objects:	press **Spacebar** or **Enter**	
4.	Specify base point or displacement:	**PICK** near the *Circle* center	base point is the handle, or where to move <u>from</u>
5.	Specify second point:	**PICK** near the original location	second point is where to move <u>to</u>

Using *Copy*

The *Copy* command is similar to *Move* because you are prompted to (1) select objects, (2) specify a "base point," a point to copy <u>from</u>, and (3) specify a "second point of displacement," a point to copy <u>to</u>.

Copy

Pull-down Menu	Command (Type)	Alias (Type)	Short-cut	Screen (side) Menu	Tablet Menu
Modify Copy	*Copy*	*CO* or *CP*	(Edit Mode) *Copy Selection*	*MODIFY1 Copy*	*V,15*

1. Using the *Circle* and 2 *Lines* from the previous exercise, use the *Copy* command to practice selecting objects and to make copies of the objects as indicated in the following table.

STEPS	COMMAND PROMPT	PERFORM ACTION	COMMENTS
1.	Command:	type *CO* and press **Enter** or **Spacebar**	*CO* is the command alias for *Copy*
2.	COPY Select objects:	type *F* and press **Enter**	*F* invokes the *Fence* selection option
3.	Specify first fence point:	**PICK** a point near two *Lines*	starts the "fence"
4.	Specify next fence point:	**PICK** a second point across the two *Lines*	lines are highlighted
5.	Specify next fence point:	Press **Enter**	completes *Fence* option
6.	Select objects:	Press **Enter**	completes object selection
7.	Specify base point or displacement:	**PICK** between the *Lines*	base point is the handle, or where to copy <u>from</u>
8.	Specify second point:	enter **@2<45** and press **Enter**	copies of the lines are created 2 units in distance at 45 degrees from the original location

2. Practice removing objects from the selection set by following the steps given in the table below.

STEPS	COMMAND PROMPT	PERFORM ACTION	COMMENTS
1.	Command:	type *CO* and press **Enter** or **Spacebar**	*CO* is the command alias for *Copy*
2.	COPY Select objects:	**PICK** in an open area near the right of your drawing, drag *Window* to the left to select all objects	all objects within and crossing the *Window* are highlighted

(continued)

STEPS	COMMAND PROMPT	PERFORM ACTION	COMMENTS
3.	Select objects:	hold down **Shift** and **PICK** all highlighted objects except one *Circle*	holding down Shift while PICKing removes objects from the selection set
4.	Select objects:	press **Enter**	only one circle is highlighted
5.	Specify base point or displacement:	**PICK** near the circle's center	base point is the handle, or where to copy <u>from</u>
6.	Specify second point:	turn on *ORTHO*, move the cursor to the right, type **3** and press **Enter**	a copy of the circle is created 3 units to the right of the original circle

CHAPTER EXERCISES

Open drawing `CH3EX1` that you created in Chapter 3 Exercises. Turn off *SNAP* (**F9**) to make object selection easier.

1. **Use the pickbox to select objects**

   Invoke the *Erase* command by any method. Select the lower-left square with the pickbox (Fig. 4-9, highlighted). Each *Line* must be selected individually. Press **Enter** to complete *Erase*. Then use the *Oops* command to unerase the square. (Type *Oops*.)

   FIGURE 4-9 ——————————

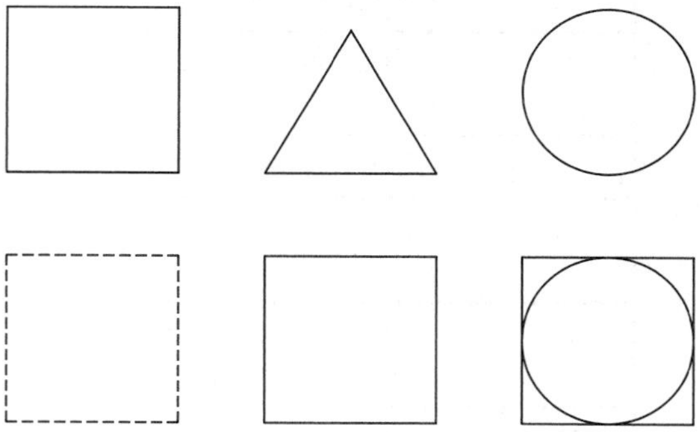

2. **Use the *AUto Window* and *AUto Crossing Window***

   Invoke *Erase*. Select the center square on the bottom row with the *AUto Window* and select the equilateral triangle with the *AUto Crossing Window*. Press **Enter** to complete the *Erase* as shown in Figure 4-10. Use *Oops* to bring back the objects.

   FIGURE 4-10 ——————————

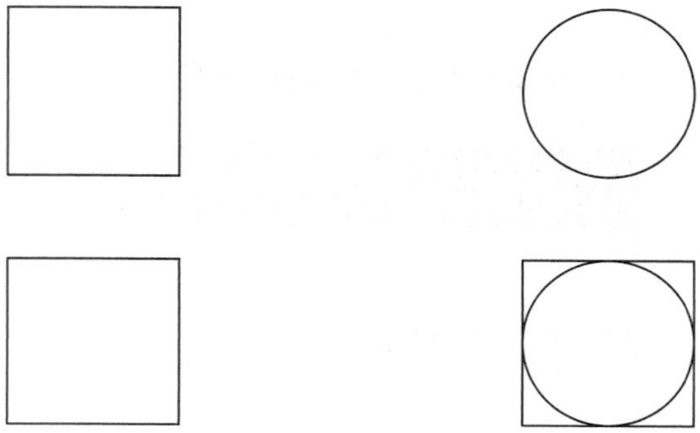

3. **Use the *Fence* selection option**

   Invoke *Erase* again. Use the **Fence** option to select all the vertical *Lines* and the *Circle* from the squares on the bottom row. Complete the *Erase* (see Fig. 4-11). Use *Oops* to unerase.

FIGURE 4-11

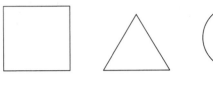

4. **Use the *ALL* option and deselect**

   Use *Erase*. Select all the objects with **ALL**. Remove the four *Lines* (shown highlighted in Fig. 4-12) from the selection set by pressing **Shift** while PICKing. Complete the *Erase* to leave only the four *Lines*. Finally, use *Oops*.

FIGURE 4-12

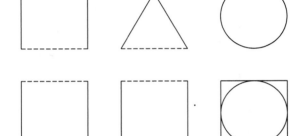

5. **Use Noun/Verb selection**

   <u>Before</u> invoking *Erase,* use the pickbox or *AUto Window* to select the triangle. (Make sure no other commands are in use.) <u>Then</u> invoke *Erase*. The triangle should disappear. Retrieve the triangle with *Oops*.

FIGURE 4-13

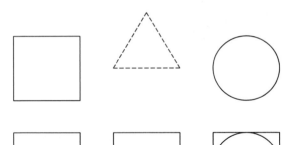

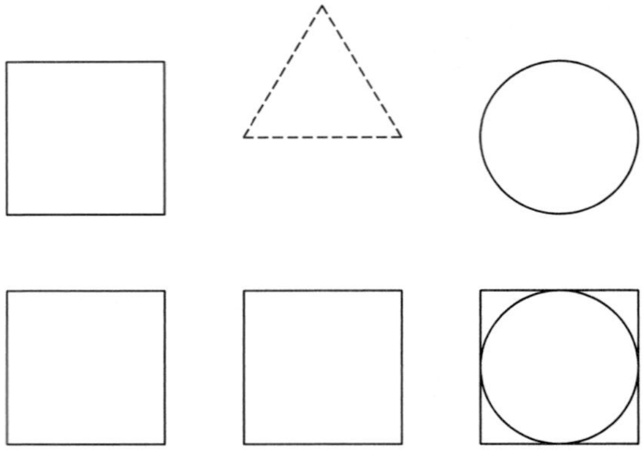

6. **Use *Move* with *Wpolygon***

   Invoke the **Move** command by any method. Use the **WP** option (*Window Polygon*) to select only the *Lines* comprising the triangle. Turn on *SNAP* and PICK the lower-left corner as the "Base point:". *Move* the triangle up 1 unit. (See Fig. 4-13.)

7. **Use *Previous* with *Move***

   Invoke *Move* again. At the "Select objects:" prompt, type *P* for *Previous*. The triangle should highlight. Using the same base point, move the triangle back to its original position.

8. ***Exit*** AutoCAD and do <u>not</u> save changes.

HELPFUL COMMANDS

CHAPTER OBJECTIVES

After completing this chapter you should:

1. be able to find *Help* for any command or system variable;

2. be able to use *Oops* to unerase objects;

3. be able to use *U* to undo one command or use *Undo* to undo multiple commands;

4. be able to *Redo* commands that were undone;

5. be able to regenerate the drawing with *Regen*.

CONCEPTS

There are several commands that do not draw or edit objects in AutoCAD, but are intended to assist you in using AutoCAD. These commands are used by experienced AutoCAD users and are particularly helpful to the beginner. The commands, as a group, are not located in any one menu, but are scattered throughout several menus.

Help

Pull-down Menu	Command (Type)	Alias (Type)	Short-cut	Screen (side) Menu	Tablet Menu
Help *Help*	*Help*	*?*	*F1*	*HELP* *Help*	*Y,7*

Help gives you an explanation for any AutoCAD command or system variable as well as help for using the menus and toolbars. *Help* displays a window that gives a variety of methods for finding the information that you need. There is even help for using *Help*!

Help can be used two ways: (1) entered as a command at the open Command: prompt or (2) used transparently while a command is currently in use.

1.   If the *Help* command is entered at an open Command: prompt (when no other commands are in use), the *Help* window appears (Fig. 5-1).

 2.   When *Help* is used transparently (when a command is in use), it is context sensitive; that is, help on the current command is given automatically. For example, Figure 5-2 displays the window that appears if *Help* is invoked during the *Line* command. (If <u>typing</u> a transparent command, an apostrophe (') symbol is typed as a prefix to the command, such as, '*HELP* or '*?*. If you PICK *Help* from the menus or press **F1**, it is automatically transparent.)

FIGURE 5-1

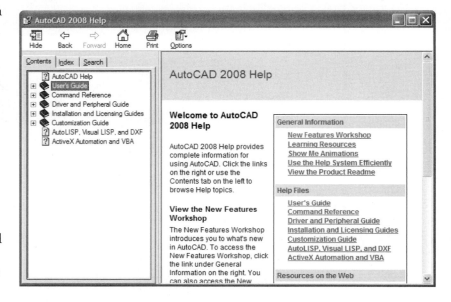

FIGURE 5-2

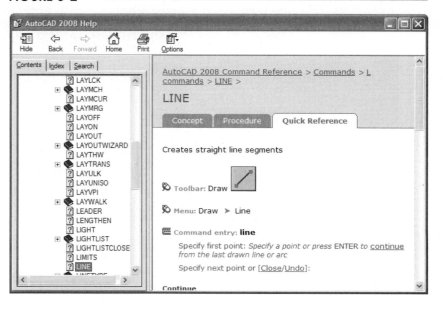

Oops

Pull-down Menu	Command (Type)	Alias (Type)	Short-cut	Screen (side) Menu	Tablet Menu
...	Oops	...	...	MODIFY1 Erase Oops:	...

The *Oops* command unerases whatever was erased with the <u>last</u> *Erase* command. *Oops* does not have to be used immediately after the *Erase,* but can be used at <u>any time after</u> the *Erase.* *Oops* is typically used after an accidental erase. However, *Erase* could be used intentionally to remove something from the screen temporarily to simplify some other action. For example, you can *Erase* a *Line* to simplify PICKing a group of other objects to *Move* or *Copy,* and then use *Oops* to restore the erased *Line.*

Oops can be used to restore the original set of objects after the *Block* or *Wblock* command is used to combine many objects into one object (explained in Chapter 20).

The *Oops* command is available only from the side menu; there is no icon button or option available in the pull-down menu. *Oops* appears on the side menu only <u>if</u> the *Erase* command is typed or selected. Otherwise, *Oops* must be typed.

U

Pull-down Menu	Command (Type)	Alias (Type)	Short-cut	Screen (side) Menu	Tablet Menu
Edit Undo	U	...	Crtl+Z or (Default Menu) Undo	EDIT Undo	T,12

The *U* command undoes only the <u>last</u> command. *U* means "undo one command." If used after *Erase,* it unerases whatever was just erased. If used after *Line,* it undoes the group of lines drawn with the last *Line* command. Both *U* and *Undo* do not undo inquiry commands (like *Help*), the *Plot* command, or commands that cause a write-to-disk, such as *Save.*

If you type the letter *U,* select the icon button, or select **Undo** from the pull-down menu or screen (side) menu, only the last command is undone. Typing **Undo** invokes the full *Undo* command. The *Undo* command is explained next.

Undo

Pull-down Menu	Command (Type)	Alias (Type)	Short-cut	Screen (side) Menu	Tablet Menu
...	Undo	...	...	ASSIST Undo	...

The full *Undo* command (as opposed to the *U* command) can be typed or can be selected using the *Undo* button drop-down list. This command has the same effect as the *U* command in that it undoes the previous command(s). However, the *Undo* command allows you to <u>undo multiple commands</u> in reverse chronological order.

Undo includes a drop-down list appearing with the *Undo* icon button. This feature allows you to select a range of commands to undo. Bring your pointer down to the last command you want to undo and select (Fig. 5-3).

FIGURE 5-3

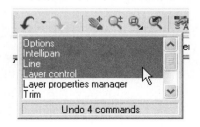

Redo

Pull-down Menu	Command (Type)	Alias (Type)	Short-cut	Screen (side) Menu	Tablet Menu
Edit Redo	Redo	...	Crtl+Y or (Default Menu) Redo	EDIT Redo	U,12

The *Redo* command undoes an *Undo*. *Redo* must be used as the <u>next</u> command after the *Undo*. The result of *Redo* is as if *Undo* was never used. *Redo* automatically reverses the action of the previous *Undo*, even when multiple commands were undone. Remember, *U* or *Undo* can be used at any time, but *Redo* has no effect unless used immediately after *U* or *Undo*.

Mredo

Pull-down Menu	Command (Type)	Alias (Type)	Short-cut	Screen (side) Menu	Tablet Menu
...	Mredo	...	...	...	...

The *Redo* command reverses the total action of the last *Undo* command. In contrast, *Mredo* (multiple redo) allows you to select how many of the commands, in reverse chronological order, you want to redo.

Mredo is useful only when the *Undo* command or drop-down list was just used (as the last command) and multiple commands were undone. Then you can immediately use *Mredo* to reverse the action of any number of the previous *Undos*. This feature is available (the icon is enabled) only if *Undo* was just used.

You can achieve a *Mredo* using the *Redo* drop-down list to select from the list of previously undone commands to reverse the action of the *Undo* (Fig. 5-74). Remember, the list includes only commands that were undone during the previous *Undo* command.

FIGURE 5-4

Typing *Mredo* yields the following prompt:

Command: **mredo**
Enter number of actions or [All/Last]:

Enter a value for the number of commands you want to redo. *All* will redo all of the undos (all the undos shown in the *Redo* drop-down list), the same as using the *Redo* command. The *Last* option generally has the same result as the *All* option; however, when the *U* command was used multiple times in succession, the *Last* option undoes only the last *U*.

Regen

Pull-down Menu	Command (Type)	Alias (Type)	Short-cut	Screen (side) Menu	Tablet Menu
View Regen	Regen	RE	...	VIEW 1 Regen	J,1

The *Regen* command reads the database and redisplays the drawing accordingly. A *Regen* is caused by some commands automatically. Occasionally, the *Regen* command is required to update the drawing to display the latest changes made to some system variables. *Regenall* is used to regenerate all viewports when several viewports are being used.

CHAPTER EXERCISES

Start AutoCAD. If the *Startup* dialog box appears, select *Start from Scratch*, choose *Imperial* as the default setting, and click the *OK* button. Alternately, select the **ACAD.DWT** template, then *Zoom All*. Complete the following exercises.

1. *Help*

   Use *Help* by any method to find information on the following commands. Use the *Contents* tab and select the *Command Reference* to locate information on each command. (Use *Help* at the open Command: prompt, not during a command in use.) Read the text screen for each command.

   New, Open, Save, SaveAs
   Oops, Undo, U

2. **Context-sensitive** *Help*

   Invoke each of the commands listed below. When you see the first prompt in each command, enter *'Help* or *'?* (transparently) or select *Help* from the menus or Standard toolbar. Read the explanation for each prompt. Select a hypertext item in each screen (underlined).

   Line, Arc, Circle, Point

3. *Oops*

   Draw 3 vertical **Lines**. **Erase** one line; then use *Oops* to restore it. Next, **Erase** two *Lines*, each with a separate use of the *Erase* command. Use *Oops*. Only the last *Line* is restored. **Erase** the remaining two *Lines*, but select both with a *Window*. Now use *Oops* to restore both *Lines* (since they were *Erased* at the same time).

4. **Delayed** *Oops*

   Oops can be used at any time, not only immediately after the *Erase*. Draw several horizontal **Lines** near the bottom of the screen. Draw a **Circle** on the **Lines**. Then **Erase** the *Circle*. Use *Move*, select the *Lines* with a *Window*, and displace the *Lines* to another location above. Now use *Oops* to make the *Circle* reappear.

5. *U*

   Press the letter *U* (make sure no other commands are in use). The *Circle* should disappear (*U* undoes the last command—*Oops*). Do this repeatedly to *Undo* one command at a time until the *Circle* and *Lines* are in their original position (when you first created them).

6. *Undo*

   Use the *Undo* command and select the *Back* option. Answer *Yes* to the warning message. This action should *Undo* everything.

   Draw a vertical **Line**. Next, draw a square with four **Line** segments (all drawn in the same *Line* command). Finally, draw a second vertical **Line**. **Erase** the first *Line*.

Now type **Undo** and enter a value of **3**. You should have only one (the first) *Line* remaining. *Undo* reversed the following three commands:

Erase	The first vertical *Line* was unerased.
Line	The second vertical *Line* was removed.
Line	The four *Lines* comprising the square were removed.

7. **Multiple Undo**

Draw two *Circles* and two more *Lines* so your drawing shows a total of two *Circles* and three *Lines*. Now use the *Multiple Undo* drop-down list. The list should contain *Line, Line, Circle, Circle, Line*. Highlight the first two lines and the two circles in the list, then left-click. Only one *Line* should remain in your drawing.

8. **Multiple Redo**

Select the *Multiple Redo* drop-down list. The list should contain *Circle, Circle, Line, Line*. Highlight the first two *Circles* in the list, then left-click. The two *Circles* should reappear in your drawing, now showing a total of two *Circles* and one *Line*.

9. **Exit** AutoCAD and answer **No** to "Save Changes to…?"

6

BASIC DRAWING SETUP

CHAPTER OBJECTIVES

After completing this chapter you should:

1. know the basic steps for setting up a drawing;

2. know how to use the *Start from Scratch* option and the *Quick Setup* and *Advanced Setup* wizards that appear in the *Startup* and *Create New Drawing* dialog boxes;

3. know how to use the *Select template* dialog box;

4. be able to specify the desired *Units*, *Angles* format, and *Precision* for the drawing;

5. be able to specify the drawing *Limits*;

6. know how to specify the *Snap* and *Grid* increments;

7. understand the basic function and use of a *Layout*.

STEPS FOR BASIC DRAWING SETUP

Assuming the general configuration (dimensions and proportions) of the geometry to be created is known, the following steps are suggested for setting up a drawing:

1. Determine and set the *Units* that are to be used.
2. Determine and set the drawing *Limits;* then *Zoom All.*
3. Set an appropriate *Snap* type and increment.
4. Set an appropriate *Grid* value to be used.

These additional steps for drawing setup are discussed also in Chapter 12, Advanced Drawing Setup.

5. Change the *LTSCALE* value based on the new *Limits.*
6. Create the desired *Layers* and assign appropriate *linetype* and *color* settings.
7. Create desired *Text Styles* (optional).
8. Create desired *Dimension Styles* (optional).
9. Activate a *Layout* tab, set it up for the plot or print device and paper size, and create a viewport (if not already existing).
10. Create or insert a title block and border in the layout.

When you start AutoCAD or use the *New* command to begin a new drawing, the *Startup, Create New Drawing,* or *Select Template* dialog box appears. These tools make available three options for setting up a new drawing. The three options represent three levels of automation/preparation for drawing setup. The *Start from Scratch* and *Use a Wizard* options are described in this chapter. *Use a Template* and creating template drawings are discussed in Chapter 12, Advanced Drawing Setup.

The *Start from Scratch* option requires you to step through each of the individual commands listed above to set up a drawing to your specifications. The *Use a Wizard* option provides two wizards, the *Quick Setup* and the *Advanced Setup* wizard. These two wizards lead you through the first two steps listed above.

STARTUP OPTIONS

When you start AutoCAD or use the *New* command, you are presented with one of three options, depending on how your system has been configured:

1. the *Startup* or *Create New Drawing* dialog box appears;
2. the *Select Template* dialog box appears; or
3. a new drawing starts with the *Qnew* template specified in the *Options* dialog box.

To control which of these actions occurs, make the following settings:

For method 1.   Set the *STARTUP* system variable to 1.
For method 2.   Set the *STARTUP* system variable to 0 and ensure the *Default Template File Name for QNEW* is set to *None.*
For method 3.   Set the *STARTUP* system variable to 0 and set the *Default Template File Name for QNEW* to the desired template file.

See "AutoCAD File Commands," "*New,*" and "*Qnew*" in Chapter 2 for more information on these settings.

DRAWING SETUP OPTIONS

There are three general options to set up a new drawing based on which of the three startup options you have configured. They are:

FIGURE 6-1

1. *Start from Scratch;*
2. *Use a Template* or the *Select Template* dialog box; and
3. *Use a Wizard.*

All three methods, *Start From Scratch, Use a Template*, and *Use a Wizard*, are available in the *Startup* or *Create New Drawing* dialog boxes (Fig. 6-1). The *Select Template* dialog (Fig. 6-2) is essentially the same as the *Use a Template* option in the *Startup* or *Create New Drawing* dialog box.

However, no matter which of the three methods you select, you can accomplish essentially the same setup to begin a drawing. Despite all the methods AutoCAD supplies, all of these startup options and drawing setup options generally start the drawing session with one of two setups, either the ACAD.DWT (inch) or the ACADISO.DWT (metric) drawing template. In fact, using *Start from Scratch* or selecting the defaults in either setup *Wizard* from any source, or selecting either of these templates by any method produces the same result—beginning with the ACAD.DWT (inch) or the ACADISO.DWT (metric) drawing template! The template used by the *Wizards* is determined by the *MEASUREINIT* system variable setting (0 for ACAD. DWT, 1 for ACADISO.DWT).

Start from Scratch

The *Start from Scratch* option is available in the *Startup* or *Create New Drawing* dialog box (see Fig. 6-1). Use this option to begin a drawing with basic drawing settings, then determine your own system variable settings using the individual setup commands such as *Units, Limits, Snap,* and *Grid.*

Imperial (feet and inches)

Use the *Imperial (feet and inches)* option if you want to begin with the traditional inch-based AutoCAD default drawing settings, such as *Limits* settings of 12 x 9. This option normally causes AutoCAD to use the ACAD.DWT template drawing. See the "Table of *Start from Scratch, Imperial* Settings (ACAD.DWT)." The same setup can be accomplished by any of the following methods.

> Select the *Start from Scratch, Imperial* option in the *Startup* or *Create New Drawing* dialog box.
> Select the ACAD.DWT by any *Template* method.
> Set the *STARTUP* system variable to 0, then set the *Default Template File Name for Qnew* in the *Files* tab of the *Options* dialog box to ACAD.DWT.
> Set the *MEASUREINIT* system variable to 0, then cancel the *Startup* dialog box that appears when AutoCAD starts.

Metric

Use the *Metric* option for setting up a drawing for use with metric units. This option causes AutoCAD to use the ACADISO.DWT template drawing. The drawing has *Limits* settings of 420 x 279, equal to a metric A3 sheet measured in mm. See the "Table of *Start from Scratch, Metric* Settings (ACADISO.DWT)." The same setup can be accomplished by any of the following methods.

> Select the *Start from Scratch, Metric* option in the *Startup* or *Create New Drawing* dialog box.
> Select the ACADISO.DWT by any *Template* method.

Set the *STARTUP* system variable to 0, then set the *Default Template File Name for Qnew* in the *Files* tab of the *Options* dialog box to ACADISO.DWT.

Set the *MEASUREINIT* system variable to 1, then cancel the *Startup* dialog box that appears when AutoCAD starts.

Template

The *Template* option is available in the *Startup* or *Create New Drawing* dialog box (see Fig. 6-1) and from the *Select Template* dialog box (see Fig. 6-2).

Use the *Template* option if you want to begin a drawing using an existing template drawing (.DWT) as a starting point. A template drawing can have many of the drawing setup steps performed but contains no geometry. Several templates are provided by AutoCAD, including the default inch (*Imperial*) template ACAD.DWT and the default metric template ACADISO.DWT. See the "Table of *Start from Scratch, Imperial* Settings (ACAD.DWT)" and the "Table of *Start from Scratch, Metric* Settings (ACADISO.DWT)." More information on this option, including creating template drawings and using templates provided by AutoCAD, is given in Chapter 12, Advanced Drawing Setup.

FIGURE 6-2

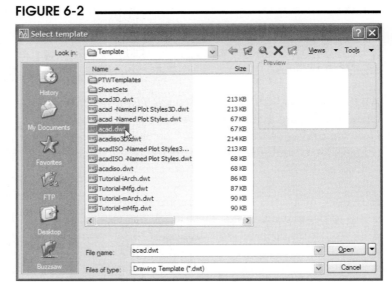

 TIP

NOTE: After selecting the ACAD.DWT template drawing, it is recommended that you perform a *Zoom* with the *All* option. Do this by selecting *Zoom, All* from the *View* pull-down menu or type Z, press Enter, then type A and Enter. This action ensures that the drawing displays all of the drawing *Limits*.

The following two tables list the AutoCAD settings for the two template drawings, ACAD.DWT and ACADISO.DWT. Many of these settings, such as *Units*, *Limits*, *Snap*, and *Grid*, are explained in the following sections.

Table of *Start from Scratch Imperial* Settings (ACAD.DWT)

Related Command	Description	System Variable	Default Setting
Units	linear units	*LUNITS*	2 (decimal)
Limits	drawing area	*LIMMAX*	12.0000,9.0000
Snap	snap increment	*SNAPUNIT*	.5000, .5000
Grid	grid increment	*GRIDUNIT*	.5000, .5000
LTSCALE	linetype scale	*LTSCALE*	1.0000
DIMSCALE	dimension scale	*DIMSCALE*	1.0000
Text, Mtext	text height	*TEXTSIZE*	.2000
Hatch	hatch pattern scale	*HPSCALE*	1.0000

Table of *Start from Scratch Metric* Settings (ACADISO.DWT)

Related Command	Description	System Variable	Default Setting
Units	linear units	*LUNITS*	2 (decimal)
Limits	drawing area	*LIMMAX*	420.0000, 297.0000
Snap	snap increment	*SNAPUNIT*	10.0000, 10.0000
Grid	grid increment	*GRIDUNIT*	10.0000, 10.0000
LTSCALE	linetype scale	*LTSCALE*	1.0000
DIMSCALE	dimension scale	*DIMSCALE*	1.0000
Text, Mtext	text height	*TEXTSIZE*	2.5000
Hatch	hatch pattern scale	*HPSCALE*	1.0000

The metric drawing setup is intended to be used with ISO linetypes and ISO hatch patterns, which are pre-scaled for these *Limits,* hence the *LTSCALE* and hatch pattern scale of 1. The individual dimensioning variables for arrow size, dimension text size, gaps, and extensions, etc. are changed so the dimensions are drawn correctly with a *DIMSCALE* of 1 with the ISO-25 dimension style.

Wizard

Selecting the *Use a Wizard* option in the *Startup* or *Create New Drawing* dialog box (see Fig. 6-1) gives a choice of using the *Quick Setup* or *Advanced Setup* wizard. The *Advanced Setup* wizard is an expanded version of the *Quick Setup* wizard.

The wizards use default settings based on either the ACAD.DWT or the ACADISO.DWT template. The template used by the wizards is determined by the *MEASUREINIT* system variable setting for your system (0 = ACAD.DWT, English template, and 1 = ACADISO.DWT, metric template). To change the setting, type *MEASUREINIT* at the Command prompt.

Quick Setup Wizard

The *Quick Setup* wizard automates only the first two steps listed under "Steps for Basic Drawing Setup" on the chapter's first page. Those functions, simply stated, are:

1. *Units*
2. *Limits*

Choosing the *Quick Setup* wizard invokes the *QuickSetup* dialog box (Fig. 6-3). There are two steps, *Units* and *Area*.

FIGURE 6-3

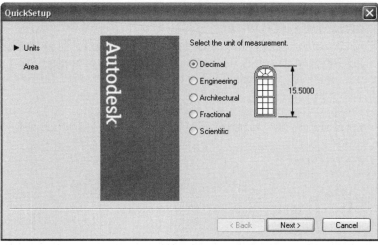

Units

Press the desired radio button to display the units you want to use for the drawing. The options are:

Decimal	Use generic decimal units with a precision 0.0000.
Engineering	Use feet and decimal inches with a precision of 0.0000.
Architectural	Use feet and fractional inches with a precision of 1/16 inch.
Fractional	Use generic fractional units with a precision of 1/16 units.
Scientific	Use generic decimal units showing a precision of 0.0000.

Use *Architectural* or *Engineering* units if you want to specify coordinate input using feet values with the apostrophe (') symbol. If you want to set additional parameters for units such as precision or system of angular measurement, use the *Units* or *-Units* command. Keep in mind the setting you select in this step changes only the display of units in the coordinate display area of the Status Bar (*Coords*) and in some dialog boxes, but not necessarily for the dimension text format. Select the *Next* button after specifying *Units*.

Area

Enter two values that constitute the X and Y measurements of the area you want to work in (Fig. 6-4). These values set the *Limits* of the drawing. The first edit box labeled *Width* specifies the X value for *Limits*. The X value is usually the longer of the two measurements for your drawing area and represents the distance across the screen in the X direction or along the long axis of a sheet of paper. The second edit box labeled *Length* specifies the Y value for *Limits* (the Y distance of the drawing area). (Beware: The terms "Width" and "Length" are misleading since "length" is normally the longer of two dimensions. When setting AutoCAD *Limits*, the Y measurement is generally the shorter of the two measurements.) The two values together specify the upper right corner of the drawing *Limits*. When you finish the two steps, you should <u>use *Zoom All* to display the entire *Limits* area in the screen</u>.

FIGURE 6-4

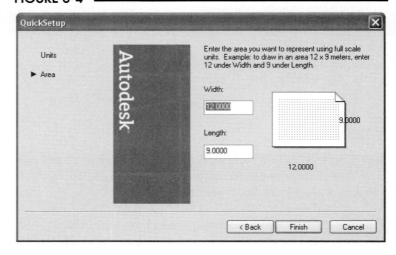

TIP

Advanced Setup Wizard

The *Advanced Setup* wizard performs the same tasks as the *Quick Setup* wizard with the addition of allowing you to select units precision and other units options (normally available in the *Units* dialog box). Selecting *Advanced Setup* produces a series of dialog boxes (not all shown). There are five steps involved in the series.

The following list indicates the "steps" in the *Advanced Setup* wizard and the related "Steps for Basic Drawing Setup" on the chapter's first page.

1.	*Units* and *Precision*	*Units* command
2.	*Angle*	*Units* command
3.	*Angle Measure*	*Units* command
4.	*Angle Direction*	*Units* command
5.	*Area*	*Limits* command

Units

You can select the units of measurement for the drawing as well as the unit's *Precision* in this first step (Fig. 6-5). These are the same options available in the *Drawing Units* dialog box (see "*Units*," this chapter). This is similar to the first step in the *Quick Setup* wizard but with the addition of *Precision*.

Similar to using the *Units* or *-Units* command, your choices in this and the next three dialog boxes determine the display of units for the coordinate display area of the Status Bar (*Coords*) and in dialog boxes. If you want to use feet units for coordinate input, select *Architectural* or *Engineering*.

FIGURE 6-5

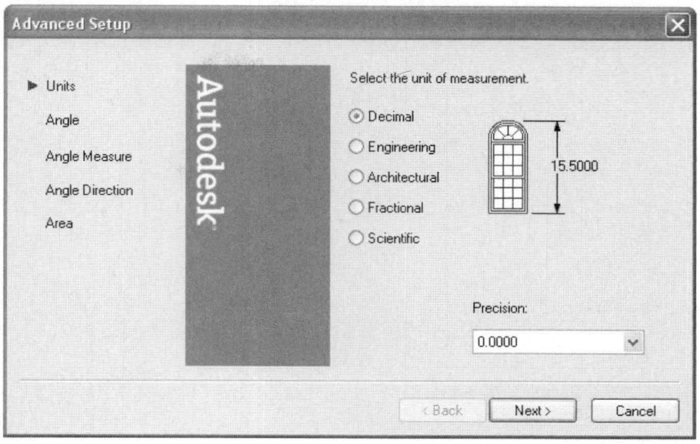

Angle

This step (not shown) provides for your input of the desired system of angular measurement. Select the drop-down list to select the angular *Precision*.

Angle Measure

This step sets the direction for angle 0. East (X positive) is the AutoCAD default. This has the same function as selecting the *Angle 0 Direction* in the *Drawing Units* dialog box.

Angle Direction

Select *Clockwise* if you want to change the AutoCAD default setting for <u>measuring</u> angles. This setting (identical to the *Drawing Units* dialog box option) affects the direction of positive and negative angles in commands such as *Rotate*, *Array Polar,* and dimension commands that measure angles, but does not affect the direction *Arcs* are drawn (always counterclockwise).

Area

Enter values to define the upper right corner for the *Limits* of the drawing. The *Width* refers to the X *Limits* component and the *Length* refers to the Y component. Generally, the *Width* edit box contains the larger of the two values unless you want to set up a vertically oriented drawing area. If you plan to print or plot to a standard scale, your input for *Area* should be based on the intended plot scale and sheet size. See the "Tables of *Limits* Settings" in Chapter 14 for appropriate values to use.

SETUP COMMANDS

If you want to set up a drawing using individual commands instead of the *Quick Setup* or *Advanced Setup* wizard, use the commands given in this section. The *Quick Setup* and *Advanced Setup* wizards use only the first two commands discussed in this section, *Units* and *Limits*.

Units

Pull-down Menu	Command (Type)	Alias (Type)	Short-cut	Screen (side) Menu	Tablet Menu
Format Units...	Units or -Units	UN or -UN	...	FORMAT Units	V,4

The *Units* command allows you to specify the type and precision of linear and angular units as well as the direction and orientation of angles to be used in the drawing. The current setting of *Units* determines the display of values by the coordinates display (*Coords*) and in some dialog boxes.

You can select *Units ...* from the *Format* pull-down or type *Units* (or command alias *UN*) to invoke the *Drawing Units* dialog box (Fig. 6-6). Type *-Units* (or *-UN*) to produce a text screen.

FIGURE 6-6

Units		Format	
1.	Scientific	1.55E + 01	Generic decimal units with an exponent.
2.	Decimal	15.50	Generic decimal usually used for applications in metric or decimal inches.
3.	Engineering	1'-3.50"	Explicit feet and decimal inches with notation, one unit equals one inch.
4.	Architectural	1'-3 1/2"	Explicit feet and fractional inches with notation, one unit equals one inch.
5.	Fractional	15 1/2	Generic fractional units.

Precision

When setting *Units,* you should also set the precision. *Precision* is the number of places to the right of the decimal or the denominator of the smallest fraction to display. The precision is set by making the desired selection from the *Precision* pop-up list in the *Drawing Units* dialog box (Fig. 6-6) or by keying in the desired selection in Command line format.

Precision displays only the display of values in the *Coords* display and in dialog boxes. The <u>actual preci-</u><u>sion</u> of the drawing database is always the same in AutoCAD, that is, 14 significant digits.

Angles

You can specify a format other than the default (decimal degrees) for expression of angles. Format options for angular display and examples of each are shown in Figure 6-6 (dialog box format).

The orientation of angle 0 can be changed from the default position (east or X positive) to other options by selecting the *Direction* tile in the *Drawing Units* dialog box. This produces the *Direction Control* dialog box (Fig. 6-7). Alternately, the *-Units* command can be typed to select these options in Command line format.

FIGURE 6-7

The direction of angular measurement can be changed from its default of counterclockwise to clockwise. The direction of angular measurement affects the direction of positive and negative angles in commands such as *Array Polar, Rotate* and dimension commands that <u>measure</u> angular values but does not change the direction *Arcs* are <u>drawn,</u> which is always counterclockwise.

Insertion Scale

This section in the *Drawing Units* dialog box specifies the units to use when you drag-and-drop *Blocks* from AutoCAD DesignCenter or Tool palettes into the current drawing. There are many choices including *Unitless, Inches, Feet, Millimeters,* and so on. If you intend to insert *Blocks* into the drawing (you are currently setting units for) using drag-and-drop, select a unit from the list that matches the units of the *Blocks* to insert. If you are not sure, select *Unitless.*

Keyboard Input of *Units* Values

When AutoCAD prompts for a point or a distance, you can respond by entering values at the keyboard. The values can be in <u>any format</u>—integer, decimal, fractional, or scientific—regardless of the format of *Units* selected.

You can <u>type in explicit feet and inch values only if</u> *Architectural* or *Engineering* units have been specified as the drawing units. For this reason, specifying *Units* is the first step in setting up a drawing. Type in explicit feet or inch values by using the apostrophe (') symbol after values representing feet and the quote (") symbol after values representing inches. If no symbol is used, the values are understood by AutoCAD to be <u>inches</u>.

Feet and inches input <u>cannot</u> contain a blank, so a hyphen (-) must be typed between inches and fractions. For example, with *Architectural* units, key in *6'2-1/2"*, which reads "six feet, two and one-half inches." The standard engineering and architectural format for dimensioning, however, places the hyphen between feet and inches (as displayed by the default setting for the *Coords* display).

Limits

Pull-down Menu	Command (Type)	Alias (Type)	Short-cut	Screen (side) Menu	Tablet Menu
Format *Drawing Limits*	*Limits*	…	…	*FORMAT* *Limits*	*V,2*

The *Limits* command allows you to set the size of the drawing area by specifying the lower-left and upper-right corners in X,Y coordinate values.

```
Command: limits
Reset Model space limits
Specify lower left corner or [ON/OFF] <0.0000,0.0000>: X,Y or Enter (Enter an X,Y value or accept
the 0,0 default—normally use 0,0 as lower-left corner.)
Specify upper right corner <12.0000,9.0000>: X,Y (Enter new values to change upper-right corner to
allow adequate drawing area.)
```

The default *Limits* values in the ACAD.DWT are 12 and 9; that is, 12 units in the X direction and 9 units in the Y direction (Fig. 6-8). Starting a drawing by the following methods results in *Limits* of 12 x 9:

Selecting the ACAD.DWT template drawing.
Selecting the *Imperial* defaults in the *Start from Scratch* option.

FIGURE 6-8

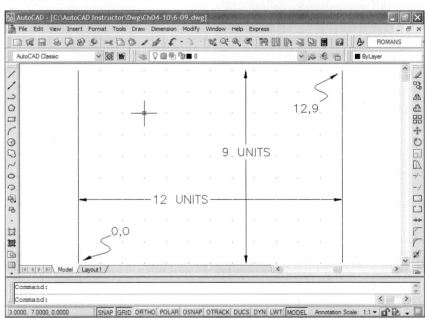

With the default settings for the ACAD.DWT, the *Grid* is displayed only over the *Limits*. The grid display can be controlled using the *Limits* option of *Grid* (see "*Grid*"). The units are generic decimal units that can be used to represent inches, feet, millimeters, miles, or whatever is appropriate for the intended drawing. Typically, however, decimal units are used to represent inches or millimeters. If the default units are used to represent inches, the default drawing size would be 12 x 9 inches.

Remember that when a CAD system is used to create a drawing, the geometry should be drawn <u>full size</u> by specifying dimensions of objects in <u>real-world units</u>. A completed CAD drawing or model is virtually an exact dimensional replica of the actual object. Scaling of the drawing occurs only when plotting or printing the file to an actual fixed-size sheet of paper.

TIP Before beginning to create an AutoCAD drawing, determine the size of the drawing area needed for the intended geometry. After setting *Units,* appropriate *Limits* should be set in order to draw the object or geometry to the <u>real-world size in the actual units</u>. There are no practical maximum or minimum settings for *Limits*.

The X,Y values you enter as *Limits* are understood by AutoCAD as values in the units specified by the *Units* command. For example, if you previously specified *Architectural units,* then the values entered are understood as inches unless the notation for feet (') is given (**240,180** or **20',15'** would define the same coordinate). Remember, you can type in explicit feet and inch values only if *Architectural* or *Engineering* units have been specified as the drawing units.

TIP If you are planning to plot the drawing to scale, *Limits* should be set to a proportion of the <u>sheet size</u> you plan to plot on. For example, setting limits to 22 x 17 (2 times 11 by 8.5) would allow enough room for drawing an object about 20" x 15" and allow plotting at 1/2 size on the 11" x 8.5" sheet. Simply stated, <u>set *Limits* to a proportion of the paper</u>.

Limits also defines the display area when a *Zoom All* is used. *Zoom All* forces the full display of the *Limits*. *Zoom All* can be invoked by typing **Z** (command alias) then **A** for the *All* option.

TIP Changing *Limits* does <u>not</u> automatically change the display. As a general rule, you should make a habit of invoking a *Zoom All* <u>immediately following</u> a change in *Limits* to display the area defined by the new limits (Fig. 6-9).

If you are already experimenting with drawing in different *Linetypes,* a change in *Limits* and *Zoom All* affects the display of the hidden and dashed lines. The *LTSCALE* variable controls the spacing of non-continuous lines. The *LTSCALE* is often changed proportionally with changes in *Limits*.

FIGURE 6-9

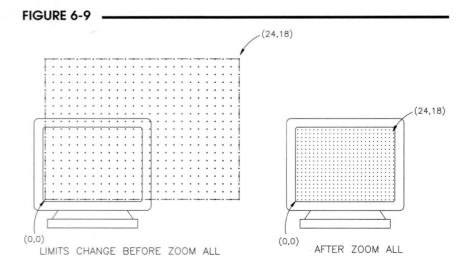

LIMITS CHANGE BEFORE ZOOM ALL   AFTER ZOOM ALL

ON/OFF

If the *ON* option of *Limits* is used, limits checking is activated. Limits checking prevents you from drawing objects outside of the limits by issuing an outside-limits error. This is similar to drawing "off the paper." Limits checking is *OFF* by default.

Snap

Pull-down Menu	Command (Type)	Alias (Type)	Short-cut	Screen (side) Menu	Tablet Menu
Tools *Drafting Settings...* *Snap and Grid*	*Snap*	*SN*	*F9 or Ctrl+B*	*TOOLS 2* *Grid* *Snap and Grid*	*W,10*

Snap in AutoCAD has two possible types, _Grid Snap_ and _Polar Snap_. When you are setting up the drawing, you should set the desired *Snap type* and increment. You may decide to use both types of Snap, so set both increments initially. You can have only <u>one of the two *Snap types* active at one time</u>. (See Chapter 3, Draw Command Concepts, for further explanation and practice using both Snap types.)

Grid Snap forces the cursor to preset positions on the screen, similar to the *Grid*. *Grid Snap* is like an invisible grid that the cursor "snaps" to. This function can be of assistance if you are drawing *Lines* and other objects to set positions on the drawing, such as to every .5 unit. The default value for *Grid Snap* is .5, but it can be changed to any value.

Polar Snap forces the cursor to move in set intervals from the previously designated point (such as the "first point" selected during the *Line* command) to the next point. *Polar Snap* operates for cursor movement at any previously set *Polar Tracking* angle, whereas *Grid Snap* (since it is rectangular) forces regular intervals only in horizontal or vertical movements. *Polar Tracking* must also be on (*POLAR* or F10) for *Polar Snap* to operate.

Set the desired snap type and increments using the *Snap* command or the *Drafting Settings* dialog box. In the *Drafting Settings* dialog box, select the *Snap and Grid* tab (Fig. 6-10).

To set the _Polar Snap_ increment, first select *Polar Snap* in the *Snap Type* section (lower left). This action causes the *Polar Spacing* section to be enabled. Next, enter the desired *Polar Distance* in the edit box.

To set the _Grid Snap_ increment, first select *Grid Snap* in the *Snap Type* section. This action causes the *Snap* section to be enabled. Next, enter the desired *Snap X Spacing* value in the edit box. If you want a non-square snap grid, remove the check by *Equal X and Y spacing*, then enter the desired values in the *Snap X spacing* and *Snap Y spacing* edit boxes.

FIGURE 6-10

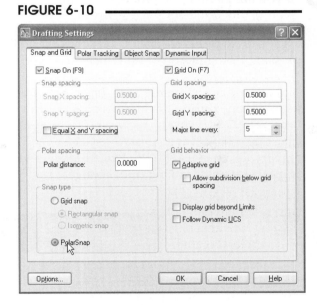

The Command line format of *Snap* is as follows:

    Command: **snap**
    Specify snap spacing or [ON/OFF/Aspect/Style/Type] <0.5000>:

Value
Entering a value at the Command line prompt sets the _Grid Snap_ increment only. You must use the *Drafting Settings* dialog box to set the *Polar Snap* distance.

ON/OFF
Selecting *ON* or *OFF* accomplishes the same action as toggling the F9 key or selecting the word *SNAP* on the Status Bar.

Aspect

The *Aspect* option allows specification of unequal X and Y spacing for the snap grid. This is identical to entering unequal *Snap X Spacing* and *Snap Y Spacing* in the dialog box.

Rotate

Although the *Rotate* option is not listed, the *Grid Snap* can be rotated about any point and set to any angle. When Snap has been rotated, the *GRID, ORTHO*, and the "crosshairs" automatically follow this alignment. This action facilitates creating objects oriented at the specified angle, for example, creating an auxiliary view of drawing part or a floor plan at an angle. To do this, use the *Rotate* option in Command line format.

Style

The *Style* option allows switching between a *Standard* snap pattern (square or rectangular) and an *Isometric* snap pattern. If using the dialog box, toggle *Isometric* snap. When *Snap Style* or *Rotate* (*Angle*) is changed, the *Grid* automatically aligns with it.

Type

This option switches between *Grid Snap* and *Polar Snap*. Remember, you can have only one of the two snap types active at one time.

Once your snap type and increment(s) are set, you can begin drawing using either snap type or no snap at all. While drawing, you can right-click on the word *SNAP* on the Status Bar to invoke the shortcut menu shown in Figure 6-11. Here you can toggle between *Grid Snap* and *Polar Snap* or turn both off (*POLAR* must also be on to use *Polar Snap*). Left-clicking on the word *SNAP* or pressing F9 toggles on or off whichever snap type is current.

FIGURE 6-11

The process of setting the *Snap Type, Grid Snap Spacing, Polar Snap Spacing*, and *Polar Tracking Angle* is usually done during the initial stages of drawing setup, although these settings can be changed at any time. The *Grid Snap Spacing* value is stored in the *SNAPUNIT* system variable and saved in the <u>drawing file</u>. The other settings (*Snap Type, Polar Snap Spacing*, and *Polar Tracking Angle*) are saved in the <u>system registry</u> (as the *SNAPTYPE, POLARDIST*, and *POLARANG* system variables) so that the settings remain in affect for any drawing until changed.

Grid

Pull-down Menu	Command (Type)	Alias (Type)	Short-cut	Screen (side) Menu	Tablet Menu
Tools *Drafting Settings...* *Snap and Grid*	*Grid*	...	F7 or Ctrl+G	*TOOLS 2* *Grid* *Snap and Grid*	*W,10*

Grid is visible on the screen, whereas *Grid Snap* is invisible. *Grid* is only a <u>visible</u> display of some regular interval. *Grid* and *Grid Snap* can be <u>independent</u> of each other. In other words, each can have separate spacing settings and the active state of each (*ON, OFF*) can be controlled independently. The *Grid* <u>follows</u> the *Snap* if *Snap* is rotated or changed to *Isometric Style*. Although the *Grid* spacing can be different than that of *Snap*, it can also be forced to follow *Snap* by using the *Snap* option. The default *Grid* setting is **0.5**.

The *Grid* <u>cannot</u> be plotted. It is <u>not</u> comprised of *Point* objects and therefore is not part of the current drawing. *Grid* is only a visual aid.

Grid can be accessed by Command line format (shown below) or set via the *Drafting Settings* dialog box (Fig. 6-10). The dialog box can be invoked by menu selection or by typing *Dsettings* or *DS*.

NOTE: When you draw in 3D (normally by beginning with a 3D template drawing, ACAD3D.DWT or ACADISO3D.DWT), <u>lines</u> rather than dots are used to designate the *Grid*.

> Command: **grid**
> Specify grid spacing(X) or [ON/OFF/Snap/Major/aDaptive/Limits/Follow/Aspect] <0.5000>:

ON/OFF

The *ON* and *OFF* options simply make the *Grid* visible or not (like toggling the F7 key, pressing Ctrl+G, or clicking *GRID* on the Status bar).

Grid Spacing

If you supply a value in Command line format or in the *Grid X spacing* and *Grid Y spacing* edit boxes, the *Grid* is displayed at the specified spacing regardless of the *Snap* spacing. If you key in an *X* as a suffix to the value (for example, 2X), the *Grid* is displayed at that value times the current *Snap* spacing.

Limits (Display grid beyond Limits)

This option determines whether or not the *Grid* is displayed beyond the area set by the *Limits* command. For example, if you chose to display the grid beyond the *Limits,* the grid pattern is displayed to the edge of the drawing area no matter how far you *Zoom* in or out. If you chose not to display the *Grid* beyond the *Limits,* when *Zoomed* out you can "see" the *Limits* by the area filled with the grid pattern. Regardless of this setting, *Zoom All* displays the area defined by the *Limits*.

Follow

This option is used only for drawing in 3D. If this option is on, the grid pattern (lines for 3D drawings) automatically follow the XY plane of the Dynamic UCS. This setting is stored in the *GRIDDISPLAY* system variable.

Aspect

The *Aspect* option of *Grid* allows different X and Y spacing (causing a rectangular rather than a square *Grid*).

Dsettings

Pull-down Menu	Command (Type)	Alias (Type)	Short-cut	Screen (side) Menu	Tablet Menu
Tools *Drafting Settings...*	*Dsettings*	*DS*	*Status Bar* (right-click) *Settings...*	*TOOLS 2* *Osnap...* *(Grid or Polar)*	*W,10*

You can access controls to *Snap* and *Grid* features using the *Dsettings* command. *Dsettings* produces the *Drafting Settings* dialog box described earlier (see Fig. 6-10). The three tabs in the dialog box are *Snap and Grid, Polar Tracking,* and *Object Snap*. Use the *Snap and Grid* tab to control settings for *Snap* and *Grid* as previously described. The *Polar Tracking* and *Object Snap* tabs are explained in Chapter 7.

INTRODUCTION TO LAYOUTS AND PRINTING

The last two steps listed in the "Steps for Basic Drawing Setup" on this chapter's first page are also listed on the next page.

9. Activate a *Layout* tab, set it up for the plot or print device and paper size, and create a viewport (if not already created).
10. Create or insert a title block and border in the layout.

Although steps 9 and 10 are often performed after the drawing is complete and just before making a print or plot, it is wise to consider these steps in the drawing setup process. Because you want hidden line dashes, text, dimensions, hatch patterns, etc. to have the correct size in the finished print or plot, it would be sensible to consider the paper size of the print or plot and the drawing scale before you create text, dimensions, and hatch patterns in the drawing. In this way, you can more accurately set the necessary sizes and system variables before you draw, or as you draw, instead of changing multiple settings upon completion of the drawing.

To put it simply, to determine the size for linetypes (*LTSCALE*), text, and dimensions, use the <u>proportion of the drawing to the paper size</u>. In other words, if the size of the drawing area (*Limits*) is 22 x 17 and the paper size you will print on is 11 x 8.5, then the drawing is 2 times the size of the paper. Therefore, set the *LTSCALE* to 2 and create the text and dimensions twice as large as you want them to appear on the print.

Model Tab and *Layout* Tabs

At the bottom of the drawing area you should see one *Model* tab and two *Layout* tabs. If the tabs have been turned off on your system, see "Setting Layout Options and Plot Options" later in this section.

 When you start AutoCAD and begin a drawing, the *Model* tab is active by default. Objects that represent the subject of the drawing (model geometry) are normally drawn in the *Model* tab, also known as "model space." Traditionally, dimensions and notes (text) are also created in model space.

 A <u>layout</u> is activated by selecting the *Layout1*, *Layout2*, or other layout tab (Fig. 6-12). A layout represents the sheet of paper that you intend to print or plot on (sometimes called paper space). The dashed line around the "paper sheet" in Figure 6-12 represents the maximum printable area for the configured printer or plotter.

 You can have multiple layouts (*Layout1*, *Layout2*, etc.), each representing a different sheet size and/or print or plot device. For example, you may have one layout to print with a laser printer on a 8.5 x 11 inch sheet and another layout set up to plot on an 24 x 18 ("C" size) sheet of paper.

With the default options set when AutoCAD is installed, a viewport is automatically created when you acti-

FIGURE 6-12

paper space   viewport

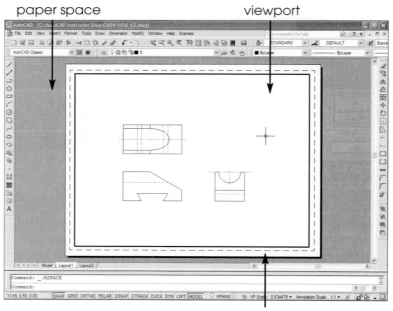

printable area

vate a layout. However, you can create layouts using the *Vports* command. A <u>viewport</u> is a "window" that looks into model space. Therefore, you first create the drawing objects in model space (the *Model* tab), then activate a layout and create a viewport to "look into" model space. Typically, only drawing objects such as a title block, border, and some text are created in a layout. Creating layouts and viewports is discussed in detail in Chapter 13, Layouts and Viewports.

Since a layout represents the actual printed sheet, you normally <u>print the layout full size (1:1)</u>. However, the view of the drawing objects appearing in the viewport is scaled to achieve the desired print scale. In other words, you can <u>control the scale of the drawing by setting the "viewport scale"</u>—the proportion of the drawing objects that appear in the viewport relative to the paper size, or simply stated, the proportion of model space to paper space. (This is the same idea as the proportion of the drawing area to the paper size, discussed earlier.)

One easy way to set the viewport scale is to use the *VP Scale* pop-up list near the lower-right corner of the Drawing Editor (Fig. 6-13). To set the scale of objects in the viewport relative to the paper size, select the desired scale from the list. Setting the scale of the drawing for printing and plotting using this and other methods is discussed in detail in Chapter 13, Layouts and Viewports, and Chapter 14, Printing and Plotting.

FIGURE 6-13

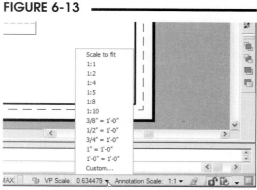

Setting the *VP Scale* (viewport scale) changes the size of the geometry in the viewport relative to the paper (*Layout*). Note in Figure 6-14 that selecting *1:1* from the *VP Scale* pop-up list changes the size of the drawing in the viewport compared to the drawing shown in Fig. 6-12.

FIGURE 6-14

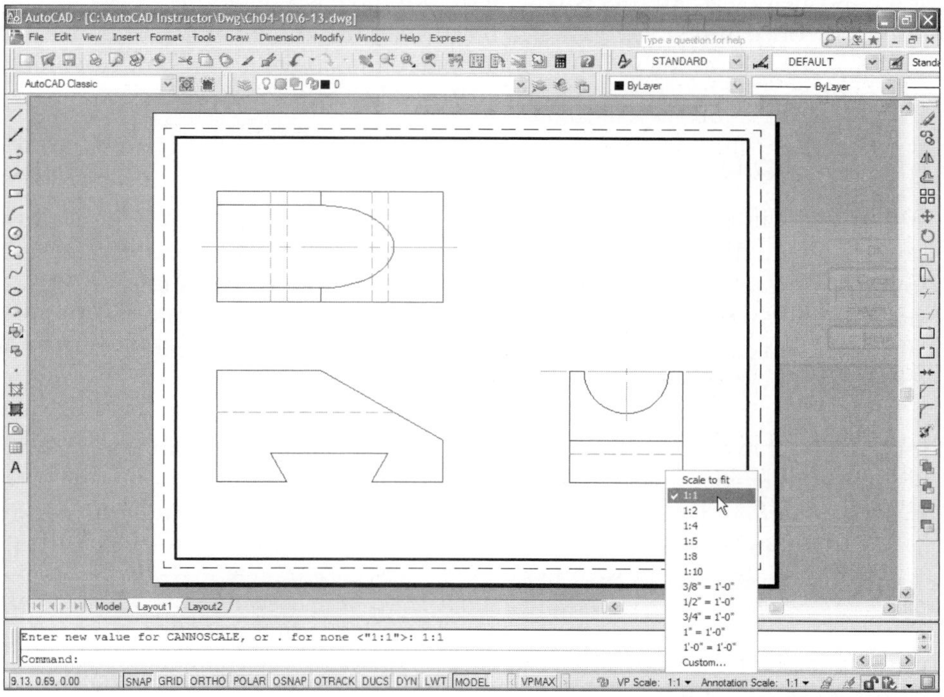

Using a *Layout* tab to set up a drawing for printing or plotting is recommended, although you can also make a print or plot directly from the *Model* tab (print from model space). This method is also discussed in Chapter 14, Printing and Plotting.

Step 10 in the "Steps for Basic Drawing Setup" is to create a titleblock and border for the layout. The titleblock and text that you want to appear only in the print but not in the drawing (in model space) are typically created in the layout (in paper space). Figure 6-15 shows the drawing after a titleblock has been added.

FIGURE 6-15

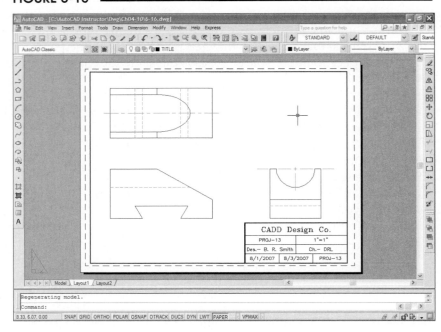

The advantage of creating a titleblock early in the drawing process is knowing how much space the title-block occupies so you can plan around it. Note that in Figure 6-15 no drawing objects can be created in the lower-right corner because the titleblock occupies that area.

Creating a titleblock is often accomplished by using the *Insert* command to bring a previously created *Block* into the layout. A *Block* in AutoCAD is a group of objects (*Lines*, *Arcs*, *Circles*, *Text*, etc.) that is saved as one object. In this way, you do not have to create the set of objects repeatedly—you need create them only once and save them as a *Block*, then use the *Insert* command to insert the *Block* into your drawing. For example, your school or office may have a standard titleblock saved as a *Block* that you can *Insert*. *Blocks* are covered in Chapter 20, Blocks and DesignCenter.

Why Set Up Layouts Before Drawing?

Knowing the intended drawing scale before completing the drawing helps you set the correct size for linetypes, text, dimensions, hatch patterns, and other size-related drawing objects. Since the text and dimensions must be readable in the final printed drawing (usually 1/8" to 1/4" or 3mm-6mm), you should know the drawing scale to determine how large to create the text and dimensions in the drawing. Although you can change the sizes when you are ready to print or plot, knowing the drawing scale early in the drawing process should save you time.

You may not have to go through the steps to create a layout and viewport to determine the drawing scale, although doing so is a very "visible" method to achieve this. The important element is knowing ahead of time the intended drawing scale. If you know the intended drawing scale, you can calculate the correct sizes for creating text and dimensions and setting the scale for linetypes. This topic is a major theme discussed in Chapters 12, 13, and 14.

AutoCAD 2008 introduced annotative objects. Annotative objects offer an alternative method for determining and specifying a static size for "annotation" in drawings such as text, dimensions, and hatch patterns. Annotative objects are dynamic in that the size of these objects can be easily changed to match the viewport scale. Using this method is particularly helpful for displaying the same text and dimensions in multiple viewports at different scales. Annotative objects are introduced in Chapters 12, 13, and 14.

Printing and Plotting

The *Plot* command allows you to print a drawing (using a printer) or plot a drawing (using a plotter). The *Plot* command invokes the *Plot* dialog box (Fig. 6-16). You can type the *Plot* command, select the *Plot* icon button from the Standard toolbar, or select *Plot* from the *File* pull-down menu, as well as use other methods (see "*Plot*," Chapter 14).

If you want to print the drawing as it appears in model space, invoke *Plot* while the *Model* tab is active. Likewise, if you want to print a layout, invoke *Plot* while the desired *Layout* tab is active.

The *Plot* dialog box allows you to specify several options with respect to printing and plotting, such as selecting the plot device, paper size, and scale for the plot.
If you are printing the drawing from the *Model* tab, you would select an appropriate scale (such as *1:1*, *1:2*, or *Scaled to Fit*) in the *Scale* drop-down list so the drawing geometry would fit appropri-

FIGURE 6-16

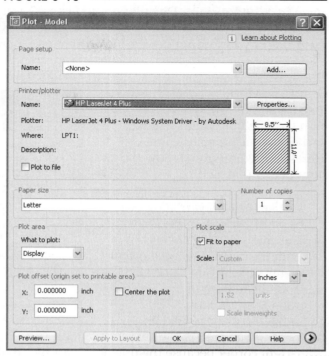

ately on the printed sheet (see Fig. 6-16). If you are printing the drawing from a layout, you would select *1:1* in the *Scale* drop-down list. In this case, you want to print the layout (already set to the sheet size for the plot device) at full size. The drawing geometry that appears in the viewport can be scaled by setting the viewport scale (using the *Viewports* drop-down list), as described earlier.

Any plot specifications you make with the *Plot* dialog box (or the *Page Setup* dialog box) are saved with each *Layout* tab and the *Model* tab; therefore, you can have several layouts, each saved with a particular print or plot setup (scale, device, etc.).

Details of printing and plotting, including all the options of the *Plot* dialog box, printing to scale, and configuring printers and plotters, are discussed in Chapter 14.

Setting Layout Options and Plot Options

In Chapters 12, 13, and 14 you will learn advanced steps in setting up a drawing, how to create layouts and viewports, and how to print and plot drawings. Until you study those chapters, and for printing drawings before that time, you may need to go through a simple process of configuring your system for a plot device and to automatically create a viewport in layouts. However, if you are at a school or office, some settings may have already been prepared for you, so the following steps may not be needed. Activating a *Layout* tab automatically creates a viewport if you use AutoCAD's default options that appear when it is first installed.

As a check, start AutoCAD and draw a *Circle* in model space. When you activate a *Layout* tab for the first time in a drawing, you see either a "blank" sheet with no viewport, or a viewport that already exists, or one that is automatically created by AutoCAD. The viewport allows you to view the circle you created in model space. (If the *Page Setup Manager* appears, select *Cancel* this time.) If a viewport exists, but no circle appears, double-click <u>inside</u> the viewport and type *Z* (for *Zoom*), press Enter, then type *A* (for *All*) and press Enter to make the circle appear.

If your system has not previously been configured for you at your school or office, follow the procedure given in "Configuring a Default Output Device" and "Creating Automatic Viewports." After doing so, you should be able to activate a *Layout* tab and AutoCAD will automatically match the layout size to the sheet size of your output device and automatically create a viewport.

Configuring a Default Output Device

If no default output device has been specified for your system or to check to see if one has already been specified for you, follow the steps below (Fig. 6-17).

FIGURE 6-17

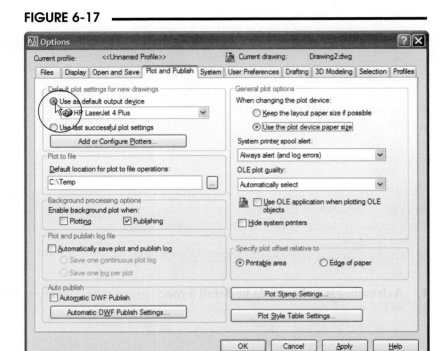

1. Invoke the *Options* dialog box by right-clicking in the drawing area and selecting *Options…* from the bottom of the shortcut menu, or selecting *Options…* from the bottom of the *Tools* pull-down menu.
2. In the *Plot and Publish* tab, locate the *Default plot settings for new drawings* cluster near the top-left corner of the dialog box. Select the desired plot device from the list (such as a laser printer), then select the *Use as default output device* button.
3. Select *OK*.

If output devices that are connected to your system do not appear in the list, see "Configuring Plotters and Printers" in Chapter 14.

Creating Automatic Viewports

If no viewport exists on your screen, configure your system to automatically create a viewport by following these steps (Fig. 6-18).

1. Invoke the *Options* dialog box again.
2. In the *Display* tab, find the *Layout elements* cluster near the lower-left corner of the dialog box. Select the *Create viewport in new layouts* checkbox from the bottom of the list.
3. Select *OK*.

FIGURE 6-18

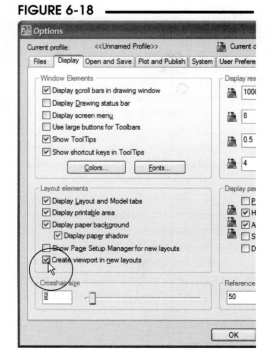

Printing the Drawing

Once an output device has been configured for your system, you are ready to print. For simplicity, you need to perform only three steps to make a print of your drawing, either from the *Model* tab or from a *Layout* tab.

1. Invoke the *Plot* dialog box. Set the desired scale in the *Plot Scale* section, *Scale* drop-down list. If you are printing from a *Layout* tab, select *1:1*. If you are printing from the *Model* tab, select *Scaled to Fit* or other appropriate scale.
2. Select the *Preview* button to ensure the drawing will be printed as you expect.
3. Assuming the preview is as you expect, select the *OK* button to produce the print or plot.

This section, "Introduction to Layouts and Printing," will give you enough information to create prints or plots of drawings for practice and for the next several Chapter Exercises. You also have learned some important concepts that will help you understand all aspects of setting up drawings, creating layouts and viewports, and printing and plotting drawings, as discussed in Chapters 12, 13, and 14.

CHAPTER EXERCISES

1. A drawing is to be made to detail a mechanical part. The part is to be manufactured from sheet metal stock; therefore, only one view is needed. The overall dimensions are 16 x 10 inches, accurate to the nearest .125 inch. Complete the steps for drawing setup:

    A. The drawing will be automatically "scaled to fit" the paper (no standard scale).

        1. Begin a *New* drawing. If a dialog box appears, either select the *ACAD.DWT* template or select *Start from Scratch*, and *Imperial* default settings.
        2. *Units* should be *Decimal*. Set the *Precision* to **0.000**.
        3. Set *Limits* in order to draw full size. Make the lower-left corner **0,0** and the upper right at **24,18**. This is a 4 x 3 proportion and should allow space for the part and dimensions or notes.
        4. *Zoom All*. (Type **Z** for *Zoom*; then type **A** for *All*.)
        5. Set the *Grid* to **1**.
        6. Set the *Grid Snap* increment to **.125**. Since the *Polar Snap* increment and *Snap Type* are not saved with the drawing (but are saved in the system registry), it is of no use to set these options at this time.
        7. Save this drawing as **CH6EX1A** (to be used again later). (When plotting at a later time, "Scale to Fit" can be specified.)

    B. The drawing will be printed from a layout to scale on engineering A size paper (11" x 8.5").

        1. Begin a *New* drawing. If a dialog box appears, either select the *ACAD.DWT* template or select *Start from Scratch*, and *Imperial* default settings.
        2. *Units* should be *Decimal*. Set the *Precision* to **0.000**.
        3. Set *Limits* to a proportion of the paper size, making the lower-left corner **0,0** and the upper-right at **22,17**. This allows space for drawing full size and for dimensions or notes.
        4. *Zoom All*. (Type **Z** for *Zoom*; then type **A** for *All*.)
        5. Set the *Grid* to **1**.
        6. Set the *Grid Snap* increment to **.125**. Since the *Polar Snap* increment and *Snap Type* are not saved with the drawing (but are saved in the system registry), it is of no use to set these options at this time.

7. Activate a **Layout** tab. Assuming your system is configured for a printer using an 11 x 8.5 sheet and is configured to automatically create a viewport (see "Setting Layout Options and Plot Options" in this chapter), a viewport should appear. If the *Page Setup Manager* appears, select **Close**.

8. Double-click inside the viewport, then select *1:2* from the **VP Scale** list (in the lower-right corner). Next, double-click outside the viewport or toggle the **MODEL** button on the Status bar to **PAPER**.

9. Activate the **Model** tab. **Save** the drawing as **CH6EX1B.**

2. A drawing is to be prepared for a house plan. Set up the drawing for a floor plan that is approximately 50' x 30'. Assume the drawing is to be automatically "Scaled to Fit" the sheet (no standard scale).

A. Begin a **New** drawing. If a dialog box appears, either select the **ACAD.DWT** template or select **Start from Scratch**, and **Imperial** default settings.

B. Set **Units** to **Architectural**. Set the **Precision** to **0'-0 1/4"**. Each unit equals 1 inch.

C. Set **Limits** to **0,0** and **80',60'**. Use the apostrophe (') symbol to designate feet. Otherwise, enter **0,0** and **960,720** (size in inch units is: 80x12=960 and 60x12=720).

D. **Zoom All**. (Type **Z** for *Zoom;* then type **A** for *All*.)

E. Set **Grid** to **24** (2 feet).

F. Set the **Grid Snap** increment to **6"**. Since the *Polar Snap* increment and *Snap Type* are not saved with the drawing (but are saved in the system registry), it is of no use to set these options at this time.

G. **Save** this drawing as **CH6EX2.**

3. A multiview drawing of a mechanical part is to be made. The part is 125mm in width, 30mm in height, and 60mm in depth. The plot is to be made on an A3 metric sheet size (420mm x 297mm). The drawing will use ISO linetypes and ISO hatch patterns, so AutoCAD's *Metric* default settings can be used.

A. Begin a **New** drawing using either the **ACADISO.DWT** template or select **Start from Scratch**, and **Metric** default settings.

B. **Units** should be **Decimal**. Set the **Precision** to **0.00**.

C. Calculate the space needed for three views. If **Limits** are set to the sheet size, there should be adequate space for the views. Make sure the lower-left corner is at **0,0** and the upper right is at **420,297**. (Since the *Limits* are set to the sheet size, a plot can be made later at 1:1 scale.)

D. Set the **Grid Snap** increment to **10**. Make **Grid Snap** current.

E. **Save** this drawing as **CH6EX3** (to be used again later).

4. Assume you are working in an office that designs many mechanical parts in metric units. However, the office uses a standard laser jet printer for 11" x 8.5" sheets. Since AutoCAD does not have a setup for metric drawings on non-metric sheets, it would help to carry out the steps for drawing setup and save the drawing as a template to be used later.

A. Begin a **New** drawing. If a dialog box appears, either select the **ACAD.DWT** template or select **Start from Scratch**, and **Imperial** default settings.

B. Set the **Units Precision** to **0.00**.

C. Change the **Limits** to match an 11" x 8.5" sheet. Make the lower-left corner **0,0** and the upper right **279,216** (11 x 8.5 times 25.4, approximately). (Since the *Limits* are set to the sheet size, plots can easily be made at 1:1 scale.)

D. **Zoom All**. (Type **Z** for *Zoom,* then **A** for *All*).

E. Change the *Grid Snap* increment to **2**. When you begin the drawing in another exercise, you may want to set the *Polar Snap* increment to 2 and make *Polar Snap* current.

F. Change *Grid* to **10**.

G. At the Command prompt, type *LTSCALE*. Change the value to **25**.

H. Activate a *Layout* tab. Assuming your system is configured for a printer using an 11 x 8.5 sheet and is configured to automatically create a viewport (see "Setting Layout Options and Plot Options" in this chapter), a viewport should appear. (If the *Page Setup Manager* appears, select *Close* and the viewport should appear.)

I. Produce the *Page Setup Manager* by selecting it from the *File* pull-down menu or typing *Pagesetup* at the Command prompt. Select *Modify* so the *Page Setup* dialog box appears. In this dialog box, find the *Printer/Plotter* section and select your configured printer if not already configured. Ensure the *Paper size* is set to *Letter*. In the *Plot Scale* section, use the drop-down list to change *inches* to *mm*. Then select *1:1* in the **Scale** drop-down list. This action causes AutoCAD to measure the 11 x 8.5 inch sheet in millimeters (as shown by the image in the *Printer/Plotter* section) and print the <u>sheet</u> at 1:1. Select *OK* in the *Page Setup* dialog box and *Close* in the *Page Setup Manager*. Using the *Page Setup Manager* saves the print settings with the layout for future use.

J. Examining the layout, it appears the viewport is extemely small in relation to the printable area. *Erase* the viewport.

K. To make a new viewport, type *-Vports* at the Command prompt (don't forget the hyphen) and press **Enter**. Accept the default (*Fit*) option by pressing **Enter**. A viewport appears to fit the printable area.

L. Double-click inside the viewport, then select *1:1* from the *VP Scale* list (in the lower-right corner). Next, double-click outside the viewport or toggle the *MODEL* button on the Status bar to *PAPER*.

M. Activate the *Model* tab. *Save* the drawing as **A-METRIC.**

5. Assume you are commissioned by the local parks and recreation department to provide a layout drawing for a major league sized baseball field. Follow these steps to set up the drawing:

A. Begin a *New* drawing. If a dialog box appears, either select the *ACAD.DWT* template or select *Start from Scratch*, and *Imperial* default settings.

B. Set the *Units* to *Architectural* and the *Precision* to 1/2".

C. Set *Limits* to an area of **512' x 384'** (make sure you key in the apostrophe to designate feet). *Zoom All*.

D. Type *DS* to invoke the *Drafting Settings* dialog box. Change the *Grid Snap* to **10'** (don't forget the apostrophe). When you are ready to draw (at a later time), you may want to set the *Polar Snap* increment to 10' and make *Polar Snap* the current *Snap Type*.

E. Use the *Grid* command and change the value to **20'**. Ensure *SNAP* and *GRID* are on.

F. Save the drawing and assign the name **BALL FIELD CH6.**

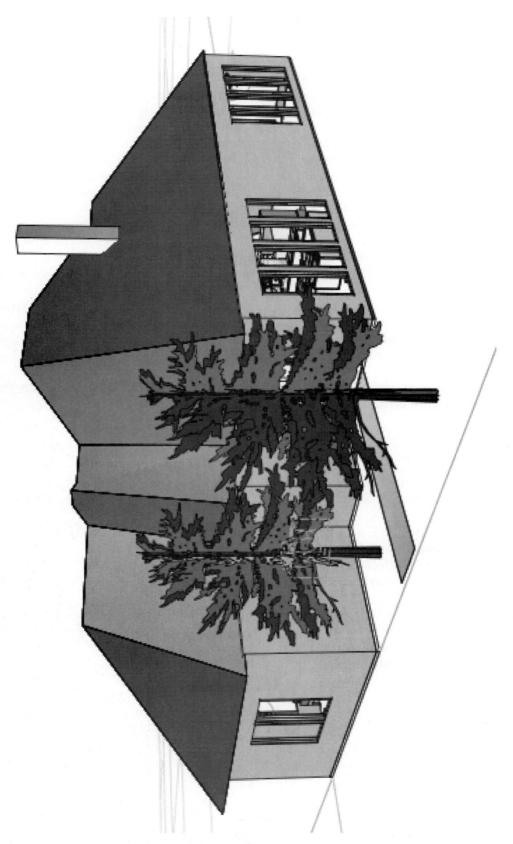

3D HOUSE.DWG, Courtesy of Autodesk, Inc.

OBJECT SNAP AND OBJECT SNAP TRACKING

CHAPTER OBJECTIVES

After completing this chapter you should:

1. understand the importance of accuracy in CAD drawings;

2. know the function of each of the Object Snap (*OSNAP*) modes;

3. be able to recognize the Snap Marker symbols;

4. be able to invoke *OSNAP*s for single point selection;

5. be able to operate Running Object Snap modes;

6. know that you can toggle Running Object Snap off to specify points not at object features;

7. be able to use Object Snap Tracking to create and edit objects that align with existing *OSNAP* points.

CAD ACCURACY

Because CAD databases store drawings as digital information with great precision (fourteen numeric places in AutoCAD), it is possible, practical, and desirable to create drawings that are 100% accurate; that is, a CAD drawing should be created as an exact dimensional replica of the actual object. For example, lines that appear to connect should actually connect by having the exact coordinate values for the matching line endpoints. Only by employing this precision can dimensions placed in a drawing automatically display the exact intended length, or can a CAD database be used to drive CNC (Computer Numerical Control) machine devices such as milling machines or lathes, or can the CAD database be used for rapid prototyping devices such as Stereo Lithography Apparatus. With CAD/CAM technology (Computer-Aided Design/Computer-Aided Manufacturing), the CAD database defines the configuration and accuracy of the finished part. <u>Accuracy is critical</u>. Therefore, in no case should you create CAD drawings with only visual accuracy such as one might do when sketching using the "eyeball method."

OBJECT SNAP

AutoCAD provides a capability called "Object Snap," or *OSNAP* for short, that enables you to "snap" to existing object endpoints, midpoints, centers, intersections, etc. When an *OSNAP* mode (*Endpoint, Midpoint, Center, Intersection*, etc.) is invoked, you can move the cursor <u>near</u> the desired object feature (endpoint, midpoint, etc.) and AutoCAD locates and calculates the coordinate location of the desired object feature. Available Object Snap modes are:

Endpoint, Midpoint, Intersection, Center, Quadrant, Tangent, Perpendicular, Parallel, Extension, Nearest, Insert, Node, Mid Between Two Points, Temporary Tracking, From, and *Apparent Intersection*

For example, when you want to draw a *Line* and connect its endpoint to an existing *Line*, you can invoke the *Endpoint OSNAP* mode at the "Specify next point or [Undo]:" prompt, then snap <u>exactly</u> to the desired line end by moving the cursor near it and PICKing (Fig. 7-1). *OSNAP*s can be used for any draw or modify operation—whenever AutoCAD prompts for a point (location).

FIGURE 7-1

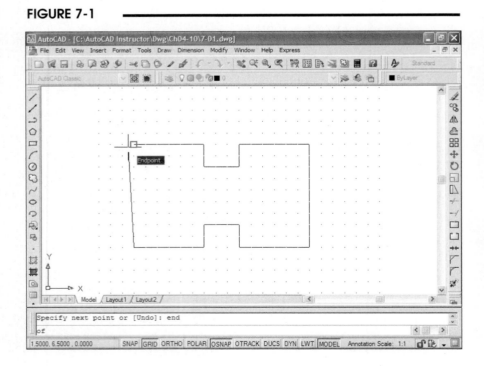

AutoCAD displays a "Snap Marker" indicating the particular object feature (endpoint, midpoint, etc.) when you move
the cursor <u>near</u> an object feature. Each *OSNAP* mode
(*Endpoint, Midpoint, Center, Intersection,* etc.) has a distinct
symbol (Snap Marker) representing the object feature. This
innovation allows you to preview and confirm the snap
points before you PICK them (Fig. 7-2).

FIGURE 7-2

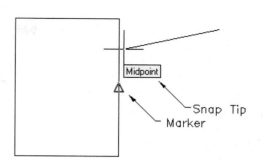

AutoCAD provides two other aids for previewing and confirming *OSNAP* points before you PICK. A
"Snap Tip" appears shortly after the Snap Marker appears (hold the cursor still and wait one second).
The Snap Tip gives the name of the found *OSNAP* point, such as an *Endpoint, Midpoint, Center,* or
Intersection, (Fig. 7-2). In addition, a "Magnet" draws the cursor to the snap point if the cursor is within
the confines of the Snap Marker. This Magnet feature helps confirm that you have the desired snap
point before making the PICK.

A visible target box, or "Aperture," <u>can be displayed</u> at the cursor (invisible by
default) whenever an *OSNAP* mode is in effect (Fig. 7-3). The Aperture is a square
box larger than the pickbox (default size of 10 pixels square). Technically, this target
box (visible or invisible) must be located <u>on an object</u> before a Snap Marker and
related Snap Tip appear. The settings for the Aperture, Snap Markers, Snap Tips, and
Magnet are controlled in the *Drafting* tab of the *Options* dialog box (discussed later).

FIGURE 7-3

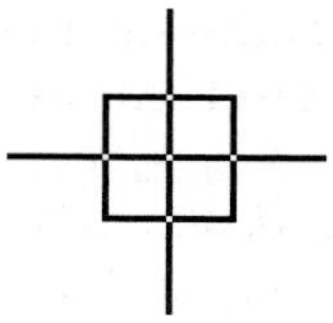

OBJECT SNAP MODES

The Object Snap modes are explained in this section. Each mode, and its relation to the AutoCAD
objects, is illustrated.

Object Snaps must be activated in order for you to "snap" to the desired object features. Two methods
for activating Object Snaps, Single Point Selection and Running Object Snaps, are discussed in the sections following Object Snap Modes. The *OSNAP* modes (*Endpoint, Midpoint, Center, Intersection,* etc.)
operate identically for either method.

Center

This *OSNAP* option finds the center of a *Circle, Arc,*
or *Donut* (Fig. 7-4). You can PICK the *Circle* <u>object</u> or
where you think the center is.

FIGURE 7-4

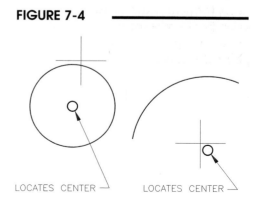

LOCATES CENTER   LOCATES CENTER

Endpoint

 The *Endpoint* option snaps to the endpoint of a *Line, Pline, Spline,* or *Arc* (Fig. 7-5). PICK the object <u>near</u> the desired end.

FIGURE 7-5

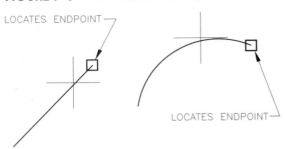

Insert

 This option locates the insertion point of *Text* or a *Block* (Fig. 7-6). PICK anywhere on the *Block* or line of *Text*.

FIGURE 7-6

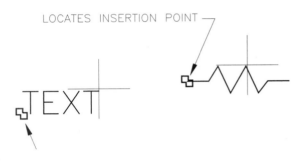

Intersection

 Using this option causes AutoCAD to calculate and snap to the intersection of any two objects (Fig. 7-7). You can locate the cursor (Aperture) so that <u>both</u> objects pass near (through) it, or you can PICK each object <u>individually</u>.

FIGURE 7-7

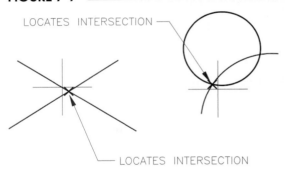

Even if the two objects that you PICK do not physically intersect, you can PICK each one individually with the *Intersection* mode and AutoCAD will find the <u>extended</u> intersection (Fig. 7-8).

FIGURE 7-8

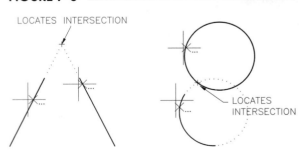

Midpoint

The *Midpoint* option snaps to the point of a *Line* or *Arc* that is <u>halfway</u> between the endpoints. PICK anywhere on the object (Fig. 7-9).

FIGURE 7-9

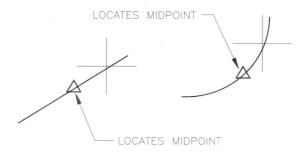

LOCATES MIDPOINT

LOCATES MIDPOINT

Nearest

The *Nearest* option locates the point on an object nearest to the <u>cursor position</u> (Fig. 7-10). Place the cursor center nearest to the desired location, then PICK.

Nearest <u>cannot</u> be used effectively with *ORTHO* or *POLAR* to draw orthogonal lines because the *Nearest* point takes precedence over *ORTHO* and *POLAR*. However, <u>the *Intersection* mode (in Running Osnap mode only) in combination with *POLAR* does allow you to construct orthogonal lines (or lines at any Polar Tracking angle) that intersect with other objects</u>.

FIGURE 7-10

LOCATES NEAREST
POINT ON OBJECT

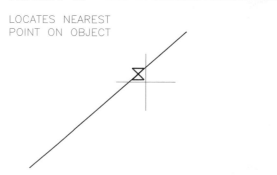

TIP

Node

This option snaps to a <u>*Point*</u> object (Fig. 7-11). The *Point* must be within the Aperture (visible or invisible Aperture).

FIGURE 7-11

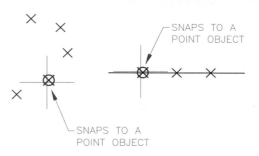

SNAPS TO A
POINT OBJECT

SNAPS TO A
POINT OBJECT

Perpendicular

Use this option to snap perpendicular to the selected object (Fig. 7-12). PICK anywhere on a *Line* or straight *Pline* segment. The *Perpendicular* option is typically used for the second point ("Specify next point:" prompt) of the *Line* command.

FIGURE 7-12

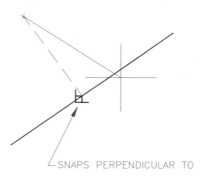

SNAPS PERPENDICULAR TO

Quadrant

The *Quadrant* option snaps to the 0, 90, 180, or 270 degree quadrant of a *Circle* (Fig. 7-13). PICK <u>nearest</u> to the desired *Quadrant*.

FIGURE 7-13

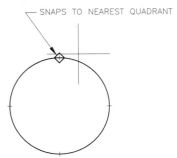

SNAPS TO NEAREST QUADRANT

Tangent

This option calculates and snaps to a tangent point of an *Arc* or *Circle* (Fig. 7-14). PICK the *Arc* or *Circle* as near as possible to the expected *Tangent* point.

FIGURE 7-14

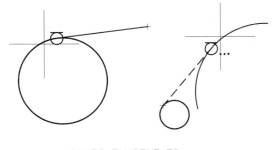

SNAPS TANGENT TO

Apparent intersection

Use this option when you are working with a 3D drawing and want to snap to a point in space where two objects appear to intersect (from your viewpoint) but do not actually physically intersect.

Acquisition Object Snap Modes

Three Object Snap modes require an "acquisition" step: *Parallel, Extension,* and *Temporary Tracking.* When you use these Object Snap modes a dotted line appears, called an <u>alignment vector</u>, that indicates a vector along which the selected point will lie. These modes require an additional step—you must "acquire" (select) a point or object. For example, using the *Parallel* mode, you must "acquire" an object to be parallel to. The process used to "acquire" objects is explained here and is similar to that used for Object Snap Tracking (see "Object Snap Tracking" later in this chapter).

To Acquire an Object

To acquire an object to use for an *Extension* or *Parallel* Object Snap mode, move the cursor over the desired object and pause briefly, <u>but do not pick the object</u>. A small plus sign (+) is displayed when AutoCAD acquires the object (Fig. 7-15). A dotted-line "alignment vector" appears as you move the cursor into a parallel or extension position (Fig. 7-16). <u>You can acquire multiple objects to generate multiple vectors</u>.

To clear an acquired object (in case you decide not to use that object), move the cursor back over the acquisition marker until the plus sign (+) disappears. Acquired points also clear automatically when another command is issued.

FIGURE 7-15

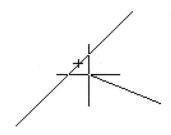

Parallel

The *Parallel* Osnap option snaps the rubberband line into a parallel relationship with any acquired object (see "To Acquire an Object," earlier this section).

Start a *Line* by picking a start point. When AutoCAD prompts to "Specify next point or [Undo]:," acquire an object to use as the parallel source (a line to draw parallel to). Next, move the cursor to within a reasonably parallel position with the acquired line. The current rubberband line snaps into an exact parallel position (Fig. 7-16). A dotted-line parallel alignment vector appears as well as a tool tip indicating the current *Line* length and angle. The parallel Osnap symbol appears on the source parallel line. Pick to specify the current *Line* length. Keep in mind multiple lines can be acquired, giving you several parallel options. The acquired source objects lose their acquisition markers when each *Line* segment is completed.

Consider the use of *Parallel* Osnap with other commands such as *Move* or *Copy*. Figure 7-17 illustrates using *Move* with a *Circle* in a direction *Parallel* to the acquired *Line*.

Extension

The *Extension* Osnap option snaps the rubberband line so that it intersects with an extension of any acquired object (see "To Acquire an Object," earlier this section).

For example, assume you use the *Line* command, then the *Extension* Osnap mode. When another existing object is acquired, the current *Line* segment intersects with an extension of the acquired object. Figure 7-18 depicts drawing a *Line* segment (upper left) to an *Extension* of an acquired *Line* (lower right). Notice the acquisition marker (plus symbol) on the acquired object.

Consider drawing a *Line* to an *Extension* of other objects, such as shown in Figure 7-19. Here a *Line* is drawn to the *Extension* of an *Arc*.

FIGURE 7-16

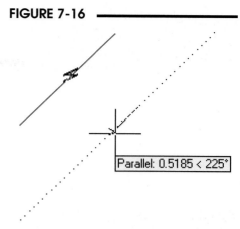

Parallel: 0.5185 < 225°

FIGURE 7-17

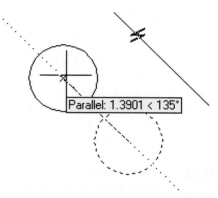

Parallel: 1.3901 < 135°

FIGURE 7-18

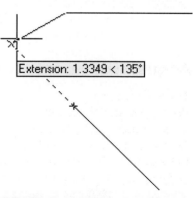

Extension: 1.3349 < 135°

FIGURE 7-19

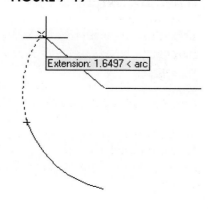

Extension: 1.6497 < arc

If you set *Extension* as a <u>Running Osnap mode</u>, each newly created *Line* segment automatically becomes acquired; therefore, you can draw an extension of the previous segment (at the same angle) easily (Fig. 7-20). (See "Osnap Running Mode" later in this chapter.)

FIGURE 7-20 ━━━━━

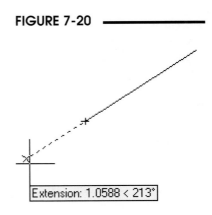

Extension: 1.0588 < 213°

Object Snap Modifiers

Three Object Snap modes can be thought of as Object Snap modifiers—that is, they are intended to be used with other Object Snap options. They are *Temporary Tracking* or *TT* (also requires an acquisition point), *From*, and *Mid Between Two Points* or *M2p*.

Temporary Tracking (TT)

Temporary Tracking sets up a temporary Polar Tracking point. This option allows you to "track" in a polar direction from any point you select. <u>Tracking</u> is the process of moving from a selected point in a preset angular direction. The preset angles are those specified in the *Increment Angle* section in the *Polar Tracking* tab of the *Drafting Settings* dialog box (see "Polar Tracking" in Chapter 3).

From

The *From* option is similar to the *Temporary Tracking* mode; however, instead of using a tracking vector to place the desired point, you specify a distance using relative Cartesian or relative polar coordinates. Like *Temporary Tracking*, you use another *Osnap* mode to specify the point to track from, except with the *From* option no tracking vector appears. There are two steps: first, select a "base point" (use another *Osnap* mode to select this point), then specify an "Offset" (enter relative Cartesian or relative polar coordinates).

Mid Between 2 Points (M2p)

M2p finds the midpoint between two points. *M2p* is intended to find a point <u>not necessarily on an object</u>—the point can be anywhere between the two selected points. Other *Osnap* modes should be used to designate the desired two points.

OSNAP SINGLE POINT SELECTION

Object Snaps (Single Point)	Pull-down Menu	Command (Type)	Alias (Type)	Short-cut	Screen (side) Menu	Tablet Menu	Cursor Menu (Shift+button 2)
	...	*End, Mid,* etc. (first three letters)	...	...	****(asterisks)	T,15 - U,22	*Endpoint Midpoint,* etc.

There are many methods for invoking *OSNAP* modes for single point selection, as shown in the command table. If you prefer to type, enter only the <u>first three letters</u> of the *OSNAP* mode at the "Specify first point: " prompt, "Specify next point or [Undo]:" prompt, or <u>any time AutoCAD prompts for a point</u>.

If desired, an *Object Snap* toolbar (Fig. 7-21) can be activated to float or dock on the screen by right-clicking on any icon button and selecting *Object Snap* from the list of toolbars.

FIGURE 7-21

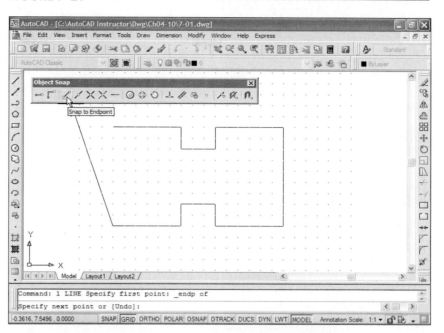

In addition to these options, a special menu called the <u>cursor menu</u> can be used. The cursor menu pops up at the <u>current location</u> of the cursor and replaces the cursor when invoked (Fig. 7-22). This menu is activated by pressing **Shift+right-click** (hold down the Shift key while clicking the right mouse button).

(If you have a wheel mouse, you can also change the *MBUTTON-PAN* system variable to 0. This action allows you to press the wheel to invoke the *Osnap* cursor menu, but disables your ability to *Pan* by holding down the wheel.)

FIGURE 7-22

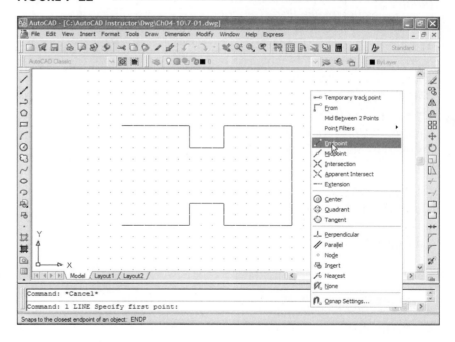

<u>With any of these methods, *OSNAP* modes are active only for selection of a single point.</u> If you want to *OSNAP* to another point, you must select an *OSNAP* mode again for the second point. With this method, the desired *OSNAP* mode is selected <u>transparently</u> (invoked during another command operation) immediately before selecting a point when prompted.

OSNAP RUNNING MODE

A more effective method for using Object Snap is called "Running Object Snap" because one or more *OSNAP* modes (*Endpoint, Center, Midpoint,* etc.) can be turned on and kept running indefinitely. Turn on Running Object Snap by selecting the *OSNAP* button on the Status Bar (Fig. 7-23) or by pressing the F3 key or Ctrl+F key sequence.

FIGURE 7-23 ───────

RTHO POLAR OSNAP OTRACK DUI

 This method can obviously be more productive because you do not have to continually invoke an *OSNAP* mode each time you need to use one. For example, suppose you have <u>several</u> *Endpoints* to connect. It would be most efficient to turn on the running *Endpoint OSNAP* mode and leave it <u>running</u> during the multiple selections. This is faster than continually selecting the *OSNAP* mode each time before you PICK.

You normally want <u>several *OSNAP* modes running at the same time</u>. A common practice is to turn on the *Endpoint, Center, Midpoint* modes simultaneously. In that way, if you move your cursor near a *Circle,* the *Center* Marker appears; if you move the cursor near the end or the middle of a *Line,* the *Endpoint,* or the *Midpoint* mode Markers appear.

Object Snaps (Running)

Pull-down Menu	Command (Type)	Alias (Type)	Short-cut	Screen (side) Menu	Tablet Menu	Cursor Menu (Shift+button 2)
Tools *Drafting Settings...* *Object Snap*	*Osnap* or *-Osnap*	*OS* or *-OS*	...	*TOOLS 2* *Osnap...*	U,22	*Osnap* *Settings...*

Running Object Snap Modes

 All features of Running Object Snaps are controlled by the *Drafting Settings* dialog box (Fig. 7-24). This dialog box can be invoked by the following methods (see the previous Command Table):

1. Type the *OSNAP* command.
2. Type *OS,* the command alias.
3. Select *Drafting Settings...* from the *Tools* pull-down menu.
4. Select *Osnap Settings...* from the bottom of the cursor menu (Shift + right-click).
5. Select the *Object Snap Settings* icon button from the *Object Snap* toolbar.
6. Right-click on the words *OSNAP* or *OTRACK* on the Status Bar, then select *Settings...* from the shortcut menu that appears.

FIGURE 7-24 ───────

Use the *Object Snap* tab to select the desired Object Snap settings. Try using three or four commonly used modes together, such as *Endpoint, Midpoint, Center,* and *Intersection.* The Snap Markers indicate which one of the modes would be used as you move the cursor near different object features. Using similar modes simultaneously, such as *Center, Quadrant,* and *Tangent,* can sometimes lead to difficulties since it requires the cursor to be placed almost in an exact snap spot, or in some cases it may not find one of the modes (*Quadrant* overrides *Tangent*). In these cases, the Tab key can be used to cycle through the options (see "Object Snap Cycling").

Running Object Snap Toggles

Another feature that makes Running Object Snap effective is Osnap Toggle. If you need to PICK a point without using *OSNAP,* use the Osnap Toggle to temporarily override (turn off) the modes. With Running Osnaps temporarily off, you can PICK any point without AutoCAD forcing your selection to an *Endpoint, Center,* or *Midpoint,* etc. When you toggle Running Object Snap on again, AutoCAD remembers which modes were previously set.

The following methods can be used to toggle Running Osnaps on and off:

1. click the word *OSNAP* that appears on the Status Bar (at the bottom of the screen)
2. press **F3**
3. press **Ctrl+F**

None

 The *None OSNAP* option is a Running Osnap <u>override effective for only one PICK</u>. None is similar to the Running Object Snap Toggle, except it is effective for one PICK, then Running Osnaps automatically come back on without having to use a toggle. If you have *OSNAP* modes running but want to deactivate them for a <u>single</u> PICK, use *None* in response to "Specify first point:" or other point selection prompt. In other words, using *None* <u>during a draw or edit command</u> overrides any Running Osnaps for that single point selection. *None* can be typed at the command prompt (when prompted for a point) or can be selected from the bottom of the Object Snap toolbar.

Object Snap Cycling

In cases when you have multiple Running Osnaps set, and it is difficult to get the desired snap Marker to appear, you can use the <u>Tab</u> key to cycle through the possible *OSNAP* modes for the highlighted object. In other words, pressing the Tab key makes AutoCAD highlight the object nearest the cursor, then cycles through the possible snap Markers (for running modes that are set) that affect the object.

For example, when the *Center* and *Quadrant* modes are both set as Running Osnaps, moving the cursor near a *Circle* makes the *Quadrant* Marker appear but not the *Center* Marker. In this case, pressing the Tab key highlights the *Circle,* then cycles through the four *Quadrant* and one *Center* snap candidates.

OBJECT SNAP TRACKING

A feature called Object Snap Tracking helps you draw objects at specific angles and in specific relationships to other objects. Object Snap Tracking works in conjunction with object snaps and displays temporary alignment paths called "tracking vectors" that help you create objects aligned at precise angular positions relative to other objects. You can toggle Object Snap Tracking on and off with the *OTRACK* button on the Status Bar or by toggling F11.

To practice with Object Snap Tracking, try turning the *Extension* and *Parallel* Osnap options off. Since these two new options require point acquisition and generate alignment vectors, it can be difficult to determine which of these features is operating when they are all on.

Object Snap Tracking is similar to Polar Tracking in that it displays and snaps to alignment vectors, but the alignment vectors are generated from <u>other existing objects</u>, not from the current object. These other objects are acquired by Osnapping to them. Once a point (*Endpoint, Midpoint,* etc.) is acquired, alignment vectors generate from them in proximity to the cursor location. This process allows you to construct geometry that has orthogonal or angular relationships to other existing objects.

Using Object Snap Tracking is essentially the same as using the *Temporary Tracking* Osnap option, then using another Osnap mode to acquire the tracking point. Object Snap Tracking can be used with either Single Point or Running Osnap mode.

For example, Figure 7-25 displays an alignment vector generated from an acquired *Endpoint* (top of the left *Line*). The current *Line* can then be drawn to a point that is horizontally aligned with the acquired *Endpoint*. Note that in Figure 7-25 and the related figures following, the Osnap Marker (*Endpoint* marker in this case) is anchored to the acquired point. The alignment vector always rotates about, and passes through, the acquired point (at the *Endpoint* marker). The current *Line* being constructed is at the cursor location.

FIGURE 7-25

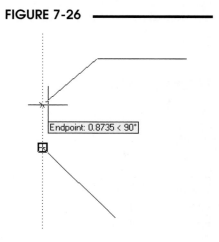

Endpoint: 0.6867 < 0°

Moving the cursor to another location causes a different alignment vector to appear; this one displays vertical alignment with the same *Endpoint* (Fig. 7-26).

FIGURE 7-26

Endpoint: 0.8735 < 90°

In addition, moving the cursor around an acquired point causes an array of alignment vectors to appear based on the current angular increment set in the *Polar Tracking* tab of the *Drafting Settings* dialog box (see Chapter 3). In Figure 7-27, alignment vectors are generated from the acquired point in 30-degree increments.

FIGURE 7-27

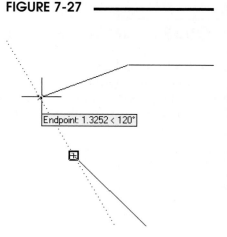

Endpoint: 1.3252 < 120°

To acquire a point to use for Object Snap Tracking (when AutoCAD prompts to specify a point), move the cursor over the object point and pause briefly when the Osnap Marker appears, but <u>do not pick the point</u>. A small plus sign (+) is displayed when AutoCAD acquires the point (Fig. 7-28). The alignment vector appears as you move the cursor away from the acquired point. <u>You can acquire multiple points to generate multiple alignment vectors</u>.

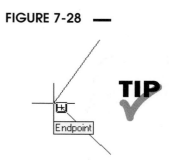

In Figure 7-28, the *Endpoint* object snap is on. Start a *Line* by picking its start point, move the cursor over another line's endpoint to acquire it, and then move the cursor along the horizontal, vertical, or polar alignment vector that appears (not shown in Fig. 7-28) to locate the endpoint you want for the line you are drawing.

To clear an acquired point (in case you decide not to use an alignment vector from that point), move the cursor back over the point's acquisition marker until the plus sign (+) disappears. Acquired points also clear automatically when another command is issued.

To Use Object Snap Tracking:

1. Turn on both Object Snap and Object Snap Tracking (press F3 and F11 or toggle *OSNAP* and *OTRACK* on the Status Bar). You can use *OSNAP* single point selection when prompted to specify a point instead of using Running Osnap.

2. Start a *Draw* or *Modify* command that prompts you to specify a point.

3. Move the cursor over an Object Snap point to temporarily acquire it. <u>Do not PICK the point</u> but only pause over the point briefly to acquire it.

4. Move the cursor away from the acquired point until the desired vertical, horizontal, or polar alignment vector appears, then PICK the desired location for the line along the alignment vector.

Object Snap Tracking with Polar Tracking

For some cases you may want to use Object Snap Tracking in conjunction with Polar Tracking. This combination allows you to track from the last point specified (on the current *Line* or other operation) as well as to connect to an alignment vector from an existing object.

Figure 7-29 displays both Polar Tracking and Object Snap Tracking in use. The current *Line* (dashed vertical line) is Polar Tracking along a vertical vector from the last point specified (on the horizontal line above). The new *Line's* endpoint falls on the horizontal alignment vector (Object Snap Tracking) acquired from the *Endpoint* of the existing diagonal *Line*.

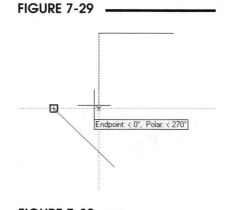

If the cursor is moved from the previous location (in the previous figure), additional Polar Tracking options and Object Snap Tracking options appear. For example, in Figure 7-30, the current *Line* endpoint is tracking at 210 degrees (Polar Tracking) and falls on a vertical alignment vector from the *Endpoint* of the diagonal line (Object Snap Tracking).

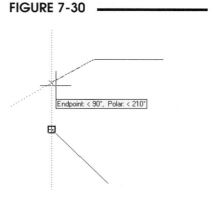

You can accomplish the same capabilities available with the combination of Polar Tracking and Object Snap Tracking by using only Object Snap Tracking and multiple acquired points. For example, Figure 7-31 illustrates the same situation as in Figure 7-29 but only Object Snap tracking is on (Polar Tracking is off). Notice that <u>two Endpoints</u> have been acquired to generate the desired vertical and horizontal alignment vectors.

FIGURE 7-31

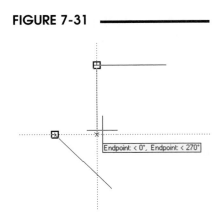

Object Snap Tracking Settings

You can set the *Object Snap Tracking Settings* in the *Drafting Settings* dialog box (Fig. 7-32). The only options are to *Track orthogonally only* or to *Track using all polar angle settings*. The Object Snap Tracking vectors are determined by the *Increment angle* and *Additional angles* set for Polar Tracking (left side of dialog box). If you want to track using these angles, select *Track using all polar angle settings*.

FIGURE 7-32

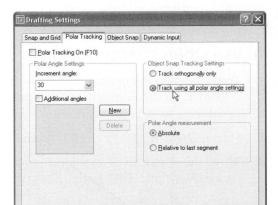

OSNAP APPLICATIONS

OSNAP can be used any time AutoCAD prompts you for a point. This means that you can invoke an *OSNAP* mode during any draw or modify command as well as during many other commands. *OSNAP* provides you with the potential to create 100% accurate drawings with AutoCAD. Take advantage of this feature whenever it will improve your drawing precision. Remember, any time you are prompted for a point, use *OSNAP* if it can improve your accuracy.

CHAPTER EXERCISES

1. *OSNAP* **Single Point Selection**

    Open the **CH6EX1A** drawing and begin constructing the sheet metal part. Each unit in the drawing represents one inch.

    A. Create four **Circles**. All *Circles* have a radius of **1.685**. The *Circles'* centers are located at **5,5**, **5,13**, **19,5**, and **19,13**.

    B. Draw four **Lines**. The *Lines* (highlighted in Fig. 7-33) should be drawn on the outside of the *Circles* by using the **Quadrant** *OSNAP* mode as shown for each *Line* endpoint.

FIGURE 7-33

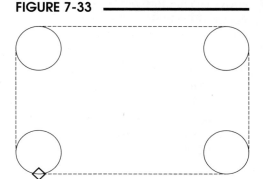

C. Draw two *Lines* from the *Center* of the existing *Circles* to form two diagonals as shown in Figure 7-34.

D. At the *Intersection* of the diagonals create a *Circle* with a **3** unit radius.

FIGURE 7-34

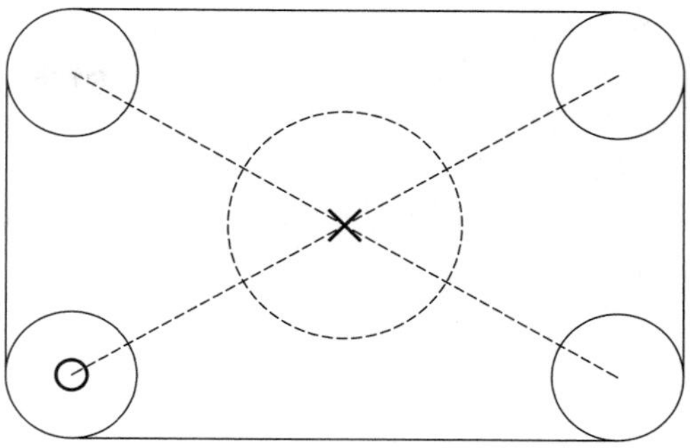

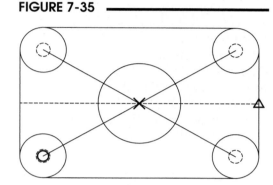

E. Draw two *Lines,* each from the *Intersection* of the diagonals to the *Midpoint* of the vertical *Lines* on each side. Finally, construct four new *Circles* with a radius of **.25**, each at the *Center* of the existing ones (Fig. 7-35).

F. *SaveAs* **CH7EX1.**

FIGURE 7-35

2. *OSNAP* **Single Point Selection**

A multiview drawing of a mechanical part is to be constructed using the **A-METRIC** drawing. All dimensions are in millimeters, so each unit equals one millimeter.

A. *Open* **A-METRIC** from Chapter 6 Exercises. Draw a *Line* from **60,140** to **140,140**. Create two *Circles* with the centers at the *Endpoints* of the *Line,* one *Circle* having a <u>diameter</u> of **60** and the second *Circle* having a diameter of **30**. Draw two *Lines Tangent* to the *Circles* as shown in Figure 7-36. *SaveAs* **PIVOTARM CH7.**

FIGURE 7-36

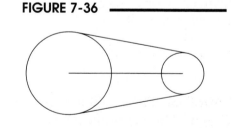

B. Draw a vertical *Line* down from the far left *Quadrant* of the *Circle* on the left. Specify relative polar coordinates using any input method to make the *Line* 100 units at 270 degrees. Draw a horizontal *Line* **125** units from the last *Endpoint* using relative polar or direct distance entry coordinates. Draw another *Line* between that *Endpoint* and the *Quadrant* of the *Circle* on the right. Finally, draw a horizontal *Line* from point **30,70** and *Perpendicular* to the vertical *Line* on the right (Fig. 7-37).

FIGURE 7-37

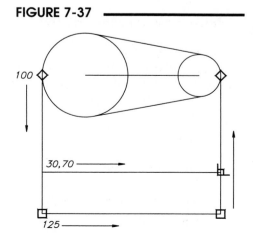

C. Draw two vertical *Lines* from the
    Intersections of the horizontal *Line*
    and *Circles* and *Perpendicular* to the
    Line at the bottom. Next, draw two
    Circles concentric to the previous
    two and with diameters of **20** and **10**
    as shown in Figure 7-38.

FIGURE 7-38 ────────────

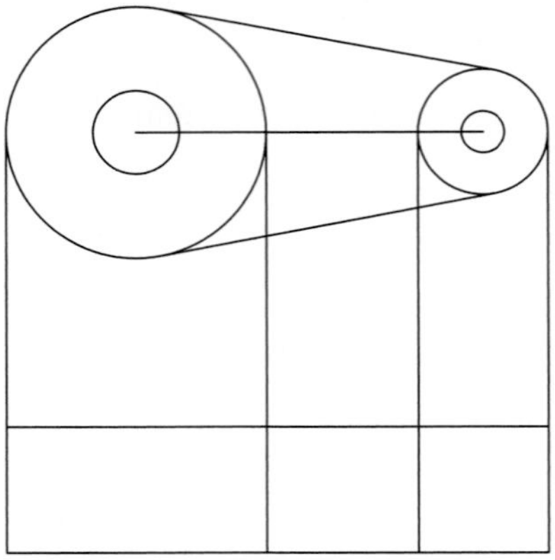

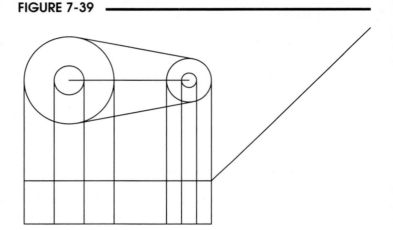

D. Draw four more vertical *Lines* as
    shown in Figure 7-39. Each *Line* is
    drawn from the new *Circles'*
    Quadrant and *Perpendicular* to the
    bottom line. Next, draw a miter
    Line from the *Intersection* of the
    corner shown to **@150<45**. *Save* the
    drawing for completion at a later
    time as another chapter exercise.

FIGURE 7-39 ────────────

3. **Running Osnap**

    Create a cross-sectional view of a door header composed of two 2 x 6
    wooden boards and a piece of 1/2" plywood. (The dimensions of a
    2 x 6 are actually 1-1/2" x 5-3/8".)

    A. Begin a *New* drawing and assign the name **HEADER**. Draw four
        vertical lines as shown in Figure 7-40.

    B. Use the *OSNAP* command or select *Drafting Settings…* from the
        Tools pull-down menu and turn on the *Endpoint* and
        Intersection modes.

FIGURE 7-40 ────────────

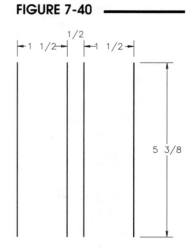

C. Draw the remaining lines as shown in Figure 7-41 to complete the header cross-section. *Save* the drawing.

FIGURE 7-41 ——

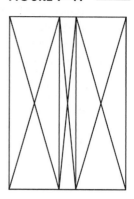

4. **Running Object Snap and *OSNAP* toggle**

Assume you are commissioned by the local parks and recreation department to provide a layout drawing for a major league sized baseball field. Lay out the location of bases, infield, and outfield as follows.

A. *Open* the **BALL FIELD CH6.DWG** that you set up in Chapter 6. Make sure *Limits* are set to 512',384', *Snap* is set to 10', and *Grid* is set to 20'. Ensure *SNAP* and *GRID* are on. Use *SaveAs* to save and rename the drawing to **BALL FIELD CH7**.

B. Begin drawing the baseball diamond by using the *Line* command and using Dynamic Input. **PICK** the "Specify first point:" at **20',20'** (watch *Coords*). Draw the foul line to first base by turning on **POLAR**, move the cursor to the right (along the X direction) and enter a value of **90'** (don't forget the apostrophe to indicate feet). At the "Specify next point or [Undo]:" prompt, continue by drawing a vertical *Line* of **90'**. Continue drawing a square with **90'** between bases (Fig. 7-42).

FIGURE 7-42 ——————

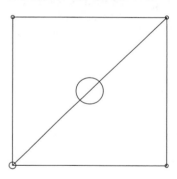

C. Invoke the *Drafting Settings* dialog box by any method. Turn on the *Endpoint, Midpoint,* and *Center* object snaps.

D. Use the *Circle* command to create a "base" at the lower-right corner of the square. At the "Specify center point for circle or [3P/2P/Ttr]:" prompt, **PICK** the *Line Endpoint* at the lower-right corner of the square. Enter a *Radius* of **1'** (don't forget the apostrophe). Since Running Object Snaps are on, AutoCAD should display the Marker (square box) at each of the *Line Endpoints* as you move the cursor near; therefore, you can easily "snap" the center of the bases (*Circles*) to the corners of the square. Draw *Circles* of the same *Radius* at second base (upper-right corner), and third base (upper-left corner of the square). Create home plate with a *Circle* of a **2'** *Radius* by the same method (see Fig. 7-42).

E. Draw the pitcher's mound by first drawing a *Line* between home plate and second base. **PICK** the "Specify first point:" at home plate (*Center* or *Endpoint*), then at the "Specify next point or [Undo]:" prompt, PICK the *Center* or the *Endpoint* at second base. Construct a *Circle* of **8'** *Radius* at the *Midpoint* of the newly constructed diagonal line to represent the pitcher's mound (see Fig. 7-42).

F. *Erase* the diagonal *Line* between home plate and second base. Draw the foul lines from first and third base to the outfield. For the first base foul line, construct a *Line* with the "Specify first point:" at the *Endpoint* of the existing first base line or *Center* of the base. Move the cursor (with *POLAR* on) to the right (X positive) and enter a value of **240′** (don't forget the apostrophe). Press **Enter** to complete the *Line* command. Draw the third base foul line at the same length (in the positive Y direction) by the same method (see Fig. 7-43).

FIGURE 7-43 ───────

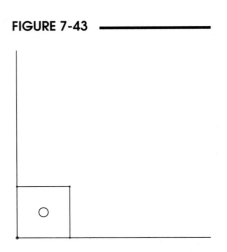

G. Draw the home run fence by using the *Arc* command from the *Draw* pull-down menu. Select the *Start, Center, End* method. **PICK** the end of the first base line (*Endpoint* object snap) for the *Start* of the *Arc* (Fig. 7-44, point 1), **PICK** the pitcher's mound (*Center* object snap) for the *Center* of the *Arc* (point 2), and the end of the third base line (*Endpoint* object snap) for the *End* of the *Arc* (point 3). Don't worry about *POLAR* in this case because *OSNAP* overrides *POLAR*.

FIGURE 7-44 ───────

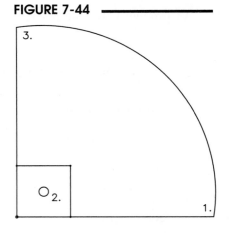

H. Now turn off **POLAR**. Construct an *Arc* to represent the end of the infield. Select the *Arc Start, Center, End* method from the *Draw* pull-down menu. For the *Start* point of the *Arc*, toggle Running Osnaps <u>off</u> by pressing **F3**, **Ctrl+F**, or clicking the word *OSNAP* on the Status Bar and **PICK** location **140′, 20′, 0′** on the first base line (watch *Coords* and ensure *SNAP* and *GRID* are on). (See Fig. 7-45, point 1.) Next, toggle Running Osnaps back on, and **PICK** the *Center* of the pitcher's mound as the *Center* of the *Arc* (point 2). Third, toggle Running Osnaps off again and **PICK** the *End* point of the *Arc* on the third base line (point 3).

FIGURE 7-45 ───────

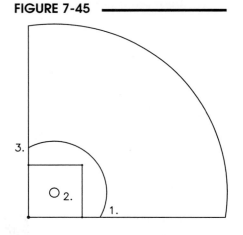

I. Lastly, the pitcher's mound should be moved to the correct distance from home plate. Type *M* (the command alias for *Move*) or select *Move* from the *Modify* pull-down menu. When prompted to "Select objects:" **PICK** the *Circle* representing pitcher's mound. At the "Specify base point or displacement:" prompt, toggle Running Osnaps <u>on</u> and **PICK** the *Center* of the mound. At the "Specify second point of displacement:" prompt, enter **@3′2″<225**. This should reposition the pitcher's mound to the regulation distance from home plate (60′-6″). Compare your drawing to Figure 7-45. *Save* the drawing (as BALL FIELD CH7).

J. Activate a *Layout* tab. Assuming your system is configured for a printer using an 11 x 8.5 sheet and is configured to automatically create a viewport (see "Setting Layout Options and Plot Options" in Chapter 6), a viewport should appear. (If the *Page Setup Manager* automatically appears, select *Close*.) If a viewport appears, proceed to step L. If no viewport appears, complete step K.

K. To make a viewport, type *-Vports* at the Command prompt (don't forget the hyphen) and press **Enter**. Accept the default (*Fit*) option by pressing **Enter**. A viewport appears to fit the printable area.

L. Double-click inside the viewport, then type *Z* for *Zoom* and press **Enter**, and type *A* for *All* and press **Enter**. From the *VP Scale* list (near the lower-right corner of the Drawing Editor), select *1/64"=1'*. Make a print of the drawing on an 11 x 8.5 inch sheet.

M. Activate the *Model* tab. *Save* the drawing as BALL FIELD CH7.

5. **Polar Tracking and Object Snap Tracking**

A. In this exercise, you will create a table base using Polar Tracking and Object Snap Tracking to locate points for construction of holes and other geometry. Begin a *New* drawing and use the **ACAD.DWT** template. *Save* the drawing and assign the name TABLE-BASE. Set the drawing limits to **48** x **32**.

B. Invoke the *Drafting Settings* dialog box. In the *Snap and Grid* tab, set *Polar Distance* to **1.00** and make *Polar Snap* current. In the *Polar Tracking* tab, set the *Increment Angle* to **45**. In the *Object Snap* tab, turn on the *Endpoint, Midpoint,* and *Center Osnap* options. Select *OK*. On the Status Bar ensure *SNAP, POLAR, OSNAP,* and *OTRACK* are on.

C. Use a *Line* and create the three line segments representing the first leg as shown in Figure 7-46. Begin at the indicated location. Use Polar Tracking and Polar Snap to assist drawing the diagonal lines segments. (Do not create the dimensions in your drawing.)

FIGURE 7-46 ━━━━━

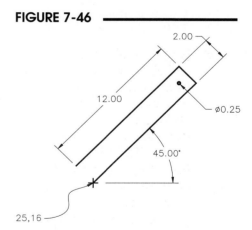

D. Place the first drill hole at the end of the leg using *Circle* with the *Center, Radius* option. Use Object Snap Tracking to indicate the center for the hole (*Circle*). Track vertically and horizontally from the indicated corners of the leg in Figure 7-47. Use *Endpoint Osnap* to snap to the indicated corners. (See Fig. 7-46 for hole dimension.)

FIGURE 7-47 ━━━━━

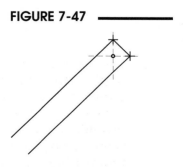

E. Create the other 3 legs in a similar fashion. You can track from the
    Endpoints on the bottom of the existing leg to identify the "first
    point" of the next *Line* segment as shown in Figure 7-48.

FIGURE 7-48 ───

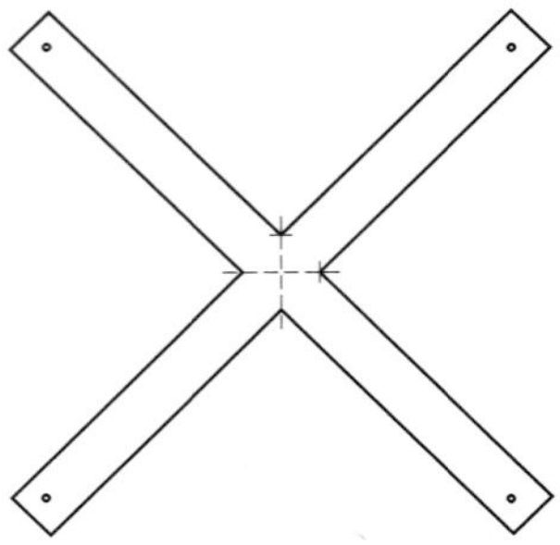

F. Use *Line* to create a 24" x 24" square
    table top on the right side of the
    drawing as shown in Figure 7-49.
    Use Object Snap Tracking and Polar
    Tracking with the *Move* command to
    move the square's center point to the
    center point of the 4 legs (HINT: At
    the "Specify base point or displace-
    ment:" prompt, Osnap Track to the
    square's *Midpoints*. At the "Specify
    second point of displacement:"
    prompt, use *Endpoint* Osnaps to
    locate the legs' center.

FIGURE 7-49 ───

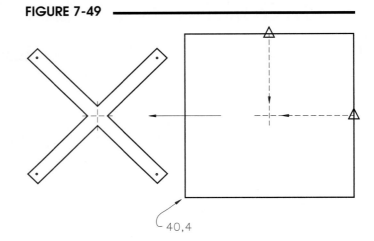

40,4

G. Create a smaller square (*Line*) in the center of the table which
    will act as a support plate for the legs. Track from the *Midpoint*
    of the legs to create the lines as indicated (highlighted) in
    Figure 7-50.

H. *Save* the drawing. You will finish the table base in Chapter 9.

FIGURE 7-50 ───

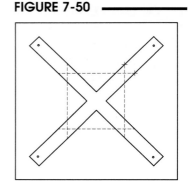

DRAW COMMANDS I

CHAPTER OBJECTIVES

After completing this chapter you should:

1. know where to locate and how to invoke the draw commands;

2. be able to draw *Lines*;

3. be able to draw *Circles* by each of the five options;

4. be able to draw *Arcs* by each of the eleven options;

5. be able to create *Point* objects and specify the *Point Style*;

6. be able to create *Plines* with width and combined of line and arc segments.

CONCEPTS

Draw Commands—Simple and Complex

Draw commands create objects. An object is the smallest component of a drawing. The draw commands listed immediately below create simple objects and are discussed in this chapter. Simple objects <u>appear</u> as one entity.

> *Line, Circle, Arc,* and *Point*

Other draw commands create more complex shapes. Complex shapes appear to be composed of several components, but each shape is usually <u>one object</u>. An example of an object that is one entity but usually appears as several segments is listed below and is also covered in this chapter:

> *Pline*

Other draw commands discussed in Chapter 15 (listed below) are a combination of simple and complex shapes:

> *Xline, Ray, Polygon, Rectangle, Donut, Spline, Ellipse, Divide, Mline, Measure, Sketch, Solid, Region,* and *Boundary*

Draw Command Access

As a review from Chapter 3, Draw Command Concepts, remember that there are many methods you can use to access the draw commands. For example, you can enter the command names or command aliases at the keyboard. Alternately, you can select the command from the *Draw* pull-down menu (Fig. 8-1). If you are using the *AutoCAD Classic* workspace, access to drawing commands is provided on the *Draw* toolbar (Fig. 8-1, left side). If you prefer the *2D Drafting and Annotation* workspace, you can select the desired tool (icon button) from the *2D Draw* control panel on the dashboard (Fig. 8-2). Note that many of the draw commands are only accessible through flyouts. Draw commands can also be accessed using the *Draw* tool palette. In addition, you can use the *DRAW1* and *DRAW2* screen menus or the digitizing tablet, if available.

FIGURE 8-1

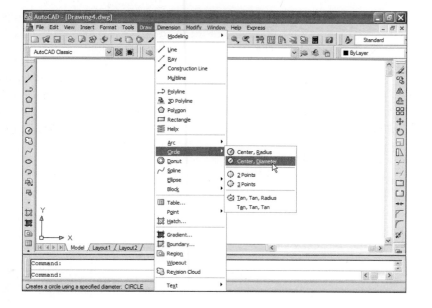

FIGURE 8-2

Coordinate Entry Input Methods

When creating objects with draw commands, AutoCAD always prompts you to indicate points (such as endpoints, centers, radii) to describe the size and location of the objects to be drawn. An example you are familiar with is the *Line* command, where AutoCAD prompts for the "Specify first point:". Indication of these points, called <u>coordinate entry</u> can be accomplished by three input methods:

1.	Mouse Input	PICK (press the left <u>mouse</u> button to specify a point on the screen).
2.	Dynamic Input	When *DYN* is on, you can <u>key values into the edit boxes</u> near the cursor and press Tab to change boxes.
3.	Command Line Input	You can <u>key values</u> in any format <u>at the Command prompt</u>.

Any of these methods can be used <u>whenever</u> AutoCAD prompts you to specify points. (For practice with these methods, see Chapter 3, Draw Command Concepts.)

Drawing Aids

Also keep in mind that you can specify points interactively using the following AutoCAD features individually or in combination: Grid Snap, Polar Snap, Ortho, Polar Tracking, Object Snap, and Object Snap Tracking. Use these drawing tools <u>whenever</u> AutoCAD prompts you to select points. (See Chapters 3 and 7 for details on these tools.)

COMMANDS

Line

Pull-down Menu	Command (Type)	Alias (Type)	Short-cut	Screen (side) Menu	Tablet Menu
Draw *Line*	*Line*	L	...	*DRAW 1* *Line*	J,10

This is the fundamental drawing command. The *Line* command creates straight line segments; each segment is an object. One or several line segments can be drawn with the *Line* command.

Command: **Line**
Specify first point: **PICK** or (**coordinates**) (A point can be designated by interactively selecting with the input device or by entering coordinates. Use any of the drawing tools to assist with interactive entry.)
Specify next point or [Undo]: **PICK** or (**coordinates**) (Again, device input or keyboard input can be used.)
Specify next point or [Undo]: **PICK** or (**coordinates**)
Specify next point or [Close/Undo]: **PICK** or (**coordinates**) or **C**
Specify next point or [Close/Undo]: press **Enter** to finish command
Command:

FIGURE 8-3

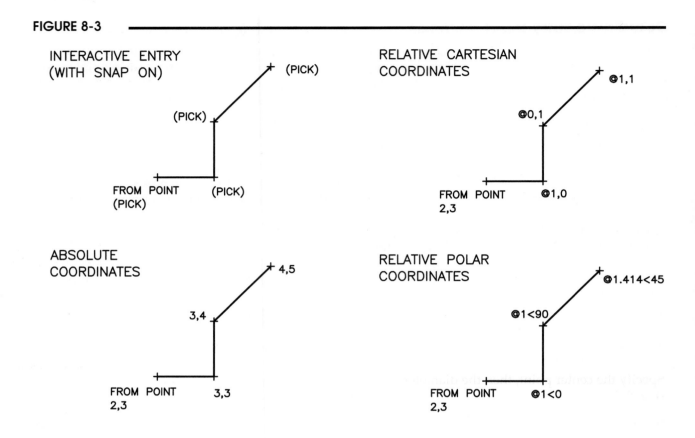

Figure 8-3 shows four examples of creating the same *Line* segments using different formats of coordinate entry. Refer to Chapter 3, Draw Command Concepts, for examples of drawing vertical, horizontal, and inclined lines using the formats for coordinate entry.

Circle

Pull-down Menu	Command (Type)	Alias (Type)	Short-cut	Screen (side) Menu	Tablet Menu
Draw *Circle >*	*Circle*	*C*	...	*DRAW 1* *Circle*	*J,9*

The *Circle* command creates one object. Depending on the option selected, you can provide two or three points to define a *Circle*. As with all commands, the Command line prompt displays the possible options:

> Command: **Circle**
> Specify center point for circle or [3P/2P/Ttr (tan tan radius)]: **PICK** or (**coordinates**) or (**option**).
> (PICKing or entering coordinates designates the center point for the circle. You can enter 3P, 2P, or T for another option.)

As with most commands, the default and other options are displayed on the Command line. The default option always appears first. The other options can be invoked by typing the numbers and/or uppercase letter(s) that appear in brackets [ ].

All of the options for creating *Circles* are available from the *Draw* pull-down menu and from the *DRAW 1* screen (side) menu. The tool (icon button) for only the *Center, Radius* method is included in the *Draw* toolbar. Although the *Circle* options are not available on the *Draw* toolbar, you can use the *Circle* button, then <u>right-click for a shortcut menu</u> showing all the options.

The options, or methods, for drawing *Circles* are listed below. Each figure gives several possibilities for each option, with and without *OSNAPs*.

Center, Radius
Specify a center point, then a radius (Fig. 8-4).

The *Radius* (or *Diameter*) can be specified by entering values or by indicating a length inter-actively (PICK two points to specify a length when prompted). As always, points can be specified by PICKing or entering coordinates. Watch *Coords* for coordinate or distance (polar format) display. Grid Snap, Polar Snap, Polar Tracking, Object Snap, and Object Snap Tracking can be used for interactive point specification.

FIGURE 8-4

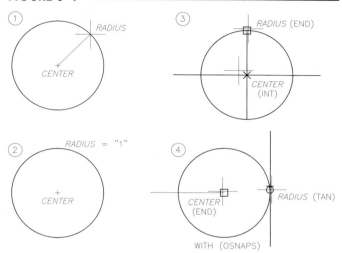

Center, Diameter
Specify the center point, then the diameter (Fig. 8-5).

FIGURE 8-5

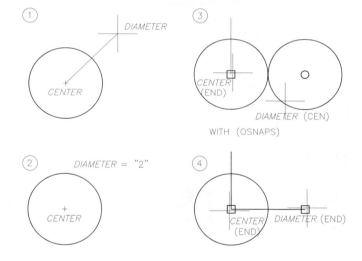

2 Points
The two points specify the location and diameter.

The *Tangent OSNAPs* can be used when selecting points with the *2 Point* and *3 Point* options, as shown in Figures 8-6 and 8-7.

FIGURE 8-6

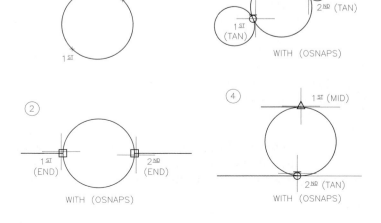

3 Points

The *Circle* passes through all three points specified (Fig. 8-7).

FIGURE 8-7

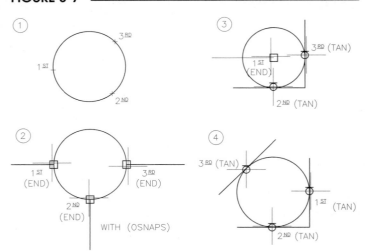

Tangent, Tangent, Radius

Specify two objects for the *Circle* to be tangent to; then specify the radius (Fig. 8-8).

TIP The *TTR* (Tangent, Tangent, Radius) method is extremely efficient and productive. The *OSNAP Tangent* modes are automatically invoked. This is the <u>only</u> draw command option that automatically calls *OSNAPs*.

FIGURE 8-8

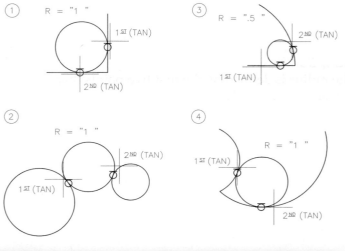

Arc

Pull-down Menu	Command (Type)	Alias (Type)	Short-cut	Screen (side) Menu	Tablet Menu
Draw *Arc >*	*Arc*	*A*	...	DRAW 1 *Arc*	*R,10*

An arc is part of a circle; it is a regular curve of <u>less</u> than 360 degrees. The *Arc* command in AutoCAD provides eleven options for creating arcs. An *Arc* is one object. *Arcs* are always drawn by default in a <u>counterclockwise</u> direction. This occurrence forces you to decide in advance which points should be designated as *Start* and *End* points (for options requesting those points). For this reason, it is often easier to create arcs by another method, such as drawing a *Circle* and then using *Trim* or using the *Fillet* command. (See "Use *Arcs* or *Circles*?" at the end of this section on *Arcs*.) The *Arc* command prompt is:

Command: **Arc**
Specify start point of arc or [Center]: **PICK** or (**coordinates**) or **C** (Interactively select or enter coordinates in any format for the start point. Type C to use the Center option.)

The prompts displayed by AutoCAD are different depending on which option is selected. At any time while using the command, you can select from the options listed on the Command line by typing in the capitalized letter(s) for the desired option.

Alternately, to use a particular option of the *Arc* command, you can select from the *Draw* pull-down menu or from the *DRAW 1* screen (side) menu. The tool (icon button) for only the *3 Points* method is included in the *Draw* toolbar. However, you can select the *Arc* button, then <u>right-click for a shortcut menu</u> at any time during the *Arc* command to show other possible options at that point. These options require coordinate entry of <u>points in a specific order</u>.

3 Points
Specify three points through which the *Arc* passes (Fig. 8-9).

FIGURE 8-9

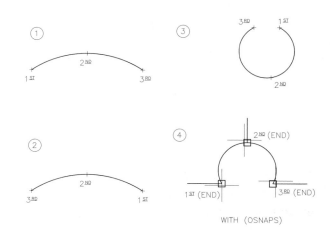

Start, Center, End
The radius is defined by the first two points that you specify (Fig. 8-10).

FIGURE 8-10

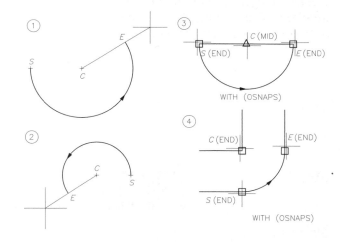

Start, Center, Angle
The angle is the <u>included</u> angle between the sides from the center to the endpoints (Fig. 8-11). A <u>negative</u> angle can be entered to generate an *Arc* in a <u>clockwise</u> direction.

FIGURE 8-11

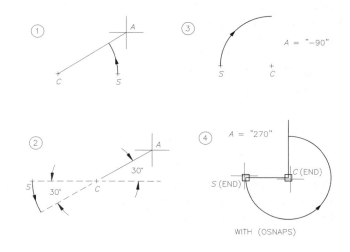

Start, Center, Length

Length means length of chord. The length of chord is between the start and the other point specified (Fig. 8-12). A negative chord length can be entered to generate an *Arc* of 180+ degrees.

FIGURE 8-12

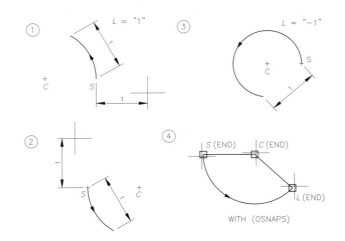

Start, End, Angle

The included angle is between the sides from the center to the endpoints (Fig. 8-13). Negative angles generate clockwise *Arcs*.

FIGURE 8-13

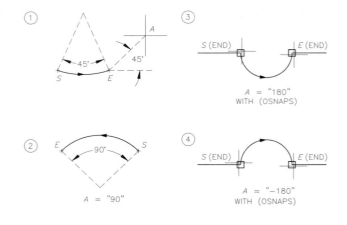

Start, End, Radius

The radius can be PICKed or entered as a value (Fig. 8-14). A negative radius value generates an *Arc* of 180+ degrees.

FIGURE 8-14

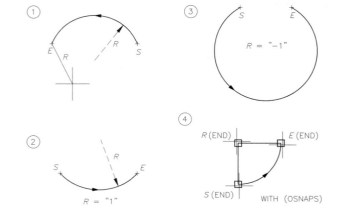

Start, End, Direction

The direction is tangent to the start point (Fig. 8-15).

FIGURE 8-15

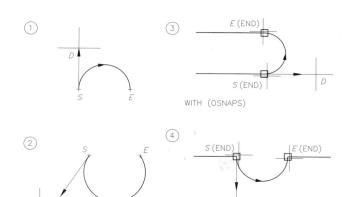

Center, Start, End

This option is like *Start, Center, End* but in a different order (Fig. 8-16).

FIGURE 8-16

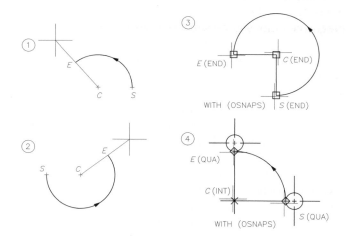

Center, Start, Angle

This option is like *Start, Center, Angle* but in a different order (Fig. 8-17).

FIGURE 8-17

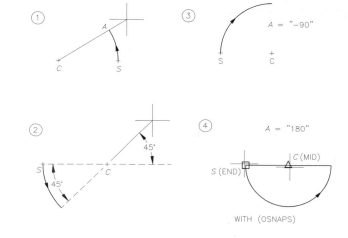

Center, Start, Length

This is similar to the *Start, Center, Length* option but in a different order (Fig. 8-18). *Length* means length of chord.

FIGURE 8-18

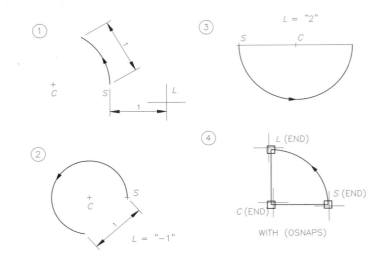

Continue

The new *Arc* continues from and is tangent to the last point (Fig. 8-19). The only other point required is the endpoint of the *Arc*. This method allows drawing *Arcs* tangent to the preceding *Line* or *Arc*.

 Arcs are always created in a <u>counterclockwise</u> direction. This fact must be taken into consideration when using any method <u>except</u> the *3-Point*, the *Start, End, Direction,* and the *Continue* options. The direction is explicitly specified with *Start, End, Direction,* and *Continue* methods, and direction is irrelevant for *3-Point* method.

FIGURE 8-19

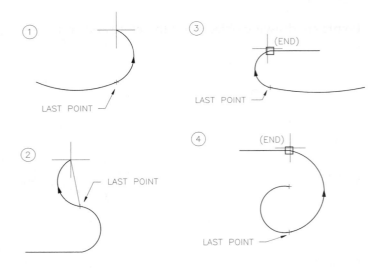

As usual, points can be specified by PICKing or entering coordinates. Watch *Coords* to display coordinate values or distances. Grid Snap, Polar Snap, Polar Tracking, Object Snap, and Object Snap Tracking can be used when PICKing. The *Endpoint, Intersection, Center, Midpoint,* and *Quadrant OSNAP* options can be used with great effectiveness. The *Tangent OSNAP* option <u>cannot</u> be used effectively with most of the *Arc* options. The *Radius, Direction, Length,* and *Angle* specifications can be given by entering values or by PICKing with or without *OSNAPs*.

Use *Arcs* or *Circles*?

 Although there are sufficient options for drawing *Arcs*, <u>usually it is easier to use the *Circle* command</u> followed by *Trim* to achieve the desired arc. Creating a *Circle* is generally an easier operation than using *Arc* because the counterclockwise direction does not have to be considered. The unwanted portion of the circle can be *Trimmed* at the *Intersection* of or *Tangent* to the connecting objects using *OSNAP*. The *Fillet* command can also be used instead of the *Arc* command to add a fillet (arc) between two existing objects (see Chapter 9, Modify Commands I).

Point

Pull-down Menu	Command (Type)	Alias (Type)	Short-cut	Screen (side) Menu	Tablet Menu
Draw Point > Single Point or Multiple Point	Point	PO	...	DRAW 2 Point	O,9

A *Point* is an object that has no dimension; it only has location. A *Point* is specified by giving only one coordinate value or by PICKing a location on the screen.

Figure 8-20 compares *Points* to *Line* and *Circle* objects.

```
Command: point
Current point modes:  PDMODE=0  PDSIZE=0.0000
Specify a point: PICK or (coordinates)
(Select a location for the Point object.)
```

Points are useful in construction of drawings to locate points of reference for subsequent construction or locational verification. The *Node OSNAP* option is used to snap to *Point* objects.

Points are drawing objects and therefore appear in prints and plots. The default "style" for points is a tiny dot. The *Point Style* dialog box can be used to define the format you choose for *Point* objects (the *Point* type [PDMODE] and size [PDSIZE]).

The *Draw* pull-down menu offers the *Single Point* and the *Multiple Point* options. The *Single Point* option creates one *Point*, then returns to the command prompt. This option is the same as using the *Point* command by any other method. Selecting *Multiple Point* continues the *Point* command until you press the Escape key.

FIGURE 8-20

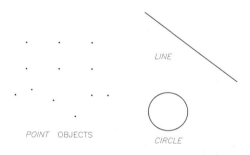

Pline

Pull-down Menu	Command (Type)	Alias (Type)	Short-cut	Screen (side) Menu	Tablet Menu
Draw Polyline	Pline	PL	...	DRAW 1 Pline	N,10

A *Pline* (or *Polyline*) has special features that make this object more versatile than a *Line*. Three features are most noticeable when first using *Plines*:

1. A *Pline* can have a specified *width,* whereas a *Line* has no width.
2. Several *Pline* segments created with one *Pline* command are treated by AutoCAD as <u>one</u> object, whereas individual line segments created with one use of the *Line* command are individual objects.
3. A *Pline* can contain arc segments.

Figure 8-21 illustrates *Pline* versus *Line* and *Arc* comparisons.

FIGURE 8-21

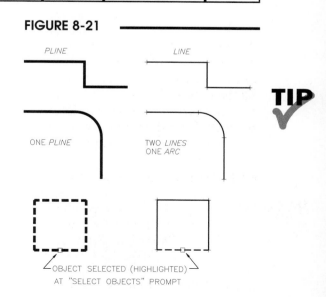

TIP

The *Pline* command begins with the same prompt as *Line;* however, <u>after</u> the "start point:" is established, the *Pline* options are accessible.

>   Command: **Pline**
>   Specify start point: **PICK** or (**coordinates**)
>   Current line-width is 0.0000
>   Specify next point or [Arc/Close/Halfwidth/Length/Undo/Width]:

Similar to the other drawing commands, you can invoke the *Pline* command by any method, then right-click for a shortcut menu at any time during the command to show other possible options at that point.

The options and descriptions follow.

Width

You can use this option to specify starting and ending widths. Width is measured perpendicular to the centerline of the *Pline* segment (Fig. 8-22). *Plines* can be tapered by specifying different starting and ending widths. See NOTE at the end of this section.

Halfwidth

This option allows specifying half of the *Pline* width. *Plines* can be tapered by specifying different starting and ending widths (Fig. 8-22).

FIGURE 8-22

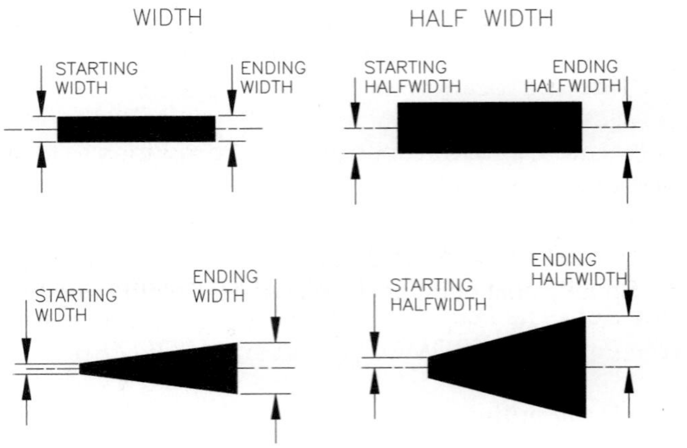

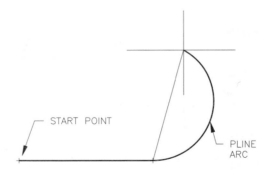

Arc

This option (by default) creates an arc segment in a manner similar to the *Arc Continue* method (Fig. 8-23). Any of several other methods are possible (see "*Pline Arc* Segments").

FIGURE 8-23

Close

The *Close* option creates the closing segment connecting the first and last points specified with the current *Pline* command as shown in Figure 8-24.

This option can also be used to close a group of connected *Pline* segments into one continuous *Pline*. (A *Pline* closed by PICKing points has a specific start and endpoint.) A *Pline Closed* by this method has special properties if you use *Pedit* for *Pline* editing or if you use the *Fillet* command with the *Pline* option (see *Fillet* in Chapter 9 and *Pedit* in Chapter 16).

FIGURE 8-24

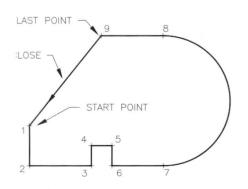

Length

Length draws a *Pline* segment at the same angle as and connected to the previous segment and uses a length that you specify. If the previous segment was an arc, *Length* makes the current segment tangent to the ending direction (Fig. 8-25).

FIGURE 8-25

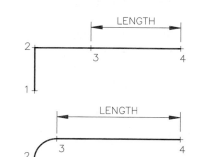

Undo

Use this option to *Undo* the last *Pline* segment. It can be used repeatedly to undo multiple segments.

NOTE: If you change the *Width* of a *Pline,* be sure to respond to <u>both</u> <u>prompts</u> for width ("Specify starting width:" and "Specify ending width:") before you draw the first *Pline* segment. It is easy to hastily PICK the endpoint of the *Pline* segment after specifying the "Specify starting width:" instead of responding with a value (or Enter) for the "Specify ending width:". In this case (if you PICK at the "Specify ending width:" prompt), AutoCAD understands the line length that you interactively specified to be the ending width you want for the next line segment (you can PICK two points in response to the "Specify starting width:" or "Specify ending width:" prompts).

```
Command: Pline
Specify start point: PICK
Current line-width is 0.0000
Specify next point or [Arc/Close/Halfwidth/Length/Undo/Width]: w
Specify starting width <0.0000>: .2
Specify ending width <0.2000>:  Enter a value or press Enter—do not PICK the "next point."
```

Pline Arc Segments

When the *Arc* option of *Pline* is selected, the prompt changes to provide the various methods for construction of arcs:

Specify endpoint of arc or [Angle/CEnter/CLose/Direction/Halfwidth/Line/Radius/Second pt/Undo/Width]:

Angle

You can draw an arc segment by specifying the included angle (a negative value indicates a clockwise direction for arc generation).

CEnter

This option allows you to specify a specific center point for the arc segment.

CLose

This option closes the *Pline* group with an arc segment.

Direction

Direction allows you to specify an explicit starting direction rather than using the ending direction of the previous segment as a default.

Line

This switches back to the line options of the *Pline* command.

Radius

You can specify an arc radius using this option.

Second pt

Using this option allows specification of a 3-point arc.

TIP Because a shape created with one *Pline* command is <u>one object</u>, manipulation of the shape is generally easier than with several objects. For some applications, one *Pline* shape can have advantages over shapes composed of several objects (see "*Offset*," Chapter 9). Editing *Plines* is accomplished by using the *Pedit* command. As an alternative, *Plines* can be "broken" back down into individual objects with *Explode*.

Drawing and editing *Plines* can be somewhat involved. As an alternative, you can draw a shape as you would normally with *Line, Circle, Arc, Trim*, etc., and then <u>convert</u> the shape to one *Pline* object using *Pedit*. (See Chapter 16 for details on converting *Lines* and *Arcs* to *Plines*.)

CHAPTER EXERCISES

Create a *New* drawing. Use the *ACAD.DWT* template, then *Zoom All*, or select *Start from Scratch* and select *Imperial* settings. Set *Polar Snap On* and set the *Polar Distance* to **.25**. Turn on *SNAP, POLAR, OSNAP, OTRACK*, and *DYN*. *Save* the drawing as **CH8EX**. For each of the following problems, *Open* **CH8EX**, complete one problem, then use *SaveAs* to give the drawing a new name.

1. *Open* **CH8EX**. Create the geometry shown in Figure 8-26. Start the first *Circle* center at point **4,4.5** as shown. Do not copy the dimensions. *SaveAs* **LINK** (HINT: Locate and draw the two small *Circles* first. Use *Arc, Start, Center, End* or *Center, Start, End* for the rounded ends.)

FIGURE 8-26

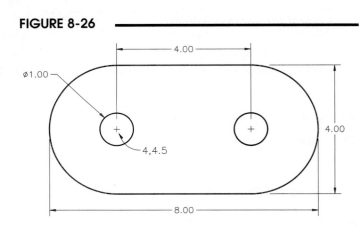

2. *Open* **CH8EX**. Create the geometry as shown in Figure 8-27. Do not copy the dimensions. Assume symmetry about the vertical axis.

    SaveAs **SLOTPLATE CH8.**

FIGURE 8-27

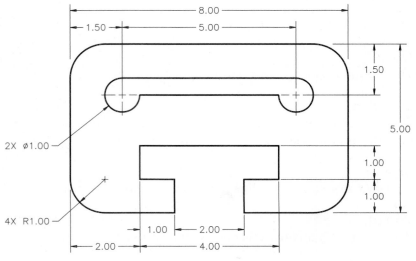

3. **Open CH8EX**. Create the shapes shown in Figure 8-28. Do not copy the dimensions. **SaveAs CH8EX3**.

   Draw the **Lines** at the bottom first, starting at coordinate **1,3**. Then create **Point** objects at **5,7**, **5.4,7**, **5.8,7**, etc. Change the **Point Style** to an X and **Regen**. Use the **NODe OSNAP** mode to draw the inclined **Lines**. Create the **Arc** on top with the **Start, End, Direction** option.

FIGURE 8-28

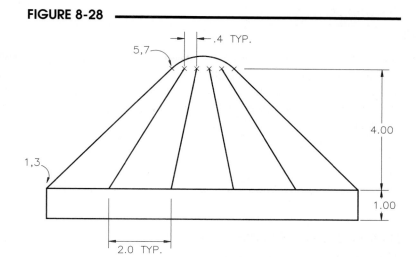

4. **Open CH8EX**. Create the shape shown in Figure 8-29. Draw the two horizontal **Lines** and the vertical **Line** first by specifying the endpoints as given. Then create the **Circle** and **Arcs**. **SaveAs CH8EX4**.

   HINT: Use the **Circle 2P** method with **Endpoint OSNAP**s. The two upper **Arcs** can be drawn by the **Start, End, Radius** method.

FIGURE 8-29

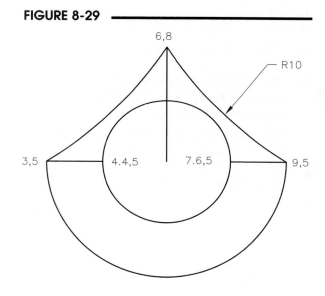

5. **Open CH8EX**. Complete the geometry in Figure 8-30. Use the coordinates to establish the **Lines**. Draw the **Circles** using the **Tangent, Tangent, Radius** method. **SaveAs CH8EX5**.

FIGURE 8-30

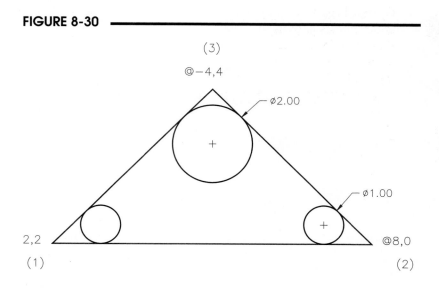

6. *Pline*

    Create the shape shown in
    Figure 8-31. Draw the outside
    shape with <u>one continuous</u>
    Pline (with 0.00 width).
    When finished, *SaveAs*
    PLINE1.

FIGURE 8-31

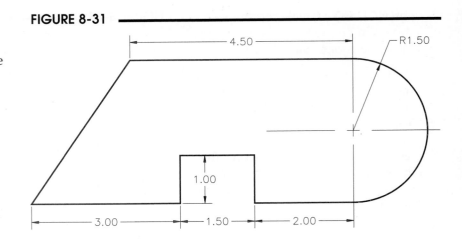

MODIFY COMMANDS I

CHAPTER OBJECTIVES

After completing this chapter you should:

1. be able to *Erase* objects;

2. be able to *Move* objects from a base point to a second point;

3. know how to *Rotate* objects about a base point;

4. be able to enlarge or reduce objects with *Scale*;

5. be able to *Stretch* objects and change the length of *Lines* and *Arcs* with *Lengthen*;

6. be able to *Trim* away parts of objects at cutting edges and *Extend* objects to boundary edges;

7. know how to use the four *Break* options;

8. be able to *Copy* objects and make *Mirror* images of selected objects;

9. know how to create parallel copies of objects with *Offset*;

10. be able to make *rectangular* and *Polar Arrays* of objects;

11. be able to create a *Fillet* and a *Chamfer* between two objects.

CONCEPTS

Draw commands are used to create new objects. *Modify* commands are used to change existing objects or to use existing objects to create new and similar objects. The *Modify* commands covered first in this chapter (listed below) only <u>change existing objects</u>.

FIGURE 9-1

>   *Erase, Move, Rotate, Scale, Stretch, Lengthen, Trim, Extend,* and *Break*

The *Modify* commands covered near the end of this chapter (listed below) <u>use existing objects to create new and similar objects</u>. For example, the *Copy* command prompts you to select an object (or set of objects), then creates an identical object (or set).

>   *Copy, Mirror, Offset, Array, Fillet,* and *Chamfer*

Several commands are found in the *Edit* pull-down menu such as *Cut, Copy,* and *Paste*. Although these command names appear similar (or the same in the case of *Copy*) to some of those in the *Modify* menu, these AutoCAD command names are actually *Cutclip, Copyclip,* and *Pasteclip*. These commands are for OLE operations (cutting and pasting) between two different AutoCAD drawings or other software applications.

Remember that there are many methods you can use to access the modify commands. If you are using the *AutoCAD Classic* workspace, select tools (icon buttons) for modify commands from the *Modify* toolbar that appears on the right side of the Drawing Editor (Fig. 9-1).

You can also select the modify commands from the *Modify* pull-down menu available from either the *AutoCAD Classic* or the *2D Drafting & Annotation* workspace (Fig. 9-2).

Alternately, you can enter the command names or command aliases at the keyboard. Modify commands can also be accessed using the *Modify* tool palette, the *MODIFY1* and *MODIFY2* screen menus, and the digitizing tablet, if available (not shown).

FIGURE 9-2

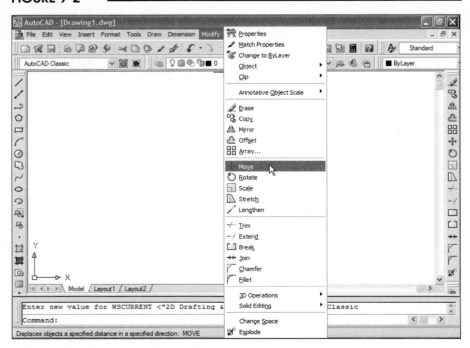

If you prefer the *2D Drafting and Annotation* workspace, you can select the desired tool from the *2D Draw* control panel on the dashboard since the *2D Draw* control panel contains both draw and modify commands. Note that you may have to locate the desired commands by selecting a flyout as shown in Figure 9-3.

Since all *Modify* commands affect or use existing geometry, the first step in using most *Modify* commands is to construct a selection set (see Chapter 4). This can be done by one of two methods:

1. Invoking the desired command and then creating the selection set in response to the "Select Objects:" prompt (Verb/Noun syntax order) using any of the select object options;

2. Selecting the desired set of objects with the pickbox or *Auto Window* or *Crossing Window* <u>before</u> invoking the edit command (Noun/Verb syntax order).

The first method allows use of any of the selection options (*Last, All, WPolygon, Fence,* etc.), while the latter method allows <u>only</u> the use of the pickbox and *Auto Window* and *Crossing Window*.

FIGURE 9-3

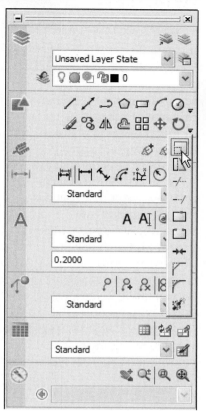

COMMANDS

Erase

Pull-down Menu	Command (Type)	Alias (Type)	Short-cut	Screen (side) Menu	Tablet Menu
Modify *Erase*	*Erase*	E	(Edit Mode) *Erase*	*MODIFY1* *Erase*	V,14

The *Erase* command deletes the objects you select from the drawing. Any of the object selection methods can be used to highlight the objects to *Erase*. The only other required action is for you to press *Enter* to cause the erase to take effect.

```
Command: erase
Select objects: PICK (Use any object selection method.)
Select objects: PICK (Continue to select desired objects.)
Select objects: Enter (Confirms the object selection process and causes Erase to take effect.)
Command:
```

If objects are erased accidentally, *U* can be used immediately following the mistake to undo one step, or *Oops* can be used to bring back into the drawing whatever was *Erased* the last time *Erase* was used (see Chapter 5). If only part of an object should be erased, use *Trim* or *Break*.

Move

Pull-down Menu	Command (Type)	Alias (Type)	Short-cut	Screen (side) Menu	Tablet Menu
Modify Move	Move	M	(Edit Mode) Move	MODIFY2 Move	V,19

Move allows you to relocate one or more objects from the existing position in the drawing to any other position you specify. After selecting the objects to *Move*, you must specify the "base point" and "second point of displacement." You can use any of the five coordinate entry methods to specify these points. Examples are shown in Figure 9-4.

FIGURE 9-4

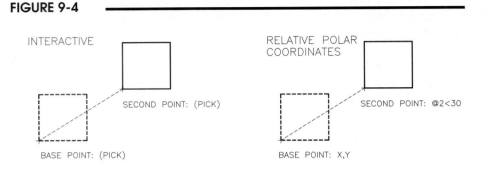

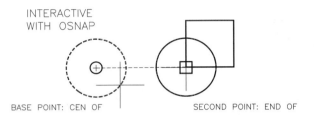

Command: **move**
Select objects: **PICK**   (Use any of the object selection methods.)
Select objects: **PICK**   (Continue to select other desired objects.)
Select objects: **Enter**   (Press Enter to indicate selection of objects is complete.)
Specify base point or displacement: **PICK** or (**coordinates**) (This is the point to <u>move from</u>. Select a point to use as a "handle." An *Endpoint* or *Center*, etc., can be used.)
Specify second point or <use first point as displacement>: **PICK** or (**coordinates**) (This is the point to <u>move to</u>. *OSNAP*s can also be used here.)
Command:

Keep in mind that *OSNAP*s can be used when PICKing any point. It is often helpful to toggle *ORTHO* or *POLAR ON* to force the *Move* in a horizontal or vertical direction.

If you know a specific distance and an angle that the set of objects should be moved, Polar Tracking or relative polar coordinates can be used. In the following sequence, relative polar coordinates are used to move objects 2 units in a 30 degree direction (see Fig. 9-4, relative polar coordinates).

Command: **move**
Select objects: **PICK**
Select objects: **PICK**
Select objects: **Enter**
Specify base point or displacement: **X,Y (coordinates)**
Specify second point or <use first point as displacement>: **@2<30**
Command:

Dynamic Input and direct distance entry can be used effectively with *Move*. For example, assume you wanted to move the right side view of a multiview drawing 20 units to the right (Fig. 9-5). First, invoke the *Move* command and select all of the objects comprising the right side view in response to the "Select Objects:" prompt. The command sequence is as follows:

FIGURE 9-5

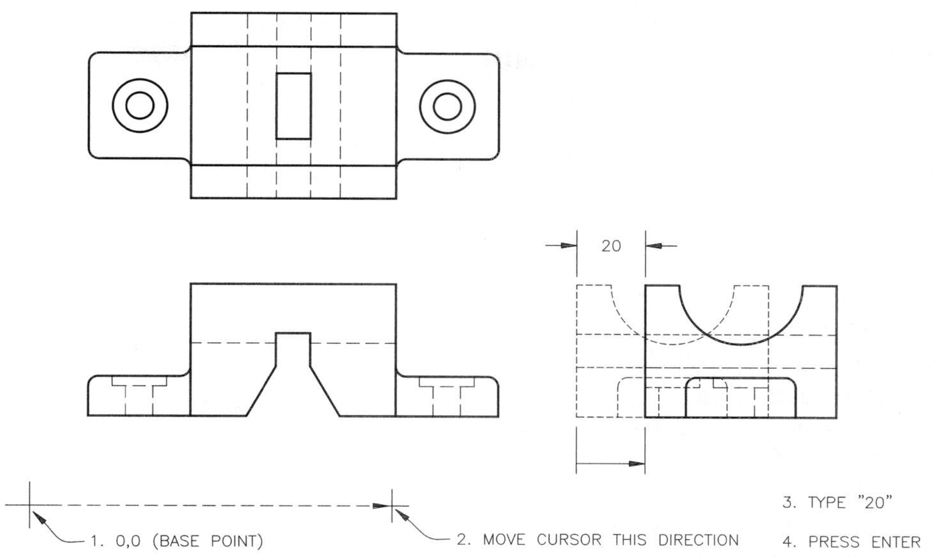

Command: **move**
Select objects: **PICK**
Select objects: **Enter**
Specify base point or displacement: **0,0** (or any value)
Specify second point or <use first point as displacement>: **20** (with *ORTHO* or *POLAR* on, move cursor to right), then press **Enter**

In response to the "Specify base point or displacement:" prompt, PICK a point or enter any coordinate pairs or single value. At the "Specify second point:" prompt, move the cursor to the right any distance (using Polar Tracking or *ORTHO* on), then type "20" and press Enter. The value specifies the distance, and the cursor location from the last point indicates the direction of movement.

In the previous example, note that <u>any value</u> can be entered in response to the "Specify base point or displacement:" prompt. If a single value is entered ("3," for example), AutoCAD recognizes it as direct distance entry. The point designated is 3 units from the last PICK point in the direction specified by wherever the cursor is at the time of entry.

Rotate

Pull-down Menu	Command (Type)	Alias (Type)	Short-cut	Screen (side) Menu	Tablet Menu
Modify *Rotate*	*Rotate*	*RO*	(Edit Mode) *Rotate*	*MODIFY2* *Rotate*	*V,20*

Selected objects can be rotated to any position with this command. After selecting objects to *Rotate*, you select a "base point" (a point to rotate about) then specify an angle for rotation. AutoCAD rotates the selected objects by the increment specified from the original position.

For example, specifying a value of **45** would *Rotate* the selected objects 45 degrees counterclockwise from their current position; a value of **-45** would *Rotate* the objects 45 degrees in a clockwise direction (Fig. 9-6).

FIGURE 9-6

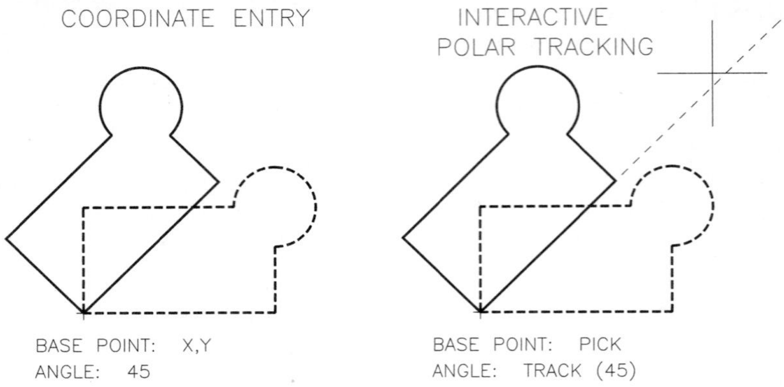

COORDINATE ENTRY

INTERACTIVE POLAR TRACKING

BASE POINT: X,Y
ANGLE: 45

BASE POINT: PICK
ANGLE: TRACK (45)

```
Command: rotate
Current positive angle in UCS: ANGDIR=counterclockwise  ANGBASE=0
Select objects: PICK
Select objects: PICK
Select objects: Enter (Indicates completion of object selection.)
Specify base point: PICK or (coordinates) (Select a point to rotate about.)
Specify rotation angle or [Copy/Reference]: PICK or (coordinates) (Interactively rotate the set or
enter a value for the number of degrees to rotate the object set.)
Command:
```

The base point is often selected interactively with *OSNAPs*. When specifying an angle for rotation, you can enter an angular value, use Polar Tracking, or turn on *ORTHO* for 90-degree rotation. Note the status of the two related system variables is given: *ANGDIR* (counterclockwise or clockwise rotation) and *ANGBASE* (base angle used for rotation).

The **Copy** option allows you to make a copy of the selected objects—that is, the original set of objects remains in its original orientation while you create rotated copies. After selecting a "base point," type "C" or right-click and select *Copy* at the following prompt.

```
Specify rotation angle or [Copy/Reference] <0>: C
```

Next, enter a rotation angle value or PICK a point. Using this option would result in, using Figure 9-6 for example, a drawing with both sets of objects: the original set (shown with dashed lines) and the new set.

TIP The **Reference** option can be used to specify a vector as the original angle before rotation. This vector can be indicated interactively (*OSNAPs* can be used) or entered as an angle using keyboard entry. Angular values that you enter in response to the "New angle:" prompt are understood by AutoCAD as <u>absolute</u> angles for the *Reference* option only.

Scale

Pull-down Menu	Command (Type)	Alias (Type)	Short-cut	Screen (side) Menu	Tablet Menu
Modify *Scale*	*Scale*	SC	(Edit Mode) *Scale*	*MODIFY2* *Scale*	V,21

The *Scale* command is used to increase or decrease the size of objects in a drawing. The *Scale* command does not normally have any relation to plotting a drawing to scale.

After selecting objects to *cale*, AutoCAD prompts you to select a "Base point:", which is the <u>stationary point</u>. You can then scale the size of the selected objects interactively or enter a scale factor (Fig. 9-7).

FIGURE 9-7

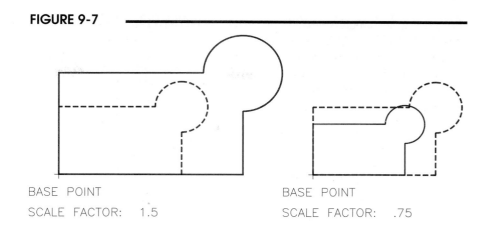

BASE POINT
SCALE FACTOR: 1.5

BASE POINT
SCALE FACTOR: .75

Using interactive input, you are presented with a rubberband line connected to the base point. Making the rubberband line longer or shorter than 1 unit increases or decreases the scale of the selected objects by that proportion; for example, pulling the rubberband line to two units length increases the scale by a factor of two.

> Command: **scale**
> Select objects: **PICK** or (**coordinates**) (Select the objects to scale.)
> Select objects: **Enter** (Indicates completion of the object selection.)
> Specify base point: **PICK** or (**coordinates**) (Select the stationary point.)
> Specify scale factor or [Copy/Reference]: **PICK** or (**value**) or **R** (Interactively scale the set of objects or enter a value for the scale factor.)
> Command:

The *Copy* option allows you to make a copy of the selected objects—that is, the original set of objects remains in its original orientation while you create scaled copies. Type "C" or right-click and select *Copy* in response to "Specify rotation angle or [Copy/Reference] <0>:". Similar to the *Copy* option of the *Rotate* command, using this option would result in two sets of objects: the original selection set and the new set.

It may be desirable in some cases to use the *Reference* option to specify a value or two points to use as the reference length. This length can be indicated interactively (*OSNAP*s can be used) or entered as a value. This length is used for the subsequent reference length that the rubberband uses when interactively scaling. For example, if the reference distance is 2, then the "rubberband" line must be stretched to a length greater than 2 to increase the scale of the selected objects.

Scale normally should <u>not</u> be used to change the scale of an entire drawing in order to plot on a specific size sheet. CAD drawings should be created <u>full size</u> in <u>actual units</u>.

Stretch

Pull-down Menu	Command (Type)	Alias (Type)	Short-cut	Screen (side) Menu	Tablet Menu
Modify *Stretch*	*Stretch*	S	...	*MODIFY2* *Stretch*	V,22

Objects can be made longer or shorter with *Stretch*.

The power of this command lies in the ability to *Stretch* groups of objects while retaining the connectivity of the group. When *Stretched, Lines* and *Plines* become longer or shorter and *Arcs* change radius to become longer or shorter. *Circles* do not stretch; rather, they move if the center is selected within the Crossing Window.

Objects to *Stretch* should be selected by a <u>Crossing Window or Crossing Polygon</u> only. The *Crossing Window* or *Polygon* should be created so the objects to *Stretch* <u>cross</u> through the window. *Stretch* actually moves the object endpoints that are located within the *Crossing Window* (Fig. 9-8).

FIGURE 9-8

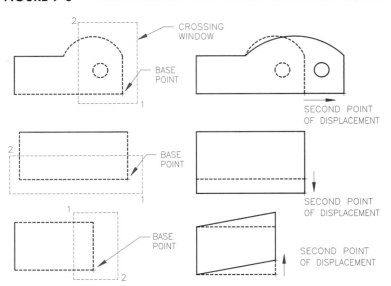

```
Command: stretch
Select objects to stretch by cross-
ing-window or crossing-polygon...
Select objects: PICK
Specify opposite corner: PICK
Select objects: Enter
Specify base point or [Displacement] <Displacement>: (Select a point to stretch from.)
Specify second point or <use first point as displacement>: (Select a point to stretch to.)
```

TIP

Stretch can be used to lengthen one object while shortening another. Application of this ability would be repositioning a door or window on a wall (Fig. 9-9).

In summary, the *Stretch* command <u>stretches objects that cross</u> the selection Window and <u>moves objects that are completely within</u> the selection Window, as shown in Figure 9-9.

The *Stretch* command allows you to select objects with other selection methods, such as a pickbox or normal (non-crossing) *Window*. Doing so, however, results in a *Move* since <u>only objects crossing through a crossing window get stretched</u>.

FIGURE 9-9

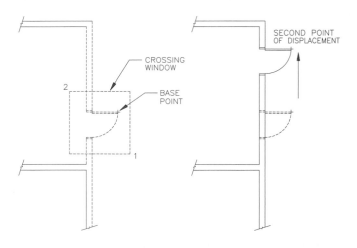

Lengthen

Pull-down Menu	Command (Type)	Alias (Type)	Short-cut	Screen (side) Menu	Tablet Menu
Modify *Lengthen*	*Lengthen*	*LEN*	...	*MODIFY2* *Lengthen*	*W,14*

Lengthen changes the length (longer or shorter) of linear objects and arcs. No additional objects are required (as with *Trim* and *Extend*) to make the change in length. Many methods are provided as displayed in the command prompt.

Command: **lengthen**
Select an object or [DElta/Percent/Total/DYnamic]:

Select an object

Selecting an object causes AutoCAD to report the current length of that object. If an *Arc* is selected, the included angle is also given.

DElta

Using this option returns the prompt shown next. You can change the current length of an object (including an arc) by any increment that you specify. Entering a positive value increases the length by that amount, while a negative value decreases the current length. The end of the object that you select changes while the other end retains its current end-point (Fig. 9-10).

> Enter delta length or [Angle] <0.0000>:
> (**value**) or **a**

FIGURE 9-10

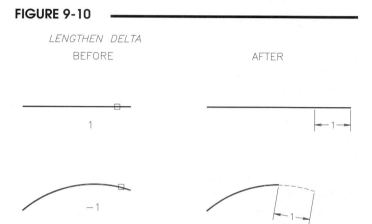

LENGTHEN DELTA
BEFORE   AFTER

The *Angle* option allows you to change the <u>included angle</u> of an arc (the length along the curvature of the arc can be changed with the *Delta* option). Enter a positive or negative value (degrees) to add or subtract to the current included angle, then select the end of the object to change (Fig. 9-11).

FIGURE 9-11

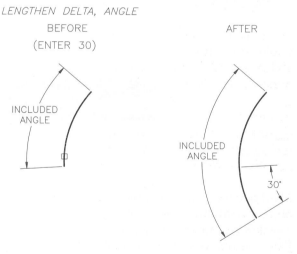

LENGTHEN DELTA, ANGLE
BEFORE   AFTER
(ENTER 30)

INCLUDED
ANGLE

INCLUDED
ANGLE

30°

Percent

Use this option if you want to change the length by a percentage of the current total length (Fig. 9-12). For arcs, the percentage applied affects the length and the included angle equally, so there is no *Angle* option. A value of greater than 100 increases the current length, and a value of less than 100 decreases the current length. Negative values are not allowed. The end of the object that you select changes.

FIGURE 9-12

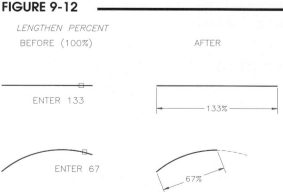

LENGTHEN PERCENT
BEFORE (100%)   AFTER

ENTER 133   133%

ENTER 67   67%

Total

This option lets you specify a value for the new total length (Fig. 9-13). Simply enter the value and select the end of the object to change. The angle option is used to change the total included angle of a selected arc.

> Specify total length or [Angle] <1.0000)>: (**value**) or **A**

FIGURE 9-13

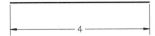

LENGTHEN TOTAL
BEFORE (3 UNITS)   AFTER

ENTER 4

ENTER 2

DYnamic

This option allows you to change the length of an object by dynamic dragging (Fig. 9-14). Select the end of the object that you want to change. Object Snaps and Polar Tracking can be used.

FIGURE 9-14

LENGTHEN DYNAMIC
BEFORE   AFTER

Trim

Pull-down Menu	Command (Type)	Alias (Type)	Short-cut	Screen (side) Menu	Tablet Menu
Modify *Trim*	*Trim*	*TR*	...	*MODIFY2* *Trim*	*W,15*

The *Trim* command allows you to trim (shorten) the end of an object back to the intersection of another object (Fig. 9-15). The middle section of an object can also be *Trimmed* between two intersecting objects. There are two steps to this command: first, PICK one or more "cutting edges" (existing objects); then PICK the object or objects to *Trim* (portion to remove). The cutting edges are highlighted after selection. Cutting edges themselves can be trimmed if they intersect other cutting edges, but lose their highlight when trimmed.

FIGURE 9-15

BEFORE *TRIM*   AFTER ALL *TRIMS*

CUTTING EDGES

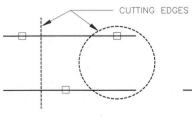

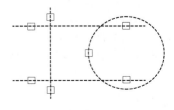
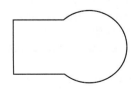

Command: **Trim**
Current settings: Projection=UCS, Edge=None
Select cutting edges ...
Select objects or <select all>: **PICK** or press **Enter** to use all objects as cutting edges
Select objects: **Enter**
Select object to trim or shift-select to extend or [Fence/Crossing/Project/Edge/eRase/Undo]: **PICK** (Select the part of an object to trim away.)
Select object to trim or shift-select to extend or [Fence/Crossing/Project/Edge/eRase/Undo]: **PICK**
Select object to trim or shift-select to extend or [Fence/Crossing/Project/Edge/eRase/Undo]: **Enter**

Select all

If you need to select many objects as <u>cutting edges</u>, you can press Enter at the "Select objects or <select all>:" prompt. In this case, all objects in the drawing become cutting edges; however, the objects are not highlighted.

Fence

The *Fence* and *Crossing Window* options are available for selecting the <u>objects you want to trim</u>. Rather than selecting each object that you want to trim individually with the pickbox, the *Fence* option allows you to select multiple objects using a crossing line. The *Fence* can have several line segments (see "*Fence*" in Chapter 3). This option is especially useful when you have many objects to trim away that are in close proximity to each other.

Crossing Window

You can also use a *Crossing Window* to select the parts of <u>objects you want to trim</u>. This option may be more efficient than using the pickbox to select objects individually.

Edge

The *Edge* option can be set to *Extend* or *No extend*. In the *Extend* mode, objects that are selected as trimming edges will be <u>imaginarily extended</u> to serve as a cutting edge. In other words, lines used for trimming edges are treated as having infinite length (Fig. 9-16). The *No extend* mode considers only the actual length of the object selected as trimming edges.

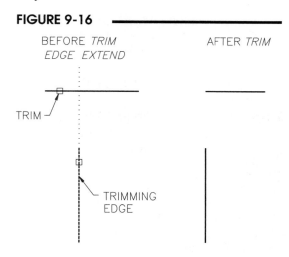

FIGURE 9-16

Select object to trim or shift-select to extend or [Fence/Crossing/Project/Edge/eRase/Undo]: **Edge**
Enter an implied edge extension mode [Extend/No extend] <No extend>:

Projection

The *Projection* switch controls how *Trim* and *Extend* operate in 3D space. *Projection* affects the projection of the cutting edge and boundary edge.

eRase

The *eRase* option actually allows you to erase entire objects during the *Trim* command. This option operates just like the *Erase* command. When you are finished erasing, you are returned to the "Select object to trim..." prompt.

Undo

The *Undo* option allows you to undo the last *Trim* in case of an accidental trim.

Shift-Select

See "*Trim* and *Extend* Shift-Select Option."

Extend

Pull-down Menu	Command (Type)	Alias (Type)	Short-cut	Screen (side) Menu	Tablet Menu
Modify Extend	Extend	EX	...	MODIFY2 Extend	W,16

Extend can be thought of as the opposite of *Trim*. Objects such as *Lines, Arcs,* and *Plines* can be *Extended* until intersecting another object called a boundary edge (Fig. 9-17). The command first requires selection of <u>existing</u> objects to serve as boundary edge(s) which become highlighted; then the objects to extend are selected. Objects extend until, and only if, they eventually intersect a boundary edge. An *Extended* object acquires a new endpoint at the boundary edge intersection.

FIGURE 9-17

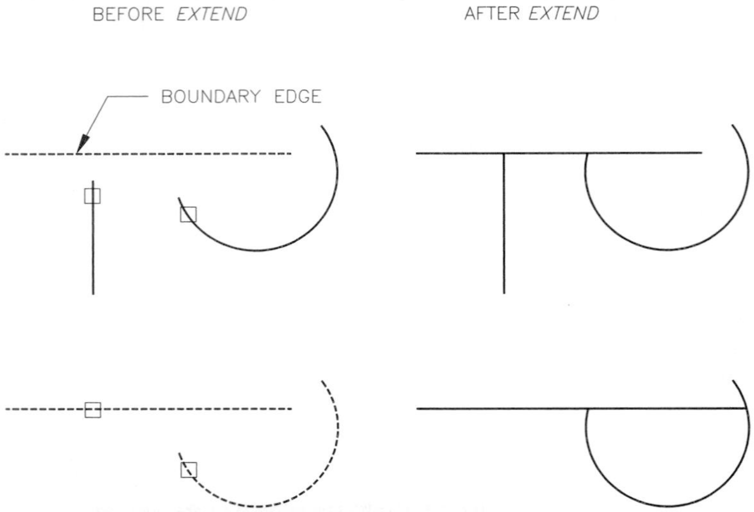

BEFORE *EXTEND*   AFTER *EXTEND*

BOUNDARY EDGE

```
Command: extend
Current settings: Projection=UCS,
Edge=None
Select boundary edges ...
Select objects or <select all>: PICK
(Select boundary edge.)
Select objects: Enter
Select object to extend or shift-select to trim or [Fence/Crossing/Project/Edge/Undo]: PICK (Select object
to extend.)
Select object to extend or shift-select to trim or [Fence/Crossing/Project/Edge/Undo]: Enter
Command:
```

Select all
Pressing Enter at the "Select objects or <select all>:" prompt automatically makes all objects in the drawing <u>boundary edges</u>. The objects are not highlighted.

Fence
Use the *Fence* option (a crossing line) to select multiple <u>objects that you want to extend</u>. This operates similar to the *Fence* option of *Trim*.

Crossing Window
Similar to the same option of the *Trim* command, you can use a *Crossing Window* to select multiple <u>objects to extend</u>.

FIGURE 9-18

BEFORE *EXTEND*   AFTER *EXTEND*
EDGE EXTEND

Edge/Projection
The *Edge* and *Projection* switches operate identically to their function with the *Trim* command. Use *Edge* with the *Extend* option if you want a boundary edge object to be imaginarily extended (Fig. 9-18).

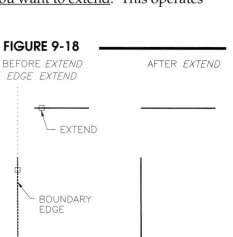

EXTEND

BOUNDARY EDGE

Shift-Select
See "*Trim* and *Extend* Shift-Select Option."

Trim and *Extend* **Shift-Select Option**

With this feature, you can toggle between *Trim* and *Extend* by holding down the Shift key. For example, if you invoked *Trim* and are in the process of trimming lines but decide to then *Extend* a line, you can simply hold down the Shift key to change to the *Extend* command without leaving *Trim*.

```
Command: trim
Current settings: Projection=UCS, Edge=None
Select cutting edges ...
Select objects or <select all>: PICK
Select objects: Enter
Select object to trim or shift-select to extend or [Fence/Crossing/Project/Edge/eRase/Undo]: PICK (to trim)
Select object to trim or shift-select to extend or [Fence/Crossing/Project/Edge/eRase/Undo]: Shift, then PICK to extend
```

Not only does holding down the Shift key toggle between *Trim* and *Extend*, but objects you selected as *cutting edges* become *boundary edges* and vice versa. Therefore, if you want to use this feature effectively, during the first step of either command (*Trim* or *Extend*), you must anticipate and select edges you might potentially use as both cutting edges and boundary edges.

Break

Pull-down Menu	Command (Type)	Alias (Type)	Short-cut	Screen (side) Menu	Tablet Menu
Modify *Break*	*Break*	*BR*	...	*MODIFY2* *Break*	*W,17*

Break allows you to break a space in an object or break the end off an object. You can think of *Break* as a partial erase. If you choose to break a space in an object, the space is created between two points that you specify (Fig. 9-19). In this case, the *Break* creates two objects from one.

FIGURE 9-19

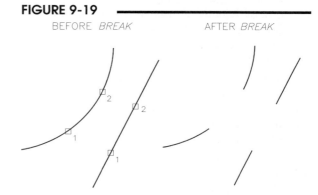

BEFORE *BREAK*   AFTER *BREAK*

If *Break*ing a circle (Fig. 9-20), the break is in a <u>counterclockwise</u> direction from the first to the second point specified.

FIGURE 9-20

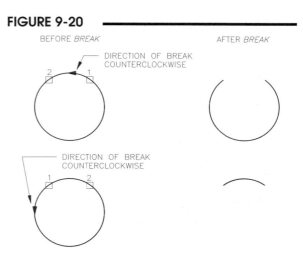

BEFORE *BREAK*   AFTER *BREAK*

If you want to *Break* the end off a *Line* or *Arc*, the first point should be specified at the point of the break and the second point should be just off the end of the *Line* or *Arc* (Fig. 9-21).

The *Modify* toolbar includes icons for only the default option (see *Select, Second*) and the *Break at Point* option. Four methods are available only from the screen menu. If you are typing, the desired option is selected by keying the letter *F* (for *First point*) or symbol @ (for "last point").

Select, Second

This method (the default method) has two steps: select the object to break; then select the second point of the break. The first point used to select the object is also the first point of the break (see Fig. 9-19, Fig. 9-20, Fig. 9-21).

> Command: **break**
> Select object: **PICK** (This is the first point of the break.)
> Specify second break point or [First point]: **PICK** (This is the second point of the break.)
> Command:

Select, 2 Points

This method uses the first selection only to indicate the object to *Break*. You then specify the point that is the first point of the *Break,* and the next point specified is the second point of the *Break*. This option can be used with *OSNAP Intersection* to achieve the same results as *Trim*. The command sequence is as follows:

> Command: **break**
> Select object: **PICK** (This selects only the object to break.)
> Specify second break point or [First point]: **f** (Indicates respecification of the first point.)
> Specify first break point: **PICK** (This is the first point of the break.)
> Specify second break point: **PICK** (This is the second point of the break.)
> Command:

Break at Point

This option creates a break with no space; however, you can select the object you want to *Break* first and the point of the *Break* next (Fig. 9-22).

> Command: **break**
> Select object: **PICK** (This selects only the object to break.)
> Specify second break point or [First point]: **f** (Indicates respecification of the first point.)
> Specify first break point: **PICK** (This is the first point of the break.)
> Specify second break point: **@** (The second point of the break is the last point.)

FIGURE 9-21

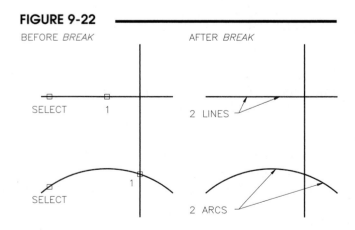
FIGURE 9-22

Join

Pull-down Menu	Command (Type)	Alias (Type)	Short-cut	Screen (side) Menu	Tablet Menu
Modify *Join*	*Join*	*J*	...	...	...

The *Join* command allows you to combine two of the same objects (for example, two collinear *Lines*) into one object (for example, one *Line*). You can also use *Join* to *Close* an *Arc* to form a complete *Circle* or close an elliptical arc to form a complete *Ellipse*. You can *Join* the following objects:

    Arcs, Elliptical arcs, *Lines, Polylines,* and *Splines*

Objects to be joined must be located in the same plane. Sample prompts and restrictions for each type of object are given.

Lines

The *Lines* to join must be collinear, but they can have gaps between them (Fig. 9-23).

    Command: **Join**
    Select source object: **PICK**
    Select lines to join to source: **PICK**
    Select lines to join to source: **Enter**
    1 line joined to source
    Command:

FIGURE 9-23

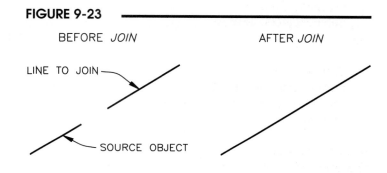

Arc

The *Arc* objects must lie on the same imaginary circle but can have gaps between them. The *Close* option converts the source arc into a *Circle* (Fig. 9-24).

    Command: **Join**
    Select source object: **PICK**
    Select arcs to join to source or [cLose]: **1**
    Arc converted to a circle.
    Command:

FIGURE 9-24

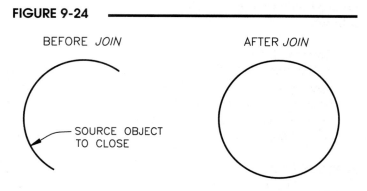

Polyline

Objects to *Join* to *Plines* can be *Plines, Lines,* or *Arcs*. The objects cannot have gaps between them and must lie on the UCS XY plane or on a plane parallel to it. If *Arcs* and *Lines* are joined to the *Pline*, they are converted to one *Pline*.

Elliptical Arc

The elliptical arcs must lie on the same ellipse shape but can have gaps between them. The *Close* option closes the source elliptical arc into a complete *Ellipse*.

Spline

The *Spline* objects must lie in the same plane, and the end points to join must be connected.

Copy

	Pull-down Menu	Command (Type)	Alias (Type)	Short-cut	Screen (side) Menu	Tablet Menu
	Modify Copy	Copy	CO or CP	(Edit Mode) Copy Selection	MODIFY1 Copy	V,15

Copy creates a duplicate set of the selected objects and allows placement of those copies. The *Copy* operation is like the *Move* command, except with *Copy* the original set of objects remains in its original location. You specify a "base point:" (point to copy <u>from</u>) and a "specify second point of displacement or <use first point as displacement>:" (point to copy <u>to</u>). See Figure 9-25. The command syntax for *Copy* is as follows.

> Command: **Copy**
> Select objects: **PICK**
> Select objects: **Enter**
> Specify base point or [Displacement/mOde] <Displacement>: **PICK**
> (This is the point to copy <u>from</u>. Select a point, usually on an object, as a "handle.")
> Specify second point or <use first point as displacement>: **PICK**
> (This is the point to copy <u>to</u>. *OSNAPs* or coordinate entry should be used.)
> Specify second point or [Exit/Undo] <Exit>: **Enter**

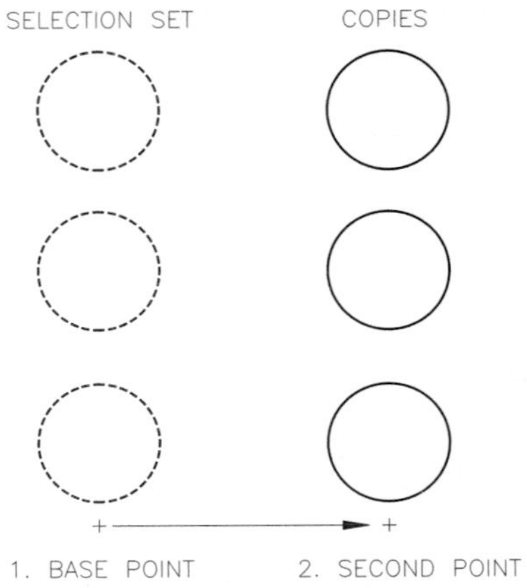

In many applications it is desirable to use *OSNAP* options to PICK the "base point:" and the "second point of displacement (Fig. 9-26).

Alternately, you can PICK the "base point:" and use Polar Tracking or enter relative Cartesian coordinates, relative polar coordinates, or direct distance entry to specify the "second point of displacement:" (Fig. 9-27).

FIGURE 9-25

SELECTION SET   COPIES

1. BASE POINT   2. SECOND POINT

FIGURE 9-26

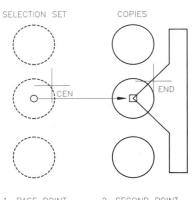

SELECTION SET   COPIES

1. BASE POINT   2. SECOND POINT

FIGURE 9-27

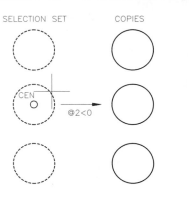

SELECTION SET   COPIES

1. BASE POINT   2. SECOND POINT

mOde: (Single/Multiple)

Copy has a *Multiple* mode. Activate the multiple mode by typing an *O* at the "Specify base point or [Displacement/mOde] <Displacement>:" prompt, then type *M* for *Multiple*. Next, continue to specify a "second point" to create additional copies. The *Copy* command will continue to operate in the *Multiple* mode until you specify the *Single* mode (Fig. 9-28).

FIGURE 9-28

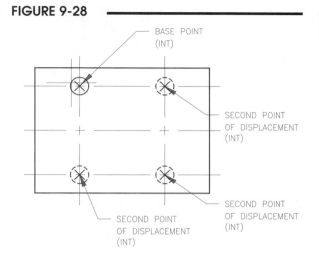

Command: **copy**
Select objects: **PICK**
Select objects: **Enter**
Current settings: Copy mode = Multiple
Specify base point or [Displacement/mOde]
<Displacement>: *o* (for *mOde* option)
Enter a copy mode option [Single/Multiple]
<Multiple>: *m* (for *Multiple* option)
Specify base point or [Displacement/mOde] <Displacement>: **PICK**
Specify second point or <use first point as displacement>: (This creates the first copy)
Specify second point or [Exit/Undo] <Exit>: (This creates the second copy)
Specify second point or [Exit/Undo] <Exit>: **Enter** (completes the command)

Mirror

Pull-down Menu	Command (Type)	Alias (Type)	Short-cut	Screen (side) Menu	Tablet Menu
Modify *Mirror*	*Mirror*	*MI*	...	*MODIFY1* *Mirror*	*V,16*

This command creates a mirror image of selected existing objects. You can retain or delete the original objects ("source objects"). After selecting objects, you create two points specifying a "rubberband line," or "mirror line," about which to *Mirror*.

The length of the mirror line is unimportant since it represents a vector or axis (Fig. 9-29).

FIGURE 9-29

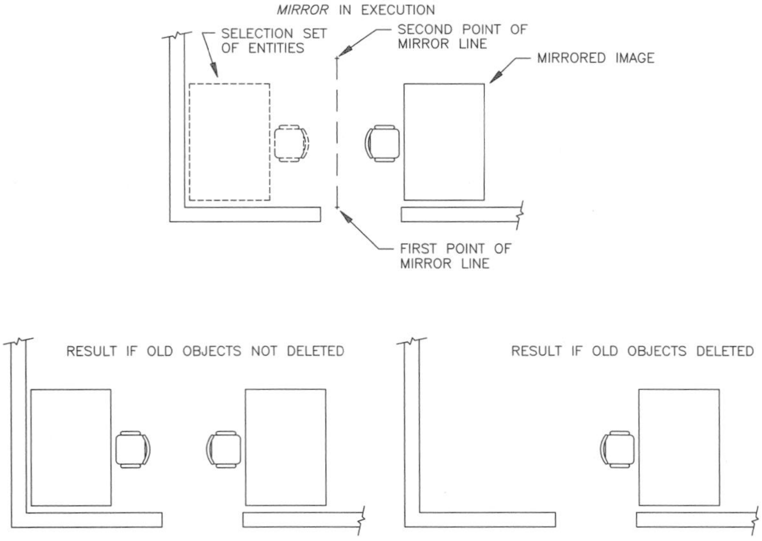

Command: **mirror**
Select objects: **PICK** (Select object or group of objects to mirror.)
Select objects: **Enter** (Press Enter to indicate completion of object selection.)
Specify first point of mirror line: **PICK** or (**coordinates**) (Draw first endpoint of line to represent mirror axis by PICKing or entering coordinates.)
Specify second point of mirror line: **PICK** or (**coordinates**) (Select second point of mirror line by PICKing or entering coordinates.)
Erase source objects? [Yes/No] <N>: **Enter** or **Y** (Press Enter to yield both sets of objects or enter Y to keep only the mirrored set.)
Command:

If you want to *Mirror* only in a vertical or horizontal direction, toggle *ORTHO* or *POLAR On* before selecting the "second point of mirror line."

Mirror can be used to draw the other half of a symmetrical object, thus saving some drawing time (Fig. 9-30).

FIGURE 9-30

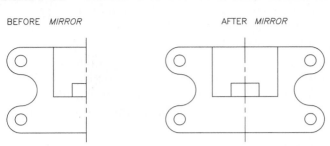

BEFORE *MIRROR*   AFTER *MIRROR*

Offset

Pull-down Menu	Command (Type)	Alias (Type)	Short-cut	Screen (side) Menu	Tablet Menu
Modify *Offset*	*Offset*	*O*	...	*MODIFY1* *Offset*	*V,17*

Offset creates a <u>parallel copy</u> of selected objects. Selected objects can be *Lines*, *Arcs*, *Circles*, *Plines*, or other objects. *Offset* is a very useful command that can increase productivity greatly, particularly with *Plines*.

Depending on the object selected, the resulting *Offset* is drawn differently (Fig. 9-31). *Offset* creates a parallel copy of a *Line* equal in length and perpendicular to the original. *Arcs* and *Circles* have a concentric *Offset*. *Offsetting* closed *Plines* or *Splines* results in a complete parallel shape.

FIGURE 9-31

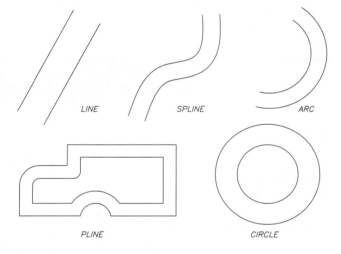

LINE   SPLINE   ARC

PLINE   CIRCLE

Distance

The default option (*distance*) prompts you to specify a distance between the original object and the newly offset object (Fig. 9-32, left).

FIGURE 9-32

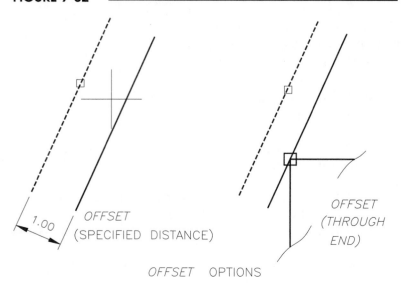

1.00   OFFSET (SPECIFIED DISTANCE)

OFFSET (THROUGH END)

OFFSET OPTIONS

Command: **offset**
Current settings: Erase source=No Layer=Source OFFSETGAPTYPE=0
Specify offset distance or [Through/Erase/Layer] <1.0000>: **(value)** or **PICK** (Specify an offset distance by entering a value or picking two points.)
Select object to offset or [Exit/Undo] <Exit>: **PICK** (Select the object to offset.)
Specify point on side to offset or [Exit/Multiple/Undo] <Exit>: **PICK** (Specify which side of original object to create the offset object.)
Select object to offset or [Exit/Undo] <Exit>: **Enter**
Command:

Through

Use the *Through* option to offset an object so it passes through a specified point (Fig. 9-32, right).

Command: **offset**
Current settings: Erase source=No Layer=Source OFFSETGAPTYPE=0
Specify offset distance or [Through/Erase/Layer] <1.0000>: **t**
Select object to offset or [Exit/Undo] <Exit>: **PICK**
Specify through point or [Exit/Multiple/Undo] <Exit>: **PICK** (Select a point for the object to be offset through. Normally, an *OSNAP* mode is used to specify the point.)

Erase

Note that the default *Erase source* setting is *No*. With this setting, the original object to offset remains in the drawing. Changing the *Erase* setting to *Yes* causes the selected (source) object to be erased automatically when the offset is created.

Specify offset distance or [Through/Erase/Layer] <1.0000>: **e**
Erase source object after offsetting? [Yes/No] <No>: **y**

Layer

You can use this setting to create an offset object on a different layer than the original layer of the source object. First, using the *Layer* command (see Chapter 11) change the current layer to the desired layer, next change the *Layer* setting of the *Offset* command to *Current*, then create the offset.

Specify offset distance or [Through/Erase/Layer] <1.0000>: **l**
Enter layer option for offset objects [Current/Source] <Source>: **c**

Notice the power of using *Offset* with closed *Pline* or *Spline* shapes (Fig. 9-33). Because a *Pline* or a *Spline* is one object, *Offset* creates one complete smaller or larger "parallel" object. Any closed shape composed of *Lines* and *Arcs* can be converted to one *Pline* object (see "*Pedit*," Chapter 16).

FIGURE 9-33

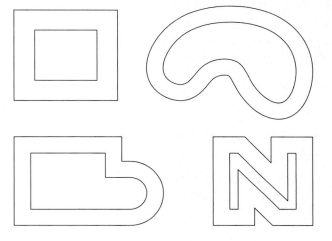

OFFSETGAPTYPE

When you create an offset of a "closed" *Pline* shape with sharp corners, such as the rectangle shown in Figure 9-33, the corners of the new offset object are also sharp (with the default setting of 0). The *OFFSETGAPTYPE* system variable controls how the corners (potential gaps) between the line segments are treated when closed *Plines* are offset (Fig. 9-34).

FIGURE 9-34

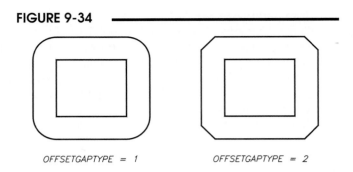

OFFSETGAPTYPE = 1   OFFSETGAPTYPE = 2

0   Fills the gaps by extending the polyline segments
1   Fills the gaps with filleted arc segments (the radius of each arc segment is equal to the offset distance)
2   Fills the gaps with chamfered line segments (the perpendicular distance to each chamfer is equal to the offset distance)

Array

Pull-down Menu	Command (Type)	Alias (Type)	Short-cut	Screen (side) Menu	Tablet Menu
Modify *Array*	*Array*	*AR*	...	*MODIFY1* *Array*	*V,18*

The *Array* command creates either a *Rectangular* or a *Polar* (circular) pattern of existing objects that you select. The pattern could be created from a single object or from a group of objects. *Array* copies a duplicate set of objects for each "item" in the array.

The *Array* command produces a dialog box to enter the desired array parameters and to preview the pattern (Fig. 9-35). The first step is to select the type of array you want (at the top of the box): *Rectangular Array* or *Polar Array*.

FIGURE 9-35

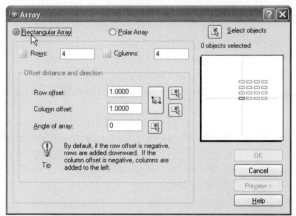

Rectangular Array

A rectangular array is a pattern of objects generated into rows and columns. Selecting a *Rectangular Array* produces the central area of the box and preview image as shown in Figure 9-36. Follow these steps to specify the pattern for the *Rectangular Array*.

1. Pick the *Select objects* button in the upper-right corner of the dialog box to select the objects you want to array.
2. Enter values in the *Rows:* and *Columns:* edit boxes to indicate how many rows and columns you want. The image area indicates the number of rows and columns you request, but does not indicate the shape of the objects you selected.

FIGURE 9-36

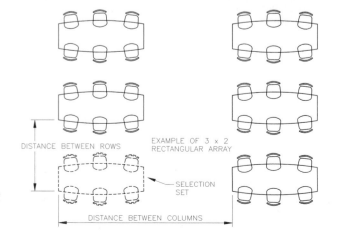

DISTANCE BETWEEN ROWS

EXAMPLE OF 3 x 2 RECTANGULAR ARRAY

SELECTION SET

DISTANCE BETWEEN COLUMNS

3. Enter values in the *Row offset:* and *Column offset:* edit boxes to indicate the distance between the rows and columns (see Fig. 9-36). Enter positive values to create an array to the right (+X) and upward (+Y) from the selected set, or enter negative values to create the array in the –X and –Y directions. You can also use the small select buttons to the far right of these edit boxes to interactively PICK two points for the *Row offset* and *Column offset*. Alternately, select the large button to the immediate right of the *Row offset* and *Column offset* edit boxes to use the "unit cell" method. With this method, you select diagonal corners of a window to specify both directions and both distances for the array (see Fig. 9-37).

FIGURE 9-37

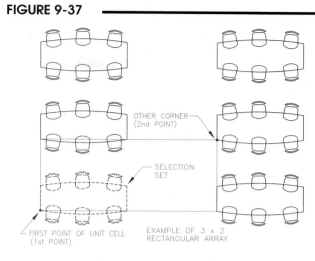

4. If you want to create an array at an angle (instead of generating the rows vertically and columns horizontally), enter a value in the *Angle of array* edit box or use the select button to PICK two points to specify the angle of the array.
5. Select the *Preview* button to temporarily close the dialog box and return to the drawing where you can view the array and choose to *Accept* or *Modify* the array. Selecting *Modify* returns you to the *Array* dialog box.

Polar Array

The *Polar Array* option generates a circular pattern of the selection set. Selecting the *Polar Array* button produces the central area of the box and preview image similar to that shown in Figure 9-38. Typical steps are as follows.

FIGURE 9-38

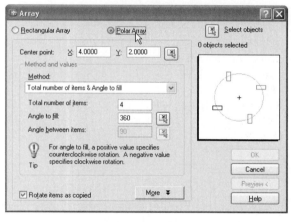

1. Pick the *Select objects* button in the upper-right corner of the dialog box to select the objects you want to array.
2. Enter values for the *Center point* of the circular pattern in the X and Y edit boxes, or use the select button to PICK a center point interactively.
3. Select the *Method* from the drop-down list. The three possible methods are:
    Total number of items & Angle to fill
    Total number of items & Angle between items
    Angle to fill & Angle between items
    Whichever option you choose enables the two applicable edit boxes below.
4. You must supply two of the three possible parameters following to specify the configuration of the array based on your selection in the *Method* drop-down list.
    Total number of items (the total number <u>includes</u> the selected object set)
    Angle to fill
    Angle between items
    Arrays are generated counterclockwise by default. To produce a clockwise array, enter a negative value in *Angle to fill*; or enter a negative value in *Angle to fill*, then switch to the *Total number of items & Angle between items*. The image area indicates the configuration of the array according to the options you selected and values you specified in the *Method and values* section. The image area does not indicate the shape of the object(s) you selected to array.
5. If you want the set of objects to be rotated but remain in the same orientation, remove the check in the *Rotate items as copied* box.

6.   Select the *Preview* button to temporarily close the dialog box and return to the drawing where you can view the array and chose to *Accept* or *Modify* the array. Selecting *Modify* returns you to the *Array* dialog box.

Figure 9-39 illustrates a *Polar Array* created using 8 as the *Total number of items*, 360 as the *Angle to fill*, and selecting *Rotate items as copied*.

FIGURE 9-39

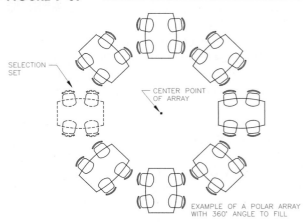

Figure 9-40 illustrates a *Polar Array* created using 5 as the *Total number of items*, 180 as the *Angle to fill*, and selecting *Rotate items as copied*.

FIGURE 9-40

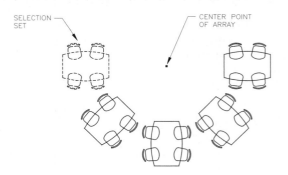

Occasionally an unexpected pattern may result if you remove the check from *Rotate items as copied*. In such a case, the objects may not seem to rotate about the specified center, as shown in Figure 9-41. The reason is that AutoCAD must select a single base point from the set of objects to use for the array such as the end of a line. This base point is not necessarily at the center of the set of objects.

The solution to this problem is to use the *More* button near the bottom of the *Array* dialog box to open the lower section of the dialog box. In this area, remove the check for *Set to object's default*, then specify a *Base point* at the center of the selected set of objects by either entering values or PICKing a point interactively.

FIGURE 9-41

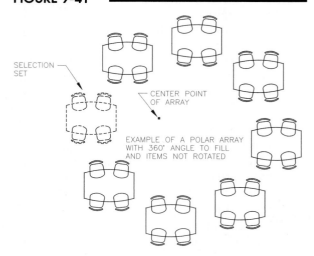

Fillet

Pull-down Menu	Command (Type)	Alias (Type)	Short-cut	Screen (side) Menu	Tablet Menu
Modify Fillet	Fillet	F	...	MODIFY2 Fillet	W,19

The *Fillet* command automatically rounds a sharp corner (intersection of two *Lines, Arcs, Circles,* or *Pline* vertices) with a radius (Fig. 9-42). You specify only the radius and select the objects' ends to be *Filleted.* The objects to fillet do <u>not</u> have to completely intersect or can overlap. You can specify whether or not the objects are automatically extended or trimmed as necessary (see Fig. 9-43).

FIGURE 9-42

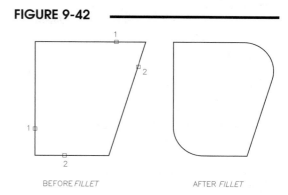

BEFORE *FILLET*   AFTER *FILLET*

Command: **Fillet**
Current settings: Mode = TRIM, Radius = 0.5000
Select first object or [Undo/Polyline/Radius/Trim/Multiple]: **PICK** (Select one *Line, Arc,* or *Circle* near the point where the fillet should be created.)
Select second object or shift-select to apply corner: **PICK** (Select the second object near the desired fillet location.)
Command:

The fillet is created at the corner selected.

Treatment of *Arcs* and *Circles* with *Fillet* is shown in Figure 9-43. Note that the objects to *Fillet* do not have to intersect or can overlap.

FIGURE 9-43

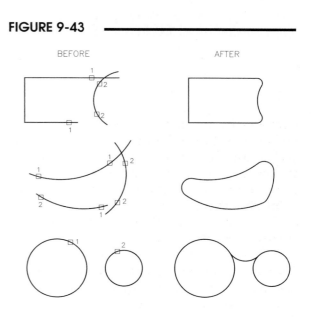

Radius

Use the *Radius* option to specify the desired radius for the fillets. *Fillet* uses the specified radius value for all new fillets until the value is changed.

TIP

If parallel objects are selected, the *Fillet* is automatically created to the correct radius (Fig. 9-44). Therefore, parallel objects can be filleted at any time <u>without specifying a radius value</u>.

FIGURE 9-44

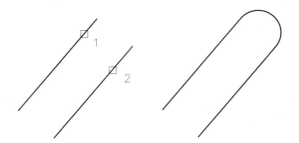

Polyline

Fillets can be created on *Polylines* in the same manner as two *Line* objects. Use *Fillet* as you normally would for *Lines*. However, if you want a fillet equal to the specified radius to be added to <u>each vertex</u> of the *Pline* (except the endpoint vertices), use the *Polyline* option; then select anywhere on the *Pline* (Fig. 9-45).

FIGURE 9-45

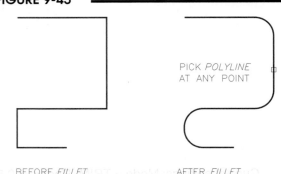

Closed Plines created by the *Close* option of *Pedit* react differently with *Fillet* than *Plines* connected by PICKing matching endpoints. Figure 9-46 illustrates the effect of *Fillet Polyline* on a *Closed Pline* and on a connected *Pline*.

FIGURE 9-46

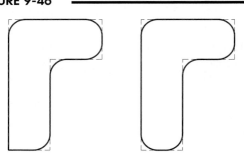

Trim/Notrim

In the previous figures, the objects are shown having been filleted in the *Trim* mode—that is, with automatically trimmed or extended objects to meet the end of the new fillet radius. The *Notrim* mode creates the fillet <u>without</u> any extending or trimming of the involved objects (Fig. 9-47). Note that the command prompt indicates the current mode (as well as the current radius) when *Fillet* is invoked.

FIGURE 9-47

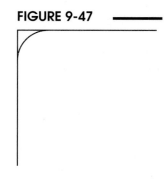

```
Command: Fillet
Current settings: Mode = TRIM, Radius = 0.5000
Select first object or [Undo/Polyline/Radius/Trim/Multiple]: T (Invokes the
Trim/No trim option)
Enter Trim mode option [Trim/No trim] <Trim>: n (Sets the option to No trim)
Select first object or [Undo/Polyline/Radius/Trim/Multiple]:
```

Multiple

Normally the *Fillet* command allows you to create one fillet, then the command ends. Use the *Multiple* option if you want to create several filleted corners. For example, if you wanted to *Fillet* the four corners of a rectangular shape you could use the *Multiple* option instead of using the *Fillet* command four separate times.

Undo

The *Undo* option operates only in conjunction with the *Multiple* option. *Undo* will remove the most recent of multiple fillets.

Shift-select

If you want to create a sharp corner instead of a rounded (radiused) corner, you can use the Shift-select option. After selecting the first line, hold down the Shift key and select the second line. Because *Fillet* can automatically trim and extend as needed, a sharp corner is created (Fig. 9-48). This action is identical to setting a *Radius* of 0 and selecting two lines normally. The *Trim* mode setting does not affect using Shift-select to create sharp corners.

FIGURE 9-48

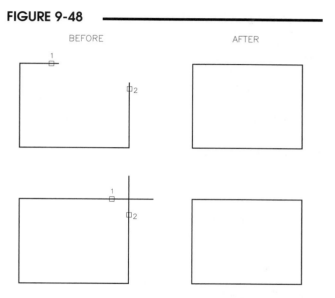

```
Command: Fillet
Current settings: Mode = TRIM, Radius = 0.5000
Select first object or
[Undo/Polyline/Radius/Trim/Multiple]: PICK
(Select the first line.)
Select second object or shift-select to apply
corner: Shift and PICK (Hold down the Shift
key and PICK the second line.)
```

Chamfer

Pull-down Menu	Command (Type)	Alias (Type)	Short-cut	Screen (side) Menu	Tablet Menu
Modify *Chamfer*	*Chamfer*	*CHA*	...	*MODIFY2* *Chamfer*	*W,18*

Chamfering is a manufacturing process used to replace a sharp corner with an angled surface. In AutoCAD, *Chamfer* is commonly used to change the intersection of two *Lines* or *Plines* by adding an angled line. The *Chamfer* command is similar to *Fillet*, but rather than rounding with a radius or "fillet," an angled line is automatically drawn at the distances (from the existing corner) that you specify.

Chamfers can be created by two methods: *Distance* (specify two distances) or *Angle* (specify a distance and an angle). The current method and the previously specified values are displayed at the Command prompt along with the options:

```
Command: chamfer
(TRIM mode) Current chamfer Dist1 = 0.0000, Dist2 = 0.0000
Select first line or [Undo/Polyline/Distance/ Angle/Trim/mEthod/Multiple]:
```

mEthod

Use this option to indicate which of the two methods you want to use: *Distance* (specify 2 distances) or *Angle* (specify a distance and an angle).

Distance

The *Distance* option is used to specify the two values applied <u>when the *Distance Method* is used</u> to create the chamfer. The values indicate the distances from the corner (intersection of two lines) to each chamfer endpoint (Fig. 9-49).

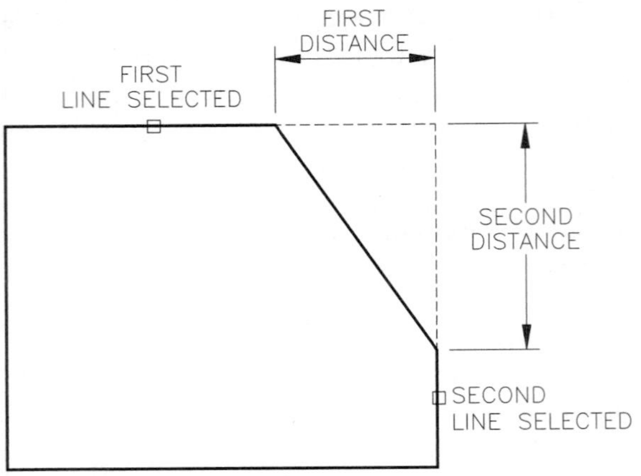

> Command: **chamfer**
> (TRIM mode) Current chamfer Dist1 = 0.0000, Dist2 = 0.0000
> Select first line or [Undo/Polyline/Distance/Angle/Trim/mEthod/Multiple]: **d**
> Specify first chamfer distance <0.0000>: **(value)** (Enter a value for the distance from the existing corner to the endpoint of the chamfer on the first line.)
> Specify second chamfer distance <0.7500>: **Enter** or **(value)** (Press Enter to use the same value as the first, or enter another value.)
> Select first line or [Undo/Polyline/Distance/Angle/Trim/mEthod/Multiple]:

FIGURE 9-49

FIRST DISTANCE
FIRST LINE SELECTED
SECOND DISTANCE
SECOND LINE SELECTED

Angle

The *Angle* option allows you to specify the values that are <u>used for the *Angle Method*</u>. The values specify a distance <u>along the first line</u> and an <u>angle from the first line</u> (Fig. 9-50).

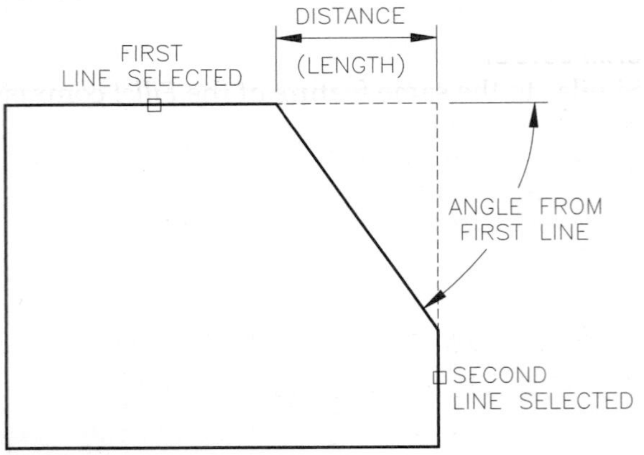

> Select first line or [Undo/Polyline/Distance/Angle/Trim/mEthod/Multiple]: **a**
> Specify chamfer length on the first line <0.0000>: **(value)**
> Specify chamfer angle from the first line <0>: **(value)**

FIGURE 9-50

DISTANCE
FIRST LINE SELECTED   (LENGTH)
ANGLE FROM FIRST LINE
SECOND LINE SELECTED

Trim/Notrim

The two lines selected for chamfering do not have to intersect but can overlap or not connect. The *Trim* setting automatically trims or extends the lines selected for the chamfer (Fig. 9-51), while the *Notrim* setting adds the chamfer without changing the length of the selected lines. These two options are the same as those in the *Fillet* command (see "*Fillet*"). Changing these options sets the *TRIMMODE* system variable to 0 (*Notrim*) or 1 (*Trim*). The variable controls trimming for both *Fillet* and *Chamfer*.

FIGURE 9-51

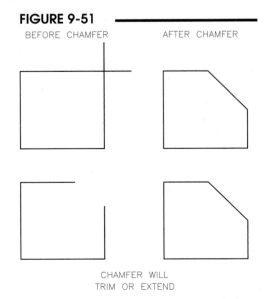

BEFORE CHAMFER   AFTER CHAMFER

CHAMFER WILL TRIM OR EXTEND

Polyline

The *Polyline* option of *Chamfer* creates chamfers on all vertices of a *Pline*. All vertices of the *Pline* are chamfered with the supplied distances (Fig. 9-52). The first end of the *Pline* that was <u>drawn</u> takes the first distance. Use *Chamfer* without this option if you want to chamfer only one corner of a *Pline*. This is similar to the same option of *Fillet* (see "*Fillet*").

FIGURE 9-52

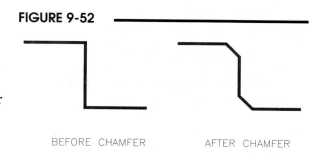

BEFORE CHAMFER   AFTER CHAMFER

A POLYLINE CHAMFER EXECUTES
A CHAMFER AT EACH VERTEX

Multiple

Normally the *Chamfer* command ends after you create one chamfer, similar to the operation of the *Fillet* command. Instead of using the *Chamfer* command several times to create several chamfered corners, use the *Multiple* option.

Undo

The *Undo* option operates only in conjunction with the *Multiple* option. *Undo* will remove the most recent of multiple chamfers.

Shift-select

Similar to the same feature of the *Fillet* command, use the Shift-select option if you want to create a sharp corner instead of a chamfered corner. After selecting the first line to chamfer, hold down the Shift key and select the second line.

CHAPTER EXERCISES

1. *Move*

   Begin a *New* drawing and create the geometry in Figure 9-53A using *Lines* and *Circles*. If desired, set *Polar Distance* to **.25** to make drawing easy and accurate.

   For practice, <u>turn *SNAP OFF*</u> (**F9**). Use the *Move* command to move the *Circles* and *Lines* into the positions shown in Fig. 9-53B. *OSNAPs* are required to *Move* the geometry accurately (since *SNAP* is off). *Save* the drawing as **MOVE1**.

FIGURE 9-53

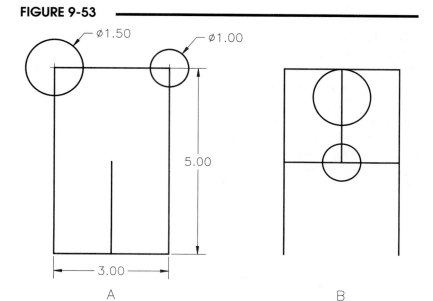

A   B

2. *Rotate*

Begin a *New* drawing and create the geometry in Figure 9-54A.

Rotate the shape into position shown in step B. *SaveAs* **ROTATE1**.

Use the *Reference* option to *Rotate* the box to align with the diagonal *Line* as shown in C. *SaveAs* **ROTATE2**.

FIGURE 9-54

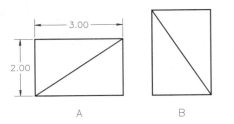
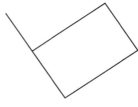

3. *Scale*

Open **ROTATE1** to again use the shape shown in Figure 9-55B and *SaveAs* **SCALE1**. *Scale* the shape by a factor of **1.5**.

Open **ROTATE2** to again use the shape shown in C. Use the *Reference* option of *Scale* to increase the scale of the three other *Lines* to equal the length of the original diagonal *Line* as shown. (HINT: *OSNAPs* are required to specify the *Reference length* and *New length*.) *SaveAs* **SCALE2**.

FIGURE 9-55

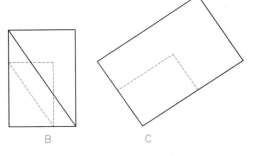

4. *Stretch*

A design change has been requested. *Open* the SLOTPLATE CH8 drawing and make the following changes.

A. The top of the plate (including the slot) must be moved upward. This design change will add 1" to the total height of the Slot Plate. Use *Stretch* to accomplish the change, as shown in Figure 9-56. Draw the *Crossing Window* as shown.

FIGURE 9-56

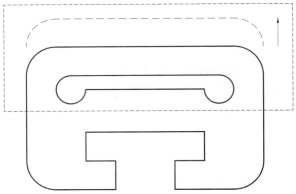

B. The notch at the bottom of the plate must be adjusted slightly by relocating it .50 units to the right, as shown in Figure 9-57. Draw the *Crossing Window* as shown. Use *SaveAs* to reassign the name to SLOTPLATE 2.

FIGURE 9-57

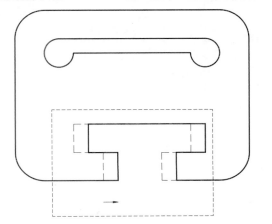

5. *Trim*

   A. Create the shape shown in
      Figure 9-58A. *SaveAs* **TRIM-EX**.

   B. Use *Trim* to alter the shape as shown in B.
      SaveAs **TRIM1**.

   C. *Open* **TRIM-EX** to create the shapes
      shown in C and D using *Trim*. *SaveAs*
      TRIM2 and **TRIM3**.

6. *Extend*

   Open each of the drawings created as solutions for Figure 9-58 (**TRIM1, TRIM2,** and
   TRIM3). Use *Extend* to return each of the
   drawings to the original form shown in
   Figure 9-58A. *SaveAs* **EXTEND1, EXTEND2,**
   and **EXTEND3**.

FIGURE 9-58

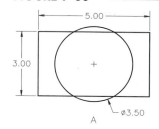
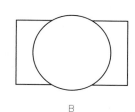

A   B

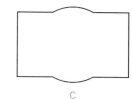
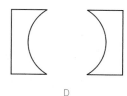

C   D

7. *Break*

   A. Create the shape shown in Figure 9-59A.
      SaveAs **BREAK-EX**.

   B. Use *Break* to make the two breaks as shown
      in B. *SaveAs* **BREAK1**. (HINT: You may
      have to use *OSNAP*s to create the breaks at
      the *Intersections* or *Quadrants* as shown.)

   C. Open **BREAK-EX** each time to create the
      shapes shown in C and D with the *Break*
      command. *SaveAs* **BREAK2** and **BREAK3**.

FIGURE 9-59

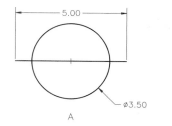
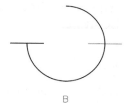

A   B

C   D

8. **Complete the Table Base**

   A. Open the **TABLE-BASE** drawing you created in the
      Chapter 7 Exercises. The last step in the previous exercise
      was the creation of the square at the center of the table base.
      Now, use *Trim* to edit the legs and the support plate to look
      like Figure 9-60.

FIGURE 9-60

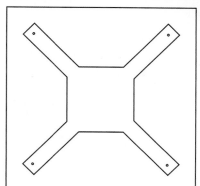

B. Use *Circle* with the ***Center, Radius*** option to create a 4″ diameter circle to represent the welded support for the center post for this table. Finally, add the other 4 drill holes (.25″ diameter) on the legs at the intersection of the vertical and horizontal lines representing the edges of the base. Use **Polar Tracking** or *Intersection* OSNAP for this step. The table base should appear as shown in Figure 9-61. *Save* the drawing.

FIGURE 9-61

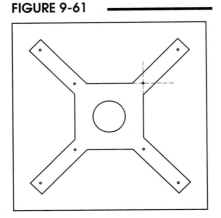

9. *Lengthen*

Five of your friends went to the horse races, each with $5.00 to bet. Construct a simple bar graph (similar to Fig. 9-62) to illustrate how your friends' wealth compared at the beginning of the day.

Modify the graph with *Lengthen* to report the results of their winnings and losses at the end of the day. The reports were as follows: friend 1 made $1.33 while friend 2 lost $2.40; friend 3 reported a 150% increase and friend 4 brought home 75% of the money; friend 5 ended the day with $7.80 and friend 6 came home with $5.60. Use the *Lengthen* command with the appropriate options to change the line lengths accordingly.

Who won the most? Who lost the most? Who was closest to even? Enhance the graph by adding width to the bars and other improvements as you wish (similar to Fig. 9-63). *SaveAs* **FRIENDS GRAPH**.

FIGURE 9-62

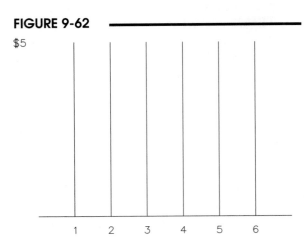

FIGURE 9-63

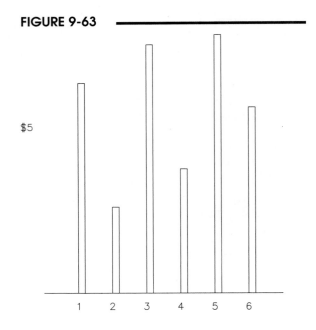

10. *Trim, Extend*

A. *Open* the **PIVOTARM CH7** drawing from the Chapter 7 Exercises. Use *Trim* to remove the upper sections of the vertical *Lines* connecting the top and front views. Compare your work to Figure 9-64.

FIGURE 9-64

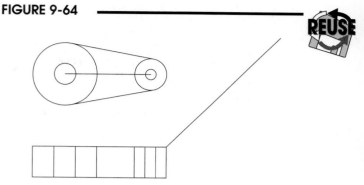

B. Next, draw a horizontal *Line* in the front view between the *Midpoints* of the vertical *Line* on each end of the view, as shown in Figure 9-65. *Erase* the horizontal *Line* in the top view between the *Circle* centers.

FIGURE 9-65

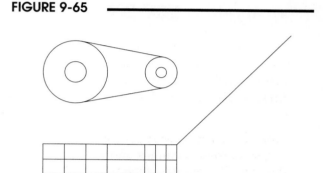

C. Draw vertical *Lines* from the *Endpoints* of the inclined *Line* (side of the object) in the top view down to the bottom *Line* in the front view, as shown in Figure 9-66. Use the two vertical lines as *Cutting edges* for *Trimming* the *Line* in the middle of the front view, as shown highlighted.

FIGURE 9-66

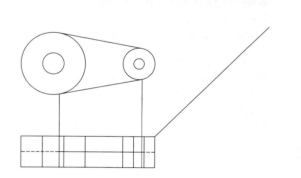

D. Finally, *Erase* the vertical lines used for *Cutting edges* and then use *Trim* to achieve the object as shown in Figure 9-67. Use *SaveAs* and name the drawing **PIVOTARM CH9.**

FIGURE 9-67

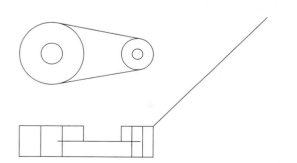

11. **GASKETA**

Begin a *New* drawing. Set *Limits* to **0,0** and **8,6**. Set *SNAP* and *GRID* values appropriately. Create the Gasket as shown in Figure 9-68. Construct only the gasket shape, not the dimensions or centerlines. (HINT: Locate and draw the four .5" diameter *Circles*, then create the concentric .5" radius arcs as full *Circles*, then *Trim*. Make use of *OSNAPs* and *Trim* whenever applicable.) *Save* the drawing and assign the name   **GASKETA.**

FIGURE 9-68

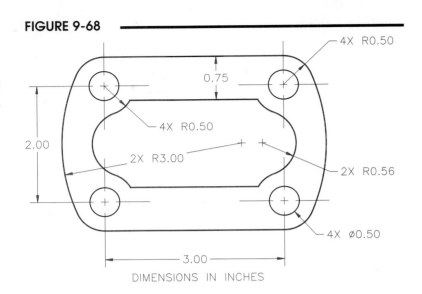

DIMENSIONS IN INCHES

12. **Chemical Process Flow Diagram**

Begin a *New* drawing and re-create the chemical process flow diagram shown in Figure 9-69. Use *Line*, *Circle*, *Arc*, *Trim*, *Extend*, *Scale*, *Break*, and other commands you feel necessary to complete the diagram. Because this is diagrammatic and will not be used for manufacturing, dimensions are not critical, but try to construct the shapes with proportional accuracy. *Save* the drawing as **FLOWDIAG**.

FIGURE 9-69

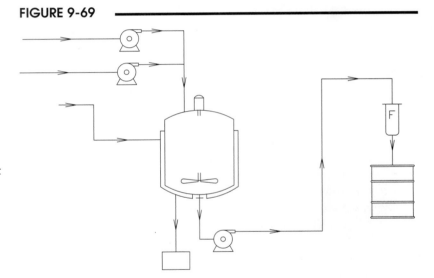

13. *Copy*

Begin a *New* drawing. Set the *Limits* to **24,18**. *Set Polar Snap* to **.5** and set your desired running *OSNAPs*. Turn On *SNAP, GRID, OSNAP,* and *POLAR*. Create the sheet metal part composed of four *Lines* and one **Circle** as shown in Figure 9-70. The lower-left corner of the part is at coordinate 2,3.

FIGURE 9-70

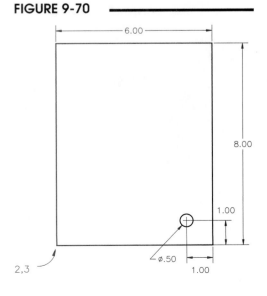

Use *Copy* to create two copies of the rectangle and hole in a side-by-side fashion as shown in Figure 9-71. Allow 2 units between the sheet metal layouts. *SaveAs* **PLATES.**

FIGURE 9-71

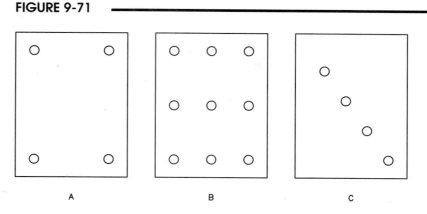

A   B   C

A. Use *Copy* to create the additional 3 holes equally spaced near the other corners of the part as shown in Figure 9-71A. *Save* the drawing.

B. Use *Copy* to create the hole configuration as shown in Figure 9-71B. *Save*.

C. Use *Copy* to create the hole placements as shown in C. Each hole <u>center</u> is **2** units at **125** degrees from the previous one (use relative polar coordinates or set an appropriate *Polar Angle*). *Save* the drawing.

14. *Mirror*

A manufacturing cell is displayed in Figure 9-72. The view is from above, showing a robot <u>centered</u> in a work station. The production line requires 4 cells. Begin by starting a *New* drawing, setting *Units* to *Engineering* and *Limits* to **40′ x 30′**. It may be helpful to set *Polar Snap* to **6″**. Draw one cell to the dimensions indicated. Begin at the indicated coordinates of the lower-left corner of the cell. *SaveAs* **MANFCELL.**

FIGURE 9-72

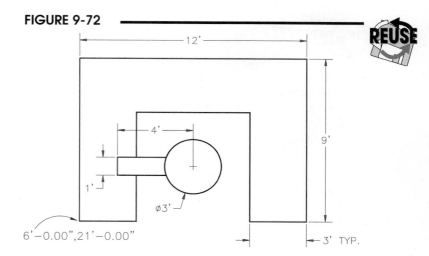

Use *Mirror* to create the other three manufacturing cells as shown in Figure 9-73. Ensure that there is sufficient space between the cells as indicated. Draw the two horizontal *Lines* representing the walkway as shown. *Save*.

FIGURE 9-73

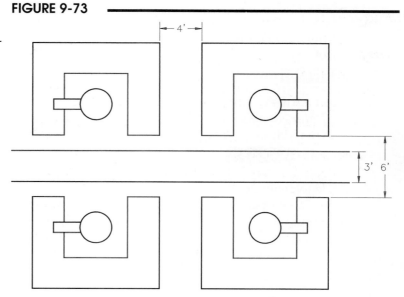

15. *Offset*

Create the electrical schematic as shown in Figure 9-74. Draw *Lines* and *Plines*, then *Offset* as needed. Use *Point* objects with an appropriate (circular) *Point Style* at each of the connections as shown. Check the *Set Size to Absolute Units* radio button (in the *Point Style* dialog box) and find an appropriate size for your drawing. Because this is a schematic, you can create the symbols by approximating the dimensions. Omit the text. *Save* the drawing as **SCHEMATIC1.** Create a *Plot* and check *Scaled to Fit*.

FIGURE 9-74

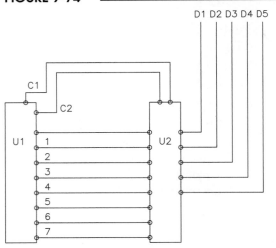

16. *Array, Polar*

Begin a *New* drawing. Create the starting geometry for a Flange Plate as shown in Figure 9-75. *SaveAs* **ARRAY**.

FIGURE 9-75

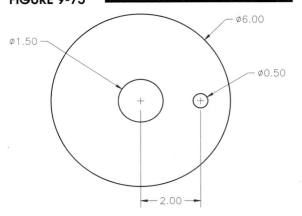

A. Create the *Polar Array* as shown in Figure 9-76A. *SaveAs* **ARRAY1.**

B. *Open* **ARRAY**. Create the *Polar Array* as shown in Figure 9-76B. *SaveAs* **ARRAY2.** (HINT: Use a negative angle to generate the *Array* in a clockwise direction.)

FIGURE 9-76

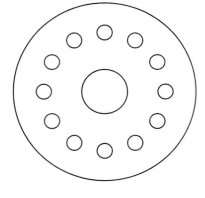

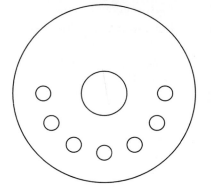

A

B

17. *Array, Rectangular*

Begin a *New* drawing. Use *Save* and assign the name **LIBRARY DESKS**. Create the *Array* of study carrels (desks) for the library as shown in Figure 9-77. The room size is 36′ x 27′ (set *Units* and *Limits* accordingly). Each carrel is **30″ x 42″**. Design your own chair. Draw the first carrel (highlighted) at the indicated coordinates. Create the *Rectangular Array* so that the carrels touch side to side and allow a 6′ aisle for walking between carrels (not including chairs).

FIGURE 9-77

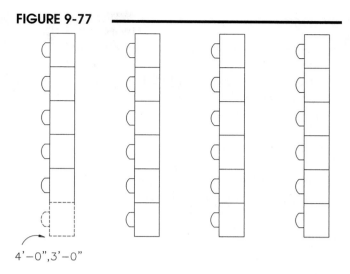

4′−0″,3′−0″

18. *Array, Rectangular*

Create the bolt head with one thread as shown in Figure 9-78A. Create the small line segment at the end (0.40 in length) as a *Pline* with a *Width* of .02. *Array* the first thread (both crest and root lines, as indicated in B) to create the schematic thread representation. There is **1** row and **10** columns with **.2** units between each. Add the *Lines* around the outside to complete the fastener. *Erase* the *Pline* at the small end of the fastener (Fig. 9-78B) and replace it with a *Line*. *Save* as **BOLT.**

FIGURE 9-78

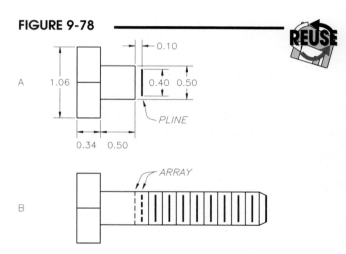

19. *Fillet*

Create the "T" Plate shown in Figure 9-79. Use *Fillet* to create all the fillets and rounds as the last step. When finished, *Save* the drawing as **T-PLATE** and make a plot.

FIGURE 9-79

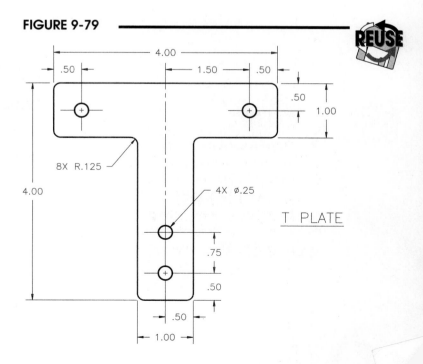

T PLATE

20. *Copy, Fillet*

Create the Gasket shown in Figure 9-80. Include center-lines in your drawing.

Save the drawing as **GASKETB.**

FIGURE 9-80

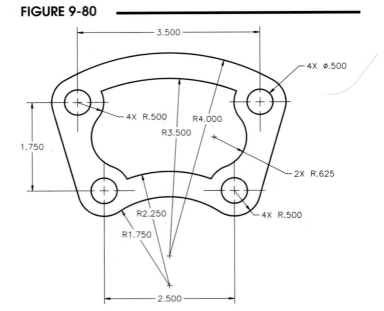

21. *Array*

Create the perforated plate as shown in Figure 9-81. *Save* the drawing as **PERF-PLAT**. (HINT: For the *Rectangular Array* in the center, create 100 holes, then *Erase* four holes, one at each corner for a total of 96. The two circular hole patterns contain a total of 40 holes each.)

FIGURE 9-81

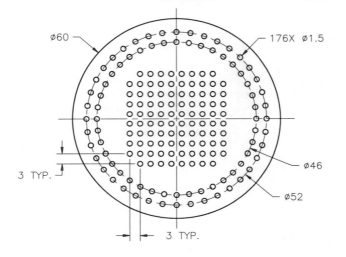

22. *Chamfer*

Begin a *New* drawing. Set the *Limits* to **279,216**. Next, type *Zoom* and use the *All* option. Set *Snap* to **2** and *Grid* to **10**. Create the Catch Bracket shown in Figure 9-82. Draw the shape with all vertical and horizontal *Lines*, then use *Chamfer* to create the six chamfers. *Save* the drawing as **CBRACKET** and create a plot.

FIGURE 9-82

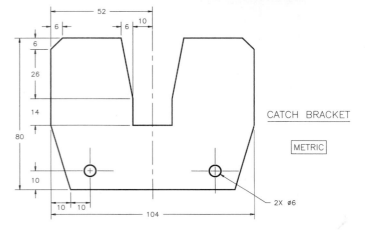

VIEWING COMMANDS

CHAPTER OBJECTIVES

After completing this chapter you should:

1. understand the relationship between the drawing objects and the display of those objects;

2. be able to use all of the *Zoom* options to view areas of a drawing;

3. be able to *Pan* the display about your screen;

4. be able to save and restore *Views*;

5. be able to turn on and off the *UCS Icon*;

6. be able to create, save, and restore model space viewports using the *Vports* command.

CONCEPTS

The accepted CAD practice is to draw full size using actual units. Since the drawing is a virtual dimensional replica of the actual object, a drawing could represent a vast area (several hundred feet or even miles) or a small area (only millimeters). The drawing is created full size with the actual units, but it can be displayed at any size on the screen. Consider also that CAD systems provide for a very high degree of dimensional precision, which permits the generation of drawings with great detail and accuracy.

Combining those two CAD capabilities (great precision and drawings representing various areas), a method is needed to view different and detailed segments of the overall drawing area. In AutoCAD the commands that facilitate viewing different areas of a drawing are *Zoom, Pan,* and *View*.

FIGURE 10-1

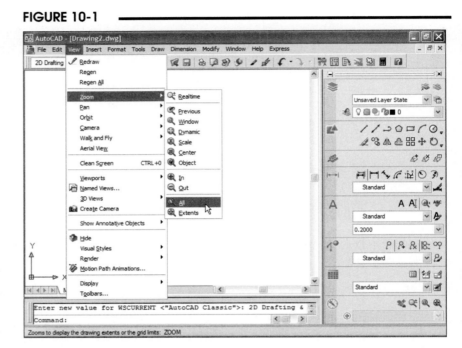

The viewing commands can be found in the *View* pull-down menu (Fig. 10-1) which is available from either of the 2D workspaces. If you are using the *2D Drafting & Annotation* workspace, the control panel at the bottom of the dashboard, *2D Navigate*, contains the viewing commands (Fig. 10-1, lower right).

The Standard toolbar contains a group of tools (icon buttons) for the viewing commands located near the right end of the toolbar (Fig. 10-2). The *Realtime* options of *Pan* and *Zoom,* and *Zoom Previous* have icons permanently displayed on the toolbar, whereas the other *Zoom* options are located on flyouts.

FIGURE 10-2

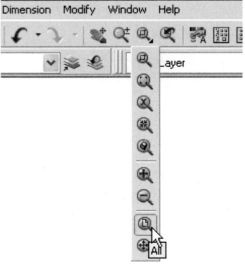

Like other commands, the viewing commands can also be invoked by typing the command or command alias, using the screen menu (*VIEW1*), or using the digitizing tablet menu (if available).

The commands discussed in this chapter are:

> *Zoom, Pan, View, Ucsicon, Syswindows,* and *Vports*

The *Realtime* options are the default options for both the *Pan* and *Zoom* commands. *Realtime* options of *Pan* and *Zoom* allow you to <u>interactively</u> change the drawing display by moving the cursor in the drawing area.

ZOOM AND *PAN* WITH THE MOUSE WHEEL

Zoom and *Pan*

To zoom in means to magnify a small segment of a drawing and to zoom out means to display a larger area of the drawing. <u>Zooming does not change the size of the drawing objects; zooming changes only the display of those objects</u>—the area that is displayed on the screen. All objects in the drawing have the same dimensions before and after zooming. Only your display of the objects changes.

To pan means to move the display area slightly without changing the size of the current view window. Using the pan function, you "drag" the drawing across the screen to display an area outside of the current view in the Drawing Editor.

Although there are many ways to change the area of the drawing you want to view using either the *Zoom* or *Pan* commands, the fastest and easiest method for simple zoom and pan operations is to use the mouse wheel (the small wheel between the two mouse buttons) if you have one. Using the mouse wheel to zoom or pan does not require you to invoke the *Zoom* or *Pan* commands. Additionally, the mouse wheel zoom and pan functions are transparent, meaning that you can use this method while another command is in use. So, if you have a mouse wheel, you can zoom and pan at any time without using any commands.

Zoom with the Mouse Wheel

If you have a mouse wheel, you can zoom by simply turning the wheel. Turn the wheel forward to zoom <u>in</u> and turn the wheel backward to zoom <u>out</u>. <u>The current location of the cursor is the center for zooming</u>. In other words, if you want to zoom in to an area, simply locate your cursor on the spot and turn the wheel forward. This type of zooming can be done transparently (during another command operation). You can also pan using the wheel (see "*Pan* with the Mouse Wheel or Third Button").

FIGURE 10-3

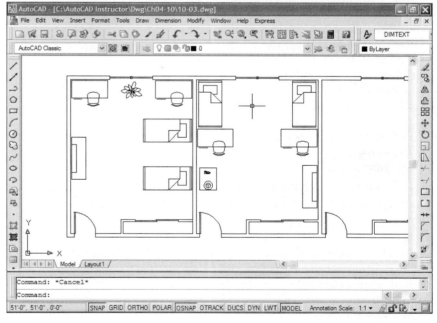

For example, Figure 10-3 displays a floor plan of a dorm room. Notice the location of the cursor in the upper-right corner of the drawing area. By turning the mouse wheel forward, the display changes by zooming in (enlarging) to the area designated by the cursor location. The resulting display (after turning the mouse wheel forward) is shown in Figure 10-4.

Using the same method, you could zoom out, or change the display from Figure 10-4 to Figure 10-3, by turning the wheel backward. When zooming out, the cursor location also controls the center for zooming.

FIGURE 10-4

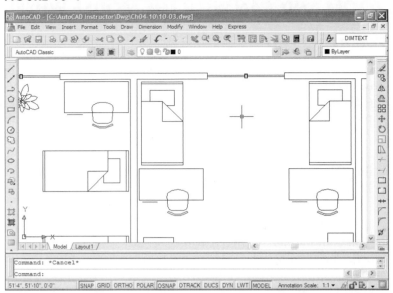

Pan with the Mouse Wheel or Third Button

If you have a mouse wheel or third mouse button, you can pan by <u>holding down</u> the wheel or button so the "hand" appears (Fig. 10-5), then move the hand cursor in any direction to "drag" the drawing around on the screen. Panning is typically used when you have previously zoomed in to an area but then want to view a different area that is slightly out of the current viewing area (off the screen). Panning with the mouse wheel or third button, similar to zooming with the mouse wheel, does not require you to invoke any commands. In addition, the pan feature is transparent, so you can pan with the wheel or third button while another command is in operation. Using the mouse wheel or third button to pan is essentially the same as using the *Pan* command with the *Realtime* option, except you do not have to invoke the *Pan* command.

FIGURE 10-5

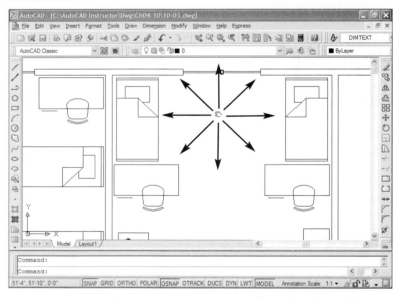

COMMANDS

Zoom

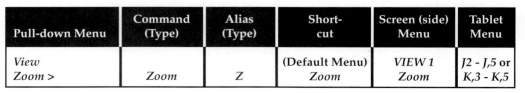

Pull-down Menu	Command (Type)	Alias (Type)	Short-cut	Screen (side) Menu	Tablet Menu
View Zoom >	Zoom	Z	(Default Menu) Zoom	VIEW 1 Zoom	J2 - J,5 or K,3 - K,5

The *Zoom* options described next can be typed or selected by any method. Unlike most commands, AutoCAD provides a tool (icon button) for each <u>option</u> of the *Zoom* command. If you type *Zoom*, the options can be invoked by typing the first letter of the desired option or pressing Enter for the *Realtime* option.

The Command prompt appears as follows.

Command: **zoom**
Specify corner of window, enter a scale factor (nX or nXP), or
[All/Center/Dynamic/Extents/Previous/Scale/Window/Object] <real time>:

Realtime (Rtzoom)

FIGURE 10-6

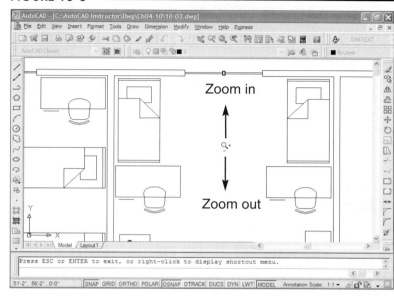

Realtime is the default option of *Zoom*. If you type *Zoom*, just press Enter to activate the *Realtime* option. Alternately, you can type *Rtzoom* to invoke this option directly.

With the *Realtime* option you can interactively zoom in and out with <u>vertical cursor motion</u>. When you activate the *Realtime* option, the cursor changes to a magnifying glass with a plus (+) and a minus (-) symbol. Move the cursor to any location on the drawing area and hold down the PICK (left) mouse button. Move the cursor up to zoom in and down to zoom out (Fig. 10-6). Horizontal movement has no effect. You can zoom in or out repetitively. Pressing Esc or Enter exits *Realtime Zoom* and returns to the Command prompt.

The current drawing window is used to determine the zooming factor. Moving the cursor half the window height (from the center to the top or bottom) zooms in or out to a zoom factor of 100%. Starting at the bottom or top allows zooming in or out (respectively) at a factor of 200%.

If you have zoomed in or out repeatedly, you can reach a zoom limit. In this case the plus (+) or minus (-) symbol disappears when you press the left mouse button. To zoom further with *Rtzoom*, first type *Regen*.

Pressing the right mouse button (or button #2 on digitizing pucks) produces a small cursor menu with other viewing options (Fig. 10-7). This same menu and its options can also be displayed by right-clicking during *Realtime Pan*. The options of the cursor menu are described below.

FIGURE 10-7

Exit
Select *Exit* to exit *Realtime Pan* or *Zoom* and return to the Command prompt. The Escape or Enter key can be used to accomplish the same action.

Pan
This selection switches to the *Realtime* option of *Pan*. A check mark appears here if you are currently using *Realtime Pan*. See "*Pan*" next in this chapter.

Zoom
A check mark appears here if you are currently using *Realtime Zoom*. If you are using the *Realtime* option of *Pan*, check this option to switch to *Realtime Zoom*.

Zoom Window

Select this option if you want to display an area to zoom in to by specifying a *Window*. This feature operates differently but accomplishes the same action as the *Window* option of *Zoom*. With this option, window selection is made with one PICK. In other words, PICK for the first corner, hold down the left button and drag, then release to establish the other corner. With the *Window* option of the *Zoom* command, PICK once for each of two corners. See "*Window.*"

Zoom Original

Selecting this option automatically displays the area of the drawing that appeared on the screen immediately before using the *Realtime* option of *Zoom* or *Pan*. Using this option successively has no effect on the display. This feature is different from the *Previous* option of the *Zoom* command, in which ten successive previous views can be displayed.

Zoom Extents

Use this option to display the entire drawing in its largest possible form in the drawing area. This option is identical to the *Extents* option of the *Zoom* command. See "*Extents.*"

Window

 To *Zoom* with a *Window* is to draw a rectangular window around the desired viewing area. You PICK a first and a second corner (diagonally) to form the rectangle. The windowed area is magnified to fill the screen (Fig. 10-8). It is suggested that you draw the window with a 4 x 3 (approximate) proportion to match the screen proportion. If you type *Zoom* or *Z*, *Window* is an <u>automatic</u> option so you can begin selecting the first corner of the window after issuing the *Zoom* command <u>without</u> indicating the *Window* option as a separate step.

FIGURE 10-8

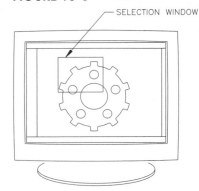
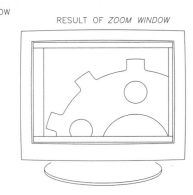

SELECTION WINDOW

RESULT OF *ZOOM WINDOW*

All

 This option displays <u>all of the objects</u> in the drawing <u>and all of the *Limits*</u>. In Figure 10-9, notice the effects of *Zoom All*, based on the drawing objects and the drawing *Limits*.

Extents

 This option results in the largest possible display of <u>all of the objects</u>, <u>disregarding</u> the *Limits*. (*Zoom All* includes the *Limits*.) See Figure 10-9.

FIGURE 10-9

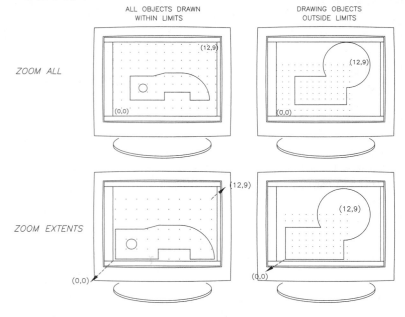

ALL OBJECTS DRAWN WITHIN LIMITS

DRAWING OBJECTS OUTSIDE LIMITS

ZOOM ALL

ZOOM EXTENTS

Zoom to Object

You can think of this option as "zoom to extents of object." *Zoom to Object* allows you to select any object, such as a *Line*, *Circle*, or *Pline*, then AutoCAD automatically zooms to display the entire object as large as possible on the screen.

Scale (X/XP)

This option allows you to enter a scale factor for the desired display. The value that is entered can be relative to the full view (*Limits*) or to the current display (Fig. 10-10). A value of **1, 2,** or **.5** causes a display that is 1, 2, or .5 times the size of the *Limits*, centered on the current display. A value of **1X, 2X,** or **.5X** yields a display 1, 2, or .5 times the size of the <u>current display</u>. If you are using paper space and viewing model space, a value of **1XP, 2XP,** or **.5XP** yields a model space display scaled 1, 2, or .5 times paper space units.

If you type *Zoom* or *Z,* you can enter a scale factor at the *Zoom* command prompt without having to type the letter *S*.

FIGURE 10-10

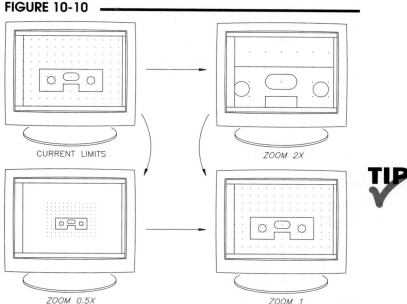

CURRENT LIMITS

ZOOM 2X

ZOOM 0.5X

ZOOM 1

In

Zoom In magnifies the current display by a factor of 2X (2 times the current display). Using this option is the same as entering a *Zoom Scale* of 2X.

Out

Zoom Out makes the current display smaller by a factor of .5X (.5 times the current display). Using this option is the same as entering a *Zoom Scale* of .5X.

Center

First, specify a location as the center of the zoomed area; then specify either a *Magnification factor* (see *Scale X/XP*), a *Height* value for the resulting display, or PICK two points forming a vertical to indicate the height for the resulting display (Fig. 10-11).

FIGURE 10-11

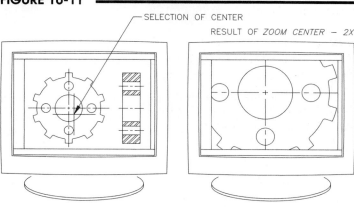

SELECTION OF CENTER

RESULT OF *ZOOM CENTER — 2X*

Previous

Selecting this option automatically changes to the previous display. AutoCAD saves the previous ten displays changed by *Zoom, Pan,* and *View.* You can successively change back through the previous ten displays with this option.

Dynamic

FIGURE 10-12

With this option you can change the display from one windowed area in a drawing to another without using *Zoom All* (to see the entire drawing) as an intermediate step. *Zoom Dynamic* causes the screen to display the drawing *Extents* bounded by a box. The current view or window is bounded by a box in a broken-line pattern. The view box is the box with an "X" in the center which can be moved to the desired location (Fig. 10-12). The desired location is selected by pressing **Enter** (not the PICK button as you might expect).

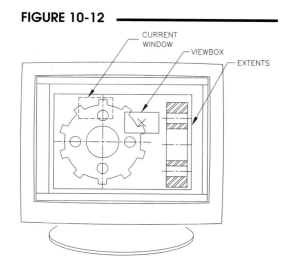

Pressing the left button allows you to resize the view box (displaying an arrow instead of the "X") to the desired size. Move the mouse or puck left and right to increase and decrease the window size. Press the left button again to set the size and make the "X" reappear.

Zoom Is Transparent

Zoom is a <u>transparent</u> command, meaning it can be invoked while another command is in operation. You can, for example, begin the *Line* command and at the "Specify next point or [Undo]:" prompt *'Zoom* with a *Window* to better display the area for selecting the endpoint. This transparent feature is automatically entered if *Zoom* is invoked by the screen, pull-down, or tablet menus or toolbar icons, but if typed it must be prefixed by the apostrophe (') symbol, e.g., "Specify next point or [Undo]: *'Zoom.*" If *Zoom* has been invoked transparently, the **>>** symbols appear at the command prompt before the listed options as follows:

```
Command: line
Specify first point: 'zoom
>>Specify corner of window, enter a scale factor (nX or nXP), or
[All/Center/Dynamic/Extents/Previous/Scale/Window/Object] <real time>:
```

Pan

Pull-down Menu	Command (Type)	Alias (Type)	Short-cut	Screen (side) Menu	Tablet Menu
View *Pan*	*Pan* or -*Pan*	*P* or -*P*	(Default Menu) *Pan*	*VIEW 1* *Pan*	*N,10-P,11*

Using the *Pan* command by typing or selecting from the tool, screen, or digitizing tablet menu produces the *Realtime* version of *Pan.* The *Point* option is available from the pull-down menu or by typing -*Pan.* The other "automatic" *Pan* options (*Left, Right, Up, Down*) can only be selected from the pull-down menu.

The *Pan* command is useful if you want to move (pan) the display area slightly without changing the <u>size</u> of the current view window. With *Pan*, you "drag" the drawing across the screen to display an area outside of the current view in the drawing area.

Realtime (RTPAN)

Realtime is the default option of *Pan*. It is invoked by selecting *Pan* by any method, including typing *Pan* or *Rtpan*. *Realtime Pan* is essentially the same as using the mouse wheel or third button to pan. That is, *Realtime Pan* allows you to interactively pan by "pulling" the drawing across the screen with the cursor motion. After activating the command, the cursor changes to a hand cursor. Move the hand to any location on the drawing, then hold the PICK (left) mouse button down and drag the drawing around on the screen to achieve the desired view (see Fig. 10-5). When you release the mouse button, panning is discontinued. You can move the hand to another location and pan again without exiting the command.

You must press Escape or Enter or use *Exit* from the pop-up cursor menu to exit *Realtime Pan* and return to the Command prompt. The following Command prompt appears during *Realtime Pan*.

Command: **pan**
Press Esc or Enter to exit, or right-click to display shortcut menu.

If you press the right mouse button (#2 button), a small cursor menu pops up. This is the same menu that appears during *Realtime Zoom* (see Fig. 10-7). This menu provides options for you to *Exit, Zoom* or *Pan* (realtime), *Zoom Window, Zoom Original,* or *Zoom Extents.* See "*Zoom*" earlier in this chapter.

Point

The *Point* option is available only through the pull-down menu or by typing "-*Pan*" (some commands can be typed with a hyphen [-] prefix to invoke the command line version of the command without dialog boxes, etc.). Using this option produces the following Command prompt.

Command: **-pan**
Specify base point or displacement: **PICK** or (**value**)
Specify second point: **PICK** or (**value**)
Command:

FIGURE 10-13

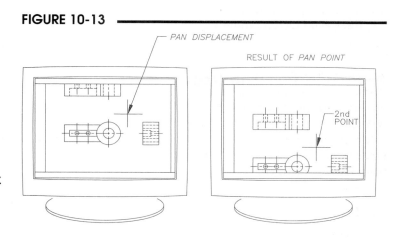

The "base point:" can be thought of as a "handle," or point to move <u>from</u>, and the "second point:" as the new location to move <u>to</u> (Fig. 10-13). You can PICK each of these points with the cursor.

You can also enter coordinate values rather than interactively PICKing points with the cursor. The command syntax is as follows:

Command: **-pan**
Specify base point or displacement: **0,-2**
Specify second point: **Enter**
Command:

Entering coordinate values allows you to *Pan* to a location outside of the current display. If you use the interactive method (PICK points), you can *Pan* only within the range of whatever is visible on the screen.

The following *Pan* options are available from the *View* pull-down menu only.

Left

Automatically pans to the left, equal to about 1/8 of the current drawing area width.

Right

Pans to the right—similar to but opposite of *Left*.

Up

Automatically pans up, equal to about 1/8 of the current drawing area height.

Down

Pans down—similar to but opposite of *Up*.

Pan Is Transparent

Pan, like *Zoom*, can be used as a transparent command. The transparent feature is automatically entered if *Pan* is invoked by the screen, pull-down, or tablet menus or icons. However, if you <u>type</u> *Pan* during another command operation, it must be prefixed by the apostrophe (') symbol.

```
Command: line
Specify first point: 'pan
>>Press ESC or ENTER to exit, or right-click to display shortcut menu. (Pan, then) Enter
Resuming LINE command.
Specify next point or [Undo]:
```

Zoom and *Pan* with *Undo* and *Redo*

FIGURE 10-14

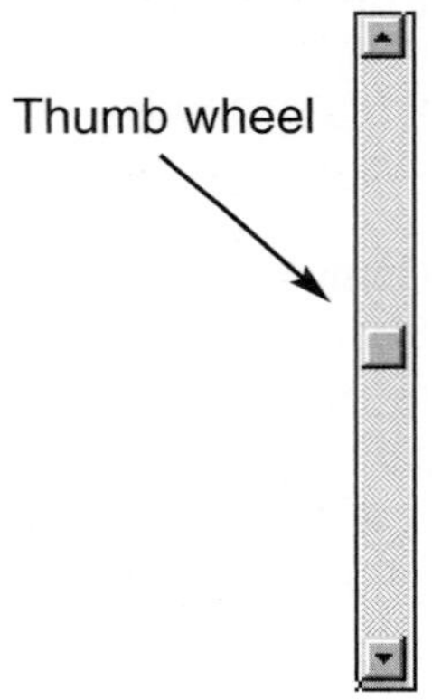

By default, consecutive uses of *Zoom* and *Pan* are treated as one command for *Undo* and *Redo* operations. For example, if you *Zoomed* three times consecutively but then wanted to return to the original display, you would have to *Undo* only once. You can change this setting (*Combine zoom and pan commands*) in the lower-right corner of the *User Preference* tab of the *Options* dialog box.

Thumb wheel

Scroll Bars

You can also use the horizontal and vertical scroll bars directly below and to the right of the graphics area to pan the drawing (Fig. 10-14). These scroll bars pan the drawing in one of three ways: (1) click on the arrows at the ends of the scroll bar, (2) move the thumb wheel, or (3) click inside the scroll bar. The scroll bars can be turned on or off by using the *Display* tab in the *Options* dialog box.

View

Pull-down Menu	Command (Type)	Alias (Type)	Short-cut	Screen (side) Menu	Tablet Menu
View *Named Views...*	*View* or *-View*	*V* or *-V*	...	*VIEW 1* *Ddview*	*M,5*

The *View* command provides a dialog box called the *View Manager* for you to create *New* views of a specified display window and restore them (make *Current*) at a later time. For typical applications you would use *View, New* to save that display under an assigned name. Later in the drawing session, the named *View* can be made *Current* any number of times. This method is preferred to continually *Zooming* in and out to display several of the same areas repeatedly. Making a named *View* the *Current* view requires no regeneration time.

Using *View* invokes the *View Manager* (Fig. 10-15) and *-View* produces the Command line format. The options of the *View* command are described below.

The left side of the *View Manager* lists the saved views contained in the current drawing. The central area gives several input fields that allow you to change *Camera* and *Target* settings, *Perspective*, *Field of view*, and *Clipping* plane settings. These features are used almost exclusively for 3D views (see Chapter 27, 3D Basics, Navigation, and Visual Styles).

FIGURE 10-15

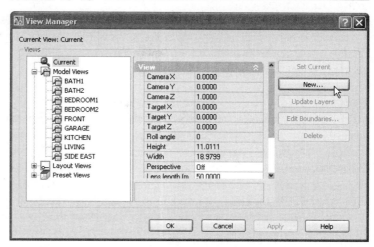

New

Using the *New* button in the *View Manager* box produces the *New View* dialog box (Fig. 10-16) that you should use to create and specify options for a new named view. Enter a *View Name* and (optional) *View Category*, a description field (such as "Elevation" or "Plan").

The *Boundary* section allows you to specify the area of the drawing you want to display when the view is restored. If already zoomed into the desired view area, select *Current Display*. Otherwise, select *Define window* or press the *Define View Window* button to return to the drawing to specify a window.

Under *Settings* you can choose to *Save layer snapshot with view* so that when the view is later restored the layer visibility settings return to those when the view was defined or later updated using the *Update Layers* button. In this way, when the view is restored, AutoCAD resets the layer visibility back to when the view was created no matter what the current visibility settings may be at the time you restore the view. (Layers are discussed in detail in Chapter 11.)

FIGURE 10-16

Set Current

To *Restore* a named view, double-click on the name from the list or highlight the name and select *Set Current*, then press *OK*. This action makes the selected view current in the drawing area. If the view you saved was in another *Layout tab* or *Model tab*, then using *Set Current* will return to the correct layout and display the desired view.

Update Layers

If the layer settings were stored when the view was created (the *Store Current Settings with View* option was used) and you want to reset the layer visibility for the view, use this option. Pressing this button causes the layer visibility information saved with the selected named view to update to the current settings (match the layer visibility in the current model space or layout viewport).

Edit Boundaries

The *Edit Boundaries* button returns to the drawing with the selected view highlighted. With this unique feature, the screen displays the current view (boundary) in the central area of the screen with the normal background color and the additional area surrounding the old view boundary in a grayed-out representation. Select the new boundary with a window.

Delete

Highlight any named view from the list, then select *Delete* to remove the named view from the list of saved views.

General

Selecting a saved view from the *Views* list on the left changes the central area to list the features of the view that were specified when the view was created (using the *New* option), such as *Name*, *Category*, and *layer snapshot*. You can change any of these specifications by entering the new information in the input fields.

Ucsicon

Pull-down Menu	Command (Type)	Alias (Type)	Short-cut	Screen (side) Menu	Tablet Menu
View *Display >* *UCS Icon >*	*Ucsicon*	...	...	*VIEW 2* *UCSicon*	*L,2*

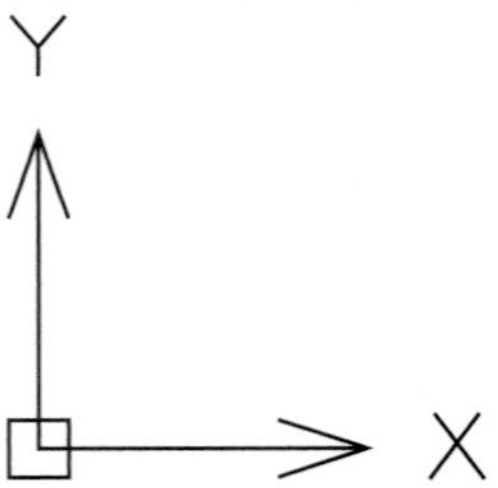

The icon that may appear in the lower-left corner of the AutoCAD Drawing Editor is the Coordinate System Icon (Fig. 10-17). The icon is sometimes called the "UCS icon" because it automatically orients itself to the new location of a coordinate system that you create, called a "UCS" (User Coordinate System).

FIGURE 10-17

The *Ucsicon* command controls the appearance and positioning of the Coordinate System Icon. It is important to display the icon when working with 3D drawings so that you can more easily visualize the orientation of the X, Y, and Z coordinate system. However, when you are working with 2D drawings, it is not necessary to display this icon. (By now you are accustomed to normal orientation of the positive X and Y axes—X positive is to the right and Y positive is up.)

Command: **ucsicon**
Enter an option [ON/OFF/All/Noorigin/ORigin/Properties] <OFF>:

ON/OFF

Use the *ON* and *OFF* options to turn the display of the *Uscicon* on and off. For example, if you draw primarily in 2D, you may want to turn the icon off since it is not necessary to constantly indicate the origin and the XY directions.

NOTE: The setting for the *Display UCS Icon* section of the *Options* dialog box, *3D Modeling* tab, <u>overrides</u> the *ON* and *OFF* options of the *Ucsicon* command. In other words, if the check is removed for *Display in 2D model space* in the dialog box (Fig. 10-18), the icon is permanently off and cannot be turned on by using the *Ucsicon* command.

FIGURE 10-18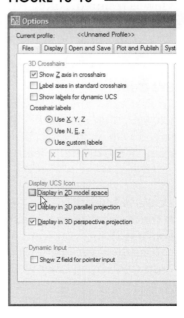

See Chapter 27 for more information and 3D applications of the UCS icon.

Vports

Pull-down Menu	Command (Type)	Alias (Type)	Short-cut	Screen (side) Menu	Tablet Menu
View Viewports >	Vports or -Vports	...	...	VIEW 1 Vports	M,3 and M,4

A "viewport" is one of several simultaneous views of a drawing on the screen. The *Vports* command invokes the *Viewports* dialog box. With this dialog box you can create model space viewports or paper space viewports, depending on which space is current when you invoke *Vports*. *Model* tab (model space) viewports are sometimes called "tiled" viewports because they fit together like tiles with no space between. *Layout* tab (paper space) viewports can be any shape and configuration and are used mainly for setting up several views on a sheet for plotting. Only an introduction to *Model* tab viewports is given in this chapter. Using model space viewports for constructing 3D models is discussed in Chapter 27. *Layout* tab (paper space) viewports are discussed in Chapter 13.

If the *Model* tab is active and you use the *Vports* command, you will create model space viewports. Several viewport configurations are available. *Vports* in this case simply divides the <u>screen</u> into multiple sections, allowing you to view different parts of a drawing. Seeing several views of the same drawing simultaneously can make construction and editing of complex drawings more efficient than repeatedly using other display commands to view detailed areas. Model space *Vports* affect <u>only the screen display</u>. The viewport configuration <u>cannot be plotted</u>. If the *Plot* command is used from Model space, only the <u>current</u> viewport is plotted.

Figure 10-19 displays the AutoCAD drawing editor after the *Vports* command was used to divide the screen into viewports. Tiled viewports always fit together like tiles with no space between. The shape and location of the viewports are not variable as with paper space viewports.

FIGURE 10-19

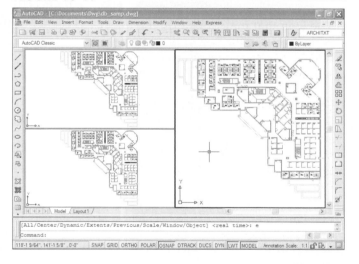

The *View* pull-down menu (see Fig. 10-1) can be used to display the *Viewports* dialog box (Fig. 10-20). The dialog box allows you to PICK the configuration of the viewports that you want. If you are working on a 2D drawing, make sure you select *2D* in the *Setup* box.

FIGURE 10-20

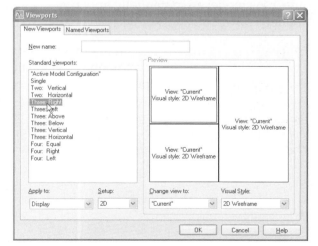

After you select the desired layout, the previous display appears in <u>each</u> of the viewports. For example, if a full view of the office layout is displayed when you use the *Vports* command, the resulting display in <u>each</u> viewport would be the same full view of the office (see Fig. 10-19). It is up to you then to use viewing commands (*Zoom*, *Pan*, etc.) in each viewport to specify what areas of the drawing you want to see in each viewport. There is no automatic viewpoint configuration option for *Vports* for 2D drawings.

A popular arrangement of viewpoints for construction and editing of 2D drawings is a combination of an overall view and one or two *Zoomed* views (Fig. 10-21). You cannot draw or project from one view-port to another. Keep in mind that there is only <u>one model</u> (drawing) but several views of it on the screen. Notice that the active or current viewport displays the cursor, while moving the pointing device to another viewport displays only the pointer (small arrow) in that viewport.

FIGURE 10-21

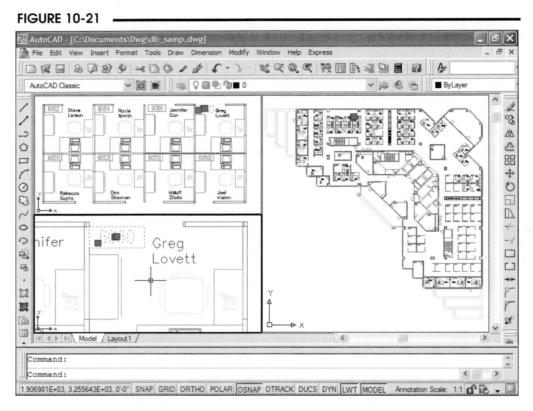

A viewport is made active by PICKing in it. Any display commands (*Zoom*, *Pan*, *Redraw*, etc.) and some drawing aids (*SNAP, GRID*) used affect only the <u>current viewport</u>. Draw and modify commands that affect the model are potentially apparent in all viewports (for every display of the affected part of the model). *Redrawall* and *Regenall* can be used to redraw and regenerate all viewports.

You can begin a drawing command in one viewport and finish in another. In other words, you can toggle viewports within a command. For example, you can use the *Line* command to PICK the "first point:" in one viewport, then make another viewport current to PICK the "next point:".

You can save a viewport configuration (including the views you prepare) by entering a name in the *New Name* box. If you save several viewport configurations for a drawing, you can then select from the list in the *Named Viewports* tab.

CHAPTER EXERCISES

1. *Zoom Extents, Realtime, Window, Previous, Center*

   Open the sample drawing supplied with AutoCAD called **DB_SAMP.DWG**. If your system has the standard installation of AutoCAD, the drawing is located in the **C:\Program Files\AutoCAD 2008\Sample** directory. The drawing shows an office building layout (Fig. 10-22). Use *SaveAs* and rename the drawing to **ZOOM TEST** and locate it in your working folder (directory). (When you view sample drawings, it is a good idea to copy them to another name so you do not accidentally change the original drawings.)

FIGURE 10-22

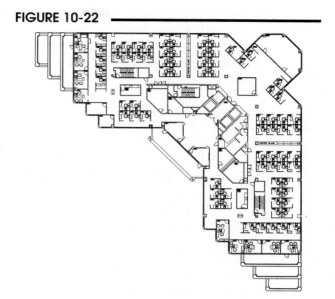

A. First, use *Zoom Extents* to display the office drawing as large as possible on your screen. Next, use *Zoom Realtime* and zoom in to the center of the layout (the center of your screen). You should be able to see rooms 6002 and 6198 located against the exterior wall (Fig. 10-23). Right-click and use *Zoom Extents* from the cursor pop-up menu to display the entire layout again. Press **Esc, Enter,** or right-click and select *Exit* to exit *Realtime*.

FIGURE 10-23

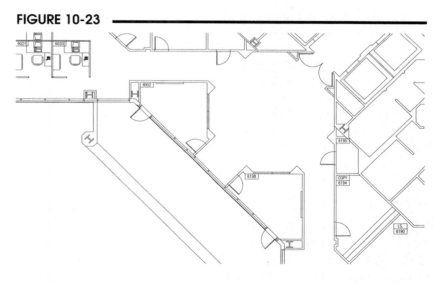

B. Use *Zoom Window* to closely examine room 6050 at the top left of the building. Whose office is it? What items on the desk are magenta in color? What item is cyan (light blue)?

C. Use *Zoom Center*. **PICK** the telephone on the desk next to the computer as the center. Specify a height of **24** (2′). You should see a display showing only the telephone. How many keys are on the telephone's number pad (*Point* objects)?

D. Next, use *Zoom Previous* repeatedly until you see the same display of rooms 6002 and 6198 as before (Fig. 10-23). Use *Zoom Previous* repeatedly until you see the office drawing *Extents*.

2. *Pan Realtime*

   A. Using the same drawing as exercise 1 (ZOOM TEST.DWG, originally DB_SAMP.DWG), use *Zoom Extents* to ensure you can view the entire drawing. Next, *Zoom* with a *Window* to an area about 1/4 the size of the building.

   B. Invoke the *Real Time* option of *Pan*. *Pan* about the drawing to find a coffee room. Can you find another coffee room? How many coffee rooms are there?

3. *Pan* and *Zoom* with the Mouse Wheel

   In the following exercise, *Zoom* and *Pan* using the wheel on your mouse (turn the wheel to *Zoom* and press and drag the wheel to *Pan*). If you do not have this capability, use *Realtime Pan* and *Zoom* in concert to examine specific details of the office layout.

   A. Find your new office. It is room 6100. (HINT: It is centrally located in the building.) Your assistant is in the office just to the right. What is your assistant's name and room number?

   B. Although you have a nice office, there are some disadvantages. How far is it from your office door to the nearest coffee room (nearest corner of the coffee room)? HINT: Use the *Dist* command with *Endpoint OSNAP* to select the two nearest doors.

   C. *Pan* and/or *Zoom* to find the copy room 6006. How far is it to the copy room (direct distance, door to door)?

   D. Perform a *Zoom Extents*. Use the wheel to zoom in to Kathy Ragerie's office (room 6150) in the lower-right corner of the building. Remember to locate the cursor at the spot in the drawing that you want to zoom in to. If you need to pan, hold down the wheel and "drag" the drawing across the screen until you can view all of room 6150 clearly. How many chairs are in Kathy's office? Finally, use *Zoom Extents* to size the drawing to the screen.

4. *Pan Point*

   A. *Zoom* in to your new office (room 6100) so that you can see the entire room and room number. Assume you were listening to music on your computer with the door open and calculated it could be heard about 100' away. Naturally, you are concerned about not bothering the company CEO who has an office in room 6048. Use *Pan Point* to see if the CEO's office is within listening range. (HINT: enter 100'<0 at the "Specify base point or displacement:" prompt.) Should you turn down the system?

   B. Do not *Exit* the **ZOOM TEST** drawing.

5. *Zoom Dynamic*

   A. *Open* the **TABLET** drawing from the **Sample** folder (C:\Program Files\AutoCAD 2008\Sample). Use *SaveAs*, name the new drawing **TABLET TEST**, and locate it in your working folder.

   B. Use *Zoom* with the *Window* option (or you can try *Realtime Zoom*). Zoom in to the first several command icons located in the middle left section of the tablet menu. Your display should reveal icons for *Zoom* and other viewing commands. *Pan* or *Zoom* in or out until you see the *Zoom* icons clearly.

C. Use *Zoom Dynamic*, then try moving the view box immediately. Can you move the box before the drawing is completely regenerated? Change the view box size so it is approximately equal to the size of one icon. Zoom in to the command in the lower-right corner of the tablet menu. What is the command? What is the command in the lower-left corner? For a drawing of this complexity and file size, which option of *Zoom* is faster and easier for your system—*Realtime* or *Dynamic*?

D. If you wish, you can *Exit* the **TABLET TEST** drawing.

6. *View*

A. Make the **ZOOM TEST** drawing that you used in previous exercises the current drawing. *Zoom* in to the office that you will be moving into (room 6100). Make sure you can see the entire office. Use the *View* command and create a *New* view named **6100**. Next, *Zoom* or *Pan* to room 6048. *Save* the display as a *View* named **6048**.

B. Use the *View* dialog box again and make view **6100** *Current*.

C. In order for the CEO to access information on your new computer, a cable must be stretched from your office to room 6048. Find out what length of cable is needed to connect the upper left corner of office 6100 to the upper right corner of office 6048. (HINT: Use *Dist* to determine the distance. Type the '-*View* command <u>transparently</u> [prefix with an apostrophe and hyphen] during the *Dist* command and use *Running Endpoint OSNAP* to select the two indicated office corners. The V*iew* dialog box <u>is not transparent</u>.) What length of cable is needed? Do not *Exit* the drawing.

7. *Zoom All, Extents*

FIGURE 10-24

A. Begin a *New* drawing using the **ACAD.DWT** or using the *Start from Scratch, Imperial* options. *Zoom All*. Right-click on the **GRID** button on the Status bar, select *Settings*, and remove all checks from the *Grid behavior* section in the *Drafting Settings* dialog box. Turn on the *SNAP* (**F9**) and *GRID* (**F7**). Draw two *Circles*, each with a **1.5** unit *radius*. The *Circle* centers are at **3,5** and at **5,5**. See Figure 10-24.

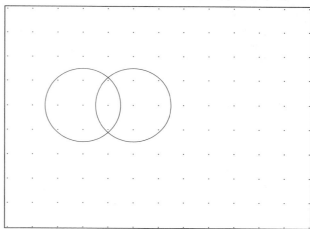

B. Use *Zoom All*. Does the display change? Now use *Zoom Extents*. What happens? Now use *Zoom All* again. Which option <u>always</u> shows all of the *Limits*?

C. Draw a *Circle* with the center at **10,10** and with a *radius* of **5**. Now use *Zoom All*. Notice the GRID appears only on the area defined by the *Limits*. Can you move the cursor to 0,0? Now use *Zoom Extents*. What happens? Can you move the cursor to 0,0?

D. *Erase* the large *Circle*. Use *Zoom All*. Can you move the cursor to 0,0? Use *Zoom Extents*. Can you find point 0,0?

E. *Exit* the drawing and discard changes.

8. *Vports*

A. *Open* the **ZOOM TEST** drawing again. Perform a *Zoom Extents*. Invoke the *Viewports* dialog box by any method. When the dialog box appears, choose the *Three: Right* option. The resulting viewport configuration should appear as Figure 10-19, shown earlier in the chapter.

B. Using *Zoom* and *Pan* in the individual viewports, produce a display with an overall view on the top and two detailed views as shown in Figure 10-21.

C. Type the *Vports* command. Use the *New* option to save the viewport configuration as **3R**.

D. Click in the bottom-left viewport to make it the current viewport. Use *Vports* again and change the display to a *Single* screen. Now use *Realtime Zoom* and *Pan* to locate and zoom to room 6100.

E. Use the *Viewports* dialog box again. Select the *Named Viewports* tab and select **3R** as the viewport configuration to restore. Ensure the **3R** viewport configuration was saved as expected, including the detail views you prepared.

F. Invoke a *Single* viewport again. Finally, use the *Vports* dialog box to create another viewport configuration of your choosing. *Save* the viewport configuration and assign an appropriate name. *Save* the **ZOOM TEST** drawing.

LAYERS AND OBJECT PROPERTIES

CHAPTER OBJECTIVES

After completing this chapter you should:

1. understand the strategy of grouping related geometry with *Layers*;

2. be able to create *Layers*;

3. be able to assign *Color, Linetype,* and *Lineweight* to *Layers*;

4. be able to control a layer's properties and visibility settings (*On, Off, Freeze, Thaw, Lock, Unlock*);

5. be able to set *LTSCALE* to adjust the scale of linetypes globally;

6. understand the concept of object properties;

7. be able to change an object's properties with the *Object Properties* toolbar, the *Properties* palette, and with *Match Properties;*

8. be able to use the *Layer Tools* included with AutoCAD.

CONCEPTS

In a CAD drawing, layers are used to group related objects in a drawing. Objects (*Lines*, *Circles*, *Arcs*, etc.) that are created to describe one component, function, or process of a drawing are perceived as related information and, therefore, are typically drawn on one layer. A single CAD drawing is generally composed of several components and, therefore, several layers. Use of layers provides you with a method to control visible features of the components of a drawing. For each layer, you can control its color on the screen, the linetype and lineweight it will be displayed with, and its visibility setting (on or off). You can also control if the layer is plotted or not.

Layers in a CAD drawing can be compared to clear overlay sheets on a manual drawing. For example, in a CAD architectural drawing, the floor plan can be drawn on one layer, electrical layout on another, plumbing on a third layer, and HVAC (heating, ventilating, and air conditioning) on a fourth layer (Fig. 11-1). Each layer of a CAD drawing can be assigned a different color, linetype, lineweight, and visibility setting similar to the way clear overlay sheets on a manual drawing can be used. Layers can be temporarily turned *Off* or *On* to simplify drawing and editing, like overlaying or

FIGURE 11-1

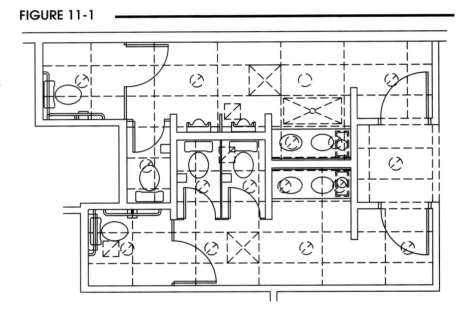

removing the clear sheets. For example, in the architectural CAD drawing, only the floor plan layer can be made visible while creating the electrical layout, but can later be cross-referenced with the HVAC layout by turning its layer on. Layers can also be made "non-plottable." Final plots can be made of specific layers for the subcontractors and one plot of all layers for the general contractor by controlling the layers' *Plot/No Plot* icon before plotting.

AutoCAD allows you to create a practically unlimited number of layers. You should assign a name to each layer when you create it. The layer names should be descriptive of the information on the layer.

Assigning Colors, Linetypes, and Lineweights

There are two strategies for assigning colors, linetypes, and lineweights in a drawing: assign these properties to layers or assign them to objects.

Assign colors, linetypes, and lineweights to layers
For most drawings, layers are assigned a color, linetype, and lineweight so that all objects drawn on a single layer have the same color, linetype, and lineweight. Assigning colors, linetypes, and lineweights to layers is called *ByLayer* color, linetype, and lineweight setting. Using the *ByLayer* method makes it visually apparent which objects are related (on the same layer). All objects on the same layer have the same linetype and color.

Assign colors, linetypes, and lineweights to individual objects

Alternately, you can assign colors, linetypes, and lineweights to specific objects, overriding the layer's color, linetype, and lineweight setting. This method is fast and easy for small drawings and works well when layering schemes are not used or for particular applications. However, using this method makes it difficult to see which layers the objects are located on.

Object Properties

Another way to describe the assignment of color, linetype, and lineweight properties is to consider the concept of Object Properties. Each object has properties such as a layer, a color, a linetype, and a lineweight. The color, linetype, and lineweight for each object can be designated as *ByLayer* or as a specific color, linetype, or lineweight. Object properties for the two drawing strategies (schemes) are described as follows.

Object Properties

	1. *ByLayer* Drawing Scheme	2. Object-Specific Drawing Scheme
Layer Assignment	Layer name descriptive of geometry on the layer	Layer name descriptive of geometry on the layer
Color Assignment	*ByLayer*	*Red, Green, Blue, Yellow,* or other specific color setting
Linetype Assignment	*ByLayer*	*Continuous, Hidden, Center,* or other specific linetype setting
Lineweight Assignment	*ByLayer*	0.05 mm, 0.15 mm, 0.010", or other specific lineweight setting

It is recommended that beginners use only one method for assigning colors, linetypes, and lineweights. After gaining some experience, it may be desirable to combine the two methods only for specific applications. Usually, the *ByLayer* method is learned first, and the object-specific color, linetype, and lineweight assignment method is used only when layers are not needed or when complex applications are needed. (The *ByBlock* color, linetype, and lineweight assignment has special applications for *Blocks* and is discussed in Chapter 20.)

LAYERS AND LAYER PROPERTIES CONTROLS

Layer Control Drop-Down List

The Object Properties toolbar and *Layers* control panel (on the dashboard) contain a drop-down list for making layer control quick and easy (Fig. 11-2). The window normally displays the current layer's name, visibility setting, and properties. When you pull down the list, all the layers (unless otherwise specified in the *Named Layer Filters* dialog box) and their settings are displayed. Selecting any layer name makes it current. Clicking on any of the visibility/properties icons changes the layers' setting as described in the following section. Several layers can be changed in one "drop." You cannot change a layer's color, linetype, or line-weight, nor can you create new layers using this drop-down list.

FIGURE 11-2

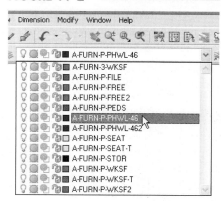

Layer

Pull-down Menu	Command (Type)	Alias (Type)	Short-cut	Screen (side) Menu	Tablet Menu
Format Layer	Layer or -Layer	LA or -LA	...	FORMAT Layer	U,5

The way to gain complete layer control is through the *Layer Properties Manager* (Fig. 11-3). The *Layer Properties Manager* is invoked by using the icon button (shown above), typing the *Layer* command or *LA* command alias, or selecting *Layer* from the *Format* pull-down or screen menu.

As an alternative to the *Layer Properties Manager*, the *Layer* command can be used in Command line format by typing *Layer* with a "-" (hyphen) prefix. The Command line format shows most of the options available through the *Layer Properties Manager*.

FIGURE 11-3

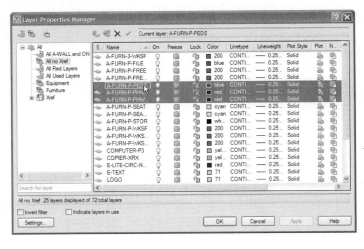

```
Command: -LAYER
Current layer: "0"
Enter an option
```

[?/Make/Set/New/ON/OFF/Color/Ltype/LWeight/MATerial/Plot/PStyle/Freeze/Thaw/LOck/Unlock/stAte]:

Layer Properties Manager Layer List

The right-central area of the *Layer Properties Manager* (see Fig. 11-3) is the layer list that displays the list of layers in the drawing. Here you can control the current visibility settings and the properties assigned to each layer. This list can display all the layers in the drawing or a subset of layers if a layer filter is applied. Layer filters are created and applied in the Filter Tree View on the left side of the *Layer Properties Manager*. For new drawings such as those created from the ACAD.DWT template, only one layer may exist in the drawing—Layer "0" (zero). New layers can be created by selecting the *New Layer* button or the *New Layer* option of the right-click menu (see *New Layer*).

All properties and visibility settings of layers can be controlled by highlighting the layer name and then selecting one of the icons for the layer such as the light bulb icon (*On, Off*), sun/snowflake icon (*Thaw/Freeze*), padlock icon (*Lock/Unlock*), *Color* tile, *Linetype*, *Lineweight*, or *Plot/No plot* icon.

The column widths can be changed by moving the pointer to the "crack" between column headings until double arrows appear (Fig. 11-4). You can also right-click on a column heading to produce a list of columns to display. Uncheck any columns from the list that you do not want displayed in the *Layer Properties Manager*.

FIGURE 11-4

A particularly useful feature is the ability to sort the layers in the list by any one of the headings (*Name, On, Freeze, Linetype*, etc.) by clicking on the heading tile. For example, you can sort the list of names in alphabetical order (or reverse order) by clicking once (or twice) on the *Name* heading above the list of names. Or you may want to sort the *Frozen* and *Thawed* layers or sort layers by *Color* by clicking on the column heading.

Right-clicking in the layer list of the *Layer Properties Manager* produces a shortcut menu (Fig. 11-5). You have a choice of selecting icon buttons or selecting shortcut menu options for most of the layer controls available in the *Layer Properties Manager*, as noted in the following sections.

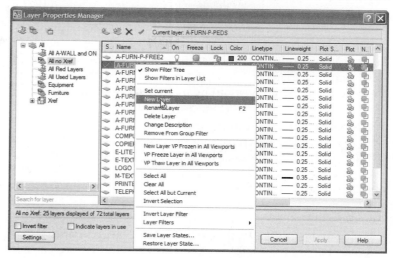

New Layer

The *New Layer* option allows you to make new layers. There is only one layer in the AutoCAD templates (ACAD.DWT and ACADISO.DWT) as they are provided to you "out of the box." That layer is Layer 0. Layer 0 is a part of every AutoCAD drawing because it cannot be deleted. You can, however, change the *Color, Linetype,* and *Lineweight* of Layer 0 from the defaults (*Continuous* linetype, *Default* lineweight, and color #7 *White*). Layer 0 is generally used as a construction layer or for geometry not intended to be included in the final draft of the drawing. Layer 0 has special properties when creating *Blocks* (see Chapter 20).

A new layer named "Layer1" (or other number) then appears in the list with the default color (*White*), linetype (*Continuous*), and lineweight (*Default*). If an existing layer name is highlighted when you make a new layer, the highlighted layer's properties (*color, linetype, lineweight,* and *plot style*) are used as a template for the new layer. The new layer name initially appears in the rename mode so you can immediately assign a more appropriate and descriptive name for the layer. You can create many new names quickly by typing (renaming) the first layer name, then typing a comma before other names. A comma forces a new "blank" layer name to appear. Colors, linetypes, and lineweights should be assigned as the next step.

You should create layers for each group of related objects and assign appropriate layer names for that geometry. You can use up to 256 characters (including spaces) for layer names.

If you want to create layers by typing, the *New* and *Make* options of the -*Layer* command can be used. *New* allows creation of one or more new layers. *Make* allows creation of one layer (at a time) and sets it as the current layer.

New Layer VP

Use this option to create a new layer that is frozen in all viewports. See Chapters 13 for concepts of viewport-specific layer settings.

Delete

The *Delete Layer* option allows you to delete layers. <u>Only layers with no geometry can be deleted</u>. You cannot delete a layer that has objects on it, nor can you delete Layer 0, the current layer, or layers that are part of externally referenced (*Xref*) drawings. If you attempt to *Delete* such a layer, accidentally or intentionally, a warning appears.

Set Current

To *Set* a layer as the *Current* layer is to make it the active drawing layer. Any objects created with draw commands are created on the *Current* layer. You can, however, edit objects on any layer, but draw only on the current layer. Therefore, if you want to draw on the FLOORPLAN layer (for example), use the *Set Current* option or double-click on the layer name. If you want to draw with a certain *Color* or *Linetype*, set the layer with the desired *Color*, *Linetype*, and *Lineweight* as the *Current* layer. Any layer can be made current, but only <u>one layer at a time</u> can be current.

Status and Indicate Layers in Use

Status is not an option; however, the *Status* column indicates whether or not the related layer contains objects. The *Status* icon is a lighter shade of gray if no objects exist on the layer. The *Indicate Layers in Use* option below the layer list <u>must be checked</u> for the *Status* icon to indicate if objects do not exist, otherwise all *Status* icons are dark gray.

On, Off

If a layer is *On*, it is visible. Objects on visible layers can be edited or plotted. Layers that are *Off* are not visible. Objects on layers that are *Off* will not plot and cannot be edited (unless the *ALL* selection option is used, such as *Erase, All*). It is not advisable to turn the current layer *Off*.

Freeze, Thaw

Freeze and *Thaw* override *On* and *Off*. *Freeze* is a more protected state than *Off*. Like being *Off*, a frozen layer is not visible, nor can its objects be edited or plotted. Objects on a frozen layer cannot be accidentally *Erased* with the *ALL* option. *Freezing* also prevents the layer from being considered when *Regen*s occur. *Freezing* unused layers speeds up computing time when working with large and complex drawings. *Thawing* reverses the *Freezing* state. Layers can be *Thawed* and also turned *Off*. Frozen layers are not visible even though the light bulb icon is on.

Lock, Unlock

Layers that are *Locked* are protected from being edited but are still visible and can be plotted. *Locking* a layer prevents its objects from being changed even though they are visible. Objects on *Locked* layers cannot be selected with the *ALL* selection option (such as *Erase, All*). Layers can be *Locked* and *Off*.

Color, Linetype, Lineweight, and Other Properties

Layers have properties of *Color, Linetype,* and *Lineweight* such that (generally) an object that is drawn on, or changed to, a specific layer assumes the layer's linetype and color. Using this scheme (*ByLayer*) enhances your ability to see what geometry is related by layer. It is also possible, however, to assign specific color, linetype, and lineweight to objects which will override the layer's color, linetype, and lineweight (see "*Color, Linetype,* and *Lineweight* Commands" and "Changing Object Properties").

Color

green

Selecting one of the small color boxes in the list area of the *Layer Properties Manager* causes the *Select Color* dialog box to pop up (Fig. 11-6). The desired color can then be selected or the name or color number (called the ACI—AutoCAD Color Index) can be typed in the edit box. This action retroactively changes the color assigned to a layer. Since the color setting is assigned to the layer, all objects on the layer that have the *ByLayer* setting change to the new layer color. Objects with specific color assigned (not *ByLayer*) are not affected. Alternately, the *Color* option of the *-Layer* command (hyphen prefix) can be typed to enter the color name or ACI number.

FIGURE 11-6

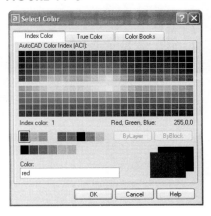

The *Select Color* dialog box also contains tabs for the *Index Color* (ACI), *True Color*, and *Color Books*.

Linetype

To set a layer's linetype, select the *Linetype* (word such as *Continuous* or *Hidden*) in the layer list, which in turn invokes the *Select Linetype* dialog box (Fig. 11-7). Select the desired linetype from the list. Similar to changing a layer's color, all objects on the layer with *ByLayer* linetype assignment are retroactively displayed in the selected layer linetype while non-*ByLayer* objects remain unchanged.

The ACAD.DWT and ACADISO.DWT template drawings as supplied by Autodesk have only one linetype available (*Continuous*). Before you can use other linetypes in a drawing, you must load the linetypes by selecting the *Load* tile (see the *Linetype* command) or by using a template drawing that has the desired linetypes already loaded.

FIGURE 11-7

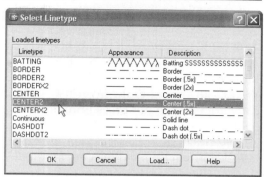

Lineweight

The lineweight for a layer can be set by selecting the *Lineweight* (word such as *Default* or *0.20 mm, 0.50 mm, 0.010"* or *0.020"*, etc.) in the layer list. This action produces the *Lineweight* dialog box (Fig. 11-8). Like the *Color* and *Linetype* properties, all objects on the layer with *ByLayer* lineweight assignment are retroactively displayed in the lineweight assigned to the layer while non-*ByLayer* objects remain unchanged. You can set the units for lineweights (mm or inches) using the *Lineweight* command.

FIGURE 11-8

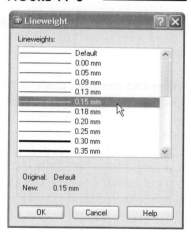

Plot Style

This column of the *Layer Properties Manager* designates the plot style assigned to the layer. If the drawing has a color-dependent Plot Style Table attached, this section is disabled since the plot styles are automatically assigned to colors. If a named Plot Style Table is attached to the drawing you can select this section to assign plot styles from the table to layers.

Plot/No Plot

You can prevent a layer from plotting by clicking the printer icon so a red line appears over the printer symbol. There are actually three methods for preventing a layer from appearing on the plot: *Freeze* it, turn it *Off*, or change the *Plot/No Plot* icon. The *Plot/No Plot* icon is generally preferred since the other two options prevent the layer from appearing in the drawing (screen) as well as in the plot.

Current VP Freeze, VP Freeze

These options are used and are displayed in the *Layer Properties Manager* only when paper space viewports exist in the current layout tab. Using these options, you can control what geometry (layers) appears in specific viewports. See Chapter 13 for more information on these options.

VP Color, VP Lineweight, VP Linetype, VP Plot Style

These options allow you to specify viewport-specific property overrides. That is, you can stipulate that layers appear in different colors, linetypes, etc. for different viewports. See Chapters 13 for concepts of viewport-specific properties.

Description

You can add an optional *Description* for each layer that appears in the *Layer Properties Manager*. Select *Change Description* from the right-click menu and add descriptive information about the objects that reside on the layer.

Search for Layer

This edit box below the Filter Tree View allows you to use wildcards to search for matching names in a long list of layers. For example, entering "A-WALL*" would cause only layers beginning with "A-WALL" to appear in the layer list. Clicking in the edit box makes the "*" (asterisk) wildcard automatically appear. <u>Do not press Enter</u> after typing in your entry.

Apply and OK

When you are finished making all the desired changes to the layers in the *Layer Properties Manager*, press the *OK* button (at the bottom of the dialog box) to apply the new settings and dismiss the dialog box. You can also use the *Apply* button to immediately apply any layer settings you have made to the drawing without dismissing the *Layer Properties Manager*.

Displaying an Object's Properties and Visibility Settings

When the Layer Control drop-down list in the Object Properties toolbar is in the normal position (not "dropped down"), it can be used to display an <u>object's</u> layer properties and visibility settings. Do this by selecting an object (with the pickbox, window, or crossing window) when no commands are in use.

FIGURE 11-9

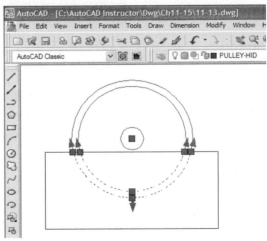

When an object is selected, the Object Properties toolbar displays the layer, color, linetype, lineweight, and plot style <u>of the selected object</u> (Fig. 11-9). The Layer Control box displays the selected object's layer name and layer settings rather than that of the current layer. The Color Control, Linetype Control, Lineweight Control, and Plot Style Control boxes in the Object Properties toolbar (just to the right) also temporarily reflect the color, linetype, lineweight, and plot style properties of the selected object. Pressing the Escape key causes the list boxes to display the current layer name and settings again, as would normally be displayed. If more than one object is selected, and the objects are on different layers or have different color, linetype, and lineweight properties, the list boxes go blank until the objects become unhighlighted or a command is used.

These sections of the Object Properties toolbar have another important feature that allows you to change a highlighted object's (or set of objects) properties. See "Changing Object Properties" later in this chapter.

Laymcur

Pull-down Menu	Command (Type)	Alias (Type)	Short-cut	Screen (side) Menu	Tablet Menu
Format, Layer Tools > Make Object's Layer Current	*Laymcur*	...	...	...	...

This productive feature can be used to make a desired layer current simply by selecting <u>any object on the layer</u>.

his option is generally faster than using the *Layer Properties Manager* or *Layer* drop-down list to set a layer current. The *Make Object's Layer Current* feature is particularly useful when you want to draw objects on the same layer as other objects you see, but are not sure of the layer name.

There are two steps in the procedure to make an object's layer current: (1) select the icon and (2) select any object on the desired layer. The selected object's layer becomes current and the new current layer name immediately appears in the Layer Control list box in the Object Properties toolbar.

OBJECT-SPECIFIC PROPERTIES CONTROLS

The commands in this section are used to control the *color*, *linetype*, and *lineweight* properties of individual objects. This method of object property assignment is used only in special cases—when you want the objects' *color*, *linetype*, and *lineweight* to override those properties assigned to the layers on which the objects reside. Using this method makes it difficult to see which objects are on which layers. In most cases, the *ByLayer* property is assigned instead to individual objects so the objects assume the properties of their layers. In the case that *color*, *linetype* and/or *lineweight* properties are accidentally set for individual objects or you want to change object properties to a *ByLayer* setting, you can use the *Setbylayer* command (see "*Setbylayer*") to retroactively change object properties to *ByLayer* properties.

Linetype

Pull-down Menu	Command (Type)	Alias (Type)	Short-cut	Screen (side) Menu	Tablet Menu
Format *Linetype...*	*Linetype* or *-Linetype*	*LT* or *-LT*	...	*FORMAT* *Linetype*	*U,3*

Invoking this command presents the *Linetype Manager* (Fig. 11-10). Even though this looks similar to the dialog box used for assigning linetypes to layers (shown earlier in Fig. 11-7), beware!

When linetypes are selected and made *Current* using the *Linetype Manager*, they are assigned to objects—not to layers. That is, selecting a linetype by this manner (making it *Current*) causes all objects from that time on to be drawn using that linetype, regardless of the layer that they are on (unless the *ByLayer* type is selected). In contrast, selecting linetypes during the *Layer* command (using the *Layer Properties Manager*) results in assignment of linetypes to layers.

FIGURE 11-10

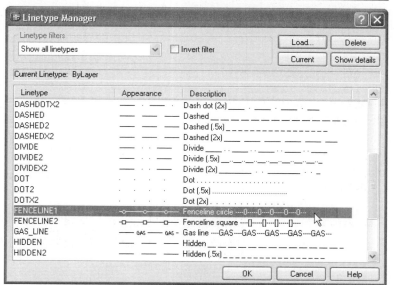

Linetype Control Drop-Down List

FIGURE 11-11

The Object Properties toolbar contains a drop-down list for selecting linetypes (Fig. 11-11). Although this appears to be quick and easy, you can <u>only assign linetypes to objects</u> by this method unless *ByLayer* is selected to use the layers' assigned linetypes. Any linetype you select from this list becomes the current object linetype. If you want to select linetypes for layers, <u>make sure this list displays the *ByLayer* setting</u>, then use the *Layer Properties Manager* to select linetypes for layers.

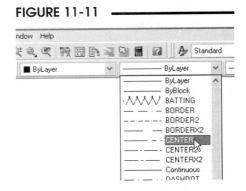

Keep in mind that the Linetype Control drop-down list as well as the Layer Control, Color Control, and Lineweight Control drop-down lists display the current settings for selected objects (objects selected when no commands are in use). When linetypes are assigned to layers rather than objects, the layer drop-down list reports a *ByLayer* setting.

Lweight

Pull-down Menu	Command (Type)	Alias (Type)	Short-cut	Screen (side) Menu	Tablet Menu
Format Lineweight...	*Lweight*	*LW*	...	...	...

The *Lweight* command (short for lineweight) produces the *Lineweight Settings* dialog box (Fig. 11-12). This dialog box is the lineweight <u>equivalent</u> of the *Linetype Manager*—that is, this dialog box assigns lineweights <u>to objects, not layers</u> (unless the *ByLayer* or *Default* lineweight is selected). See the previous discussion under *"Linetype."* The *ByBlock* setting is discussed in Chapter 20.

FIGURE 11-12

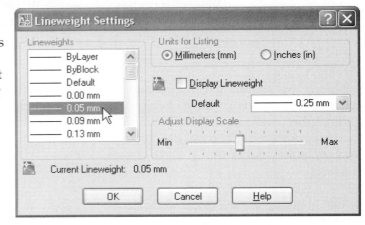

Any lineweight selected in the *Lineweight Settings* dialog box automatically becomes current (without having to select a *Current* button or double-click). The current lineweight is assigned to all subsequently drawn objects. That is, selecting a lineweight by this manner (making it current) causes all objects <u>from that time on</u> to be drawn using that lineweight, regardless of the layer that they are on (unless *ByLayer* or *Default* is selected). In contrast, selecting lineweights during the *Layer* command (using the *Layer Properties Manager*) results in assignment of lineweights to <u>layers</u>.

As a reminder, this type of drawing method (object-specific property assignment) can be difficult for beginning drawings and for complex drawings. If you want to draw objects in the <u>layer's</u> assigned lineweight, select *ByLayer*. If you want all layers to have the same lineweight, select *Default*.

Lineweight Control Drop-Down List

The Object Properties toolbar contains a drop-down list for selecting lineweights (Fig. 11-13). Similar to the function of the Linetype Control drop-down list, you can <u>only assign lineweights to objects</u> by this method unless *ByLayer* is selected to use the layers' assigned lineweights. Any lineweight you select from this list becomes the current object lineweight. If you want to select lineweights for layers, make sure this list displays the *ByLayer* setting, then use the *Layer Properties Manager* to select lineweights for layers.

FIGURE 11-13

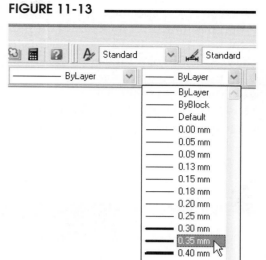

Remember that the Lineweight Control drop-down list as well as the Layer Control, Color Control, and Linetype Control drop-down lists display the current settings for selected objects (objects selected when no commands are in use). When lineweights are assigned to layers rather than objects, the layer drop-down list reports a *ByLayer* setting.

Color

Pull-down Menu	Command (Type)	Alias (Type)	Short-cut	Screen (side) Menu	Tablet Menu
Format Color...	*Color*	*COL*	*...*	*FORMAT Color*	*U,4*

Similar to linetypes and lineweights, colors can be assigned to layers or to objects. Using the *Color* command assigns a color for all newly created <u>objects</u>, regardless of the layer's color designation (unless the *ByLayer* color is selected). This color setting <u>overrides</u> the layer color for any newly created objects so that all new objects are drawn with the specified color no matter what layer they are on. This type of color designation prohibits your ability to see which objects are on which layers by their color; however, for some applications object color setting may be desirable. Use the *Layer Properties Manager* to set colors for layers.

Invoking this command by the menus or by typing *Color* presents the *Select Color* dialog box shown in Figure 11-14. This is essentially the same dialog box used for assigning colors to layers; however, the *ByLayer* and *ByBlock* tiles are accessible. (The buttons are grayed-out when this dialog is invoked from the *Layer Properties Manager* because, in that case, any setting is a *ByLayer* setting.) *ByBlock* color assignment is discussed in Chapter 20.

FIGURE 11-14

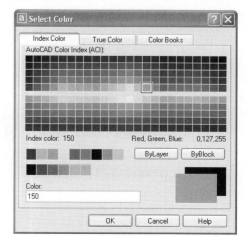

Color Control Drop-Down List

When an object-specific color has been set, the top item in the Object Properties toolbar Color Control drop-down list displays the current color (Fig. 11-15). Beware—using this list to select a color assigns an object-specific color unless *ByLayer* is selected. Any color you select from this list becomes the current object color. Choosing *Select Other...* from the bottom of the list invokes the *Select Color* dialog box (see Fig. 11-14). If you want to assign colors to layers, make sure this list displays the *ByLayer* setting, then use the *Layer Properties Manager* to select colors for layers.

FIGURE 11-15

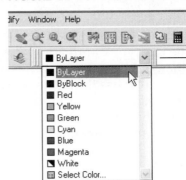

Setbylayer

Pull-down Menu	Command (Type)	Alias (Type)	Short-cut	Screen (side) Menu	Tablet Menu
Modify > Change to ByLayer	*Setbylayer*	...	...	...	...

You can change object-specific (*ByBlock*) property settings for selected objects to *ByLayer* settings using this command. *Setbylayer* prompts you to select objects. Properties for the selected objects are automatically changed to the properties of the layer that the objects are on. Properties that can be changed are *Color, Linetype, Lineweight, Material,* and *Plot Style*.

```
Command: setbylayer
Current active settings: Color Linetype Lineweight Material
Select objects or [Settings]: PICK
Select objects or [Settings]: Enter
Change ByBlock to ByLayer? [Yes/No] <Yes>: Y or N
Include blocks? [Yes/No] <Yes>: Y or N
nn objects modified, nn objects did not need to be changed.
Command:
```

If *Settings* is selected, the *SetByLayer Settings* dialog box is displayed (Fig. 11-16). Here you can specify which object properties are set to *ByLayer*.

FIGURE 11-16

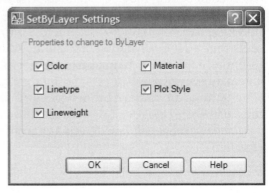

Occasionally you may be required to edit a drawing created by another person or company. This command is especially useful when you want to change such a drawing (that has *ByBlock* properties) to a *ByLayer* setting.

CONTROLLING LINETYPE SCALE

LTSCALE

Pull-down Menu	Command (Type)	Alias (Type)	Short-cut	Screen (side) Menu	Tablet Menu
Format Linetype... Show details >> Global scale factor	*Ltscale*	*LTS*	...	*FORMAT Linetype Show details >> Global scale factor*	...

Hidden, dashed, dotted, and other linetypes that have spaces are called <u>non-continuous</u> linetypes. When drawing objects that have non-continuous linetypes (either *ByLayer* or object-specific linetype designations), the linetype's dashes or dots are automatically created and spaced. The *LTSCALE* (Linetype Scale) system variable controls the length and spacing of the dashes and/or dots. The value that is specified for *LTSCALE* affects the drawing <u>globally and retroactively</u>. That is, all existing non-continuous lines in the drawing as well as new lines are affected by *LTSCALE*. You can therefore adjust the drawing's linetype scale for all lines at any time with this one command.

If you choose to make the dashes of non-continuous lines smaller and closer together, reduce *LTSCALE*; if you desire larger dashes, increase *LTSCALE*. The *Hidden* linetype is shown in Figure 11-17 at various *LTSCALE* settings. Any positive value can be specified.

FIGURE 11-17 ——————

LTSCALE

4 —————— —————— —————— ——————

2 —— —— —— —— —— —— ——

1 — — — — — — — — — — — —

0.5 =

TIP

LTSCALE can be set in the *Details* section of the *Linetype Manager* or in Command line format. In the *Linetype Manager*, select the *Details* button to allow access to the *Global scale factor* edit box (Fig. 11-18). Changing the value in this edit box sets the *LTSCALE* variable. Changing the value in the *Current object scale* edit box sets the linetype scale for the current object only (see "*CELTSCALE*"), but does not affect the global linetype scale (*LTSCALE*).

FIGURE 11-18 ——————

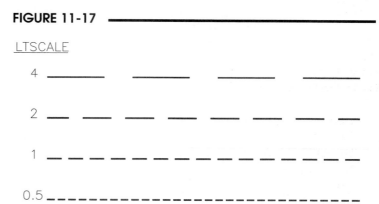

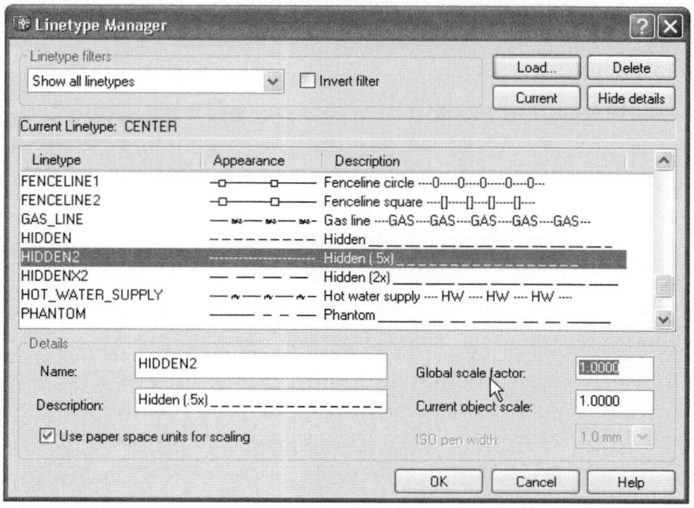

LTSCALE can also be used in the Command line format.

    Command: **ltscale**
    Enter new linetype scale factor <1.0000>: (**value**) (Enter any positive value.)

The *LTSCALE* for the default template drawing (ACAD.DWT) is 1. This value represents an appropriate *LTSCALE* for objects drawn within the default *Limits* of 12 x 9.

As a general rule, you should change the *LTSCALE* proportionally when *Limits* are changed (more specifically, when *Limits* are changed to other than the intended plot sheet size). For example, if you increase the drawing area defined by *Limits* by a factor of 2 (to 24 x 18) from the default (12 x 9), you might also change *LTSCALE* proportionally to a value of 2. Since *LTSCALE* is retroactive, it can be changed at a later time or repeatedly adjusted to display the desired spacing of linetypes.

If you load the *Hidden* linetype, it displays dashes (when plotted 1:1) of 1/4" with a *LTSCALE* of 1. The ANSI and ISO standards state that hidden lines should be displayed on drawings with dashes of approximately 1/8" or 3mm. To accomplish this, you can use the *Hidden* linetype and change *LTSCALE* to .5. It is recommended, however, that you use the *Hidden2* linetype that has dashes of 1/8" when plotted 1:1. Using this strategy, *LTSCALE* remains at a value of 1 to create the standard linetype sizes for *Limits* of 12 x 9 and can easily be changed in proportion to the *Limits*. (For more information on *LTSCALE*, see Chapter 12 and Chapter 13.)

CELTSCALE

CELTSCALE stands for Current Entity Linetype Scale. *CELTSCALE* is actually a system variable for linetype scale stored with each object—an object property. This setting changes the object-specific linetype scale proportional to the *LTSCALE*. The *LTSCALE* value is global and retroactive, whereas *CELTSCALE* sets the linetype scale for all newly created objects and is not retroactive. *CELTSCALE* is object-specific. Using *CELTSCALE* to set an object linetype scale is similar to setting an object color, linetype, and lineweight in that the properties are assigned to the specific object.

Using *CELTSCALE* is not the recommended method for adjusting individual lines' linetype scale. Setting this variable each time you wanted to create a new entity with a different linetype scale would be too time consuming and confusing. The recommended method for adjusting linetypes in the drawing to different scales is not to use *CELTSCALE*, but to use the following strategy.

1.   Set *LTSCALE* to an appropriate value for the drawing.
2.   Create all objects in the drawing with the desired linetypes.
3.   Adjust the *LTSCALE* again if necessary to globally alter the linetype scale.
4.   Use *Properties* and *Matchprop* to retroactively change the linetype scale for selected objects (see "*Properties*" and "*Matchprop*").

CHANGING OBJECT PROPERTIES

Often it is desirable to change the properties of an object after it has been created. For example, an object's *Layer* property could be changed. This can be thought of as "moving" the object from one layer to another. When an object is changed from one layer to another, it assumes the new layer's color, linetype, and lineweight, provided the object was originally created with color, linetype, and lineweight assigned *ByLayer*, as is generally the case. In other words, if an object was created on the wrong layer (possibly with the wrong linetype or color), it could be "moved" to the desired layer, therefore assuming the new layer's linetype and color.

Another example is to change an individual object's linetype scale. In some cases an individual object's linetype scale requires an adjustment to other than the global linetype scale (*LTSCALE*) setting. One of several methods can be used to adjust the individual object's linetype scale to an appropriate value.

Several methods can be used to retroactively change properties of selected objects. The Object Properties toolbar, the *Properties* palette, and the *Match Properties* command can be used to <u>retroactively</u> change the properties of individual objects. Properties that can be changed by these three methods are *Layer*, *Linetype*, *Lineweight*, *Color*, (object-specific) *Linetype scale*, *Plot Style*, and other properties. Although some of the commands discussed in this section have additional capabilities, the discussion is limited to changing the object properties covered in this chapter—specifically layer, color, linetype, and lineweight.

Object Properties Toolbar

The five drop-down lists in the Object Properties toolbar (when not "dropped down") generally show the current layer, color, linetype, lineweight, and plot style. However, if an object or set of objects is selected, the information in these lists changes to display the current objects' settings. <u>You can change the selected objects' settings</u> by "dropping down" any of the lists and making another selection.

First, select (highlight) an object when no commands are in use. Use the pickbox (that appears on the crosshairs), window or crossing window to select the desired object or set of objects. The entries in the five lists (Layer Control, Color Control, Linetype Control, Lineweight Control, and Plot Style Control) then change to display the settings for the <u>selected object or objects</u>. If several objects are selected that have different properties, the boxes display no information (go "blank"). Next, use any of the drop-down lists to make another selection (Fig. 11-19). The highlighted object's properties are changed to those selected in the lists. Press the Escape key to complete the process.

FIGURE 11-19

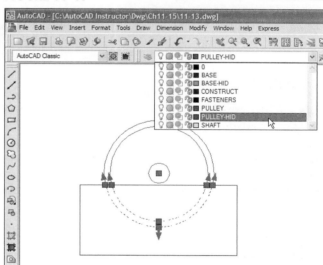

Remember that in most cases color, linetype, and lineweight settings are assigned *ByLayer*. In this type of drawing scheme, to change the linetype or color properties of an object, you would change the object's <u>layer</u> (see Fig. 11-19). If you are using this type of drawing scheme, refrain from using the Color Control, Linetype Control, and Lineweight Control drop-down lists for changing properties.

Properties

Pull-down Menu	Command (Type)	Alias (Type)	Short-cut	Screen (side) Menu	Tablet Menu
Modify *Properties*	*Properties*	*PR* or *CH*	(Edit Mode) *Properties* or *Ctrl+1*	*MODIFY1* *Property*	Y,12 to Y,13

The *Properties* palette (Fig. 11-20) gives you complete access to one or more objects' properties. The contents of the palette change based on what type and how many objects are selected.

You can use the palette two ways: you can invoke the palette, then select (highlight) one or more objects, or you can select objects first and then invoke the palette. Once opened, this palette remains on the screen until dismissed by clicking the "X" in the upper corner. You can even toggle the palette on and off with Ctrl+1 (to appear and disappear). When multiple objects are selected, a drop-down list in the top of the palette appears. Select the object(s) whose properties you want to change.

To change an object's layer, linetype, lineweight, or color properties, highlight the desired objects in the drawing and select the properties you want to change from the right side of the palette. In the *General* list, the layer, linetype, lineweight, and color properties are located in the top half of the list.

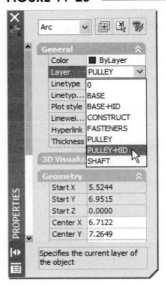

FIGURE 11-20

 TIP

If you use the *ByLayer* strategy of linetype, lineweight, and color properties assignment, you can change an object's linetype, lineweight, and color simply by changing its layer (see Fig. 11-20). This method is recommended for most applications.

Matchprop

	Pull-down Menu	Command (Type)	Alias (Type)	Short-cut	Screen (side) Menu	Tablet Menu
	Modify Match Properties	*Matchprop*	*MA*	...	*MODIFY1 Matchprp*	**Y,14** and **Y,15**

Matchprop is used to "paint" the properties of one object to another. The process is simple. After invoking the command, select the object that has the desired properties (source object), then select the object you want to "paint" the properties to (destination object). The command prompt is as follows.

```
Command: matchprop
Select source object: PICK
Current active settings:  Color Layer Ltype Ltscale Lineweight Thickness
    PlotStyle Dim Text Hatch Polyline Viewport Table Material Shadow display Multileader
Select destination object(s) or [Settings]: PICK
Select destination object(s) or [Settings]: Enter
Command:
```

Only one "source object" can be selected, but its properties can be painted to several "destination objects." The "destination object(s)" assume all of the properties of the "source object" (listed as "Current active settings").

FIGURE 11-21

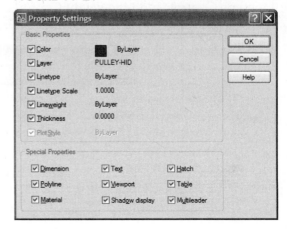

Use the *Settings* option to control which of several possible properties and other settings are "painted" to the destination objects. At the "Select destination object(s) or [Settings]:" prompt, type *S* to display the *Property Settings* dialog box (Fig. 11-21). In the dialog box, designate which of the *Basic Properties* or *Special Properties* are to be painted to the "destination objects." The following *Basic Properties* correspond to properties discussed in this chapter.

Color	Paints the object-specific or *ByLayer* color.
Layer	Moves selected objects to Source Object layer.
Linetype	Paints the object-specific or *ByLayer* linetype.
Lineweight	Paints the object-specific or *ByLayer* lineweight.
Linetype Scale	Changes the individual object's linetype scale, not global (*LTSCALE*).

The *Special Properties* of the *Property Settings* dialog box are discussed in Chapter 16, Modify Commands II. Also see Chapter 16 for a full explanation of the *Properties* palette (*Properties*).

CHAPTER EXERCISES

1. *Layer Properties Manager,* **Layer Control drop-down list, and** *Make Object's Layer Current*

   Open the **DB_SAMP.DWG** sample drawing located in the C:\Program Files\Acad2008\Sample\ directory. Next, use *Saveas* to save the drawing <u>in your working directory</u> as **DB_SAMP2**.

   A. Invoke the *Layer* command in command line format by typing *-LAYER* or *-LA*. Use the *?* option to yield the list of layers in the drawing. Notice that all the layers have the same linetype except one. Which layer has a different linetype and what is the linetype?

   B. Use the Layer Control drop-down list to indicate the layers of selected items. On the upper-left side of the drawing, **PICK** one or two cyan (light blue) lines. What layer are these lines on? Press the **Esc** key twice to cancel the selection. Next, **PICK** the gray staircase just above the center of the floor plan. What layer is the staircase on? Press the **Esc** key twice to cancel the selection.

   C. Invoke the *Layer Properties Manager* by selecting the icon button or using the *Format* pull-down menu. Turn *Off* the **CPU** layer (click the light bulb icon). Select *OK* to return to the drawing and check the new setting. Are the computers displayed? Next, use the **Layer Control** drop-down list to turn the **CPU** layer back *On*.

   D. Use the *Layer Properties Manager* and determine if any layers are *Frozen*. Which layers are *Frozen*? Next, right-click for the shortcut menu and select the *Select All* option. *Freeze* all layers. Which layer will not *Freeze*? Select *OK* in the warning dialog box, then *Apply*. Notice the light bulb icon indicates the layers are still *On* although the layers do not appear in the drawing (*Freeze* overrides *On*). Use the same procedure to select all layers and *Thaw* them. Then *Freeze* the two layers that were previously frozen.

E. Use the *Layer Properties Manager* again and select *All Used Layers* in the Filter Tree View area. Are there any layers in the drawing that are unused (examine the text just below the Filter Tree View)? How many layers are in the drawing? Select the *All* group in the Filter Tree View.

F. Now we want to *Freeze* all the *Cyan* layers. Using the *Layer Properties Manager* again, sort the list by color. **PICK** the first layer <u>name</u> in the list of the four cyan ones. Now hold down the **Shift** key and **PICK** the last name in the list. All four layers should be selected. *Freeze* all four by selecting any one of the sun/snowflake icons. Select *OK* to return to the drawing to examine the changes. Next, use the *LayerP* command to undo the last action in the *Layer Properties Manager*.

G. Next, you will "move" objects from one layer to another using the Object Properties toolbar. Make layer **0** (zero) the current layer by using the Layer Control drop-down list and selecting the name. Then use the *Layer Properties Manager*, right-click to *Select All* names. *Freeze* all layers (except the current layer). Then *Thaw* layer **CHAIRS**. Select *OK* and return to the drawing to examine the chairs (cyan in color). Use a crossing window to select all of the objects in the drawing (the small blue Grips may appear). Next, list the layers using the **Layer Control** drop-down list and **PICK** layer **0**. Press **Esc** twice to disable the Grips. The objects should be black and on layer 0. You actually "moved" the objects to layer 0. Finally, type *U* to undo the last action so the chairs return to layer CHAIRS (cyan). Also, *Thaw* all layers.

H. Let's assume you want to draw some additional objects on the GRIDLN layer and the E-F-TERR layer. Select the *Make Object's Layer Current* icon button. At the "Select object whose layer will become current" prompt, select one of the horizontal or vertical grid lines (gray color). The **GRIDLN** layer name should appear in the Layer Control box as the current layer. Draw a *Line* and it should be on layer GRIDLN and gray in color. Next, use *Make Object's Layer Current* again and select one of the cyan lines at the upper-left corner of the drawing. The **E-F-TERR** layer should appear as the current layer. Draw another *Line*. It should be on layer E-F-TERR and cyan in color. Keep in mind that this method of setting a layer current works especially well when you do not know the layer names or know what objects are on which layers.

I. *Close* the drawing. Do not save the changes.

2. *Layer Properties Manager* **and** *Linetype Manager*

A. Begin a *New* drawing and use *Save* to assign the name **CH11EX2**. Set *Limits* to **11 x 8.5**, then *Zoom All*. Use the *Linetype Manager* to list the loaded linetypes. Are any linetypes already loaded? **PICK** the *Load* button then right-click to *Select All* linetypes. Select *OK*, then close the *Linetype Manager*.

B. Use the *Layer Properties Manager* to create 3 new layers named **OBJ**, **HID**, and **CEN**. Assign the following colors, linetypes, and lineweights to the layers by clicking on the *Color*, *Linetype*, and *Lineweight* columns.

OBJ	*Red*	*Continuous*	*Default*
HID	*Yellow*	*Hidden2*	*Default*
CEN	*Green*	*Center2*	*Default*

C. Make the **OBJ** layer
Current and PICK the
OK tile. Verify the
current layer by looking
at the *Layer Control*
drop-down list, then
draw the visible object
Lines and the *Circle* only
(not the dimensions) as
shown in Figure 11-22.

FIGURE 11-22

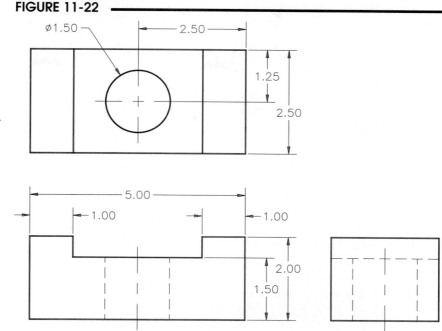

D. When you are finished drawing the visible object lines, create the necessary hidden lines by
making layer **HID** the *Current* layer, then drawing *Lines*. Notice that you only specify the *Line*
endpoints as usual and AutoCAD creates the dashes.

E. Next, create the center lines for the holes by making layer **CEN** the *Current* layer and drawing
Lines. Make sure the center lines extend <u>slightly beyond</u> the *Circle* and beyond the horizontal
Lines defining the hole. *Save* the drawing.

F. Now create a *New* layer
named **BORDER** and draw a
border and title block of your
design on that layer. The
final drawing should appear
as that in Figure 11-23.

FIGURE 11-23

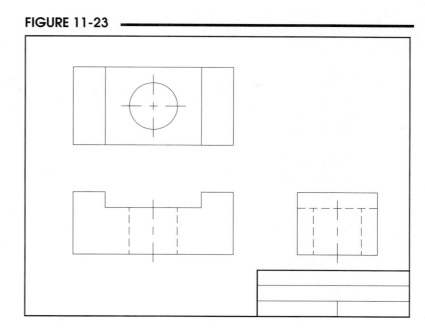

G. Open the *Linetype Manager* again and right-click to *Select All*, then select the *Delete* button. All unused linetypes should be removed from the drawing.

H. *Save* the drawing, then *Close* the drawing.

3. *LTSCALE, Properties* **palette,**
    Matchprop

A. Begin a *New* drawing and use *Save* to assign the name **CH11EX3.** Create the same four *New* layers that you made for the previous exercise (**OBJ, HID, CEN,** and **BORDER**) and assign the same linetypes, lineweights, and colors. Set the *Limits* equal to a "C" size sheet, **22 x 17**, then *Zoom All*.

B. Create the part shown in Figure 11-24. Draw on the appropriate layers to achieve the desired line-types.

FIGURE 11-24

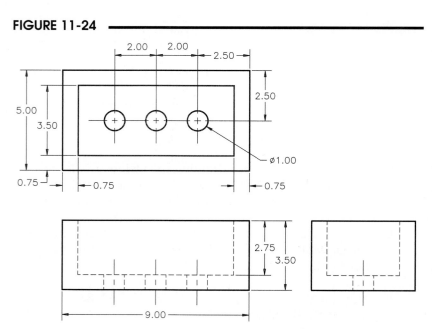

C. Notice that the *Hidden* and *Center* linetype dashes are very small. Use *LTSCALE* to adjust the scale of the non-continuous lines. Since you changed the *Limits* by a factor of slightly less than 2, try using **2** as the *LTSCALE* factor. Notice that all the lines are affected (globally) and that the new *LTSCALE* is retroactive for existing lines as well as for new lines. If the dashes appear too long or short because of the small hidden line segments, *LTSCALE* can be adjusted. Remember that you cannot control where the multiple <u>short</u> centerline dashes appear for one line segment. Try to reach a compromise between the hidden and center line dashes by adjusting the *LTSCALE*.

D. The six short vertical hidden lines in the front view and two in the side view should be adjusted to a smaller linetype scale. Invoke the *Properties* palette by selecting *Properties* from the *Modify* pull-down menu or selecting the *Properties* icon button. Select <u>only the two short lines in the side view</u>. Use the palette to retroactively adjust the two individual objects' *Linetype Scale* to **0.6000**.

E. When the new *Linetype Scale* for the two selected lines is adjusted correctly, use *Matchprop* to "paint" the *Linetype Scale* to the several other short lines in the front view. Select **Match Properties** from the **Modify** pull-down menu or select the "paint brush" icon.   When prompted for the "source object," select one of the two short vertical lines in the side view. When prompted for the "destination object(s)," select <u>each</u> of the short lines in the front view. This action should "paint" the *Linetype Scale* to the lines in the front view.

F. When you have the desired linetype scales, **Exit** AutoCAD and **Save Changes**. Keep in mind that the drawing size may appear differently on the screen than it will on a print or plot. Both the plot scale and the paper size should be considered when you set *LTSCALE*. This topic will be discussed further in Chapter 12, Advanced Drawing Setup.

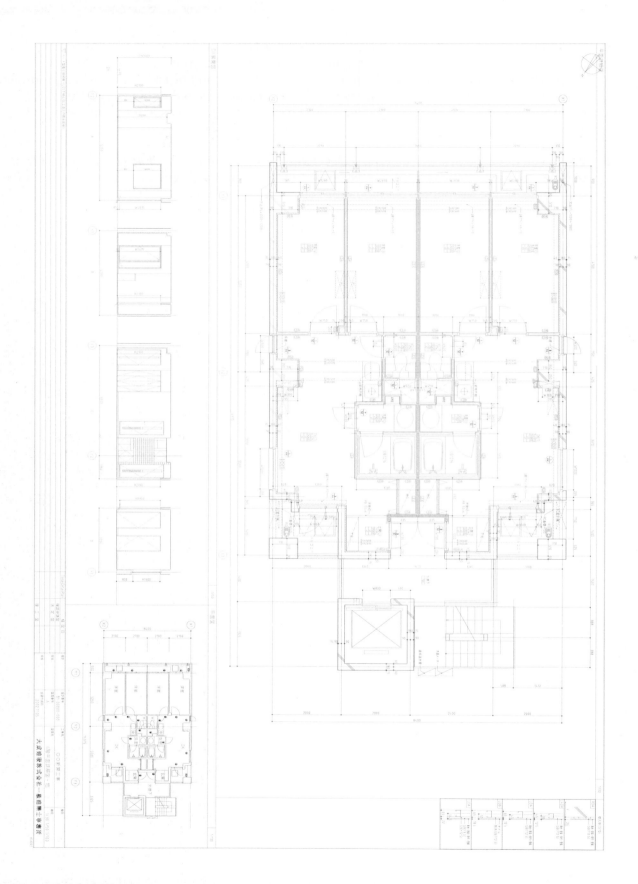

TAISEI DETAIL PLAN.DWG, Courtesy of Autodesk, Inc.

12

ADVANCED
DRAWING SETUP

CHAPTER OBJECTIVES

After completing this chapter you should:

1. know the steps for setting up
   a drawing;

2. be able to determine an appropriate Limits
   setting for the drawing;

3. be able to calculate and apply the "drawing
   scale factor";

4. be able to access existing and create new
   template drawings;

5. know what setup steps can be
   considered for creating
   template drawings.

CONCEPTS

When you begin a drawing, there are several steps that are typically performed in preparation for creating geometry, such as setting *Units, Limits,* and creating *Layers* with *linetypes, lineweights,* and *colors.* Some of these basic concepts were discussed in Chapters 6 and 11. This chapter discusses setting *Limits* for correct plotting as well as other procedures, such as layer creation and variables settings, that help prepare a drawing for geometry creation.

To correctly set up a drawing for printing or plotting to scale, <u>any two</u> of the following three variables must be known. The third can be determined from the other two.

> Drawing *Limits*
> Print or plot sheet (paper) size
> Print or plot scale

The method given in this text for setting up a drawing (Chapters 6 and 12) is intended for plotting from both the *Model* tab (model space) and from *Layout* tabs (paper space). The method applies when you plot from the *Model* tab or plot with <u>one viewport</u> in a *Layout* tab. If multiple viewports are created in one layout, the same general steps would be taken; however, the plot scale might vary for each viewport based on the size and number of viewports. (See Chapters 13 and 14 for more information.)

 Rather than performing the steps for drawing setup each time you begin, you can use "template drawings." Template drawings have many of the setup steps performed but contain no geometry. AutoCAD provides several template drawings and you can create your own template drawings. A typical engineering, architectural, design, or construction office generally produces drawings that are similar in format and can benefit from creation and use of individualized template drawings. The drawing similarities may be subject of the drawing (geometry), plot scale, sheet size, layering schemes, dimensioning styles, and/or text styles. Template drawing creation and use are discussed in this chapter.

STEPS FOR DRAWING SETUP

Assuming that you have in mind the general dimensions and proportions of the drawing you want to create, and the drawing will involve using layers, dimensions, and text, the following steps are suggested for setting up a drawing:

1. Determine and set the *Units* that are to be used.
2. Determine and set the drawing *Limits;* then *Zoom All.*
3. Set an appropriate *Snap Type* (*Polar Snap* or *Grid Snap*), *Snap* spacing, and *Polar* spacing if useful.
4. Set an appropriate *Grid* value if useful.
5. Change the *LTSCALE* value based on the new *Limits.* Set *PSLTSCALE* and *MSLTSCALE* to 0.
6. Create the desired *Layers* and assign appropriate *linetype, lineweight,* and *color* settings.
7. Create desired *Text Styles* (optional, discussed in Chapter 18).
8. Create desired *Dimension Styles* (optional, discussed in Chapter 26).
9. Activate a *Layout* tab, set it up for the plot or print device and paper size, and create a viewport (if not already existing).
10. Create or insert a title block and border in the layout.

Each of the steps for drawing setup is explained in detail here.

1. **Set *Units***
    This task is accomplished by using the *Units* command or the *Quick Setup* or *Advanced Setup* wizard. Set the linear units and precision desired. Set angular units and precision if needed. (See Chapter 6 for details on the *Units* command and setting *Units* using the wizards.)

2. **Set *Limits***

Before beginning to create an AutoCAD drawing, determine the size of the drawing area needed for the intended geometry. Using the actual *Units*, appropriate *Limits* should be set in order to draw the object or geometry to the <u>real-world size</u>. *Limits* are set with the *Limits* command by specifying the lower-left and upper-right corners of the drawing area. Always *Zoom All* after changing *Limits*. *Limits* can also be set using the *Quick Setup* or *Advanced Setup* wizard. (See Chapter 6 for details on the *Limits* command and the setup wizards.)

If you are planning to plot the drawing to scale, *Limits* should be set to a proportion of the sheet size you plan to plot on. For example, if the sheet size is 11" x 8.5", set *Limits* to 11 x 8.5 if you want to plot full size (1"=1"). Setting *Limits* to 22 x 17 (2 times 11 x 8.5) provides 2 times the drawing area and allows plotting at 1/2 size (1/2"=1") on the 11" x 8.5" sheet. Simply stated, set *Limits* to a <u>proportion</u> of the paper size.

Setting *Limits* to the <u>paper size</u> allows plotting at 1=1 scale. Setting *Limits* to a <u>proportion</u> of the sheet size allows plotting at the <u>reciprocal</u> of that proportion. For example, setting *Limits* to 2 times an 11" x 8.5" sheet allows you to plot 1/2 size on that sheet. Or setting *Limits* to 4 times an 11" x 8.5" sheet allows you to plot 1/4 size on that sheet. (Standard paper sizes are given in Chapter 14.)

Even if you plan to use a layout tab (paper space) and create one viewport, setting model space *Limits* to a proportion of the paper size is recommended. In this way you can easily calculate the <u>viewport scale</u> (the proportion of paper space units to model space units); therefore, the model space geometry can be plotted to a standard scale.

If you plan to plot from a layout, *Limits* in <u>paper space</u> should be set to the actual paper size; however, setting the paper space *Limits* values is automatically done for you when you select a *Plot Device* and *Paper Size* from the *Page Setup* or *Plot* dialog box (see Chapter 13, Layouts and Viewports, and Chapter 14, Printing and Plotting).

To set model space *Limits*, you can use the individual commands for setting *Units* and *Limits* (*Units* command and *Limits* command) or use the *Quick Setup* wizard or *Advanced Setup* wizard.

Drawing Scale Factor

<u>The proportion of the *Limits* to the intended print or plot sheet size is the "drawing scale factor."</u> This factor can be used as a general scale factor for other size-related drawing variables such as *LTSCALE*, dimension *Overall Scale* (*DIMSCALE*), and *Hatch* pattern scale. The drawing scale factor can also be used to determine the viewport scales.

Most size-related AutoCAD drawing variables are set to 1 by default. This means that variables (such as *LTSCALE*) that control sizing and spacing of objects are set appropriately for creating a drawing plotted full size (1=1).

FIGURE 12-1

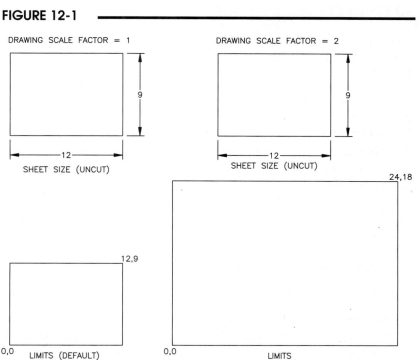

Therefore, when you set *Limits* to the sheet size and print or plot at 1=1, the sizing and spacing of linetypes and other variable-controlled objects are correct. When *Limits* are changed to some proportion of the sheet size, the size-related variables should also be changed proportionally. For example, if you intend to plot on a 12 x 9 sheet and the default *Limits* (12 x 9) are changed by a factor of 2 (to 24 x 18), then 2 becomes the drawing scale factor. Then, as a general rule, the values of variables such as *LTSCALE, DIMSCALE (Overall Scale)*, and other scales should be multiplied by a factor of 2 (Fig. 12-1, on the previous page). When you then print the 24 x 18 area onto a 12" x 9" sheet, the plot scale would be 1/2, and all sized features would appear correct.

Limits should be set to the paper size or to a proportion of the paper size used for plotting or printing. In many cases, "cut" paper sizes are used based on the 11" x 8.5" module (as opposed to the 12" x 9" module called "uncut sizes"). Assume you plan to print on an 11" x 8.5" sheet. Setting *Limits* to the sheet size provides plotting or printing at 1=1 scale. Changing the *Limits* to a proportion of the sheet size <u>by some multiplier</u> makes that value the drawing scale factor. The <u>reciprocal</u> of the drawing scale factor is the scale for plotting on that sheet (Fig. 12-2).

FIGURE 12-2

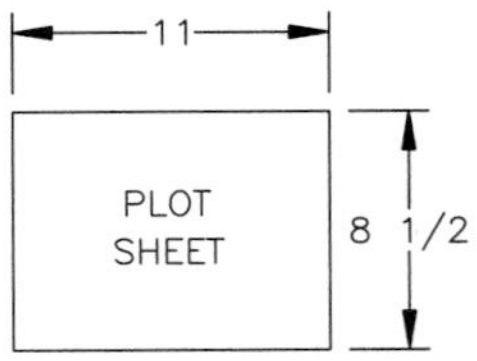

PLOT SHEET — 11 — 8 1/2

(CUT PAPER SIZE)

LIMITS	SCALE FACTOR	PLOT SCALE	
11 x 8.5	1	1=1	(FULL SIZE)
22 x 17	2	1=2	(HALF SIZE)
44 x 34	4	1=4	(1/4 SIZE)
110 x 85	10	1=10	(1/10 SIZE)

Since the drawing scale factor (DSF) is the proportion of *Limits* to the sheet size, you can use this formula:

$$DSF = \frac{Limits}{Sheet\ size}$$

Because plot/print scale is the reciprocal of the drawing scale factor:

$$Plot\ scale = \frac{1}{DSF}$$

The term "drawing scale factor" is not a variable or command that can be found in AutoCAD or its official documentation. This concept has been developed by AutoCAD users to set system variables appropriately for printing and plotting drawings to standard scales.

3. **Set *Snap***

Use the *Snap* command or *Snap and Grid* tab of the *Drafting Settings* dialog box to set the *Snap* type (*Grid Snap* and/or *Polar Snap*) and appropriate *Snap Spacing* and/or *Polar Spacing* values. The *Snap Spacing* and *Polar Spacing* values are dependent on the <u>interactive</u> drawing accuracy that is desired.

The accuracy of the drawing and the size of the *Limits* should be considered when setting the *Snap Spacing* and *Polar Spacing* values. On one hand, to achieve accuracy and detail, you want the values to be the smallest dimensional increments that would commonly be used in the drawing. On the other hand, depending on the size of the *Limits*, the *Snap* value should be large enough to make interactive selection (PICKing) fast and easy. As a starting point for determining appropriate *Snap Spacing* and *Polar Spacing* values, the default values can be multiplied by the "drawing scale factor."

If you are creating a template drawing (.DWT file), it is of no help to set the *Snap* type (*Grid Snap* or *Polar Snap*), *Polar Spacing* value, or *Polar Tracking* angles since these values are stored in the system registry, not in the drawing. The *Grid Snap Spacing* value, however, is stored in the drawing file.

TIP

4. **Set *Grid***

The *Grid* value setting is usually set equal to or proportionally larger than that of the *Grid Snap* value. *Grid* should be set to a proportion that is <u>easily visible</u> and to some value representing a regular increment (such as .5, 1, 2, 5, 10, or 20). Setting the value to a proportion of *Grid Snap* gives visual indication of the *Grid Snap* increment. For example, a *Grid* value of 1X, 2X, or 5X would give visual display of every 1, 2, or 5 *Grid Snap* increments, respectively. If you are not using *Snap* because only extremely small or irregular interval lengths are needed, you may want to turn *Grid* off. (See Chapter 6 for details on the *Grid* command.)

5. **Set the *LTSCALE*, *MSLTSCALE*, and *PSLTSCALE***

A change in *Limits* (then *Zoom All*) affects the display of non-continuous (hidden, dashed, dotted, etc.) lines. The *LTSCALE* variable controls the spacing of non-continuous lines. The drawing scale factor can be used to determine the *LTSCALE* setting. For example, if the default *Limits* (based on sheet size) have been changed by a factor of 2, the drawing scale factor=2; so set the *LTSCALE* to 2. In other words, if the DSF=2 and plot scale=1/2, you would want to double the linetype dashes to appear the correct size in the plot, so *LTSCALE*=2. If you prefer a *LTSCALE* setting of other than 1 for a drawing in the default *Limits,* multiply that value by the drawing scale factor.

American National Standards Institute (ANSI) requires that <u>hidden lines be plotted with 1/8"</u> <u>dashes</u>. An AutoCAD drawing plotted full size (1 unit=1") with the default *LTSCALE* value of 1 plots the *Hidden* linetype with 1/4" dashes. Therefore, it is recommended for English drawings that you <u>use the *Hidden2, Center2,* and other *2 linetypes</u> with dash lengths .5 times as long so an *LTSCALE* of 1 produces the ANSI standard dash lengths. Multiply your *LTSCALE* value by the drawing scale factor when plotting to scale. If individual lines need to be adjusted for linetype scale, use *Properties* when the drawing is nearly complete. (See Chapter 11 for details on the *LTSCALE* variable.)

Assuming you are planning to create only one viewport in a *Layout* tab, set the *PSLTSCALE* (paper space linetype scale) and *MSLTSCALE* (model space linetype scale) values to 0 (zero). AutoCAD's default setting is 1. The setting of 0 forces the linetype scale to appear the same in *Layout* viewports as in the *Model* tab. Set the *PSLTSCALE* and *MSLTSCALE* values at the command prompt.

6. **Create *Layers*; Assign *Linetypes*, *Lineweights*, and *Colors***

Use the *Layer Properties Manager* to create the layers that you anticipate needing. Multiple layers can be created with the *New* option. You can type in several new names separated by commas. Assign a descriptive name for each layer, indicating its type of geometry, part name, or function. Include a *linetype* designator in the layer name if appropriate; for example, PART1-H and PART1-V indicate hidden and visible line layers for PART1.

Once the *Layers* have been created, assign a *Color*, *Linetype,* and *Lineweight* to each layer. *Colors* can be used to give visual relationships to parts of the drawing. Geometry that is intended to be plotted with different pens (pen size or color) or printed with different plotted appearances should be drawn in different screen colors (especially when color-dependent plot styles are used). Use the *Linetype* command to load the desired linetypes. You should also select *Lineweights* for each layer (if desired) using the *Layer Properties Manager*. (See Chapter 11 for details on creating layers and assigning linetypes, lineweights, and colors.)

7. **Create Text Styles**

AutoCAD has only one text style as part of the standard template drawings (ACAD.DWT and ACADISO.DWT) and one or two text styles for most other template drawings. If you desire other *Text Styles,* they are created using the *Style* command or the *Text Style…* option from the *Format* pull-down menu. (See Chapter 18, Text.) If you desire engineering standard text, create a *Text Style* using the ROMANS.SHX or .TTF font file.

8. Create Dimension Styles

If you plan to dimension your drawing, Dimension Styles can be created at this point; however, they are generally created during the dimensioning process. Dimension Styles are names given to groups of dimension variable settings. (See Chapter 26 for information on creating Dimension Styles.)

Although you do not have to create Dimension Styles until you are ready to dimension the geometry, it is helpful to create Dimension Styles as part of a template drawing. If you produce similar drawings repeatedly, your dimensioning techniques are probably similar. Much time can be saved by using a template drawing with previously created Dimension Styles. Several of the AutoCAD-supplied template drawings have prepared Dimension Styles.

9. Activate a *Layout* Tab, Set the Plot Device, and Create a Viewport

This step can be done just before making a print or plot, or you can complete this and the next step before creating your drawing geometry. Completing these steps early allows you to see how your printed drawing might appear and how much area is occupied by the titleblock and border.

Activate a *Layout* tab. If your *Options* are set as explained in Chapter 6, Basic Drawing Setup, the layout is automatically set up for your default print/plot device and an appropriate-sized viewport is created. Otherwise, use the *Page Setup Manager* (*Pagesetup* command) to set the desired print/plot device and paper size. Then use the *Vports* command to create one or more viewports. This topic is discussed briefly in Chapter 6, Basic Drawing Setup, and is discussed in more detail in Chapter 13, Layouts and Viewports.

10. Create a Titleblock and Border

For 2D drawings, it is helpful to insert a titleblock and border early in the drawing process. This action gives a visual drawing boundary as well as reserves the space needed by the titleblock.

Rather than create a new titleblock and border for each new drawing, a common practice is to use the *Insert* command to insert an existing titleblock and border as a *Block*.

You can draw or *Insert* a titleblock and border in the *Model* tab (model space) or into a layout (paper space), depending on how you want to plot the drawing. If you plan to plot from the *Model* tab, multiply the actual size of the titleblock and border by the drawing scale factor to determine their sizes for the drawing. If you want to plot from a layout, draw or insert the titleblock and border at the actual size (1:1) since a layout represents the plot sheet. (See Chapter 20 for information on *Block* creation and *Insertion*.)

USING AND CREATING TEMPLATE DRAWINGS

Instead of going through the steps for setup <u>each time</u> you begin a new drawing, you can create one or more template drawings or use one of the AutoCAD-supplied template drawings. AutoCAD template drawings are saved as a .DWT file format. A template drawing has the initial setup steps (*Units, Limits, Layers, Linetypes, Lineweights, Colors,* etc.) completed and saved, but no geometry has been created yet. For some templates, *Layout* tabs have been set up with titleblocks, viewports, and plot settings. Template drawings are used as a <u>template</u> or <u>starting point</u> each time you begin a new drawing. AutoCAD actually makes a copy of the template you select to begin the drawing. The creation and use of template drawings can save hours of preparation.

Using Template Drawings

To use a template drawing, select the *Template* option in the *Startup* or *Create New Drawing* dialog box (Fig. 12-3, on the next page) or in the *Select Template* dialog box. This option allows you to select the desired template (.DWT) drawing from a list including all templates in the Template folder.

Any template drawings that you create using the *SaveAs* command (as described in the next section) appear in the list of available templates. Select the *Browse...* option to browse other folders.

FIGURE 12-3

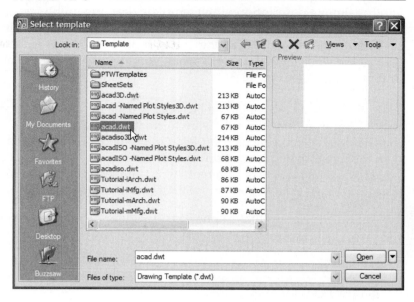

Creating Template Drawings

To make a template drawing, begin a *New* drawing using a default template drawing (ACAD.DWT) or other template drawing (*.DWT), then make the initial drawing setups. Alternately, *Open* an existing drawing that is set up as you need (layers created, linetypes loaded, *Limits* and other settings for plotting or printing to scale, etc.), and then *Erase* all the geometry. Next, use *SaveAs* to save the drawing under a different descriptive name with a .DWT file extension. Do this by selecting the *AutoCAD Drawing Template File (*.dwt)* option in the *Save Drawing As* dialog box (Fig. 12-4).

FIGURE 12-4

AutoCAD automatically (by default) saves the drawing in the folder where other template (*DWT) files are found. Use the Save in: drop-down list at the top of the Save Drawing As dialog box to specify another location for saving the .DWT file.

Location of the Template folder aligns with the Microsoft Windows 2000 and XP conventions. That is, the templates are stored by default (for a stand-alone installation) in a folder multiple levels under the user's profile (C:\Documents and Settings\user's name\Local Settings\Application Data\AutoCAD 2008\... and so on). To find the complete path, use the *Options* dialog box, *Files* tab, and expand *Drawing Template Settings*.

After you assign the file name in the *Save Drawing As* dialog box, you can enter a description of the template drawing in the *Template Description* dialog box that appears (Fig. 12-5). The description is saved with the template drawing and appears in the *Startup* and *Create New Drawing* dialog boxes when you highlight the template drawing name (see Fig. 12-3).

FIGURE 12-5

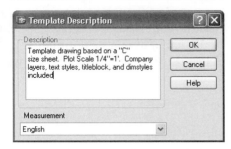

Multiple template drawings can be created, each for a particular situation. For example, you may want to create several templates, each having the setup steps completed but with different *Limits* and with the intention to plot or print each in a different scale or on a different size sheet. Another possibility is to create templates with different layout and layering schemes. There are many possibilities, but the specific settings used depend on your applications (geometry), typical scales, or your plot/print devices.

Typical drawing steps that can be considered for developing template drawings are listed below:

> Set *Units*
> Set *Limits*
> Set *Snap*
> Set *Grid*
> Set *LTSCALE, MSLTSCALE,* and *PSLTSCALE*
> Create *Layers* with color and linetypes assigned
> Create *Text Styles* (see Chapter 18)
> Create Dimension Styles (see Chapter 26)
> Create *Layout* tabs, each with titleblock, border and plot settings saved (plot device, paper size, plot scale, etc.). (See Chapter 13, Layouts and Viewports.)

You can create template drawings for different sheet sizes and different plot scales. If you plot from the *Model* tab, a template drawing should be created for each case (scale and sheet size). When you learn to create *Layouts*, you can save one template drawing that contains several layouts, each layout for a different sheet size and/or scale.

Additional Advanced Drawing Setup Concepts

The drawing setup described in this chapter is appropriate for learning the basic concepts of setting up a drawing in preparation for creating layouts and viewports and plotting to scale. In Chapter 13, Layouts and Viewports, you will learn the basics of creating layouts and viewports—particularly for drawings that use one large viewport in a layout.

In many cases, you may want to create several viewports in one or more layouts. The primary purpose of this action is to print the same drawing, or parts of the same drawing, in different scales. Annotative objects (such as annotative text, annotative dimensions, and annotative hatch patterns) can be used to facilitate this objective. Annotative objects automatically adjust for size when the *Annotative Scale* for the viewport is changed. That is, annotative objects can be automatically resized to appear "readable" in different viewports at different scales. If you plan to create multiple viewports and intend to use annotative objects, the steps for setting up a drawing are given next.

Steps for Drawing Setup Using Annotative Objects

The first four steps are the same as those given previously (see "Steps for Drawing Setup"):

1. Determine and set the *Units* that are to be used.
2. Determine and set the drawing *Limits*; then *Zoom All*.
3. Set an appropriate *Snap Type*, *Snap* spacing, and *Polar* spacing if useful.
4. Set an appropriate *Grid* value if useful.

These steps are for using annotative objects specifically:

5. Change *MSLTSCALE* to 0 and *PSLTSCALE* to 1.
6. Set the *Annotation Scale* pop-up list in the *Model* tab (lower-right corner of the Drawing Editor) to the scale you intend to print the drawing for the primary viewport (use the reciprocal of the Drawing Scale Factor). In other words, use the following formula where AS is the *Annotation Scale* and DSF is the Drawing Scale Factor:

AS = 1/DSF

The remaining steps are the same as previously given (see "Steps for Drawing Setup").

CHAPTER EXERCISES

For the first three exercises, you will set up several drawings that will be used in other chapters for creating geometry. Follow the typical steps for setting up a drawing given in this chapter.

1. **Create a Metric Template Drawing**
   You are to create a template drawing for use with metric units. The drawing will be printed full size on an "A" size sheet (not on an A4 metric sheet), and the dimensions are in millimeters. Start a *New* drawing, use the **ACAD.DWT** template or select *Start from Scratch*, and select the *Imperial* default settings. Follow these steps.

   A. Set *Units* to *Decimal* and *Precision* to **0.00**.
   B. Set *Limits* to a metric sheet size, **279.4 x 215.9**; then *Zoom All* (scale factor is approximately 25).
   C. Set *Snap Spacing* (for *Grid Snap*) to **1**. (*Polar Tracking* and *Polar Snap* settings do not have to be preset since they are saved in the system registry.)
   D. Set *Grid* to **10**.
   E. Set *LTSCALE* to **25** (drawing scale factor of 25.4). Set *PSLTSCALE* and *MSLTSCALE* to **0**.
   F. *Load* the *Center2* and *Hidden2 Linetypes*.
   G. Create the following *Layers* and assign the *Colors*, *Linetypes*, and *Lineweights* as shown:

OBJECT	red	continuous	0.40 mm
CONSTR	white	continuous	0.25 mm
CENTER	green	center2	0.25 mm
HIDDEN	yellow	hidden2	0.25 mm
DIM	cyan	continuous	0.25 mm
TITLE	white	continuous	0.25 mm
VPORTS	white	continuous	0.25 mm

   H. Use the *SaveAs* command. In the dialog box under *Save as type:*, select *AutoCAD Drawing Template File (.dwt)*. Assign the name **A-METRIC.** Enter **Metric A Size Drawing** in the *Template Description* dialog box.

2. **Barguide Setup**
A drawing of a mechanical part is to be made and plotted full size (1″=1″) on an 11″ x 8.5″ sheet.

A. Begin a *New* drawing using the **ACAD.DWT** template.
B. Set the *Units* to *Decimal* with a *Precision* of *0.000*.
C. Set the *Limits* to **11 x 8.5**.
D. Set the *Grid* to *.500*.
E. Change the *Snap* to *.250*.
F. Set *LTSCALE* to **1**. Set *PSLTSCALE* and *MSLTSCALE* to **0**.
G. *Load* the *Center2* and *Hidden2 Linetypes*.
H. Create the following *Layers* and assign the given *Colors, Linetypes,* and *Lineweights*:

OBJECT	*red*	*continuous*	*0.40 mm*
CONSTR	*white*	*continuous*	*0.25 mm*
CENTER	*green*	*center2*	*0.25 mm*
HIDDEN	*yellow*	*hidden2*	*0.25 mm*
DIM	*cyan*	*continuous*	*0.25 mm*
TITLE	*white*	*continuous*	*0.25 mm*
VPORTS	*white*	*continuous*	*0.25 mm*

I. Save the drawing as **BARGUIDE** for use in another chapter exercise.

3. **Apartment setup**
A floor plan of an apartment has been requested. Dimensions are in feet and inches. The drawing will be plotted at 1/2″=1′ scale on an 24″ x 18″ sheet.

A. Begin a *New* drawing using the **ACAD.DWT** template.
B. Set the *Units* to *Architectural* with a *Precision* of *1/8*.
C. Set the *Limits* to **48′ x 36′** (make sure you use the apostrophe to designate feet).
D. Set the *Snap* value to **1** (inch).
E. Set the *Grid* value to **12** (inches).
F. Set the *LTSCALE* to **12**. Set *PSLTSCALE* and *MSLTSCALE* to **0**.
G. Create the following *Layers* and assign the *Colors* and *Lineweights*:

FLOORPLN	*red*	*0.016″*
CONSTR	*white*	*0.010″*
TEXT	*green*	*0.010″*
TITLE	*yellow*	*0.010″*
DIM	*cyan*	*0.010″*
VPORTS	*white*	*0.010″*

H. *Save* the drawing and assign the name **APARTMENT** to be used later.

4. **Mechanical setup**
    A drawing is to be created and printed in 1/4"=1" scale on an A size (11" x 8.5") sheet.

    A. Begin a *New* drawing using the **ACAD.DWT** template.
    B. Set the *Units* to *Decimal* with a *Precision* of *0.000*.
    C. Set the *Limits* to **44** x **34**.
    D. Set *Snap* to **1**.
    E. Set *Grid* to **2**.
    F. Type *LTS* at the command prompt and enter **4**.
    G. Set *PSLTSCALE* and *MSLTSCALE* to **0**.
    H. Save the drawing and name it **CH12EX4**.

5. **Create a Template Drawing**
    In this exercise, you will create a "generic" template drawing that can be used at a later time. Creating templates now will save you time later when you begin new drawings.

    A. Create a template drawing for use with decimal dimensions and using standard paper "A" size format. Begin a *New* drawing and use the **ACAD.DWT** template or select *Start from Scratch* with the *Imperial* default settings.
    B. Set *Units* to *Decimal* and *Precision* to **0.00**.
    C. Set *Limits* to **11** x **8.5**.
    D. Set *Snap* to **.25**.
    E. Set *Grid* to **1**.
    F. Set *LTSCALE* to **1**. Set *PSLTSCALE* and *MSLTSCALE* to **0**.
    G. Create the following *Layers*, assign the *Linetypes* and *Lineweights* as shown, and assign your choice of *Colors*. Create any other layers you think you may need or any assigned by your instructor.

OBJECT	*continuous*	*0.016"*
CONSTR	*continuous*	*0.010"*
TEXT	*continuous*	*0.010"*
TITLE	*continuous*	*0.010"*
VPORTS	*continuous*	*0.010"*
DIM	*continuous*	*0.010"*
HIDDEN	*hidden2*	*0.010"*
CENTER	*center2*	*0.010"*
DASHED	*dashed2*	*0.010"*

    H. Use *SaveAs* and name the template drawing **ASHEET.** Make sure you select *AutoCAD Drawing Template File (*.dwt)* from the *Save as Type:* drop-down list in the *Save Drawing As* dialog box. Also, from the *Save In:* drop-down list on top, select your working directory as the location to save the template.

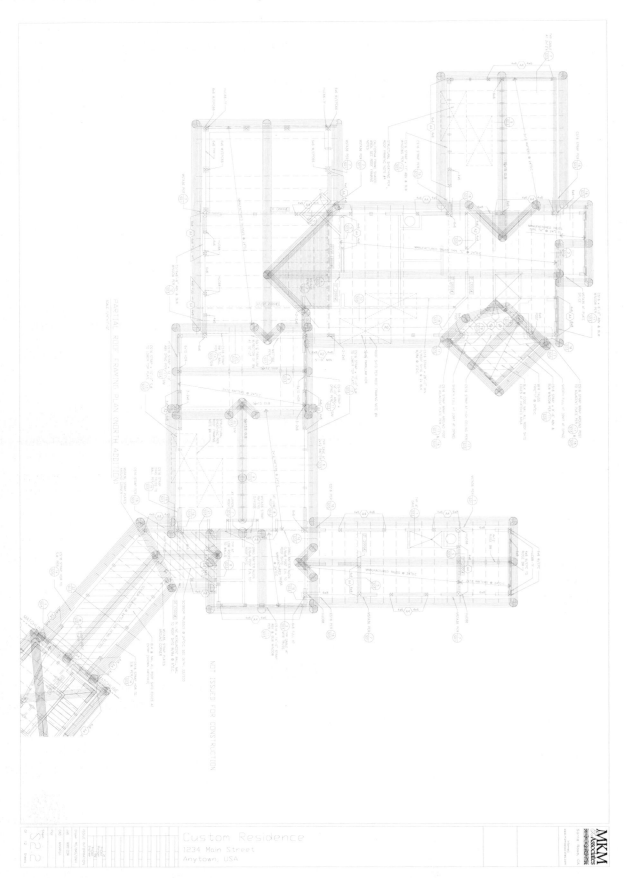

LAYOUTS AND VIEWPORTS

CHAPTER OBJECTIVES

After completing this chapter you should:

1. know the difference between paper space and model space and between tiled viewports and paper space viewports;

2. know that the purpose of a layout is to prepare model space geometry for plotting;

3. know the "Guidelines for Using Layouts and Viewports";

4. be able to create layouts using the *Layout Wizard* and the *Layout* command;

5. know how to set up a layout for plotting using the *Page Setup Manager* and the *Page Setup* dialog box;

6. be able to use the options of *Vports* and *-Vports* to create and control viewports in paper space;

7. know how to scale the model geometry displayed in paper space viewports.

CONCEPTS

You are already somewhat familiar with model space (the *Model* tab) and paper space (*Layout* tabs). Model space (the *Model* tab) is used for construction of geometry, whereas paper space (*Layout* tabs) is used for preparing to print or plot the model geometry. In order to view the model geometry in a layout, one or more viewports must be created. This can be done automatically by AutoCAD when you activate a *Layout* tab for the first time, or you can use the *Vports* command to create viewports in a layout. An introduction to these concepts is given in Chapter 6, Basic Drawing Setup, "Introduction to Layouts and Printing." However, this chapter (Chapter 13) gives a full explanation of paper space, layouts, and creating paper space viewports.

Paper Space and Model Space

The two drawing spaces that AutoCAD provides are model space and paper space. Model space is activated by selecting the *Model* tab and paper space is activated by selecting the *Layout1*, *Layout2*, or other layout tab. When you start AutoCAD and begin a drawing, model space is active by default. Objects that represent the subject of the drawing (model geometry) are normally drawn in model space. Dimensioning is traditionally performed in model space because it is associative—directly associated to the model geometry. The model geometry is usually completed before using paper space. It is possible to create dimensions in paper space that are attached to objects in model space; however, this practice is recommended only for certain applications.

Paper space represents the <u>paper that you plot or print on</u>. When you enter paper space (by activating a *Layout* tab) for the first time, you may see only a blank "sheet," unless your system is configured to automatically generate a viewport as explained in Chapter 6. Normally the only geometry you would create in paper space is the title block, drawing border, and possibly some other annotation (text). In order to see any model geometry from paper space, viewports must be created (like cutting rectangular holes) so you can "see" into model space. Viewports are created either automatically by AutoCAD or by using the *Vports* command. Any number or size of rectangular or non-rectangular viewports can be created in paper space. Since there is only <u>one model space</u> in a drawing, you see the same model space geometry in each viewport initially. You can, however, control the size and area of the geometry displayed and which layers are *Frozen* and *Thawed* <u>in each viewport</u>. Since paper space represents the actual paper used for plotting, you should plot from paper space at a scale of 1:1.

The *TILEMODE* system variable controls which space is active—paper space (*TILEMODE=0*) or model space (*TILEMODE=1*). *TILEMODE* is automatically set when you activate the *Model* tab or a *Layout* tab.

Layouts

By default there is a *Layout1* tab and *Layout2* tab. You can produce any configuration of layouts by creating new layouts or copying, renaming, deleting, or moving existing layouts. A shortcut menu is available (right-click while your pointer is on a layout tab) that allows you to *Rename* the selected layout as well as create a *New Layout* from scratch or *From Template, Delete, Move or Copy* a layout, *Select all Layouts*, or set up for plotting (Fig. 13-1).

The primary function of a layout is to prepare the model geometry for plotting. A layout simulates the sheet of paper you will plot on and allows you to specify plotting parameters and see the changes in settings that you make, similar to a plot preview. Plot specifications are made using the *Page Setup* dialog box to select such parameters as plot device, paper size and orientation, scale, and Plot Style Tables to attach.

Since you can have multiple layouts in AutoCAD, you can set up multiple plot schemes. For example, assume you have a drawing (in model space) of an office

FIGURE 13-1

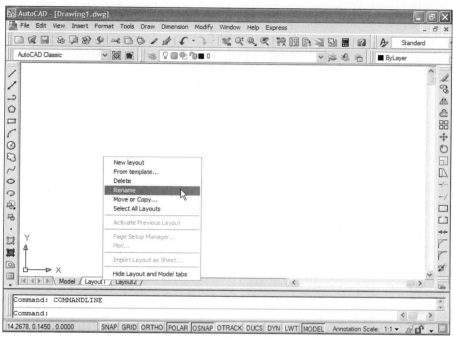

complex. You can set up one layout to plot the office floor plan on a "C" size sheet with an electrostatic plotter in 1/4"=1' scale and set up a second layout to plot the same floor plan on a "A" size sheet with a laser jet printer at 1/8"=1' scale. Since the plotting setup is saved with each layout, you can plot any layout again without additional setup.

NOTE: If the *Model* and *Layout* tabs are not visible at the bottom left of the drawing area, you can make them appear by right-clicking on the layout icon just to the right of the Status bar, then select *Display Layout and Model Tabs* (Fig. 13-2).

FIGURE 13-2

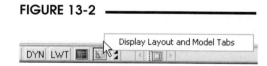

Viewports

When you activate a layout tab (enter paper space) for the first time, you may see a "blank sheet." However, if *Create Viewport in New Layouts* is checked in the *Display* tab of the *Options* dialog box (see Fig. 13-3, lower-left corner), you may see a viewport that is automatically created. If no viewport exists, you can create one or more with the *Vports* command. A viewport in paper space is a window into model space. Without a viewport, no model geometry is visible in a layout.

FIGURE 13-3

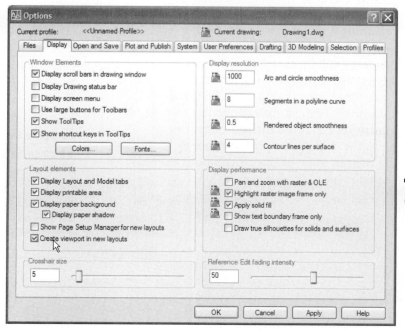

There are two types of viewports in AutoCAD, (1) viewports in model space that divide the screen like "tiles" described in Chapter 10 (known as model space viewports or tiled viewports) and (2) viewports in paper space described in this chapter (known as paper space viewports or floating viewports). The term "floating viewports" is used because these viewports can be moved or resized and can take on any shape. Floating viewports are objects that can be *Erased*, *Moved*, *Copied*, stretched with grips, or the viewport's layer can be turned *Off* or *Frozen* so no viewport "border" appears in the plot.

The command generally used to create viewports, *Vports*, can create either tiled viewports or floating viewports, but only one type at a time depending on the active space—*Model* tab or *Layout* tab. When the *Model* tab is active (*TILEMODE*=1), *Vports* creates tiled viewports. When a *Layout* tab is active (*TILE-MODE*=0), *Vports* creates paper space viewports.

Typically there is only one viewport per layout; however, you can create multiple viewports in one layout. This capability gives you the flexibility to display two or more areas of the same model (model space) in one layout. For example, you may want to show the entire floor plan in one viewport and a detail (enlarged view) in a second viewport, both in the same layout (Fig. 13-4).

FIGURE 13-4

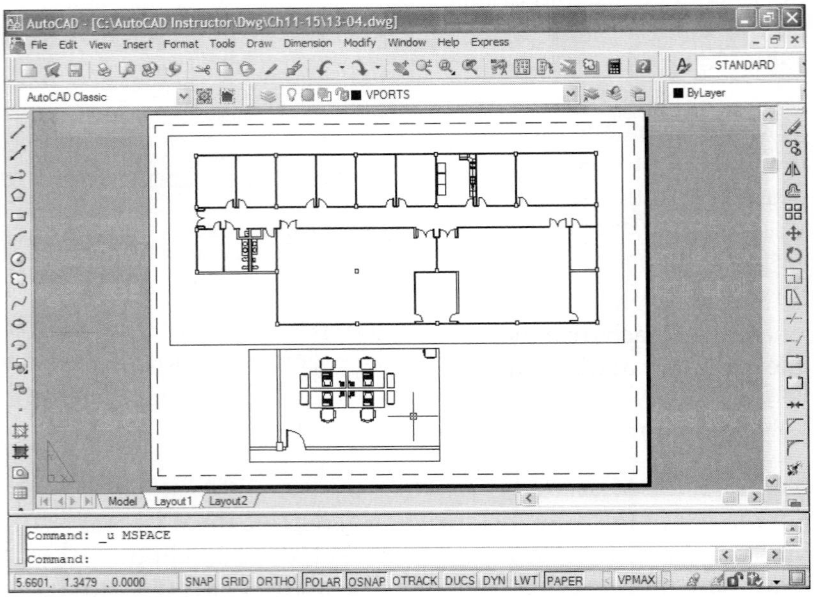

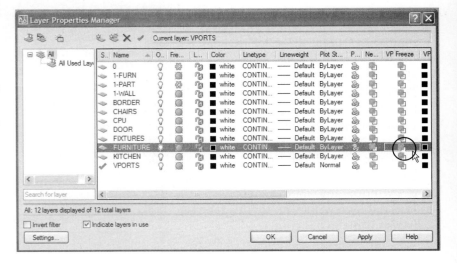

You can control the layer visibility in each viewport. In other words, you can control which layers are visible in which viewports. Do this by using the *VP Freeze* button and *New VP Freeze* button in the *Layer Properties Manager* (Fig. 13-5, last two columns). For example, notice in Figure 13-4 that the furniture layer is on in the detail viewport, but not in the overall view of the office floor plan above.

FIGURE 13-5

You can also scale the display of the geometry that appears in a viewport. You can set the viewport scale by using the *VP Scale* pop-up list (Fig. 13-6), the *Viewports* toolbar drop-down list, the *Properties* palette, or entering a *Zoom XP* factor. This action sets the proportion (or scale) of paper space units to model space units.

FIGURE 13-6

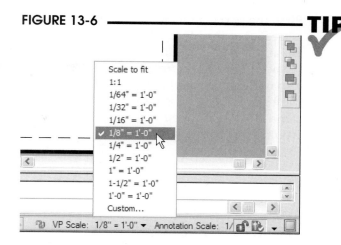

Since a layout represents the plotted sheet, you normally <u>plot the layout at 1:1 scale</u>. However, the geometry in each viewport is scaled to achieve the scale for the drawing objects. For example, you could have two or more viewports in one layout, each displaying the geometry at different scales, as is the case for Figure 13-7. Or, you can set up one layout to display model geometry in one scale and set up a similar layout to display the same geometry at a different scale.

FIGURE 13-7

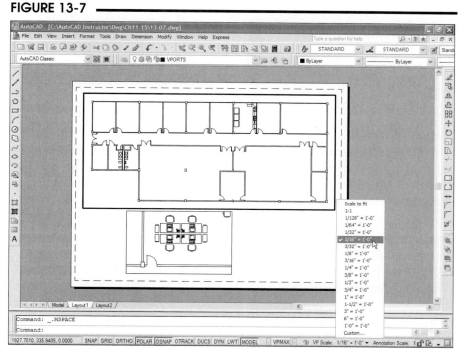

An important concept to remember when using layouts is that <u>the *Plot Scale* that you select in the *Page Setup* and *Plot* dialog boxes should almost always be 1:1</u>. This is because the layout is the actual paper size. Therefore, <u>the geometry displayed in the viewports must be scaled to the desired plot scale</u> (the reciprocal of the "drawing scale factor," which can be set by using the *Viewports* toolbar, the *Properties* palette for the viewport, or a *Zoom XP* factor.

Layouts and Viewports Example

Consider this brief example to explain the basics of using layouts and viewports. In order to keep this example simple, only one viewport is created in the layout to set up a drawing for plotting. This example assumes that no automatic viewports or page setups are created. Your system may be configured to automatically create viewports when a *Layout* tab is activated (as recommended in Chapter 6 to simplify viewport creation). Disabling automatic setups is accomplished by ensuring *Show Page Setup Manager for New Layouts* and *Create Viewport in New Layouts* is unchecked in the *Display* tab in the *Options* dialog box.

First, the part geometry is created in the *Model* tab as usual (Fig. 13-8). Associative dimensions are also normally created in the *Model* tab. This step is the same method that you would have normally used to create a drawing.

FIGURE 13-8

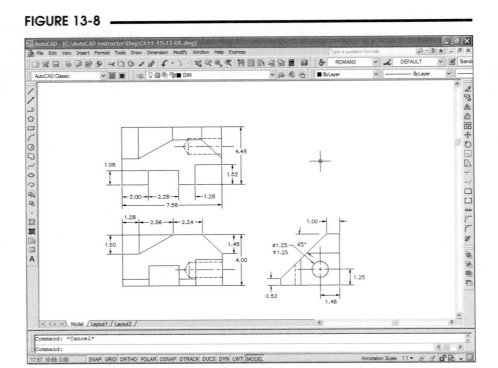

When the part geometry is complete, enable paper space by double-clicking a *Layout* tab. AutoCAD automatically changes the setting of the *TILEMODE* variable to 0. When you enable paper space for the first time in a drawing, a "blank sheet" appears (assuming no automatic viewports or page setups are created as a result of settings in the *Options* dialog box).

By activating a *Layout* tab, you are automatically switched to paper space. When that occurs, the paper space *Limits* are set automatically based on the selected plot device and sheet size previously selected for plotting. (The default plot device is set in the *Plotting* tab of the *Options* dialog box.) Objects such as a title block and border should be created <u>in the paper space layout</u> (Fig. 13-9). Normally, only objects that are <u>annotations</u> for the drawing (title blocks, tables, border, company logo, etc.) are drawn in the layout.

FIGURE 13-9

Next, create a viewport in the "paper" with the *Vports* command in order to "look" into model space. All of the model geometry in the drawing initially appears in the viewport. At this point, there is no specific scale relation between paper space units and the size of the geometry in the viewport (Fig. 13-10).

You can control what part of model space geometry you see in a viewport and control the scale of model space to paper space. Two methods are used to control the geometry that is visible in a particular viewport.

FIGURE 13-10

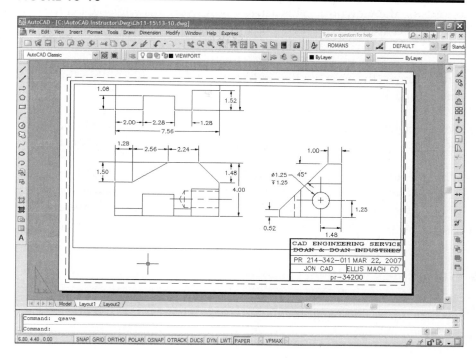

1. Display commands and Viewport Scale

The *Zoom, Pan, View, 3Dorbit* and other display commands allow you to specify what area of model space you want to see in a viewport. In addition, the scale of the model geometry in the viewport can be set by using the *VP Scale* pop-up list, the *Viewports* toolbar, the *Properties* palette for the viewport, or a *Zoom XP* factor.

2. Viewport-specific layer visibility control

You can control what <u>layers</u> are visible in specific viewports. This function is often used for displaying different model space geometry in separate viewports. You can control which layers appear in which viewports by using the *Layer Properties Manager*.

For example, the *VP Scale* pop-up list could be used to scale the model geometry in a viewport (Fig. 13-11). In this example, double-click inside the viewport and select *1:2* to scale the model geometry to 1/2 size (model space units equal 1/2 times paper space units).

FIGURE 13-11

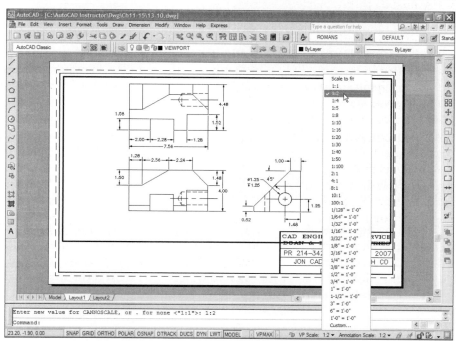

While in a layout with a viewport created to display model space geometry, you can double-click inside or outside the viewport. This action allows you to switch between model space (inside a viewport) and paper space (outside a viewport) so you can draw or edit in either space. You could instead use the *Mspace* (Model Space) and *Pspace* (Paper Space) commands or single-click the words *MODEL* or *PAPER* on the Status Bar to switch between inside and outside the viewport. The "crosshairs" can move completely across the screen when in paper space, but only within the viewport when in model space (model space inside a viewport).

An object <u>cannot be in both spaces</u>. You can, however, draw in paper space and <u>*OSNAP* to objects in</u> <u>model space</u>. Commands that are used will affect the objects or display of the current space.

When drawing is completed, activate paper space and plot at 1:1 since the layout size (paper space *Limits*) is set to the <u>actual paper size</u>. Typically, the *Page Setup Manager* or the *Plot* dialog box is used to prepare the layout for the plot. The plot scale to use for the layout is almost always 1:1 since the layout represents the plotted sheet and the model geometry is already scaled.

Guidelines for Using Layouts and Viewports

Although there are other alternatives, the typical steps used to set up a layout and viewports for plotting are listed here. (These guidelines assume that no automatic viewports or page setups are created, as described in Chapter 6, "Introduction to Layouts and Printing.")

1. Create the part geometry in model space. Associative dimensions are typically created in model space.

2. Click on a *Layout* tab. When this is done, AutoCAD automatically sets up the size of the layout (*Limits*) for the default plot device, paper size, and orientation. Use the *Page Setup Manager* and *Page Setup* dialog box to ensure the correct device, paper size, and orientation are selected. If not, make the desired selections. (If the layout does not automatically reflect the desired settings, check the *Plotting* tab of the *Options* dialog box to ensure that *Use the Plot Device Paper Size* is checked.)

3. Set up a title block and border as indicated.

    A. Make a layer named BORDER or TITLE and *Set* that layer as current.
    B. Draw, *Insert*, or *Xref* a border and a title block.

4. Make viewports in paper space.

    A. Make a layer named VIEWPORT (or other descriptive name) and set it as the *Current* layer (it can be turned *Off* later if you do not want the viewport objects to appear in the plot).
    B. Use the *Vports* command to make viewports. Each viewport contains a view of model space geometry.

5. Control the display of model space geometry in each viewport. Complete these steps for each viewport.

    A. Double-click inside the viewport or use the *MODEL/PAPER* toggle to "go into" model space (in the viewport). Move the cursor and PICK to activate the desired viewport if several viewports exist.
    B. Use the *VP Scale* pop-up list, the *Viewport Scale Control* drop-down list, the *Properties* palette for the viewport, or *Zoom XP* to scale the display of the paper space units to model units. This action dictates the plot scale for the model space geometry. The viewport scale is the same as the plot scale factor that would otherwise be used (reciprocal of the "drawing scale factor").
    C. Control the layer visibility for each viewport by using the icons in the *Layer Properties Manager*.

6. Plot from paper space at a scale of 1:1.

   A. Use the *MODEL/PAPER* toggle or double-click outside the viewport to switch to paper space.

   B. If desired, turn *Off* the VIEWPORT layer so the viewport borders do not plot.

   C. Invoke the *Plot* dialog box or the *Page Setup* dialog box to set the plot scale and other options. Normally, set the plot scale for the layout to *1:1* since the layout is set to the actual paper size. The paper space geometry will plot full size, and the resulting model space geometry will plot to the <u>scale</u> selected in the *VP Scale* pop-up list, *Viewport Scale* drop-down list, *Properties* palette, or by the designated *Zoom XP* factor.

   D. Make a *Full Preview* to check that all settings are correct. Make changes if necessary. Finally, make the plot.

CREATING AND SETTING UP LAYOUTS

Listed below are several ways to create layouts.

Shortcut menu	Right-click on a *Layout* tab and select *New Layout, From Template,* or *Move or Copy* (see Fig. 13-1).
Layout command	Type or select from the *Insert* pull-down menu. Use the *New Layout, Layout from Template,* or *Create Layout Wizard* option.
Layoutwizard command	Type or select from the *Insert* pull-down menu.

There are several steps involved in correctly setting up a layout, as listed previously in the "Guidelines for Using Layouts and Viewports." The steps involve specifying a plot device, setting paper size and orientation, setting up a new *Layout* tab, and creating the viewports and scaling the model geometry. One of the best methods for accomplishing all these tasks is to use the *Layout Wizard.*

Layoutwizard

Pull-down Menu	Command (Type)	Alias (Type)	Short-cut	Screen (side) Menu	Tablet Menu
Insert Layout > Create Layout Wizard	*Layoutwizard*	...	...	...	...

The *Layoutwizard* command produces a wizard to automatically lead you through eight steps to correctly set up a layout tab for plotting. The steps are essentially the same as those listed earlier in the "Guidelines for Using Layouts and Viewports."

Begin

The first step is intended for you to enter a name for the layout (Fig. 13-12). The default name is whatever layout number is next in the sequence, for example, *Layout3*. Enter any name that has not yet been used in this drawing. You can change the name later if you want using the *Rename* option of the *Layout* command.

FIGURE 13-12

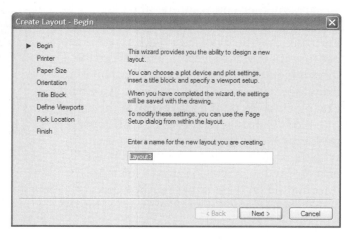

Printer

Because the *Limits* in paper space are automatically set based on the size and orientation of the selected paper, choosing a print or plot device is essential at this step (Fig. 13-13). The list includes all previously configured printer and plotter devices. If you intend to use a different device, it may be wise to *Cancel*, use *Plotter Manager* to configure the new device, then begin *Layoutwizard* again (see Chapter 14, "Configuring Plotters and Printers").

FIGURE 13-13

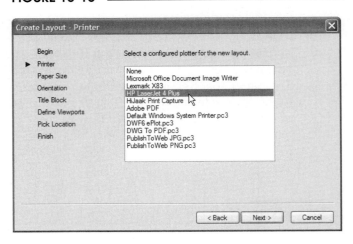

Paper Size

Select the paper size you expect to use for the layout (Fig. 13-14). Make sure you also select the units for the paper (*Drawing Units*) since the layout size (*Limits* in paper space) is automatically set in the selected units.

Orientation

Select either *Portrait* or *Landscape* (not shown). *Landscape* plots horizontal lines in the drawing along the long axis of the paper, whereas *Portrait* plots horizontal lines in the drawing along the short axis of the paper.

FIGURE 13-14

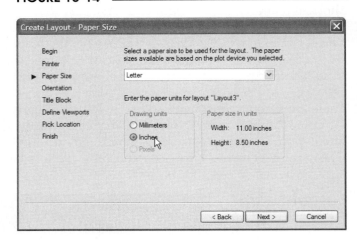

Title Block

In this step, you can select *None* or select from the AutoCAD-supplied title blocks (Fig. 13-15). The selected title block is inserted into the paper space layout. You can also specify whether you want the title block to be inserted as a *Block* or as an *Xref*. Normally inserting the title block as a *Block* object is safer because it becomes a permanent part of the drawing although it occupies a small bit of drawing space (increased file size). An *Xrefed* title block saves on file size because the actual title block is only referenced, but can cause problems if you send the drawing to a client who does not have access to the same referenced drawing. (See Chapter 20 for more information on these subjects.) You can also create your own title block drawings and save them in the Template folder so they will appear in this list.

FIGURE 13-15

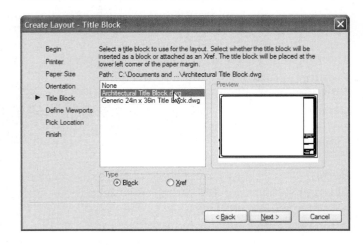

Define Viewports

This step has two parts (Fig. 13-16). First select the type and number of viewports to set up. Use *None* if you want to set up viewports at a later time or want to create a non-rectangular viewport. Normally, use *Single* if you need only one rectangular viewport. Use *Std. 3D Engineering Views* if you have a 3D model. The *Array* option creates multiple viewports with the number of *Rows* and *Columns* you enter in the edit boxes below. Second, determine the *Viewport Scale*. This setting determines the scale that the model space geometry will be displayed in the viewport (ratio of paper space units to model space units).

FIGURE 13-16

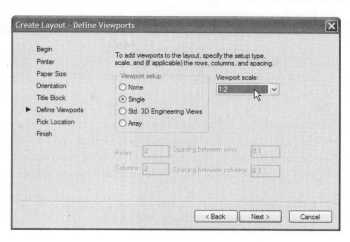

Pick Location

In this step, pick *Select Location* to temporarily exit the wizard and return to the layout to pick diagonal corners for the viewport(s) to fit within (Fig. 13-17). Normally, you do not want the corners of the viewport(s) to overlap the title block or border. You can bypass this step to have AutoCAD automatically draw the viewport border at the extents of the printable area.

FIGURE 13-17

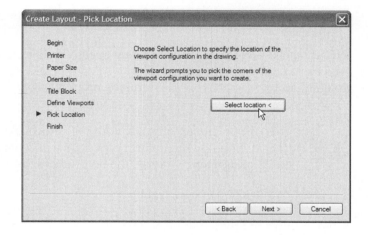

Finish

This is not really a step, rather a confirmation (not shown). In all other steps, you can use *Back* to go backward in the process to change some of the specifications. In this step, *Back* is disabled, so you can only select *Finish* or *Cancel*.

At this time, the new layout is activated so you can see the setup you specified. If you want to make changes to the plot device, paper size, orientation, or viewport scale, you can do so by using the *Page Setup* dialog box. You can also change the viewport sizes or configuration using *Vports* or grips (explained later in this chapter).

Layout

Pull-down Menu	Command (Type)	Alias (Type)	Short-cut	Screen (side) Menu	Tablet Menu
Insert *Layout >* *New Layout*	*Layout*	*LO* or *-LO*	...	...	...

If you prefer to create a layout without using the *Layout Wizard*, use the *Layout* command by typing it or selecting it from the *Insert* pull-down menu or *Layout* toolbar. Using the wizard leads you through all the steps required to create a new layout including selecting plot device, paper size and orientation, and creating viewports. Use the *Layout* command to create viewports "manually" (if your system is not set to automatically create viewports). You will then have to use *Vports* to create viewports in the layout.

It is a good idea to ensure the correct default plot device is specified in the *Options* dialog box before you use *Layout* to create a new layout. See the *New* option for more information. Besides creating a new layout, using a template, and copying an existing layout, you can rename, save, and delete layout.

> Command: **layout**
> Enter layout option [Copy/Delete/New/Template/Rename/SAveas/Set/?] <set>:

Copy

Use this option if you want to copy an existing layout including its plot specifications. If you do not provide the name of a layout to copy, AutoCAD assumes you want to copy the active layout tab. If you do not assign a new name, AutoCAD uses the name of the copied layout (*Floor Plan*, for example) and adds an incremental number in parentheses, such as *Floor Plan (2)*.

> Enter layout to copy <current>:
> Enter layout name for copy <default>:

Delete

Use this option to *Delete* a layout. If you do not enter a name, the most current layout is deleted by default.

> Enter name of layout to delete <current>:

Wildcard characters can be used when specifying layout names to delete. You can select several *Layout* tabs to delete by holding down Shift while you pick. If you select all layouts to delete, all the layouts are deleted, and a single layout tab remains named *Layout1*. The *Model* tab cannot be deleted.

New

 This option creates a new layout tab. AutoCAD creates a new layout from scratch based on settings defined by the default plot device and paper size (in the *Options* dialog box). If you choose not to assign a new name, the layout name is automatically generated (*Layout3*, for example).

> Enter new layout name <Layout#>:

Keep in mind that when the new layout is created, its size (*Limits*) and shape are determined by the *Use as Default Output Device* setting specified in the *Plot and Publish* tab of the *Options* dialog. Therefore, it is a good idea to ensure the correct plot device is specified before you use the *New* option. You can, however, change the layout at a later time using the *Page Setup* dialog box, assuming that *Use Plot Device Paper Size* is checked in the *Plot and Publish* tab of the *Options* dialog box (see *Options* in this chapter).

Template

 You can create a new template based on an existing layout in a template file (.DWT) or drawing file (.DWG) with this option. The layout, viewports, associated layers, and all the paper space geometry in the layout (from the specified template or drawing file) are inserted into the current drawing. No dimension styles or other objects are imported. The standard file navigation dialog box (not shown) is displayed for you to select a file to use as a template. Next, the *Insert Layout* dialog box appears for you to select the desired layout from the drawing (Fig. 13-18).

FIGURE 13-18

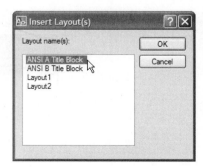

Rename

Use *Rename* to change the name of an existing layout. The last current layout is used as the default for the layout to rename.

    Enter layout to rename <current>:
    Enter new layout name:

Layout names can contain up to 255 characters and are not case sensitive. Only the first 32 characters are displayed in the tab.

Save

This option creates a .DWT file. The file contains all of the (paper space) geometry in the layout and all of the plot settings for the layout, but no model space geometry. All *Block* definitions appearing in the layout, such as a titleblock, are also saved to the .DWT file, but not unused *Block* definitions. All layouts are stored in the template folder as defined in the *Options* dialog box. The last current layout is used as the default for the layout to save.

    Enter layout to save to template <current>:

The standard file selection dialog box is displayed in which you can specify a file name for the .DWT file. When you later create new layout from a template (using the *Template* option of *Layout*), all the saved information (plot settings, layout geometry and *Block* definitions) are imported into the new layout.

Set

This option simply makes a layout current. This option has identical results as picking a layout tab.

? (List Layouts)

Use this option to list all the layouts when you have turned off the layout tabs in the *Display* tab of the *Options* dialog box. Otherwise, layout tabs are visible at the bottom of the drawing area.

Inserting Layouts with AutoCAD DesignCenter

A feature in AutoCAD, called DesignCenter, allows you to insert content from any drawing into another drawing. Content that can be inserted includes drawings, *Blocks*, *Dimstyles*, *Layers*, *Layouts*, *Linetypes*, *Tablestyles*, *Textstyles*, *Xrefs*, raster images, and URLs (web site addresses). If you have created a layout in a drawing and want to insert it into the current drawing, you can use the *Template* option of *Layout* (as previously explained) or use DesignCenter to drag and drop the layout name into the drawing. With DesignCenter, only layouts from .DWG files (not .DWT files) can be imported. See Chapter 20 Blocks, DesignCenter, and Tool Palettes for more information.

Setting Up the Layout

When you do not use the *Layout Wizard*, the steps involved in setting up the layout must be done individually. In cases when you need flexibility, such as creating non-rectangular viewports or when some factors are not yet known, it is desirable to take each step individually. These steps are listed in "Guidelines for Using Layouts and Viewports."

Options

Pull-down Menu	Command (Type)	Alias (Type)	Short-cut	Screen (side) Menu	Tablet Menu
Tools Options...	Options	OP	(Default) Options...	TOOLS2 Options	Y,10

When you create *New* layouts using the *Layout* command, the size and shape of the layout is determined by the selected plot device, paper size, and paper size. Default settings are made in the *Plot and Publish* tab of the *Options* dialog box (Fig. 13-19). Although most settings concerned with plotting a layout can later be changed using the *Page Setup* dialog box, it may be efficient to make the settings in the *Options* dialog box <u>before</u> creating layouts from scratch. Four settings in this dialog box affect layouts.

FIGURE 13-19

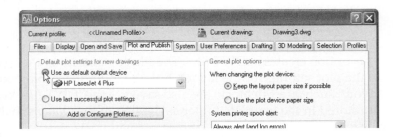

Default Plot Settings for New Drawings

Your choice in this section determines what plot device is used to automatically set the paper size (paper space *Limits*) when new layouts are created. Your setting here applies to <u>new drawings and the current drawing</u>. (Paper size information for each device is stored either in the plotter configuration file [.PC3] or in the default system settings if the output device is a system printer.)

Use as Default Output Device
The selected device is used to determine the paper size (paper space *Limits*) that are set automatically when a new layout is created in the current drawing and for new drawings. This drop-down list contains all configured plot or print devices (any plotter configuration files [PC3] and any system printers that are configured in the system).

Use Last Successful Plot Settings
This button uses the plotting settings (device and paper size) according to those of the last successful plot that was made on the system (a plot must be made from the layout for this option to be valid) and applies them to new layouts that are created. This setting affects new layouts created for the current drawing as well as new drawings. This option overrides the output device appearing in the *Use as Default Output Device* drop-down list.

General Plot Options

These options apply only when plot devices are changed for existing layouts <u>and</u> the layout has a viewport created. Plot devices for layouts can be changed using the *Page Setup* or *Plot* dialog boxes. When you change the plot device for an existing layout, the paper size (*Limits* setting) of the layout is determined by your choice in this section.

Keep the Layout Paper Size if Possible
If this button is pressed, AutoCAD attempts to use the paper size initially specified for the layout if you decide later to change the plot device for the layout. This option is useful if you have two devices that can use the same size paper as specified in the layout, so you can select either device to use without affecting the layout. However, if the selected output device cannot plot to the previously specified paper size, AutoCAD displays a warning message and uses the paper size specified by the plot device you select in the *Page Setup* or *Plot* dialog box. If no viewports have been created in the layout, you can change the plot device and automatically reset the layout limits without consequence. This button sets the *PAPERUPDATE* system variable to 0 (layout paper size is not updated).

Use the Plot Device Paper Size
As the alternative to the previous button, this one sets the paper size (*Limits*) for an existing layout to the paper size of whichever device is selected for that layout in the *Page Setup* or *Plot* dialog box. This option sets *PAPERUPDATE* to 1.

Pagesetup

Pull-down Menu	Command (Type)	Alias (Type)	Short-cut	Screen (side) Menu	Tablet Menu
File *Page Setup Manager...*	*Pagesetup*	...	...	...	V,25

Plot specifications that you make and save for a layout are called "page setups." There are two separate dialog boxes that are used for saving a page setup. Using the *Pagesetup* command produces the *Page Setup Manager* (Fig. 13-20) which is a "front end" for the *Page Setup* dialog box. The *Page Setup* dialog box (see Fig. 13-21) is used to select the plot specifications for the layout such as plot device, paper size, plot scale, etc. All settings made in the *Page Setup Manager* and related dialog box are saved with the layout. Therefore, the plot settings are prepared when you are ready to make a plot or print.

If you have created a layout using the *Layout* command and the desired settings were made in the *Options* dialog box as explained in the previous section, the next step in setting up a layout is using the *Page Setup Manager* and the related *Page Setup* dialog box. If you used the *Layout Wizard*, you can optionally use the *Page Setup Manager* and related dialog box to alter some of those settings.

To save a page setup for a layout, first activate the desired layout tab, then use *Pagesetup* to produce the *Page Setup Manager*. The active layout tab's name appears highlighted in the *Page setups* list. Next, select the *Modify* option as shown in Figure 13-20 to produce the *Page Setup* dialog box.

When the *Page Setup* dialog box appears (Fig. 13-21), specify the following plotting options:

1. Select a plotter or printer in the *Printer Plotter* section.

FIGURE 13-20

FIGURE 13-21

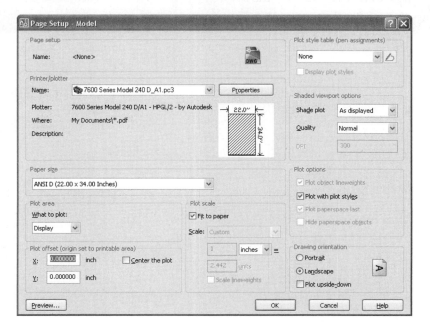

2. Select the desired *Paper size*.
3. Ensure the *Plot scale* is set to 1:1.
4. Select any other desired options such as *Drawing orientation, Plot options*, etc.
5. Select the *OK* button, then *Close* the *Page Setup Manager* to save the settings for the layout.

Remember that the size of the layout (paper space *Limits*) is automatically set based on *Paper Size* and *Drawing Orientation* you select here. Depending on your selection of *Keep the Layout Size if Possible* or *Use the Plot Device Paper Size* in the *Plotting* tab of the *Options* dialog box (see previous section), the layout size and shape may or may not change to the new plot device settings. You can, however, select a new plot device, paper size, or orientation <u>after creating the layout, but before creating viewports</u>.

TIP

Since the layout size and shape may change with a new device, viewports that were previously created may no longer fit on the page. In that case, the viewports must be changed to a new size or shape. Alternately, you can *Erase* the viewports and create new ones. Either of these choices is not recommended. Instead, ensure you have the desired plot device and paper settings <u>before creating viewports</u>. (See the "Guidelines for Using Layouts and Viewports.")

Plot Scale is almost always set to *1:1* since the layout represents the sheet you will be plotting on. The model geometry that appears in the layout (in a viewport) will be scaled by setting a viewport scale (see "Scaling Viewport Geometry").

For more information on the *Page Setup Manager* and the *Page Setup* dialog box, such as how to apply an existing page setup to the active layout, see Chapter 14.

USING VIEWPORTS IN PAPER SPACE

If you use the *Layout Wizard*, viewports can be created during that process. If you do not use the wizard, viewports should be created for a layout only after specifying the plot device, paper size, and orientation in the *Page Setup* dialog box or through the default plot options set in the *Options* dialog box. You can create viewports with the *Vport* or *-Vport* commands. The Command line version (*-Vports*) produces several options not available in the dialog box, including options to create non-rectangular viewports.

Vports

Pull-down Menu	Command (Type)	Alias (Type)	Short-cut	Screen (side) Menu	Tablet Menu
View Viewports >	*Vports* or *-Vports*	...	...	*VIEW 1 Vports*	M,3 and M,4

The *Vports* command produces the *Viewports* dialog box. Here you can select from several configurations to create rectangular viewports as shown in Figure 13-22. Making a selection in the *Standard Viewports* list on the left of the dialog box in turn displays the selected *Preview* on the right.

After making the desired selection from the dialog box, AutoCAD issues the following prompt.

    Command: **vports**
    Specify first corner or [Fit] <Fit>:

FIGURE 13-22

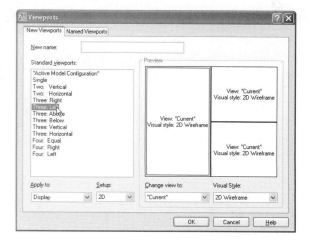

You can pick two diagonal corners to specify the area for the viewports to fill. Generally, pick just inside the titleblock and border if one has been inserted. If you use the *Fit* option, AutoCAD automatically fills the printable area with the specified number of viewports.

The *Setup: 3D* and *Change View to:* options are intended for use with 3D models.

You may notice that the *Viewports* dialog box is the same dialog box that appears when you use *Vports* while in the *Model* tab. However, when *Vports* dialog is used in the *Model* tab, <u>tiled</u> viewports are created, whereas when the *Vports* is used in a layout (paper space), paper space viewports are created.

Paper space viewports differ from tiled viewports. AutoCAD treats paper space viewports created with *Vports* as <u>objects</u>. Like other objects, paper space viewports can be affected by most editing commands. For example, you could use *Vports* to create one viewport, then use *Copy* or *Array* to create other viewports. You could edit the size of viewports with *Stretch* or *Scale*. You can use Grips to change paper space viewports. Additionally, you can use *Move* (Fig. 13-23) to relocate the position of viewports. Delete a viewport using *Erase*. To edit a paper space viewport, you must be in paper space to PICK the viewport objects (borders).

Paper Space viewports can also overlap. This feature makes it possible for geometry appearing in different viewports to occupy the same area on the screen or on a plot.

NOTE: Avoid creating one viewport completely within another's border because visibility and selection problems may result.

Another difference between tiled (*Model* tab) viewports and paper space viewports is that viewports in layouts can have space between them. The option in the bottom left of the *Viewports* dialog box called *Viewport Spacing* allows you to create space between the viewports (if you choose any option other than *Single*) by entering a value in the edit box. Figure 13-24 illustrates a layout with three viewports after entering .5 in the *Viewport Spacing* edit box. Also note several of the *-Vports* command options are available in the right-click shortcut menu that appears when a viewport object is selected.

FIGURE 13-23

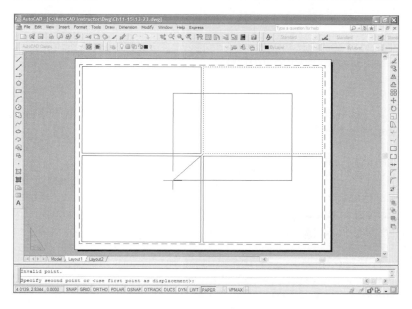

FIGURE 13-24

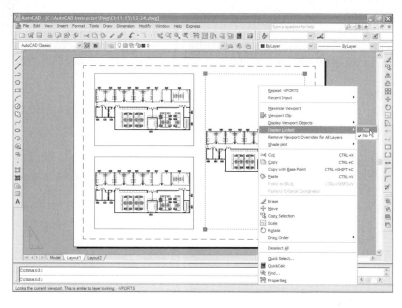

Using the -*Vports* command (with a hyphen prefix) invokes the Command line version and produces the following options.

> Command: -**vports**
> Specify corner of viewport or [ON/OFF/Fit/Shadeplot/Lock/Object/Polygonal/Restore/Layer/2/3/4] <Fit>:

The default option (Specify corner of viewport) allows you to pick two diagonal corners for one viewport. The other options are as follows. Remember to select the viewport objects (borders) while in paper space when AutoCAD prompts you to select a viewport.

ON

On turns on the display of model space geometry in the selected viewport. Select the desired viewport to turn on.

OFF

Turn off the display of model space geometry in the selected viewport with this option. Select the desired viewport to turn off.

Fit

Fit creates a new viewport and fits it to the size of the printable area. The new viewport becomes the current viewport.

Shadeplot

If you are displaying a 3D surface or solid model, *Shadeplot* allows you to select how you want to print the current viewport: *As Displayed*, *Wireframe*, *Hidden*, *Visual Styles*, or *Rendered*. (See Chapter 28, 3D Basics, Navigation, and Visual Styles, for more information.)

Lock

Use this option to <u>lock the scale</u> of the current viewport. Turn viewport locking *On* or *Off*. Double-clicking inside a viewport and then using *Zoom* normally changes the scale of a viewport. If *Lock* is used on the viewport, the specified scale of the viewport cannot be changed (unless viewport locking is turned *Off*).

Object

You can use *Object* to convert a closed *Polyline*, *Ellipse*, *Spline*, *Region*, or *Circle* into a viewport (border). The polyline you specify must be closed and contain at least three vertices. It can be self-intersecting and can contain arcs as well as line segments.

Polygonal

Use this option to create an irregularly shaped viewport defined by specifying points. You can define the viewport by straight line or arc segments.

Restore

Use this option to restore a previously created named viewport configuration.

LAyer

The *LAyer* option allows you to reset viewport layer property overrides back to global properties.

2/3/4

These options create a number of viewports within a rectangular area that you specify. The 2 option allows you to arrange 2 viewports either *Vertically* or *Horizontally*. The 4 option automatically divides the specified area into 4 equal viewports. The possible configurations using the 3 option are displayed if you use the *Viewports* dialog box. Using the 3 option yields the following prompt:

Horizontal/Vertical/Above/Below/Left/<Right>:

Pspace

Pull-down Menu	Command (Type)	Alias (Type)	Short-cut	Screen (side) Menu	Tablet Menu
...	*Pspace*	*PS*	...	*VIEW 1* *Pspace*	*L,5*

The *Pspace* command switches from model space (inside a viewport) to paper space (outside a viewport). The cursor is displayed in paper space and can move <u>across the entire screen</u> when in paper space, while the cursor appears only inside the current viewport if the *Mspace* command is used. This command accomplishes the same action as double-clicking outside the viewport in a layout or using the *MODEL/PAPER* toggle on the Status Bar.

If you type this command while in the *Model* tab, the following message appears:

Command: **pspace**
PSPACE
** Command not allowed in Model Tab **

Mspace

Pull-down Menu	Command (Type)	Alias (Type)	Short-cut	Screen (side) Menu	Tablet Menu
...	*Mspace*	*MS*	...	*VIEW* *Mspace*	*L,4*

The *Mspace* command switches from paper space to <u>model space inside a viewport</u>, similar to double-clicking inside a viewport. If several viewports exist in the layout when you use the command, you are switched to the last active viewport. The cursor appears only in that viewport. The current viewport displays a heavy border. You can switch to another viewport (make another current) by PICKing in it. To use *Mspace*, there must be at least one viewport in the layout, and it must be *On* (not turned *Off* by the *-Vports* command) or AutoCAD issues a message and cancels the command.

Model

Pull-down Menu	Command (Typo)	Alias (Type)	Short-cut	Screen (side) Menu	Tablet Menu
...	*Model*	...	...	...	...

The *Model* command accomplishes the same action as selecting the *Model* tab; that is, *Model* makes the *Model* tab active. If you issue this command while you are in the *Model* tab, nothing happens.

Scaling the Display of Viewport Geometry

Paper space (layout) objects such as title block and border should correspond to the paper at a 1:1 scale. A paper space layout is intended to represent the print or plot sheet. *Limits* in a <u>layout</u> are automatically set to the exact paper size, and the finished drawing is plotted from the layout to a scale of 1:1. The model space geometry, however, should be true scale in real-world units, and *Limits* in <u>model space</u> are generally set to accommodate that geometry.

When model space geometry appears in a viewport in a layout, the size of the <u>displayed</u> geometry can be controlled so that it appears and plots in the correct scale. There are three ways to scale the display of model space geometry in a paper space viewport. You can (1) use the *VP Scale* pop-up list, (2) use the *Viewport Scale Control* drop-down list, (3) use the *Properties* palette, or (4) use the *Zoom* command and enter an *XP* factor. Once you set the scale for the viewport, use the *Lock* option of *-Vports* or the *Properties* palette to lock the scale of the viewport so it is not accidentally changed when you *Zoom* or *Pan* inside the viewport.

When you use layouts, the *Plot Scale* specified in the *Plot* dialog box and *Page Setup* dialog box dictates the scale for the layout, which <u>should normally be set to 1:1</u>. Therefore, <u>the viewport scale actually determines the plot scale of the geometry in a finished plot</u>. Set the viewport scale to the reciprocal of the "drawing scale factor," as explained in Chapter 12, Advanced Drawing Setup.

VP Scale Pop-up List

The *VP Scale* pop-up list is located in the lower-right corner of the screen (see Figs. 13-6 and 13-7). To set the viewport scale (scale of the model geometry in the viewport), first make the viewport current by double-clicking in it or typing *Mspace*. When the crosshairs appear in the viewport and the viewport border is highlighted (a wide border appears), select the desired scale from the list. The entries in the list can be deleted, added, or edited by selecting *Custom...*, which produces the *Edit Scale List* dialog box (see "*Scalelistedit*" near the end of this chapter).

Viewport Scale Control Drop-Down List

The *Viewport Scale Control* drop-down list is located in the *Viewports* toolbar. Invoke the *Viewports* toolbar using the *Toolbars...* option at the bottom of the *View* pull-down menu.

The *Viewport Scale Control* drop-down list provides only standard scales for decimal, metric, and architectural scales. <u>You can select from the list or enter a value.</u> Values can be entered as a decimal, a fraction (proper or improper), a proportion using a colon symbol (:), or an equation using an equal symbol (=). Feet (') and inch (") unit symbols can also be entered. For example, if you wanted to scale the viewport geometry at 1/2"=1', you could enter any of the following, as well as other, values.

1/2"=1'	1:24
.5"=1'	1=24
1/2=12	.0417
1/24	

Remember that the viewport scale is actually the desired plot scale for the geometry in the viewport, which is the reciprocal of the "drawing scale factor" (see Chapter 12).

Properties

The *Properties* palette can also be used to specify the scale for the display of model space geometry in a viewport. Do this by first double-clicking <u>in paper space</u> or issuing the <u>*Pspace*</u> command, then invoking *Properties*. When (or before) the dialog box appears, select the viewport object (border). Ensure the word "Viewport" appears in the top of the box (Fig. 13-25).

Under *Standard Scale*, select the desired scale from the drop-down list. This method offers only standard scales. Alternately, you can enter any values in the *Custom Scale* box. (See *Properties*, Chapter 16.) Similar to using the *Viewport Scale Control* edit box, you can enter values as a decimal or fraction and use a colon symbol or an equal symbol. Feet and inch unit symbols can also be entered. You can also *Lock* the viewport scale using the *Display Locked* option in the palette. (See *Properties*, Chapter 16.)

FIGURE 13-25

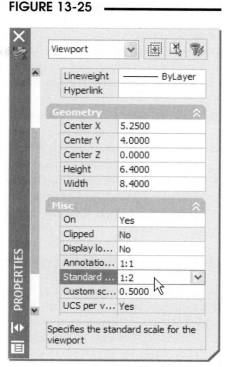

ZOOM XP Factors

To scale the display of model space geometry relative to paper space (in a viewport), you can also use the *XP* option of the *Zoom* command. *XP* means "times paper space." Thus, model space geometry is *Zoomed* to some factor "times paper space."

Since paper space is set to the actual size of the paper, <u>the *Zoom XP* factor that should be used for a view-port</u> <u>is equivalent to the plot scale</u> that would otherwise be used for plotting that geometry in model space. The *Zoom XP* factor is the reciprocal of the "drawing scale factor." *Zoom XP* <u>only</u> while you are "in" the desired model space viewport. Fractions or decimals are accepted.

For example, if the model space geometry would normally be plotted at 1/2"=1" or 1:2, the *Zoom* factor would be .5XP or 1/2XP. If a drawing would normally be plotted at 1/4"=1', the *Zoom* factor would be 1/48XP. Other examples are given in the following list.

1:5	*Zoom* .2XP or 1/5 XP
1:10	*Zoom* .1XP or 1/10XP
1:20	*Zoom* .05XP or 1/20XP
1/2"=1"	*Zoom* 1/2XP
3/8"=1"	*Zoom* 3/8XP
1/4"=1"	*Zoom* 1/4XP
1/8"=1"	*Zoom* 1/8XP
3"=1'	*Zoom* 1/4XP
1"=1'	*Zoom* 1/12XP
3/4'=1'	*Zoom* 1/16XP
1/2"=1'	*Zoom* 1/24XP
3/8"=1'	*Zoom* 1/32XP
1/4"=1'	*Zoom* 1/48XP
1/8"=1'	*Zoom* 1/96XP

Refer to the "Tables of Limits Settings," Chapter 14, for other plot scale factors.

Locking Viewport Geometry

Once the viewport scale has been set to yield the correct plot scale, you should lock the scale of the view-port so it is not accidentally changed when you *Zoom* or *Pan* inside the viewport. If viewport lock is on and you use *Zoom* or *Pan* when the viewport is active, AutoCAD automatically switches to paper space (*Pspace*) for zooming and panning.

You can lock the viewport scale by the following methods.

1. Toggle the small *Lock/Unlock Viewport* icon located left of the *VP Scale* pop-up list (see Fig. 13-6).
2. Use the *Lock* option of the *-Vports* command (see *-Vports*).
3. Use the *Display Locked* setting in the *Properties* palette (see Fig. 13-25).
4. Use the *Display Locked* option in the right-click shortcut edit menu that appears when a viewport object is selected (see Fig. 13-24).

Linetype Scale in Viewports—*PSLTSCALE*

The *LTSCALE* (linetype scale) setting controls how hidden, center, dashed, and other non-continuous lines appear in the *Model* tab. *LTSCALE* can be set to any value. The *PSLTSCALE* (paper space linetype scale) setting determines if those lines appear and plot the same in *Layout* tabs as they do in the *Model* tab. *PSLTSCALE* can be set to 1 or 0 (on or off).

If *PSLTSCALE* is set to 0, the non-continuous line scaling that appears in *Layout* tabs (and in viewports) is the same as that in the *Model* tab. In this way, the linetype spacing always looks the same relative to model space units, no matter whether you view it from the *Model* tab or from a viewport in a *Layout*. For example, if in the *Model* tab a particular center line shows one short dash and two long dashes, it will look the same when viewed in a viewport in a layout—one short dash and two long dashes. When *PSLTSCALE* is 0, linetype scaling is controlled exclusively by the *LTSCALE* setting. A *PSLTSCALE* setting of 0 is recommended for most drawings unless they contain multiple viewports or layouts.

If *PSLTSCALE* is set to 1 (the default setting for all AutoCAD template drawings), the linetype scale for non-continuous lines in viewports is automatically changed relative to the viewport scale; therefore, the scale of those lines in *Layout* tabs can appear differently than they do in the *Model* tab. A *PSLTSCALE* setting of 1 is recommended for drawings that contain multiple viewports or layouts, especially when they are at different scales. In such a situation, the line dashes would appear the same size in different viewports relative to paper space, even though the viewport scales were different. Technically speaking, if *PSLTSCALE* is set to 1, *LTSCALE* controls linetype scaling globally for the drawing, but lines that appear in viewports are automatically scaled to the *LTSCALE* times the viewport scale.

Until you gain more experience with layouts and viewports, and in particular creating multiple view-ports, a *PSLTSCALE* setting of 0 is the simplest strategy. This setting also agrees with the strategy used in this text discussed earlier (in Chapter 6 and Chapter 12) for drawing setup. You can change the default *PSLTSCALE* setting of 1 to 0 by typing *PSLTSCALE* at the Command prompt.

CHAPTER EXERCISES

1. *Pagesetup, Vports*

   A. *Open* the **A-METRIC.DWT** template drawing that you created in the Chapter 12
      Exercises. Activate a *Layout* tab. If your *Options* are set as explained in Chapter 6, the
      layout is automatically set up for your default print/plot device. Otherwise, use the
      Page Setup Manager and select the *Modify* button. Then use the *Page Setup* dialog box
      to set the desired print/plot device and to select an 11 x 8.5 paper size.

   B. If a viewport already exists in the layout, *Erase* it. Make **VPORTS** the *Current* layer.
      Then use the *Vports* command and select a *Single* viewport and accept the default (*Fit*)
      option to create one viewport at the maximum size for the printable area. Ensure that
      PSLTSCALE is set to **0** by typing it at the Command prompt. Finally, make the *Model*
      tab active. *Save* and *Close* the drawing.

2. *Pagesetup, Vports*

   A. *Open* the **BARGUIDE** drawing that you set up in the Chapter 12 Exercises. Activate a
      Layout tab. If your *Options* are set as explained in Chapter 6, the layout is automatically
      set up for your default print/plot device. Otherwise, use the *Page Setup Manager* and
      select the *Modify* button. Then use the *Page Setup* dialog box that appears to set the
      desired print/plot device and to select an **11 x 8.5** paper size.

   B. If a viewport already exists in the layout, *Erase* it. Make **VPORTS** the *Current* layer.
      Then use the *Vports* command and select a *Single* viewport and accept the default (*Fit*)
      option to create one viewport at the maximum size for the printable area. Type
      PSLTSCALE and set the value to **0**. Finally, make the *Model* tab active. *Save* and *Close*
      the drawing.

3. **Create a Layout and Viewport for the ASHEET Template Drawing**

   A. *Open* the **ASHEET.DWT** template drawing you created in the Chapter 12 Exercises.
      Activate a *Layout* tab. Unless already set up, use *Pagesetup* to set the desired
      print/plot device and to select an **11 x 8.5** paper size.

   B. If a viewport already exists in the layout, *Erase* it. Make **VPORTS** the *Current* layer.
      Then use the *Vports* command and select a *Single* viewport and accept the default (*Fit*)
      option to create one viewport at the maximum size for the printable area. Ensure that
      PSLTSCALE is set to **0**. Finally, make the *Model* tab active. *Save* and *Close* the ASHEET
      template drawing.

4. **Create a New BSHEET Template Drawing**
    Complete this exercise if your system is configured with a "B" size printer or plotter.

    A. Using the template drawing in the previous exercise, create a template for a standard engineering "B" size sheet. First, use the *New* command. When the *Select a Template File* dialog box appears, locate and select the **ASHEET.DWT.** When the drawing opens, set the model space *Limits* to **17** x **11** and *Zoom All*.

    B. Activate the *Layout 1* tab. *Erase* the existing viewport. Activate the *Page Setup Manager* and select the *Modify* button. In the *Page Setup* dialog box select a print or plot device that can use a "B" size sheet (17 x 11). In the *Paper Size* drop-down list, select *ANSI B (11 x 17 inches)*. Close the dialog box and the *Page Setup Manager*.

    C. Make **VPORTS** the *Current* layer. Then use the *Vports* command and select a *Single* viewport and accept the default (*Fit*) option to create one viewport at the maximum size for the printable area. Finally, make the *Model* tab active. All other settings and layers are okay as they are. Use *Saveas* and save the new drawing as a template (.**DWT**) drawing in your working directory. Assign the name **BSHEET.**

5. **Create Multiple Layouts for Plotting on C and D Size Sheets**
    Complete this exercise if your system is configured with a "C" and "D" sized plotter.

    A. Using the ASHEET.DWT template drawing from an earlier exercise, create a template for standard engineering "C" and "D" size sheets. First, use the *New* command and select the *Template* option from the *Create New Drawing* dialog box. Select the *Browse* button. When the *Select a Template File* dialog box appears, select the **ASHEET.DWT.** When the drawing opens, set the model space *Limits* to **34** x **22** and *Zoom All*.

    B. Activate the *Layout1* tab. *Erase* the existing viewport. Activate the *Page Setup Manager* and select the *Modify* button. In the *Page Setup* dialog box select a plot device that can use a "C" size sheet (22 x 17). In the *Paper Size* drop-down list, select *ANSI C (22 x 17 inches)*. Close the dialog box and the *Page Setup Manager*.

    C. Make **VPORTS** the *Current* layer. Then use the *Vports* command and select a *Single* viewport and use the *Fit* option to create one viewport at the maximum size for the printable area. Right-click on the *Layout 1* tab and select *Rename* from the shortcut menu. Rename the layout to "**C Sheet**."

    D. Activate the *Layout2* tab. *Erase* the existing viewport if one exists. Activate the *Page Setup Manager*, select the *Modify* button, then select a plot device that can use a "D" size sheet (34 x 22). (This can be the same device you selected in step B, as long as it supports a D size sheet.) In the *Paper Size* drop-down list, select *ANSI D (34 x 22 inches)*. Close this dialog box and the *Page Setup Manager*.

    E. Make **VPORTS** the *Current* layer. Then use the *Vports* command and select *Single* to create one viewport. *Rename* the layout "**D Sheet**."

    F. Finally, make the *Model* tab active. All other settings and layers are okay as they are. Use *Saveas* and save the new drawing as a template (.**DWT**) drawing in your working directory. Assign the name **C-D-SHEET.**

6. **Layout Wizard**

In this exercise, you will open an existing drawing and use the *Layout Wizard* to set up a layout for plotting.

A. **Open** the **DB_SAMP** drawing that is located in the AutoCAD 2008\Sample folder. Use the *Saveas* command and assign the name **VP_SAMP** and specify your working folder as the location to save.

B. Create a new layer named **VPORTS** and make it the *Current* layer. Invoke the *Layout Wizard*. In the *Begin* step, enter a name for the new layout, such as **Layout 3**. Select *Next*.

C. In the *Printer* step, the list of available printers in your lab or office appears. The default printer should be highlighted. Choose any device that you know will operate for your lab or office and select *Next*. If you are unsure of which devices are usable, keep the default selection and select *Next*.

D. For *Paper Size*, select the largest paper size supported by your device. Ensure **Inches** is selected unless you are using a standard metric sheet. Select *Next*.

E. Select **Landscape** for the *Orientation*.

F. Select a *Title Block* that matches the paper size you specified if available. Insert the title block as a **Block**. Select *Next*.

G. In the *Define Viewports* step, select a **Single** viewport. Under *Viewport Scale*, select **Scaled to Fit**. Proceed to the next step.

H. Bypass the next step in the wizard, *Pick Location*, by selecting *Next*. Also, when the *Finish* step appears, select *Next*. This action causes AutoCAD to create a viewport that fits the extents of the printable area. The resulting layout should look similar to that shown in Figure 13-26. As you can see, the model space geometry extends beyond the title block. Your layout may appear slightly different depending on the sheet size and title block you selected. **Save** the drawing.

FIGURE 13-26

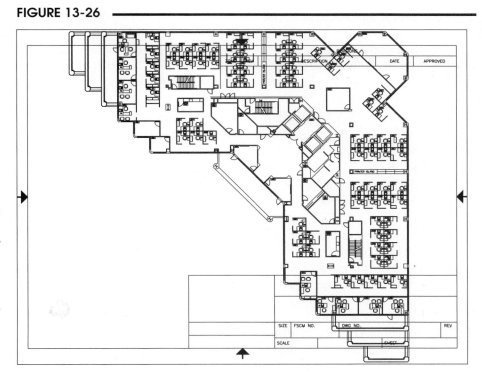

I. Use the *Mspace* command or double-click inside the viewport to activate model space. To specify a standard scale for the geometry in the viewport, use the *VP Scale* pop-up list and select a scale (such as *1/64"=1'*) until the model geometry appears completely within the title block border (Fig. 13-27).

FIGURE 13-27

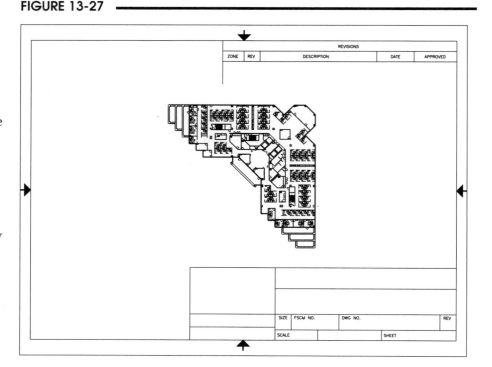

J. *Freeze* layer **VPORTS** so the viewport border does not appear. *Save* the drawing. With the layout active, access the *Plot* dialog box and ensure the *Plot Scale* is set to **1:1**. Select *Plot*.

7. ***Page Setup, Vports, and Zoom XP***

In this exercise, you will use an existing metric drawing, access *Layout1*, draw a border and title block, create one viewport, and use *Zoom XP* to scale the model geometry to paper space. Finally, you will *Plot* the drawing to scale on an "A" size sheet.

A. *Open* the **CBRACKET** drawing you worked on in Chapter 9 Exercises. If you have already created a title block and border, *Erase* them.

B. Invoke the *Options* dialog box and access the *Display* tab. In the *Layout Elements* section (lower left), ensure that *Show Page Setup Manager for New Layouts* is checked and *Create Viewport in New Layouts* is not checked.

C. Select the *Layout1* tab. The *Page Setup Manager* should appear. Select the *Modify* button. In the *Page Setup* dialog box select a print or plot device that will allow you to print or plot on an "A" size sheet (11 x 8.5). In the *Paper Size* drop-down list, select the correct sheet size, then select *Landscape* orientation. Select *OK* in the *Page Setup* dialog box and *Close* the *Page Setup Manager*.

D. Set layer **TITLE** *Current*. Draw a title block and border (in paper space). HINT: Use *Pline* with a *width* of .02, and draw the border with a .5 unit margin within the edge of the *Limits* (border size of 10 x 7.5). Provide spaces in the title block for your school or company name, your name, part name, date, and scale as shown in the following figure. Use *Saveas* and assign the name **CBRACKET-PS**.

E. Create layer **VPORTS** and make it
    Current. Use the *Vports* command to
    create a *Single* viewport. Pick diago-
    nal corners for the viewport at **.5,1.5**
    and **9.5,7**. The CBRACKET drawing
    should appear in the viewport at no
    particular scale, similar to Figure
    13-28.

FIGURE 13-28

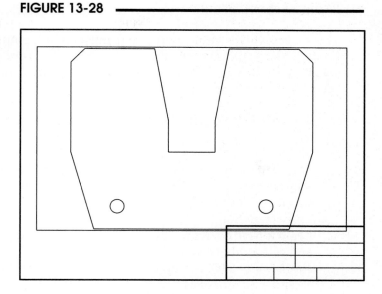

F. Next, use the *Mspace* command, the *MODEL/PAPER* toggle on the Status Bar, or
    double-click inside the viewport to bring the cursor into model space. Use *Zoom* with a
    .03937XP factor to scale the model space geometry to paper space. (The conversion
    from inches to millimeters is 25.4 and from millimeters to inches is .03937. Model space
    units are millimeters and paper space units are inches; therefore, enter a *Zoom XP* factor
    of .03937.)

G. Finally, activate paper space by using
    the *Pspace* command, the
    MODEL/PAPER toggle, or double-
    clicking outside the viewport. Use
    the *Layer Properties Manager* to
    make layer **VPORTS** *non-plottable*.
    Your completed drawing should look
    like that in Figure 13-29. *Save* the
    drawing. *Plot* the drawing from the
    layout at **1:1** scale.

FIGURE 13-29

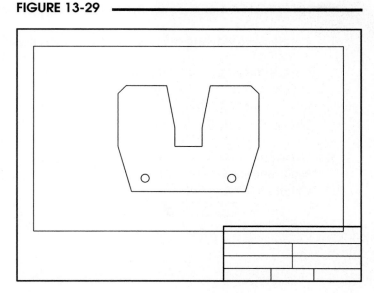

8. **Using a Layout Template**

This exercise gives you experience using a layout template already set up with a title block and border. You will draw the hammer (see Fig. 13-30) and use a layout template to plot on a B size sheet. It is necessary to have a plot or print device configured for your system that can use a "B" size sheet.

A. Start AutoCAD. Begin a *New* drawing by any method. Change the *STARTUP* system variable to **1**.

B. Now access the *Display* tab in the *Options* dialog box. In the *Layout Elements* section (lower left), <u>remove the check for both</u> *Show Page Setup Manager for New Layouts* and *Create Viewport in New Layouts*. Select *OK*.

C. Begin a *New* drawing. In the *Create New Drawing* dialog box that appears, select **Use a Wizard** and select the **Advanced Setup Wizard**. In the first step, *Units*, select *Decimal* units and a *Precision* of **0.00**. Accept the defaults for the next three steps. In the *Area* step, enter a *Width* of **17** and a *Length* of **11**.

D. Right-click on an existing layout tab and use the **From Template** option from the shortcut menu. Locate and select the **ANSI B -Color Dependent Plot Styles** <u>template drawing</u> (.DWT) in the *Select Template From File* dialog box, then select the **ANSI B Title Block** in the *Insert Layout(s)* dialog box. Access the new tab. Note that the template loads a layout with a title block and border in paper space and has one viewport created. Access the **Page Setup Manager** box and select the matching plot device and sheet size for the selected ANSI B title block.

E. Pick the *Model* tab. Set up appropriate *Snap* and *Grid* (for the *Limits* of 17 x 11) if desired. Also set *LTSCALE* and any running *Osnaps* you want to use. Type *PSLTSCALE* and set the value to **0**.

F. Create the following layers and assign linetypes and line weights. Assign colors of your choice.

GEOMETRY	*Continuous*
CENTER	*Center2*

Notice the *Title Block* and *Viewport* layers already exist as part of the template drawing.

G. Draw the hammer according to the dimensions given in Figure 13-30. *Save* the drawing as **HAMMER.**

FIGURE 13-30

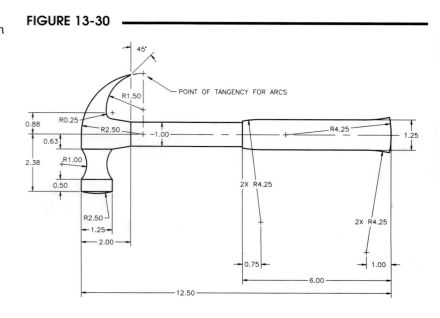

H. When you are finished with the drawing, select the *ANSI B Title Block* tab. Now, scale the hammer to plot to a standard scale. You can use either the *Viewports* toolbar or the *VP Scale* pop-up list to set the viewport scale to **1:1**. Alternately, use *Zoom 1XP* to correctly size model space units to paper space units.

I. Your completed drawing should look like that in Figure 13-31. Use the *Layer Properties Manager* to make layer **Viewports** *non-plottable*. *Save* the drawing. Make a *Plot* from the layout at a scale of **1:1**.

FIGURE 13-31

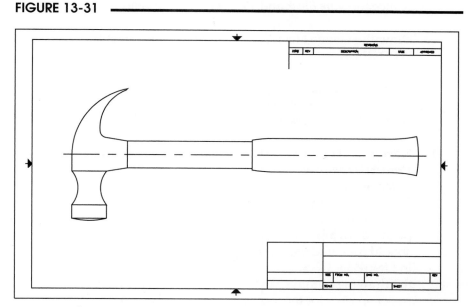

9. Using a Template Drawing

This exercise gives you experience using a template drawing that includes a title block and border. You will construct the wedge block in Figure 13-32 and plot on a B size sheet. It is necessary to have a plot or print device configured for your system that can use a "B" size sheet.

FIGURE 13-32

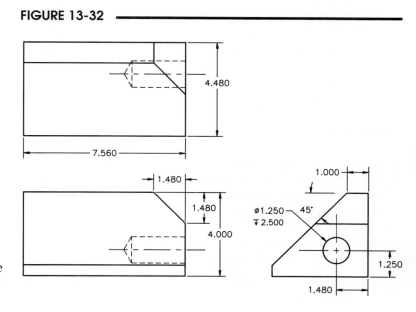

A. Complete steps 8.A. and 8.B. from the previous exercise.

B. Begin a *New* drawing. In the *Create New Drawing* dialog box that appears, select **Use a Template**. Locate and select the *ANSI B -Color Dependent Plot Styles.Dwt*.

C. Select the *ANSI B Title Block* tab. Then invoke the *Page Setup Manager* box and select a matching plot or print device and sheet size for the ANSI B title block.

D. Pick the *Model* tab. Set up model space with *Limits* of **17 x 11**. Set up appropriate *Snap* and *Grid* if desired. Also set *LTSCALE* and any running *Osnaps* you want to use. Type *PSLTSCALE* and set the value to **0**.

E. Create the following layers and assign linetypes and line weights. Assign colors of your choice.

GEOMETRY   *Continuous*
CENTER   *Center2*
HIDDEN   *Hidden2*

Notice the *Title Block* and *Viewport* layers already exist as part of the template drawing.

F. Draw the wedge block according to the dimensions given (Fig. 13-32). *Save* the drawing as **WEDGEBK**.

G. When you are finished with the views, select the *ANSI B Title Block* tab. To scale the model space geometry to paper space, you can use either the *Viewports* toolbar or the *VP Scale* pop-up list to set the viewport scale to **1:2**. You may also need to use *Pan* to center the geometry.

H. Next, use the *Layer Properties Manager* to make layer **Viewports** *non-plottable*. Your completed drawing should look like that in Figure 13-33. *Save* the drawing. From paper space, plot the drawing on a B size sheet and enter a scale of **1:1**.

FIGURE 13-33

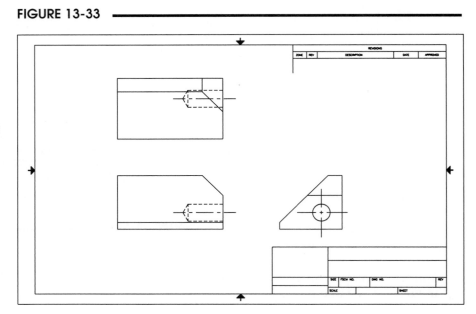

10. **Multiple Layouts**

Assume a client requests that a copy of the HAMMER drawing be faxed to him immediately; however, your fax machine cannot accommodate the B size sheet that contains the plotted drawing. In this exercise you will create a second layout using a layout template to plot the same model geometry on an A size sheet.

A. Open the drawing you created in a previous exercise named **HAMMER**. Access the *Display* tab in the *Options* dialog box. In the *Layout Elements* section (lower left), remove the check for both *Show Page Setup Manager for New Layouts* and *Create Viewport in New Layouts*. Select *OK*.

B. Right-click on an existing layout tab and use the *From Template* option from the shortcut menu. Locate and select the *ANSI A -Color Dependent Plot Styles* template drawing (.DWT) to import. When the layout is imported, click the new *ANSI A Title Block* tab. Note that the template loads a layout with a title block and border in paper space and has one viewport created. Access the *Page Setup Manager* and select the matching print or plot device and sheet size for the selected ANSI A title block.

C. Double-click inside the viewport or use *Mspace* or the *MODEL/PAPER* toggle. Now scale the hammer to plot to a standard scale by using *Zoom 3/4XP* or enter *.75* in the appropriate edit box in the *Viewports* toolbar or *Properties* palette to correctly scale the model space geometry. Use *Pan* in the viewport to center the geometry within the viewport.

D. Your new layout should look like that in Figure 13-34. Access the *ANSI B Title Block* tab you created earlier. Note that you now have two layouts, each layout has settings saved to plot the same geometry using different plot devices and at different scales.

E. Access the new layout tab. *Save* the drawing. Make a *Plot* from the layout at a scale of **1:1** and send the fax.

FIGURE 13-34

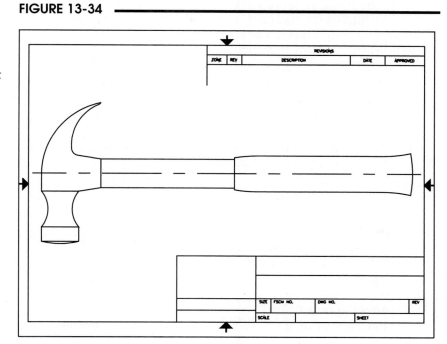

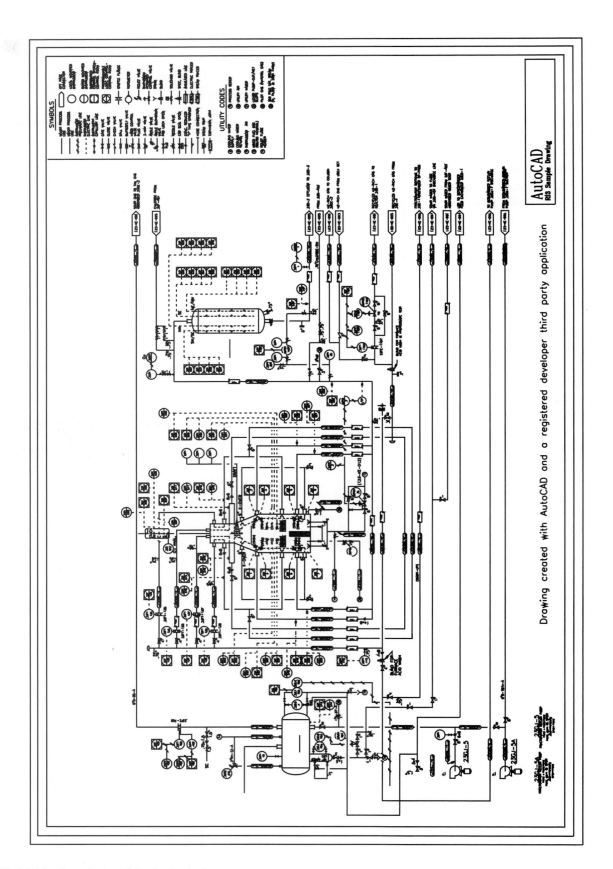

AutoCAD
R13 Sample Drawing

Drawing created with AutoCAD and a registered developer third party application

PNID.DWG, Courtesy of Autodesk, Inc.

PRINTING AND PLOTTING

CHAPTER OBJECTIVES

After completing this chapter you should:

1. know the typical steps for printing or plotting;

2. be able to invoke and use the *Plot* dialog box and *Page Setup Manager*;

3. be able to select from available plotting devices and set the paper size and orientation;

4. be able to specify what area of the drawing you want to plot;

5. be able to preview the plot before creating a plotted drawing;

6. be able to specify a scale for plotting a drawing;

7. know how to set up a drawing for plotting to a standard scale on a standard size sheet;

8. be able to use the Tables of *Limits* Settings to determine *Limits*, scale, and paper size settings;

9. know how to configure plot and print devices in AutoCAD.

CONCEPTS

Several concepts, procedures, and tools in AutoCAD relate to printing and plotting. Many of these topics have already been discussed, such as setting up a drawing to draw true size, creating layouts and viewports, specifying the viewport scale, and specifying the plot or print options. This chapter explains the typical steps to plotting from either the *Model* tab or *Layout* tabs, using the *Plot* dialog box and the *Page Setup Manager*, configuring print and plot devices, and plotting to scale.

In AutoCAD, the term "plotting" can refer to plotting on a plot device (such as a pen plotter or electrostatic plotter) or printing with a printer (such as a laser jet printer). The *Plot* command is used to create a plot or print by producing the *Plot* dialog box. You can plot only by using the *Plot* command (Fig. 14-1).

FIGURE 14-1

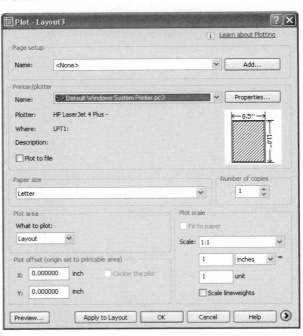

Before making a plot, you specify the parameters of how the drawing will appear on the plot, such as what type of printer or plotter you want to use, the paper size and orientation, what area of the drawing to plot, plot scale, and other options. You can save the plot specifications for each layout. These saved plot specifications are called "page setups." In this way, each layout can have its own page setup—set up to plot or print with a specific device, paper size, orientation, scale, and so on. Therefore, one drawing can be plotted and printed in multiple ways.

Although you can create a plot only by using the *Plot* dialog box, two methods can be used to create and save the plot specifications for a layout:

1. Use the *Plot* dialog box to specify the plot settings and select the *Apply to Layout* button.
2. Use the *Page Setup Manager* and related *Page Setup* dialog box.

These two methods are almost identical since the *Plot* dialog box and the *Page Setup* dialog box have almost the same features and functions except that you cannot plot from the *Page Setup* dialog box. Therefore, for most applications, it is suggested that you use the *Plot* dialog box for saving the page setups and for plotting. This chapter explains the features and distinctions of both methods.

TYPICAL STEPS TO PLOTTING

Assuming your CAD system and plotting devices have been properly configured, the typical basic steps to plotting the *Model* tab or a *Layout* tab are listed below.

1. Use *Save* to ensure the drawing has been saved in its most recent form before plotting (just in case some problem arises while plotting).

2. Make sure the plotter or printer is turned on, has paper (and pens for some devices) loaded, and is ready to accept the plot information from the computer.

3. Invoke the *Plot* dialog box.

4. Select the intended plot device from the *Printer/Plotter* drop-down list. The list includes all devices currently configured for your system.

5. Ensure the desired *Drawing Orientation* and *Paper Size* are selected.

6. Determine and select the desired *Plot Area* for the drawing: *Layout, Limits, Extents, Display, Window,* or *View*.

7. Select the desired scale from the *Scale* drop-down list or enter a *Custom* scale. If no standard scale is needed, select *Scaled to Fit*. If you are plotting from a *Layout*, normally set the scale to 1:1.

8. If necessary, specify a *Plot Offset* or *Center the Plot* on the sheet.

9. If necessary, specify the *Plot Options*, such as plotting with lineweights or plot styles.

10. Always preview the plot to ensure the drawing will plot as you expect. Select a *Preview* to view the drawing objects as they will plot. If the preview does not display the plot as you intend, make the appropriate changes. Otherwise, needless time and media could be wasted.

11. If all settings are acceptable and you want to save the page setup (plot specifications) with the layout (recommended), select the *Apply to Layout* button.

12. If you want to make the plot at this time, select the *OK* button. If you do not want to plot, select *Cancel* and the settings will be saved (if you used *Apply to Layout*).

PRINTING, PLOTTING, AND SAVING PAGE SETUPS

Plot

Pull-down Menu	Command (Type)	Alias (Type)	Short-cut	Screen (side) Menu	Tablet Menu
File *Plot...*	*Plot* *or -Plot*	*PRINT*	*Ctrl+P*	*FILE* *Plot*	*W,25*

Using *Plot* invokes the *Plot* dialog box (see Fig. 14-2 on the next page). This dialog box has a collapsed and expanded mode controlled by the arrow button in the lower-right corner. The *Plot* dialog box allows you to set plotting parameters such as plotting device, paper size, orientation, and scale. Settings made to plotting options can be saved for the *Model* tab and each *Layout* tab in the drawing so they do not have to be re-entered the next time you want to make a plot.

Page Setup

The *Name* drop-down list at the top left corner of the *Plot* dialog box gives the page setups saved in the drawing. You can select a previously saved page setup from the list to apply those settings to the current layout. Although the default page setup is listed as *<None>*, it will be listed as the current tab name (such as *Layout1*) once you save it. You can also assign a specific name using the *Add* button. The list can contain saved page setups (such as *Layout1* or *Layout2*) and "named" (assigned) setups (such as *Setup1, Setup2*, or other assigned names). You can also *Import* page setups from other drawings that have assigned names.

Add

This option produces the *Add Page Setup* dialog box (not shown) where you can assign a name for the current page setup. You can accept the default assigned name (*Setup1*) or enter a name of your choice. You must assign a name if you want to *Import* the page setup into another drawing.

Printer/plotter

Many devices (printers and plotters) can be configured for use with AutoCAD. Once configured, those devices appear in the *Printer/plotter* list. For example, you may configure an "A" size plotter, a "D" size plotter, and a laser printer. Any one of those devices could then be used to plot the current drawing.

FIGURE 14-2

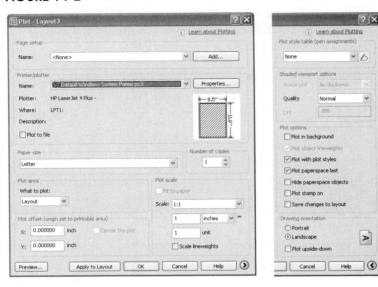

When you install AutoCAD, it automatically configures itself to use the Windows system print devices, a *DWF6 ePlot*, two PDF devices, and two *PublishToWeb* devices. Additional devices must be configured within AutoCAD by using the *Plotter Manager* (see "Configuring Plotters and Printers" near the end of this chapter).

Name:
Select the desired device from the list. The list is composed of all previously configured devices for your system.

Properties...
This button invokes the *Plotter Configuration Editor*. In the editor, you can specify a variety of characteristics for the specific plot device shown in the *Name:* drop-down list. The *Plotter Configuration Editor* is also accessible through the *Plotter Manager*.

Plot to File
Choosing this option writes the plot to a file instead of making a plot. This action generally creates a .PLT file type; however, the format of the file (for example, PCL or HP/GL language) depends on the device brand and model that is configured. When you use the *OK* button to create the plot, the *Browse for Plot File* dialog box (not shown) appears for you to specify the name and location for the plot file. The plot file can be printed or plotted later <u>without</u> AutoCAD, assuming the correct interpreter for the device is available.

Paper Size and Number of Copies

The options that appear in this list depend on the selected printer/plotter device. Normally, several paper sizes are available. Select the desired *Paper Size* from the drop-down list, then select the *Number of Copies* you want to create.

Plot Area

In the *What to Plot:* list section you can select what aspect of the drawing you want to plot, described next.

Limits or Layout
This option changes depending on whether you invoke the *Plot* dialog box while the *Model* tab or a *Layout* tab is current. If the model tab is current, this selection plots the area defined by the *Limits* command. If a *Layout* tab is current, the layout is plotted as defined by the selected paper size.

Extents
Plotting *Extents* is similar to *Zoom Extents*. This option plots the entire drawing (all objects), disregarding the *Limits*. Use the *Zoom Extents* command in the drawing to make sure the extents have been updated before using this plot option.

Display
This option plots the current display on the screen. If using viewports, it plots the current viewport display.

View
With this option you can plot a named view previously saved using the *View* command.

Window
Window allows you to plot any portion of the drawing. You must specify the window by picking or supplying coordinates for the lower-left and upper-right corners.

Plot Offset
Plotting and printing devices cannot plot or print to the edge of the paper. Therefore, the lower left corner of your drawing will not plot at exactly the lower left corner of the paper. You can choose to reposition, or offset, the drawing on the paper by entering positive or negative values in the X and Y edit boxes. Home position for plotters is the <u>lower-left</u> corner (landscape orientation) and for printers it is the <u>upper-left</u> corner of the paper (portrait orientation). If plotting from the *Model* tab, you can choose to *Center the Plot* on the paper.

Plot Scale
You can select a standard scale from the drop-down list or enter a custom scale.

Fit to Paper
The *Fit to Paper* option is disabled if a layout tab is active. If the *Model* tab is active, the *Fit to Paper* option is checked and the *Scale* section is disabled. The drawing is automatically sized to fit on the sheet based on the specified area to plot (*Display*, *Limits*, *Extents*, etc.). Changing the *inches* or *mm* specification changes only the units of measure for other areas of the *Plot* dialog box.

Scale
This section provides the scale drop-down list for you to select a specific scale for the drawing to be plotted. These options are the same as those appearing in all other scale drop-down lists (such as the *VP Scale* pop-up list, *Viewport* toolbar, and *Properties* palette). First ensure the correct paper units are selected (*inches* or *mm*), then select the desired scale from the list.

If you are plotting from a *Layout* tab, select 1:1 from the *Scale* drop-down list since you almost always want to plot paper space at 1:1 scale. (To set the scale for the viewport geometry in a layout, use the *Viewport* toolbar, *Zoom XP* factor, or *Properties* palette as explained in Chapter 13.) If you are plotting the *Model* tab, select an option directly from the *Scale* drop-down list to specify the desired plot scale for the drawing. In either case, the plot scale for the drawing is equal to the reciprocal of the drawing scale factor and can be determined by referencing the Tables of *Limits* Settings. (See "Plotting to Scale.")

Custom Edit Boxes
Instead of selecting from the drop-down list, you can enter the desired ratio in the edit boxes. Decimals or fractions can be entered. The first edit box represents paper units and the second represents drawing units. For example, to prepare for a plot of 3/8 size (3/8"=1"), the following ratios can be entered: 3=8, 3/8=1, or .375=1 (paper units =drawing units). For guidelines on plotting a drawing to scale, see "Plotting to Scale" in this chapter.

Scale Lineweights
This option scales the lineweights for plots. When checked, lineweights are scaled proportionally with the plot scale, so a 1mm lineweight, for example, would plot at .5mm when the drawing is plotted at 1:2 scale. Normally (no check in this box) lineweights are plotted at their absolute value (0.010", 0.016", 0.25 mm, etc.) at any scale the drawing is plotted.

The following options are available from the expanded mode of the *Plot* dialog box (Fig. 14-3).

FIGURE 14-3

Plot Style Table (pen assignments)
If desired, select a plot style table to use with your plot device. A plot style table can contain several plot styles. Each plot style can assign specific line character-istics (lineweight, screening, dithering, end and joint types, fill patterns, etc.). Normally, a plot style table can be assigned to the *Model* tab or to *Layout* tabs. The plot style tables appearing in the list are managed by the *Plot Style Manager*. Selecting the *New* option invokes the *Add Plot Style Table* wizard.

Shaded Viewport Options
These options are <u>useful only if you are printing from the *Model* tab</u> and you are printing a 3D surface or solid model. These options <u>do not affect the display of 3D objects in viewports</u> of a *Layout* tab. You should have a laser jet, inkjet, or electrostatic type printer configured as the plot device (one that can print solid areas) since a pen plotter only creates line plots.

Shade Plot
Use these options if you are printing from the *Model* tab. You can shade the 3D object on the screen using any Visual Style, then select *As Displayed* in the *Shade Plot* section to make a shaded print, or you can select any option from the drop-down list for printing the drawing no matter which display is visible on screen.

Quality and *DPI*
You can also select from *Draft* (wireframe), *Preview* (shaded at 150 dpi), *Normal* (shaded at 300 dpi), *Presentation* (shaded at 600 dpi), or *Maximum* (for your printer) to specify the resolution quality of the print. Select *Custom* if you want to supply a different value in the *DPI* (dots per inch) edit box.

NOTE: The label for this section is misleading since it cannot be used to control printing for viewports. If you are printing from a *Layout* tab and want to print shaded images of 3D objects in viewports, you must control *Shadeplot* for each viewport using the *-Vports* command or the *Properties* palette.

Plot Options
These checkboxes control the following options.

Plot in Background
When this box is not checked, the *Plot Job Progress* window appears while the drawing is processed to the print or plot device so you must wait some time (depending on the complexity of the drawing) before continuing to work. Checking this option allows you to return to the drawing immediately while plot-ting continues in the background. Plotting in the background generally requires more time to produce the plot.

NOTE: The *Plot in Background* setting is valid only for the current plot and is <u>not saved in the page setup</u>. The default setting for this box (checked or unchecked) is controlled by the *Background processing options* section in the *Plot and Publish* tab of the *Options* dialog box.

Plot Object Lineweights
Checking this option plots the lineweights as they are assigned in the drawing. Lineweights can be assigned to layers or to objects. Checking *Plot with Plot Styles* disables this option.

Plot with Plot Styles
You can use plot styles to override lineweights assigned in the drawing. You must have a plot style table attached (see *Plot Device Tab, plot style table*) for this option to have effect.

Plot Paperspace Last
Checking this option plots paper space geometry last. Normally, paper space geometry is plotted before model space geometry.

Hide Paperspace Objects
This option is useful <u>only if you have 3D surface or solid objects in paper space</u> and you want to remove hidden lines (edges obscured from view). 3D objects are almost always created in model space. In that case, use the *Shade Plot* options (see *Shaded Viewport Options*) if you are printing from the *Model* tab. If you are printing from a *Layout* tab, you must control *Shadeplot* for each viewport using the *-Vports* command or the *Properties* palette.

Plot Stamp On
You have the capability of adding an informational text "stamp" on each plot. For example, the drawing name, date, and time of plot can be added to each drawing that you plot for tracking and verification purposes. To specify the plot stamp, use the *Plotstamp* command. Select this option to apply the stamp you specified to the plotted drawing.

Save Changes to Layout
Check this box if you want settings you make (the page setup) to be saved with the drawing for the selected *Model* tab or *Layout* tab. Selecting this option and making a plot has the same effect as pressing the *Apply to Layout* button. If you want to save changes without plotting, use the *Apply to Layout* button.

Drawing Orientation

Landscape is the typical horizontal orientation (similar to the drawing area orientation) and *Portrait* is a vertical orientation. If plotting from a layout tab, this setting affects the orientation of the paper; therefore, set this option before creating your viewport arrangement and title block to ensure they match the paper orientation. If plotting from the *Model* tab, *Portrait* positions horizontal lines in the drawing along the short axis of the paper.

Preview

Selecting the *Preview* button displays a simulated plot sheet to ensure the drawing will be plotted as you expect (Fig. 14-4). This feature is helpful particularly when plot style tables are attached since the resulting plots may appear differently than the *Model* tab or *Layout* tabs display. The *Zoom* feature is on by default so you can check specific areas of the drawing before making the plot. Right-clicking produces a shortcut menu, allowing you to use *Pan* and other *Zoom* options. Changing the display during a *Preview* <u>does not</u> change the area of the drawing to be plotted or printed.

FIGURE 14-4

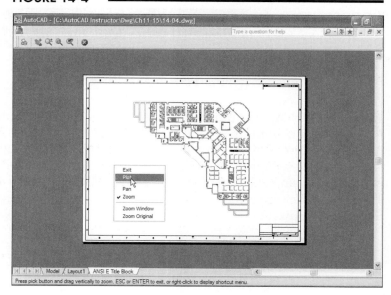

TIP

Apply to Layout

Using this button saves the current *Plot* dialog box settings to the active layout as a page setup. If no changes have been made to the default settings, this button is disabled. If this button is disabled and you want to save the current settings, assign a specific page setup name using the *Add* button. If you do not want to plot after saving the page setup, then use *Cancel*.

NOTE: Using this option (in the *Plot* dialog box) accomplishes essentially the same results as using the *Page Setup Manager* and *Page Setup* dialog box to save a page setup.

OK

Press *OK* to create the print or plot as assigned by the current settings in the *Plot* dialog box. The settings are not saved with the layout as a page setup unless the *Save changes to layout* box is checked or you used the *Apply to Layout* option.

Cancel

Cancel the plot and the current settings with this option. If you used the *Apply to Layout* option, *Cancel* will not undo the saved settings.

-Plot

Entering *-Plot* at the Command line invokes the Command line version of *Plot*. This is an alternate method for creating a plot if you want to use the current settings or if you enter a page setup name to define the plot settings.

```
Command:  -plot
Detailed plot configuration? [Yes/No] <No>:
Enter a layout name or [?] <ANSI E Title Block>:
Enter a page setup name <>:
Enter an output device name or [?] <HP 7586B.pc3>:
Write the plot to a file [Yes/No] <N>:
Save changes to page setup [Yes/No]? <N>
Proceed with plot [Yes/No] <Y>:
Effective plotting area:  33.54 wide by 42.60 high
Plotting viewport 2.
Command:
```

Pagesetup

Pull-down Menu	Command (Type)	Alias (Type)	Short-cut	Screen (side) Menu	Tablet Menu
File *Page Setup...*	*Pagesetup*	...	...	...	V,25

As you already know, AutoCAD allows you to set and save plotting specifications, called page setups, for each layout. In this way, one drawing can contain several layouts, each layout can display the same or different aspects of the drawing, and each layout can be set up to print or plot with the same or different plot devices, paper sizes, scales, plot styles, and so on. Oddly enough, you can create and manage page setups using either the *Plot* dialog box or the *Page Setup Manager* and related *Page Setup* dialog box. The main difference is that you cannot plot from the *Page Setup Manager* or *Page Setup* dialog box. See the following discussion, "Should you use the *Plot* dialog box or the *Page Setup Manager*?"

The *Pagesetup* command produces the *Page Setup Manager* (Fig. 14-5). Produce the *Page Setup Manager* by any method shown in the command table above or right-click on a *Layout* or *Model* tab and select *Page Setup Manager*. To cause the *Page Setup Manager* to appear automatically when you create a new layout, check the box in the lower-left corner of this dialog box or check *Show Page Setup Manager for new layouts* in the *Display* tab of the *Options* dialog box.

The *Page Setup Manager* has two basic functions: 1) apply existing page setups appearing in the *Page Setups* list to any layout or to a sheet set, and 2) provide access to the *Page Setup* dialog box where you can *Modify* or *Import* existing page setups or create *New* page setups. The *Page Setup* dialog box (see Fig. 14-7) allows you to specify and save the individual settings such as plot device, paper size, and so on.

FIGURE 14-5

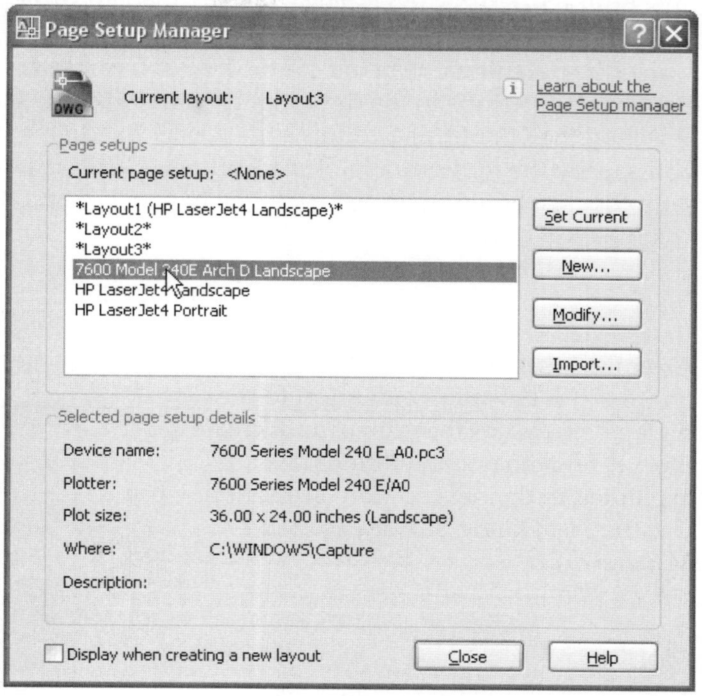

To apply a page setup to a layout, first activate the desired layout tab, then use *Pagesetup* to produce the *Page Setup Manager*. The *Current layout* note at the top of the manager displays the name of the current layout.

Page Setups List

This area lists the saved page setups that are available in the drawing (or sheet set). There are two types of page setups: those that are <u>unnamed</u> but applied to layouts, such as *Layout1*, and page setups that have an assigned name, or <u>named</u> page setups, such as *HP LaserJet4 Landscape*. "Applied" or "attached" page setups are denoted by asterisks.

By default, every initialized layout (initialized by clicking the layout tab) has a default unnamed page setup automatically applied to it, such as *Layout1*. Layouts that have a named page setup applied to them are enclosed in asterisks with the named page setup in parentheses, such as *Layout 2 (HP LaserJet4 Landscape)*. Therefore, the *Page Setups* list can contain 1) unnamed applied page setups such as *Layout1*, 2) named but unapplied page setups such as *HP LaserJet4 Landscape*, and 3) named page setups that have been applied to a layout such as *Layout 2 (HP LaserJet4 Landscape)*.

Set Current

To <u>apply</u> or attach a page setup from the list to the current layout, highlight the desired page setup, then use *Set Current*. You can select any named page setup, such as *HP LaserJet4 Landscape*, to apply to the current layout.

You can also select and apply another layout's unnamed page setup, such as *Layout3*; however, doing so does <u>not</u> change the name of the current layout's page setup. Once you assign a page setup to the current layout, you cannot "un-apply" or "detach" it. You can only assign another named page setup. *Set Current* is disabled for sheet sets or when you select the current page setup from the list.

New

This button produces the *New Page Setup* dialog box (Fig. 14-6) where you can assign a name for the new page setup. Assign a descriptive name (the default name being "Setup1"). You can select from the existing page setups in the drawing to "start with" as a template. Selecting the *OK* button produces the *Page Setup* dialog box where you specify the individual plotting options. Creating a *New* page setup does not automatically apply the new page setup to the current layout—you must use *Set Current* to do so.

FIGURE 14-6

As you can see, the *Page Setup* dialog box (Fig. 14-7) is essentially identical to the *Plot* dialog box, except that you cannot create a plot. A few other small differences are explained in the next section (see "Should You Use the *Plot* Dialog Box or *Page Setup Manager*?").

FIGURE 14-7

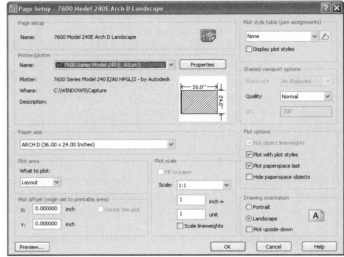

Modify

This button in the *Page Setup Manager* (see previous Fig. 14-5) produces the *Page Setup* dialog box where you can edit the existing settings for the selected page setup. Select any page set up in the *Page Setup* list to modify, including the current layout's page setup.

Import

Selecting this button invokes a standard file selection dialog box (not shown) from which you can import one or more page setups. You can select from any .DWG, .DWT, or .DXF file. Next, the *Import Page Setups* dialog box appears (not shown) for you to select the individual page setups in that specific drawing to import. The imported page setups then appear in the *Page Setup Manager* list, but are not yet applied to any layout.

Should You Use the *Plot* Dialog Box or the *Page Setup Manager*?

As you probably noticed, the *Page Setup Manager* and related *Page Setup* dialog box are similar and have essentially the same functions as the *Plot* dialog box, including managing page setups, with a few exceptions. The main differences are noted below.

1. You cannot plot from the *Page Setup Manager* or *Page Setup* dialog box.
2. The *Plot* dialog box includes all features in one dialog box compared to two for the *Page Setup Manager* and *Page Setup* dialog box.
3. Only the *Plot* dialog box allows you to set options to *Plot to a file*, *Plot in the background*, and turn *Plot Stamp on*.
4. You can modify any page setup from the *Page Setup Manager*, whereas you can modify only the current layout's page setup in the *Plot* dialog box.

So it has probably occurred to you, "Why do I need the *Page Setup Manager*?" Autodesk designed the *Page Setup Manager* interface primarily for use with sheet sets. A "sheet set" is a group of sheets related to one project, and therefore all "sheets" are plotted with the same plot settings. With the *Page Setup Manager*, you can apply one page setup to the entire sheet set. Without the *Page Setup Manager* and only the *Plot* dialog, you would have to open every sheet in the set (therefore every drawing file) and apply the same page setup to each layout in each drawing. Also, the desired page setup may have to be imported into each sheet (.DWG) individually before applying it.

So which method should you use? If you want to accomplish typical plotting and printing tasks, use the *Plot* dialog box for simplicity. You can also create, import, and assign page setups easily using the top section of the *Plot* dialog box. However, if you want to manage a large group of page setups, including making modifications to them, use the *Page Setup Manager*.

Preview

Pull-down Menu	Command (Type)	Alias (Type)	Short-cut	Screen (side) Menu	Tablet Menu
File *Plot Preview*	*Preview*	*PRE*	...	...	X,24

The *Preview* command accomplishes the same action as selecting *Preview* in the *Plot* dialog box. The advantage to using this command is being able to see a full print preview directly without having to invoke the *Plot* dialog box first. All functions of this preview (right-click for shortcut menu to *Pan* and *Zoom*, etc.) operate the same as a preview from the *Plot* dialog box including one important option, *Plot*.

You can print or plot during *Preview* directly from the shortcut menu (Fig. 14-8). Using the *Plot* option creates a print or plot based on your settings in the *Plot* or *Page Setup* dialog box. If you use this feature, ensure you first set the plotting parameters in the *Plot* or *Page Setup* dialog box.

FIGURE 14-8

PLOTTING TO SCALE

When you create a manual drawing, a scale is determined before you can begin drawing. The scale is determined by the proportion between the size of the object on the paper and the actual size of the object. You then complete the drawing in that scale so the actual object is proportionally reduced or enlarged to fit on the paper.

With a CAD drawing you are not restricted to drawing on a sheet of paper, so the geometry can be created full size. Set *Limits* to provide an appropriate amount of drawing space; then the geometry can be drawn using the actual dimensions of the object. The resulting drawing on the CAD system is a virtual full-size replica of the actual object. Not until the CAD drawing is plotted on a fixed size sheet of paper, however, is it scaled to fit on the sheet.

You can specify the plot scale in one of two ways. If you are plotting a *Layout* tab, specify the scale of the geometry that appears in the viewport by using the *VP Scale* pop-up list, *Viewport* toolbar, the *Properties* palette, or the *Zoom XP* factor, and then select 1:1 in the *Plot* or *Page Setup* dialog box. If you plot the *Model* tab, select the desired scale directly in the *Plot* or *Page Setup* dialog box. In either case, the value or ratio that you specify is the proportion of paper space units to model space units—the same as the proportion of the paper size to the *Limits*. This ratio is also equal to the reciprocal of the "drawing scale factor," as explained in Chapters 12 and 13. Exceptions to this rule are cases in which multiple viewports are used.

For example, if you want to draw an object 15" long and plot it on an 11" x 8.5" sheet, you can set the model space *Limits* to 22 x 17 (2 x the sheet size). The value of **2** is then the drawing scale factor, and the plot scale is **1:2** (or **1/2**, the reciprocal of the drawing scale factor). If, in another case, the calculated drawing scale factor is **4**, then **1:4** (or **1/4**) would be the plot scale to achieve a drawing plotted at 1/4 actual size. In order to calculate *Limits* and drawing scale factors, it is helpful to know the standard paper sizes.

Standard Paper Sizes

Size	Engineering (") (ANSI)	Architectural (")
A	8.5 x 11	9 x 12
B	11 x 17	12 x 18
C	17 x 22	18 x 24
D	22 x 34	24 x 36
E	34 x 44	36 x 48

Size	Metric (mm) (ISO)
A4	210 x 297
A3	297 x 420
A2	420 x 594
A1	594 x 841
A0	841 x 1189

Calculating the Drawing Scale Factor

Assuming you are planning to plot from the *Model* tab or from a *Layout* tab using one viewport that fills almost all of the printable area, use the following method to determine the "drawing scale factor" and, therefore, the plot or print scale to specify.

1. In order to provide adequate space to create the drawing geometry full size, it is recommended to set the model space *Limits*, then *Zoom All*. *Limits* should be set to the intended sheet size used for plotting times a factor, if necessary, that provides enough area for drawing. This factor, <u>the proportion of the *Limits* to the sheet size, is the "drawing scale factor."</u>

$$\text{DSF} = \frac{\text{Limits}}{\text{sheet size}}$$

    You should also use a proportion that will yield a standard drawing scale (1/2"=1", 1/8"=1', 1:50, etc.), instead of a scale that is not a standard (1/3"=1", 3/5"=1', 1:23, etc.). See the "Tables of *Limits* Settings" for common standard drawing scales.

2. The "drawing scale factor" is used as the value (at least a starting point) for changing all size-related variables (*LTSCALE*, *Overall Scale* for dimensions, text height, *Hatch* pattern scale, etc.). The <u>reciprocal of the drawing scale factor is the plot scale</u> to use for plotting or printing the drawing.

$$\text{Plot Scale} = \frac{1}{\text{DSF}}$$

3. If you using millimeter dimensions, set *Units* to *Decimal* and use metric values for the sheet size. In this way, the reciprocal of the drawing scale factor is the plot scale, the same as feet and inch drawings. However, multiply the drawing's scale factor by 25.4 (25.4mm = 1") to determine the factor for changing all size-related variables (*LTSCALE*, *Overall Scale* for dimensions, etc.).

$$\text{DSF(mm)} = \frac{\text{Limits}}{\text{sheet size}} \times 25.4$$

If drawing or inserting a border on the sheet, its maximum size cannot exceed the printable area. The printable area is displayed in a layout by the dashed line border just inside the "paper" edge. Since plotters or printers do not draw or print all the way to the edge of the paper, the border should not be drawn outside of the printable area. Generally, approximately 1/4" to 1/2" (6mm to 12mm) offset from each edge of the paper (margin) is required. When plotting a *Layout*, the printable area is denoted by a dashed border around the sheet. If plotting the *Model* tab, you should multiply the margin (1/2 of the difference between the *Paper Size* and printable area) by the "drawing scale factor" to determine the margin size in model space drawing units.

Guidelines for Plotting the *Model* Tab to Scale

Even though model space *Limits* can be changed and plot scale can be reset at <u>any time</u> in the drawing process, it is helpful to begin the process during the initial drawing setup (also see Chapter 12, Advanced Drawing Setup).

1. Set *Units* (*Decimal, Architectural, Engineering*, etc.) and *Precision* to be used in the drawing.
2. Set model space *Limits*. Set the *Limits* values to a proportion of the desired sheet size that provides enough area for drawing geometry as described in "Calculating the Drawing Scale Factor." The resulting value is the drawing scale factor.
3. Create the drawing geometry in the *Model* tab. Use the DSF as the initial value (or multiplier) for linetype scale, text size, dimension scale, hatch scale, etc.
4. Use the DSF to determine the scale factor to create, *Insert* or *Xref* the title block and border in model space, if one is needed.
5. Access the *Plot* dialog box. Select the desired scale in the *Scale* drop-down list or enter a value in the *Custom* edit boxes. The value to enter is 1/DSF. (Values are also given in the "Tables of *Limits* Settings.") Make a plot *Preview*, then make the plot or make the needed adjustments.

Guidelines for Plotting *Layouts* to Scale

Assuming you are using one viewport that fills almost all of the printable area, plot from the *Layout* tab using this method.

1. Set *Units* (*Decimal, Architectural, Engineering*, etc.) and *Precision* to be used in the drawing.
2. Set model space *Limits* and create the geometry in the *Model* tab. Use the DSF to determine values for linetype scale, text size, dimension scale, hatch scale, etc. Complete the drawing geometry in model space.
3. Click a *Layout* tab to begin setting up the layout. You can also use the *Layout Wizard* to create a new layout and complete steps 2 through 5.
4. Use *Page Setup* to select the desired *Plot Device* and *Paper Size*.
5. Create, *Insert*, or *Xref* the title block and border in the layout. If you are using one of AutoCAD's template drawings or another template, the title block and border may already exist.
6. Create a viewport using *Vports*.
7. Click inside the viewport to set the viewport scale. You can set the scale using the *VP Scale* pop-up list, *Viewports* toolbar, the *Properties* palette, or use *Zoom* and enter an *XP* factor. Use 1/DSF as the viewport scale.
8. Activate the *Plot* dialog box and ensure the scale in the *Scale* drop-down list is 1:1. Make a plot *Preview*, then make the plot or make the needed adjustments.

If you want to print the same drawing using different devices and different scales, you can create multiple layouts. With each layout you can save the specific plot settings. Beware—a drawing printed at different scales may require changing or setting different values for linetype scale, text size, dimension variables, hatch scales, etc. for each layout.

2008

Annotative Objects

Annotative objects (annotative text and dimensions, etc.) automatically change size when you change the viewport scale. However, calculating *Limits* and a drawing scale factor is a valid strategy for using annotative objects effectively. For example, the drawing scale factor should be used to set the *Annotation Scale* in the *Model* tab.

Template Drawings

Simplifying the process of plotting to scale can be accomplished by preparing template drawings. One method is to create a template for each sheet size that is used in the lab or office. In this way, the CAD operator begins the session by selecting the template drawing representing the sheet size and then multiplies <u>those</u> *Limits* by some factor to achieve the desired *Limits* and drawing scale factor. Another method is to create multiple layouts in the template drawing, one for each device and sheet size that you have available. In this way, a template drawing can be selected with the final layouts, title blocks, plot devices, and sheet sizes already specified, then you need only set the model space *Limits* and plot scale(s).

TABLES OF *LIMITS* SETTINGS

Rather than making calculations of *Limits*, drawing scale factor, and plot scale for each drawing, the Tables of *Limits* Settings on the following pages can be used to make calculating easier. There are five tables, one for each of the following applications:

 Mechanical Engineering
 Architectural
 Metric (using ISO standard metric sheets)
 Metric (using ANSI standard engineering sheets)
 Civil Engineering

To use the tables correctly, you must know <u>any two</u> of the following three variables. The third variable can be determined from the other two.

 Approximate drawing <u>*Limits*</u>
 Desired print or plot <u>paper size</u>
 Desired print or plot <u>scale</u>

1. If you know the *Scale* and **paper size**:
   Assuming you know the scale you want to use, look along the top row to find the desired <u>scale</u> that you eventually want to print or plot. Find the desired <u>paper size</u> by looking down the left column. The intersection of the row and column yields the model space <u>*Limits*</u> settings to use to achieve the desired plot scale.

2. If you know the approximate model space ***Limits*** and **paper size**:
   Calculate how much space (minimum *Limits*) you require to create the drawing actual size. Look down the left column of the appropriate table to find the <u>paper size</u> you want to use for plotting. Look along that row to find the next larger <u>*Limits*</u> settings than your required area. Use these values to set the drawing *Limits*. The <u>scale</u> to use for the plot is located on top of that column.

3. If you know the *Scale* and approximate model space ***Limits***:
   Calculate how much space (minimum *Limits*) you need to create the drawing objects actual size. Look along the top row of the appropriate table to find the desired <u>scale</u> that you eventually want to plot or print. Look down that column to find the next larger <u>*Limits*</u> than your required minimum area. Set the model space *Limits* to these values. Look to the left end of that row to give the <u>paper size</u> you need to print or plot the *Limits* in the desired scale.

NOTE: Common scales are given in the Tables of *Limits* Settings. If you need to create a print or plot in a scale that is not listed in the tables, you may be able to use the tables as a guide by finding the nearest table and standard scale to match your needs, then calculating <u>proportional</u> settings.

MECHANICAL TABLE OF *LIMITS* SETTINGS
For ANSI Sheet Sizes
(X axis x Y axis)

Paper Size (Inches)	Scale Factor 1 / Scale 1"=1" / Proportion 1:1	1.33 / 3/4"=1" / 3:4	2 / 1/2"=1" / 1:2	2.67 / 3/8"=1" / 3:8	4 / 1/4"=1" / 1:4	5.33 / 3/16"=1" / 3:16	8 / 1/8"=1" / 1:8
A 11 x 8.5 In.	11.0 x 8.5	14.7 x 11.3	22.0 x 17.0	29.3 x 22.7	44.0 x 34.0	58.7 x 45.3	88.0 x 68.0
B 17 x 11 In.	17.0 x 11.0	22.7 x 14.7	34.0 x 22.0	45.3 x 29.3	68.0 x 44.0	90.7 x 58.7	136.0 x 88.0
C 22 x 17 In.	22.0 x 17.0	29.3 x 22.7	44.0 x 34.0	58.7 x 45.3	88.0 x 68.0	117.0 x 90.7	176.0 x 136.0
D 34 x 22 In.	34.0 x 22.0	45.3 x 29.3	68.0 x 44.0	90.7 x 58.7	136.0 x 88.0	181.3 x 117.3	272.0 x 176.0
E 44 x 34 In.	44.0 x 34.0	58.7 x 45.3	88.0 x 68.0	117.3 x 90.7	176.0 x 136.0	235.7 x 181.3	352.0 x 272.0

ARCHITECTURAL TABLE OF *LIMITS* SETTINGS
For Architectural Sheet Sizes
(X axis x Y axis)

Paper Size (Inches)		Drawing Scale Factor 12 Scale 1" = 1' Proportion 1:12	16 3/4" = 1' 1:16	24 1/2" = 1' 1:24	32 3/8" = 1' 1:32	48 1/4" = 1' 1:48	64 3/16" = 1' 1:64	96 1/8" = 1' 1:96
A 12 x 9	Ft	12 x 9	16 x 12	24 x 18	32 x 24	48 x 36	64 x 48	96 x 72
	In.	144 x 108	192 x 144	288 x 216	384 x 288	576 x 432	768 x 576	1152 x 864
B 18 x 12	Ft	18 x 12	24 x 16	36 x 24	48 x 32	64 x 48	96 x 64	128 x 96
	In.	216 x 144	288 x 192	432 x 288	576 x 384	768 x 576	1152 x 768	1536 x 1152
C 24 x 18	Ft	24 x 18	32 x 24	48 x 36	64 x 48	96 x 72	128 x 96	192 x 144
	In.	288 x 216	384 x 288	576 x 432	768 x 576	1152 x 864	1536 x 1152	2304 x 1728
D 36 x 24	Ft	36 x 24	48 x 32	72 x 48	96 x 64	144 x 96	192 x 128	288 x 192
	In.	432 x 288	576 x 384	864 x 576	1152 x 768	1728 x 1152	2304 x 1536	3456 x 2304
E 48 x 36	Ft	48 x 36	64 x 48	96 x 72	128 x 96	192 x 144	256 x 192	384 x 288
	In.	576 x 432	768 x 576	1152 x 864	1536 x 1152	2304 x 1728	3072 x 2304	4608 x 3456

METRIC TABLE OF *LIMITS* SETTINGS
For Metric (ISO) Sheet Sizes
(X axis x Y axis)

Paper Size (mm)		Drawing Scale Factor 25.4 Scale 1:1 Proportion 1:1	50.8 1:2 1:2	127 1:5 1:5	254 1:10 1:10	508 1:20 1:20	1270 1:50 1:50	2540 1:100 1:100
A4 297 x 210	mm	297 x 210	594 x 420	1485 x 1050	2970 x 2100	5940 x 4200	14,850 x 10,500	29,700 x 21,000
	m	.297 x .210	.594 x .420	1.485 x 1.050	2.97 x 2.10	5.94 x 4.20	14.85 x 10.50	29.70 x 21.00
A3 420 x 297	mm	420 x 297	840 x 594	2100 x 1485	4200 x 2970	8400 x 5940	21,000 x 14,850	42,000 x 29,700
	m	.420 x .297	.840 x .594	2.100 x 1.485	4.20 x 2.97	8.40 x 5.94	21.00 x 14.85	42.00 x 29.70
A2 594 x 420	mm	594 x 420	1188 x 840	2970 x 2100	5940 x 4200	11,880 x 8400	29,700 x 21,000	59,400 x 42,000
	m	.594 x .420	1.188 x .840	2.97 x 2.10	5.94 x 4.20	11.88 x 8.40	29.70 x 21.00	59.40 x 42.00
A1 841 x 594	mm	841 x 594	1682 x 1188	4205 x 2970	8410 x 5940	16,820 x 11,880	42,050 x 29,700	84,100 x 59,400
	m	.841 x .594	1.682 x 1.188	4.205 x 2.970	8.41 x 5.94	16.82 x 11.88	42.05 x 29.70	84.10 x 59.40
A0 1189 x 841	mm	1189 x 841	2378 x 1682	5945 x 4205	11,890 x 8410	23,780 x 16,820	59,450 x 42,050	118,900 x 84,100
	m	1.189 x .841	2.378 x 1.682	5.945 x 4.205	11.89 x 8.41	23.78 x 16.82	59.45 x 42.05	118.90 x 84.10

METRIC TABLE OF *LIMITS* SETTINGS
For ANSI Sheet Sizes
(X axis x Y axis)

Paper Size (mm)		Drawing Scale Factor 25.4 / Scale 1:1 / Proportion 1:1	50.8 / 1:2 / 1:2	127 / 1:5 / 1:5	254 / 1:10 / 1:10	508 / 1:20 / 1:20	1270 / 1:50 / 1:50	2540 / 1:100 / 1:100
A 279.4 x 215.9	mm	279.4 x 215.9	558.8 x 431.8	1397 x 1079.5	2794 x 2159	5588 x 4318	13,970 x 10,795	27,940 x 21,590
	m	0.2794 x 0.2159	0.5588 x 0.4318	1.397 x 1.0795	2.794 x 2.159	5.588 x 4.318	13.97 x 10.795	27.94 x 21.59
B 431.8 x 279.4	mm	431.8 x 279.4	863.6 x 558.8	2159 x 1397	4318 x 2794	8636 x 5588	21,590 x 13,970	43,180 x 27,940
	m	0.4318 x 0.2794	0.8636 x 0.5588	2.159 x 1.397	4.318 x 2.794	8.636 x 5.588	21.59 x 13.97	43.18 x 27.94
C 558.8 x 431.8	mm	558.8 x 431.8	1117.6 x 863.6	2794 x 2159	5588 x 4318	11,176 x 8636	27,940 x 21,590	55,880 x 43,180
	m	0.5588 x 0.4318	1.1176 x 0.8636	2.794 x 2.159	5.588 x 4.318	11.176 x 86.36	27.94 x 21.59	55.88 x 43.18
D 863.6 x 558.8	mm	863.6 x 558.8	1727.2 x 1117.6	4318 x 2794	8636 x 5588	17,272 x 11,176	43,180 x 27,940	86,360 x 55,880
	m	0.8636 x 0.5588	1.7272 x 1.1176	4.318 x 2.794	8.636 x 5.588	17.272 x 11.176	43.18 x 27.94	86.36 x 55.88
E 1117.6 x 863.6	mm	1117.6 x 863.6	2235.2 x 1727.2	5588 x 4318	11,176 x 8636	22,352 x 17,272	55,880 x 43,180	111,760 x 86,360
	m	1.1176 x 0.8636	2.2352 x 1.7272	5.588 x 4.318	11.176 x 8.636	22.352 x 17.272	55.88 x 43.18	111.76 x 86.36

CIVIL TABLE OF *LIMITS* SETTINGS
For ANSI Sheet Sizes
(X axis x Y axis)

Paper Size (Inches)	Drawing Scale Factor 120 / 1" = 10' / 1:120	240 / 1" = 20' / 1:240	360 / 1" = 30' / 1:360	480 / 1" = 40' / 1:480	600 / 1" = 50' / 1:600
A. 11 x 8.5					
In.	1320 x 1020	2640 x 2040	3960 x 3060	5280 x 4080	6600 x 5100
Ft	110 x 85	220 x 170	330 x 255	440 x 340	550 x 425
B. 17 x 11					
In.	2040 x 1320	4080 x 2640	6120 x 3960	8160 x 5280	10,200 x 6600
Ft	170 x 110	340 x 220	510 x 330	680 x 440	850 x 550
C. 22 x 17					
In.	2640 x 2040	5280 x 4080	7920 x 6120	10,560 x 8160	13,200 x 10,200
Ft	220 x 170	440 x 340	660 x 510	880 x 680	1100 x 850
D. 34 x 22					
In.	4080 x 2640	8160 x 5280	12,240 x 7920	16,320 x 10,560	20,400 x 13,200
Ft	340 x 220	680 x 440	1020 x 660	1360 x 880	1700 x 1100
E. 44 x 34					
In.	5280 x 4080	10,560 x 8160	15,840 x 12,240	21,120 x 16,320	26,400 x 20,400
Ft	440 x 340	880 x 680	1320 x 1020	1760 x 1360	2200 x 1700

Examples for Plotting to Scale

Following are several hypothetical examples of drawings that can be created using the "Guidelines for Plotting to Scale." As you read the examples, try to follow the logic and check the "Tables of *Limits* Settings."

A. A one-view drawing of a mechanical part that is 40" in length is to be drawn requiring an area of approximately 40" x 30", and the drawing is to be plotted on an 8.5" x 11" sheet. The template drawing *Limits* are preset to 0,0 and 11,8.5. (The ACAD.DWT template *Limits* are set to 0,0 and 12,9, which represents an uncut sheet size. Template *Limits* of 11 x 8.5 are more practical in this case.) The expected plot scale is 1/4"=1". The following steps are used to calculate the new *Limits*.

    1. ***Units*** are set to ***decimal***. Each unit represents 1.00 inches.
    2. Multiplying the *Limits* of 11 x 8.5 by a factor of **4**, the new **Limits** should be set to **44 x 34**, allowing adequate space for the drawing. The "drawing scale factor" is **4**.
    3. All size-related variable default values (***LTSCALE, Overall Scale*** for dimensions, etc.) are multiplied by **4**. The plot scale (for *Layout* tabs, entered in the *Viewport* toolbar, etc., or for the *Model* tab, entered in the *Plot* or *Page Setup* dialog box) is **1:4** to achieve a plotted drawing of 1/4"=1".

B. A floorplan of a residence will occupy 60' x 40'. The drawing is to be plotted on a "D" size architectural sheet (24" x 36"). No prepared template drawing exists, so the ACAD.DWT template drawing (12 x 9) *Limits* are used. The expected plot scale is 1/2"=1'.

    1. ***Units*** are set to ***Architectural***. Each unit represents 1".
    2. The floorplan size is converted to inches (60' x 40' = 720" x 480"). The sheet size of 36" x 24" (or 3' x 2') is multiplied by **24** to arrive at **Limits** of **864" x 576"** (72' x 48'), allowing adequate area for the floor plan. The *Limits* are changed to those values.
    3. All default values of size-related variables (***LTSCALE, Overall Scale*** for dimensions, etc.) are multiplied by **24**, the drawing scale factor. The plot scale (for *Layout* tabs, entered in the *VP Scale* list, etc., or for the *Model* tab, entered in the *Plot* or *Page Setup* dialog box) is **1/2"=1'** (or 1/24, which is the reciprocal of the drawing scale factor).

C. A roadway cloverleaf is to be laid out to fit in an acquired plot of land measuring 1500' x 1000'. The drawing will be plotted on "D" size engineering sheet (34" x 22"). A template drawing with *Limits* set equal to the sheet size is used. The expected plot scale is 1"=50'.

    1. ***Units*** are set to ***Engineering*** (feet and decimal inches). Each unit represents 1.00".
    2. The sheet size of 34" x 22" (or 2.833' x 1.833') is multiplied by **600** to arrive at **Limits** of **20400" x 13200"** (1700' x 1100'), allowing enough drawing area for the site. The *Limits* are changed to **1700' x 1100'**.
    3. All default values of size-related variables (***LTSCALE, Overall Scale*** for dimensions, etc.) are multiplied by **600**, the drawing scale factor. The plot scale (for *Layout* tabs, entered in the *VP Scale* list, etc., or for the *Model* tab, entered in the *Plot* or *Page Setup* dialog box) is **1/600** (***inches = drawing units***) to achieve a drawing of 1"=50' scale.

NOTE: Many civil engineering firms use one unit in AutoCAD to represent one foot. This simplifies the problem of using decimal feet (10 parts/foot rather than 12 parts/foot); however, problems occur if architectural layouts are inserted or otherwise combined with civil work.

CONFIGURING PLOTTERS AND PRINTERS

Plotter-manager

Pull-down Menu	Command (Type)	Alias (Type)	Short-cut	Screen (side) Menu	Tablet Menu
File *Plotter Manager...*	*Plottermanager*	...	...	...	...

The AutoCAD *Plotter Manager* can be invoked by several methods as shown in the command table above. In addition to those methods, you can select the *Add or Configure Plotters* button in the *Plotting* tab of the *Options* dialog box.

In any case, the *Plotter Manager* is a separate system window which can run as an independent application to AutoCAD (Fig. 14-9). By default, the window displays icons of all plot devices configured specifically for AutoCAD. The default Windows configured device is included in this group. (When you install AutoCAD, the Windows configured system printer is automatically configured if you have a previously installed Windows printer.) You can add or configure a plotter with the *Plotter Manager*.

FIGURE 14-9

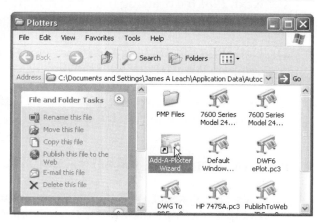

Add-A-Plotter Wizard

The *Add-A-Plotter Wizard* makes the process of configuring a new device very simple. Several choices are available (Fig. 14-10). You can configure a new device to be connected directly to and managed by your system or to be shared and managed by a network server. You can also reconfigure printers that were previously set up as Windows system printers. In this case, icons for the printers will appear in the *Plotter Manager* and you can access and control configuration settings to use specifically for AutoCAD.

FIGURE 14-10

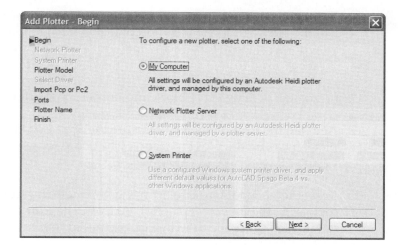

CHAPTER EXERCISES

1. **Print GASKETA from the *Model* Tab**

    A. *Open* the **GASKETA** drawing that you created in Chapter 9 Exercises. Activate the *Model* tab. What are the model space *Limits*?

    B. From the *Model* tab, **Plot** the drawing **Extents** on an 11″ x 8.5″ sheet. Select the **Fit to paper** option from the *Plot Scale* section in the *Plot* dialog box.

    C. Next, **Plot** the drawing from the *Model* tab again. Plot the **Limits** on the same size sheet. **Scale to Fit**.

    D. Now, **Plot** the drawing **Limits** as before but plot the drawing at **1:1**. Measure the drawing and compare the accuracy with the dimensions given in the exercise in Chapter 9.

    E. Compare the three plots. What are the differences and why did they occur? (When you finish, there is no need to *Save* the changes.)

2. **Print GASKETA from a *Layout* Tab in Two Scales**

    A. Using the **GASKETA** drawing again, activate the *Layout1* tab. If a viewport exists in the layout, *Erase* it. Activate the *Plot* or *Page Setup* dialog box and configure the layout to use a printer to print on a *Letter* (or "A" size) sheet.

    B. Next, make layer **0** *Current*. Use *Vports* to create a *Single* viewport using the *Fit* option. Double-click inside the viewport and *Zoom All* to see all of the model space *Limits*. Next, set the viewport scale to **1:1** using any method (*Zoom XP*, the *Viewport* toolbar, or the *Properties* palette). You may need to *Pan* to center the geometry, but do <u>not</u> *Zoom*.

    C. (Optional) Create a title block and border for the drawing in paper space.

    D. Activate the *Plot* dialog box. Ensure the *Scale* edit box is set at **1:1**. Plot the *Layout*. When complete, measure the plot for accuracy by comparing with the dimensions given for the exercise in Chapter 9. *Save* the drawing.

    E. Reset the viewport scale to **2:1** using any method (*Zoom XP*, the *VP Scale* list, the *Viewport* toolbar, or the *Properties* palette). Activate the *Plot* dialog box and make a second plot at the new scale. *Close* the drawing, but <u>do not save</u> it.

3. **Print the MANFCELL Drawing in Two Standard Scales**

    A. *Open* the **MANFCELL** drawing from Chapter 9. Check to make sure that the *Limits* settings are at **0,0** and **40′,30′**.

    B. Activate the *Layout1* tab. If a viewport exists, *Erase* it. Use *Pagesetup* to select a printer to print on a letter ("A") sheet size. Next, create a *Single* viewport with the *Fit* option on layer **0**.

C. In this step you will make one print of the drawing at the largest possible size using a standard architectural scale. Use any method to set the viewport scale. Use the "Architectural Table of *Limits* Settings" in the chapter for guidance on an appropriate scale to set. (HINT: Look in the "A" size sheet row and find the next larger *Limits* settings that will accommodate your current *Limits*, then use the scale shown above at the top of the column.) *Save* the drawing. Make a print using the *Plot* dialog box.

D. Now make a plot of the drawing using the next smaller standard architectural scale. This time determine the appropriate scale to set using either the "Architectural Table of *Limits* Settings" or by selecting it from the *VP Scale* pop-up list or *Viewport Scale* drop-down list.

4. **Use *Plotter Manager* to Configure a "D" Size Plot Device**
Even if your system currently has a "D" size plot device configured, this exercise will give you practice with *Plotter Manager*. This exercise does not require that the device actually exist in your lab or office or that it be physically attached to your system.

A. Invoke the *Plotter Manager*. In *Plotter Manager*, double-click on *Add-A-Plotter Wizard*.

B. Select the *Next* button in the *Introduction* page. In the *Begin* page, select *My Computer*. In the *Plotter Model* page, select any *Manufacturer* and *Model* that can plot on a "D" size sheet, such as *CalComp 68444 Color Electrostatic, Hewlett-Packard Draft-Pro EXL (7576A)*, or *Oce 9800 FBBS R3.x*.

C. Select *Next* in both the *Import PCP or PC2* and *Ports* pages. In the *Plotter Name* step, select the default name but insert the manufacturer's name as a prefix, such as "HP" or "CalComp."

D. Finally, activate the *Plot* or *Page Setup* dialog box and ensure your new device is listed in the *Name* drop-down list.

5. **Create Multiple Layouts for Different Plot Devices, Set *PSLTSCALE***
In this exercise you will create two layouts—one for plotting on an engineering "A" size sheet and one on a "C" size sheet. Check the table of "Standard Paper Sizes" in the chapter and find the correct size in inches for both engineering "A" and "C" size sheets. You will also adjust *LTSCALE* and *PSLTSCALE* appropriately for the layouts.

A. *Open* drawing CH11EX3. Check to ensure the model space *Limits* are set at **22 x 17**. If you plan to print on a "A" size sheet, what is the drawing scale factor? Is the *LTSCALE* set appropriately?

B. Activate the *Layout1* tab. Right-click and use *Rename* to rename the tab to **A Sheet**. If a viewport exists in the layout, *Erase* it. Activate the *Page Setup* dialog box and configure the layout to use a printer to print on a *Letter* ("A" size) sheet. Next, use *Vports* to create a viewport on layer **0** using the *Fit* option. Double-click inside the viewport and set the viewport scale to **1:2**.

C. Toggle between *Model* tab and the *A Sheet* tab to examine the hidden and center lines. Do the dashes appear the same in both tabs? If the dashes appear differently in the two tabs, type *PSLTSCALE* and ensure the value is set to **0**. (Make sure you always use *Regenall* after using *PSLTSCALE*.)

D. Use the *Plot* dialog box to make the plot. Measure the plot for accuracy by comparing it with the dimensions given for the exercise in Chapter 11. To refresh your memory, that exercise involved adjusting the *LTSCALE* factor. Does the *LTSCALE* in the drawing yield hidden line dashes of 1/8" (for the long lines) on the plot? If not, make the *LTSCALE* adjustment and plot again. Use *SaveAs* to save and rename the drawing as **CH14EX5**.

E.   Next, activate the *Layout2* tab. Right-click and use *Rename* to rename the tab to **C Sheet**. If a viewport exists in the layout, *Erase* it. Activate the *Page Setup* dialog box and configure the layout to use a plotter for a *ANSI C (22 x 17 Inches)* sheet. Next, on layer **0**, use *Vports* to create a viewport using the *Fit* option. Double-click inside the viewport and set the viewport scale. Since the model space limits are set to 22 x 17 (same as the sheet size), set the viewport scale to **1:1**.

F.   Do the hidden lines appear the same in all tabs? *PSLTSCALE* set to 0 ensures that the hidden and center lines appear with the same *LTSCALE* in the viewports as in the *Model* tab. If everything looks okay, make the plot. Does the plot show hidden line dashes of 1/8"? *Save* the drawing.

6.   **Create an Architectural Template Drawing**
In this exercise you will create a new template for architectural applications to plot at 1/8"=1' scale on a "D" size sheet.

A.   Begin a *New* drawing using the template **ASHEET.DWT** you worked on in Chapter 13 Exercises. Use *Save* to create a new template (.DWT) named **D-8-AR** in your working folder (not where the AutoCAD-supplied templates are stored).

B.   Set *Units* to *Architectural*. Use the "Architectural Table of *Limits* Settings" to determine and set the new model space *Limits* for an architectural "D" size sheet to plot at **1/8"=1'**. Multiply the existing *LTSCALE* (**1.0**) times the drawing scale factor shown in the table. Turn *Snap* to *Polar Snap* and turn *Grid* off.

C.   Activate the *Layout2* tab. Right-click to produce the shortcut menu and select *Rename*. Rename the tab to **D Sheet**. *Erase* the viewport if one exists.

D.   Activate the *Page Setup* dialog box. Select a plot device that can use an architectural D size sheet (36 x 24), such as the device you configured for your system in Exercise 4. In the *Layout Settings* tab, select *ARCH D (24 x 36 Inches)* in the *Paper Size* drop-down list. Select *OK* to dismiss the *Page Setup* dialog box and save the settings.

E.   Make **VPORTS** the current layer. Use the *Vports* command to create a *Single* viewport using the *Fit* option. Double-click inside the viewport and set the viewport scale to **1/8"=1'**.

F.   Make the *Model* tab active and set the **OBJECT** layer current. Finally, *Save* as a template (.DWT) drawing in your working folder.

7.   **Create a Civil Engineering Template Drawing**
Create a new template for civil engineering applications to plot at 1"=20' scale on a "C" size sheet.

A.   Begin a *New* drawing using the template **C-D-SHEET.** Use *Save* to create a new template (.DWT) named **C-CIVIL-20** in your working folder (not where the AutoCAD-supplied templates are stored).

B.   Set *Units* to *Engineering*. Use the "Civil Table of *Limits* Settings" to determine and set the new *Limits* for a "C" size sheet to plot at **1"=20'**. Multiply the existing *LTSCALE* (.5) times the scale factor shown. Turn *Snap* to *Polar* and *Grid* Off.

C.   Activate the *D Sheet* layout tab. Use any method to set the viewport scale to **1:240**. *Save* the drawing as a template (.DWT) drawing in your working folder.

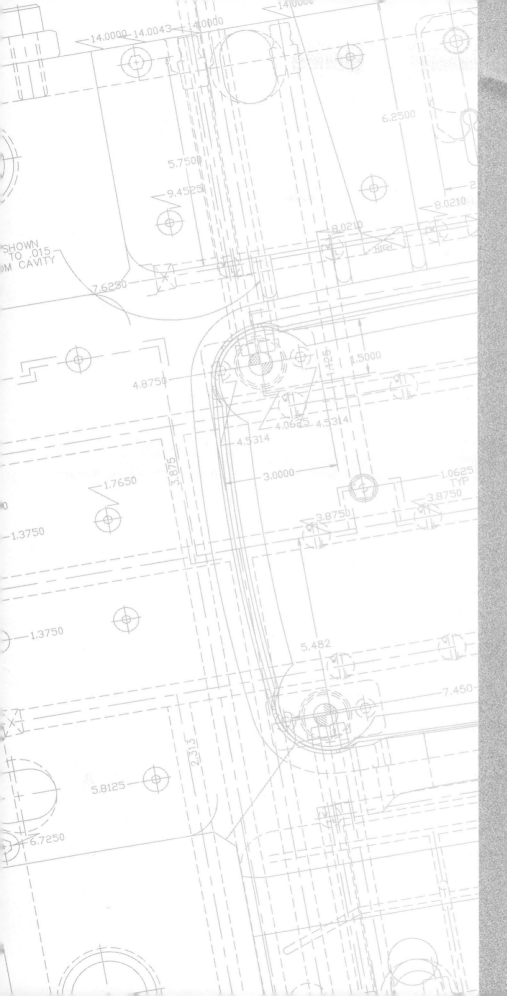

15

DRAW
COMMANDS II

CHAPTER OBJECTIVES

After completing this chapter you should:

1. be able to create construction lines using
   the *Xline* and *Ray* commands;

2. be able to create *Polygons* by the
   Circumscribe, Inscribe, and *Edge* methods;

3. be able to create *Rectangles*;

4. be able to use *Donut* to create circles with
   width;

5. be able to create *Spline* curves passing
   exactly through the selected points;

6. be able to create *Ellipses* using the *Axis
   End* method, the *Center* method, and the
   Arc method;

7. be able to use *Divide* to add points at equal
   parts of an object and *Measure* to add
   points at specified segment lengths along
   an object;

8. be able to create a *Boundary* by PICKing
   inside a closed area;

9. know that objects forming a closed shape
   can be combined into one *Region* object;

10. be able to use the *Revcloud* command to
    draw revision clouds.

CONCEPTS

Remember that *Draw* commands create AutoCAD objects. The draw commands addressed in this chapter create more complex objects than those discussed in Chapter 8, Draw Commands I. The draw commands covered previously are *Line, Circle, Arc, Point,* and *Pline.* The shapes created by the commands covered in this chapter are more complex. Most of the objects <u>appear</u> to be composed of several simple objects, but each shape is actually treated by AutoCAD as <u>one object</u>. Only the *Divide, Measure* and *Sketch* commands create multiple objects. The following commands are explained in this chapter.

Xline, Ray, Polygon, Rectangle, Donut, Spline, Ellipse, Divide, Measure, Boundary, and *Region*

These draw commands can be accessed using any of the command entry methods, including the menus and icon buttons, as illustrated in Chapter 8. Buttons that appear by the Command tables in this chapter are available in the default *Draw* toolbar.

COMMANDS

Xline

Pull-down Menu	Command (Type)	Alias (Type)	Short-cut	Screen (side) Menu	Tablet Menu
Draw *Construction Line*	*Xline*	*XL*	...	*DRAW 1* *Xline*	*L,10*

When you draft with pencil and paper, light "construction" lines are used to lay out a drawing. These construction lines are not intended to be part of the finished object lines but are helpful for preliminary layout such as locating intersections, center points, and projecting between views.

There are two types of construction lines—*Xline* and *Ray.* An *Xline* is a line with infinite length, therefore having no endpoints. A *Ray* has one "anchored" endpoint and the other end extends to infinity. Even though these lines extend to infinity, they do not affect the drawing *Limits* or *Extents* or change the display or plot area in any way.

An *Xline* has no endpoints (*Endpoint Osnap* cannot be used) but does have a <u>root</u>, which is the theoretical <u>midpoint</u> (*Midpoint Osnap* can be used). If *Trim* or *Break* is used with *Xlines* or *Rays* such that two end-points are created, the construction lines become *Line* objects. *Xlines* and *Rays* are drawn on the current layer and assume the current linetype and color (object-specific or *ByLayer*).

There are many ways that you can create *Xlines* as shown by the options appearing at the command prompt. All options of *Xline* automatically repeat so you can easily draw multiple lines.

    Command: **xline**
    Specify a point or [Hor/Ver/Ang/Bisect/Offset]:

Specify a point

The default option only requires that you specify two points to construct the *Xline* (Fig. 15-1). The first point becomes the root and anchors the line for the next point specification. The second point, or "through point," can be PICKed at any location and can pass through any point (*Polar Snap, Polar Tracking, Osnaps*, and *Objects Snap Tracking* can be used). If horizontal or vertical *Xlines* are needed, *ORTHO* can be used in conjunction with the "Specify a point:" option.

> Command: **xline**
> Specify a point or [Hor/Ver/Ang/Bisect/Offset]:
> Specify through point:

FIGURE 15-1

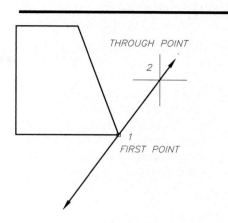

Hor

This option creates a horizontal construction line. Type the letter "H" at the first prompt. You only specify one point, the through point or root (Fig. 15-2).

FIGURE 15-2

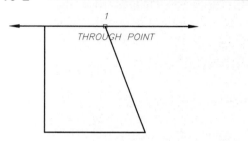

Ver

Ver creates a vertical construction line. You only specify one point, the through point (root) (Fig. 15-3).

FIGURE 15-3

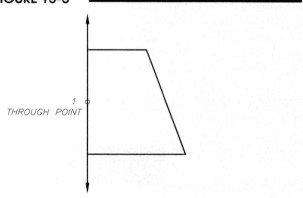

Ang

The *Ang* option provides two ways to specify the desired angle. You can (1) *Enter angle* or (2) select a *Reference* line (*Line, Xline, Ray*, or *Pline*) as the starting angle, then specify an angle from the selected line (in a counterclockwise direction) for the *Xline* to be drawn (Fig. 15-4).

> Command: **xline**
> Specify a point or [Hor/Ver/Ang/Bisect/Offset]: **a**
> Enter angle of xline (0) or [Reference]:

FIGURE 15-4

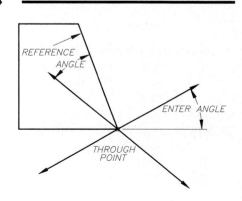

Bisect

This option draws the *Xline* at an angle between two selected points. First, select the angle vertex, then two points to define the angle (Fig. 15-5).

> Command: **xline**
> Specify a point or [Hor/Ver/Ang/Bisect/Offset]: **b**
> Specify angle vertex point: **PICK**
> Specify angle start point: **PICK**
> Specify angle end point: **PICK**
> Specify angle end point: **Enter**
> Command:

Offset

Offset creates an *Xline* parallel to another line. This option operates similarly to the *Offset* command. You can (1) specify a *Distance* from the selected line or (2) PICK a point to create the *Xline Through*. With the *Distance* option, enter the distance value, select a line (*Line, Xline, Ray,* or *Pline*), and specify on which side to create the offset *Xline* (Fig. 15-6).

> Command: **xline**
> Specify a point or [Hor/Ver/Ang/Bisect/Offset]: **o**
> Specify offset distance or [Through] <1.0000>:
> **(value)** or **PICK**
> Select a line object: **PICK**
> Specify side to offset: **PICK**
> Select a line object: **Enter**
> Command:

Using the *Through* option, select a line (*Line, Xline, Ray,* or *Pline*); then specify a point for the *Xline* to pass through. In each case, the anchor point of the *Xline* is the "root." (See "*Offset*," Chapter 9.)

FIGURE 15-5

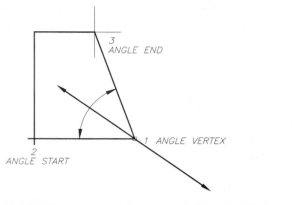

FIGURE 15-6

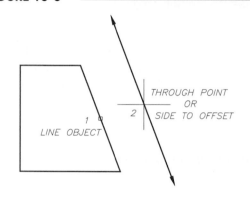

Ray

Pull-down Menu	Command (Type)	Alias (Type)	Short-cut	Screen (side) Menu	Tablet Menu
Draw Ray	Ray	...	...	DRAW 1 Ray	K,10

A *Ray* is also a construction line (see *Xline*), but it extends to infinity in only <u>one direction</u> and has one "anchored" endpoint. Like an *Xline*, a *Ray* extends past the drawing area but does not affect the drawing *Limits* or *Extents*. The construction process for a *Ray* is simpler than for an *Xline*, only requiring you to establish a "start point" (endpoint) and a "through point" (Fig. 15-7). Multiple *Rays* can be created in one command.

FIGURE 15-7

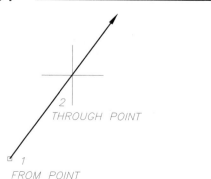

Command: **ray**
Specify start point: **PICK** or (**coordinates**)
Specify through point: **PICK** or (**coordinates**)
Specify through point: **PICK** or (**coordinates**)
Specify through point: **Enter**
Command:

Rays are especially helpful for construction of geometry about a central reference point or for construction of angular features. In each case, the geometry is usually constructed in only one direction from the center or vertex. If horizontal or vertical *Rays* are needed, just toggle on *ORTHO*. To draw a *Ray* at a specific angle, use relative polar coordinates or Polar Tracking. *Endpoint* and other appropriate *Osnaps* can be used with *Rays*. A *Ray* has one *Endpoint* but no *Midpoint*.

Polygon

Pull-down Menu	Command (Type)	Alias (Type)	Short-cut	Screen (side) Menu	Tablet Menu
Draw Polygon	*Polygon*	*POL*	...	*DRAW 1 Polygon*	*P,10*

The *Polygon* command creates a regular polygon (all angles are equal and all sides have equal length). A *Polygon* object appears to be several individual objects but, like a *Pline*, is actually <u>one</u> object. In fact, AutoCAD uses *Pline* to create a *Polygon*. There are two basic options for creating *Polygons*: you can specify an *Edge* (length of one side) or specify the size of an imaginary circle for the *Polygon* to *Inscribe* or *Circumscribe*.

Inscribe/Circumscribe
The command sequence for this default method follows:

Command: **polygon**
Enter number of sides <4>: (**value**)
Specify center of polygon or [Edge]: **PICK** or (**coordinates**)
Enter an option [Inscribed in circle/Circumscribed about circle] <I>: **I** or **C**
Specify radius of circle: **PICK** or (**coordinates**)
Command:

The orientation of the *Polygon* and the imaginary circle are shown in Figure 15-8. Note that the *Inscribed* option allows control of one-half of the distance <u>across the corners</u>, and the *circumscribed* option allows control of one-half of the distance <u>across the flats</u>.

Using *ORTHO ON* with specification of the *radius of circle* forces the *Polygon* to a 90 degree orientation.

Edge
The *Edge* option only requires you to indicate the number of sides desired and to specify the two endpoints of one edge (Fig. 15-8).

FIGURE 15-8

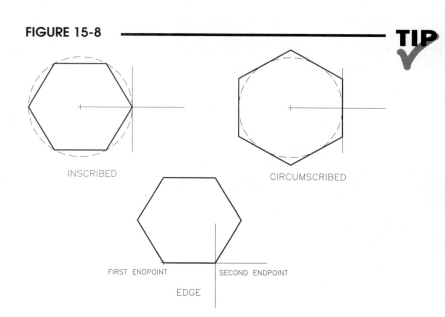

INSCRIBED

CIRCUMSCRIBED

FIRST ENDPOINT   SECOND ENDPOINT

EDGE

TIP

Command: **polygon**
Enter number of sides <4>: (**value**)
Specify center of polygon or [Edge]: **e**
Specify first endpoint of edge: **PICK** or (**coordinates**)
Specify second endpoint of edge: **PICK** or (**coordinates**)
Command:

Because *Polygon*s are created as *Pline*s, *Pedit* can be used to change the line width or edit the shape in some way (see "*Pedit*," Chapter 16). *Polygon*s can also be *Exploded* into individual objects similar to the way other *Pline*s can be broken down into component objects (see "*Explode*," Chapter 16).

Rectang

Pull-down Menu	Command (Type)	ALIAS (Type)	Short-cut	Screen (side) Menu	Tablet Menu
Draw Rectangle	Rectang	REC	...	DRAW 1 Rectang	Q,10

The *Rectang* command only requires the specification of two diagonal corners for construction of a rectangle, identical to making a selection window (Fig. 15-9). The corners can be PICKed or dimensions can be entered. The resulting <u>rectangle is actually a *Pline* object</u>. Therefore, a <u>rectangle is one object</u>, not four separate objects.

Specify first corner point or
[Chamfer/Elevation/Fillet/Thickness/Width]:
PICK, (coordinates) or option
Specify other corner point or [Area/Dimensions/Rotation]:
PICK, (coordinates) or option

FIGURE 15-9

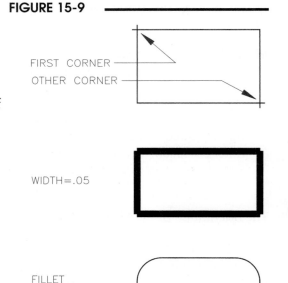

FIRST CORNER
OTHER CORNER

Several options of the *Rectangle* command affect the shape as if it were a *Pline* (see *Pline* and *Pedit*). For example, a *Rectangle* can have *width*:

Specify first corner point or
[Chamfer/Elevation/Fillet/Thickness/Width]: **w**
Specify line width for rectangles <0.0000>: (**value**)

WIDTH=.05

The *Fillet* and *Chamfer* options allow you to specify values for the fillet radius or chamfer distances:

Specify first corner point or
[Chamfer/Elevation/Fillet/Thickness/Width]: **f**
Specify fillet radius for rectangles <0.0000>: (**value**)

FILLET
RADIUS=.5

Specify first corner point or
[Chamfer/Elevation/Fillet/Thickness/Width]: **c**
Specify first chamfer distance for rectangles <0.0000>:
(**value**)
Specify second chamfer distance for rectangles <0.5000>:
(**value**)

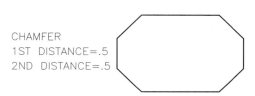

CHAMFER
1ST DISTANCE=.5
2ND DISTANCE=.5

Thickness and *Elevation* are 3D properties.

After specifying the first corner of the rectangle, you can enter the other corner or use these options (Fig. 15-10). The *Dimensions* option prompts for the length and width rather than picking two diagonal corners.

FIGURE 15-10

LENGTH=4
WIDTH=2.5

```
Specify other corner point or [Area/Dimensions/Rotation]: d
Specify length for rectangles <4.0000>: (value)
Specify width for rectangles <2.5000>: (value)
```

The *Area* option prompts for an area value and either the length or width to use for calculating the area.

AREA=10
LENGTH=4

```
Enter area of rectangle in current units <100.0000>:
(value)
Calculate rectangle dimensions based on [Length/Width]
<Length>: l
Enter rectangle length <10.0000>: (value)
```

The *Rotation* option allows you to specify a rotational value for the orientation of the *Rectangle*.

ROTATION=15

```
Specify other corner point or [Area/Dimensions/Rotation]: r
Specify rotation angle or [Pick points] <0>: (value)
```

Donut

Pull-down Menu	Command (Type)	Alias (Type)	Short-cut	Screen (side) Menu	Tablet Menu
Draw *Donut*	*Donut*	*DO*	...	*DRAW 1* *Donut*	*K,9*

A *Donut* is a circle with width (Fig. 15-11). Invoking the command allows changing the inside and outside diameters and creating multiple *Donuts*:

FIGURE 15-11

```
Command: donut
Specify inside diameter of donut
<0.5000>: (value)
Specify outside diameter of donut
<1.0000>: (value)
Specify center of donut or <exit>:
PICK
Specify center of donut or <exit>:
Enter
```

FILL ON

INSIDE DIA = 0.5
OUTSIDE DIA = 1

INSIDE DIA = 0
OUTSIDE DIA = 1

INSIDE DIA = 1.8
OUTSIDE DIA = 2

FILL OFF

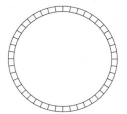

Donuts are actually solid filled circular *Plines* with width. The solid fill for *Donuts*, *Plines*, and other "solid" objects can be turned off with the *Fill* command or *FILLMODE* system variable.

Spline

Pull-down Menu	Command (Type)	Alias (Type)	Short-cut	Screen (side) Menu	Tablet Menu
Draw Spline	Spline	SPL	...	DRAW 1 Spline	L,9

The *Spline* command creates a NURBS (non-uniform rational bezier spline) curve. The non-uniform feature allows irregular spacing of selected points to achieve sharp corners, for example. A *Spline* can also be used to create regular (rational) shapes such as arcs, circles, and ellipses. Irregular shapes can be combined with regular curves, all in one spline curve definition.

Spline is the newer and more functional version of a *Spline*-fit *Pline* (see "*Pedit*," Chapter 16). The main difference between a *Spline*-fit *Pline* and a *Spline* is that a *Spline* curve passes through the points selected, while the points selected for construction of a *Spline*-fit *Pline* only have a "pull" on the curve. Therefore, *Splines* are more suited to accurate design because the curve passes exactly through the points used to define the curve (data points) (Fig. 15-12).

FIGURE 15-12

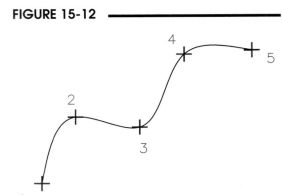

SPLINE
(DEFAULT TANGENTS)

The construction process involves specifying points that the curve will pass through and determining tangent directions for the two ends (for non-closed *Splines*) (Fig. 15-13).

FIGURE 15-13

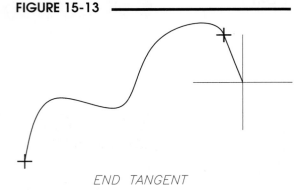

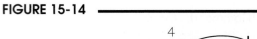

END TANGENT

The *Close* option allows creation of closed *Splines* (these can be regular curves if the selected points are symmetrically arranged) (Fig. 15-14).

FIGURE 15-14

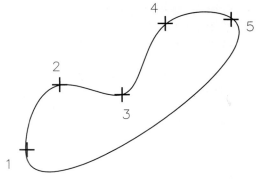

CLOSED SPLINE

Command: *spline*
Specify first point or [Object]: **PICK** or (**coordinates**)
Specify next point: **PICK** or (**coordinates**)
Specify next point or [Close/Fit tolerance] <start tangent>: **PICK** or (**coordinates**)
Specify next point or [Close/Fit tolerance] <start tangent>: **PICK** or (**coordinates**)
Specify next point or [Close/Fit tolerance] <start tangent>: **Enter**
Specify start tangent: **PICK** or **Enter** (Select direction for tangent or Enter for default)
Specify end tangent: **PICK** or **Enter** (Select direction for tangent or Enter for default)
Command:

The *Object* option allows you to convert *Spline*-fit *Plines* into NURBS *Splines*. Only *Spline*-fit *Plines* can be converted (see *"Pedit,"* Chapter 16).

Command: *spline*
Specify first point or [Object]: *o*
Select objects to convert to splines ...
Select objects: **PICK**
Select objects: **Enter**
Command:

FIGURE 15-15

A *Fit Tolerance* applied to the *Spline* "loosens" the fit of the curve. A tolerance of 0 (default) causes the *Spline* to pass exactly through the data points. Entering a positive value allows the curve to fall away from the points to form a smoother curve (Fig. 15-15).

Specify next point or [Close/Fit tolerance] <start tangent>: *f*
Specify fit tolerance <0.0000>: (Enter a positive value)

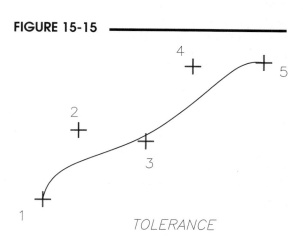

TOLERANCE

Ellipse

Pull-down Menu	Command (Type)	Alias (Type)	Short-cut	Screen (side) Menu	Tablet Menu
Draw *Ellipse*	*Ellipse*	*EL*	...	DRAW 1 *Ellipse*	*M,9*

An *Ellipse* is one object. AutoCAD *Ellipses* are (by default) NURBS curves (see *Spline*). There are three methods of creating *Ellipses* in AutoCAD: (1) specify one <u>axis</u> and the <u>end</u> of the second, (2) specify the <u>center</u> and the ends of each axis, and (3) create an elliptical <u>arc</u>. Each option also permits supplying a rotation angle rather than the second axis length.

Command: *ellipse*
Specify axis endpoint of ellipse or [Arc/Center]: **PICK** or (**coordinates**)
(This is the first endpoint of either the major or minor axis.)
Specify other endpoint of axis: **PICK** or (**coordinates**)
(Select a point for the other endpoint of the first axis.)
Specify distance to other axis or [Rotation]: **PICK** or (**coordinates**)
(This is the distance measured perpendicularly from the established axis.)
Command:

Axis End

This default option requires PICKing three points as indicated in the command sequence above (Fig. 15-16).

FIGURE 15-16

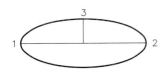
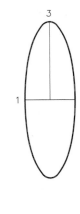
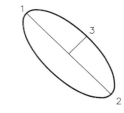

Rotation

If the *Rotation* option is used with the *Axis End* method, the following syntax is used:

> Specify distance to other axis or [Rotation]: **r**
> Specify rotation around major axis: **PICK** or (**value**)

The specified angle is the number of degrees the shape is rotated <u>from the circular position</u> (Fig. 15-17).

FIGURE 15-17

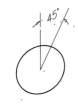

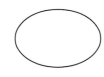

ROTATION = 0   ROTATION = 45   ROTATION = 70

TIP Center

With many practical applications, the center point of the ellipse is known, and therefore the *Center* option should be used (Fig. 15-18):

> Command: **ellipse**
> Specify axis endpoint of ellipse or [Arc/Center]: **c**
> Specify center of ellipse: **PICK** or (**coordinates**)
> Specify endpoint of axis: **PICK** or (**coordinates**) (Select a point for the other endpoint of the first axis.)
> Specify distance to other axis or [Rotation]: **PICK** or (**coordinates**) (This is the distance measured perpendicularly from the established axis.)
> Command:

FIGURE 15-18

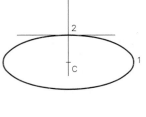
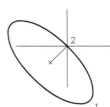
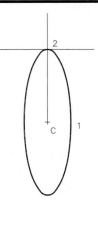

Arc

Use this option to construct an elliptical arc (partial ellipse). The procedure is identical to the *Center* option with the addition of specifying the start- and endpoints for the arc.

Divide

Pull-down Menu	Command (Type)	Alias (Type)	Short-cut	Screen (side) Menu	Tablet Menu
Draw Point > Divide	*Divide*	*DIV*	...	DRAW 2 *Divide*	*V,13*

The *Divide* and *Measure* commands add *Point* objects to existing objects. Both commands are found in the *Draw* pull-down menu under *Point* because they create *Point* objects.

The *Divide* command finds equal intervals along an object such as a *Line, Pline, Spline,* or *Arc* and adds a *Point* object at each interval. The object being divided is <u>not</u> actually broken into parts—it remains as <u>one</u> object. *Point* objects are automatically added to display the "divisions."

The point objects that are added to the object can be used for subsequent construction by allowing you to *OSNAP* to equally spaced intervals (*Nodes*).

The command sequence for the *Divide* command is as follows:

```
Command: divide
Select object to divide: PICK (Only one object can be selected.)
Enter the number of segments or [Block]: (value)
Command:
```

Point objects are added to divide the object selected into the desired number of parts. Therefore, there is <u>one less</u> *Point* added than the number of segments specified.

After using the *Divide* command, the *Point* objects may not be visible unless the point style is changed with the *Point Style* dialog box (*Format* pull-down menu) or by changing the *PDMODE* variable by command line format. Figure 15-19 shows *Points* displayed using a *PDMODE* of 3.

FIGURE 15-19

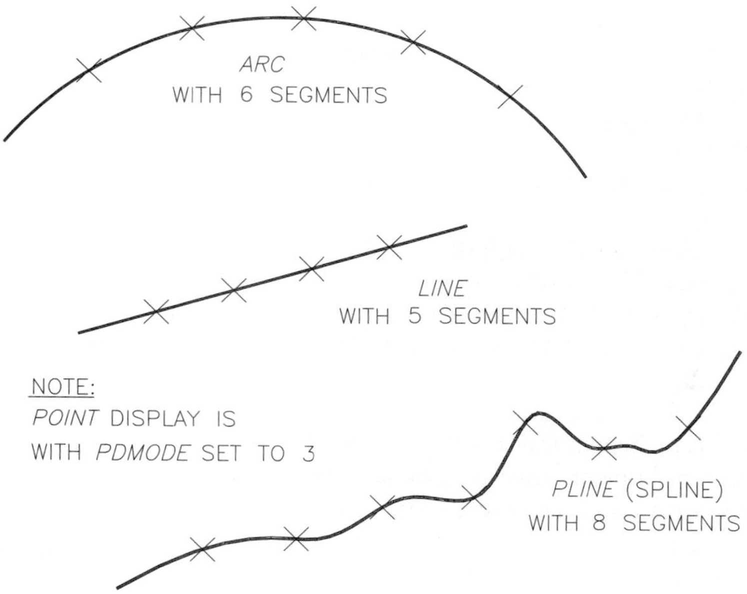

ARC
WITH 6 SEGMENTS

LINE
WITH 5 SEGMENTS

NOTE:
POINT DISPLAY IS
WITH *PDMODE* SET TO 3

PLINE (SPLINE)
WITH 8 SEGMENTS

You can request that *Blocks* be inserted rather than *Point* objects along equal divisions of the selected object. Figure 15-20 displays a generic rectangular-shaped block inserted with *Divide*, both aligned and not aligned with a *Line, Arc,* and *Pline.* In order to insert a *Block* using the *Divide* command, the name of an <u>existing</u> *Block* must be given. (See Chapter 20.)

FIGURE 15-20

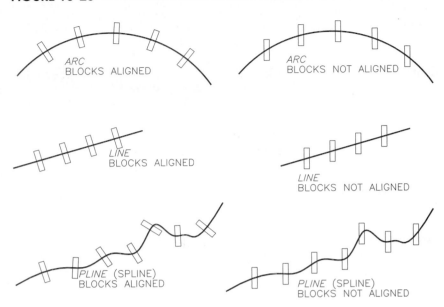

Measure

Pull-down Menu	Command (Type)	Alias (Type)	Short-cut	Screen (side) Menu	Tablet Menu
Draw *Point >* *Measure*	*Measure*	*ME*	...	DRAW 2 *Measure*	V,12

The *Measure* command is similar to the *Divide* command in that *Point* objects (or *Blocks*) are inserted along the selected object. The *Measure* command, however, allows you to designate the <u>length</u> of segments rather than the <u>number</u> of segments as with the *Divide* command.

    Command: **measure**
    Select object to measure: **PICK** (Only one object can be selected.)
    Specify length of segment or [Block]: (**value**) (Enter a value for length of one segment.)
    Command:

Point objects are added to the selected object at the designated intervals (lengths). One *Point* is added for each interval <u>beginning at the end nearest</u> the end used for object selection (Fig. 15-21). The intervals are of equal length except possibly the last segment, which is whatever length is remaining.

You can request that *Blocks* be inserted rather than *Point* objects at the designated intervals of the selected object. Inserting *Blocks* with *Measure* requires that an <u>existing</u> *Block* be used.

FIGURE 15-21

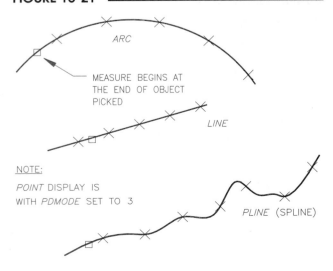

Boundary

Pull-down Menu	Command (Type)	Alias (Type)	Short-cut	Screen (side) Menu	Tablet Menu
Draw *Boundary...*	*Boundary* or *-Boundary*	*BO* or *-BO*	...	*DRAW 2* *Boundary*	Q,9

Boundary finds and draws a boundary from a group of connected or overlapping shapes forming an enclosed area. The shapes can be any AutoCAD objects and can be in any configuration, as long as they form a totally enclosed area. *Boundary* creates either a *Polyline* or *Region* object forming the boundary shape (see *Region* next). The resulting *Boundary* does not affect the existing geometry in any way. *Boundary* finds and includes internal closed areas (called "islands") such as circles (holes) and includes them as part of the *Boundary*.

To create a *Boundary*, select an internal point in any enclosed area (Fig. 15-22). *Boundary* finds the boundary (complete enclosed shape) surrounding the internal point selected. You can use the resulting *Boundary* to construct other geometry or use the generated *Boundary* to determine the *Area* for the shape (such as a room in a floor plan). The same technique of selecting an internal point is also used to determine boundaries for sectioning using *Hatch*.

FIGURE 15-22

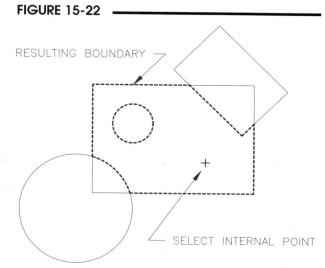

RESULTING BOUNDARY

SELECT INTERNAL POINT

Boundary operates in dialog box mode (Fig. 15-23), or type *-Boundary* for Command line mode. After setting the desired options, select the *Pick Points* tile to select the desired internal point (as shown in Fig. 15-22).

Object Type
Construct either a *Region* or *Polyline* boundary (see *Region* next). If islands are found, two or more *Plines* are formed but only one *Region*.

Boundary Set
The *Current Viewport* option is sufficient for most applications; however, for large drawings you can make a new boundary set by selecting a smaller set of objects to be considered for the possible boundary.

Island Detection
This options tells AutoCAD whether or not to include interior enclosed objects in the Boundary.

Pick Points
Use this option to PICK an enclosed area in the drawing that is anywhere <u>inside</u> the desired boundary.

FIGURE 15-23

Boundary Creation

Pick Points

☑ Island detection

Boundary retention
☑ Retain boundaries
Object type: Polyline

Boundary set
Current viewport | New

OK   Cancel   Help

Region

	Pull-down Menu	Command (Type)	Alias (Type)	Short-cut	Screen (side) Menu	Tablet Menu
	Draw Region...	Region	REG	...	DRAW 2 Region	R,9

The *Region* command converts one object or a set of objects forming a closed shape into one object called a *Region*. This is similar to the way in which a set of objects (*Lines, Arcs, Plines,* etc.) forming a closed shape and having matching endpoints can be converted into a closed *Pline*. A *Region,* however, has special properties:

1. Several *Regions* can be combined with Boolean operations known as *Union, Subtract,* and *Intersect* to form a "composite" *Region*. This process can be repeated until the desired shape is achieved.

2. A *Region* is considered a planar surface. The surface is defined by the edges of the *Region* and no edges can exist within the *Region* perimeter. *Regions* can be used with surface modeling.

In order to create a *Region,* a closed shape must exist. The shape can be composed of one or more objects such as a *Line, Arc, Pline, Circle, Ellipse,* or anything composed of a *Pline (Polygon, Rectangle, Boundary)*. If more than one object is involved, the <u>endpoints must match (having no gaps or overlaps)</u>. Simply invoke the *Region* command and select all objects to be converted to the *Region*.

```
Command: region
Select objects: PICK
Select objects: PICK
Select objects: PICK
Select objects: Enter
1 loop extracted.
1 Region created.
Command:
```

Consider the shape shown in Figure 15-24 composed of four connecting *Lines*. Using *Region* combines the shape into one object, a *Region*. The appearance of the object does not change after the conversion, even though the resulting shape is one object.

Although the *Region* appears to be no different than a closed *Pline,* it is more powerful because several *Regions* can be combined to form complex shapes (composite *Regions*) using the three Boolean operations explained in Chapter 16.

As an example, a set of *Regions* (converted *Circles*) can be combined to form the sprocket with only <u>one</u> *Subtract* operation (Fig. 15-25). Compare the simplicity of this operation to the process of using *Trim* to delete <u>each</u> of the unwanted portions of the small circles.

FIGURE 15-24

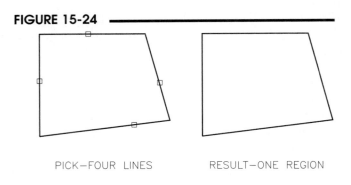

PICK—FOUR LINES   RESULT—ONE REGION

FIGURE 15-25

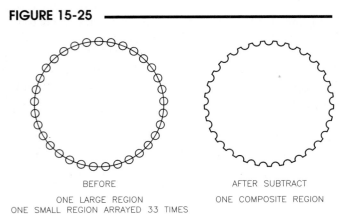

BEFORE   AFTER SUBTRACT
ONE LARGE REGION   ONE COMPOSITE REGION
ONE SMALL REGION ARRAYED 33 TIMES

The Boolean operators, *Union, Subtraction,* and *Intersection,* can be used with *Regions* as well as solids. Any number of these commands can be used with *Regions* to form complex geometry (see Chapter 16, Modify Commands II).

Wipeout

Pull-down Menu	Command (Type)	Alias (Type)	Short-cut	Screen (side) Menu	Tablet Menu
Draw *Wipeout*	*Wipeout*	...	...	...	...

Wipeout creates an "opaque" object that is used to "hide" other objects. The *Wipeout* command creates a closed *Pline* or uses an existing closed *Pline*. All objects behind the *Wipeout* become "invisible." For example, assume a *Pline* shape was created so as to cross other objects. You could use *Wipeout* to make the shape appear to be "on top of" the other objects (Fig. 15-26, left).

FIGURE 15-26

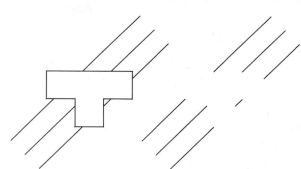

```
Command: wipeout
Specify first point or [Frames/Polyline]
<Polyline>: PICK
Specify next point: PICK
Specify next point or [Undo]: PICK
Specify next point or [Close/Undo]: Enter
Command:
```

You can use the *Frames* option to make the *Wipeout* object disappear. Doing so would result in a display similar to Figure 15-26, right.

```
Specify first point or [Frames/Polyline] <Polyline>: f
Enter mode [ON/OFF] <ON>: off
```

If you have an existing *Pline* and want to use it to create a *Wipeout* of the same shape, use the *Polyline* option.

```
Command: wipeout
Specify first point or [Frames/Polyline] <Polyline>: p
Select a closed polyline: PICK
Erase polyline? [Yes/No] <No>:
```

Selecting *No* to the "Erase polyline?" prompt results in two objects that occupy the same space—the existing *Pline* and the new *Wipeout*. To create an "invisible" *Wipeout* you must specify the *Erase Polyline* option when you create the *Wipeout*, then use the *Frame Off* option to make the *Wipeout* object disappear.

Revcloud

Pull-down Menu	Command (Type)	ALIAS (Type)	Short-cut	Screen (side) Menu	Tablet Menu
Draw *Revision Cloud*	*Revcloud*	...	...	...	...

Use this command to draw a revision cloud on the current layer. Revision clouds are the customary method of indicating an area of a drawing (usually an architectural application) that contains a revision to the original design. Revision clouds should be created on a separate layer so they can be controlled for plotting.

Revcloud is an automated routine that simplifies drawing revision clouds. All you have to do is pick a point and move the cursor in a circular direction (Fig. 15-27). You do not have to "pick" an endpoint. The cloud automatically closes when your cursor approaches the start point and the command ends. A *Revcloud* is a series of *Pline* arc segments.

FIGURE 15-27

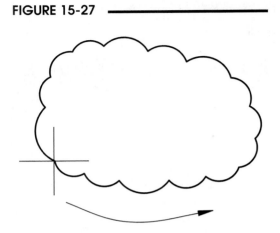

```
Command: revcloud
Minimum arc length: 0.5000   Maximum arc length:
0.5000  Style: Normal
Specify start point or [Arc length/Object/Style] <Object>:
Guide crosshairs along cloud path...
Revision cloud finished.
Command:
```

Arc length

Use the *Arc length* option to specify a value in drawing units for the arc segments. Entering different values for the minimum and maximum arc lengths draws *Revclouds* with different size arcs, similar to that shown in Figure 15-27.

Object

You can convert some closed objects into *Revclouds* using the *Object* option. *Rectangles, Polygons, Circles,* and *Ellipses* can be converted to *Revclouds*.

Style

Use this option to specify the line style that the *Revcloud* is drawn with, either *Normal* or *Calligraphy*. The *Normal* option creates arcs with a constant line width of 0, such as shown in Figure 15-27. The *Calligraphy* option creates a "stroke" similar to a calligraphy pen. With this option each arc is drawn starting with a thin lineweight and ending with a thick lineweight.

CHAPTER EXERCISES

1. *Polygon*

   Open the **PLINE1** drawing you created in Chapter 8 Exercises. Use *Polygon* to construct the polygon located at the *Center* of the existing arc, as shown in Figure 15-28. When finished, *SaveAs* **POLYGON1.**

FIGURE 15-28

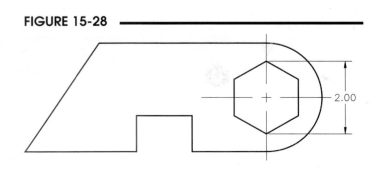

2. *Xline, Ellipse, Rectangle*

FIGURE 15-29

Open the **APARTMENT** drawing that you created in Chapter 12. Draw the floor plan of the efficiency apartment on layer FLOORPLAN as shown in Figure 15-29. Use *SaveAs* to assign the name **EFF-APT.** Use *Xline* and *Line* to construct the floor plan with 8" width for the exterior walls and 5" for interior walls. Use the *Ellipse* and *Rectangle* commands to design and construct the kitchen sink, tub, wash basin, and toilet. *Save* the drawing but do not exit AutoCAD. Continue to Exercise 3.

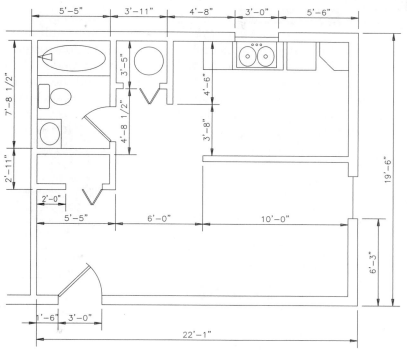

3. *Pline, Page Setup, Vports, and Viewport Scale*

This exercise gives you experience creating a border and plotting the efficiency apartment to scale. To complete the exercise as instructed, you should have a device configured to plot on a "C" size sheet.

A. Invoke the *Options* dialog box and access the *Display* tab. In the *Layout Elements* section (lower left), ensure that *Show Page Setup Manager for New Layouts* and *Create Viewport in New Layouts* <u>are not checked</u>.

B. Select the *Layout1* tab. Activate the *Page Setup Manager*. Select *Modify*, then select a plot device from the list that will allow you to plot on an architectural "C" size sheet (**24" x 18"**). Next, select the correct sheet size. Pick *OK* to finish the setup.

C. Make layer **TITLE** the *Current* layer. Draw a border (in paper space) using a *Pline* with a *width* of **.04**, and draw the border with a **.75** unit margin within the edge of the *Limits* (border size of 22.5 x 16.5). *Save* the drawing as **EFF-APT-PS.**

D. Make layer **VPORTS** *Current*. Use *Vports* to create a *Single* viewport. Pick diagonal corners for the viewport at **1,1** and **23,17**. The apartment drawing should appear in the viewport at no particular scale. Use any method to set the viewport scale to *1/2"=1'*.

FIGURE 15-30

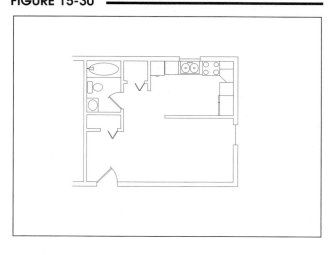

E. Next, *Freeze* layer **VPORTS**. Your completed drawing should look like that in Figure 15-30. *Save* the drawing. *Plot* the drawing from the layout at **1:1** scale.

4. *Donut*

FIGURE 15-31

Open **SCHEMATIC1** drawing you created in Chapter 9
Exercises. Use the *Point Style* dialog box (*Format* pull-down
menu) to change the style of the *Point* objects to dots instead of
small circles. Turn on the *Node Running Osnap* mode. Use the
Donut command and set the *Inside diameter* and the *Outside
diameter* to an appropriate value for your drawing such that the
Inside diameter is 1/2 the value of the *Outside diameter*. Create a
Donut at each existing *Node*. *SaveAs* **SCHEMATIC1B**. Your finished drawing should look like that in Figure 15-31.

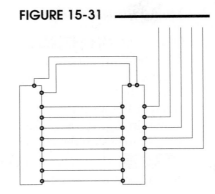

5. *Ellipse, Polygon*

FIGURE 15-32

Begin a *New* drawing using the
template drawing that you
created in Chapter 12 named
A-METRIC Use *SaveAs* and
assign the name **WRENCH.**

Complete the construction of
the wrench shown in Figure
15-32. Center the drawing
within the *Limits*. Use *Line*,
Circle, Arc, Ellipse, and
Polygon to create the shape.
For the "break lines," create one

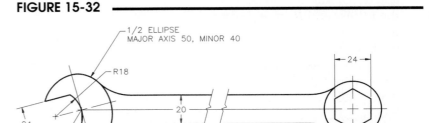

connected series of jagged *Lines*, then *Copy* for the matching set. Utilize *Trim, Rotate*, and other
edit commands where necessary. HINT: Draw the wrench head in an orthogonal position; then
rotate the entire shape 15 degrees. Notice the head is composed of ½ of a *Circle* (radius=25) and ½
of an *Ellipse*. *Save* the drawing when completed. Activate a *Layout* tab, configure a print or plot
device using *Pagesetup*, make one *Viewport*, and *Plot* the drawing at a standard scale.

FIGURE 15-33

6. *Xline, Spline*

Create the handle shown in Figure 15-33.
Construct multiple *Horizontal* and *Vertical Xlines* at
the given distances on a separate layer (shown as
dashed lines in Figure 15-33). Construct two
Splines that make up the handle sides by PICKing
the points indicated at the *Xline* intersections.
Note the *End Tangent* for one *Spline*. Connect the
Splines at each end with horizontal *Lines. Freeze*
the construction (*Xline*) layer. *Save* the drawing as
HANDLE.

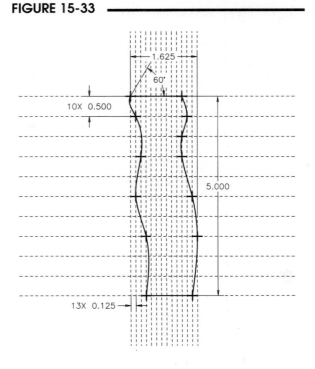

7. *Divide, Measure*

Use the **ASHEET** template drawing, use *SaveAs,* and assign the name **BILLMATL.**   Create the table in Figure 15-34 to be used as a bill of materials. Draw the bottom *Line* (as dimensioned) and a vertical *Line.* Use *Divide* along the bottom *Line* and *Measure* along the vertical *Line* to locate *Points* as desired. Create *Offsets Through* the *Points* using *Node OSNAP.* (*POLAR* and *Trim* may be of help.)

FIGURE 15-34

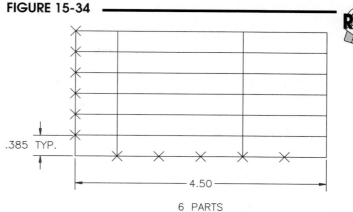

.385 TYP.

4.50

6 PARTS

8. *Boundary*

Open the **GASKETA** drawing that you created in Chapter 9 Exercises. Use the *Boundary* command to create a boundary of the <u>outside shape only</u> (no islands). Use *Move Last* to displace the new shape to the right of the existing gasket. Use *SaveAs* and rename the drawing to **GASKET-AREA.**   Keep this drawing to determine the *Area* of the shape after reading Chapter 17.

9. *Region*

Create a gear, using *Regions.* Begin a *New* drawing, using the **ACAD.DWT** template or *Start from Scratch* with the **English** defaults settings. Use *Save* and assign the name **GEAR-REGION.**

A. Set *Limits* at 8 x 6 and *Zoom All.* Create a *Circle* of **1** unit *diameter* with the center at **4,3.** Create a second concentric *Circle* with a *radius* of **1.557.** Create a closed *Pline* (to represent one gear tooth) by entering the following coordinates:

From point:	**5.571,2.9191**
To point:	**@.17<160**
To point:	**@.0228<94**
To point:	**@.0228<86**
To point:	**@.17<20**
To point:	**c**

The gear at this stage should look like that in Figure 15-35.

B. Use the *Region* command to convert the three shapes (two *Circles* and one "tooth") to three *Regions.*

FIGURE 15-35

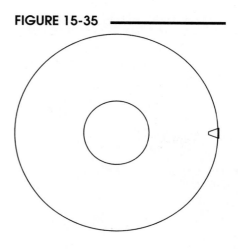

C. **Array** the small *Region* (tooth) in a **Polar** array about the center of the gear. There are **40** items that are rotated as they are copied. This action should create all the teeth of the gear (Fig. 15-36).

D. **Save** the drawing. The gear will be completed in Chapter 16 Exercises by using the *Subtract* Boolean operation.

FIGURE 15-36

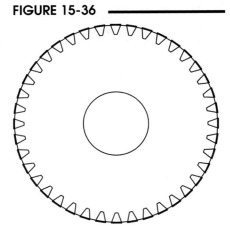

10. *Revcloud*

FIGURE 15-37

Assume you are working with the city engineering office checking the new apartment design submitted to your office for approval. **Open** **EFF-APT** drawing from Exercise 2. You notice that the location of the front door has some issues concerning passing the safety codes. Use **Revcloud** to denote the area for review. Use the **Arc length** option and specify a **Minimum arc length** of **1′** and a **Maximum arc length** of **2′**. Draw a revision cloud similar to that shown in Figure 15-37. Use *SaveAs* to save the drawing as **EFF-APP-REV**.

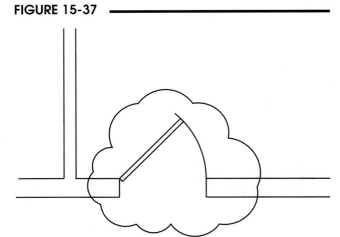

16

MODIFY COMMANDS II

CHAPTER OBJECTIVES

After completing this chapter you should:

1. be able to use *Properties* to modify any type of object;

2. know that you can double-click on any object to produce the *Properties* palette or a more specific editing tool;

3. be able to use *Matchprop* and the *Object Properties* toolbar to change properties of an object;

4. be able to use *Chprop* to change an object's properties;

5. know how to use *Change* to change points or properties;

6. be able to *Explode* a *Pline* into its component objects;

7. be able to use all the *Pedit* options to modify *Plines* and to convert *Lines* and *Arcs* to *Plines*;

8. be able to modify *Splines* with *Splinedit*;

9. know that composite *Regions* can be created with *Union*, *Subtract*, and *Intersect*.

CONCEPTS

This chapter examines commands that are similar to, but generally more advanced and powerful than, those discussed in Chapter 9, Modify Commands I. None of the commands in this chapter create new duplicate objects from existing objects but instead modify the properties of the objects or convert objects from one type to another. Several commands are used to modify specific types of objects such as *Pedit* (modifies *Plines*), *Splinedit* (modifies *Splines*), and *Union, Subtract,* and *Intersect* (modify *Regions*). Several of the commands and features discussed in this chapter were mentioned in Chapter 11 (*Properties, MatchProp,* and *Object Properties* toolbar) but are explained completely here.

Only some commands in this chapter that modify general object properties have icon buttons that appear in the AutoCAD Drawing Editor by default. For example, you can access *Properties* and *Matchprop* from the *Object Properties* toolbar and *Explode* by using its icon from the *Modify* toolbar.

Other commands that modify specific objects such as *Pedit* and *Splinedit* appear in the *Modify II* toolbar (Fig. 16-1). The Boolean operators (*Union, Subtract,* and *Intersect*) appear in the *Solids Editing* toolbar. Activate a toolbar by right-clicking on any tool (icon button) and selecting from the list.

If you are using the *2D Drafting & Annotation* workspace, the modify commands are available from the *2D Draw* control panel.

FIGURE 16-1

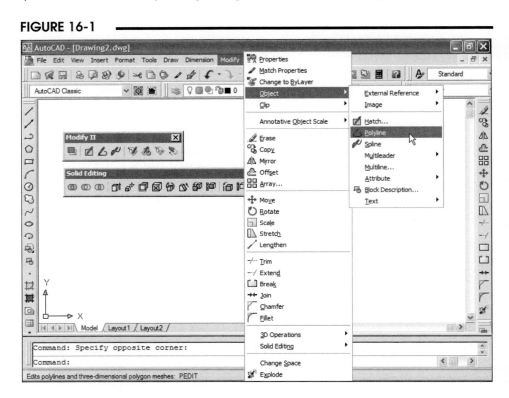

COMMANDS

You can use several methods to change properties of an object or of several objects. If you want to change an object's layer, color, or linetype only, the quickest method is by using the *Object Properties* toolbar (see "*Object Properties*" toolbar). If you want to change many properties of one or more objects (including layer, color, linetype, linetype scale, text style, dimension style, or hatch style) to match the properties of another existing object, use *Matchprop*. If you want to change any property (including coordinate data) for any object, use *Properties*.

You can double-click on any object (assuming the *Dblclkedit* command is set to *On*) to produce the *Properties* palette or a more specific editing tool, such as the *Hatch Edit* dialog box or the *Multiline Text Editor*, depending on the type of object you select.

Considerations for Changing Basic Properties of Objects

The *Properties* palette, the *Object Properties* toolbar, and the *Matchprop, Chprop,* and *Change* commands can be used to change the basic properties of objects. Here are several basic properties that can be changed and considerations when doing so.

Layer

By changing an object's *Layer*, the selected object is effectively <u>moved</u> to the designated layer. In doing so, if the object's *Color, Linetype,* and *Lineweight* are set to *BYLAYER*, the object assumes the color, linetype, and lineweight of the layer to which it is moved.

Color

It may be desirable in some cases to assign explicit *Color* to an object or objects independent of the layer on which they are drawn. The properties editing commands allow changing the color of an existing object from one object color to another, or from *BYLAYER* assignment to an object color. <u>An object drawn with an object-specific color can also be changed to *BYLAYER* with this option.</u>

Linetype

An object can assume the *Linetype* assigned *BYLAYER* or can be assigned an object-specific *Linetype*. The *Linetype* option is used to change an object's *Linetype* to that of its layer (*BYLAYER*) or to any object-specific linetype that has been loaded into the current drawing. <u>An object drawn with an object-specific linetype can also be changed to *BYLAYER* with this option.</u>

Linetype Scale

When an individual object is selected, the <u>object's individual linetype scale</u> can be changed with this option, but <u>not the global</u> linetype scale (*LTSCALE*). <u>This is the recommended method to alter an individual object's linetype scale.</u> First, draw all the objects in the current global *LTSCALE*. One of the properties editing commands could then be used with this option to retroactively adjust the linetype scale of <u>specific</u> objects. The result would be similar to setting the *CELTSCALE* before drawing the specific objects. Using this method to adjust a specific object's linetype scale <u>does not reset</u> the global *LTSCALE* or *CELTSCALE* variables.

Thickness

An object's *Thickness* can be changed by this option. *Thickness* is a three-dimensional quality (Z dimension) assigned to a two-dimensional object.

Lineweight

An object can assume the *Lineweight* assigned *ByLayer* or can be assigned an object-specific *Linetype*. The *Lineweight* option is used to change an object's *Lineweight* to that of its layer (*ByLayer*) or to any object-specific linetype that has been loaded into the current drawing. <u>An object drawn with an object-specific linetype can also be changed to *ByLayer* with this option.</u>

Plotstyle

Use this option to change an object's *Plotstyle*. Plot styles assigned as *ByLayer* or to individual objects can change the way the objects appear in plots, such as having certain screen patterns, line end joints, plotted colors, plotted lineweights, and so on.

Properties

	Pull-down Menu	Command (Type)	Alias (Type)	Short-cut	Screen (side) Menu	Tablet Menu
	Modify *Properties…*	*Properties*	*PR* or *CH*	**(Edit Mode)** *Properties* or *Ctrl+1*	MODIFY1 *Property*	*Y,12* to *Y,13*

The *Properties* palette (Fig. 16-2) gives you complete access to one or more objects' properties. You can edit the properties simply by changing the entries in the right column.

Once opened, this window remains on the screen until dismissed by clicking the "X" in the upper corner. The *Properties* palette can be "docked" on the side of the screen (Fig. 16-2) or can be "floating." You can toggle the palette on and off with Ctrl+1 (to appear and disappear).

You can also set the palette to *Auto-hide* so only the title bar appears on your screen until you point to it to produce the full palette. If you click on the bottom icon on the title bar (*Properties*), a shortcut menu displays options to *Move*, *Size*, *Close*, *Allow Docking*, toggle *Auto-hide*, and toggle the *Description* at the bottom of the palette (Fig. 16-3). Changes you make to this palette are "persistent" (remain until changed).

The contents of the palette change based on what type and how many objects are selected. The power of this feature is apparent because the contents of the palette are specific to the type of object that you select. For example, if you select a *Line*, entries specific to that *Line* appear, allowing changes to any properties that the *Line* possesses (see Fig. 16-2). Or, if you select a *Circle* or some *Text*, a window appears specific to the *Circle* or *Text* properties.

FIGURE 16-2

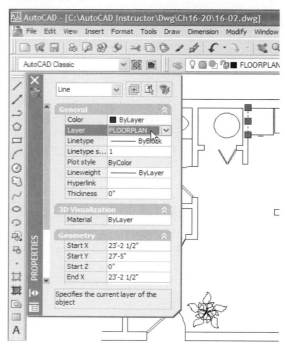

FIGURE 16-3

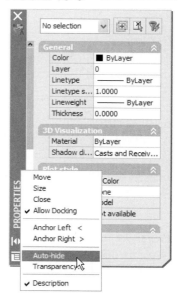

TIP You can use the palette three ways: you can invoke the palette, then select (highlight) one or more objects, you can select objects first and then invoke the palette, or you can double-click on an object (for most objects) to produce the *Properties* palette (see *DBLCLKEDIT*). When multiple objects are selected, a drop-down list in the top of the palette appears. Select the listed object(s) whose properties you want to change.

If you prefer to leave the palette on your screen, select the objects you want to change when no commands are in use with the pickbox or *Auto* window (see Fig. 16-2, highlighted line). <u>After</u> changing the objects' properties, press Escape to deselect the objects. Pressing Escape does not dismiss the *Properties* palette, nor does it undo the changes as long as the cursor is outside the palette.

If no objects are selected, the *Properties* palette displays "No selection" in the drop-down list at the top (Fig. 16-3) and gives the current settings for the drawing. Drawing-wide settings can be changed such as the *LTSCALE* (see Fig. 16-3).

When objects are selected, the dialog box gives access to properties of the selected objects. Figure 16-2 displays the palette after selecting a *Line*. Notice the selection (*Line*) in the drop-down box at the top of the palette. The properties displayed in the central area of the window are specific to the object or group of objects highlighted in the drawing and selected from the drop-down list. In this case (see Fig. 16-2), any aspect of the *Line* can be modified.

If a *Pline* is selected, for example, any property of the *Pline* can be changed. For example, the *Width* of the *Pline* can be changed by entering a new value in the *Properties* palette (Fig. 16-4).

FIGURE 16-4

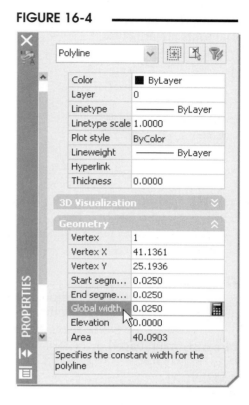

If a dimension is selected, for example, properties specific to a dimension, or to a group of dimensions if selected, can be changed. Note the list of variables that can be changed for a single dimension in Figure 16-5.

The *General* group lists basic properties for an object such as *Layer*, *Color*, or *Linetype*. To change an object's layer, linetype, lineweight, or color properties, highlight the desired objects in the drawing and select the properties you want to change from the right side of the palette. If you use the *ByLayer* strategy of linetype, lineweight, and color properties assignment, you can change an object's linetype, lineweight, and color simply by changing its layer (see Fig. 16-2). This method is recommended for most applications and is described in Chapter 11.

FIGURE 16-5

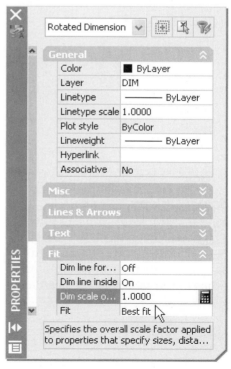

2007

The *Geometry* group lists and allows changing any values that control the selected object's geometry. For example, to change the diameter of a *Circle*, highlight the property and change the value in the right side of the palette (Fig. 16-6).

The *3D Visualization* group lists only *Material* by default for most objects. The *Bylayer*, *Byblock*, and *Global* properties determine how materials are attached.

FIGURE 16-6

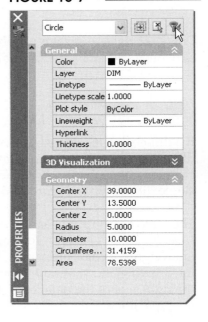

Quick Select

The *Quick Select* button appears near the upper-right corner of the *Properties* palette (Fig. 16-7). Selecting this button produces the *Quick Select* dialog box in which a selection set can be constructed based on criteria you choose from the dialog box.

FIGURE 16-7

Select Objects

Normally you can select objects by the typical methods any time the *Properties* palette is open. When objects are selected, AutoCAD searches its database and presents the properties of each object, one at a time, in the *Properties* palette. Instead, you can use the *Select Objects* button to save time posting information on each object individually to the window. Using *Select objects*, the properties of the total set of selected objects are not posted to the *Properties* palette until you press Enter.

Toggle Value of Pickadd Sysvar

This button toggles the *PICKADD* system variable on or off (1 or 0). This variable affects object selection at all times, not only for use with the *Properties* palette. *PICKADD* determines whether objects you PICK are added to the current selection set (normal setting, *PICKADD*=1) or replace the previous object or set (*PICKADD*=0). Changing the *PICKADD* variable to 0 works well with the *Properties* palette because selecting an object replaces the previous set of properties appearing in the palette with the new object's properties.

The button position of "+" (plus symbol) indicates a current setting of 1, or on, for *PICKADD* (PICKed objects are added). Confusing as it appears, a button position of "1" indicates a <u>current setting of 0</u>, or off, for *PICKADD* (PICKed objects replace the previous ones).

NOTE: Since the *PICKADD* setting affects object selection anytime, <u>make sure you set this variable back to your desired setting (usually on, or "+")</u> before dismissing the *Properties* palette.

DBLCLKEDIT

You can double-click on an object to produce the *Properties* palette or other similar dialog box to edit the object. The *DBLCLKEDIT* variable (double-click edit) controls whether double-clicking an object produces a dialog box.

```
Command: dblclkedit
Enter new value for DBLCLKEDIT <ON>:
```

DBLCLKEDIT is a system variable; however, you can change the setting simply by typing the variable at the Command line. (*DBLCLKEDIT* was originally a considered a command until recent versions of AutoCAD.)

If double-click editing is turned on, the *Properties* palette or other dialog box is displayed when an object is double-clicked. When you double-click most objects, the *Properties* palette is displayed.

However, double-clicking some types of objects displays editing tools that are specific to the type of object. For example, double-clicking a line of *Text* produces the *Text Edit* dialog box. The object types (and resulting editing tool) <u>that do not produce the *Properties* palette</u> when the object is double-clicked are listed below. These objects and the related editing tools are discussed fully in upcoming chapters of this text.

Attribute	Displays the *Edit Attribute Definition* dialog box (*Ddedit*).
Attribute within a block	Displays the *Enhanced Attribute Editor* (*Eattedit*).
Block	Displays the *Reference Edit* dialog box (*Refedit*).
Hatch	Displays the *Hatch Edit* dialog box (*Hatchedit*).
Leader text	Displays the *Multiline Text Editor* dialog box (*Ddedit*).
Mline	Displays the *Multiline Edit Tools* dialog box (*Mledit*).
Mtext	Displays the *Multiline Text Editor* dialog box (*Ddedit*).
Text	Displays the *Edit Text* dialog box (*Ddedit*).
Xref	Displays the *Reference Edit* dialog box (*Refedit*).

NOTE: The *PICKFIRST* system variable must be on (set to 1) for the *Properties* palette or other editing tool to appear when an object is double-clicked.

Matchprop

Pull-down Menu	Command (Type)	Alias (Type)	Short-cut	Screen (side) Menu	Tablet Menu
Modify *Match Properties*	*Matchprop*	*MA*	...	*MODIFY1* *Matchprp*	**Y,14 and Y,15**

Matchprop is explained briefly in Chapter 11 but is explained again in this chapter with the full details of the *Special Properties* palette.

Matchprop is used to "paint" the properties of one object to another. Simply invoke the command, select the object that has the desired properties ("source object"), then select the object you want to "paint" the properties to ("destination object"). The Command prompt is as follows:

```
Command: matchprop
Select source object: PICK
Current active settings:  Color Layer Ltype LTSCALE Lineweight Thickness PlotStyle Text Dim Hatch
    Polyline Viewport
Select destination object(s) or [Settings]: PICK
Select destination object(s) or [Settings]: PICK
Select destination object(s) or [Settings]: Enter
Command:
```

You can select several destination objects. The destination object(s) assume all of the "Current active settings" of the source object.

The *Property Settings* dialog box can be used to set which of the *Basic Properties* palette and *Special Properties* palette are to be painted to the destination objects (Fig. 16-8). At the "Select destination object(s) or [Settings]:" prompt, type *S* to display the dialog box. You can control the following *Basic Properties* palette. Only the checked properties are painted to the destination objects.

FIGURE 16-8

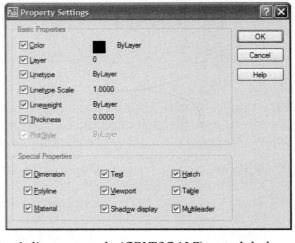

Color	This option paints the object-specific or *ByLayer* color.
Layer	Move selected objects to the source object layer with this option.
Linetype	Paint the object-specific or *ByLayer* linetype of the source object to the destination object.
Linetype Scale	This option changes the individual object's linetype scale (*CELTSCALE*) not global linetype scale (*LTSCALE*).
Thickness	Thickness is a 3-dimensional quality.
Lineweight	You can paint the object-specific or *ByLayer* lineweight of the source object to the destination object.
PlotStyle	The plot style of the source object is painted to the destination object when this setting is checked.

The *Special Properties* palette section allows you to specify features of dimensions, text, hatch patterns, viewports, and polylines to match, as explained next.

Dimension	This setting paints the *Dimension Style.* A *Dimension Style* defines the appearance of a dimension such as text style, size of arrows and text.
Text	This setting paints the source object's text *Style.* The text *Style* defines the text font and many other parameters that affect the appearance of the text.
Hatch	Checking this box paints the hatch properties such as *Pattern, Angle, Scale,* and other characteristics.
Polyline	If you use *Matchprop* with *Plines*, the *Width* and *Linetype Generation* properties in addition to basic object properties of the first *Pline* are painted to the second. If the source polyline has variable *Width*, it is not transferred.
Viewport	If you match properties from one paper space viewport to another, the following properties are changed in addition to the basic object properties: *On/Off, Display Locking*, standard or custom *Scale, Shadeplot, Snap, Grid,* and *UCSicon* settings.
Table	Use this option to change the table style of the destination object to that of the source object.
Material	This setting changes the material applied to the destination object.
Shadow Display	A 3D object can cast shadows, receive shadows, or both, or it can ignore shadows. This property changes the shadow display.
Multileader	This option paints the *Multileader Style.* A *Multileader Style* determines the appearance of the leaders such as arrowheads, text, and line.

Object Properties Toolbar

The five drop-down lists in the Object Properties toolbar (when not "dropped down") generally show the <u>current</u> layer, color, linetype, lineweight, and plot style. However, if an object or set of objects is selected, the information in these boxes changes to display the current <u>object's</u> settings. You can change an object's settings by picking an object (when no commands are in use), then dropping down any of the three lists and selecting a different layer, linetype, or color, etc.

Make sure you select (highlight) an object when no commands are in use. Use the pickbox (that appears on the cursor), *Window,* or *Crossing Window* to select the desired object or set of objects. The entries in the five boxes (Layer Control, Color Control, Linetype Control, etc.) then change to display the settings for the <u>selected object or objects</u>. If several objects are selected that have different properties, the boxes display no information (go "blank").

Next, use any of the drop-down lists to make another selection (Fig. 16-9). The highlighted object's properties are changed to those selected in the lists. Press the Escape key to complete the process.

FIGURE 16-9

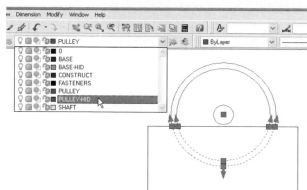

Remember that in most cases color, linetype, lineweight, and plot style settings are assigned *ByLayer.* In this type of drawing scheme, to change the linetype or color properties of an object, you would change only the object's <u>layer</u> (see Fig. 16-9). If you are using this type of drawing scheme, refrain from using the Color Control and Linetype Control dropdown lists for changing properties.

This method for changing object properties is about as quick and easy as *Matchprop*; however, only the layer, linetype, color, lineweight, or plot style can be changed with the *Object Properties* toolbar. The *Object Properties* toolbar method works well if you do not have other existing objects to match or if you want to change <u>only</u> layer, linetype, color, lineweight, and plot style <u>without</u> matching text, dimension, and hatch styles, and viewport or polyline properties.

TIP

Chprop

Pull-down Menu	Command (Type)	Alias (Type)	Short-cut	Screen (side) Menu	Tablet Menu
...	Chprop	...	...	...	...

Chprop allows you to change basic properties of one or more objects using Command line format. If you want to change properties of several objects, pick several objects or select with a window or crossing window at the "Select objects:" prompt.

```
Command: chprop
Select objects: PICK
Select objects: Enter
Enter property to change [Color/LAyer/LType/ltScale/LWeight/Thickness/Material/Annotative]:
```

Change

Pull-down Menu	Command (Type)	Alias (Type)	Short-cut	Screen (side) Menu	Tablet Menu
...	Change	-CH	...	...	...

The *Change* command allows changing three options: *Points, Properties,* or *Text.*

Point

This option allows changing the endpoint of an object or endpoints of several objects to one new position:

```
Command: change
Select objects: PICK
Select objects: Enter
Specify change point or [Properties]: PICK (Select a point to establish as new endpoint of all objects.
OSNAPs can be used.)
```

The endpoint(s) of the selected object(s) <u>nearest</u> the new point selected at the "Specify change point or [Properties]:" prompt is changed to the new point (Fig. 16-10).

Properties

These options are discussed previously. The *Elevation* property (a 3D property) of an object can also be changed only with *Change.*

```
Specify change point or [Properties]: p
Enter property to change
[Color/Elev/LAyer/LType/ltScale/
    LWeight/Thickness/Material/Annotative]:
```

FIGURE 16-10

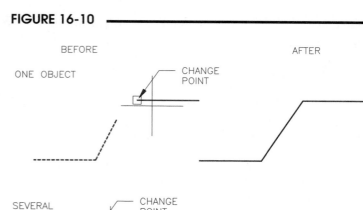
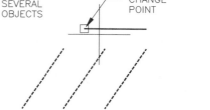

BEFORE   AFTER

ONE OBJECT   CHANGE POINT

SEVERAL OBJECTS   CHANGE POINT

Text

Although the word *"Text"* does <u>not</u> appear as an option, the *Change* command recognizes text created with *Text* if selected. <u>*Change* does not change *Mtext*</u>. (See Chapter 18 for information on *Text* and *Mtext*.)

You can change the following characteristics of *Text* objects:

    Text insertion point
    Text style
    Text height
    Text rotation angle
    Textual content

To change text, use the following command syntax:

    Command: **change**
    Select objects: **PICK** (Select one or several lines of text)
    Select objects: **Enter**
    Specify change point or [Properties]: **Enter**
    Specify new text insertion point <no change>: **PICK** or **Enter**
    Enter new text style <Standard>: **(text style name)** or **Enter**
    Specify new height <0.2000>: **(value)** or **Enter**
    Specify new rotation angle <0>: **(value)** or **Enter**
    Enter new text <text>: **(new text)** or **Enter** (Enter the complete new line of text)

Explode

Pull-down Menu	Command (Type)	Alias (Type)	Short-cut	Screen (side) Menu	Tablet Menu
Modify *Explode*	*Explode*	*X*	...	*MODIFY2* *Explode*	*Y,22*

Many graphical shapes can be created in AutoCAD that are made of several elements but are treated as one object, such as *Plines, Polygons, Blocks, Hatch* patterns, and dimensions. The *Explode* command provides you with a means of breaking down or "exploding" the complex shape from one object into its many component segments (Fig. 16-11). Generally, *Explode* is used to allow subsequent editing of one or more of the component objects of a *Pline, Polygon,* or *Block*, etc., which would otherwise be impossible while the complex shape is considered one object.

The *Explode* command has no options and is simple to use. You only need to select the objects to *Explode*.

FIGURE 16-11

BEFORE EXPLODE

1 PLINE SPLINE

1 PLINE

1 POLYGON

1 BLOCK

EACH SHAPE IS ONE OBJECT

AFTER EXPLODE

16 LINES

4 LINES

6 LINES

7 LINES

EACH SHAPE IS EXPLODED INTO SEVERAL OBJECTS

Command: **explode**
Select objects: **PICK** (Select one or more *Plines*, *Blocks*, etc.)
Select objects: **Enter** (Indicates selection of objects is complete.)

 TIP When *Plines*, *Polygons*, *Blocks*, or hatch patterns are *Exploded*, they are transformed into *Line*, *Arc*, and *Circle* objects. Beware, *Plines* having *width* lose their width information when *Exploded* since *Line*, *Arc*, and *Circle* objects cannot have width. *Exploding* objects can have other consequences such as losing "associativity" of dimensions and hatch objects and increasing file sizes by *Exploding Blocks*.

Pedit

Pull-down Menu	Command (Type)	Alias (Type)	Short-cut	Screen (side) Menu	Tablet Menu
Modify *Object >* *Polyline*	*Pedit*	*PE*	...	*MODIFY1* *Pedit*	*Y,17*

This command provides numerous options for editing *Polylines* (*Plines*). As an alternative, *Properties* can be used to change many of the *Pline's* features in dialog box form (see Fig. 16-4). The list of options below emphasizes the great flexibility possible with *Polylines*. The first step after invoking *Pedit* is to select the *Pline* to edit.

Command: **pedit**
Select polyline or [Multiple]: **PICK**
Enter an option [Close/Join/Width/Edit vertex/Fit/Spline/Decurve/Ltype gen/Undo]:

 TIP *Multiple*
The *Multiple* option allows multiple *Plines* to be edited simultaneously. The selected *Plines* can be totally separate objects and do not have to be connected in any way. Once the *Multiple* option is invoked and the objects are selected, any *Pline* option, such as *Close*, *Open*, *Join*, *Width*, *Fit*, *Spline*, *Decurve*, or *Ltype gen*, operates on the selected *Plines*.

Command: **Pedit**
Select polyline or [Multiple]: **m**
Select objects: **PICK**
Select objects: **PICK**
Select objects: **Enter**
Enter an option [Close/Open/Join/Edit vertex/Width/Fit/Spline/Decurve/Ltype gen/Undo]:

For example, you can change the width of all *Plines* in a drawing simultaneously using the *Multiple* option, selecting all *Plines*, and using the *Width* option.

Close/Open

Close connects the last segment with the first segment of an existing "open" *Pline,* resulting in a "closed" *Pline* (Fig. 16-12). A closed *Pline* is one continuous object having no specific start or endpoint, as opposed to one closed by PICKing points. A *Closed Pline* reacts differently to the *Spline* option and to some commands such as *Fillet, Pline* option (see "*Fillet,*" Chapter 9). *Open* removes the closing segment if the *Close* option was used previously (Fig. 16-12).

FIGURE 16-12

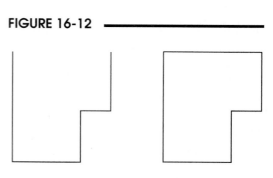

OPEN   CLOSE

Join

The *Join* option combines two or more objects into one *Pline*. The line segments do not have to meet exactly for *Join* to work. You must first use the *Multiple* option, then *Join*.

Command: **Pedit**
Select polyline or [Multiple]: ***m***
Select objects: **PICK**
Select objects: **Enter**
Enter an option [Close/Open/Join/Width/Fit/Spline/Decurve/Ltype gen/Undo]: ***j***
Join Type = Extend
Enter fuzz distance or [Jointype] <0.0000>: **.5**
1 segments added to polyline
Enter an option [Close/Open/Join/Width/Fit/Spline/Decurve/Ltype gen/Undo]:

If the ends of the line segments do not touch but are within a distance that you can set, called the *fuzz distance*, the ends can be joined by *Pedit*. *Pedit* handles this automatically by either extending and trimming the line segments or by adding a new line segment based on your setting for *Jointype*.

Select objects: **PICK**
Enter an option [Close/Open/Join/Width/Fit/Spline/Decurve/Ltype gen/Undo]: ***j***
Join Type = Extend
Enter fuzz distance or [Jointype] <2.0000>: ***j***
Enter join type [Extend/Add/Both] <Extend>:

Extend
This option causes AutoCAD to join the selected polylines by extending or trimming the segments to the nearest endpoints (see Fig. 16-13).

Add
Use this option to add a straight segment between the nearest endpoints (see Fig. 16-13).

Both
If you use this option, the selected polylines are joined by extending or trimming if possible. If not, as in the case of near parallel lines when an extension would be outside the fuzz distance, a straight segment is added between the nearest endpoints.

FIGURE 16-13

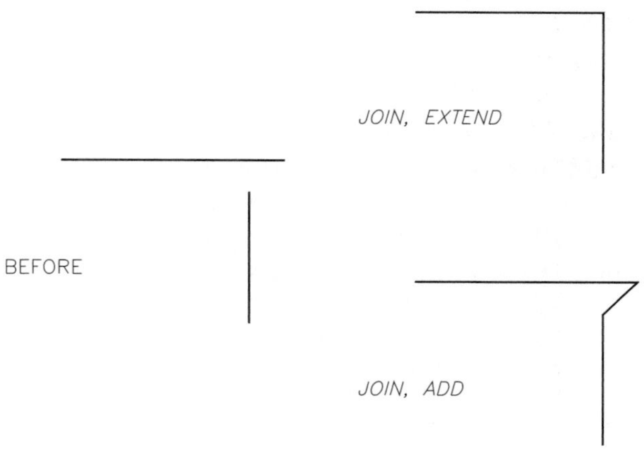

If the properties of several objects being joined into a polyline differ, the resulting polyline inherits the properties of the first object you select.

Width

Width allows specification of a uniform width for *Pline* segments (Fig. 16-14). Non-uniform width can be specified with the *Edit vertex* option.

Edit vertex

This option is covered in the next section.

FIGURE 16-14

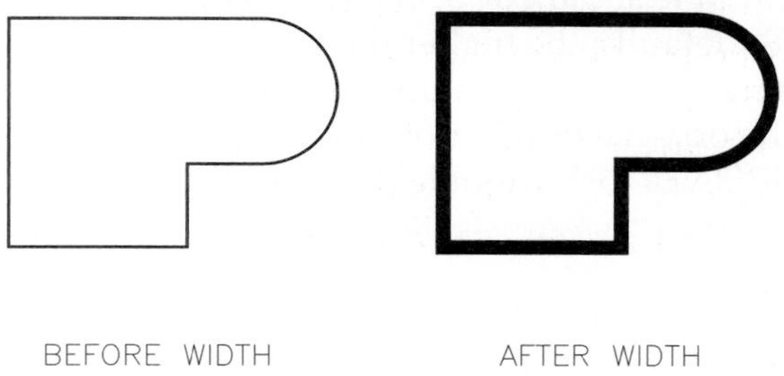

Fit

This option converts the *Pline* from straight line segments to arcs. The curve consists of two arcs for each pair of vertices. The resulting curve can be radical if the original *Pline* consists of sharp angles. The resulting curve passes <u>through all</u> vertices (Fig. 16-15).

FIGURE 16-15

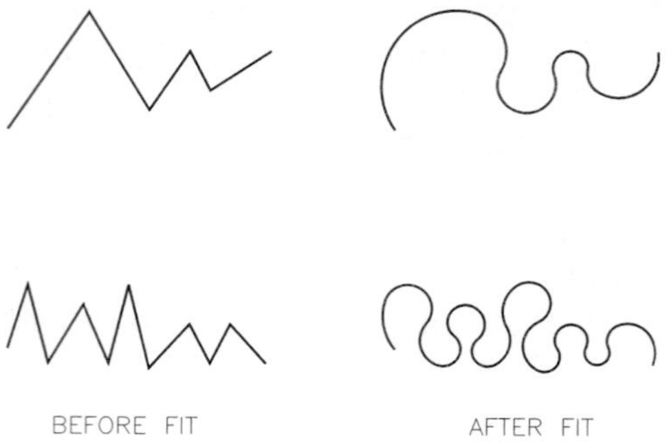

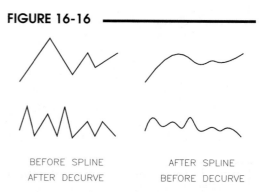

BEFORE FIT   AFTER FIT

Spline

This option converts the *Pline* to a B-spline (Bezier spline) (Fig. 16-16). The *Pline* vertices act as "control points" affecting the shape of the curve. The resulting curve passes through <u>only the end</u> vertices. A *Spline*-fit *Pline* is <u>not the same as a spline curve created with the *Spline*</u> command. This option produces a less versatile version of the newer *Spline* object.

Decurve

Decurve removes the *Spline* or *Fit* curve and returns the *Pline* to its original straight line segments state (Fig. 16-16).

FIGURE 16-16

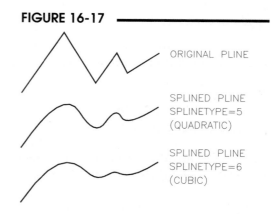

BEFORE SPLINE   AFTER SPLINE
AFTER DECURVE   BEFORE DECURVE

When you use the *Spline* option of *Pedit,* the amount of "pull" can be affected by setting the *SPLINETYPE* system variable to either 5 or 6 <u>before</u> using the *Spline* option. *SPLINETYPE* applies either a quadratic (5=more pull) or cubic (6=less pull) B-spline function (Fig. 16-17).

FIGURE 16-17

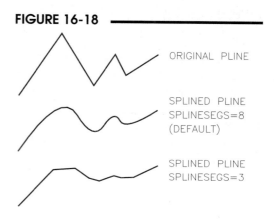

ORIGINAL PLINE

SPLINED PLINE
SPLINETYPE=5
(QUADRATIC)

SPLINED PLINE
SPLINETYPE=6
(CUBIC)

The *SPLINESEGS* system variable controls the number of line segments created when the *Spline* option is used. The variable should be set <u>before</u> using the option to any value (8=default): the higher the value, the more line segments. The actual number of segments in the resulting curve depends on the original number of *Pline* vertices and the value of the *SPLINETYPE* variable (Fig. 16-18).

FIGURE 16-18

ORIGINAL PLINE

SPLINED PLINE
SPLINESEGS=8
(DEFAULT)

SPLINED PLINE
SPLINESEGS=3

TIP ✓ Changing the *SPLFRAME* variable to 1 causes the *Pline* frame (the original straight segments) to be displayed for *Splined* or *Fit Plines*. *Regen* must be used after changing the variable to display the original *Pline* "frame" (Fig. 16-19).

FIGURE 16-19

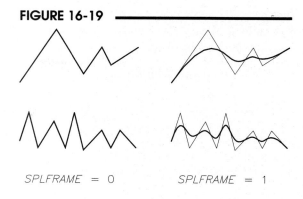

SPLFRAME = 0   SPLFRAME = 1

Ltype gen

This setting controls the generation of non-continuous linetypes for *Plines*. If *Off*, non-continuous linetype dashes start and stop at each vertex, as if the *Pline* segments were individual *Line* segments. For dashed linetypes, each line segment begins and ends with a full dashed segment (Fig. 16-20). If *On*, linetypes are drawn in a consistent pattern, disregarding vertices. In this case, it is possible for a vertex to have a space rather than a dash. Using the *Ltype gen* option <u>retroactively</u> changes *Plines* that have already been drawn. *Ltype gen* affects objects composed of *Plines* such as *Polygons*, *Rectangles*, and *Boundaries*.

FIGURE 16-20

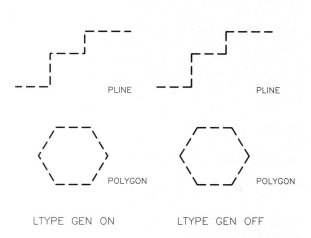

LTYPE GEN ON   LTYPE GEN OFF

Undo

Undo reverses the most recent *Pedit* operation.

eXit

This option exits the *Pedit* options, keeps the changes, and returns to the Command prompt.

Vertex Editing

Upon selecting the *Edit Vertex* option from the *Pedit* options list, the group of suboptions is displayed on the screen menu and Command line:

    Command: **pedit**
    Select polyline or [Multiple]: **PICK**
    Enter an option [Close/Join/Width/Edit vertex/Fit/Spline/Decurve/Ltype gen/Undo]: **e**
    Enter a vertex editing option
    [Next/Previous/Break/Insert/Move/Regen/Straighten/Tangent/Width/eXit] <N>:

Next

AutoCAD places an **X** marker at the first endpoint of the *Pline*. The *Next* and *Previous* options allow you to sequence the marker to the desired vertex.

Previous

See *Next* above.

Break

This selection causes a break between the marked vertex and another vertex you then select using the *Next* or *Previous* option.

> Enter an option [Next/Previous/Go/eXit] <N>:

Selecting *Go* causes the break. An endpoint vertex cannot be selected.

Insert

Insert allows you to insert a new vertex at any location <u>after</u> the vertex that is marked with the **X**. Place the marker before the intended new vertex, use *Insert*, then PICK the new vertex location.

Move

You are prompted to indicate a new location to *Move* the marked vertex.

Straighten

You can *Straighten* the *Pline* segments between the current marker and the other marker that you then place by one of these options:

> Enter an option [Next/Previous/Go/eXit] <N>:

Selecting *Go* causes the straightening to occur.

Tangent

Tangent allows you to specify the direction of tangency of the current vertex for use with curve *Fitting*.

Width

This option allows changing the *Width* of the *Pline* segment immediately following the marker, thus achieving a specific width for one segment of the *Pline*. *Width* can be specified with different starting and ending values.

eXit

This option exits from vertex editing, saves changes, and returns to the main *Pedit* prompt.

Grips

Plines can also be edited easily using Grips (see Chapter 19). A Grip appears on each vertex of the *Pline*. Editing *Plines* with Grips is sometimes easier than using *Pedit* because Grips are more direct and less dependent on the command interface.

Converting *Lines* and *Arcs* to *Plines*

A very important and productive feature of *Pedit* is the ability to convert *Lines* and *Arcs* to *Plines* and closed *Pline* shapes. Potential uses of this option are converting a series of connected *Lines* and *Arcs* to a closed *Pline* for subsequent use with *Offset* or for inquiry of the area (*Area* command) or length (*List* command) of a single shape. The only requirement for conversion of *Lines* and *Arcs* to *Plines* is that the selected objects must have <u>exact</u> matching endpoints.

To accomplish the conversion of objects to *Plines,* simply select a *Line* or *Arc* object and request to turn it into one:

> Command: **pedit**
> Select polyline or [Multiple]: **PICK** (Select only one *Line* or *Arc*)
> Object selected is not a polyline
> Do you want to turn it into one? <Y> **Enter**
> Enter an option [Close/Join/Width/Edit vertex/Fit/Spline/Decurve/Ltype gen/Undo]: **j** (Use the *Join* option)

Select objects: **PICK**
Select objects: **Enter**
1 segments added to polyline
Enter an option [Close/Join/Width/Edit vertex/Fit/Spline/Decurve/Ltype gen/Undo]: **Enter**
Command:

The resulting conversion is a one *Polyline* shape.

Splinedit

Pull-down Menu	Command (Type)	Alias (Type)	Short-cut	Screen (side) Menu	Tablet Menu
Modify Object > Splinedit	*Splinedit*	*SPE*	...	*MODIFY1 Splinedt*	*Y,18*

Splinedit is an extremely powerful command for changing the configuration of existing *Splines*. You can use multiple methods to change *Splines*. All of the *Splinedit* methods fall under two sets of options.

The two groups of options that AutoCAD uses to edit *Splines* are based on two sets of points: data points and control points. Data points are the points that were specified when the *Spline* was created—the points that the *Spline* actually passes through (Fig. 16-21).

FIGURE 16-21

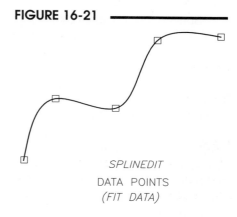

SPLINEDIT
DATA POINTS
(FIT DATA)

Control points are other points outside of the path of the *Spline* that only have a "pull" effect on the curve (Fig. 16-22).

FIGURE 16-22

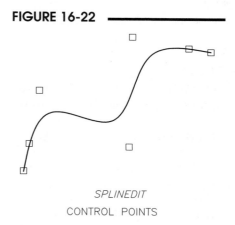

SPLINEDIT
CONTROL POINTS

Editing *Spline* Data Points

The command prompt displays several levels of options. The top level of options uses the control points method for editing. Select *Fit Data* to use data points for editing. The *Fit Data* methods are recommended for most applications. Since the curve passes directly through the data points, these options offer direct control of the curve path.

TIP

Command: **splinedit**
Select spline: **PICK**
Enter an option [Fit data/Close/Move vertex/Refine/rEverse/Undo]: **f**
Enter a fit data option [Add/Close/Delete/Move/Purge/Tangents/toLerance/eXit] <eXit>:

Add

You can add points to the *Spline*. The *Spline* curve changes to pass through the new points. First, PICK an existing point on the curve. That point and the next one in sequence (in the order of creation) become highlighted. The new point will change the curve between those two highlighted data points (Fig. 16-23).

> Specify control point: **PICK** (Select an existing point on the curve before the intended new point.)
> Specify new point: **PICK** (PICK a new point location between the two marked points.)

FIGURE 16-23

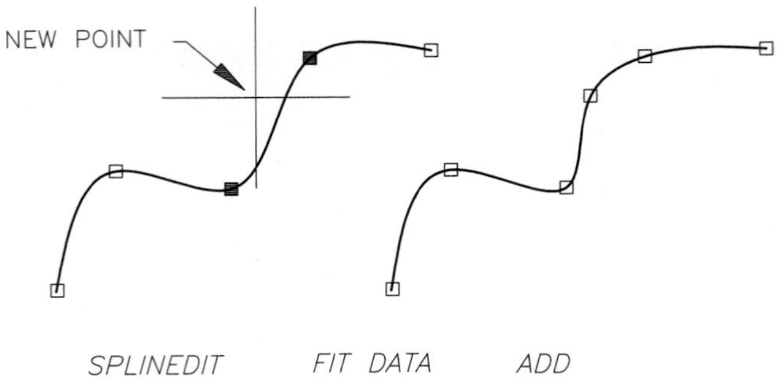

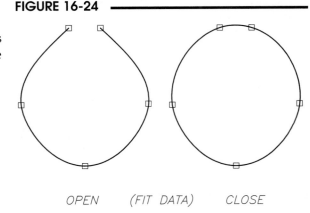

SPLINEDIT   FIT DATA   ADD

Close/Open

The *Close* option appears only if the existing curve is open, and the *Open* prompt appears only if the curve is closed. Selecting either option automatically forces the opposite change. *Close* causes the two ends to become tangent, forming one smooth curve (Fig. 16-24). This tangent continuity is characteristic of *Closed Splines* only. *Splines* that have matching endpoints do not have tangent continuity unless the *Close* option of *Spline* or *Splinedit* is used.

FIGURE 16-24

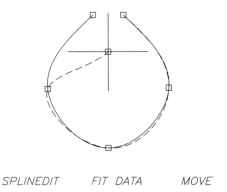

OPEN   (FIT DATA)   CLOSE

Move

You can move any data point to a new location with this option (Fig. 16-25). The beginning endpoint (in the order of creation) becomes highlighted. Type *N* for next or *S* to select the desired data point to move; then PICK the new location.

> Specify new location or [Next/Previous/Select point/eXit] <N>:

Purge

Purge deletes all data points and renders the *Fit Data* set of options unusable. You are returned to the control point options (top level). To reinstate the points, use *Undo*.

FIGURE 16-25

SPLINEDIT   FIT DATA   MOVE

Tangents

You can change the directions for the start and endpoint tangents with this option. This action gives the same control that exists with the "Enter start tangent" and "Enter end tangent" prompts of the *Spline* command used when the curves were created (see Fig. 15-13, *Spline, End Tangent*).

toLerance

Use *toLerance* to specify a value, or tolerance, for the curve to "fall" away from the data points. Specifying a tolerance causes the curve to smooth out, or fall, from the data points. The higher the value, the more the curve "loosens." The *toLerance* option of *Splinedit* is identical to the *Fit Tolerance* option available with *Spline* (see Fig. 15-15, *Spline, Tolerance*).

Editing *Spline* <u>Control Points</u>

Use the top level of command options (except *Fit Data*) to edit the *Spline's* control points. These options are similar to those used for editing the data points; however, the results are different since the curve does not pass through the control points.

> Command: **splinedit**
> Select spline: **PICK**
> Enter an option [Fit data/Close/Move vertex/Refine/rEverse/Undo]:

Fit Data

Discussed previously.

Close/Open

These options operate similar to the *Fit Data* equivalents; however, the resulting curve falls away from the control points (Fig. 16-26; see also Fig. 16-24, *Fit Data, Close*).

FIGURE 16-26

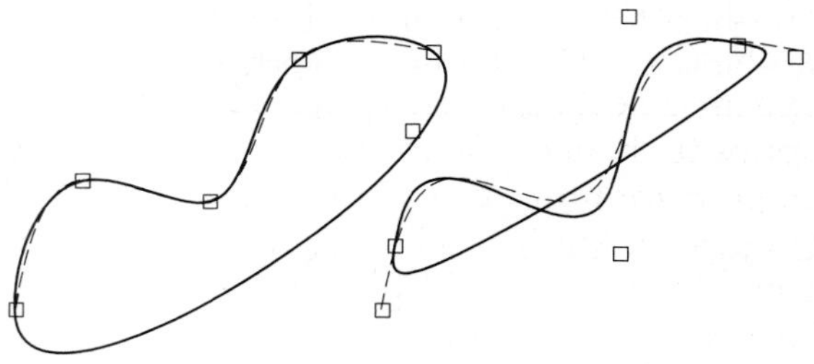

FIT DATA
CLOSE

CLOSE
(CONTROL POINTS)

Move Vertex

Move Vertex allows you to move the location of any control points. This is the control points' equivalent to the *Move* option of *Fit Data* (see Fig. 16-25, *Fit Data, Move*). The method of selecting points (*Next/Previous/Select point/eXit/*) is the same as that used for other options.

Refine

Selecting the *Refine* option reveals another level of options.

> Enter a refine option [Add control point/Elevate order/Weight/eXit] <eXit>:

Add control points is the control points' equivalent to *Fit Data Add* (see Fig. 16-23). At the "Select a point on the Spline" prompt, simply PICK a point near the desired location for the new point to appear. Once the *Refine* option has been used, the *Fit Data* options are no longer available.

Elevate order allows you to <u>increase the number of control points</u> uniformly along the length of the *Spline*. Enter a value from n to 26, where n is the current number of points + one. Once a *Spline* is elevated, it cannot be reduced.

Weight is an option that you use to assign a value to the <u>amount of "pull"</u> that a <u>specific control point</u> has on the *Spline* curve. The higher the value, the more "pull," and the closer the curve moves toward the control point. The typical method of selecting points (*Next/Previous/Select point/eXit/*) is used.

rEverse

The *rEverse* option reverses the direction of the *Spline*. The first endpoint (when created) then becomes the last endpoint. Reversing the direction may be helpful for selection during the *Move* option.

TIP

Grips

Splines can also be edited easily using *Grips* (see Chapter 19). The *Grip* points that appear on the *Spline* are identical to the *Fit Data* points. Editing with *Grips* is a bit more direct and less dependent on the command interface.

Boolean Commands

Region combines one or several objects forming a closed shape into one object, a *Region*. The appearance of the object(s) does not change after the conversion, even though the resulting shape is one object (see "*Region,*" Chapter 15).

Although the *Region* appears to be no different than a closed *Pline,* it is more powerful because several *Regions* can be combined to form complex shapes (called "composite *Regions*") using the three Boolean operations explained next. As an example, a set of *Regions* (converted *Circles*) can be combined to form the sprocket with only <u>one</u> *Subtract* operation.

The Boolean operators, *Union, Subtract,* and *Intersect,* can be used with *Regions* as well as solids. Any number of these commands can be used with *Regions* to form complex geometry. To illustrate each of the Boolean commands, consider the shapes shown in Figure 16-27. The *Circle* and the closed *Pline* are <u>first</u> converted to *Regions*; then *Union, Subtract,* or *Intersection* can be used.

FIGURE 16-27

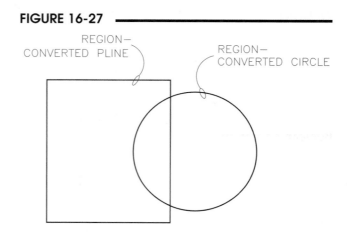

Union

Pull-down Menu	Command (Type)	Alias (Type)	Short-cut	Screen (side) Menu	Tablet Menu
Modify *Solid Editing >* *Union*	*Union*	*UNI*	...	*MODIFY2* *Union*	*X,15*

Union combines <u>two or more</u> *Regions* (or solids) into <u>one</u> *Region* (or solid). The resulting composite *Region* has the encompassing perimeter and area of the original *Regions*. Invoking *Union* causes AutoCAD to prompt you to select objects. You can select only existing *Regions* (or solids).

Command: **union**
Select objects: **PICK** (*Region*)
Select objects: **PICK** (*Region*)
Select objects: **Enter**
Command:

The selected *Regions* are combined into one composite *Region* (Fig. 16-28). Any number of Boolean operations can be performed on the *Region(s)*.

FIGURE 16-28

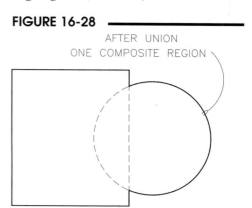

Several *Regions* can be selected in response to the "Select objects:" prompt. For example, a composite *Region* such as that in Figure 16-29 can be created with one *Union*.

Two or more *Regions* can be *Unioned* even if they do not overlap. They are simply combined into one object although they still appear as two.

FIGURE 16-29

BEFORE— 10 REGIONS

AFTER UNION— ONE REGION

Subtract

Pull-down Menu	Command (Type)	Alias (Type)	Short-cut	Screen (side) Menu	Tablet Menu
Modify *Solid Editing >* *Subtract*	*Subtract*	*SU*	...	*MODIFY2* *Subtract*	*X,16*

Subtract enables you to remove one *Region* (or set of *Regions*) from another. The *Regions* must be created before using *Subtract* (or another Boolean operator). *Subtract* also works with solids (as do the other Boolean operations).

There are two steps to *Subtract*. First, you are prompted to select the *Region* or set of *Regions* to "subtract from" (those that you wish to <u>keep</u>), then to select the *Regions* "to subtract" (those you want to <u>remove</u>). The resulting shape is one composite *Region* comprising the perimeter of the first set minus the second (sometimes called "difference").

```
Command: subtract
Select solids and regions to subtract from...
Select objects: PICK
Select objects: Enter
Select solids and regions to subtract...
Select objects: PICK
Select objects: Enter
Command:
```

Consider the two shapes previously shown in Figure 16-27. The resulting *Region* shown in Figure 16-30 is the result of *Subtracting* the circular *Region* from the rectangular one.

FIGURE 16-30

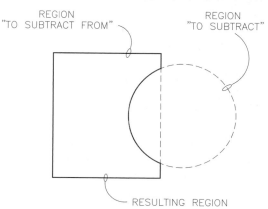

REGION "TO SUBTRACT FROM"

REGION "TO SUBTRACT"

RESULTING REGION

Intersect

Pull-down Menu	Command (Type)	Alias (Type)	Short-cut	Screen (side) Menu	Tablet Menu
Modify *Solid Editing >* *Intersect*	*Intersect*	*IN*	...	*MODIFY2* *Intrsect*	*X,17*

Intersect is the Boolean operator that finds the common area from two or more *Regions*.

Consider the rectangular and circular *Regions* previously shown (Fig. 16-27). Using the *Intersect* command and selecting both shapes results in a *Region* comprising only that area that is shared by both shapes (Fig. 16-31):

```
Command: intersect
Select objects: PICK
Select objects: PICK
Select objects: Enter
Command:
```

FIGURE 16-31 ────────────

COMPOSITE REGION—
RESULT OF INTERSECTION

If more than two *Regions* are selected, the resulting *Intersection* is composed of only the common area from all shapes (Fig. 16-32). If all of the shapes selected do not overlap, a null *Region* is created (all shapes disappear because no area is common to all).

FIGURE 16-32 ────────────

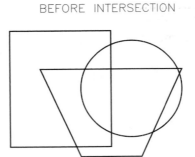
BEFORE INTERSECTION

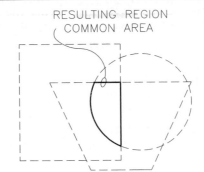
RESULTING REGION
COMMON AREA

Intersect is a powerful operation when used with solid modeling. Techniques for saving time using Boolean operations are discussed in Chapter 30, Solid Modeling Construction.

CHAPTER EXERCISES

1. *Chprop* or *Properties*

    Open the **PIVOTARM CH9** drawing that you worked on in Chapter 9 Exercises. *Load* the *Hidden2* and *Center2 Linetypes*. Make two *New Layers* named **HID** and **CEN** and assign the matching linetypes and yellow and green colors, respectively. Check the *Limits* of the drawing; then calculate and set an appropriate *LTSCALE*. Use *Chprop* or *Properties* palette to change the *Lines* representing the holes in the front view to the **HID** layer as shown in Figure 16-33. Use *SaveAs* and name the drawing **PIVOTARM CH16.**

FIGURE 16-33

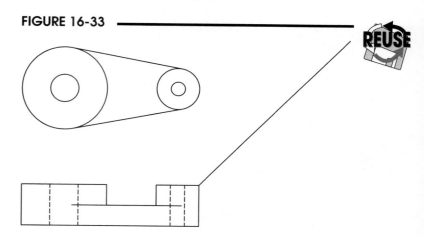

2. *Change*

    Open **CH8EX3.** *Erase* the *Arc* at the top of the object. *Erase* the *Points* with a window. Invoke the *Change* command. When prompted to *Select objects*, **PICK** all of the inclined *Lines* near the top. When prompted to "Specify change point," enter coordinate **6,8**. The object should appear as that in Figure 16-34. Use *SaveAs* and assign the name **CH16EX2.**

FIGURE 16-34

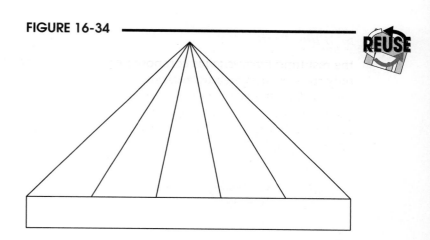

3. *Properties* **Palette**

    A design change is required for the bolt holes in **GASKETA** (from Chapter 9 Exercises). *Open* **GASKETA** and invoke the *Properties* palette. Change each of the four bolt holes to **.375** diameter (Fig. 16-35). *Save* the drawing.

FIGURE 16-35

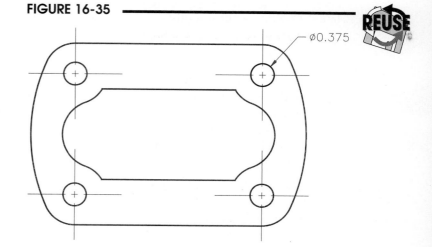

4. *Explode*

    Open the **POLYGON1** drawing that you completed in Chapter 15 Exercises. You can quickly create the five-sided shape shown (continuous lines) in Figure 16-36 by *Exploding* the *Polygon*. First, *Explode* the *Polygon* and *Erase* the two *Lines* (highlighted). Draw the bottom *Line* from the two open *Endpoints*. Do not exit the drawing.

FIGURE 16-36

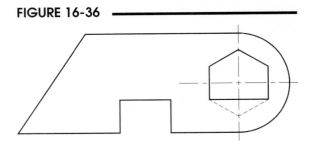

5. *Pedit*

    Use *Pedit* with the *Edit vertex* options to alter the shape as shown in Figure 16-37. For the bottom notch, use *Straighten*. For the top notch, use *Insert*. Use *SaveAs* and change the name to **PEDIT1.**

FIGURE 16-37

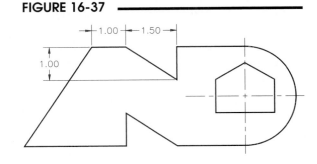

6. *Pline, Pedit*

    A. Create a line graph as shown in Figure 16-38 to illustrate the low temperatures for a week. The temperatures are as follows:

X axis	Y axis
Sunday	20
Monday	14
Tuesday	18
Wednesday	26
Thursday	34
Friday	38
Saturday	27

    Use equal intervals along each axis. Use a *Pline* for the graph line. *Save* the drawing as **TEMP-A**. (You will label the graph at a later time.)

    B. Use *Pedit* to change the *Pline* to a *Spline*. Note that the graph line is no longer 100% accurate because it does not pass through the original vertices (see Fig. 16-39). Use the *SPLFRAME* variable to display the original "frame" (*Regen* must be used after).

    Use the *Properties* palette and try the *Cubic* and *Quadratic* options. Which option causes the vertices to have more pull? Find the most accurate option. Set *SPLFRAME* to **0** and *SaveAs* **TEMP-B**.

FIGURE 16-38

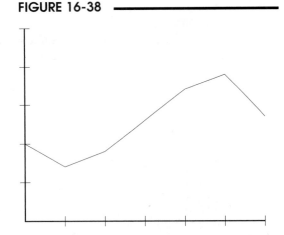

FIGURE 16-39

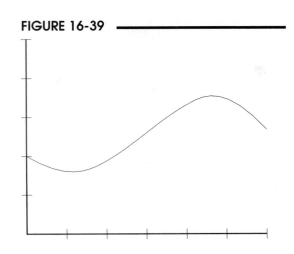

C. Use *Pedit* to change the curve from *Pline* to *Fit Curve*. Does the graph line pass through the vertices? *Saveas* TEMP-C.

D. *Open* drawing **TEMP-A**. *Erase* the *Splined Pline* and construct the graph using a *Spline* instead (Fig. 16-40, see Exercise 6A for data). *SaveAs* **TEMP-D.** Compare the *Spline* with the variations of *Plines*. Which of the four drawings (A, B, C, or D) is smoothest? Which is the most accurate?

FIGURE 16-40
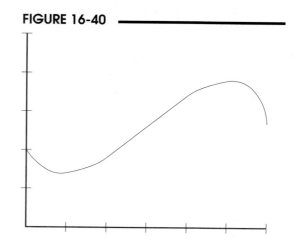

E. It was learned that there was a mistake in reporting the temperatures for that week. Thursday's low must be changed to 38 degrees and Friday's to 34 degrees. *Open* **TEMP-D** (if not already open) and use *Splinedit* to correct the mistake. Use the *Move* option of *Fit Data* so that the exact data points can be altered as shown in Figure 16-41. *SaveAs* **TEMP-E**.

FIGURE 16-41
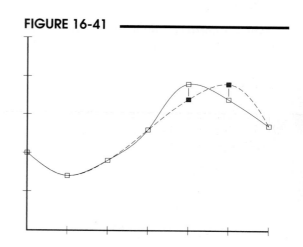

7. **Converting *Lines*, *Circles*, and *Arcs* to *Plines***

A. Begin a *New* drawing and use the **A-METRIC** template (that you worked on in Chapter 13 Exercises). Use *Save* and assign the name **GASKETC.** Change the *Limits* for plotting on an A sheet at 2:1 (refer to the Metric Table of *Limits* Settings and set *Limits* to 1/2 x *Limits* specified for 1:1 scale). Change the *LTSCALE* to **12**. First, draw only the <u>inside</u> shape using *Lines* and *Circles* (with *Trim*) or *Arcs* (Fig. 16-42). Then convert the *Lines* and *Arcs* to one closed *Pline* using *Pedit*. Finally, locate and draw the 3 bolt holes. *Save* the drawing.

FIGURE 16-42

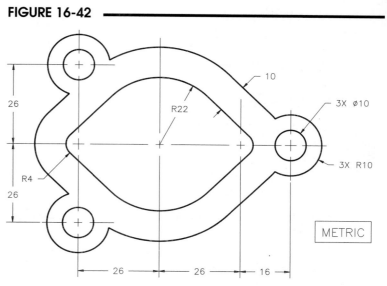

B. *Offset* the existing inside shape to create the outside shape. Use *Offset* to create concentric circles around the bolt holes. Use *Trim* to complete the gasket. *Save* the drawing and create a plot at 2:1 scale.

8. *Splinedit*

Open the HANDLE drawing that you created in Chapter 15 Exercises. During the testing and analysis process, it was discovered that the shape of the handle should have a more ergonomic design. The finger side (left) should be flatter to accommodate varying sizes of hands, and the thumb side (right) should have more of a protrusion on top to prevent slippage.

First, add more control points uniformly along the length of the left side with *Spinedit, Refine*. *Elevate* the *Order* from 4 to **6**, then use *Move Vertex* to align the control points as shown in Figure 16-43.

On the right side of the handle, *Add* two points under the *Fit Data* option to create the protrusion shown in Figure 16-44. You may have to *Reverse* the direction of the *Spline* to add the new points between the two high-lighted ones. *SaveAs* **HANDLE2**.

FIGURE 16-43 —   **FIGURE 16-44** —

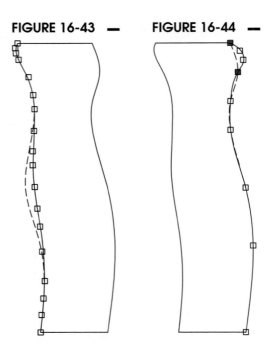

9. *Line, Pline, Stretch*

Draw the floor plan of the storage room shown in Figure 16-45. Use *Line* objects for the walls and *Plines* for the windows and doors. When your drawing is complete according to the given specifications, use *Stretch* to center the large window along the top wall. Save the drawing as STORE ROOM.

FIGURE 16-45

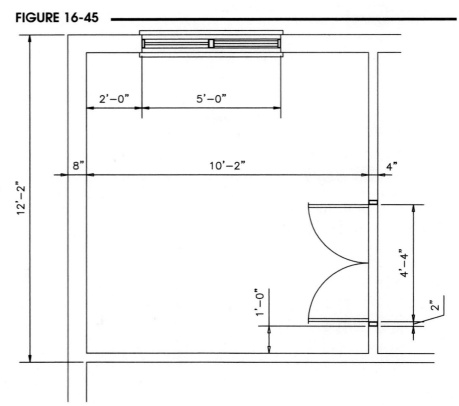

10. **Gear Drawing**

FIGURE 16-46

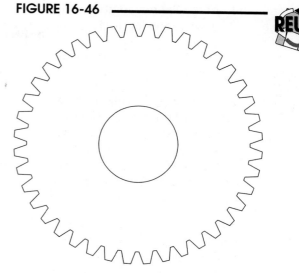

Complete the drawing of the gear you began in Chapter 15 Exercises called **GEAR-REGION.** If you remember, three shapes were created (two *Circles* and one "tooth") and each was converted to a *Region*. Finally, the "tooth" was *Arrayed* to create the total of 40 teeth.

A. To complete the gear, *Subtract* the small circular *Region* and all of the teeth from the large circular *Region*. First, use *Subtract*. At the "Select solids and regions to subtract from…" prompt, PICK the large circular *Region*. At the "Select solids and regions to subtract…" prompt, use a window to select <u>everything</u> (the large circular *Region* is automatically filtered out). The resulting gear should resemble Figure 16-46. *Save* the drawing as **GEAR-REGION 2**.

B. Consider the steps involved if you were to create the gear (as an alternative) by using *Trim* to remove 40 small sections of the large *Circle* and all unwanted parts of the teeth. *Regions* are clearly easier in this case.

11. **Wrench Drawing**

FIGURE 16-47

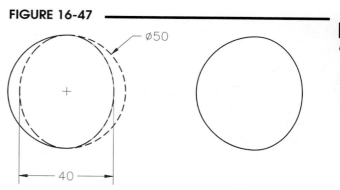

Create the same wrench you created in Chapter 15 again, only this time use region modeling. Refer to Chapter 15, Exercise 5 for dimensions (exclude the break lines). Begin a *New* drawing and use the **A-METRIC** template. Use *SaveAs* and assign the name **WRENCH-REG**.

A. Set *Limits* to **372,288** to prepare the drawing for plotting at 3:4 (the drawing scale factor is 33.87). Set the *GRID* to **10**.

B. Draw a *Circle* and an *Ellipse* as shown on the left in Figure 16-47. The center of each shape is located at **60,150**. *Trim* half of each shape as shown (highlighted). Use *Region* to convert the two remaining halves into a *Region* as shown on the right of Figure 16-47.

C. Next, create a *Circle* with the center at **110,150** and a diameter as shown in Figure 16-48. Then draw a closed *Pline* in a rectangular shape as shown. The height of the rectangle must be drawn as specified; however, the width of the rectangle can be drawn <u>approximately</u> as shown on the left. Convert each shape to a *Region*; then use *Intersect* to create the region as shown on the right side of the figure.

FIGURE 16-48

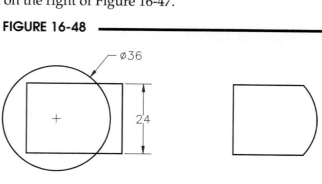

D. *Move* the rectangular-shaped region **68** units to the left to overlap the first region as shown in Figure 16-49. Use *Subtract* to create the composite region on the right representing the head of the wrench.

E. Complete the construction of the wrench in a manner similar to that used in the previous steps. Refer to Chapter 15 Exercises for dimensions of the wrench. Complete the wrench as one *Region*. *Save* the drawing as **WRENCH-REG**.

FIGURE 16-49

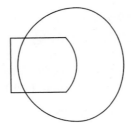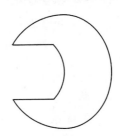

17

INQUIRY COMMANDS

CHAPTER OBJECTIVES

After completing this chapter you should;

1. be able to list the *Status* of a drawing;

2. be able to *List* the AutoCAD database information about an object;

3. know how to list the entire database of all objects with *Dblist*;

4. be able to calculate the *Area* of a closed shape with and without "islands";

5. be able to find the distance between two points with *Dist*;

6. be able to report the coordinate value of a selected point using the *ID* command;

7. know how to list the *Time* spent on a drawing or in the current drawing session.

CONCEPTS

AutoCAD provides several commands that allow you to find out information about the current drawing status and specific objects in a drawing. These commands as a group are known as "*Inquiry* commands" and are grouped together in the menu systems. The *Inquiry* commands are located in the *Inquiry* toolbar (Fig. 17-1). You can also use the *Tools* pull-down menu to access the *Inquiry* commands (Fig. 17-2).

FIGURE 17-1 ————

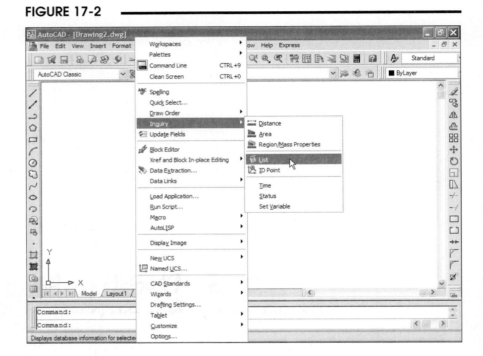

Using *Inquiry* commands, you can find out such information as the amount of time spent in the current drawing, the distance between two points, the area of a closed shape, the database listing of properties for specific objects (coordinates of endpoints, lengths, angles, etc.), and other information. The *Inquiry* commands are:

   Status, List, Dblist, Area, Dist, ID, and *Time*

FIGURE 17-2 ————

COMMANDS

Status

Pull-down Menu	Command (Type)	Alias (Type)	Short-cut	Screen (side) Menu	Tablet Menu
Tools Inquiry > Status	Status	...	...	TOOLS 1 Status	...

The *Status* command gives many pieces of information related to the current drawing. Typing or **PICK**ing the command from the icon or one of the menus causes a text screen to appear similar to that shown in Figure 17-3, on the next page. The information items are:

   Total number of objects in the current drawing.
   Paper space limits: values set by the *Limits* command in Paper Space. (*Limits* in Paper Space are set by selecting a *Paper size* in the *Page Setup* or *Plot* dialog box.)
   Paper space uses: area used by the objects (drawing extents) in Paper Space.
   Model space limits: values set by the *Limits* command in Model Space.
   Model space uses: area used by the objects (drawing extents) in Model Space.

Display shows: current display or windowed area.

Insertion basepoint: point specified by the *Base* command or default (0,0).

Snap resolution: value specified by the *Snap* command.

Grid spacing: value specified by the *Grid* command.

Current space: Paper Space or Model Space

Current layer: name.

Current color: current color assignment

Current linetype: current linetype assignment

Current lineweight: current lineweight assignment.

Current plot style: current plot style assignment.

Current elevation, thickness: 3D properties—current height above the XY plane and Z dimension.

On or off status: *FILL, GRID, ORTHO, QTEXT, SNAP, TABLET.*

Object Snap Modes: current *Running OSNAP* modes.

Free dwg disk: space on the current drawing hard disk drive.

Free temp disk: space on the current temporary files hard disk drive.

Free physical memory: amount of free RAM (total RAM).

Free swap file space: amount of free swap file space (total allocated swap file).

FIGURE 17-3

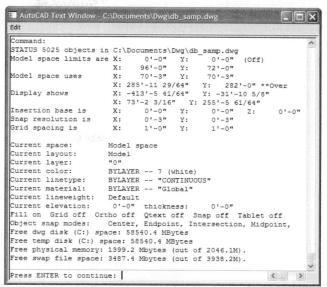

List

Pull-down Menu	Command (Type)	Alias (Type)	Short-cut	Screen (side) Menu	Tablet Menu
Tools *Inquiry >* *List*	*List*	*LS or LI*	...	*TOOLS 1* *List*	*U,8*

The *List* command displays the database list of information in text window format for one or more specified objects. The information displayed depends on the <u>type</u> of object selected. Invoking the *List* command causes a prompt for you to select objects. AutoCAD then displays the list for the selected objects (see Fig. 17-4, right and Fig. 17-5, on the next page). A *List* of a *Line* and an *Arc* is given in Figure 17-4.

For a *Line,* coordinates for the endpoints, line length and angle, current layer, and other information are given.

For an *Arc,* the center coordinate, radius, start and end angles, and length are given.

FIGURE 17-4

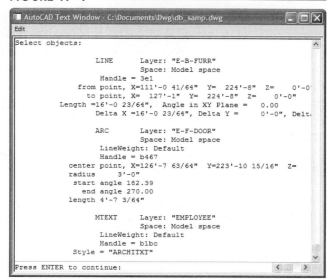

The *List* for a *Pline* is shown in Figure 17-5. The location of each vertex is given, as well as the length and perimeter of the entire *Pline*.

Plines are created and listed as *Lwpolylines,* or "lightweight polylines." The data common to all vertices are stored only once, and only the coordinate data are stored for each vertex. Because this data structure saves file space, the *Plines* are known as lightweight *Plines*.

FIGURE 17-5

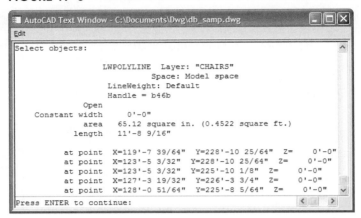

Dblist

Pull-down Menu	Command (Type)	Alias (Type)	Short-cut	Screen (side) Menu	Tablet Menu
...	*Dblist*	...	...	...	...

The *Dblist* command is similar to the *List* command in that it displays the database listing of objects; however, *Dblist* gives information for <u>every</u> object in the current drawing! This command is generally used when you desire to copy the list to a file or when only a few objects are in the drawing. If you use this command in a complex drawing, be prepared to page through many screens of information. Press Escape to cancel *Dblist* and return to the Command: prompt. Press F2 to open and close the text window.

Area

Pull-down Menu	Command (Type)	Alias (Type)	Short-cut	Screen (side) Menu	Tablet Menu
Tools Inquiry > Area	*Area*	*AA*	...	*TOOLS 1 Area*	*T,7*

The *Area* command is helpful for many applications. With this command AutoCAD calculates the area and the perimeter of any enclosed shape in a matter of milliseconds. You specify the area (shape) to consider for calculation by PICKing the *Object* (if it is a closed *Pline, Polygon, Circle, Boundary, Region* or other closed object) or by PICKing points (corners of the outline) to define the shape. The options are given below.

Specify first corner point
The command sequence for specifying the area by PICKing points is shown below. This method should be used only for shapes with <u>straight</u> sides. An example of the *Point* method (PICKing points to define the area) is shown in Figure 17-6.

FIGURE 17-6

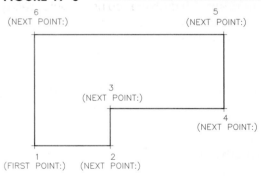

Command: **area**
Specify first corner point or [Object/Add/Subtract]: **PICK** (Locate the first corner to define the shape.)
Specify next corner point or press ENTER for total: **PICK** (Locate the second corner on the shape.)
Specify next corner point or press ENTER for total: **PICK** (Locate the next corner.)
Specify next corner point or press ENTER for total: **PICK** (Continue selecting points until all corners have been defined.)
Specify next corner point or press ENTER for total: **Enter**
Area = *nn.nnn* perimeter = *nn.nnn*
Command:

Object

If the shape for which you want to find the area and perimeter is a *Circle, Polygon, Rectangle, Ellipse, Boundary, Region,* or closed *Pline,* the *Object* option of the *Area* command can be used. Select the shape with one PICK (since all of these shapes are considered as one object).

The ability to find the area of a closed *Pline, Region,* or *Boundary* is extremely helpful. Remember that <u>any</u> closed shape, even if it includes *Arcs* and other curves, can be converted to a closed *Pline* with the *Pedit* command (as long as there are no gaps or overlaps) or can be used with the *Boundary* command. This method provides you with the ability to easily calculate the area of any shape, curved or straight. In short, convert the shape to a closed *Pline, Region,* or *Boundary* and find the *Area* with the *Object* option.

Add, Subtract

Add and *Subtract* provide you with the means to find the area of a closed shape that has islands, or negative spaces. For example, you may be required to find the surface area of a sheet of material that has several punched holes. In this case, the area of the holes is subtracted from the area defined by the perimeter shape. The *Add* and *Subtract* options are used specifically for that purpose. The following command sequence displays the process of calculating an area and subtracting the area occupied by the holes.

Command: **area**
Specify first corner point or [Object/Add/Subtract]: **a** (Use the *Add* option to begin a running total.)
Specify first corner point or [Object/Subtract]: **o** (Use the *Object* option to select the outside shape.)
(ADD mode) Select objects: **PICK** (Select the closed object.)
Area = 13.31, Perimeter = 14.39
Total area = 13.31

(ADD mode) Select objects: **Enter** (Completion of *Add* mode.)
Specify first corner point or [Object/Subtract]: **s** (Switch to *Subtract* mode.)
Specify first corner point or [Object/Add]: **o** (Use *Object* mode.)
(SUBTRACT mode) Select objects: **PICK** (Select the first *Circle* to subtract.)
Area = 0.69, Length = 2.95
Total area = 12.62

(SUBTRACT mode) Select objects: **PICK** (Select the second *Circle* to subtract.)
Area = 0.69, Length = 2.95
Total area = 11.93

(SUBTRACT mode) Select objects: **Enter** (Completion of *Subtract* mode.)
Specify first corner point or [Object/Add]: **Enter** (Completion of *Area* command.)
Command:

Make sure that you press Enter between the *Add* and *Subtract* modes.

An example of the last command sequence used to find the area of a shape minus the holes is shown in Figure 17-7. Notice that the object selected in the first step is a closed *Pline* shape, including an *Arc*.

FIGURE 17-7

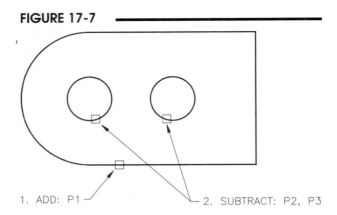

1. ADD: P1   2. SUBTRACT: P2, P3

Dist

Pull-down Menu	Command (Type)	Alias (Type)	Short-cut	Screen (side) Menu	Tablet Menu
Tools Inquiry > Distance	Dist	DI	...	TOOLS 1 Dist	T,8

 The *Dist* command reports the distance between any two points you specify. *OSNAPs* can be used to snap to the existing points. This command is helpful in many engineering or architectural applications, such as finding the clearance between two mechanical parts, finding the distance between columns in a building, or finding the size of an opening in a part or doorway. The command is easy to use.

```
Command: dist
Specify first point: PICK  (Use Osnaps if needed.)
Specify second point: PICK  (Use Osnaps if needed.)
Distance = 3.63,  Angle in XY Plane = 165,  Angle from XY Plane = 0
Delta X = -3.50,  Delta Y = 0.97,  Delta Z = 0.00
Command:
```

AutoCAD reports the absolute and relative distances as well as the angle of the line between the points.

Id

Pull-down Menu	Command (Type)	Alias (Type)	Short-cut	Screen (side) Menu	Tablet Menu
Tools Inquiry > ID Point	Id	...	...	TOOLS 1 ID	U,9

The *ID* command reports the coordinate value of any point you select with the cursor. If you require the location associated with a specific object, an *OSNAP* mode (*Endpoint, Midpoint, Center,* etc.) can be used.

```
Command: id
Specify point: PICK  (Use Osnaps if needed.)
X = 7.63    Y = 6.25    Z = 0.00
Command:
```

 NOTE: *ID* also sets AutoCAD's "last point." The last point can be referenced in commands by using the @ (at) symbol with relative rectangular or relative polar coordinates.

Time

Pull-down Menu	Command (Type)	Alias (Type)	Short-cut	Screen (side) Menu	Tablet Menu
Tools Inquiry > Time	Time	...	...	TOOLS 1 Time	...

This command is useful for keeping track of the time spent in the current drawing session or total time spent on a particular drawing. Knowing how much time is spent on a drawing can be useful in an office situation for bidding or billing jobs.

The *Time* command reports the information shown in Figure 17-8. The *Total editing time* is automatically kept, starting from when the drawing was first created until the current time. Plotting and printing time is not included in this total, nor is the time spent in a session when changes are discarded.

FIGURE 17-8

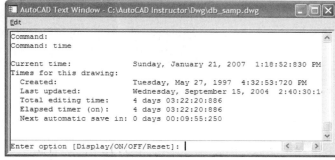

Display

The *Display* option causes *Time* to repeat the display with the updated times.

ON/OFF/Reset

The *Elapsed timer* is a separate compilation of time controlled by the user. The *Elapsed timer* can be turned *ON* or *OFF* or can be *Reset*.

Time also reports when the next automatic save will be made. The time interval of the Automatic Save feature is controlled by the *SAVETIME* system variable. To set the interval between automatic saves, type *SAVETIME* at the Command line and specify a value for time (in minutes).

CHAPTER EXERCISES

1. *List*

    A. *Open* the **GASKETA** drawing from Chapter 16 Exercises. Assume that a laser will be used to cut the holes from the gaskets, and you must program the coordinates. Use the *List* command to give information on the *Circles*. Determine and write down the coordinates for the centers of the 4 holes. CENTER PT   X 9.8354 Y 9.5049 Z = Ø

    B. *Open* the **EFF-APT** drawing from Chapter 15 Exercises. Use *List* to determine the area of the inside of the tub. If the tub were filled with 10" of water, what would be the volume of water in the tub? 611.0 SQ IN (4.2 SQ FT)

2. *Area*

    A. *Open* the **EFF-APT** drawing. The entry room is to be carpeted at a cost of $12.50 per square yard. Use the *Area* command (with the PICK points option) to determine the cost for carpeting the room, not including the closet. 194.184 SQ FT

B. *Open* the **GASKETA** drawing from Chapter 16 Exercises. Use *SaveAs* to create a file named **GASKET-AREA**. Using the *Area* command, calculate the wasted material (the center section and the four holes after the gaskets have been cut or stamped). HINT: Use *Boundary* to create objects from the waste areas for determining the *Area*.

C. Create a *Boundary* (with islands) of **GASKETA**. *Move* the (6) new boundary objects **10** units to the right. Using the *Add* and *Subtract* options of *Area*, calculate the surface area for painting a sealer on 100 gaskets, both sides. (Remember to press Enter between the *Add* and *Subtract* operations.) *Save* the drawing.

3. *Dist*

A. *Open* the **EFF-APT** drawing. Use the *Dist* command to determine the best location for installing a wall-mounted telephone in the apartment. Where should the telephone be located in order to provide the most equal access from all corners of the apartment? What is the farthest distance that you would have to walk to answer the phone?   8'5" approx

B. Using the *Dist* command, determine what length of pipe would be required to connect the kitchen sink drain (use the center of the far sink) to the tub drain (assume the drain is at the far end of the tub). Calculate only the direct distance (under the floor).

15'-3 5/8"

4. *ID*

Open the **GASKETA** drawing once again. In order to program the laser to cut the 4 holes, you must confirm the coodinates for the centers. Use the *ID* command (with *OSNAP*s) to determine the coordinates for the 4 corners and the hole centers and compare them with the coordinates from Exercise 1A.

x6.8354 y 9.5049 z - 0

5. *Time*

4 CORNER - BL x 6.5415 y 9.0866 BR 10.1093x 9.0866 y

Using the *Time* command, what is the total amount of editing time you spent with the **GASKETA** drawing? How much time have you spent in this session? How much time until the next automatic save?

40:30:720

NO MODIFICATIONS YET

CONT 4. TR - x 10.1093
              y 11.9233

TL - x 6.5415
      y 11.9233

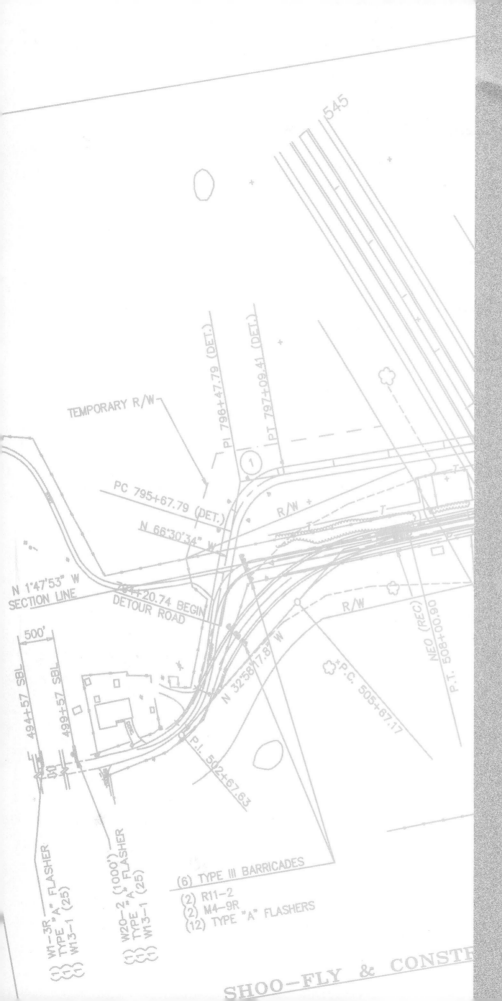

18

TEXT

CHAPTER OBJECTIVES

After completing this chapter you should:

1. be able to create lines of text in a drawing using *Text;*

2. be able to create and format paragraph text using *Mtext;*

3. know the options available for creating and editing text using the *Text Formatting* Editor;

4. be able to create text styles with the *Style* command including annotative styles;

5. know how to edit text using *Ddebit, Mtedit,* and *Properties;*

6. be able to use *Spell* to check spelling and be able to *Find and Replace* text in a drawing.

CONCEPTS

The *Text* and *Mtext* commands provide you with a means of creating text in an AutoCAD drawing. "Text" in CAD drawings usually refers to sentences, words, or notes created from alphabetical or numerical characters that appear in the drawing. The numeric values that are part of specific dimensions are generally not considered "text," since dimensional values are a component of the dimension created automatically with the use of dimensioning commands.

Text in technical drawings is typically in the form of notes concerning information or descriptions of the objects contained in the drawing. For example, an architectural drawing might have written descriptions of rooms or spaces, special instructions for construction, or notes concerning materials or furnishings (Fig. 18-1). An engineering drawing may contain, in addition to the dimensions, manufacturing notes, bill of materials, schedules, or tables (Fig. 18-2). Technical illustrations may contain part numbers or assembly notes. Title blocks also contain text.

FIGURE 18-1

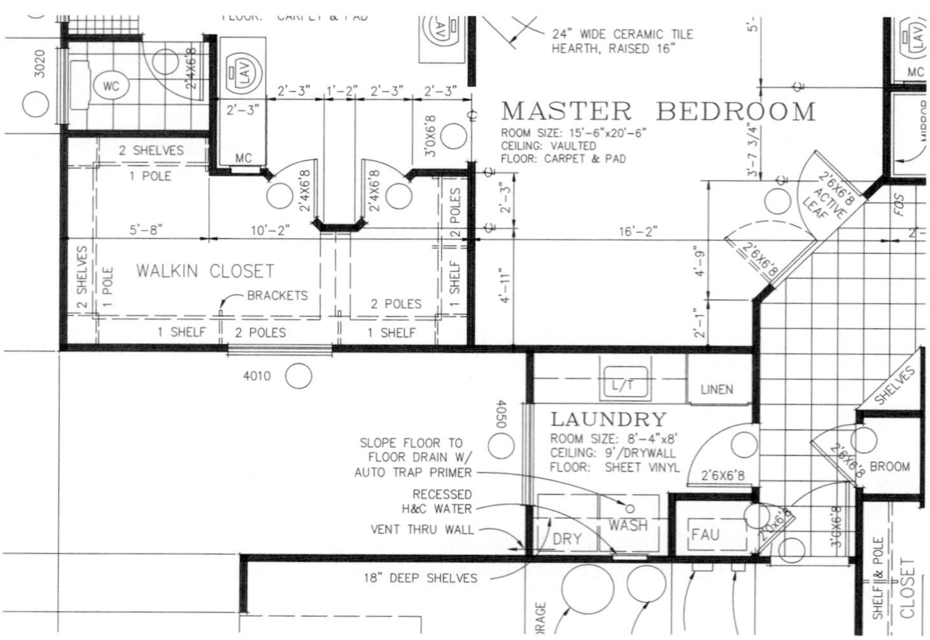

A line of text or paragraph of text in an AutoCAD drawing is treated as an object, just like a *Line* or a *Circle*. Each text object can be *Erased, Moved, Rotated,* or otherwise edited as any other graphical object. The letters themselves can be changed individually with special text editing commands. A spell checker is available by using the *Spell* command. Since text is treated as a graphical element, the use of many lines of text in a drawing can slow regeneration time and increase plotting time significantly.

FIGURE 18-2

#1 DETOUR CURVE DATA
PI STA. 796+47.81
Δ=66°30'34"
D=46°57'49"
T=80'
L=141.62'
R=122.0'
E=23.89

#2 DETOUR CURVE DATA
PI STA. 807+69.32
Δ=27°36'09"
D=18°46'12"
T=75'
L=147.09'
R=305.25'
E=9.07

#3 DETOUR CURVE DATA
PI STA. 809+28.49
Δ=30°11'20"
D=20°36'17"
T=75'
L=146.51'
R=278.07'
E=9.94

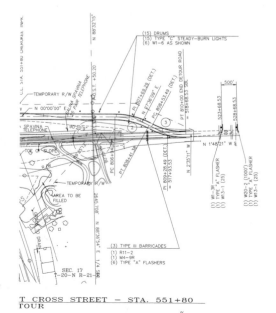

ROUTE	POINT	STATION	COORDINATES	
			NORTH	EAST
℄ BASE LINE	P.I.	502+67.63	452555.90	2855805.56
℄ BASE LINE	P.C.	505+67.16	453207.20	2855642.55
℄ BASE LINE	P.I.	506+86.85	453307.61	2855577.41
℄ BASE LINE	P.T.	508+00.90	453427.17	2855572.01
℄ BASE LINE	P.T.	510+00.00	453626.07	2855563.02
℄ BASE LINE	P.T.	516+50.00	454275.41	2855533.69
℄ BASE LINE	P.O.T.	519+00.00	454264.13	2855783.44
DETOUR @ STA 551+80				
℄ DETOUR	P.O.T.	794+20.74	453044.48	2855659.62
℄ DETOUR	P.C.	795+67.79	453103.09	2855524.75
℄ DETOUR	P.I	796+47.79	453134.98	2855451.38
℄ DETOUR	P.T.	797+09.41	453214.98	2855451.38
℄ DETOUR	P.C.	806+94.29	454199.87	2855451.38
℄ DETOUR	P.I.	807+69.29	454274.87	2855451.38
℄ DETOUR	P.T.	808+41.38	454341.30	2855486.12
℄ DETOUR	P.C.	808+53.49	454352.03	2855491.73
℄ DETOUR	P.I.	809+28.49	454418.50	2855526.48
℄ DETOUR	P.C.	810+00.00	454493.42	2855523.09

The *Text* and *Mtext* commands perform basically the same function; they create text in a drawing. *Text* creates a single line of text but allows entry of multiple lines with one use of the command. *Mtext* is the more sophisticated method of text entry. With *Mtext* (multiline text) you can create a paragraph of text that "wraps" within a text boundary (rectangle) that you specify. You can also use *Mtext* to import text content from an external file. An *Mtext* paragraph is treated as one AutoCAD object.

The form or shape of the individual letters is partly determined by the text *Style*. Create a style with the *Style* command, select any Windows standard TrueType font (.TTF) or AutoCAD-supplied font file (.TTF or .SHX shape file), then specify other parameters such as width or angle to create a style to suit your needs. When you use the *Text* or *Mtext* commands, you can select any previously created *Style* and apply one of many options for text justification used for aligning multiple lines of text, such as right, center, or left justified. Only two simple text styles, called *Standard* and *Annotative*, have been created as part of the ACAD.DWT and ACADISO.DWT template drawings. The *Standard* text style should be used as the "template" for most text objects. The *Annotative* style allows you to create annotative text, which is useful if you want to display the same text objects in multiple viewports at different scales.

Once the text has been placed in the drawing, many options are available for changing, or "editing," the text. For *Text* objects you can edit the text "in place" (directly in the drawing), and for *Mtext* objects you can use the "text editor" you used to create the text, complete with all the formatting options. Options for editing text include changing the actual content of the text, changing the text style, scale, or justification method, spell checking, and using a find and replace tool. You can also use the *Properties* palette to change any aspect of the text. When text is printed you can specify that the text be unfilled (outline text) or that each line be printed as a rectangular block instead of individual letters.

Commands related to creating or editing text and tables in an AutoCAD drawing include:

Text	Places individual lines of text in a drawing.
Mtext	Places text in paragraph form (with word-wrap) within a text boundary and allows many methods of formatting the appearance of the text using the *Text Formatting* Editor.
Style	Creates text styles for use with any of the text creation commands. You can select from font files and specify other parameters to design the appearance of the letters.
Spell	Checks spelling of existing text in a drawing.
Find	Used to find or replace a text string globally in the drawing.
Ddedit	Allows you to edit *Text* objects in place.
Mtedit	Invokes the *Text Formatting* Editor for editing any aspect of individual *Mtext* characters or the entire *Mtext* paragraph(s).
Qtext, Textfill	Changes the appearance of text to speed up regenerations and/or plots.
Scaletext	Allows you to change the text size without changing the insertion point.
Justifytext	Allows you to change the text justification method without moving the text.

TEXT CREATION COMMANDS

FIGURE 18-3

The commands for creating text are formally named *Text* and *Mtext* (these are the commands used for typing). The *Draw* pull-down and screen (side) menus provide access to the two commonly used text commands, *Multiline Text...* (*Mtext*) and *Single-Line Text* (*Text*) (Fig. 18-3).

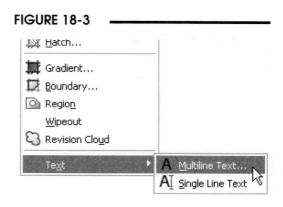

Only the *Mtext* command has an icon button (by default) near the bottom of the *Draw* toolbar (Fig. 18-4). If you are using the dashboard, both text commands are available in the *Text* control panel (Fig. 18-5).

FIGURE 18-4 —— **FIGURE 18-5** ——

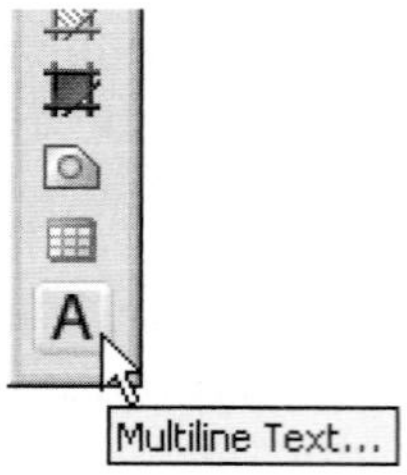

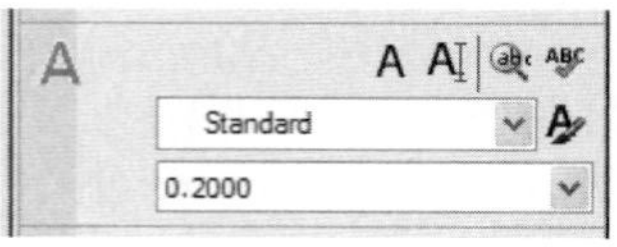

Multiline Text...

Text

Pull-down Menu	Command (Type)	Alias (Type)	Short-cut	Screen (side) Menu	Tablet Menu
Draw *Text >* *Single Line Text...*	*Text*	*DT*	...	DRAW 2 Dtext	K,8

Text lets you insert single lines of text into an AutoCAD drawing. *Text* displays each character in the drawing as it is typed. You can enter multiple lines of text without exiting the *Text* command. The lines of text do not wrap. The options are presented below:

> Command: **text**
> Current text style: "Standard" Text height: 0.2000 Annotative: No
> Specify start point of text or [Justify/Style]:

Start Point

The *Start point* for a line of text is the <u>left end</u> of the baseline for the text (Fig. 18-6). *Height* is the distance from the baseline to the top of upper case letters. Additional lines of text are automatically spaced below and left justified. The *rotation angle* is the angle of the baseline (Fig. 18-7).

The command sequence for this option is:

> Command: **text**
> Current text style: "Standard" Text height: 0.2000
> Specify start point of text or [Justify/Style]: **PICK** or (**coordinates**)
> Specify height <0.2000>: **Enter** or (**value**)
> Specify rotation angle of text <0>: **Enter** or (**value**)
> Enter text: (Type the desired line of text and press **Enter**)
> (Press **Enter** once to add an additional line of text. Press **Enter** twice to complete the command.)
> Command:

FIGURE 18-6 ——————

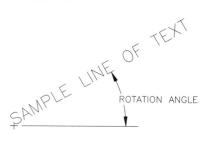

FIGURE 18-7 ——————

Justify

If you want to use one of the justification methods, invoking this option displays the choices at the prompt:

> Command: **text**
> Current text style: "Standard" Text height: 0.2000 Annotative: No
> Specify start point of text or [Justify/Style]: **J** (*Justify* option)
> Enter an option [Align/Fit/Center/Middle/Right/TL/TC/TR/ML/MC/MR/BL/BC/BR]:

After specifying a justification option, you can enter the desired text in response to the "Enter text:" prompt. The text is justified <u>as you type</u>.

Align

Aligns the line of text between the two points specified (P1, P2). The text height is adjusted automatically (Fig. 18-8).

Fit

Fits (compresses or extends) the line of text between the two points specified (P1, P2). The text height does not change (Fig. 18-8).

Center

Centers the baseline of the first line of text at the specified point. Additional lines of text are centered below the first (Fig. 18-9).

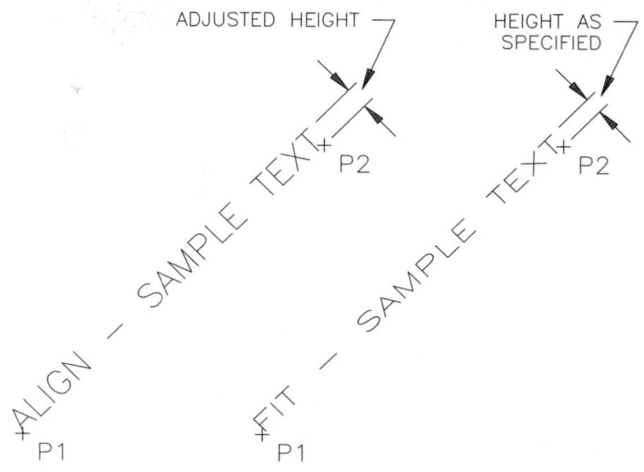

Middle

Centers the first line of text both vertically and horizontally about the specified point. Additional lines of text are centered below it (Fig. 18-9).

Right

Creates text that is right justified from the specified point (Fig. 18-9).

FIGURE 18-8

ADJUSTED HEIGHT — — HEIGHT AS SPECIFIED —

ALIGN — SAMPLE TEXT   P2

FIT — SAMPLE TEXT   P2

P1   P1

FIGURE 18-9

LEFT JUSTIFIED TEXT (START POINT) SAMPLE

RIGHT JUSTIFIED TEXT (R OPTION) SAMPLE

CENTER JUSTIFIED TEXT (C OPTION) SAMPLE

MIDDLE JUSTIFIED TEXT (M OPTION) SAMPLE

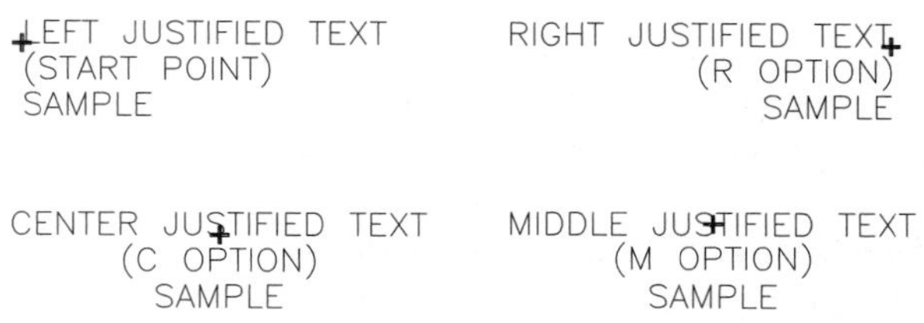

TL

Top Left. Places the text in the drawing so the top line (of the first line of text) is at the point specified and additional lines of text are left justified below the point. The top line is defined by the upper case and tall lower case letters (Fig. 18-10).

FIGURE 18-10

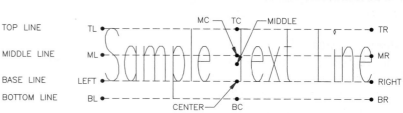

TOP LINE   TL

MIDDLE LINE   ML

BASE LINE   LEFT

BOTTOM LINE   BL

MC   TC   MIDDLE   TR

MR

RIGHT

BR

CENTER   BC

TC

Top Center. Places the text so the top line of text is at the point specified and the line(s) of text are centered below the point (Fig. 18-10).

TR

Top Right. Places the text so the top right corner of the text is at the point specified and additional lines of text are right justified below that point (Fig. 18-10).

ML

Middle Left. Places text so it is left justified and the middle line of the first line of text aligns with the point specified. The middle line is half way between the top line and the baseline, not considering the bottom (extender) line (Fig. 18-10).

MC
Middle Center. Centers the first line of text both vertically and horizontally about the midpoint of the middle line. Additional lines of text are centered below that point (Fig. 18-10).

MR
Middle Right. Justifies the first line of text at the right end of the middle line. Additional lines of text are right justified (Fig. 18-10).

BL
Bottom Left. Attaches the bottom (extender) line of the first line of text to the specified point. The bottom line is determined by the lowest point of lower case extended letters such as y, p, q, j, and g. If only upper-case letters are used, the letters appear to be located above the specified point. Additional lines of text are left justified (Fig. 18-10).

BC
Bottom Center. Centers the first line of text horizontally about the bottom (extender) line (Fig. 18-10).

BR
Bottom Right. Aligns the bottom (extender) line of the first line of text at the specified point. Additional lines of text are right justified (Fig. 18-10).

NOTE: Because there is a separate baseline and bottom (extender) line, the *MC* and *Middle* points do not coincide and the *BL*, *BC*, *BR* and *Left, Center, Right* options differ. Also, because of this feature, when all uppercase letters are used, they ride above the bottom line. This can be helpful for placing text in a table because selecting a horizontal *Line* object for text alignment with *BL*, *BC*, or *BR* options automatically spaces the text visibly above the *Line*.

Style   (option of *Text* or *Mtext*)

The *style* option of the *Text* or *Mtext* command allows you to select from the existing text styles that have been previously created as part of the current drawing. The style selected from the list becomes the current style and is used when placing text with *Text* or *Mtext*. Alternately, before you use *Text* or *Mtext*, you can select the current text style from the *Text Style Control* drop-down list (in the Object Properties toolbar and *Text* control panel) to make it the current text style.

Since only one text *style, Standard,* is available in the traditional (inch) template drawing (ACAD.DWT) and the metric template drawing (ACADISO.DWT), other styles must be created before the *style* option of *Text* is of any use. Various text styles are created with the *Style* command (this topic is discussed later).

Use the *Style* option of *Text* or *Mtext* to list existing styles and specifications. An example listing is shown below:

```
Command: text
Current text style: "Standard" Text height: 0.2000 Annotative: No
Specify start point of text or [Justify/Style]: S  (Style option)
Enter style name or [?] <STANDARD>: ?  (list option)
Enter text style(s) to list <*>: Enter
Text styles:
```

Style name: "Annotative"   Font files: txt.shx
Height: 0.0000 Width factor: 1.0000 Obliquing angle: 0
Generation: Normal

Style name: "ROMANS"   Font files: romans
Height: 1.50 Width factor: 1.00 Obliquing angle: 0.000
Generation: Normal

Style name: "STANDARD"   Font files: txt
Height: 0.00 Width factor: 1.00 Obliquing angle: 0.000
Generation: Normal

Mtext

Pull-down Menu	Command (Type)	Alias (Type)	Short-cut	Screen (side) Menu	Tablet Menu
Draw Text > Multiline Text...	*Mtext* or *-Mtext*	*T, -T or MT*	...	*DRAW 2 Mtext*	*J,8*

Multiline Text (*Mtext*) has more editing options than other text commands. You can apply underlining, color, bold, italic, font, and height changes to individual characters or words within a paragraph or multiple paragraphs of text.

Mtext allows you to create paragraph text defined by a text boundary. The <u>text boundary</u> is a reference rectangle that specifies the paragraph width. The *Mtext* object that you create can be a line, one paragraph, or several paragraphs. AutoCAD references *Mtext* (created with one use of the *Mtext* command) as one object, regardless of the amount of text supplied. Like *Text,* several justification methods are possible.

Command: **mtext**
Current text style: "Standard" Text height: 0.2000 Annotative: No
Specify first corner: **PICK**
Specify opposite corner or [Height/Justify/Line spacing/Rotation/Style/Width/Columns]: **PICK** or **(option)**

You can PICK two corners to invoke the *Text Formatting* Editor, or enter the first letter of one of these options: *Height, Justify, Rotation, Style, Line Spacing,* or *Width.* Other options can also be accessed <u>within the *Text Formatting* Editor</u>.

Using the default option the *Mtext* command, you supply a "first corner" and "opposite corner" to define the diagonal corners of the text boundary (like a window). Although this boundary confines the text on two or three sides, one or two arrows indicate the direction text flows if it "spills" out of the boundary (Fig. 18-11). (See "*Justification.*")

FIGURE 18-11

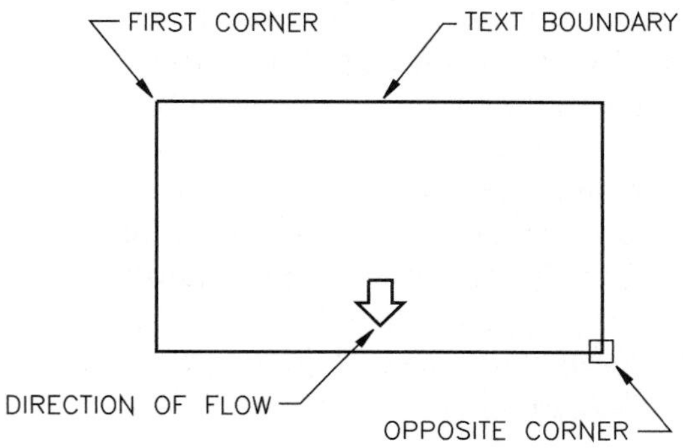

FIRST CORNER   TEXT BOUNDARY

DIRECTION OF FLOW   OPPOSITE CORNER

After you PICK the two points defining the text boundary, the *Text Formatting* Editor appears ready for you to enter the desired text (Fig. 18-12). The text wraps based on the width you defined for the text boundary. Select the *OK* button to have the text entered into the drawing.

FIGURE 18-12

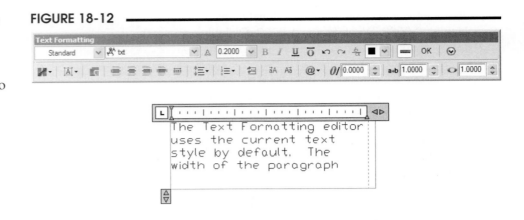

Four ways you can format and edit text using *Mtext* are: 1) the command line options, 2) the *Toolbar* (top toolbar in Fig. 18-12), 3) the *Options* toolbar (second row toolbar in Fig. 18-12), and *Options* menu (same as the right-click shortcut menu).

Text Formatting Toolbar and Command Line Options

Style
This option is the first (far left) drop-down list on the *Text Formatting* toolbar. The box (not dropped down) displays the current text *Style* (*Standard* for most template drawings), and the list displays all existing text styles in the drawing. Select the style you want to use (make current) from this list. Alternately, before you use *Text* or *Mtext,* you can select the current text style from the *Text Style Control* drop-down list (in the Object Properties toolbar or *Text* control panel) to make it the current text style. (See the *Style* command for information on creating new text styles.)

Font
Choose from any font in the drop-down list. The list includes all fonts registered on your system. Use this feature to change the appearance of selected words in the paragraph. Your selection here overrides the font originally assigned to the text *Style* for the selected or newly created words or paragraphs. Even though you can change the font for the entire *Mtext* object (paragraph) using this feature, it is recommended that you set the paragraph to the desired *Style* (in the first drop-down list), rather than change the characters in the *Style* to a different font for the current *Mtext* object. See following NOTE.

Text Color
Select individual text, then use this drop-down list from the toolbar to select a color for the selected text. This selection overrides the layer color. See following NOTE.

NOTE: The *Font, Height,* and *Color* options in the *Text Formatting* toolbar override the properties of the text *Style* and layer color. For example, changing the font in the *Font* list overrides the text style's font used for the paragraph so it is possible to have one font used for the paragraph and others for individual characters within the paragraph. This is analogous to object-specific color and linetype assignment in that you can have a layer containing objects with different linetypes and colors than the linetype and color assigned to the layer. To avoid confusion, it is recommended for global formatting to create text *Styles* to be used for the paragraphs, layer color to determine color for the paragraphs, then if needed, use the *Font* and *Color* options in the *Text Formatting* toolbar to change fonts and colors for selected text rather than for the entire paragraph.

Annotative
Use this toggle to set the *Annotative* feature of the selected text. See "Annotative Text."

Height

Select from the list or enter a new value for the height to be used globally for the paragraph or for selected words or letters. If *Height* was defined when the selected text *Style* was created for the drawing, this option overrides the *Style's Height*.

Bold, Italic, Underline, Overline

Select (highlight) the desired letters or words then PICK the desired button. Only authentic TrueType fonts (not the AutoCAD-supplied .SHX equivalents) can be bolded or italicized.

Undo, Redo

Use *Undo* to reverse the action of the last formatting option used. You can *Redo* the last *Undo*, but only immediately after the *Undo*.

Stack/Unstack

This option is a toggle to *Stack* or *Unstack* specific text. Highlight the fraction or text, then use this option to stack or unstack the fraction or text.

Ruler

Use the *Ruler* button or *Show Ruler* option from the shortcut menu to make the ruler on top of the text area to appear (Fig. 18-13).

FIGURE 18-13

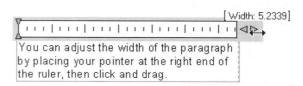

Width

Even though this option does not appear on the *Text Formatting* toolbar, you can interactively change the width of the existing text boundary. Do this by placing your pointer at the right border of the ruler above the text so the double arrows appear, then adjust the paragraph width (Fig. 18-13). The lines of text will automatically "wrap" to adjust to the new text boundary. You can also use the *Width* option from the Command line to set the width to a specific value. Also see "*Justification*" for changing the width of the text boundary with Grips.

Rotation

The *Rotation* option specifies the rotation angle of the entire *Mtext* object (paragraph) including the text boundary. This option is available only by command line; however, the Grips Rotate option can be used after the *Mtext* object has been created. Using the Command line method, you can see the rotated text boundary as you specify the corners (Fig. 18-14).

FIGURE 18-14

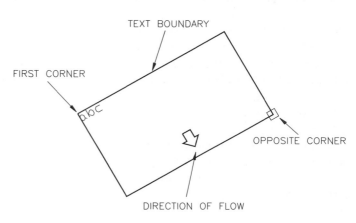

Options Toolbar and *Options* Menu (or Right-Click Shortcut Menu)

The *Options* toolbar (second row toolbar in Fig. 18-15) can be toggled on or off by selecting *Show Options* from the *Options* menu. The *Options* menu (Fig. 18-15) is the same as the right-click shortcut menu that appears when you right-click inside the text editing area.

FIGURE 18-15

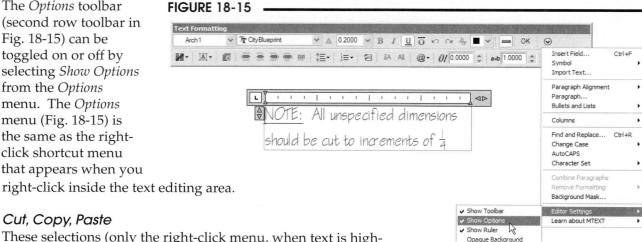

Cut, Copy, Paste

These selections (only the right-click menu, when text is high-lighted) allow you to *Cut* (erase) highlighted text from the para-graph, *Copy* highlighted text, and *Paste* (the *Cut* or *Copied*) text to the current cursor position. Text that you *Cut* or *Copy* is held in the Clipboard until Pasted.

Columns

Use this button to display a drop-down menu for creating text columns. See "Creating Columns."

MText Justification

The choices from the *MText Justification* button on the *Options* toolbar (Fig. 18-16) determine how the text in the paragraph is oriented (originally), how the text flows, and what the paragraph *Insertion* point is for Object Snaps. These options are not available in the right-click shortcut menus. These options are <u>not the same</u> as, and can be overridden by, the *Paragraph Alignment* settings (*Left*, *Center*, *Right*, *Justify*, and *Distribute*) available just to the right on the toolbar or from the right-click short-cut menus.

Justification is the method of aligning the text with respect to the text boundary. The options are *Top Left*, *Top Centered*, *Top Right*, *Middle Left*, *Bottom Left*, etc. The text paragraph (*Mtext* object) is effec-tively "attached" to the boundary based on the option selected. These options are illustrated in Figure 18-17. The illustra-tion shows the relationship among the selected option, the text boundary, the direction of flow, and the resulting text paragraph.

FIGURE 18-16

FIGURE 18-17

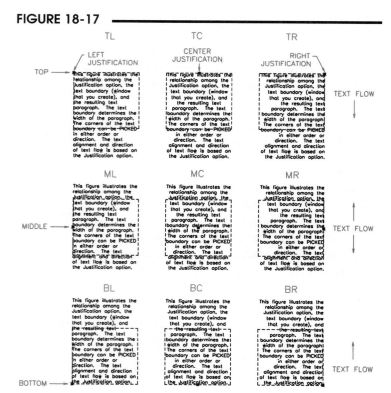

If you want to adjust the text boundary <u>after</u> creating the *Mtext* object, you can do so within the *Text Formatting* editor (see previous discussion, "Width") or you can use use grips (Chapter 19) to stretch the text boundary. If you activate the grips, four grips appear at the text boundary corners and one grip appears at the defined justification point (Fig. 18-18).

NOTE: The *MText Justification* options can be overridden by selecting one of the *Paragraph Alignment* options from the right-click shortcut menus or from the toolbar. If a *Paragraph Alignment* option is used, however, it does not change the paragraph's *Insertion* point.

FIGURE 18-18

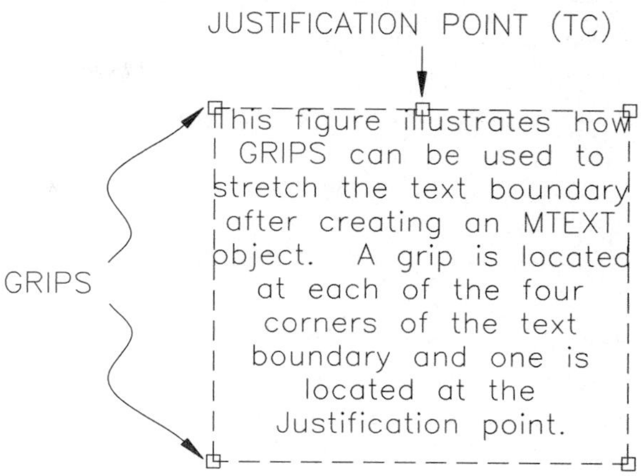

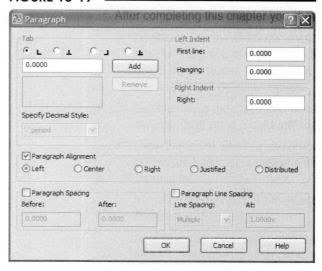

Paragraph Alignment

These options are available from the buttons on the *Options* toolbar and from the shortcut menus. The *Left, Center, Right, Justify,* and *Distribute* options automatically align the text in the paragraph accordingly and override the *Mtext Justification* setting for the paragraph (see previous NOTE).

Left, Center, Right
These options move the text in the paragraph to align flush left, centered, and flush right, respectively. The character spacing is not altered. The text does not have to be selected.

Justify, Distribute
Both of these options change the horizontal spacing such that the text in each line is adjusted to begin flush left and end flush right. The *Justify* option changes the spacing between words, whereas the *Distribute* option changes the individual character spacing.

Paragraph

This option, selected from the button on the *Options* toolbar or from shortcut menus, produces the paragraph dialog box (Fig. 18-19). The *Paragraph* dialog box is used to set tabs, indents, paragraph alignment, paragraph spacing, and line spacing.

Tab
To designate each tab, specify the increment in units and the tab type (left, center, right, decimal), then select *Add*.

Left Indent, Right Indent
Enter desired values for each indent.

Paragraph Alignment
These choices are identical to those for *Paragraph Alignment* discussed previously.

FIGURE 18-19

Paragraph Spacing
If you have multiple paragraphs created as one *Mtext* object, use the *Before* and *After* edit boxes to specify the space before and after each paragraph.

Paragraph Line Spacing

These options affect the vertical spacing between lines in the *Mtext* object. *Exactly* forces the line spacing to be the same for all lines. *At least* automatically adds space between lines based on the height of the largest character; therefore, use this option if you have different sizes of text in the paragraph. The *Multiple* option sets the spacing to a multiple of single line spacing.

Alternately, you can set the first line indent and the paragraph indent interactively by using your pointer on the ruler (just above the text boundary). Move the top indent marker to set the indent for the *First line* (Fig. 18-20) and the bottom indent marker to set an indent for the entire *Paragraph*.

FIGURE 18-20 ─────────

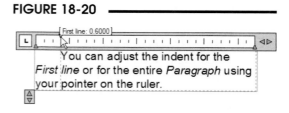

You can adjust the indent for the *First line* or for the entire *Paragraph* using your pointer on the ruler.

You can also set tab positions by clicking in the ruler appearing above the text box (Fig. 18-21). Pick the desired position to make a tab appear. Drag and drop tab stops out of the ruler to clear them. Select from left, right, center, or decimal tabs as shown by the pointer in the figure.

FIGURE 18-21 ─────────

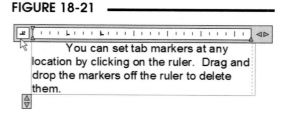

You can set tab markers at any location by clicking on the ruler. Drag and drop the markers off the ruler to delete them.

Numbering, Bullets and Lists

You can create a variety of lettered, numbered, or bulleted lists using these options in the *Options* menu or using the related buttons on the *Options* toolbar. See "Bullets and Lists" later in this chapter.

Symbol

Common symbols (*Degrees, Plus/Minus, Diameter*) can be inserted as well as many mathematical and other drawing symbols (*Almost Equal* through *Cubed*). Selecting *Other…* produces a character map to select symbols from (See *Other Symbols* next).

Other Symbols

The steps for inserting symbols from the *Character Map* dialog box (Fig. 18-22) are as follows:

1. Highlight the symbol.
2. Double-click or pick *Select* so the item appears in the *Characters to copy*: edit box.
3. Select the *Copy* button to copy the item(s) to the Windows Clipboard.
4. *Close* the dialog box.
5. In the *Text Formatting* Editor, move the cursor to the desired location to insert the symbol.
6. Finally, right-click and select *Paste* from the menu.

FIGURE 18-22 ─────────

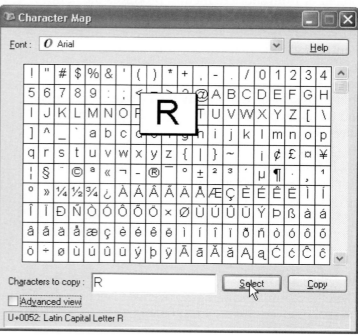

Character Set

This option (available only from the *Options* menu or right-click shortcut menu) displays a menu of code pages. If *Big Fonts* have been used for text styles in the drawing, you can select a text string, then select a code page from the menu to apply it to the text.

Oblique Angle

This option, available only from the *Options* toolbar, allows you to apply an angle (incline) to text. Enter the desired angle (in degrees) in the edit box. Apply an *Oblique Angle* to newly created text or existing text.

Tracking

This parameter, also known as kerning, determines the spacing between letters. This does not affect the width of the individual letters. The acceptable range is .75 through 4.0, with 1.0 representing normal spacing. This option is available only from the *Options* toolbar.

Width Factor

Use this option from the *Options* toolbar to increase the width of the individual letters. Apply this characteristic to newly created text or existing text.

Find and Replace

See "Editing Text" later in this chapter.

Change Case

First select the desired text, then select either *UPPERCASE* or *lowercase* from the *Options* toolbar or *Options* menu. All letters are changed; therefore, you cannot have mixed upper and lower case, such as Title Case, unless you change letters individually.

AutoCAPS

This option affects only newly typed and imported text. When checked, *AutoCAPS* turns on Caps Lock on your keyboard so only uppercase letters appear as you type. If needed, use *Change Case* to convert the text back to lowercase.

Remove Formatting

Select the desired text, then select this option to remove bold, italic, or underline formatting.

Combine Paragraphs

If you have more than one paragraph in one *Mtext* object, you can select all the text in the paragraphs, then use this option. The selected paragraphs are converted into a single paragraph and each paragraph return is replaced with a space.

Background Mask

In some cases, it is necessary to create text "on top of" other objects, such as a hatch pattern (Fig. 18-23). With this option, you can specify that the *Mtext* object you create has a background, or mask, to "cover" the objects "behind" the text. The resulting background is similar to that created by a *Wipeout*, only for text.

FIGURE 18-23

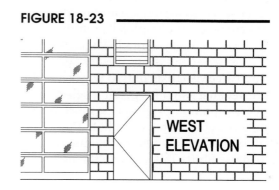

To create a background mask for an *Mtext* object, select *Background Mask* from the *Options* menu or right-click shortcut menu. This produces the *Background Mask* dialog box (Fig. 18-24) where you specify the *Fill color* for the mask and the *Border offset factor* (distance the mask boundary is offset from the actual text inserted). Selecting the *Use drawing background color* option under *Fill Color* forces AutoCAD to use the screen color as the background, therefore producing a "clear" mask.

FIGURE 18-24

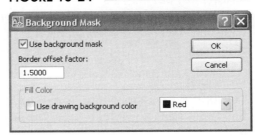

Stack/Unstack

Stack/Unstack appears in the shortcut menu if you highlight only stacked text, such as a fraction, then right-click to invoke the menu. This option is a toggle so that stacked characters become unstacked and vice versa.

Stack Properties

This option appears in the *Options* menu or right-click shortcut menu if you first highlight only stacked text such as a fraction. The *Stack Properties* dialog box appears (Fig. 18-25).

The *Text* section of the *Stack Properties* dialog box allows you to change the text characters (numbers or letters) contained in the stacked set. Although using a slash (/), pound (#), or carat (^) symbol to specify a stack normally creates horizontal, diagonal, and tolerance format, respectively, that format can be changed with the *Style* drop-down list in this dialog box. Use the *Position* option to align the fraction with the *Top*, *Center*, or *Bottom* of the other normal characters in the line of text. You can also specify the *Text size* for the stack (which should generally be smaller than normal characters since the stacked set occupies more vertical space).

FIGURE 18-25

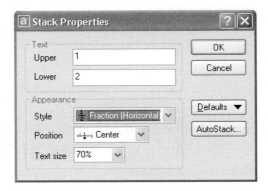

Bullets and Lists

You can use the *Text Formatting* editor or *Options* menu to create a list with numbers, letters, bullets, or other characters as separators (Fig. 18-26). Items in the list are automatically indented based on the tab stops on the ruler. You can add or delete an item or move an item up or down a level and the list numbering automatically adjusts. You can create sub-levels in the list and also remove and reapply list formatting. To turn automatic list creation off completely, remove the check from *Numbering On* in the *Options* menu or right-click shortcut menu.

FIGURE 18-26

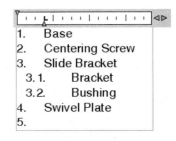

You can create lists in two ways: create the list as you type or apply list formatting to selected text. To create a list as you type, first select the *Lettered*, *Numbered*, or *Bulleted* option from the *Options* toolbar, *Options* menu or right-click shortcut, then begin typing the first item in the list. Pressing Enter adds the next number, letter, or bullet. Press Enter twice to end the list. To create a sub-list (a lower-level list), press the Tab key before the first item in the sub-list. Press Enter to start the next item at the same level, or press Shift+Tab to move the item up a level.

Creating Columns

Creating columns using *Mtext* is relatively simple, and adjusting the columns to your needs is intuitive. The *Columns* button on the *Options* toolbar or from the shortcut menus provide several ways to create the columns. To create columns, begin by using *Mtext* and <u>entering the desired text</u>. Next, either during the process of entering the text or after the paragraph(s) are complete, select the desired column option from the menu. Once the columns have been created, you can easily adjust the height, width, and spacing of the columns dynamically using the multiple sliders on the ruler (Fig. 18-27) or by using grip editing.

FIGURE 18-27

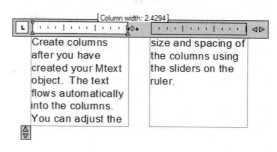

There are two different methods for creating and manipulating columns: *Static Columns* or *Dynamic Columns*. The following options are available from the *Columns* menu.

Dynamic Columns

Selecting this option sets dynamic columns mode to the current *Mtext* object. *Dynamic Columns* are text driven—that is, the number of columns varies depending on how much text is entered and by the height and/or width of the columns. Adjusting the columns affects text flow and text flow causes columns to be added or removed. *Auto height* columns allow you to set the height for all columns with one slider, whereas *Manual height* columns can be individually adjusted for height.

Static Columns

This option sets a static number of columns to the current *Mtext* object. Although you specify a static number of columns, the width, height, and spacing between the columns can be adjusted using the sliders. Text automatically flows between columns as needed. All static columns share the same height and are aligned at both sides.

Insert Column Break Alt+Enter

For *Dynamic Columns*, this option inserts a manual column break. This feature is disabled when you select *No Columns*.

Column Settings

Choosing this option produces the *Column Settings* dialog box (Fig. 18-28). This dialog box offers the same options that are available using the menus and the sliders. The dialog box is useful if you have the need for exact height and width spacing. Even though values can be entered into the edit boxes in the *Height* and *Width* clusters, the columns can be interactively adjusted later using the sliders.

FIGURE 18-28

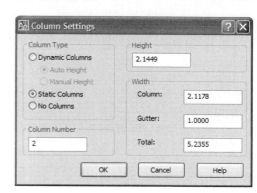

Grip Editing

Columns can be edited using grips in a fashion similar to using the sliders available within the *Mtext* command. See Chapter 19, Grip Editing.

Calculating Text Height for Scaled Drawings

To achieve a specific height in a drawing intended to be plotted to scale, multiply the desired (non-annotative) text height for the plot by the drawing scale factor. (See Chapter 14, Printing and Plotting.) For example, if the drawing scale factor is 48 and the desired text height on the plotted drawing is 1/8", enter **6** (1/8 x 48) in response to the "Height:" prompt of the *Text* or *Mtext* command.

If you know the plot scale (for example, 1/4"= 1') but not the drawing scale factor (DSF), calculate the reciprocal of the plot scale to determine the DSF, then multiply the intended text height for the plot by the DSF, and enter the value in response to the "Height:" prompt. For example, if the plot scale is 1/4"=1', then the DSF = 48 (reciprocal of 1/48).

If *Limits* have already been set and you do not know the drawing scale factor, use the following steps to calculate a text height to enter in response to the "Height:" prompt to achieve a specific plotted text height.
1. Determine the sheet size to be used for plotting (for example, 36" x 24").
2. Decide on the text height for the finished plot (for example, .125").
3. Check the *Limits* of the current drawing (for example, 144' x 96' or 1728" x 1152").
4. Divide the *Limits* by the sheet size to determine the drawing scale factor (1728"/36" = 48).
5. Multiply the desired text height by the drawing scale factor (.125 x 48 = 6).

TEXT STYLES AND ANNOTATIVE TEXT

Style

Pull-down Menu	Command (Type)	Alias (Type)	Short-cut	Screen (side) Menu	Tablet Menu
Format *Text Style...*	*Style* or *-Style*	*ST*	...	*FORMAT* *Style:*	*U,2*

Text styles can be created by using the *Style* command. A text *Style* is created by selecting a font file as a foundation and then specifying several other parameters to define the configuration of the letters.

Using the *Style* command invokes the *Text Style* dialog box (Fig. 18-29). All options for creating and modifying text styles are accessible from this device. Selecting the font file is the initial step in creating a style. The font file selected then becomes a foundation for "designing" the new style based on your choices for the other parameters (*Effects*).

FIGURE 18-29

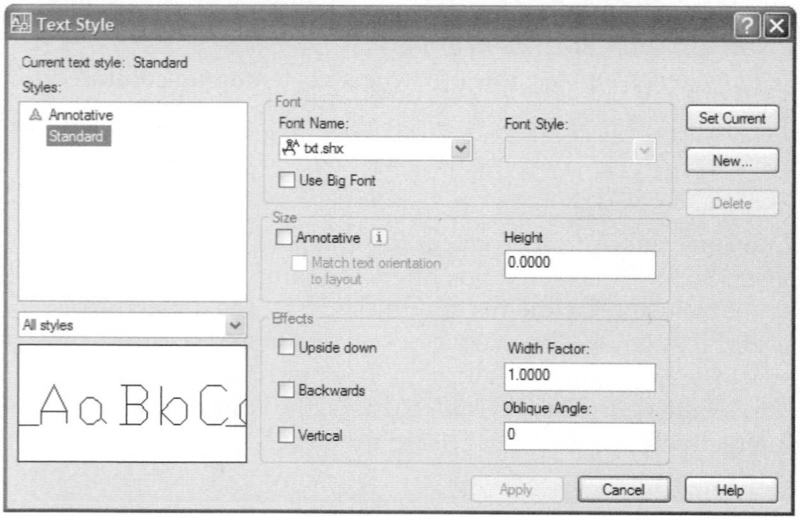

Text Style dialog box showing Current text style: Standard; Styles list with Annotative and Standard; Font Name: txt.shx; Use Big Font; Size with Annotative, Match text orientation to layout, Height 0.0000; Effects with Upside down, Backwards, Vertical, Width Factor 1.0000, Oblique Angle 0; preview showing AaBbC; buttons Set Current, New..., Delete, Apply, Cancel, Help.

Recommended steps for creating text *Styles* are as follows:

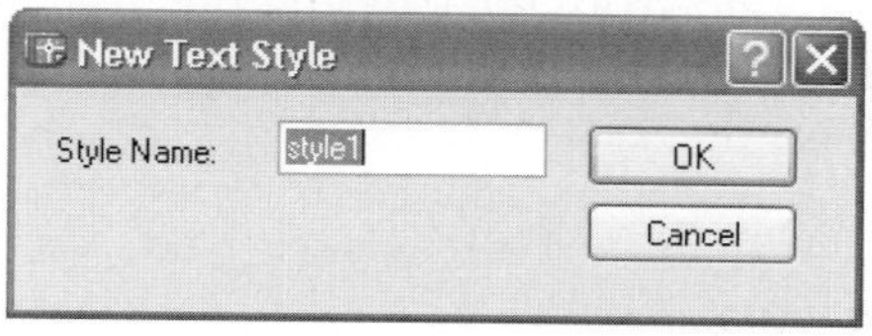

1. Use the *New* button to create a new text *Style*. This button opens the *New Text Style* dialog box (Fig. 18-30). By default, AutoCAD automatically assigns the name Style*n*, where *n* is a number that starts at 1. Enter a descriptive name into the edit box. Style names can be up to 256 characters long and can contain letters, numbers, and the special characters dollar sign ($), underscore (_), and hyphen (-).

FIGURE 18-30

2. Select a font to use for the style from the *Font Name* dropdown list (see Fig. 18-29). Authentic TrueType font files (.TTF) as well as AutoCAD equivalents of the TrueType font files (.SHX) are available.
3. If you want to create annotative text, check the *Annotative* box. Enter the desired *Paper Text Height* in the edit box. (See "Annotative Text" for more information.)
4. Select parameters in the *Effects* section of the dialog box. Changes to the *Upside down, Backwards, Width Factor,* or *Oblique Angle* are displayed immediately in the *Preview* tile.
5. Select *Apply* to save the changes you made to the new style.
6. Create other new *Styles* using the same procedure listed in steps 1 through 5.
7. Select *Close*. The new (or last created) text style that is created automatically becomes the current style inserted when *Text* or *Mtext* is used.

You can modify existing styles using this dialog box by selecting the existing *Style* from the list, making the changes, then selecting *Apply* to save the changes to the existing *Style*.

Rename
Select an existing *Style* from the list, right click, then select *Rename*. Enter a new name in the edit box.

Delete
Select an existing *Style* from the list, then select *Delete*. You cannot delete the current *Style* or *Styles* that have been used for creating text in the drawing.

Annotative
Check the *Annotative* box if you want to create annotative text. (See "Annotative Text" for more information.)

Height
The height should be 0.000 if you want to be prompted again for height each time the *Text* command is used. In this way, the height is variable for the style each time you create text in the drawing. If you want the height to be constant, enter a value other than 0. Then, *Text* will not prompt you for a height since it has already been specified. The *Mtext* command, however, allows you to change the height in the *Text Formatting* Editor, even if you specified a height when you created the text style. A specific height assignment with the *Style* command also overrides the *DIMTXT* setting (see Chapter 26).

Paper Text Height
If you check the *Annotative* box, the *Height* edit box changes to display *Paper Text Height*. Enter the desired height for the text to appear in paper space units. In other words, if you want the text to appear at 1/8" in all viewports, no matter what the model space size of the text is, enter .125 in the *Paper Text Height* edit box. (See "Annotative Text" for more information.)

Width factor
A *width factor* of 1 keeps the characters proportioned normally. A value less than 1 compresses the width of the text (horizontal dimension) proportionally; a value of greater than 1 extends the text proportionally.

Obliquing angle

An angle of 0 keeps the font file as vertical characters.
Entering an angle of 15, for example, would slant the text
forward from the existing position, or entering a negative
angle would cause a back-slant on a vertically oriented font
(Fig. 18-31).

FIGURE 18-31

0° OBLIQUING ANGLE

15° OBLIQUING ANGLE

−15° OBLIQUING ANGLE

Backwards

Backwards characters can be helpful for special applications, such as those in the printing industry.

Upside-down

Each letter is created upside-down in the order as typed
(Fig. 18-32). This is different than entering a rotation angle
of 180 in the *Text* command. (Turn this book 180 degrees to
read the figure.)

FIGURE 18-32

∩ᑭƧIDE DOMИ ⊥EX⊥

⊥EX⊥ ᴚO⊥A⊥ED 180°

Vertical

Vertical letters are shown in Figure 18-33. The normal rotation angle for vertical
text when using *Text* or *Mtext* is 270. Only .SHX fonts can be used for this option.
Vertical text does not display in the *Preview* image tile.

FIGURE 18-33

```
V   T
E   E
R   X
T   T
I
C
A
L
```

TIP Since specification of a font file is an
initial step in creating a style, it seems
logical that different styles could be
created using <u>one</u> font file but changing
the other parameters. It is possible, and
in many cases desirable, to do so. For
example, a common practice is to create
styles that reference the same font file
but have different obliquing angles
(often used for lettering on isometric
planes). This can be done by using the
Style command to assign the same font
file for each style, but assign different

FIGURE 18-34

FONT FILE	STYLE NAME (EXAMPLE)	RESULTING TEXT (BASED ON OTHER PARAMETERS)
ROMAN SIMPLEX (ROMANS.SHX)	ROMANS−VERT	ABCDEFG 1234
	ROMANS−ITAL	*ABCDEFG 1234*
ROMAN DUPLEX (ROMAND.SHX)	ROMAND−EXT	A B C 1 2 3 4
	ROMAND−COMP	ABCDEFG 1234

parameters (obliquing angle, width factor, etc.) and a unique name to each style. The relationship
between fonts, styles, and resulting text is shown in Figure 18-34.

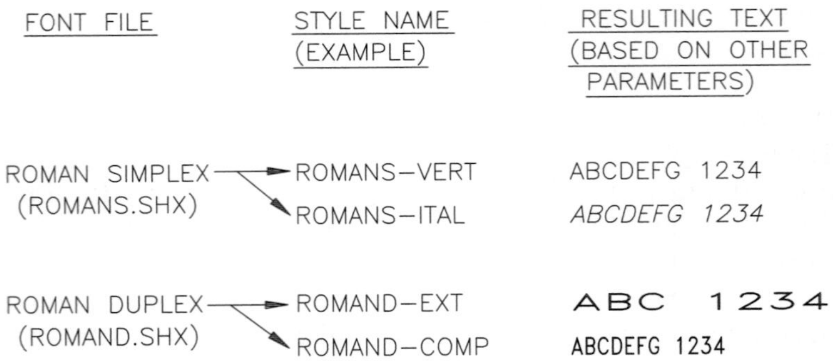

Annotative Text

For complex drawings including multiple layouts and viewports, it may be desirable to create annotative objects. Annotative objects, such as annotative text, annotative dimensions, and annotative hatch
patterns, automatically adjust for size when the *Annotation Scale* for each viewport is changed. That is,
annotative text objects can be automatically resized to appear "readable" in different viewports at different scales. The following section is helpful if you plan to create multiple viewports and want to use
annotative text. To give an overview for creating and using annotative text, follow these steps.

Steps for Creating and Viewing Annotative Text

1. Use the *Style* command to produce the *Text Style* dialog box.
2. In the *Text Style* dialog box, either create a new text style or change an existing style to *Annotative*. Next, set the desired *Paper Text Height*, or height that you want the text to appear in the viewports in paper space. Select the *Apply* or *Close* button.
3. In the *Model* tab, set the *Annotation Scale* for model space. You can use the *Annotation Scale* pop-up list or set the *CANNOSCALE* variable. Use the reciprocal of the "drawing scale factor" (DSF) as the annotation scale (AS) value, or AS = 1/DSF.
4. Create the text in the *Model* tab using *Text* or *Mtext*. Accept the default (model space) height.
5. Create the desired viewports if not already created.
6. Set the desired scale for the current viewport by double-clicking inside the viewport and using the *VP Scale* pop-up list or *Viewports* toolbar. Alternately, select the viewport object (border) and use the *Properties* palette for the viewport. Setting the *VP Scale* automatically adjusts the annotation scale for the viewport.

To simplify, create annotative text using the following formulas, where DSF represents "drawing scale factor":

Annotation Scale in the *Model* tab = 1/DSF
Paper Text Height x DSF = model space text height.

For example, assume you are drawing a small apartment. In order to draw the floor plan in the *Model* tab full size, you change the default *Limits* (12, 9) by a factor of 48 to arrive at *Limits* of 48',36' (drawing scale factor = 48). After creating the floor plan, you are ready to label the rooms.

Following steps 1 through 3, create a new *Text Style* as *Annotative*, then set the *Paper Text Height* to 1/8". Next, set the annotative scale for the *Model* tab using the *Annotation Scale* pop-up list in the lower-right corner of the Drawing Editor (see Fig. 18-35). Using the drawing scale factor (DSF) of 48, set the *Annotation Scale* to 1:48 (or AS = 1/DSF). (Also see "Steps for Drawing Setup Using Annotative Objects" in Chapter 12.)

FIGURE 18-35

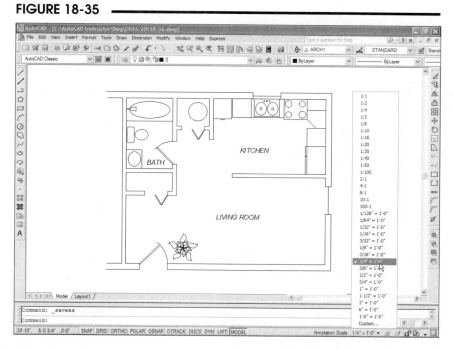

Next, following step 4, use *Text* to create the room labels in the *Model* tab as shown in Figure 18-35. The command prompt appears as shown. Note that AutoCAD automatically calculates the model space text height (1/8" x 48 = 6", or paper text height x DSF = model text height) so you do not have to specify the height during the *Text* command.

Command: **text**
Current text style: "ARCH1" Text height: 0'-6" Annotative: Yes
Specify start point of text or [Justify/Style]:

Next, following steps 5 and 6, switch to a *Layout* tab, set it up for the desired plot or print device using the *Page Setup Manager*, and create a viewport. Set the viewport scale for the viewport using the *VP Scale* pop-up list as shown in Figure 18-36. The text should appear appropriately in the layout at 1/8" (paper text height). Note that the *Annotation Scale* is locked to the *VP Scale* (by default) so its setting changes accordingly as the *VP Scale* changes.

FIGURE 18-36

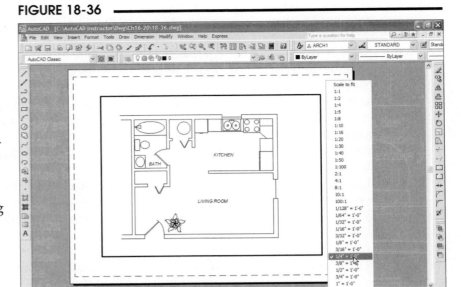

Note that at this point, using only one viewport, there is no real advantage to using annotative text as opposed to non-annotative text. The text appears in the *Model* tab and *Layout* tab at the correct size, just as if non-annotative text were used and set to a (model space) text height of 6".

Annotative Text for Multiple Viewports

Annotative objects such as annotative text are useful when multiple viewports are used and the geometry is displayed at different scales in the viewports. To display annotative text in such cases, and assuming the text is created and viewports are already prepared (as in our example), follow these steps:

1. In a layout tab when *PAPER* space is active (not in a viewport), enable the *Automatically add scales to annotative objects…* toggle in the lower-right corner of the Drawing Editor or set the *ANNOAUTOSCALE* variable to a positive value (1, 2, 3, or 4).

2. Activate one viewport by double-clicking inside the viewport. Set the viewport scale for the viewport using the *VP Scale* pop-up list, *Viewports* toolbar, or *Properties* palette. The *Annotation Scale* is locked to the *VP Scale* by default, so it changes automatically to match the *VP Scale*.

3. Repeat step 2 for each viewport.

To illustrate setting the annotative scale for each viewport, use the same apartment floor plan drawing used in the previous example. Assume one layout was created but the layout contained two viewports, one to display the entire apartment at 1/4" = 1' scale and the second viewport to display only the bath at 1/2" = 1' scale (Fig. 18-37, on the next page). Note that the scale for the bath (left) is shown much larger (approximately twice the size of the geometry in the large viewport on the right), but at this point the text object "BATH" displays at the same size relative to the geometry.

To change the display of the annotative text in the left viewport, make the viewport active and select 1/2" = 1' from the *VP Scale* pop-up list as shown. Once this is completed (not shown), the size of the geometry will be changed to exactly 1:2, and since the *Annotation Scale* is locked to the *VP Scale*, the annotative text object will be displayed at the correct size, 1/8" in paper units. Simply stated, the label "BATH" will be the same size as "LIVING ROOM" and "KITCHEN" in the other viewport.

FIGURE 18-37

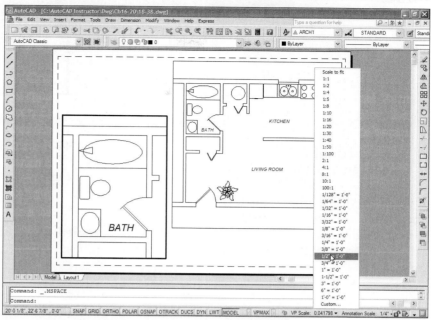

It is important to note that the original text objects do not change size when a different *VP Scale* is selected. Instead, <u>a new annotative text object is created as a copy of the original</u>, but in the new scale (Fig. 18-38). When the *Automatically add scales to annotative objects...* toggle (the *ANNOAUTOSCALE* variable) is on, new text objects are created for each new scale selected from the *VP Scale* list. Therefore, note that a single annotative object can have multiple *Annotative Scales* (or *Object Scales*), one for each time the object is displayed in a different scale.

FIGURE 18-38   **TIP** ✓

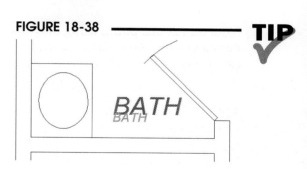

The *SELECTIONANNODISPLAY* system variable, when set to 1, allows you to see each text object when selected, as shown in Figure 18-38. The text objects can be manipulated individually using grips.

EDITING TEXT

Although several methods are available to edit text, probably the <u>simplest method to use is to double-click</u> on a *Text* or *Mtext* object (assuming *DBLCLKEDIT* is set to *On*). Doing so invokes the appropriate editing method for the specific text. If you double-click on a *Text* object, you can edit the text "in place." No real text editor appears—the text becomes reversed (in color) and you can change the text itself (see "*Ddedit*"). If you double-click on an *Mtext* object, the *Text Formatting* Editor appears (see "*Mtedit*"). You can also edit *Text* and *Mtext* objects using the *Properties* palette. The *Properties* palette is useful for editing all properties of *Text* objects; however, to gain the most control over editing *Mtext* objects, double-click, use *Ddedit*, or use *Mtedit*.

Ddedit

Pull-down Menu	Command (Type)	Alias (Type)	Short-cut	Screen (side) Menu	Tablet Menu
Modify Object > Text > Edit...	Ddedit	ED	...	MODIFY1 Ddedit	Y,21

Ddedit can be used for editing either *Text* or *Mtext* objects. The following prompt is issued:

Command: **ddedit**
Select an annotation object or [Undo]:

FIGURE 18-39

INSIDE DIAMETER TO BE HONED AFTER HEAT TREAT

If you select a *Text* object, the text becomes "reversed" and you can edit the contents of the text directly in the drawing (Fig. 18-39). Only the textual content can be changed. If you want to change other properties of the *Text* object such as *Style*, *Justification*, *Height*, *Rotation*, etc., use the *Properties* palette.

Mtedit

Pull-down Menu	Command (Type)	Alias (Type)	Short-cut	Screen (side) Menu	Tablet Menu
...	Mtedit	...	...	...	...

If you select an *Mtext* object, the *Text Formatting* Editor appears (see previous Fig. 18-15). In the *Text Formatting* Editor, you can edit the textual contents as well as any properties of the *Mtext* object. (See also "*Mtedit*.")

The *Mtedit* command is designed specifically for editing *Mtext* objects by producing the *Text Formatting* Editor. The following prompt appears.

Command: **mtedit**
Select an MTEXT object:

The *Text Formatting* Editor (see previous Fig. 18-15) allows you to change any aspect of the text content and all other properties of the *Mtext* objects. (You can also produce the *Text Formatting* Editor by double-clicking on an *Mtext* object.) *Mtedit* cannot be used to edit *Text* objects.

NOTE: If you select *Mtext* that is too small or too large to read and edit properly, you can *Zoom* with the mouse wheel <u>inside the edit boundary</u> to change the display of the text and text editing boundary.

Properties

Pull-down Menu	Command (Type)	Alias (Type)	Short-cut	Screen (side) Menu	Tablet Menu
Modify Properties...	Properties	PR or CH	(Default Menu) Find or Ctrl + 1	MODIFY1 Modify	Y,14

The *Properties* command invokes the *Properties* palette. As described in Chapter 16, the palette that appears is <u>specific</u> to the type of object that is PICKed—kind of a "smart" properties palette.

Remember that you can select objects and then invoke the palette, or you can keep the *Properties* palette on the screen and select objects in the drawing whose properties you want to change. (See Chapter 16 for more information on the *Properties* palette.)

FIGURE 18-40

If a line of *Text* is selected, the palette displays properties as shown in Figure 18-40. The *Properties* palette allows you to change almost any properties associated with the text, including the *Contents* (text), *Style*, *Justification*, *Height*, *Rotation*, *Width*, *Obliquing*, and *Text alignment*. Simply locate the property in the left column and change the value in the right column.

Note that the palette displays an entry for the *Annotative* property. You can use this section of the *Properties* palette to convert non-annotative objects to annotative by changing the *No* value to *Yes*. In this case, a new entry appears for *Annotative Scale*. A single annotative object can have multiple *Annotative Scales* (or *Object Scales*), one for each time the object is displayed in a different scale (in different viewports).

If you select an *Mtext* paragraph, the *Properties* palette appears allowing you to change the properties of the object. However, if you select the *Contents* section, a small button on the right appears. Pressing this button invokes the *Text Formatting* Editor with the *Mtext* contents and allows you to edit the text with the full capabilities of the editor.

Spell

Pull-down Menu	Command (Type)	Alias (Type)	Short-cut	Screen (side) Menu	Tablet Menu
Tools *Spelling*	*Spell*	*SP*	...	*TOOLS 1* *Spell*	*T,10*

AutoCAD has an internal spell checker that can be used to spell check and correct existing text in a drawing after using *Text* or *Mtext*. You can also select *Block* attributes and *Blocks* containing text to spell check. The *Check Spelling* dialog box (see Fig. 18-41) has options to *Ignore* the current word or *Change* to the suggested word. The *Ignore All* and *Change All* options treat every occurrence of the highlighted word.

FIGURE 18-41

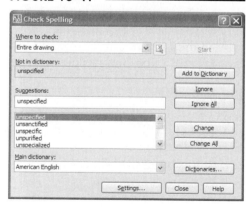

You can check the *Entire drawing*, *Current space/layout*, or *Select objects*. Press *Start* for AutoCAD to begin checking. AutoCAD matches the words to the words in the current dictionary. If the speller indicates that a word is misspelled but it is a proper name or an acronym you use often, it can be added to a custom dictionary. Choose *Add to Dictionary* if you want to leave a word unchanged but add it to the current custom dictionary.

Find

Pull-down Menu	Command (Type)	Alias (Type)	Short-cut	Screen (side) Menu	Tablet Menu
Edit *Find...*	*Find*	...	(Default menu) *Find...*	...	*X,10*

Find is used to find or replace text strings <u>globally</u> in a drawing. It can find and replace any text in a drawing, whether it is *Text* or *Mtext*. *Find* produces the *Find and Replace* dialog box (see Fig. 18-42, on the next page).

TIP The *Find* command searches a drawing globally and operates with all types of text, including dimensions, tables, and *Block* attributes. In contrast, the *Find/Replace* option in the *Text Formatting* Editor right-click menu operates only for *Mtext* and for only one *Mtext* object at a time.

FIGURE 18-42

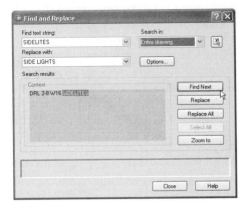

Find Text String, Find/Find Next

To find text only (not replace), enter the desired string in the *Find Text String* edit box and press *Find* (Fig. 18-42). When matching text is found, it appears in the context of the sentence or paragraph in the *Search Results* area. When one instance of the text string is located, the *Find* button changes to *Find Next*.

Replace With, Replace

You can find any text string in a drawing, or you can search for text and replace it with other text. If you want to search for a text string and replace it with another, enter the desired text strings in the *Find Text String* and *Replace With* edit boxes, then press the *Find/Find Next* button. You can verify and replace each instance of the text string found by alternately using *Find Next* and *Replace*, or you can select *Replace All* to globally replace all without verification. The status area confirms the replacements and indicates the number of replacements that were made.

Select Objects

The small button in the upper-right corner of the dialog box allows you to select objects with a pickbox or window to determine your selection set for the search. Press Enter when objects have been selected to return to the dialog box. Once a selection set has been specified, you can search the *Entire drawing* or limit the search to the *Current selection* by choosing these options in the *Search in*: drop-down list.

Select All

This options finds and selects (highlights) all objects in the current selection set containing instances of the text that you enter in *Find Text String*. This option is available only when you use *Select Objects* and set *Search In* to *Current Selection*. When you choose *Select All*, the dialog box closes and AutoCAD displays a message on the Command line indicating the number of objects that it found and selected. Note that *Select All* does not replace text; AutoCAD ignores any text in *Replace With*.

TIP ### Zoom To

This feature is extremely helpful for finding text in a large drawing. Enter the desired text string to search for in the *Find Text String* edit box, select *Find/Find Next* button, then use *Zoom to*. AutoCAD automatically locates the desired text string and zooms in to display the text (Fig. 18-43). Although AutoCAD searches model space and all layouts defined for the drawing, you can zoom only to text in the current *Model* or *Layout* tab.

FIGURE 18-43

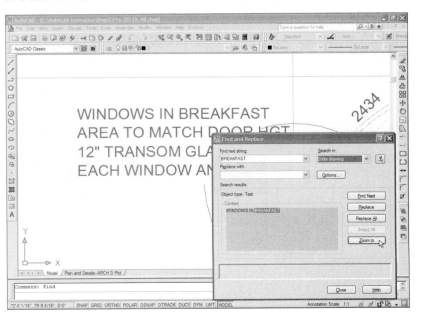

Options

Selecting the *Options* button in the *Find and Replace* dialog box produces the *Find and Replace Options* dialog box (Fig. 18-44). Note that the *Find* feature can locate and replace *Block Attribute Values, Dimension Annotation Text, Text (Mtext, Text), Table Text, Hyperlink Description,* and *Hyperlinks.* Removing any check disables the search for that text type.

FIGURE 18-44

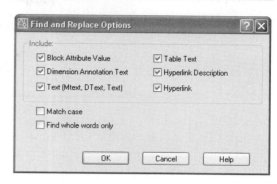

If it is important that the search exactly match the case (upper and lowercase letters) you entered in the *Find Text String* and *Replace With* edit boxes, check the *Match Case* box. If you want to search for complete words only, check the *Find whole words only* box. For example, without a check in either of these boxes, entering "door" in the *Find Text String* edit box would yield a find of "Doors" in the *Search Results* area. With a check in either *Match Case* or *Find whole words only,* a search for "door" would not find "Doors."

Scaletext

Pull-down Menu	Command (Type)	Alias (Type)	Short-cut	Screen (side) Menu	Tablet Menu
Modify Object > Text > Scale	*Scaletext*	...	...	...	...

You cannot effectively scale multiple lines of text (*Text* objects) or multiple paragraphs of text (*Mtext* objects) using the normal *Scale* command since all text objects become scaled with relation to only one base point; therefore, individual lines of text lose their original insertion point.

The text editing command *Scaletext* allows you to scale multiple text objects using this one command, and each line or paragraph is scaled relative to its individual justification point.

TIP

For example, consider the three *Text* objects shown in Figure 18-45. Assume each line of text was created using the *Start point* option (left-justified) and with a height of .20. If you wanted to change the height of all three text objects to .25, the following sequence could be used.

FIGURE 18-45

SCALETEXT scales each text object individually. Each line is scaled relative to the justification point. These sample lines are created by DTEXT.

```
Command: scaletext
Select objects: PICK
Select objects: PICK
Select objects: PICK
Select objects: Enter
Enter a base point option for scaling
[Existing/Left/Center/Middle/Right/TL/TC/TR/ML/MC/MR/BL/BC/BR] <Existing>: Enter
Specify new model height or [Paper height/Match object/Scale factor] <0.20000>: .25
Command:
```

Justifytext

Pull-down Menu	Command (Type)	Alias (Type)	Short-cut	Screen (side) Menu	Tablet Menu
Modify Object > Text > Justify	Justifytext	...	...	...	...

Justifytext changes the justification point of selected text objects without changing the text locations. Technically, this command relocates the insertion point (sometimes called attachment point) for the text object (*Mtext* objects, *Text* objects, leader text objects, and block attributes) and then justifies the text to the new insertion point.

Consider the two *Text* objects and the single *Mtext* object shown in Figure 18-46. In each case, the default justification methods were used when creating the text, resulting in left-justified text and insertion points as shown by the small "blip" (the "blip" is not created as part of the text—it is shown here only for illustration).

FIGURE 18-46 ————

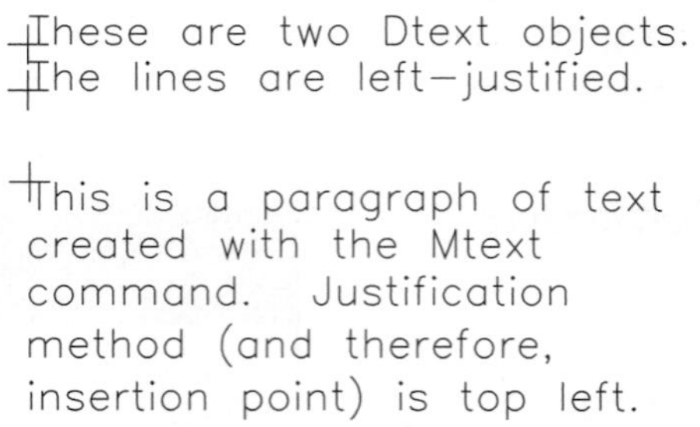

These are two Dtext objects.
The lines are left-justified.

This is a paragraph of text created with the Mtext command. Justification method (and therefore, insertion point) is top left.

Using previously available methods for changing the text justification yielded different results based on the command used or type of text object selected. For example, using the *Properties* palette to change the *Justify* field and selecting the *Bottom Right* option results in the changes shown in Figure

FIGURE 18-47 ————————————

These are two Dtext objects.
The lines are left-justified.

This is a paragraph of text created with the Mtext command. Justification method (and therefore, insertion point) is top left.

18-47. Note that for the *Text* objects, the insertion point is not changed, but the text objects are moved to justify according to the original insertion point. On the other hand, the insertion point for the *Mtext* paragraph is changed from the original top-left insertion point (shown as dashed in the figure) to the bottom right and the text is justified to the new insertion point.

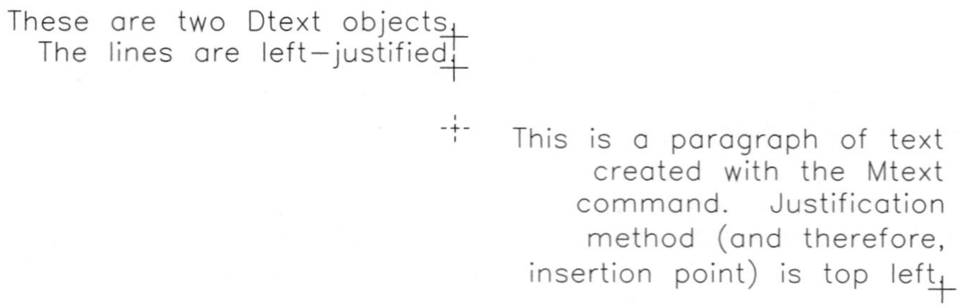

Qtext

Pull-down Menu	Command (Type)	Alias (Type)	Short-cut	Screen (side) Menu	Tablet Menu
...	Qtext	...	...	...	...

Qtext (quick text) allows you to display a line of text <u>as a box</u> in order to speed up drawing and plotting times. Because text objects are treated as graphical elements, a drawing with much text can be relatively slower to regenerate and take considerably more time plotting than the same drawing with little or no text.

When *Qtext* is turned *ON* and the drawing is regenerated, each text line is displayed as a rectangular box (Fig. 18-48). Each box displayed represents one line of text and is approximately equal in size to the associated line of text.

FIGURE 18-48

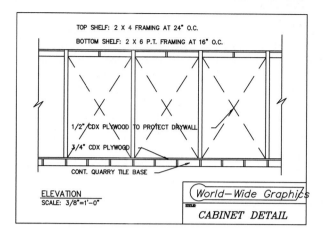

QTEXT OFF

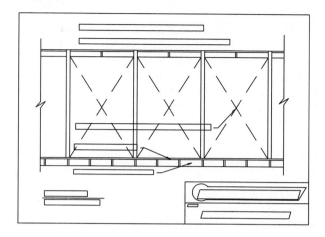

QTEXT ON

For drawings with considerable amounts of text, *Qtext ON* noticeably reduces regeneration time. For check plots (plots made during the drawing or design process used for checking progress), the drawing can be plotted with *Qtext ON*, requiring considerably less plotting time. *Qtext* is then turned *OFF* and the drawing must be *Regenerated* to make the final plot. When *Qtext* is turned *ON*, the text remains in a readable state until a *Regen* is invoked or caused. When *Qtext* is turned *OFF*, the drawing must be regenerated to read the text again.

Textfill

Pull-down Menu	Command (Type)	Alias (Type)	Short-cut	Screen (side) Menu	Tablet Menu
...	*Textfill*	...	...	...	...

The *TEXTFILL* variable controls the display of TrueType fonts for <u>printing and plotting only</u>. *TEXTFILL* does not control the display of fonts for the screen—fonts always appear filled in the Drawing Editor. If *TEXTFILL* is set to 1 (on), these fonts print and plot with solid-filled characters (Fig. 18-49). If *TEXTFILL* is set to 0 (off), the fonts print and plot as outlined text. The variable controls text display globally and retroactively.

FIGURE 18-49

TEXTFILL ON

TEXTFILL OFF

CHAPTER EXERCISES

1. *Text*

    Open the **EFF-APT-PS** drawing. Make *Layer* **TEXT** *Current* and use *Text* to label the three rooms: **KITCHEN, LIVING ROOM,** and **BATH**. Use the *Standard style* and the *Start point* justification option. When prompted for the *Height:*, enter a value to yield letters of 3/16″ on a 1/4″=1′ plot (3/16 x the drawing scale factor = text height). *Save* the drawing.

2. *Style, Text, Ddedit*

    Open the drawing of the temperature graph you created as **TEMP-D.** Use *Style* to create two styles named **ROMANS** and **ROMANC** based on the *romans.shx* and *romanc.shx* font files (accept all defaults). Use *Text* with the *Center* justification option to label the days of the week and the temperatures (and degree symbols) with the **ROMANS** style as shown in Figure 18-50. Use a *Height* of **1.6**. Label the axes as shown using the **ROMANC** style. Use *Ddedit* for editing any mistakes. Use *SaveAs* and name the drawing **TEMPGRPH.**

FIGURE 18-50

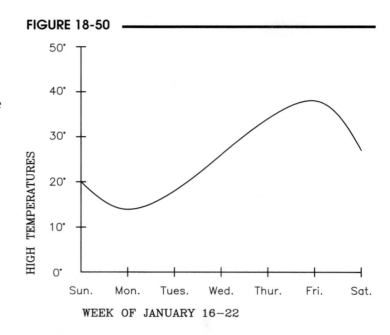

3. *Style, Text*

    Open the **BILLMATL** drawing created in the Chapter 15 Exercises. Use *Style* to create a new style using the *romans.shx* font. Use whatever justification methods you need to align the text information (not the titles) as shown in Figure 18-51. Next, type the *Style* command to create a new style that you name as **ROMANS-ITAL**. Use the *romans.shx* font file and specify a **15** degree *obliquing angle*. Use this style for the **NO., PART NAME,** and **MATERIAL**. *SaveAs* **BILLMAT2**.

FIGURE 18-51

NO.	PART NAME	MATERIAL
1	Base	Cast Iron
2	Centering Screw	N/A
3	Slide Bracket	Mild Steel
4	Swivel Plate	Mild Steel
5	Top Plate	Cast Iron

4. *Edit Text*

Open the **EFF-APT-PS** drawing. Create
a new style named **ARCH1** using the
CityBlueprint (.TTF) font file. Next, invoke
the *Properties* command. Use this dialog box
to modify the text style of each of the existing
room names to the new style as shown in
Figure 18-52. *SaveAs* **EFF-APT2.**

FIGURE 18-52

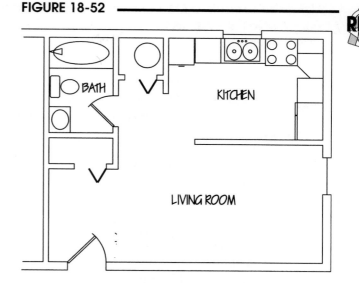

5. *Text, Mtext*

Open the **CBRACKET** drawing
from Chapter 9 Exercises. Using
romans.shx font, use *Text* to place the
part name and METRIC annotation
(Fig. 18-53). Use a *Height* of **5** and **4**,
respectively, and the *Center
Justification* option. For the notes,
use *Mtext* to create the boundary as
shown. Use the default *Justify*
method (*TL*)and a *Height* of **3**.
Use *Ddedit, Mtedit,* or *Properties* if
necessary. *SaveAs* **CBRACTXT.**

FIGURE 18-53

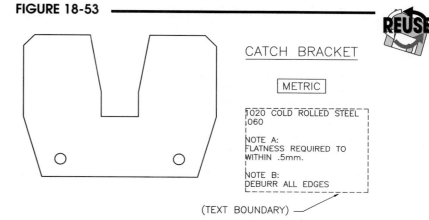

6. *Style*

Create two new styles for each of your template drawings: **ASHEET**, **BSHEET**, and
C-D-SHEET. Use the *romans.shx* style with the default options for engineering applications or
CityBlueprint (.TTF) for architectural applications. Next, design a style of your choosing to use
for larger text as in title blocks or large notes.

7. *Create a Title Block*
   A. Begin a *New* drawing and assign
      the name **TBLOCK.** Create the
      title block as shown in Figure 18-54
      or design your own, allowing
      space for eight text entries. The
      dimensions are set for an A size
      sheet.

FIGURE 18-54

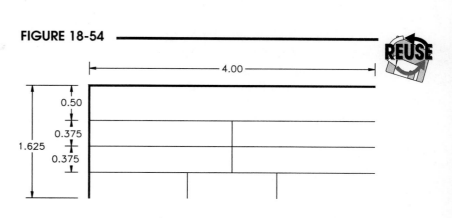

Draw on *Layer* 0. Use a *Pline* with .02 *width* for the boundary and *Lines* for the interior divisions. (No *Lines* are needed on the right side and bottom because the title block will fit against the border lines.)

B. Create two text *Styles* using *romans.shx* and *romanc.shx* font files. Insert text similar to that shown in Figure 18-55. Examples of the fields to create are:

> Company or School Name
> Part Name or Project Title
> Scale
> Designer Name
> Checker or Instructor Name
> Completion Date
> Check or Grade Date
> Project or Part Number

FIGURE 18-55

CADD Design Company		
Adjustable Mount	1/2"=1"	
Des.— B.R. Smith	Chk.—JRS	
4/1/2008	4/3/2008	42B—ADJM

Choose text items relevant to your school or office. *Save* the drawing.

8. *Mtext*

REUSE

Open the **STORE ROOM** drawing that you created in Chapter 16 Exercises. Use *Mtext* to create text paragraphs giving specifications as shown in Figure 18-56. Format the text below as shown in the figure. Use *CityBlueprint* (.TTF) as the base font file and specify a base *Height* of **3.5"**. Use a *Color* of your choice and *CountryBlueprint* (.TTF) font to emphasize the first line of the Room paragraph. All paragraphs use the *TC Justify* methods except the Contractor Notes paragraph, which is *TL*. Save the drawing as **STORE ROOM2.**

Room: <u>STORAGE ROOM</u>
    11'-2" x 10'-2"
    Cedar Lined - 2 Walls

Doors: 2 - 2268 DOORS
    Fire Type A
    Andermax

Windows: 2 - 2640 CASE-MENT WINDOWS
    Triple Pane Argon
        Filled
    Andermax

Notes: <u>Contractor Notes:</u>

    Contractor to verify
    all dimensions in
    field. Fill door and
    window roughouts
    after door and
    window placement.

FIGURE 18-56

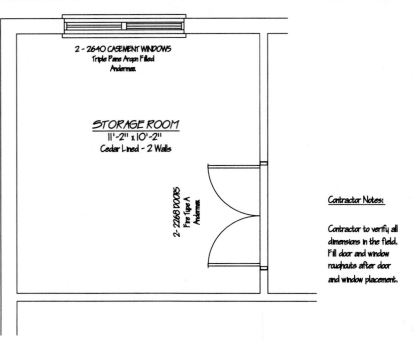

19

GRIP EDITING

CHAPTER OBJECTIVES

After completing this chapter you should:

1. be able to use the *GRIPS* variable to enable or disable object grips;

2. be able to activate the grips on any object;

3. be able to make an object's grips hot or cold;

4. be able to use each of the grip editing options, namely, STRETCH, MOVE, ROTATE, SCALE, and MIRROR;

5. be able to use the copy and Base suboptions;

6. be able to use the auxiliary grid that is automatically created when the Copy suboption is used;

7. be able to change grip variable settings using the *Options* dialog box or the Command line format.

CONCEPTS

Grips provide an alternative method of editing AutoCAD objects. The object grips are available for use by setting the *GRIPS* variable to **1**. Object grips are small squares appearing on selected objects at endpoints, midpoints, or centers, etc. The object grips are activated (made visible) by **PICK**ing objects with the cursor pickbox only <u>when no commands are in use</u> (at the open Command prompt). Grips are like small, magnetic *OSNAPs* (*Endpoint, Midpoint, Center, Quadrant*, etc.) that can be used for snapping one object to another, for example. If the cursor is moved within the small square, it is automatically "snapped" to the grip. Grips can replace the use of *OSNAP* for many applications. The grip option allows you to STRETCH, MOVE, ROTATE, SCALE, MIRROR, or COPY objects without invoking the normal editing commands or *OSNAPs*.

As an example, the endpoint of a *Line* could be "snapped" to the quadrant of a *Circle* (shown in Fig. 19-1) by the following steps:

1. Activate the grips by selecting both objects. Selection is done when no commands are in use (during the open Command prompt).
2. Select the grip at the endpoint of the *Line* (1). The grip turns **hot** (red).
3. The ** STRETCH ** option appears in place of the Command prompt.
4. STRETCH the *Line* to the quadrant grip on the *Circle* (2). **PICK** when the cursor "snaps" to the grip.
5. The *Line* and the *Circle* should then have connecting points. The Command prompt reappears. Press Escape to cancel (deactivate) the grips.

FIGURE 19-1

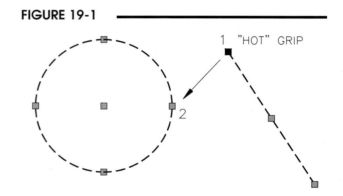

1 "HOT" GRIP

GRIPS FEATURES

Grips are enabled or disabled by changing the setting of the system variable, *GRIPS*. A setting of 1 enables or turns *ON GRIPS* and a setting of 0 disables or turns *OFF GRIPS*. This variable can be typed at the Command prompt, or Grips can be invoked from the *Selection* tab of the *Options* dialog box, or by typing *Ddgrips* (Fig. 19-2). Using the dialog box, toggle *Enable Grips* to turn *GRIPS ON*. The default setting in AutoCAD for the *GRIPS* variable is **1** (*ON*). (See "Grip Settings" near the end of the chapter for explanations of the other options in the *Grips* section of the *Options* dialog box.)

FIGURE 19-2

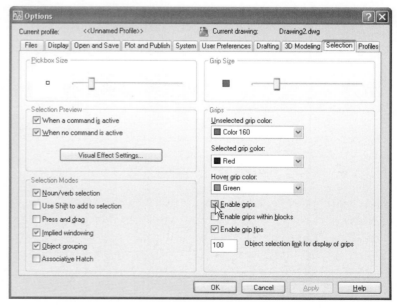

The *GRIPS* variable is saved in the user profile rather than in the current drawing as with most other system variables. Variables saved in the system registry are effective for any drawing session for that <u>particular user</u>, no matter which drawing is current. The reasoning is that grip-related variables (and selection set-related variables) are a matter of personal preference and therefore should remain constant for a particular user's CAD station.

When *GRIPS* have been enabled, a small pickbox appears at the center of the cursor crosshairs. (The pickbox also appears if the *PICKFIRST* system variable is set to **1**.) This pickbox operates in the same manner as the pickbox appearing during the "Select objects:" prompt. <u>Only</u> the pickbox, *AUTO window*, or *AUTO crossing window* methods can be used for selecting objects to activate the grips. (These three options are the only options available for Noun/Verb object selection as well.)

Activating Grips on Objects

The grips on objects are activated by selecting desired objects with the cursor pickbox, window, or crossing window. This action is done <u>when no commands are in use</u> (at the open Command prompt). When an object has been selected, two things happen: the grips appear and the object is highlighted. The grips are the small blue (default color) boxes appearing at the endpoints, midpoint, center, quadrants, vertices, insertion point, or other locations depending on the object type (Fig. 19-3). Highlighting indicates that the object is included in the selection set.

FIGURE 19-3

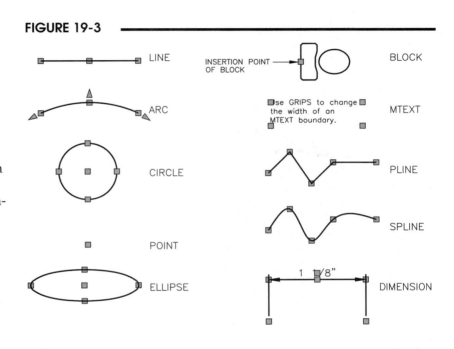

Cold and Hot Grips

Grips can have only two states: **cold** and **hot**. When *Grips* are turned on, and an object is selected (when no commands are in use), the object's grips become **cold**—its grips are displayed in blue (default color) and the object is highlighted.

One or more grips can then be made **hot**. A hot grip is created by "hovering" the cursor over a cold grip until the grip changes to a light green (default) color, then you PICK the grip so it changes to red (default color). The intermittent green grip is called a "hover" grip.

A **hot** grip is red (by default) and its object is almost always highlighted. Any grip can be changed to **hot** by selecting the grip itself. A <u>hot grip is the default base point</u> used for the editing action such as MOVE, ROTATE, SCALE, or MIRROR, or is the stretch point for STRETCH. When a **hot** grip exists, a new series of prompts appear in place of the Command prompt that displays the various grip editing options.

The grip editing options are also available from a right-click cursor menu (Fig. 19-4). <u>A grip must be changed to **hot** before the editing options appear</u>. Two or more grips can be made **hot** simultaneously by pressing Shift while selecting <u>each</u> grip.

FIGURE 19-4

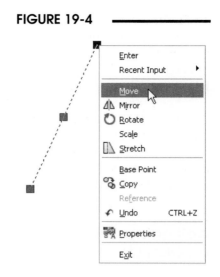

If you have made a *Grip* **hot** and want to deactivate it, possibly to make another *Grip* **hot** instead, press Escape once. This returns the grip to a **cold** state. In effect, this is an undo only for the **hot** grip. Pressing Escape again cancels all *Grips*. Pressing Escape demotes hot grips or cancels all grips:

<u>Grip State</u>	<u>Press Escape once</u>	<u>Press Escape twice</u>
only **cold**	grips are deactivated	
hot and **cold**	**hot** demoted to **cold**	grips are deactivated

FIGURE 19-5

NOTE: Beware that there are <u>two right-click (short-cut) menus available when grips are activated</u>. The Grip menu (see previous Fig. 19-4) is available only when a grip is <u>hot</u> (red). If you make any grip hot, then right-click, the <u>Grip menu</u> appears. The Grip menu contains grip editing options. However, if grips are <u>cold</u> and you right-click, the <u>Edit Mode menu</u> appears (Fig. 19-5). This menu appears any time one or more objects are selected, no commands are active, and you right-click (see Chapter 1, Edit Mode Menu). Although the commands displayed in the two menus appear to be the same, they are not. The Edit Mode menu contains the full commands. For example, *Move* in Figure 19-4 is the MOVE grip option, whereas *Move* in Figure 19-5 is the *Move* command.

TIP

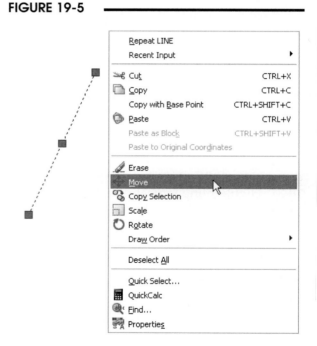

NOTE: When *PICKFIRST* is set to 0, the Edit-mode shortcut menus do not appear—the Default menu appears instead.

Grip Editing Options

When a **hot grip** has been activated, the grip editing options are available. The Command prompt is replaced by the STRETCH, MOVE, ROTATE, SCALE, or MIRROR grip editing options. You can sequentially cycle through the options by pressing the Space bar or Enter key. The editing options are displayed in Figures 19-6 through 19-10.

Alternately, you can <u>right-click when a grip is **hot**</u> to activate the grip menu (see Fig. 19-4). This menu has the same options available in Command line format with the addition of *Reference* and *Properties*. The options are described in the following figures.

** STRETCH **
Specify stretch point or
[Basepoint/Copy/Undo/eXit]:

Note that for *Arc* objects only, grips appear at the endpoints, midpoints, and center. In addition, arrowshaped grips appear at the endpoints and midpoint (Fig. 19-6). These arrow grips allow you to change the radius (midpoint arrow) or extend the arc's length (endpoint arrows) in the STRETCH mode only.

FIGURE 19-6

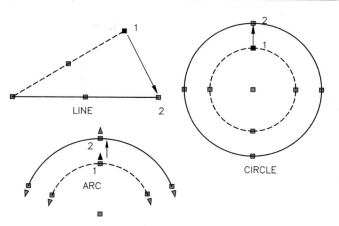

** MOVE **
Specify move point or
[Base point/Copy/Undo/eXit]:
(Fig. 19-7)

FIGURE 19-7

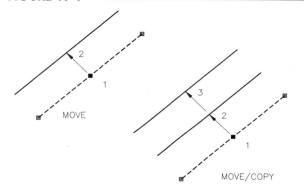

** ROTATE **
Specify rotation angle or
[Base point/Copy/Undo/Reference/eXit]:
(Fig. 19-8)

FIGURE 19-8

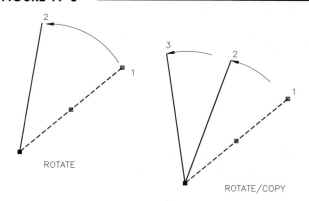

** SCALE **
Specify scale factor or
[Base point/Copy/Undo/Reference/eXit]:
(Fig. 19-9)

FIGURE 19-9

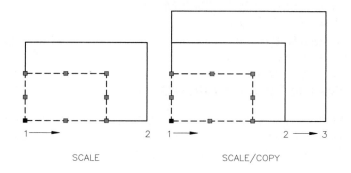

** MIRROR **
Specify second point or
[Base point/Copy/Undo/eXit]:
(Fig. 19-10)

FIGURE 19-10

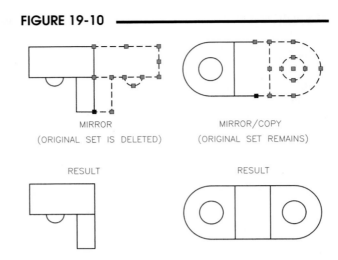

MIRROR
(ORIGINAL SET IS DELETED)

MIRROR/COPY
(ORIGINAL SET REMAINS)

RESULT

RESULT

The *Grip* options are easy to understand and use. Each option operates like the full AutoCAD command by the same name. Generally, the editing option used (except for STRETCH) affects all highlighted objects. The **hot** grip is the base point for each operation. The suboptions, Base and Copy, are explained next.

NOTE: The STRETCH option differs from other options in that STRETCH affects only the object that is attached to the **hot** grip, rather than affecting all highlighted (**cold**) objects.

Base
The Base suboption appears with all of the main grip editing options (STRETCH, MOVE, etc.). Base allows using any other grip as the base point instead of the **hot** grip. Type the letter *B* or select from the right-click cursor menu to invoke this suboption.

Copy
Copy is a suboption of every main choice. Activating this suboption by typing the letter *C* or selecting from the right-click cursor menu invokes a Multiple copy mode, such that whatever set of objects is STRETCHed, MOVEd, ROTATEd, etc., becomes the first of an unlimited number of copies (see the previous five figures). The Multiple mode remains active until exiting back to the Command prompt.

Undo
The Undo option, invoked by typing the letter *U* or selecting from the right-click cursor menu will undo the last Copy or the last Base point selection. Undo functions only after a Base or Copy operation.

Reference
This option operates similarly to the reference option of the *Scale* and *Rotate* commands. Use *Reference* to enter or PICK a new reference length (SCALE) or angle (ROTATE). (See "*Scale* and *Rotate*," Chapter 9.) With grips, *Reference* is only enabled when the SCALE or ROTATE options are active.

Properties... (Right-Click Menu Only)
Selecting this option from the right-click grip menu (see Fig. 19-4) activates the *Properties* palette. All highlighted objects become subjects of the palette. Any property of the selected objects can be changed with the palette. (See "*Properties*," Chapter 16.)

Shift to Turn on *ORTHO*

Using the Copy option allows you to make multiple copies, such as MOVE/Copy or ROTATE/Copy. If you hold down the Shift key as you make copies, *ORTHO* is turned on as long as Shift is depressed.

Auxiliary Grid

An auxiliary grid is <u>automatically</u> established on creating the first Copy (Fig. 19-11). The grid is activated by pressing Ctrl while placing the subsequent copies. The subsequent copies are then "snapped" to the grid in the same manner that *SNAP* functions. The spacing of this auxiliary grid is determined by the location of the first Copy, that is, the X and Y intervals between the base point and the second point.

FIGURE 19-11

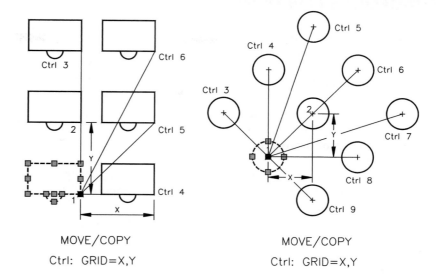

MOVE/COPY
Ctrl: GRID=X,Y

MOVE/COPY
Ctrl: GRID=X,Y

For example, a "polar array" can be simulated with grips by using ROTATE with the Copy suboption (Fig. 19-12). The "array" can be constructed by making one Copy, then using the auxiliary grid to achieve equal angular spacing. The steps for creating a "polar array" are as follows:

1. Select the object(s) to array.

2. Select a grip on the set of objects to use as the center of the array. Cycle to the ROTATE option by pressing Enter or selecting from the right-click cursor menu. Next, invoke the Copy suboption.

3. Make the first copy at any desired location.

4. After making the first copy, activate the auxiliary angular grid by holding down Ctrl while making the other copies.

5. Cancel the grips or select another command from the menus.

FIGURE 19-12

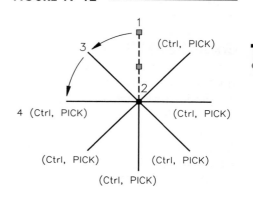

ROTATE/COPY
Ctrl — ANGLES EQUAL

TIP ✓

Editing Dimensions

One of the most effective applications of grips is as a dimension editor. Because grips exist at the dimension's extension line origins, arrowhead endpoints, and dimensional value, a dimension can be changed in many ways and still retain its associativity (Fig. 19-13). See Chapter 25, Dimensioning, for further information about dimensions, associativity, and editing dimensions with *Grips*.

FIGURE 19-13

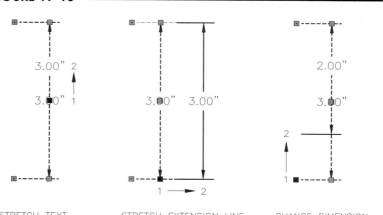

STRETCH TEXT   STRETCH EXTENSION LINE   CHANGE DIMENSION

TIP ✓

GUIDELINES FOR USING GRIPS

Although there are many ways to use grips based on the existing objects and the desired application, a general set of guidelines for using grips is given here:

1. Create the objects to edit.

2. Select objects to make **cold** grips appear.

3. Select the desired **hot** grip(s). The *Grip* options should appear in place of the Command line.

4. Press Space or Enter to cycle to the desired editing option (STRETCH, MOVE, ROTATE, SCALE, MIRROR) or select from the right-click cursor menu.

5. Select the desired suboptions, if any. If the *Copy* suboption is needed or the base point needs to be re-specified, do so at this time. *Base* or *Copy* can be selected in either order.

6. Make the desired STRETCH, MOVE, ROTATE, SCALE, or MIRROR.

7. Cancel the grips by pressing Escape or selecting a command from a menu.

GRIPS SETTINGS

FIGURE 19-14

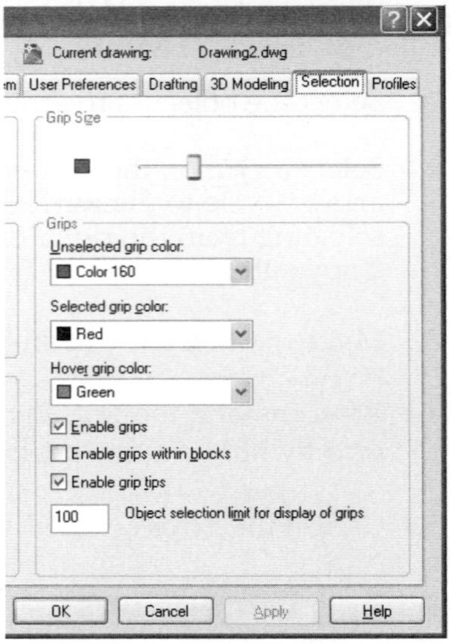

Several settings are available in the *Selection* tab of the *Options* dialog box (Fig. 19-14) that control the way grips appear or operate. The settings can also be changed by typing in the related variable name at the Command prompt.

Grip Size
Use the slider bar to interactively increase or decrease the size of the grip "box." The *GRIPSIZE* variable could alternately be used. The default size is 5 pixels square (*GRIPSIZE=5*).

Unselected Grip Color
This setting enables you to change the color of **cold** grips. Select the desired color from the *Unselected Grip Color* drop-down list. PICKing the *Select Color...* tile produces the *Select Color* dialog box (identical to that used with other color settings). The default setting is blue (ACI number 5). Alternately, the *GRIPCOLOR* variable can be typed and any ACI number from 0 to 255 can be entered.

Selected Grip Color
The color of hot grips can be specified with this option. Select the desired color from the *Selected Grip Color* drop-down list. The *Select Color...* tile produces the *Select Color* dialog box. You can also type *GRIPHOT* and enter any ACI number to make the change. The default color is red (1).

Hover Grip Color
This setting specifies the color of grips when you "hover" your cursor over a cold (blue) grip, usually just before you PICK it to make it a hot grip. The drop-down list operates identically to the other grip color options.

Enable Grips

If a check appears in the checkbox, *Grips* are enabled for the workstation. A check sets the *GRIPS* variable to 1 (on). Removing the check disables grips and sets *GRIPS*=0 (off). The default setting is on.

Enable Grips within Blocks

When no check appears in this box, only one grip at the block's insertion point is visible. This allows you to work with a block as one object. The related variable is *GRIPBLOCK*, with a setting of 0=off. This is the default setting (disabled). When the box is checked, *GRIPBLOCK* is set to 1. All grips on all objects contained in the block are visible and functional. This allows you to use any grip on any object within the block.

CHAPTER EXERCISES

1. *Open* the **TEMP-D** drawing from Chapter 16 Exercises. Use the grips on the existing *Spline* to generate a new graph displaying the temperatures for the following week. **PICK** the *Spline* to activate grips (Fig. 19-15). Use the **STRETCH** option to stretch the first data point grip to the value indicated below for Sunday's temperature. **Cancel** the grips; then repeat the steps for each data point on the graph. *Save* the drawing as **TEMP-F**.

X axis	Y axis
Sunday	29
Monday	32
Tuesday	40
Wednesday	33
Thursday	44
Friday	49
Saturday	45

FIGURE 19-15

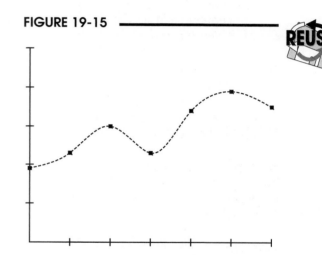

2. *Open* **CH16EX2** drawing. Activate **grips** on the *Line* to the far right. Make the top grip **hot**. Use the **STRETCH** option to stretch the top grip to the right to create a vertical *Line*. **Cancel** the grips.

    Next, activate the **grips** for all the other *Lines* (not including the vertical one on the right); as shown in Figure 19-16. **STRETCH** the top of all inclined *Lines* to the top of the vertical *Line* by making the common top grips **hot**, then *OSNAPing* to the top *ENDpoint* of the vertical *Line*. *Save* the drawing as **CH19EX2**.

FIGURE 19-16

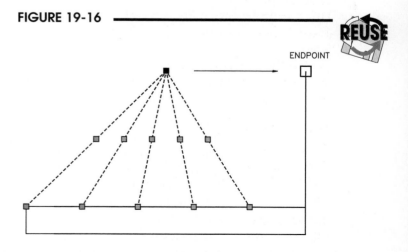

ENDPOINT

3. This exercise involves all the options of grip editing to create a Space Plate. Begin a *New* drawing or use the **ASHEET** *template* and use *SaveAs* to assign the name **SPACEPLT**.

A. Set the *Snap* value to **.125**. Set **Polar Snap** to **.125** and turn on **Polar Tracking**. Draw the geometry shown in Figure 19-17 using the *Line* and *Circle* commands.

FIGURE 19-17

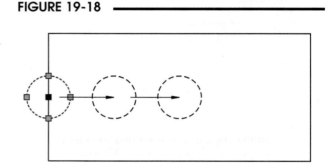

B. Activate the **grips** on the *Circle*. Make the center grip **hot**. Cycle to the **MOVE** option. Enter *C* for the **Copy** option. You should then get the prompt for **MOVE (multiple)**. Make two copies as shown in Figure 19-18. The new *Circles* should be spaced evenly, with the one at the far right in the center of the rectangle. If the spacing is off, use **grips** with the **STRETCH** option to make the correction.

FIGURE 19-18

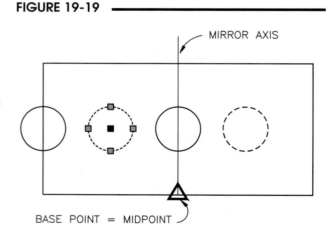

C. Activate the **grips** on the number 2 *Circle* (from the left). Make the center *Circle* grip **hot** and cycle to the **MIRROR** option. Enter *C* for the **Copy** option. (You'll see the **MIRROR (multiple)** prompt.) Then enter *B* to specify a new base point as indicated in Figure 19-19. Press Shift to turn *On ORTHO* and specify the mirror axis as shown to create the new *Circle* (shown in Fig. 19-19 in hidden linetype).

FIGURE 19-19

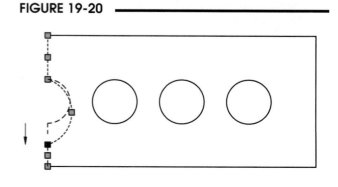

D. Use *Trim* to trim away the outer half of the *Circle* and the interior portion of the vertical *Line* on the left side of the Space Plate. Activate the **grips** on the two vertical *Lines* and the new *Arc*. Make the common grip **hot** (on the *Arc* and *Line* as shown in Fig. 19-20) and **STRETCH** it downward .5 units. (Note how you can affect multiple objects by selecting a common grip.) **STRETCH** the upper end of the *Arc* upward .5 units.

FIGURE 19-20

E. **Erase** the *Line* on the right side of the
    Space Plate. Use the same method that
    you used in step C to **MIRROR** the *Lines*
    and *Arc* to the right side of the plate (as
    shown in Fig. 19-21).

    (REMINDER: After you select the **hot** grip,
    use the **Copy** option <u>and</u> the **Base** point
    option. Use Shift to turn on *ORTHO*.)

FIGURE 19-21

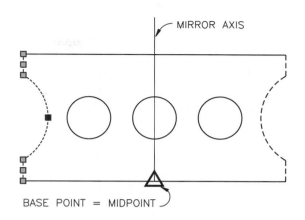

F. In this step, you will **STRETCH** the top
    edge upward one unit and the bottom edge
    downward one unit by selecting <u>multiple</u>
    hot grips.

    Select the desired horizontal <u>and</u> attached
    vertical *Lines*. Hold down Shift while
    selecting <u>each</u> of the endpoint grips, as
    shown in Figure 19-22. Although they
    appear **hot**, you must select one of the two
    again to activate the **STRETCH** option.

FIGURE 19-22

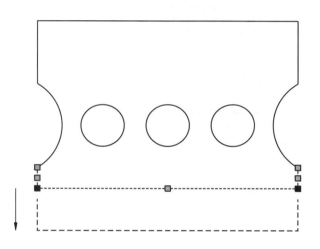

G. In this step, two more *Circles* are created by
    the **ROTATE** option (see Fig. 19-23). Select
    the two outside *Circles* to make the grips
    appear. PICK the center grip to make it
    hot. Cycle to the **ROTATE** option. Enter *C*
    for the **Copy** option. Next, enter *B* for the
    Base option and *OSNAP* to the *CENter* as
    shown. Press Shift to turn on *ORTHO* and
    create the new *Circles*.

FIGURE 19-23

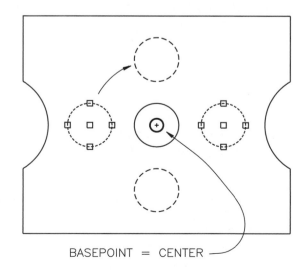

H. Select the center *Circle* and make the
center grip **hot** (Fig. 19-24). Cycle to the
SCALE option. The scale factor is **1.5**.
Since the *Circle* is a 1 unit diameter, it can be
interactively scaled (watch the *Coords*
display), or you can enter the value. The
drawing is complete. *Save* the drawing as
SPACEPLT.

FIGURE 19-24

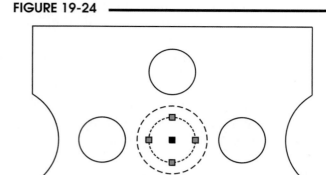

4. *Open* the **T-PLATE** drawing that
you created in Chapter 9. An order
has arrived for a modified version of
the part. The new plate requires two
new holes along the top, a 1"
increase in the height, and a .5"
increase from the vertical center to
the hole on the left (Fig. 19-25). Use
grips to **STRETCH** and **Copy** the
necessary components of the existing part. Use *SaveAs* to rename the
part to **TPLATEB**.

FIGURE 19-25

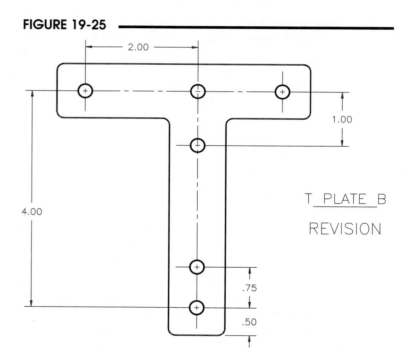

20

BLOCKS, DESIGN- CENTER, AND TOOL PALETTES

CHAPTER OBJECTIVES

After completing this chapter you should:

1. be able to use the *Block* command to transform a group of objects into one object that is stored in the current drawing's block definition table;

2. be able to use the *Insert* and *Minsert* commands to bring *Blocks* into drawings;

3. know that *color*, *linetype* and *lineweight* of *Blocks* are based on conditions when the *Block* is made;

4. be able to convert *Blocks* to individual objects with *Explode*;

5. be able to use *Wblock* to prepare .DWG files for insertion into other drawings;

6. be able to redefine and globally change previously inserted *Blocks*;

7. be able to use DesignCenter™ and Tool Palettes to drag and drop *Blocks* into the current drawing.

CONCEPTS

A *Block* is a <u>group</u> of objects that are combined into <u>one</u> object with the *Block* command. The typical application for *Blocks* is in the use of symbols. Many drawings contain symbols, such as doors and windows for architectural drawings, capacitors and resistors for electrical schematics, or pumps and valves for piping and instrumentation drawings. In AutoCAD, symbols are created first by constructing the desired geometry with objects like *Line, Arc,* and *Circle,* then transforming the set of objects comprising the symbol into a *Block.* A description of the objects comprising the *Block* is then stored in the drawing's "block definition table." The *Blocks* can then each be *Inserted* into a drawing many times and treated as a single object. Text can be attached to *Blocks* (called *Attributes*) and the text can be modified for each *Block* when inserted.

Figure 20-1 compares a shape composed of a set of objects and the same shape after it has been made into a *Block* and *Inserted* back into the drawing. Notice that the original set of objects is selected (highlighted) individually for editing, whereas, the *Block* is only one object.

FIGURE 20-1 ────────────────

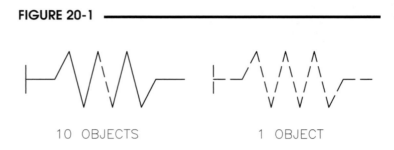

10 OBJECTS   1 OBJECT

Since an inserted *Block* is one object, it uses less file space than a set of objects that is copied with *Copy.* The *Copy* command creates a duplicate set of objects, so that if the original symbol were created with 10 objects, 3 copies would yield a total of 40 objects. If instead the original set of 10 were made into a *Block* and then *Inserted* 3 times, the total objects would be 13 (the original 10 + 3).

Upon *Inserting* a *Block,* its scale can be changed and rotational orientation specified without having to use the *Scale* or *Rotate* commands (Fig. 20-2). If a design change is desired in the *Blocks* that have already been *Inserted,* the original *Block* can be redefined and the previously inserted *Blocks* are automatically updated. *Blocks* can be made to have explicit *Linetype, Lineweight* and *Color,* regardless of the layer they are inserted onto, or they can be made to assume the *Color, Linetype,* and *Lineweight* of the layer onto which they are *Inserted.*

Similar to annotative text objects (see Chapter 18), you can create *Annotative Blocks* that can change size when drawing geometry is displayed in different scales (*VP Scale*) for different viewports. However, *Blocks* are generally used to represent parts of the drawing geometry and

FIGURE 20-2 ────────────────

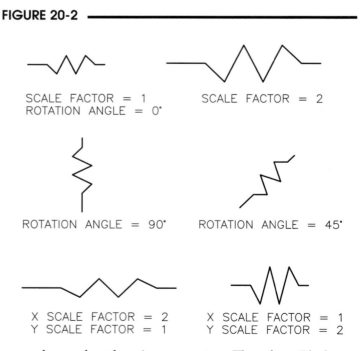

SCALE FACTOR = 1   SCALE FACTOR = 2
ROTATION ANGLE = 0°

ROTATION ANGLE = 90°   ROTATION ANGLE = 45°

X SCALE FACTOR = 2   X SCALE FACTOR = 1
Y SCALE FACTOR = 1   Y SCALE FACTOR = 2

are normally intended to be displayed in the same scale as other drawing geometry. Therefore, *Blocks* are *Annotative* only for special applications such as dynamic callouts or view labels in architectural drawings.

Blocks can be <u>nested</u>; that is, one *Block* can reference another *Block*. Practically, this means that the definition of *Block* "C" can contain *Block* "A" so that when *Block* "C" is inserted, *Block* "A" is also inserted as part of *Block* "C" (Fig. 20-3).

FIGURE 20-3

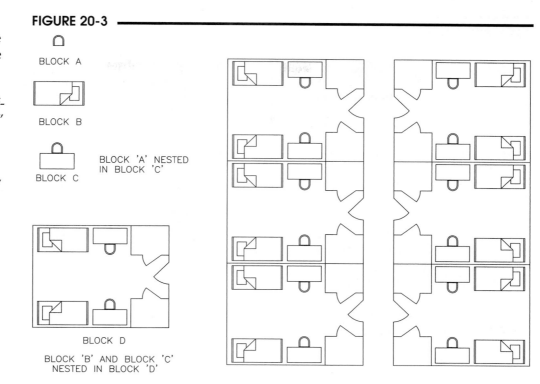

BLOCK A

BLOCK B

BLOCK C

BLOCK 'A' NESTED IN BLOCK 'C'

BLOCK D

BLOCK 'B' AND BLOCK 'C' NESTED IN BLOCK 'D'

Blocks created within the current drawing can be copied to disk as complete and separate drawing files (.DWG file) by using the *Wblock* command (Write Block). This action allows you to *Insert* the *Blocks* into other drawings.

Commands related to using *Blocks* are:

Block	Creates a *Block* from individual objects
Insert	Inserts a *Block* into a drawing
Minsert	Permits a multiple insert in a rectangular pattern
Explode	Breaks a *Block* into its original set of multiple objects
Wblock	Writes an existing *Block* or a set of objects to a file on disk
Base	Allows specification of an insertion base point
Purge	Deletes uninserted *Blocks* from the block definition table
Rename	Allows renaming *Blocks*
Bedit	Invokes the Block Editor for editing blocks

The following additional tools, discussed in this chapter, provide powerful features for creating, inserting, and editing blocks.

Tool palettes   Tool palettes allow you to drag-and-drop *Blocks* and *Hatch* patterns into a drawing. This process is similar, but easier, than using *Insert*. You can easily add blocks to create your own tool palettes.

DesignCenter   This tool allows you to drag-and-drop *Blocks* from other <u>drawings into the current drawing</u>. You can easily locate content from other drawings such as *Dimension Styles, Layers, Linetypes, Text Styles,* and *Xref* drawings to drag and drop into the current drawing.

COMMANDS

Block

Pull-down Menu	Command (Type)	Alias (Type)	Short-cut	Screen (side) Menu	Tablet Menu
Draw *Block >* *Make...*	*Block,* or *-Block*	*B* or *-B*	...	DRAW 2 *Bmake*	*N,9*

Selecting the icon button, using the pull-down or screen menu, or typing *Block* or *Bmake*, produces the *Block Definition* dialog box shown in Figure 20-4. This dialog box provides the same functions as using the *-Block* command (a hyphen prefix produces the Command line equivalent).

To make a *Block*, first create the *Lines, Circles, Arcs,* or other objects comprising the shapes to be combined into the *Block*. Next, use the *Block* command to transform the objects into one object—a *Block*.

In the *Block Definition* dialog box, enter the desired *Block* name in the *Name* edit box. Then use the *Select Objects* button (top center) to return to the drawing temporar-

FIGURE 20-4 ─────────────

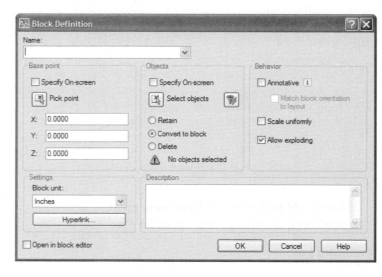

ily to select the objects you wish to comprise the *Block*. After selection of objects, the dialog box reappears. Use the *Pick Insertion Base Point* button (in the *Base Point* section of the dialog box) if you want to use a point other than the default 0,0,0 as the "insertion point" when the *Block* is later inserted. Usually select a point in the corner or center of the set of objects as the base point. When you select *OK*, the new *Block* is defined and stored in the drawing's block definition table awaiting future insertions.

If *Delete* is selected in the *Objects* section of the dialog box, the original set of "template" objects comprising the *Block* disappear even though the definition of the *Block* remains in the table. Checking *Retain* forces AutoCAD to retain the original objects (similar to using *Oops* after the *Block* command), or selecting *Convert to Block* keeps the original set of objects visible in the drawing but transforms them into a *Block*.

The *Block Unit* drop-down option and the *Description* edit box are used to specify how *Blocks* are described when DesignCenter is used to drag and drop the *Blocks* into a drawing instead of using the *Insert* command. The *Block Units* options are described later in this chapter. The *Annotative* checkbox allows you to create annotative blocks. If the *Scale uniformly* checkbox is not checked, you have the option to scale the block non-uniformly when it is *Inserted* into the drawing. If the *Allow exploding* box is checked, blocks can be "broken down" into the individual objects used to create the block (*Line, Arc, Circle,* etc.) using the *Explode* command. Use the *Hyperlink* button to produce the *Insert Hyperlink* dialog box for attaching a hyperlink to a *Block*. If *Open in block editor* is checked, the Block Editor is opened when you press the *OK* button. Normally, remove this check if you are making a block and do not want to edit it further. The *Names* drop-down list (at the top) is used to select existing *Blocks* if you want to redefine a *Block* (see "Redefining Blocks and the Block Editor" later in this chapter).

If you prefer to type, use -*Block* to produce the Command line equivalent of the *Block Definition* dialog box. The command syntax is as follows:

Command: **-Block**
Block name (or ?): (**name**) (Enter a descriptive name for the *Block* up to 255 characters.)
Specify insertion base point or [Annotative]: **PICK** or (**coordinates**) (Select a point to be used later for insertion.)
Select objects: **PICK**
Select objects: **PICK** (Continue selecting all desired objects.)
Select objects: **Enter**

The *Block* then <u>disappears</u> as it is stored in the current drawing's "block definition table." The *Oops* command can be used to restore the original set of "template" objects (they reappear), but the definition of the *Block* remains in the table. Using the *?* option of the *Block* command lists the *Blocks* stored in the block definition table.

Block Color, Linetype, and Lineweight Settings

The <u>color</u>, <u>linetype</u>, and <u>lineweight</u> of the *Block* are determined by one of the following settings when the *Block* is created:

1.   When a *Block* is inserted, it is drawn on its original layer with its original *color, linetype,* and *lineweight* (when the objects were <u>created</u>), regardless of the layer or *color, linetype* and *lineweight* settings that are current when the *Block* is inserted (unless conditions 2 or 3 exist).

2.   If a *Block* is created on <u>Layer 0</u> (Layer 0 is current when the original objects comprising the *Block* are created), then the *Block* assumes the *color, linetype,* and *lineweight* of any layer that is current when it is inserted (Fig. 20-5).

FIGURE 20-5

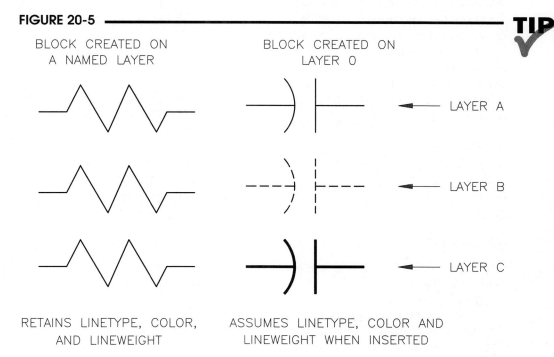

BLOCK CREATED ON A NAMED LAYER

BLOCK CREATED ON LAYER 0

LAYER A

LAYER B

LAYER C

RETAINS LINETYPE, COLOR, AND LINEWEIGHT

ASSUMES LINETYPE, COLOR AND LINEWEIGHT WHEN INSERTED

3.   If the *Block* is created with the special *BYBLOCK color, linetype,* and *lineweight* setting, the *Block* is inserted with the *color, linetype,* and *lineweight* settings that are <u>current during insertion</u>, whether the *BYLAYER* or explicit object *color, linetype,* and *lineweight* settings are current.

Insert

Pull-down Menu	Command (Type)	Alias (Type)	Short-cut	Screen (side) Menu	Tablet Menu
Insert Block...	Insert or -Insert	I or -I	...	INSERT Ddinsert	T,5

Once the *Block* has been created, it is inserted back into the drawing at the desired location(s) with the *Insert* command. The *Insert* command produces the *Insert* dialog box (Fig. 20-6) which allows you to select which *Block* to insert and to specify the *Insertion Point*, *Scale*, and *Rotation*, either interactively (*On-screen*) or by specifying values.

First, select the *Block* you want to insert. All *Blocks* located in the drawing's block definition table are listed in the *Name* drop-down list. Next, determine the parameters for *Insertion Point*, *Scale*, and *Rotation*. You can enter values in the edit boxes if you have specific parameters in mind or check *Specify On-screen* to interactively supply the parameters. For example, with the settings shown in Figure 20-6, AutoCAD would allow you to preview the *Block* as you dragged it about the screen and picked the *Insertion Point*. You would not be prompted for a *Scale* or *Rotation* angle since they are specified in the dialog box as 1.0000 and 0 degrees, respectively. Entering any other values in the *Scale* or *Rotation* edit boxes causes AutoCAD to preview the *Block* at the specified scale and rotation angle as you drag it about the drawing to pick the insertion point. Remember that *Osnaps* can be used when specifying the parameters interactively. Check *Uniform Scale* to ensure the X, Y, and Z values are scaled proportionally. *Explode* can also be toggled, which would insert the *Block* as multiple objects (see "*Explode*").

FIGURE 20-6

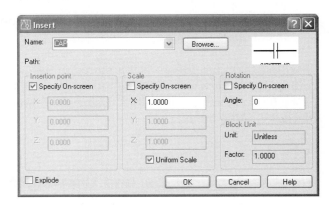

Inserting Other Drawings as *Blocks*

Selecting the *Browse* tile in the *Insert* dialog box produces the *Select Drawing* dialog box (Fig. 20-7). Here you can select <u>any</u> drawing (.DWG file) for insertion. When one drawing is *Inserted* into another, the entire drawing comes into the current drawing as a *Block*, or as <u>one object</u>. If you want to edit individual objects in the inserted drawing, you must *Explode* the object.

If you prefer the Command line equivalent, type -*Insert*.

FIGURE 20-7

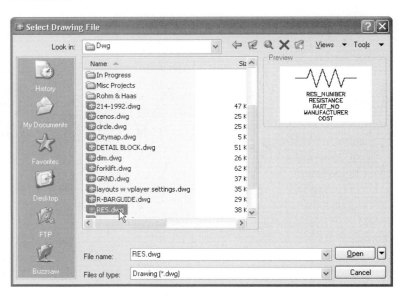

> Command: **-insert**
> Enter block name or [?] : **name**
> (Type the name of an existing block or .DWG file to insert.)
> Specify insertion point or [Basepoint/Scale/X/Y/Z/Rotate]: **PICK** or **option**
> Enter X scale factor, specify opposite corner, or [Corner/XYZ] <1>: **value**, **PICK** or **option**
> Enter Y scale factor <use X scale factor>: **value** or **PICK**
> Specify rotation angle <0>: **value** or **PICK**

Minsert

Pull-down Menu	Command (Type)	Alias (Type)	Short-cut	Screen (side) Menu	Tablet Menu
...	*Minsert*	...	...	...	...

This command allows a <u>multiple insert</u> in a rectangular pattern (Fig. 20-8). *Minsert* is actually a combination of the *Insert* and the *Array Rectangular* commands. The *Blocks* inserted with *Minsert* are associated (the group is treated as one object) and cannot be edited independently (unless *Exploded*).

Examining the command syntax yields the similarity to a *Rectangular Array*.

FIGURE 20-8

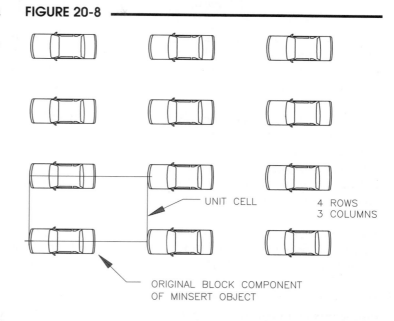

UNIT CELL

4 ROWS
3 COLUMNS

ORIGINAL BLOCK COMPONENT
OF MINSERT OBJECT

Command: **Minsert**
Enter block name [or ?] <current>: **name**
Specify insertion point or [Basepoint/Scale/X/Y/Z/Rotate]: (**value**), **PICK** or **option**
Enter X scale factor, specify opposite corner, or [Corner/XYZ] <1>: (**value**) or **Enter**
Enter Y scale factor <use X scale factor>: (**value**) or **Enter**
Specify rotation angle <0>: (**value**) or **Enter**
Enter number of rows (—-) <1>: (**value**)
Enter number of columns (|||) <1>: (**value**)
Enter distance between rows or specify unit cell (—-): (**value**) or **PICK** (Value specifies Y distance
    from *Block* corner to *Block* corner; PICK allows drawing a unit cell rectangle.)
Distance between columns: (**value**) or **PICK** (Specifies X distance between *Block* corners.)
Command:

Explode

Pull-down Menu	Command (Type)	Alias (Type)	Short-cut	Screen (side) Menu	Tablet Menu
Modify *Explode*	*Explode*	*X*	...	*MODIFY2* *Explode*	*Y,22*

Explode breaks a <u>previously</u> inserted *Block* back into its original set of objects (Fig. 20-9), which allows you to edit individual objects comprising the shape. If *Allow exploding* was unchecked in the *Block Definition* dialog box when the *Block* was created, the *Block* cannot be *Exploded*. *Blocks* that have been *Minserted* cannot be *Exploded*.

Command: **explode**
Select objects: **PICK**
Select objects: **Enter**
Command:

FIGURE 20-9

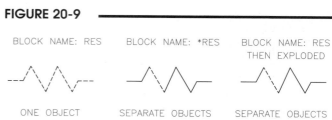

BLOCK NAME: RES   BLOCK NAME: *RES   BLOCK NAME: RES
                                                    THEN EXPLODED

ONE OBJECT   SEPARATE OBJECTS   SEPARATE OBJECTS

Wblock

Pull-down Menu	Command (Type)	Alias (Type)	Short-cut	Screen (side) Menu	Tablet Menu
File *Export...*	*Wblock*	*W*	...	*FILE* *Export*	*W,24*

The *Wblock* command writes a *Block* out to disk as a separate and complete drawing (.DWG) file. The *Block* used for writing to disk can exist in the current drawing's *Block* definition table or can be created by the *Wblock* command. Remember that the *Insert* command inserts *Blocks* (from the current drawing's block definition table) or finds and accepts .DWG files and treats them as *Blocks* upon insertion.

There are two ways to create a *Wblock* from the current drawing, (1) using an existing *Block* and (2) using a set of objects not previously defined in the current drawing as a *Block*. If you are using an existing *Block*, a copy of the *Block* is essentially transformed by the *Wblock* command to create a complete AutoCAD drawing (.DWG) file. The original block definition remains in the current drawing's block definition table. In this way, *Blocks* that were originally intended for insertion into the current drawing can be easily inserted into other drawings.

Figure 20-10 illustrates the relationship among a *Block*, the current drawing, and a *WBlock*. In the figure, SCHEM1.DWG contains several *Blocks*. The RES block is written out to a .DWG file using *Wblock* and named RESISTOR. RESISTOR is then *Inserted* into the SCHEM2 drawing.

FIGURE 20-10

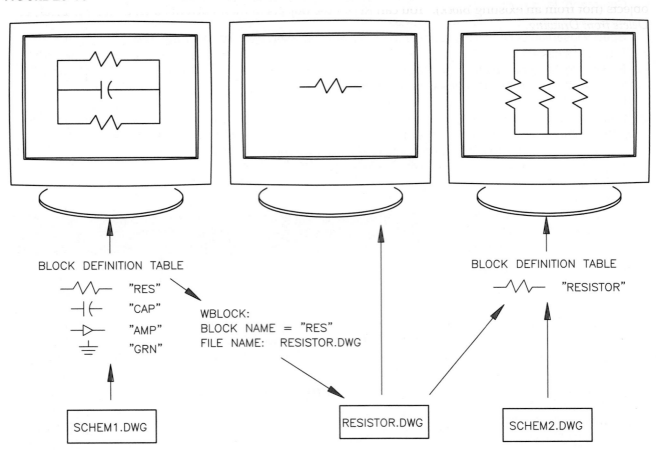

If you want to transform a set of objects into a *Block* to be used in other drawings but not in the current one, you can use *Wblock* to transform (a copy of) the objects in the current drawing into a separate .DWG file. This action does not create a *Block* in the current drawing. As an alternative, if you want to create symbols specifically to be inserted into other drawings, each symbol could be created initially as a separate .DWG file.

The *Wblock* command produces the *Write Block* dialog box (Fig. 20-11). You should notice similarities to the *Block Definition* dialog box. Under *Source*, select *Block* if you want to write out an existing *Block* and select the *Block* name from the list, or select *Objects* if you want to transform a set of objects (not a previously defined *Block*) into a separate .DWG file.

FIGURE 20-11

The *Base Point* section allows you to specify a base point to use upon insertion of the *Block*. Enter coordinate values or use the *Specify Insert Base Point* button to pick a location in the drawing. The *Objects* section allows you to specify how you want to treat selected objects if you create a new .DWG file from objects (not from an existing *Block*). You can *Retain* the objects in their current state, *Convert to Block*, or *Delete from Drawing*.

The *Destination* section defines the desired *File Name and Path*, and *Insert Units*. Your choice for *Insert Units* is applicable only when you drag and drop a *Block*, as with DesignCenter. See "DesignCenter" later in this chapter for an explanation of this subject.

If you prefer the Command line equivalent to the *Write Block* dialog box, type *-Wblock* and follow the prompt sequence shown here to create *Wblocks* (.DWG files) <u>from existing *Blocks*,</u>

> Command: **-wblock**
> (At this point, the *Create Drawing File* dialog box appears, prompting you to supply a name for the .DWG file to be created. Typically, a new descriptive name would be typed in the edit box rather than selecting from the existing names.)
> Enter name of existing block or [= (block=output file)/* (whole drawing)] <define new drawing>:
> (Enter the name of the desired existing *Block*. If the file name given in the previous step is the same as the existing *Block* name, an "=" symbol can be entered, or enter an asterisk to write out the entire drawing.)
> Command:

A copy of the existing *Block* is then created in the selected directory as a *Wblock* (.DWG file).

When *Wblocks* are *Inserted*, the *Color, Linetype,* and *Lineweight* settings of the *Wblock* are determined by the settings current when the original objects comprising the *Wblock* were created. The three possible settings are the same as those for *Blocks* (see "Block," *Color, Linetype,* and *Lineweight* Settings).

When a *Wblock* is *Inserted*, its parent (original) layer is also inserted into the current drawing. *Freezing* <u>either</u> the parent layer or the layer that was current during the insertion causes the *Wblock* to be frozen.

Base

Pull-down Menu	Command (Type)	Alias (Type)	Short-cut	Screen (side) Menu	Tablet Menu
Draw Block > Base	*Base*	...	...	*DRAW 2 Base*	...

The *Base* command allows you to specify an "insertion base point" (see the *Block* command) in the current drawing for subsequent insertions. If the *Insert* command is used to bring a .DWG file into another drawing, the insertion base point of the .DWG is 0,0 by default. The *Base* command permits you to specify another location as the insertion base point.

The *Base* command is used in the symbol drawing, that is, used in the drawing <u>to be inserted</u>. For example, while creating separate symbol drawings (.DWGs) for subsequent insertion into other drawings, the *Base* command is used to specify an appropriate point on the symbol geometry for the *Insert* command to use as a "handle" other than point 0,0 (Fig. 20-12).

FIGURE 20-12

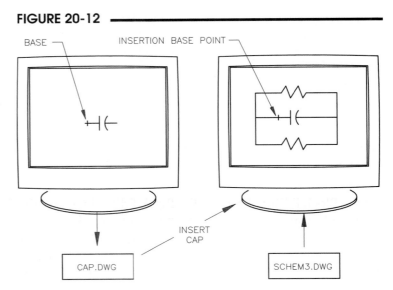

Redefining Blocks and the Block Editor

If you want to change the configuration of a *Block*, even after it has been inserted, it can be accomplished by redefining the *Block*. When you redefine a block (change a block and save it), all of the previous *Block* insertions in the drawing are automatically and globally updated (Fig. 20-13). For each *Block* insertion in the drawing, AutoCAD stores two fundamental pieces of information—the insertion point and the *Block* name. The actual block definition is stored in the block definition table. Redefining the *Block* involves changing that definition.

FIGURE 20-13

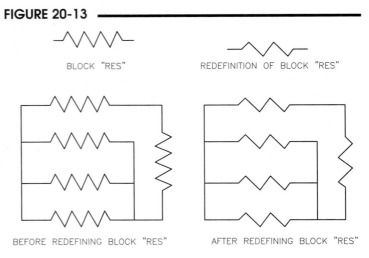

You can use three methods to redefine a *Block*: 1) use *Refedit* to change and save the block, 2) redraw the block geometry, then use the *Block* command to save the new geometry under the old block name, and 3) use the Block Editor to change the block, then *Save* the block from the editor.

To Redefine a Block Using *Refedit*

The *Refedit* command "opens" the *Block* for editing, then you make the necessary changes, and finally use *Refclose* to "close" the editing session and save the *Block*. This process has the same result as redefining the *Block*. This method is probably the least efficient method.

To Redefine a Block Using the *Block* Command

To redefine a *Block* by this method, you must first draw the new geometry or change the original "template" set of objects. Alternately, you can *Explode* one insertion of the block in the Drawing Editor, then make the desired changes. (The original unexploded block cannot be included in the new *Block* because a block definition cannot reference itself.) Next, use the *Block* command and select the new or changed geometry. The old block is redefined with the new geometry as long as the original block name is used.

To Redefine a Block Using the Block Editor

To use the Block Editor, double-click on a block or use the *Bedit* command. See "*Bedit*" next.

Bedit

Pull-down Menu	Command (Type)	Alias (Type)	Short-cut	Screen (side) Menu	Tablet Menu
Tools *Block Editor*	*Bedit*	*BE*	…	…	…

To edit and redefine existing blocks, or to create dynamic blocks, use the Block Editor. Produce the Block Editor by using the *Bedit* command by any method shown in the command table above. Alternately, double-click on any block (without attributes) in the drawing (assuming *Dblclkedit* is on) to automatically open the Block Editor.

Before the Block Editor appears, the *Edit Block Definition* dialog box appears for you to select the desired block name from the *Block to create or edit* list (Fig. 20-14).

FIGURE 20-14

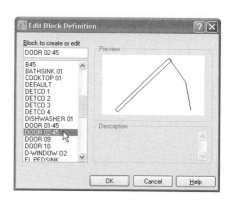

Once the desired block is selected in the *Edit Block Definition* dialog box, the Block Editor appears in place of the Drawing Editor (Fig. 20-15). The Block Editor has a light colored background by default. The Block Editor is similar to the Drawing Editor in that you can draw and edit geometry as you would normally. Most AutoCAD commands operate in the Block Editor, although a few commands, such as those related to plotting, publishing, layouts, views, and viewports, are not allowed.

FIGURE 20-15

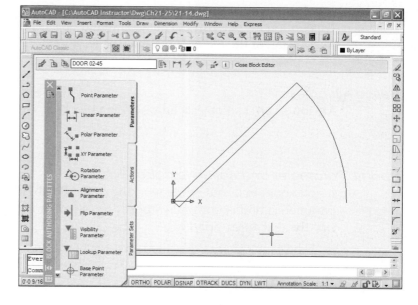

The selected block geometry also appears in the drawing editor. However, instead of the block appearing as one object, it appears in its original "exploded" state—that is, the individual *Lines, Arcs, Circles,* or other objects that make up the block can be changed.

The UCS icon appears to indicate the insertion base point for the block. Note that the *Block Authoring Palettes* appear in the Block Editor by default. If you are not creating dynamic blocks, close the *Block Authoring Palettes.*

The Block Editor toolbar at the top of the editor contains several buttons. These options, in order from left to right on the toolbar, are described next.

Edit or Create Block Definition

Use this button (far left) to work with a new block in the Block Editor. The Edit Block Definition dialog box appears for you to select the desired block name you want to edit (see previous Fig. 20-14).

Save Block Definition

This button activates the *Bsave* command. This command saves the block in the editor to the drawing's block definition table. Any previous insertions of the block in the drawing are automatically changed to the new block definition.

Save Block As

This button produces the *Save Block As* dialog box, which is essentially the same as the *Edit Block Definition* dialog box (see previous Fig. 20-14). Enter the desired name in the edit box and select *OK* to save the current block with the new name.

Authoring Palettes

If you are creating dynamic blocks, use this button to produce the *Block Authoring Palettes.* If you are redefining an existing "normal" block, close the palettes.

Parameter

This option invokes the *Bparameter* command. This is a Command line alternate to using the *Parameters* tab of the *Block Authoring Palette.*

Action

This option invokes the *Baction* command. *Baction* is a Command line alternate to using the Actions tab of the *Block Authoring Palette.*

Define Attributes

Attributes are text objects attached to a block.

Update Parameter and Action Text Size

If you zoom in or out while working in the Block Editor, the text that appears when assigning parameters and actions may appear too small or too large. Use this option to automatically size the text appropriately.

Close Block Editor

Close the Block Editor when you have completed the process of creating a new block or redefining an existing block definition and you are ready to return to the Drawing Editor. If you forget to use *Save Block Definition* first, a warning appears and allows you to save your changes.

Purge

Pull-down Menu	Command (Type)	Alias (Type)	Short-cut	Screen (side) Menu	Tablet Menu
File *Drawing Utilities >* *Purge...*	*Purge*	*PU*	...	*FILE* *Purge*	X,25

Purge allows you to selectively delete a *Block* that is not referenced in the drawing. In other words, if the drawing has a *Block* defined but not appearing in the drawing, it can be deleted with *Purge*. In fact, *Purge* can selectively delete <u>any named object</u> that is not referenced in the drawing. Examples of unreferenced named objects are:

Blocks that have been defined but not *Inserted*;
Layers that exist without objects residing on the layers;
Dimstyles that have been defined, yet no dimensions are created in the style;
Linetypes that were loaded but not used;
Shapes that were loaded but not used;
Text Styles that have been defined, yet no text has been created in the *Style*;
Table Styles that were created but not used in the drawing;
Plot Styles that were used for a page setup but not assigned to objects or layouts;
Mlstyles (multiline styles) that have been defined, yet no *Mlines* have been drawn in the style;
Materials that have been imported into the drawing but are not attached to objects;
Visual Styles that have been created for the drawing but are not currently applied;

Purge is especially useful for getting rid of unused *Blocks* because unused *Blocks* can occupy a huge amount of file space compared to other named objects (*Blocks* are the only geometry-based named objects). Although some other named objects can be deleted by other methods (such as selecting *Delete* in the *Text Style* dialog box), *Purge* is the only method for deleting *Blocks*.

The *Purge* command produces the *Purge* dialog box (Fig. 20-16). Here you can view all named objects in the drawing, but you can purge only those items that are not used in the current drawing.

View items you can purge

Generally this option is the default choice since you can remove unused named objects from your drawing only if this option is checked.

View items you cannot purge

This option is useful if you want to view all the named objects contained in the drawing. You can also get an explanation appearing at the bottom of the dialog box as to why a selected item cannot be removed.

Purge

Use the *Purge* button to remove only the items you select from the list. First, select the items, then select *Purge*. You can select more than one item using the Ctrl or Shift keys while selecting multiple items or a range, respectively.

FIGURE 20-16

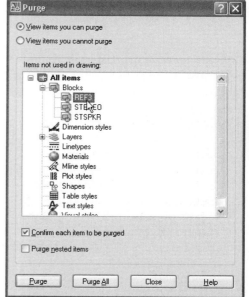

Purge All

Use this button to remove all unused named objects from the drawing. You do not have to first select items from the list. To be safe, you should also select *Confirm each item to be purged* when using this option.

Rename

Pull-down Menu	Command (Type)	Alias (Type)	Short-cut	Screen (side) Menu	Tablet Menu
Format Rename...	*Rename* or *-Rename*	*REN* or *-REN*	...	*FORMAT Rename*	*V,1*

This utility command allows you to rename a *Block* or <u>any named object</u> that is part of the current drawing. *Rename* allows you to rename the named objects listed here:

> *Blocks, Dimension Styles, Layers, Linetypes, Materials, Plot Styles, Table Styles, Text Styles, User Coordinate Systems, Views,* and *Viewport* configurations.

If you prefer to use dialog boxes, you can type *Rename* or select *Rename...* from the *Format* pull-down menu to access the dialog box shown in Figure 20-17.

You can select from the *Named Objects* list to display the related objects existing in the drawing. Then select or type the old name so it appears in the *Old Name:* edit box. Specify the new name in the *Rename To:* edit box. <u>You must then PICK *the Rename To:* tile</u> and the new names will appear in the list. Then select the *OK* tile to confirm.

FIGURE 20-17

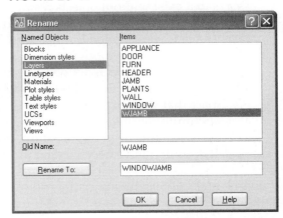

Setbylayer

Pull-down Menu	Command (Type)	Alias (Type)	Short-cut	Screen (side) Menu	Tablet Menu
Modify > Change to Bylayer	*Setbylayer*	...	...	...	...

The *Setbylayer* command allows you to change the *Byblock* property settings (*color, linetype, lineweight,* etc.) for selected objects to a *Bylayer* setting. This feature can be very helpful in collaborative environments where drawings and *Blocks* are shared. For example, you may insert *Blocks* from a library or from other drawings that have *Byblock* settings. Use the *Bylayer* command to convert the *Blocks* to assume the properties of the layers where the *Blocks* reside in the new drawing.

```
Command: setbylayer
Current active settings: Color Linetype Lineweight Material PlotStyle
Select objects or [Settings]: PICK
Select objects or [Settings]: PICK
Select objects or [Settings]: Enter
Change ByBlock to ByLayer? [Yes/No] <Yes>: Enter
Include blocks? [Yes/No] <Yes>: Enter
2 objects modified.
Command:
```

DESIGNCENTER

The DesignCenter window (Fig. 20-18) allows you to navigate, find, and preview a variety of content, including *Blocks*, located anywhere accessible to your workstation, then allows you to open or insert the content using drag-and-drop. "Content" that can be viewed and managed includes other drawings, *Blocks*, *Dimstyles*, *Layers*, *Layouts*, *Linetypes*, *Table Styles*, *Textstyles*, *Xrefs*, raster images, and URLs (Web site addresses). In addition, if you have multiple drawings open, you can streamline your drawing process by copying and pasting content, such as layer definitions, between drawings.

FIGURE 20-18

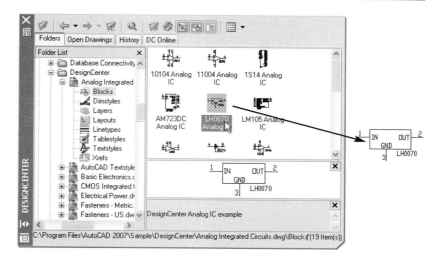

Adcenter

Pull-down Menu	Command (Type)	Alias (Type)	Short-cut	Screen (side) Menu	Tablet Menu
Tools *Palettes >* *DesignCenter*	*Adcenter*	*ADC*	*Ctrl + 2*	...	...

Accessing *Adcenter* by any method produces the DesignCenter palette. There are two sections to the window. The left side is called the Tree View and displays a Windows Explorer-type hierarchical directory (folder) structure of the local system (Fig. 20-19). The right side is called the Content Area and displays lists, icons, or thumbnail sketches of the content selected in the Tree View. The Content Area can display *Blocks*, *Dimstyles*, *Layers*, *Layouts*, *Linetypes*, *Table Styles*, *Textstyles*, *Xrefs*, and a variety of other content.

FIGURE 20-19

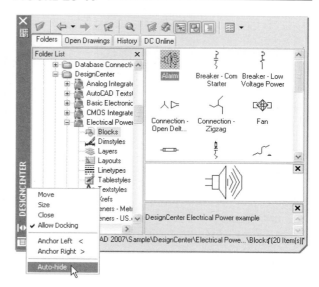

Generally, the Content Area is used to drag and drop the icons or thumbnails from the content area into the current drawing (see Fig. 20-18). However, you can also streamline a variety of tasks such as those listed here:

- Browse sources of drawing content including open drawings, other drawings, raster images, content within the drawings (*Blocks*, *Dimstyles*, *Layers*, and so on), content on network drives, or content on a Web page.

- Insert, attach, or copy and paste the content (drawings, images, *Blocks*, *Layers*, etc.) into the current drawing.

- Create shortcuts to drawings, folders, and Internet locations that you access frequently.

- Use a special search engine to find drawing content on your computer or network drives. You can specify criteria for the search based on key words, names of *Blocks*, *Dimstyles*, *Layers*, etc., or the date a drawing was last saved. Once you have found the content, you can load it into DesignCenter or drag it into the current drawing.

- Open drawings by dragging a drawing (.DWG) file from the Content Area into the drawing area.

These features of DesignCenter are explained in the following descriptions of the DesignCenter toolbar icons.

Resizing, Docking, and Hiding the DesignCenter Palette

You can change the width and height of DesignCenter by resting your pointer on one of the borders (not the title bar) until a double arrow appears, then dragging to the desired size. Alternately, rest your pointer on the lower corner (where the bevel appears) to resize both height and width. You can also move the bar between the Content Area and the Tree View area. Dock DesignCenter by clicking the title bar, then dragging it to either edge of the drawing window. You can hide the DesignCenter palette by using the *Auto-hide* option (click on the *Properties* button at the bottom of the title bar to produce the *Properties* menu as shown in Fig. 20-19). When *Auto-hide* is on, the palette is normally hidden (only the title bar is visible) and the palette appears when you bring the pointer to the title bar.

DesignCenter Options

In addition to selecting the icons from the toolbar, you can right-click the Palette background to produce the shortcut menu and choose the desired option (Fig. 20-20).

FIGURE 20-20

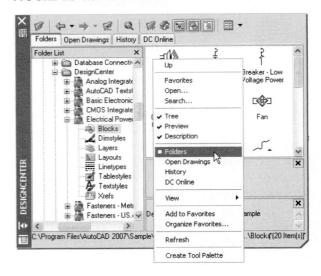

Folders Tab

Folders is the default display for the Tree View side of DesignCenter. This choice displays a hierarchical structure of the desktop (local workstation). Because this arrangement is similar to Windows Explorer, you can navigate and locate content anywhere accessible to your system, including network drives. Figure 20-20 displays a typical hierarchical structure on the Tree View side. For example, you may want to import layer definitions (including color and linetype information) from a drawing into the current drawing by dragging and dropping.

Open Drawings Tab

This option changes the Tree View to display all open drawings (Fig. 20-21). This feature is helpful when you have several drawings open and want to locate content from one drawing and import it into the current drawing. As shown in Figure 20-21, you may want to locate *Block* definitions from one drawing and *Insert* them into <u>another</u> drawing. Ensure you make the desired "target" drawing current in the drawing area before you drag and drop content from the Content Area.

FIGURE 20-21

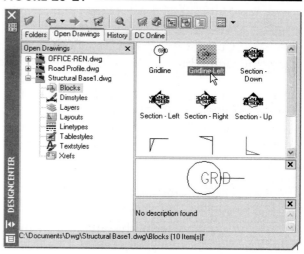

History Tab

This option displays a history (chronological list) of the last 20 file locations accessed through DesignCenter (Fig. 20-22). The purpose of this feature is simply to locate the file and load it into the Content Area. Load the file into the Content Area by double-clicking on it.

FIGURE 20-22

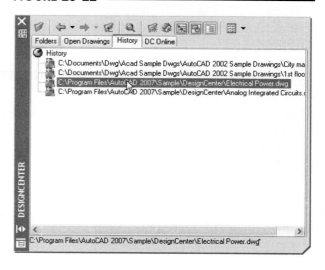

Tree View Toggle

Tree View is helpful for navigating your system for content. Once the desired folder or drawing is found and highlighted in Tree View, you may want to toggle Tree View off so only the Content Area is displayed with the desired content. The desired content may be drawings, *Blocks*, images, or a variety of other content. For example, consider previous Figure 20-21 which displays Tree View on. Figure 20-23 illustrates Tree View toggled off and only the Content Area displayed. The resulting configuration displays only the *Blocks* contained in the selected drawing. Keep in mind that you can also change the Views of the Content Area to display *Large Icons*, *Small Icons*, a *List*, or *Details*.

FIGURE 20-23

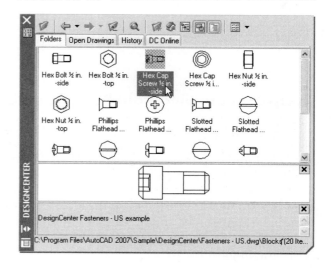

Favorites

This button displays the contents of the AutoCAD Favorites folder in the Content Area. The Tree View section displays the highlighted folder in the Desktop view.

You can add folders and files to *Favorites* by highlighting an item in Tree View or the Content Area, right-clicking on it, and selecting *Add to Favorites* from the shortcut menu (Fig. 20-24).

Load

Displays the *Load* dialog box (not shown), in which you can load the Content Area with content from anywhere accessible from your system. The *Load* dialog box is identical to the *Select File* dialog box (see "*Open*," Chapter 2, Working with Files). After selecting a file, DesignCenter automatically finds the file in Tree View and loads its content (*Blocks*, *Layers*, *Dimstyles*, etc.) into the Content Area. You can also load files into the Content Area using Windows Explorer.

FIGURE 20-24

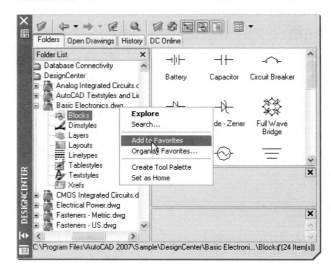

Search

This button invokes the *Search* dialog box (Fig. 20-25), in which you can specify search criteria to locate drawing files, *Blocks*, *Layers*, *Dimstyles*, and other content within drawings. <u>Once the desired content is found in the list at the bottom of the dialog box, double-click on it to load it into the Content Area.</u>

FIGURE 20-25

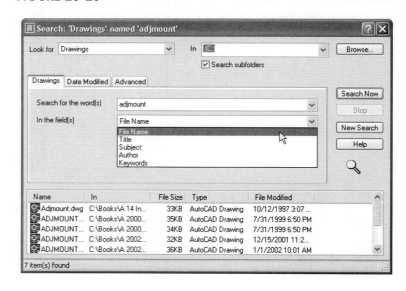

To Open a Drawing from DesignCenter

In addition to using DesignCenter to view and import content contained within drawing files, it is possible to use DesignCenter to *Open* drawings. Click on the symbol until the "i-Drop" icon (eye dropper) appears, then drag the icon of the drawing file you want to open from the <u>Content Area</u> and drop it in the drawing area. Remember, you must drag it from the Content Area, not from the Tree View area.

Usually, you would want to drop the selected drawing into a blank (*New*) drawing. Unless the selected drawing is a titleblock and border, drop it into model space, not into a layout (paper space). Make sure the background (drawing area) is visible. You may need to resize the windows displaying any currently open drawings.

A drawing file that is dropped into AutoCAD is actually *Inserted* as a *Block*. The typical Command line prompts appear for insertion point, scale factors, and rotation angle. Generally, enter 0,0 as the insertion point and accept the defaults for X and Y scale factors and rotation angle. If you want to edit the geometry, you will have to *Explode* the drawing.

To *Insert* a *Block* Using DesignCenter

One of the primary functions of DesignCenter is to insert *Block* definitions into a drawing. When you insert a *Block* into a drawing, the block definition is copied into the drawing database. Any instance of that *Block* that you *Insert* into the drawing from that time on references the original *Block* definition.

You cannot add *Blocks* to a drawing while another command is active. If you try to drop a *Block* into AutoCAD while a command is active at the Command line, the icon changes to a slash circle indicating the action is invalid.

There are two methods for inserting *Blocks* into a drawing using DesignCenter: 1) using drag-and-drop with Autoscaling, and 2) using the *Insert* dialog box with explicit insertion point, scale, and rotation value entry.

FIGURE 20-26 ────────────

Block Insertion with Drag-and-Drop

When you drag-and-drop a *Block* from DesignCenter into a drawing, you cannot key in an insertion point, X and Y scale factors, or a rotation angle. Although you specify an insertion point interactively when you "drop" the *Block* icon, AutoCAD uses Autoscaling to determine the scaling parameters. Autoscaling is a process of comparing the specified units of the *Block* definition (when the *Block* was created) and the *Insertion Scale* set in the *Drawing Units* dialog box of the target drawing. The value options are *Unitless, Inches, Feet, Miles, Millimeters, Centimeters, Kilometers,* and many other choices.

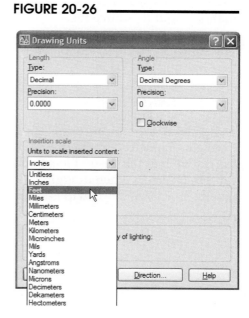

The *Units to scale inserted content* set in the *Drawing Units* dialog box (Fig. 20-26) <u>should be set in the target drawing</u> and controls how the *Block* units are scaled when the *Block* is dropped (into the target drawing). It is helpful that the <u>setting here can be changed immediately before dropping a *Block* into the drawing</u>. (Also see *"Units"* in Chapter 6.)

TOOL PALETTES

Tool palettes offer a faster and more visible process of inserting *Blocks* and *Hatch* patterns into a drawing. You can use tool palettes to insert a *Block* instead of using the *Insert* command or DesignCenter. You can also use tool palettes to insert hatch patterns (section lines or fill patterns) into a drawing. *Blocks* and hatches that reside on a tool palette are called tools, and several tool properties including scale, rotation, and layer can be set for each tool individually. The features of tool palettes, using tool palettes to insert *Blocks*, and creating tool palettes are discussed in this chapter. Using tool palettes to insert hatch patterns is discussed in Chapter 23, Section Views and Hatch Patterns.

Toolpalettes

Pull-down Menu	Command (Type)	Alias (Type)	Short-cut	Screen (side) Menu	Tablet Menu
Tools Palettes > Tool Palettes	Toolpalettes	TP	Ctrl+3	...	...

Any of the methods shown in the Command Table above produce the *Tool Palettes* window (Fig. 20-27). Tool palettes are the individual tabbed areas (like pages) of the *Tool Palettes* window. When you need to add a *Block* to a drawing, you can drag it from the tool palette into your drawing instead of using the *Insert* command or using DesignCenter. You can create your own tool palettes by placing the *Blocks* that you use often on a tool palette.

FIGURE 20-27

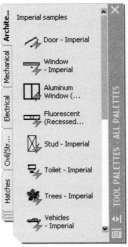

Once you add blocks from a drawing into the tool palettes, the blocks are "saved" in the tool palettes and can then be dragged and dropped into any other drawing. Therefore, blocks in the tool palettes are not usually located in the current drawing's block definition table, but are automatically located from their source drawing or file. Each block in the tool palettes contains the needed information about its source file.

The AutoCAD sample tool palettes contain many dynamic blocks (designated by the "lighting bolt" icon).

Tool palettes are simple to use. To insert a *Block* (tool) from a tool palette, simply drag it from the palette into the drawing at the desired location (Fig. 20-28). You can use *OSNAPs* when dragging *Blocks* from a tool palette; however, Grid Snap is suppressed during dragging.

FIGURE 20-28

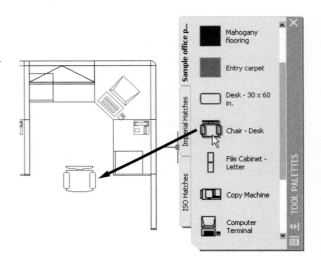

When a *Block* is dragged from a tool palette into a drawing, it is scaled automatically according to the ratio of units defined in the *Block* and units used in the current drawing. This action is similar to using DesignCenter, in that the scale and rotation angle may have to be changed after the *Block* is inserted. See "*Block* Insertion with Drag-and-Drop" earlier in this chapter.

Using the *Tool Palette* window, you can preset the *Block* insertion properties or hatch pattern properties of any tool on a tool palette. For example, you can change the insertion scale of a *Block* or the angle of a hatch pattern. Do this by right-clicking on a tool and selecting *Properties... .*

Palette *Properties*

The options and settings for tool palettes are accessible from shortcut menus in different areas on the *Tool Palettes* window. Right-clicking inside the palette produces the menu shown in Figure 20-29. You can also click on the *Properties* button (bottom of the vertical title bar) to produce essentially the same menu with the addition of *Move*, *Size*, and *Close* options. Palette properties are saved with your AutoCAD profile.

FIGURE 20-29

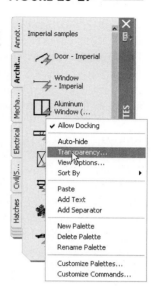

Move, Size, Close, Allow Docking, Auto-hide
These options are identical to the same features in DesignCenter. See "Resizing, Docking, and Hiding the DesignCenter Palette" earlier in this chapter.

Tool *Properties*

If you right-click on a tool (*Block* or hatch pattern), a shortcut menu appears (Fig. 20-30) allowing you to change properties <u>for that specific tool</u>. You can *Cut* any tool and *Paste* it to another palette or *Copy* a tool to another palette. Selecting *Delete* removes it from that palette only. If the source block file has been changed, you can *Update tool image*. Select *Block Editor* to edit the current block in the Block Editor.

FIGURE 20-30

You can change the properties (such as scale and rotation angle) for any specific tool. Do this by right-clicking on the tool and selecting *Properties…* from the shortcut menu (see Fig. 20-30) to produce the *Tool Properties* dialog box (Fig. 20-31).

The *Tool Properties* dialog box can have three categories of properties—*Insert* (or *Pattern*) properties, *General* properties, and *Custom* properties. You can control object-specific properties such as *Scale* and *Rotation* angle in the *Insert* list.

FIGURE 20-31

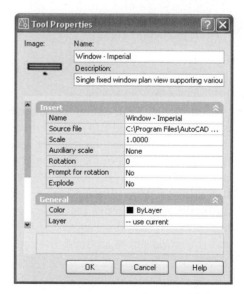

Use the *General* properties section <u>only</u> if you want to override the current drawing's property settings such as *Layer*, *Color*, and *Linetype*. (This action is sometimes referred to as setting a tool property <u>override</u>.) When a property override is set, special conditions may exist. For example, if a specified layer does not exist in the drawing, that layer is created automatically when the *Block* or hatch is inserted.

Creating Tool Palettes

Creating Tool Palettes using DesignCenter

DesignCenter provides a simple method for creating new tool palettes. Begin by opening both the DesignCenter window and the *Tool Palettes* window. Locate the desired *Blocks* or hatch patterns you want to insert into a new (not yet created) palette and ensure they appear in the DesignCenter Content Area. Next, right-click on any tool you want to move into the new palette so a shortcut menu appears (Fig. 20-32, center). Select *Create Tool Palette* from the menu. A new palette appears with the selected *Block* shown at the top of the palette. Near the new *Block*, a small edit box (not shown) appears for you to enter in the desired tool palette name. After doing so, you are presented with the new palette containing the selected *Block* (as shown in Fig. 20-32, right side).

To add additional *Blocks* or hatch patterns into your new palette, drag and drop them from the DesignCenter Content Area into the new palette (Fig. 20-33). Next, use the tool *Properties* shortcut menu, if needed, to preset the scale and rotation angle for the tool, as described earlier.

FIGURE 20-32

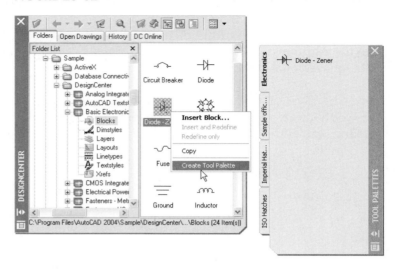

FIGURE 20-33

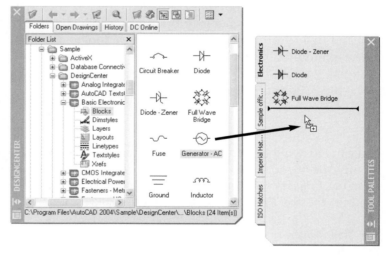

You can also create entire tool palettes fully populated with *Blocks* from all the *Blocks* contained in any drawing using DesignCenter. Do this by locating the drawing file in the Tree View area of DesignCenter, then right-clicking on the drawing file so the shortcut menu appears. Select *Create Tool Palette* (Fig. 20-34). A new tool palette is automatically added to the *Tool Palettes* window containing all the *Blocks* from the selected drawing. The new tool palette has the same name as the drawing.

FIGURE 20-34

CHAPTER EXERCISES

1. ***Block, Insert***
   In the next several exercises, you will create an office floor plan, then create pieces of furniture as *Blocks* and *Insert* them into the office. All of the block-related commands are used.

   A. Start a *New* drawing. Select the *ACAD.DWT* or *Start from Scratch* and use the *Imperial* defaults. Use *Save* and assign the name **OFF-ATT.** Set up the drawing as follows:

1. *Units*	*Architectural*	*1/2" Precision*	
2. *Limits*	48' x 36'	(1/4"=1' scale on an A size sheet), drawing scale factor = 48	
3. *Snap, Grid*	3		
4. *Grid*	12		
5. *Layers*	**FLOORPLAN**	*continuous*	*.014*	**colors of your choice**
	FURNITURE	*continuous*	*.060*	
	ELEC-HDWR	*continuous*	*.060*	
	ELEC-LINES	*hidden 2*	*.060*	
	DIM-FLOOR	*continuous*	*.060*	
	DIM-ELEC	*continuous*	*.060*	
	TEXT	*continuous*	*.060*	
	TITLE	*continuous*	*.060*	
6. *Text Style*	*CityBlueprint*	*CityBlueprint (TrueType font)*	
7. *Ltscale*	48		

   B. Create the floor plan shown in Figure 20-35. Center the geometry in the *Limits*. Draw on layer **FLOORPLAN**. Use any method you want for construction (e.g., *Line, Pline, Xline, Mline, Offset*).

FIGURE 20-35

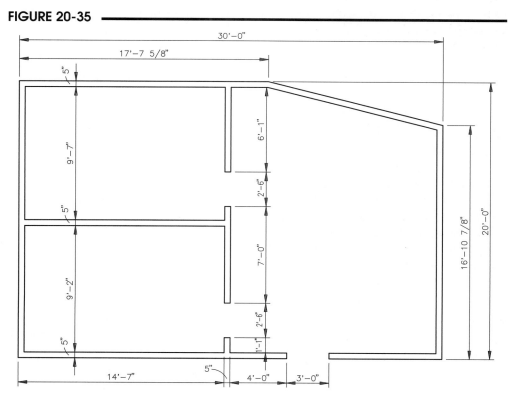

C. Create the furniture shown in Figure
20-36. Draw on layer **FURNITURE**.
Locate the pieces anywhere for now.
Do <u>not</u> make each piece a *Block*. *Save*
the drawing as **OFF-ATT**.

Now make each piece a *Block*. Use the
name as indicated and the *insertion
base point* as shown by the "blip."
Next, use the *Block* command again
but only to check the list of *Blocks*. Use
SaveAs and rename the drawing
OFFICE.

FIGURE 20-36

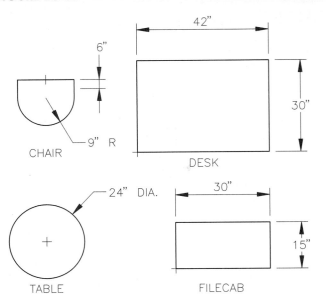

D. Use *Insert* to insert the furniture into
the drawing, as shown in Figure 20-37.
You may use your own arrangement
for the furniture, but *Insert* the same
number of each piece as shown.
Save the drawing.

FIGURE 20-37

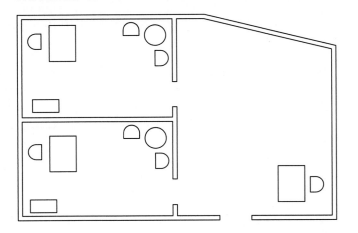

2. **Creating a .DWG file for *Insertion, Base***

Begin a *New* drawing. Assign the name **CONFTABL**.
Create the table as shown in Figure 20-38 on *Layer 0*.
Since this drawing is intended for insertion into the
OFFICE drawing, use the *Base* command to assign an
insertion base point at the lower-left corner of the table.

When you are finished, *Save* the drawing.

FIGURE 20-38

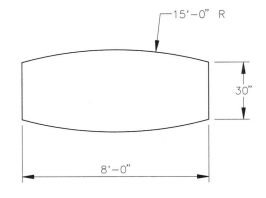

3. *Insert, Explode, Divide*

FIGURE 20-39

A. *Open* the **OFFICE** drawing. Ensure that layer **FURNITURE** is current. Use **DesignCenter** to bring the **CONFTABL** drawing in as a *Block* in the placement shown in Figure 20-39.

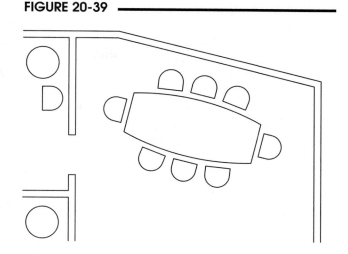

Notice that the CONFTABL assumes the linetype and color of the current layer, since it was created on layer **0**.

B. *Explode* the CONFTABL. The *Exploded* CONFTABL returns to *Layer* **0**, so use the *Properties* palette to change it back to *Layer* **FURNITURE**. Then use the *Divide* command (with the *Block* option) to insert the **CHAIR** block as shown in Figure 20-39. Also *Insert* a **CHAIR** at each end of the table. *Save* the drawing.

4. **Redefining a** *Block*

FIGURE 20-40

After a successful meeting, the client accepts the proposed office design with one small change. The client requests a slightly larger chair than that specified. Open the **Block Editor** and select the **CHAIR** block to edit. Redesign the chair to your specifications. **Save** the CHAIR block and *Close* the Block Editor. All previous insertions of the CHAIR should reflect the design change. *Save* the drawing. Your design should look similar to that shown in Figure 20-40. *Plot* to a standard scale based on your plotting capabilities.

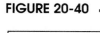
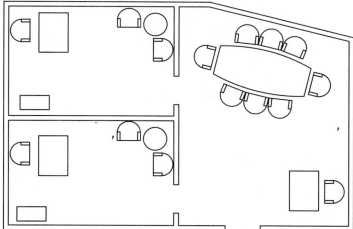

5. **Use DesignCenter to create a Tool Palette**

    A. *Close* all open drawings and begin a *New* drawing. Next, open *DesignCenter* and use the *Folders* tab to locate the **OFFICE** drawing in the Tree View area. Expand the plus (+) symbol next to OFFICE.DWG, then click on the word *Blocks* just below the drawing name. All of the *Blocks* contained in the drawing should appear in the Content Area.

    B. Use the *Toolpalettes* command or press **Ctrl+3** to produce the *Tool Palettes* window. Now, back in DesignCenter, locate the OFFICE drawing in the Tree View area and right-click on the drawing name. From the shortcut menu select *Create Tool Palette*.

    C. This action should have automatically created a new tool palette named OFFICE and populated it with all the *Blocks* contained in the drawing. Examine the new tool palette. Right-click on any tool and examine its *Properties*. Now you have a new tool palette ready to use for inserting the office furniture *Blocks* into any other drawings.

    D. Use *DesignCenter* again to locate the **HOME-SPACE PLANNER** drawing in the AutoCAD 2008 /Sample/DesignCenter folder. In the Tree View area, right-click on the drawing name, then select *Create Tool Palette* from the shortcut menu. A new palette should appear in the *Tool Palettes* window fully populated with furniture.

6. **Using Tool Palettes**

    In this exercise you will change some of the blocks in the OFFICE drawing using blocks from the *Tool Palettes* window.

    A. *Open* the **OFFICE** drawing again if not already open. Assume as a result of the meeting with the client, new desks and chairs are requested for the receptionist and the two offices. If not already available, produce the *Tool Palettes* window using the *Toolpalettes* command or pressing **Ctrl+3**. Select the new *Home-Space Planner* palette (showing the home furniture).

    B. Right-click on the *Chair-Desk* tool and select *Properties*. In the *Tool Properties* dialog box that appears, note that the scale is 1.00 and the layer is set to use current. Check the settings for the *Desk - 30 x 60 in*. With these settings, the new furniture should be usable without changes.

C. Make **FURNITURE** the current layer.
    Erase the three desks and the three desk
    chairs. Drag and drop three chairs and
    three desks from the palette window and
    place them near the desired locations in
    the office. Use *Rotate* and *Move* to posi-
    tion the new *Blocks* into suitable locations
    and orientations. Your drawing should
    look similar to that in Figure 20-41. Use
    SaveAs and name the drawing
    OFFICE-REV.

FIGURE 20-41

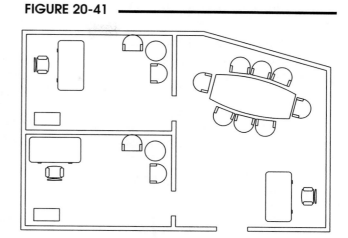

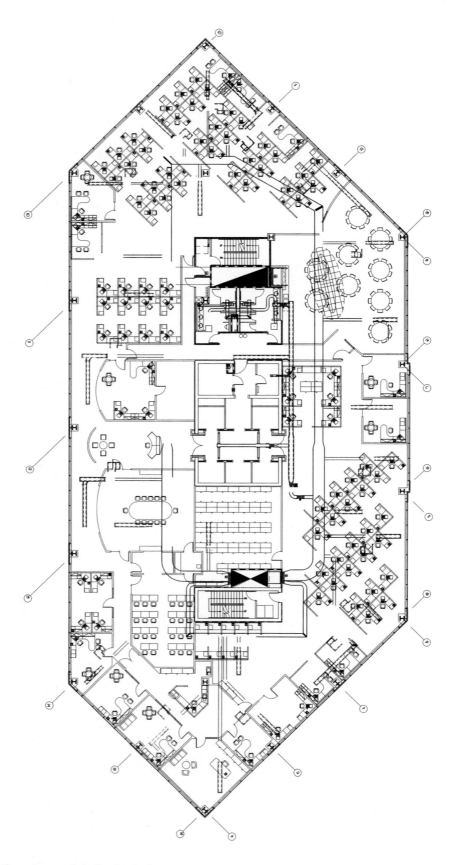

8TH FLOOR.DWG, Courtesy of Autodesk, Inc.

MULTIVIEW DRAWING

CHAPTER OBJECTIVES

After completing this chapter you should:

1. be able to draw projection lines using *ORTHO* and *SNAP*, and *Polar Tracking*;

2. be able to use *Xline* and *Ray* to create construction lines;

3. know how to use *Offset* for construction of views;

4. be able to use *Object Snap Tracking* for alignment of lines and views;

5. be able to use construction layers for managing construction lines and notes;

6. be able to use linetypes, lineweights, and layers to draw and manage ANSI standards;

7. know how to create fillets, rounds, and runouts;

8. know the typical guidelines for creating a three-view multiview drawing.

CONCEPTS

Multiview drawings are used to represent 3D objects on 2D media. The standards and conventions related to multiview drawings have been developed over years of using and optimizing a system of representing real objects on paper. Now that our technology has developed to a point that we can create 3D models, some of the methods we use to generate multiview drawings have changed, but the standards and conventions have been retained so that we can continue to have a universally understood method of communication.

This chapter illustrates methods of creating 2D multiview drawings with AutoCAD (without a 3D model) while complying with industry standards. Many techniques can be used to construct multiview drawings with AutoCAD because of its versatility. The methods shown in this chapter are the more common methods because they are derived from traditional manual techniques. Other methods are possible.

PROJECTION AND ALIGNMENT OF VIEWS

Projection theory and the conventions of multiview drawing dictate that the views be aligned with each other and oriented in a particular relationship. AutoCAD has particular features, such as *SNAP*, *ORTHO*, construction lines (*Xline, Ray*), Object Snap, Polar Tracking and Object Snap Tracking, that can be used effectively for facilitating projection and alignment of views.

Using *ORTHO* and *OSNAP* to Draw Projection Lines

ORTHO (F8) can be used effectively in concert with *OSNAP* to draw projection lines during construction of multiview drawings. For example, drawing a Line interactively with *ORTHO ON* forces the *Line* to be drawn in either a horizontal or vertical direction.

Figure 21-1 simulates this feature while drawing projection *Lines* from the top view over to a 45 degree miter line (intended for transfer of dimensions to the side view). The "first point:" of the *Line* originated from the *Endpoint* of the *Line* on the top view.

FIGURE 21-1

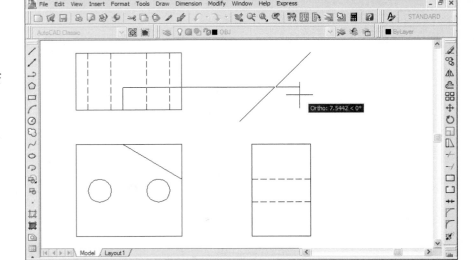

Figure 21-2 illustrates the next step. The vertical projection *Line* is drawn from the *Intersection* of the 45 degree line and the last projection line. *ORTHO* forces the *Line* to the correct vertical alignment with the side view.

Remember that any draw <u>or</u> edit command that requires PICKing is a candidate for *ORTHO* and/or *OSNAP*.

NOTE: *OSNAP* overrides *ORTHO*. If *ORTHO* is *ON* and you are using an *OSNAP* mode to PICK the "next point:" of a *Line*, the *OSNAP* mode has priority; and, therefore, the construction may not result in an orthogonal *Line*.

FIGURE 21-2

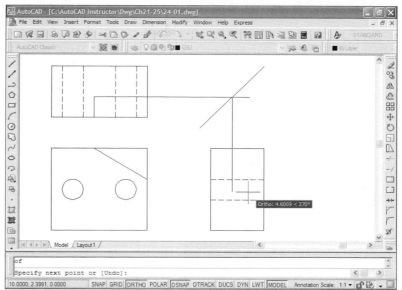

Using *Polar Tracking* to Draw Projection Lines

Similar to using *ORTHO* and *OSNAP*, you can use *Polar Tracking* and *OSNAP* options to draw projection lines at 90-degree increments. Using the previous example, *Polar Tracking* is used to project dimensions between the top and side views through the 45-degree miter line.

To use *Polar Tracking*, set the desired *Polar Angle Settings* in the *Polar Tracking* tab of the *Drafting Settings* dialog box. Ensure *POLAR* appears recessed on the Status Bar. You can also set a *Polar Snap* increment or a *Grid Snap* increment to use with *Polar Tracking* (see Chapter 3 for more information on these settings).

For example, you could use the *Endpoint OSNAP* option to snap to the *Line* in the top view, then *Polar Tracking* forces the line to a previously set polar increment (0 degrees in this case).

Note that you can use *OSNAPs* in conjunction with *Polar Tracking*, as shown in Figure 21-3, such that the current horizontal line *OSNAPs* to its *Intersection* with the 45-degree miter line. (*OSNAPs* <u>cannot</u> be used effectively with *ORTHO*, since *OSNAPs* override *ORTHO*.)

FIGURE 21-3

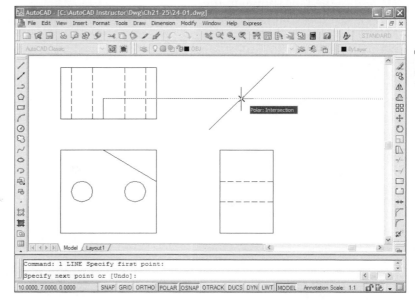

In the following step, use *Intersection OSNAP* option to snap to the intersection of the previous *Line* and the 45-degree miter line (Fig. 21-4). Again *Polar Tracking* forces the *Line* to a vertical position (270 degrees).

FIGURE 21-4

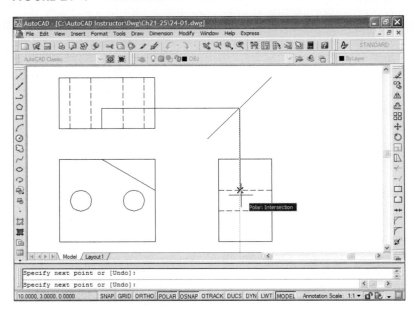

Using *Object Snap Tracking* to Draw Projection Lines Aligned with *OSNAP* Points

This feature, although the most complex, provides the greatest amount of assistance in constructing multiview drawings. The advantage is the availability of the features described in the previous method (*Polar Tracking* and *OSNAP* for alignment of vertical and horizontal *Lines*) in addition to the creation of *Lines* and other objects that align with *OSNAP* points (*Endpoint, Intersection, Midpoint,* etc.) of other objects.

To use *Object Snap Tracking*, first set the desired running *OSNAP* options (*Endpoint, Intersection,* etc.). Next, toggle on *OTRACK* at the Status Bar. Use *Polar Tracking* in conjunction with *Object Snap Tracking* by setting polar angle increments (see previous discussion) and toggling on *POLAR* on the Status Bar.

For example, to begin the *Line* in Figure 21-5, first "acquire" the *Endpoint* of the *Line* shown in the front view. Note that the "first point" of the new *Line* (in the top view) aligns vertically with the acquired *Endpoint*. At this point in time, the short vertical *Line* for the top view could be drawn from the intersection shown. *Object Snap Tracking* prevents having to draw the vertical projection line from the front up to the top view.

FIGURE 21-5

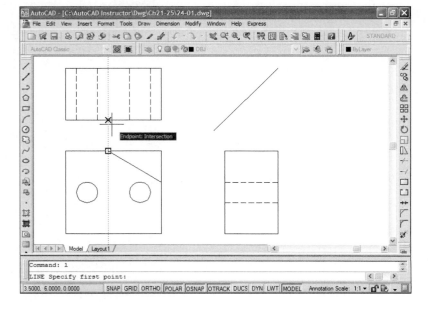

After constructing the two needed *Lines* for the top view, a tracking vector aligns with the acquired *Endpoint* of the indicated line in the top view and the *Intersection* of the 45-degree miter line (Fig. 21-6). Note that a <u>horizontal projection line is not needed</u> since *Object Snap Tracking* ensures the "first point" aligns with the appropriate point from the top view.

FIGURE 21-6

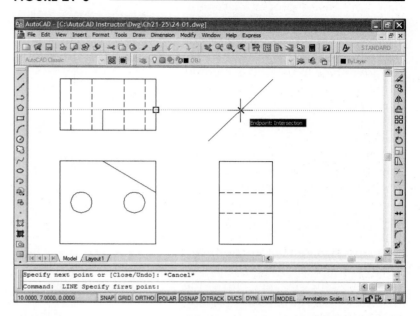

The next step is to draw the short horizontal visible line in the side view (Fig. 21-7). Coming from the "first point" (on the 45-degree miter line found in the previous step), draw the vertical projection line down to the appropriate point in the side view that "tracks" with the related *Endpoint* in the front view (see horizontal tracking vector). Since this *Line* is constructed to the correct endpoint, <u>*Trimming* is unnecessary here</u>, but it would be needed in the previous two methods.

FIGURE 21-7

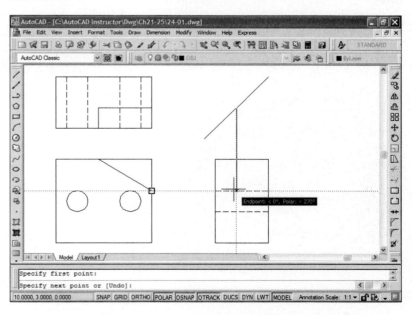

These AutoCAD features (*Object Snap Tracking* in conjunction with *Polar Tracking*) are probably the most helpful features for construction of multiview drawing since the introduction of AutoCAD in 1982.

Using *Xline* and *Ray* for Construction Lines

Another strategy for constructing multiview drawings is to make use of the AutoCAD construction line commands *Xline* and *Ray*.

Xlines can be created to "box in" the views and ensure proper alignment (Fig. 21-8). *Ray* is suited for creating the 45-degree miter line for projection between the top and side views. The advantage to using this method is that horizontal and vertical *Xlines* can be created quickly.

These lines can be *Trimmed* to become part of the finished view. Alternately, other lines could be drawn "on top of" the construction lines to create the final lines of the views. In this case, <u>construction lines should be drawn on a separate layer</u> so that the layer can be frozen before plotting. If you intend to *Trim* the construction lines so that they become part of the final geometry, draw them originally on the view layers.

TIP

FIGURE 21-8

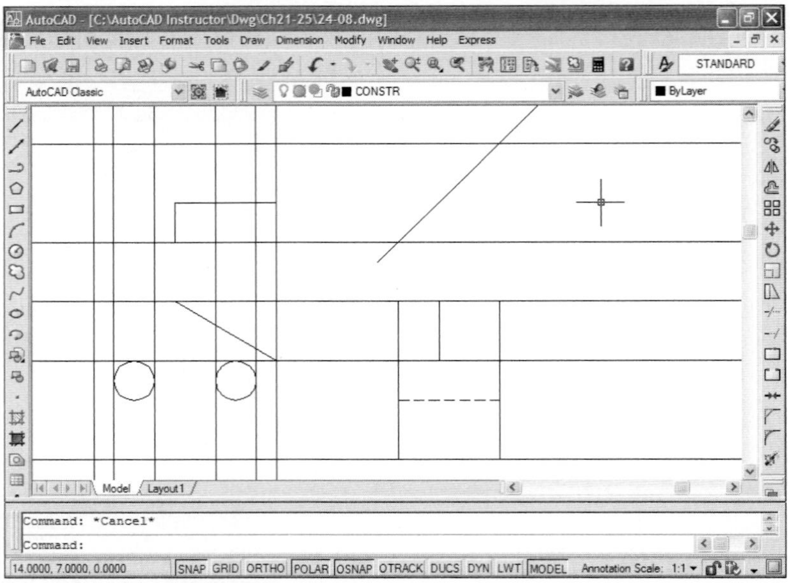

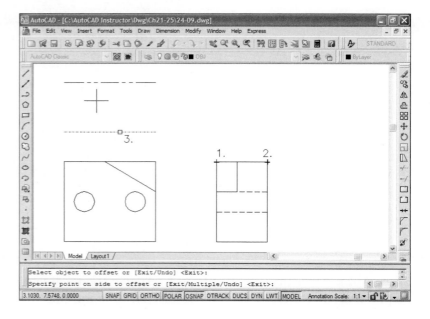

Using *Offset* for Construction of Views

An alternative to using the traditional miter line method for construction of a view by projection, the *Offset* command can be used to transfer distances from one view and to construct another. The *Distance* option of *Offset* provides this alternative.

For example, assume that the side view was completed and you need to construct a top view (Fig. 21-9). First, create a horizontal line as the inner edge of the top view (shown highlighted) by *Offset* or other method. To create the outer edge of the top view (shown in phantom linetype), use *Offset* and PICK points (1) and (2) to specify the *distance*. Select the existing line (3) as the "*Object to Offset:*", then PICK the "*side to offset*" at the current cursor position.

FIGURE 21-9

Realignment of Views Using *Polar Snap* and *Polar Tracking*

Another application of *Polar Snap* and *Polar Tracking* is the use of *Move* to change the location of an entire view while retaining its orthogonal alignment.

For example, assume that the views of a multiview (Fig. 21-10) are complete and ready for dimensioning; however, there is not enough room between the front and side views. You can invoke the *Move* command, select the entire view with a window or other option, and "slide" the entire view outward. *Polar Tracking* ensures proper orthogonal alignment. *Polar Snap* forces the movement to a regular increment so the coordinate points of the geometry retain a relationship to the original points (for example, moving exactly 1 unit).

An alternative to moving a view using mouse input is using Dynamic Input. Turn *DYN* on, then key in the desired distance value to move in the distance edit box, and enter an angular value in the angle edit box. Another alternative is moving the view using Direct Distance Entry. With this method, indicate the direction to move with the rubberband line (assuming *POLAR* is on), then key in the desired distance at the Command prompt.

FIGURE 21-10

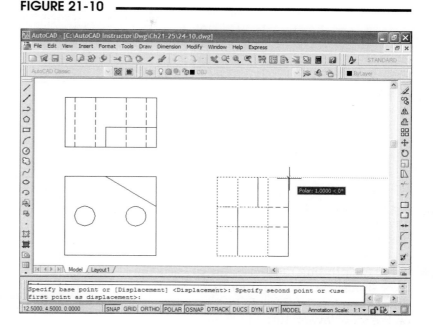

USING CONSTRUCTION LAYERS

The use of layers for isolating construction lines can make drawing and editing faster and easier. The creation of multiview drawings can involve construction lines, reference points, or notes that are not intended for the final plot. Rather than *Erasing* these construction lines, points, or notes before plotting, they can be created on a separate layer and turned *Off* or made *Frozen* before running the final plot. If design changes are required, as they often are, the construction layers can be turned *On*, rather than having to recreate the construction.

There are two strategies for creating construction objects on separate layers:

1. Use Layer 0 for construction lines, reference points, and notes. This method can be used for fast, simple drawings.

2. Create a new layer for construction lines, reference points, and notes. Use this method for more complex drawings or drawings involving use of *Blocks* on Layer 0.

For example, consider the drawing during construction in Figure 21-11. A separate layer has been created for the construction lines, notes, and reference points.

FIGURE 21-11

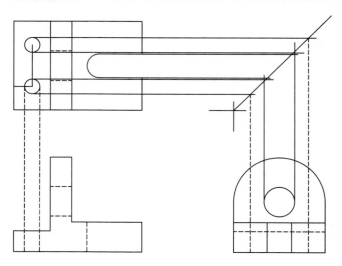

In Figure 21-12, the drawing is ready for making a print. Notice that the construction layer is *Frozen*.

FIGURE 21-12

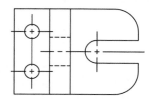

 If you are printing the *Limits*, the construction layer should be *Frozen*, rather than turned *Off*, since *Zoom All* and *Zoom Extents* are affected by geometry on layers turned *Off* and not *Frozen* (unless the objects are *Xlines* or *Rays*).

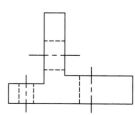
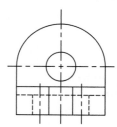

USING LINETYPES

Different types of lines are used to represent different features of a multiview drawing. Linetypes in AutoCAD are accessed by the *Linetype* command or through the *Layer Properties Manager*. *Linetypes* can be changed retroactively by the *Properties* palette. *Linetypes* can be assigned to individual objects specifically or to layers (*ByLayer*). See Chapter 11 for a full discussion on this topic.

AutoCAD complies with the ANSI and ISO standards for linetypes. The principal AutoCAD linetypes used in multiview drawings and the associated names are shown in Figure 21-13.

Many other linetypes are provided in AutoCAD. Refer to Chapter 11 for the full list and illustration of the linetypes.

FIGURE 21-13

CONTINUOUS	—————————————	DARK, WIDE
HIDDEN	– – – – – – – – – –	MEDIUM
CENTER	—— — —— — ——	MEDIUM
PHANTOM	—— — — ——	VARIES
DASHED	— —— — ——	VARIES

Other standard lines are created by AutoCAD automatically. For example, dimension lines can be automatically drawn when using dimensioning commands (Chapter 25), and section lines can be automatically drawn when using the *Hatch* command (Chapter 23).

Objects in AutoCAD can have lineweight. This is accomplished by using the *Lineweight* command or by assigning *Lineweight* in the *Layer Properties Manager* (see Chapter 11). Additionally, lineweights can be assigned by using plot styles or assigning plot device lineweights or pen thickness.

Drawing Hidden and Center Lines

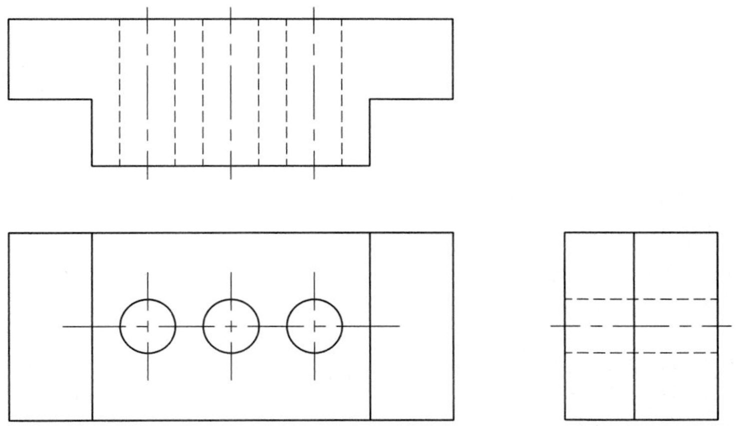

Figure 21-14 illustrates a typical application of AutoCAD *Hidden* and *Center* linetypes. Notice that the horizontal center line in the front view does not automatically locate the short dashes correctly, and the hidden lines in the right side view incorrectly intersect the center vertical line.

FIGURE 21-14 ━━━━━━━━━━

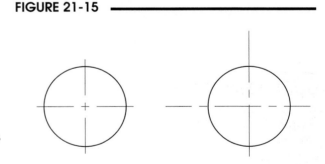

Although AutoCAD supplies ANSI standard linetypes, the application of those linetypes does not always follow ANSI standards. For example, you do <u>not</u> have control over the placement of the individual dashes of center lines and hidden lines. (You have control of only the endpoints of the lines and the *Ltscale*.) Therefore, the short dashes of center lines may not cross exactly at the circle centers, or the dashes of hidden lines may not always intersect as desired.

You do, however, have control of the endpoints of the lines. Draw lines with the *Center* linetype such that the endpoints are symmetric about the circle or group of circles. This action assures that the short dash occurs at the center of the circle (if an odd number of dashes are generated). Figure 21-15 illustrates correct and incorrect technique.

You can also control the relative size of non-continuous linetypes with the *Ltscale* variable. <u>In some cases,</u> the variable can be adjusted to achieve the desired results.

FIGURE 21-15 ━━━━━━━━━━

CORRECT SYMMETRY   INCORRECT SYMMETRY

For example, Figure 21-16 demonstrates the use of *Ltscale* to adjust the center line dashes to the correct spacing. Remember that *Ltscale* adjusts linetypes <u>globally</u> (all linetypes across the drawing).

When the *Ltscale* has been optimally adjusted for the drawing <u>globally</u>, use the *Properties* palette to adjust the linetype scale of <u>individual objects</u>. In this way, the drawing lines can originally be created to the global linetype scale without regard to the *Celtscale*. The finished drawing linetype scale can be adjusted with *Ltscale* globally; then <u>only those objects</u> that need further adjusting can be fine-tuned retroactively with *Properties*.

FIGURE 21-16 ━━━━━━━━━━

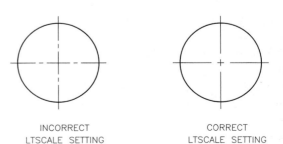

INCORRECT   CORRECT
LTSCALE SETTING   LTSCALE SETTING

TIP

TIP   The *Dimcenter* command (a dimensioning command) can be used to draw center lines automatically with correct symmetry and spacing (see Chapter 25, Dimensioning).

ANSI Standards require that multiview drawings are created with object lines having a dark lineweight, and hidden, center, dimension, and other reference lines created in a medium lineweight (see Figure 21-13).

You can assign *Lineweight* to objects using the *Lineweight* command or assign *Lineweight* to layers in the *Layer Properties Manager*. As described previously in Chapter 11, *Lineweight* can be assigned to individual objects or to layers (*ByLayer*). Use the *Lineweight Settings* dialog box to assign *Lineweight* property to objects. Use the *Layer Properties Manager* to assign *Lineweight* to layers. Generally, the *ByLayer Lineweight* assignment is preferred, similar to the *ByLayer* method of assigning *Color* and *Linetype*.

Additionally, lineweights can be assigned by using plot styles or assigning plot device lineweights or pen thickness.

Managing Linetypes, Lineweights, and Colors

There are two strategies for assigning linetypes, lineweights, and colors: *ByLayer* and object-specific assignment. In either case, thoughtful layer utilization for linetypes will make your drawings more flexible and efficient.

ByLayer Linetypes, Lineweights, and Colors

The *ByLayer* linetype, lineweight, and color settings are recommended when you want the most control over linetype visibility and plotting. This is accomplished by creating layers with the *Layer Properties Manager* and assigning *Linetype, Lineweight,* and *Color* for each layer. After you assign *Linetypes* to specific layers, you simply set the layer (with the desired linetype) as the *Current* layer and draw on that layer in order to draw objects in a specific linetype.

A template drawing for creating typical multiview drawings could be set up with the layer and linetype assignments similar to that shown in Figure 21-17. Each layer has its own *Color, Linetype,* and *Lineweight* setting, and the layer name indicates the type of lines used (hidden, center, object, etc.). This strategy is useful for simple, generic applications.

FIGURE 21-17

TIP   Using the *ByLayer* strategy (*ByLayer Linetype, Lineweight,* and *Color* assignment) gives you flexibility. You can control the linetype visibility by controlling the layer visibility (show only object layers or hidden line layers). You can also retroactively change the *Linetype, Lineweight,* and *Color* of an existing object by changing the object's *Layer* property with the *Properties* palette or *Matchprop*. Objects changed to a different layer assume the *Color, Linetype,* and *Lineweight* of the new layer.

TIP   Alternately, you can change object-specific (*ByBlock*) property settings for selected objects to *ByLayer* settings using the *Setbylayer* command. *Setbylayer* prompts you to select objects. Properties for the selected objects are automatically changed to the properties of the layer that the objects are on. Properties that can be changed are *Color, Linetype, Lineweight, Material,* and *Plot Style*. See Chapter 11 for more information on *Setbylayer*.

Another strategy for multiview drawings involving several parts, such as an assembly, is to create layers for each linetype <u>specific to each part</u>, as shown in Figure 21-18. With this strategy, each part has the complete set of linetypes, but only <u>one color per part</u> in order to distinguish the part from others in the display.

Related layer groups could be organized and managed in the *Layer Properties Manager* by creating *Group* filters and *Property* filters (Fig. 21-18). For example, *Group* filters could be created for the "Base," "Fasten," and "Mount" groups. *Property* filters could be created based on *Linetypes* for "Hidden," "Center," and "Object" lines.

FIGURE 21-18

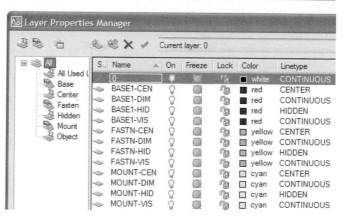

Object *Linetypes*, *Lineweights*, and *Colors*

Although this method can be complex, object-specific *Linetype*, *Lineweight*, and *Color* assignment can also be managed by skillful utilization of layers. One method is to create one layer for each part or each group of related geometry (Fig. 21-19). The *Color, Lineweight,* and *Linetype* settings should be left to the defaults. Then object-specific linetype settings can be assigned (for hidden, center, visible, etc.) using the *Linetype*, *Lineweight*, and *Color* commands.

FIGURE 21-19

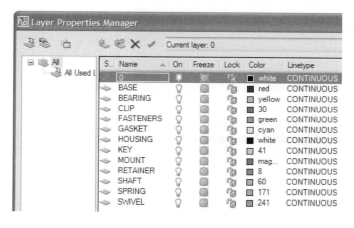

For assemblies, all lines related to one part would be drawn on one layer. Remember that you can draw everything with one linetype and color setting, then use *Properties* to retroactively set the desired color and linetype for each set of objects. Visibility of parts can be controlled by *Freezing* or *Thawing* part layers. You cannot isolate and control visibility of linetypes or colors by this method.

CREATING FILLETS, ROUNDS, AND RUNOUTS

Many mechanical metal or plastic parts manufactured from a molding process have slightly rounded corners. The otherwise sharp corners are rounded because of limitations in the molding process or for safety. A convex corner is called a <u>round</u> and a concave corner is called a <u>fillet</u>. These fillets and rounds are created easily in AutoCAD by using the *Fillet* command.

The example in Figure 21-20 shows a multiview drawing of a part with sharp corners before the fillets and rounds are drawn.

FIGURE 21-20

The corners are rounded using the *Fillet* command (Fig. 21-21). First, use *Fillet* to specify the *Radius*. Once the *Radius* is specified, just select the desired lines to *Fillet* near the end to round.

If the *Fillet* is in the middle portion of a *Line* instead of the end, *Extend* can be used to reconnect the part of the *Line* automatically trimmed by *Fillet*, or *Fillet* can be used in the *Notrim* mode.

FIGURE 21-21

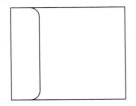

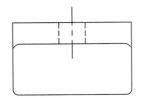

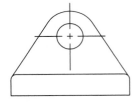

A <u>runout is a visual representation</u> of a complex fillet or round. For example, when two filleted edges intersect at less than a 90 degree angle, a runout should be drawn as shown in the top view of the multiview drawing (Fig. 21-22).

The finish marks (V-shaped symbols) indicate machined surfaces. Finished surfaces have sharp corners.

FIGURE 21-22

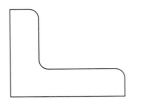

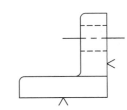

A close-up of the runouts is shown in Figure 21-23. No AutoCAD command is provided for this specific function. The *3point* option of the *Arc* command can be used to create the runouts, although other options can be used. Alternately, the *Circle TTR* option can be used with *Trim* to achieve the desired effect. As a general rule, use the same radius or slightly larger than that given for the fillets and rounds, but draw it less than 90 degrees.

FIGURE 21-23

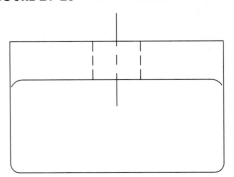

GUIDELINES FOR CREATING A TYPICAL THREE-VIEW DRAWING

Following are some guidelines for creating the three-view drawing in Figure 21-24. This object is used only as an example. The steps or particular construction procedure may vary, depending on the specific object drawn. Dimensions are shown in the figure so you can create the multiview drawing as an exercise.

FIGURE 21-24

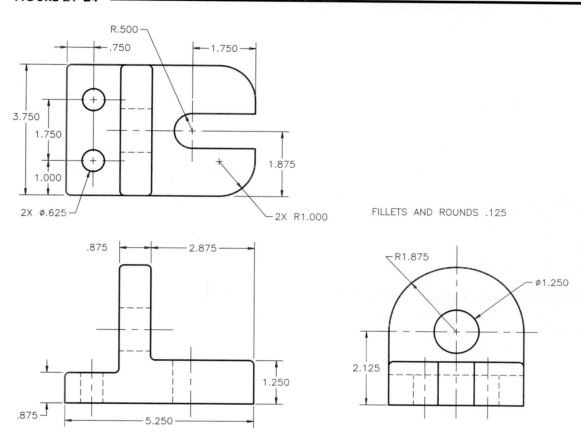

1.   Drawing Setup

Units are set to *Decimal* with 3 places of *Precision*. *Limits* of 22 x 17 are set to allow enough drawing space for both views. The finished drawing can be plotted on a B size sheet at a scale of 1"=1" or on an A size sheet at a scale of 1/2"=1". *Snap* is set to an increment of .125. *Grid* is set to an increment of .5. Although it is recommend that you begin drawing the views using *Grid Snap*, a *Polar Snap* increment of .125 is set, and *Polar Tracking* angles are set. Turn *POLAR* on. Set the desired running *OSNAPs,* such as *Endpoint, Midpoint, Intersection, Quadrant,* and *Center.* Turn on *OTRACK. Ltscale* is not changed from the default of 1. Layers are created (OBJ, HID, CEN, DIM, BORDER, and CONSTR) with appropriate *Linetypes, Lineweights,* and *Colors* assigned.

2. An outline of each view is "blocked in" by drawing the appropriate *Lines* and *Circles* on the OBJ layer similar to that shown in Figure 21-25. Ensure that *SNAP* is *ON*. *ORTHO* or *Polar Tracking* should be turned *ON* when appropriate. Use the cursor to ensure that the views align horizontally and vertically. Note that the top edge of the front view was determined by projecting from the *Circle* in the right side view.

FIGURE 21-25

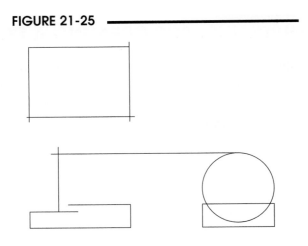

Another method for construction of a multi-view drawing is shown in Figure 21-26. This method uses the *Xline* and *Ray* commands to create construction and projection lines. Here all the views are blocked in and some of the object lines have been formed. The construction lines should be kept on a separate layer, except in the case where *Xlines* and *Rays* can be trimmed and converted to the object lines. (The following illustrations do not display this method because of the difficulty in seeing which are object and construction lines.)

FIGURE 21-26

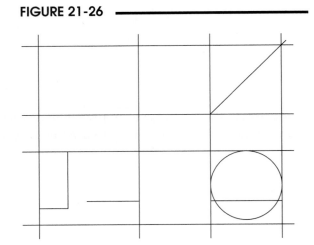

3. This drawing requires some projection between the top and side views (Fig. 21-27). The CONSTR layer is set as *Current*. Two *Lines* are drawn from the inside edges of the two views (using *OSNAP* and *ORTHO* for alignment). A 45-degree miter line is constructed for the projection lines to "make the turn." A *Ray* is suited for this purpose.

FIGURE 21-27

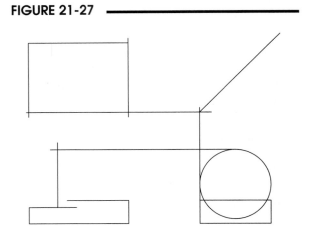

FIGURE 21-28

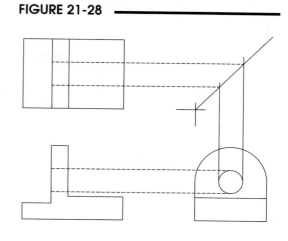

4. Details are added to the front and top views (Fig. 21-28). The projection line from the side view to the front view (previous figure) is *Trimmed*. A *Circle* representing the hole is drawn in the side view and projected up and over to the top view and to the front view. The object lines are drawn on layer OBJ and some projection lines are drawn on layer CONSTR. The horizontal projection lines from the 45-degree miter line are drawn on layer HID awaiting *Trimming*. Alternately, those two projection lines could be drawn on layer CONSTR and changed to the appropriate layer with *Properties* after *Trimming*. Keep in mind that *Object Snap Tracking* can be used here to ensure proper alignment with object features and to prevent having to actually draw some construction lines.

FIGURE 21-29

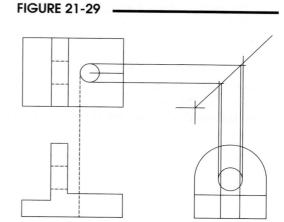

5. The hidden lines used for projection to the top view and front view (previous figure) are *Trimmed* (Fig. 21-29). The slot is created in the top view with a *Circle* and projected to the side and front views. It is usually faster and easier to draw round object features in their circular view first, then project to the other views. Make sure you use the correct layers (OBJ, CONSTR, HID) for the appropriate features. If you do not, *Properties* or *Matchprop* can be used retroactively.

FIGURE 21-30

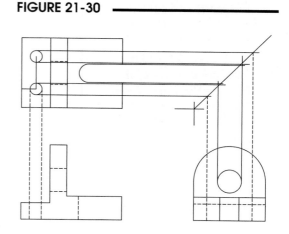

6. The lines shown in the previous figure as projection lines or construction lines for the slot are *Trimmed* or *Erased* (Fig. 21-30). The holes in the top view are drawn on layer OBJ and projected to the other views. The projection lines and hidden lines are drawn on their respective layers.

7. *Trim* the appropriate hidden lines (Fig. 21-31). *Freeze* layer CONSTR. On layer OBJ, use *Fillet* to create the rounded corners in the top view. Draw the correct center lines for the holes on layer CEN. The value for *Ltscale* should be adjusted to achieve the optimum center line spacing. The *Properties* palette can be used to adjust individual object line-type scale.

FIGURE 21-31 ──────────────

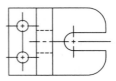

8. Fillets and rounds are added using the *Fillet* command (Fig. 21-32). The runouts are created by drawing a *3point Arc* and *Trimming* or *Extending* the *Line* ends as necessary. Use *Zoom* for this detail work.

FIGURE 21-32 ──────────────

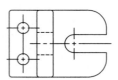

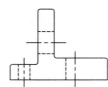
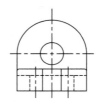

9. Activate a *Layout* tab, configure a print or plot device using *Pagesetup*, make one *Viewport*, and set the *Viewport scale* to a standard scale (Fig. 21-33). Add a border and a title block using *Pline*. Include the part name, company, draftsperson, scale, date, and drawing file name in the title block. The drawing is ready for dimensioning and manufacturing notes.

FIGURE 21-33 ──────────────

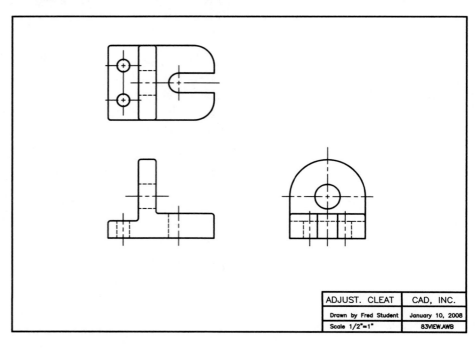

ADJUST. CLEAT	CAD, INC.
Drawn by Fred Student	January 10, 2008
Scale 1/2"=1"	83VIEW.AWB

CHAPTER EXERCISES

1. *Open* the **PIVOTARM CH16** drawing.

   A. Create the right side view. Use *OSNAP* and *ORTHO* or *Polar Tracking* to create *Lines* or *Rays* to the miter line and down to the right side view as shown in Figure 21-34. *Offset* may be used effectively for this purpose instead. Use *Extend*, *Offset*, or *Ray* to create the projection lines from the front view to the right side view. Use *Object Snap Tracking* when appropriate.

FIGURE 21-34

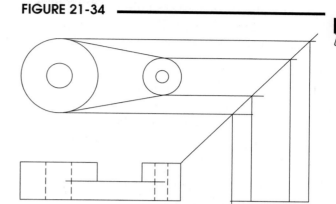

   B. *Trim* or *Erase* the unwanted projection lines, as shown in Figure 21-35. Draw a *Line* or *Ray* from the *Endpoint* of the diagonal *Line* in the top view down to the front to supply the boundary edge for *Trimming* the horizontal *Line* in the front view as shown.

FIGURE 21-35

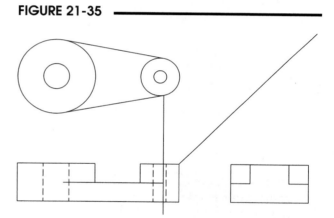

   C. Next, create the hidden lines for the holes by the same fashion as before (Fig. 21-36). Use previously created *Layers* to achieve the desired *Linetypes*. Complete the side view by adding the horizontal hidden *Line* in the center of the view.

FIGURE 21-36

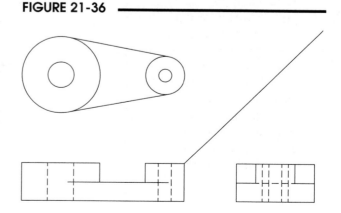

D. Another hole for a set screw must be added to the small end of the Pivot Arm. Construct a *Circle* of **4**mm diameter with its center located **8**mm from the top edge in the side view as shown in Figure 21-37. Project the set screw hole to the other views.

FIGURE 21-37

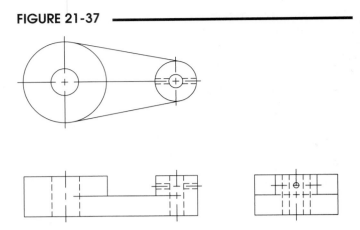

E. Make new layers **CONSTR**, **OBJ**, and **TITLE** and change objects to the appropriate layers with *Properties*. *Freeze* layer **CONSTR**. Add centerlines on the **CEN** layer as shown in Figure 21-37. Change the *Ltscale* to **18**. Activate a *Layout* tab, configure a print or plot device using *Pagesetup*, make one *Viewport*, and set the *Viewport scale* to a standard scale. To complete the PIVOTARM drawing, draw a *Pline* border (*width* **.02** x scale factor) in the layout and *Insert* the **TBLOCK** drawing that you created in Chapter 18 Exercises. *SaveAs* **PIVOTARM CH21**.

For Exercises 2 through 5, construct and plot the multiview drawings as instructed. Use an appropriate *template* drawing for each exercise unless instructed otherwise. Use conventional practices for *layers* and *linetypes*. Draw a *Pline* border with the correct *width* and *Insert* your **TBLOCK** drawing.

2. Make a two-view multiview drawing of the Clip (Fig. 21-38). *Plot* the drawing full size (**1=1**). Use the **ASHEET** template drawing to achieve the desired plot scale. *Save* the drawing as **CLIP**.

FIGURE 21-38

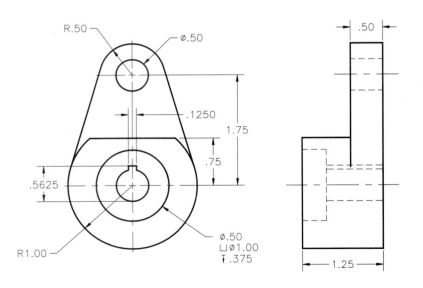

CLIP

3. Make a three-view multiview drawing of the Bar Guide and *Plot* it **1=1** (Fig. 21-39). Use the BARGUIDE drawing you set up in Chapter 12 Exercises. Note that a partial *Ellipse* will appear in one of the views.

FIGURE 21-39

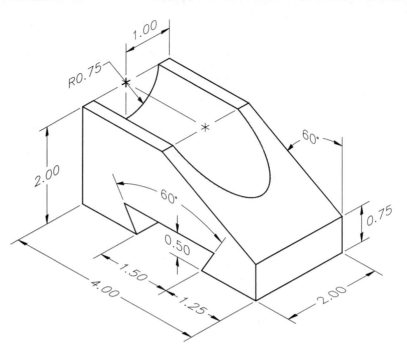

4. Construct a multi-view drawing of the Saddle (Fig. 21-40). Three views are needed. The channel along the bottom of the part intersects with the Saddle on top to create a slotted hole visible in the top view. *Plot* the drawing at **1=1**.

Save as SADDLE.

FIGURE 21-40

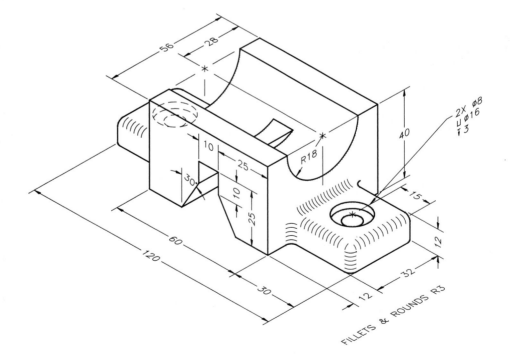

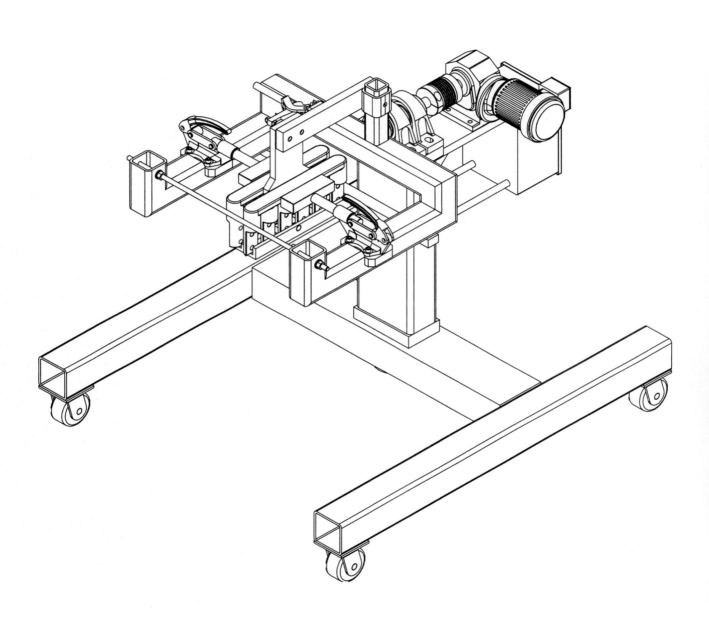

WELDING FIXTURE MODEL.DWG, Courtesy of Autodesk, Inc.

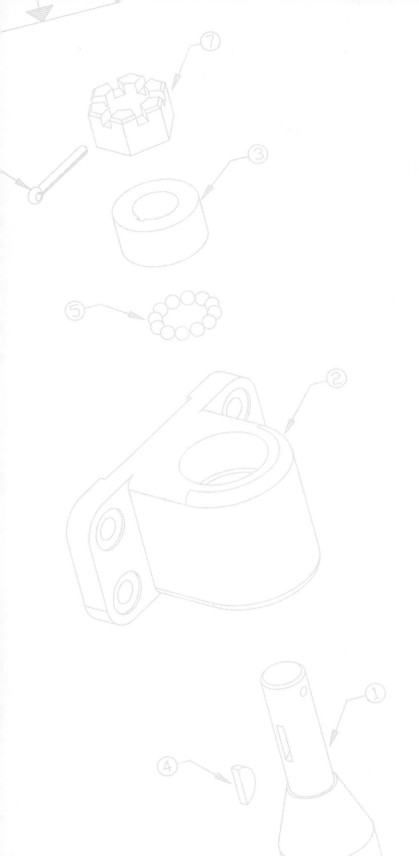

22

PICTORIAL DRAWINGS

CHAPTER OBJECTIVES

After completing this chapter you should:

1. be able to activate the *Isometric Style* of *Snap* for creating isometric drawings;

2. know how to draw on the three isometric planes by toggling *Isoplane* using Ctrl+E;

3. be able to create isometric ellipses with the *Isocircle* option of *Ellipse*;

4. be able to construct an isometric drawing in AutoCAD;

5. be able to create Oblique Cavalier and Cabinet drawings in AutoCAD.

CONCEPTS

Isometric drawings and oblique drawings are pictorial drawings. Pictorial drawings show three principal faces of the object in one view. A pictorial drawing is a drawing of a 3D object as if you were positioned to see (typically) some of the front, some of the top, and some of the side of the object. All three dimensions of the object (width, height, and depth) are visible in a pictorial drawing.

Multiview drawings differ from pictorial drawings because a multiview only shows two dimensions in each view, so two or more views are needed to see all three dimensions of the object. A pictorial drawing shows all dimensions in the one view. Pictorial drawings depict the object similar to the way you are accustomed to viewing objects in everyday life, that is, seeing all three dimensions. Figure 22-1 and Figure 22-2 show the same object in multiview and in pictorial representation, respectively. Notice that multiview drawings use hidden lines to indicate features that are normally obstructed from view, whereas <u>hidden lines are normally omitted</u> in isometric drawings (unless certain hidden features must be indicated for a particular function or purpose).

Types of Pictorial Drawings

Pictorial drawings are classified as follows:

1. Axonometric drawings
   a. Isometric drawings
   b. Dimetric drawings
   c. Trimetric drawings

2. Oblique drawings

Axonometric drawings are characterized by how the angle of the edges or axes (axon-) are measured (-metric) with respect to each other.

Isometric drawings are drawn so that each of the axes have equal angular measurement. ("Isometric" means equal measurement.) The isometric axes are always drawn at 120 degree increments (Fig. 22-3). All rectilinear lines on the object (representing horizontal and vertical edges—not inclined or oblique) are drawn on the isometric axes.

A 3D object seen "in isometric" is thought of as being oriented so that each of three perpendicular faces (such as the top, front, and side) are seen equally. In other words, the angles formed between the line of sight and each of the principal faces are equal.

FIGURE 22-1

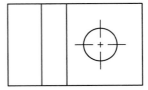

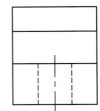

FIGURE 22-2

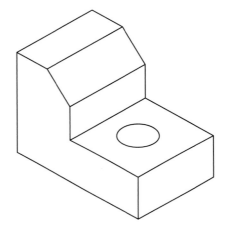

FIGURE 22-3

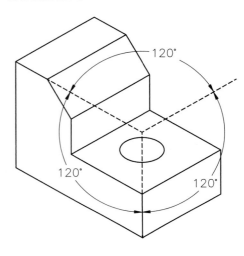

Dimetric drawings are constructed so that the angle between any two of the three axes is equal. There are many possibilities for dimetric axes. A common orientation for dimetric drawings is shown in Figure 22-4. For 3D objects seen from a dimetric viewpoint, the angles formed between the line of sight and each of two principal faces are equal.

FIGURE 22-4

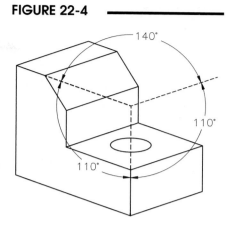

Trimetric drawings have three unequal angles between the axes. Numerous possibilities exist. A common orientation for trimetric drawings is shown in Figure 22-5.

FIGURE 22-5

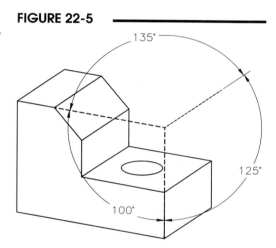

Oblique drawings are characterized by a vertical axis and horizontal axis for the two dimensions of the front face and a third (receding) axis of either 30, 45, or 60 degrees (Fig. 22-6). Oblique drawings depict the true size and shape of the front face, but add the depth to what would otherwise be a typical 2D view. This technique simplifies construction of drawings for objects that have contours in the profile view (front face) but relatively few features along the depth. Viewing a 3D object from an oblique viewpoint is not possible.

This chapter will explain the construction of isometric and oblique drawings in AutoCAD.

FIGURE 22-6

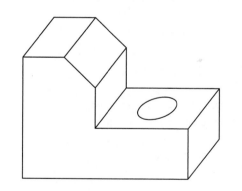

Pictorial Drawings Are 2D Drawings

Isometric, dimetric, trimetric, and oblique drawings are <u>2D drawings</u>, whether created with AutoCAD or otherwise. Pictorial drawing was invented before the existence of CAD and therefore was intended to simulate a 3D object on a 2D plane (the plane of the paper). If AutoCAD is used to create the pictorial, the geometry lies on a 2D plane—the XY plane. All coordinates defining objects have X and Y values with a Z value of 0. When the *Isometric* style of *Snap* is activated, an isometrically structured *SNAP* and *GRID* appear on the XY plane.

Figure 22-7 illustrates the 2D nature of an isometric drawing created in AutoCAD. Isometric lines are created on the XY plane. The *Isometric SNAP* and *GRID* are also on the 2D plane. (The *3Dorbit* command was used to give other than a *Plan* view of the drawing in this figure.)

Although pictorial drawings are based on the theory of projecting 3D objects onto 2D planes, it is physically possible to achieve an axonometric (isometric, dimetric, or trimetric) viewpoint of a 3D object using a 3D CAD system. In AutoCAD, the *Vpoint* and *3Dorbit* commands can be used to specify the observer's position in 3D space with respect to a 3D model. Chapter 28 discusses the specific commands and values needed to attain axonometric viewpoints of a 3D model.

FIGURE 22-7

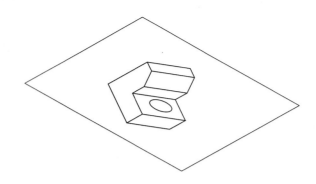

ISOMETRIC DRAWING IN AutoCAD

AutoCAD provides the capability to construct isometric drawings. An isometric *SNAP* and *GRID* are available, as well as a utility for creation of isometrically correct ellipses. Isometric lines are created with the *Line* command. There are no special options of *Line* for isometric drawing, but isometric *SNAP* and *GRID* can be used to force *Lines* to an isometric orientation. Begin creating an isometric drawing in AutoCAD by activating the *Isometric Style* option of the *Snap* command. This action can be done using any of the options listed in the following Command table.

Snap

Pull-down Menu	Command (Type)	Alias (Type)	Short-cut	Screen (side) Menu	Tablet Menu
Tools Drafting Settings... Snap and Grid	Snap	SN	F9 or Ctrl+B	TOOLS 2 Grid	W,10

Command: **snap**
Specify snap spacing or
[ON/OFF/Aspect/Rotate/Style/Type] <0.5000>: **s**
Enter snap grid style [Standard/Isometric] <S>: **i**
Specify vertical spacing <0.5000>: **Enter**
Command:

Alternately, toggling the indicated checkbox in the lower-left corner of the *Drafting Settings* dialog box activates the *Isometric snap* and *Grid snap* (Fig. 22-8).

NOTE: For AutoCAD 2007 or newer drawings, toggle *Adaptive grid* off (Fig. 22-8, right). Otherwise, the *GRID* will not display at the set increment when the *Isometric snap* style is current.

FIGURE 22-8

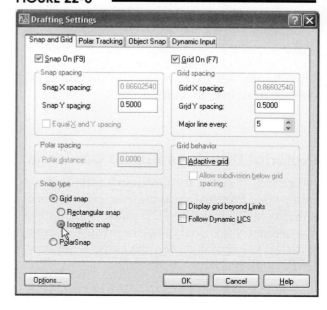

Figure 22-9 illustrates the effect of setting the *Isometric SNAP* and *GRID*. Notice the new orientation of the cursor.

FIGURE 22-9

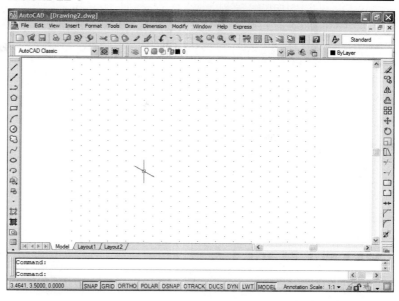

Using <u>Ctrl+E</u> (pressing the Ctrl key and the letter "E" simultaneously) toggles the cursor to one of three possible *Isoplanes* (AutoCAD's term for the three faces of the isometric pictorial). If *ORTHO* is *ON*, only <u>isometric</u> lines are drawn; that is, you can only draw *Lines* aligned with the isometric axes. *Lines* can be drawn on only two axes for each isoplane. Ctrl+E allows drawing on the two axes aligned with another face of the object. *ORTHO* is *OFF* in order to draw inclined or oblique lines (not on the isometric axis). The functions of *GRID* (F7) and *SNAP* (F9) remain unchanged.

With *SNAP ON*, toggle *Coords* (F6) several times and examine the read-out as you move the cursor. The <u>Cartesian coordinate format is of no particular assistance</u> while drawing in isometric because of the configuration of the *GRID*. The <u>relative polar</u> format, however, is <u>very helpful</u>. Use relative polar format for *Coords* while drawing in isometric (Fig. 22-10).

Alternately, use *Polar Tracking* instead of *ORTHO*. Set the *Polar Angle Settings* to 30 degrees. The advantage of using *Polar Tracking* is that the current line length is given on the polar tracking tip (see Fig. 22-10). The disadvantage is that it is possible to draw non-isoplane lines accidentally. Only one setting at a time can be used in AutoCAD—*Polar Tracking* or *ORTHO*. (The remainder of the figures illustrate the use of *ORTHO*.)

FIGURE 22-10

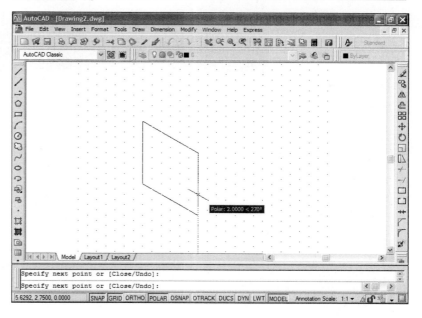

The effects of changing the *Isoplane* are shown in the following figures. Press Ctrl+E to change *Isoplane*.

With *ORTHO ON*, drawing a *Line* is limited to the two axes of the current *Isoplane*. Only one side of a cube, for example, can be drawn on the current *Isoplane*. Watch *Coords* (in a polar format) or turn on *DYN* to give the length of the current *Line* as you draw.

Toggling Ctrl+E switches the cursor and the effect of *ORTHO* to another *Isoplane*. One other side of a cube can be constructed on this *Isoplane* (Fig. 22-11).

FIGURE 22-11

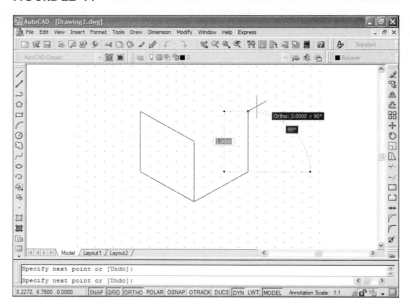

Direct Distance Entry can be of great help when drawing isometric lines. Use Ctrl+E and *ORTHO* to force the *Line* to the correct orientation, then enter the desired distance value at the Command line.

Isometric Ellipses

Isometric ellipses are easily drawn in AutoCAD by using the *Isocircle* option of the *Ellipse* command. This option appears <u>only</u> when the isometric *SNAP* is *ON*.

Ellipse

Pull-down Menu	Command (Type)	Alias (Type)	Short-cut	Screen (side) Menu	Tablet Menu
Draw *Ellipse*	*Ellipse*	*EL*	...	*DRAW 1* *Ellipse*	*M,9*

Although the *Isocircle* option does not appear in the pull-down or digitizing tablet menus, it can be invoked as an option of the *Ellipse* command. The *Isocircle* option of *Ellipse* appears only when the *Snap Type* is set to *Isometric*. You <u>must type "I"</u> to use the *Isocircle* option. The command syntax is as follows:

    Command: **ellipse**
    Specify axis endpoint of ellipse or [Arc/Center/Isocircle]: **i**
    Specify center of isocircle: **PICK** or **(coordinates)**
    Specify radius of isocircle or [Diameter]: **PICK** or **(coordinates)**
    Command:

After selecting the center point of the *Isocircle*, the isometrically correct ellipse appears on the screen on the current *Isoplane*. Use Ctrl+E to toggle the ellipse to the correct orientation. When defining the radius interactively, use *ORTHO* to force the rubberband line to an isometric axis (Fig. 22-12, next page).

Since isometric angles are equal, all isometric ellipses have the same proportion (major to minor axis). The only differences in isometric ellipses are the size and the orientation (*Isoplane*).

FIGURE 22-12

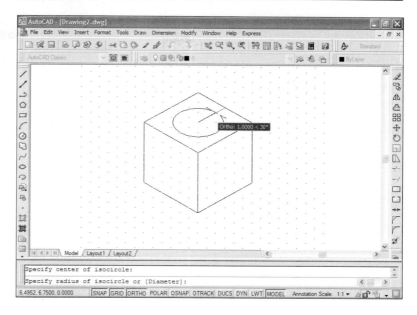

Figure 22-13 shows three ellipses correctly oriented on their respective faces. Use Ctrl+E to toggle the correct *Isoplane* orientation: *Isoplane Top*, *Isoplane Left*, or *Isoplane Right*.

FIGURE 22-13

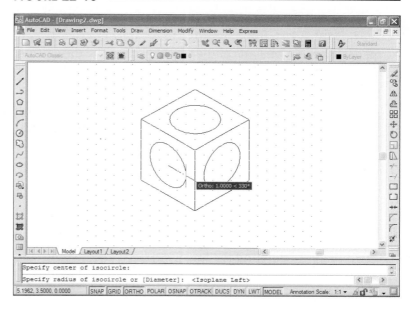

When defining the radius or diameter of an ellipse, it should always be measured in an isometric direction. In other words, an isometric ellipse is always measured on the two isometric axes (or center lines) parallel with the plane of the ellipse.

If you define the radius or diameter interactively, use *ORTHO ON*. If you enter a value, AutoCAD automatically applies the value to the correct isometric axes.

Creating an Isometric Drawing

In this exercise, the object in Figure 22-14 is to be drawn as an isometric.

The initial steps to create an isometric drawing begin with the typical setup (see Chapter 6, Drawing Setup):

1. Set the desired *Units*.

2. Set appropriate *Limits*.

3. Set the *Isometric Style* of *Snap* and specify an appropriate value for spacing. Toggle *Adaptive grid* off.

FIGURE 22-14

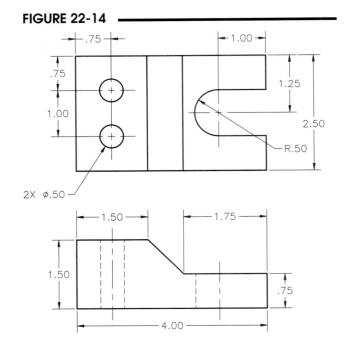

4. The next step involves creating an isometric framework of the desired object. In other words, draw an isometric box equal to the overall dimensions of the object. Using the dimensions given in Figure 22-14, create the encompassing isometric box with the *Line* command (Fig. 22-15).

   Use *ORTHO* to force isometric *Lines*. Watch the *Coords* display (in a relative polar format) to give the current lengths as you draw or use direct distance entry.

FIGURE 22-15

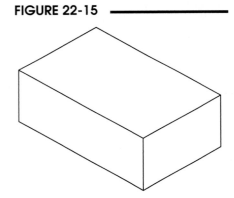

5. Add the lines defining the lower surface. Define the needed edge of the upper isometric surface as shown (Fig. 22-16).

FIGURE 22-16

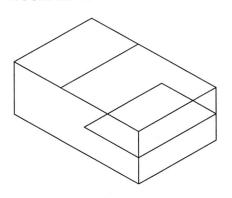

6. The <u>inclined</u> edges of the inclined surface can be drawn (with
    Line) only when *ORTHO* is *OFF*. <u>Inclined</u> lines in isometric
    cannot be drawn by transferring the lengths of the lines, but
    only by defining the <u>ends</u> of the inclined lines on <u>isometric</u>
    lines, then connecting the endpoints. Next, *Trim* or *Erase* the
    necessary *Lines* (Fig. 22-17).

FIGURE 22-17

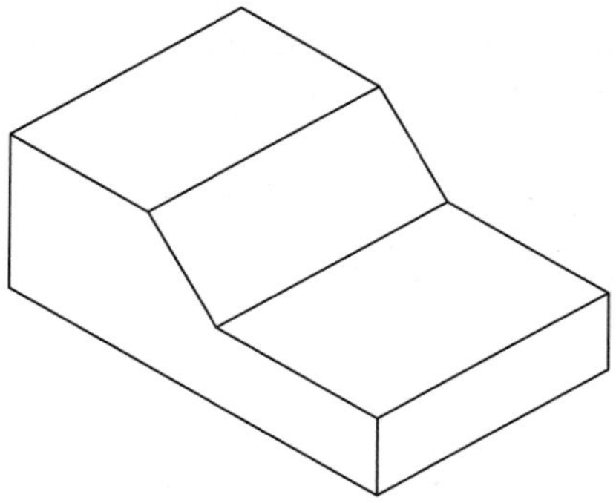

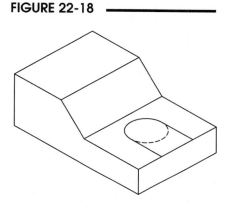

7. Draw the slot by constructing an *Ellipse* with the *Isocircle*
    option. Draw the two *Lines* connecting the circle to the right
    edge. *Trim* the unwanted part of the *Ellipse* (highlighted) using
    the *Lines* as cutting edges (Fig. 22-18).

FIGURE 22-18

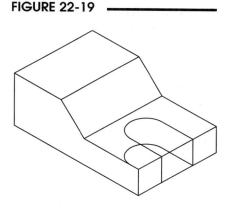

8. *Copy* the far *Line* and the *Ellipse* down to the bottom surface.
    Add two vertical *Lines* at the end of the slot (Fig. 22-19).

FIGURE 22-19

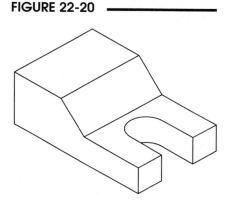

9. Use *Trim* to remove the part of the *Ellipse* that would normally
    be hidden from view. *Trim* the *Lines* along the right edge at the
    opening of the slot (Fig. 22-20).

FIGURE 22-20

10. Add the two holes on the top with *Ellipse*, *Isocircle* option (Fig. 22-21). Use *ORTHO ON* when defining the radius. *Copy* can also be used to create the second *Ellipse* from the first.

FIGURE 22-21

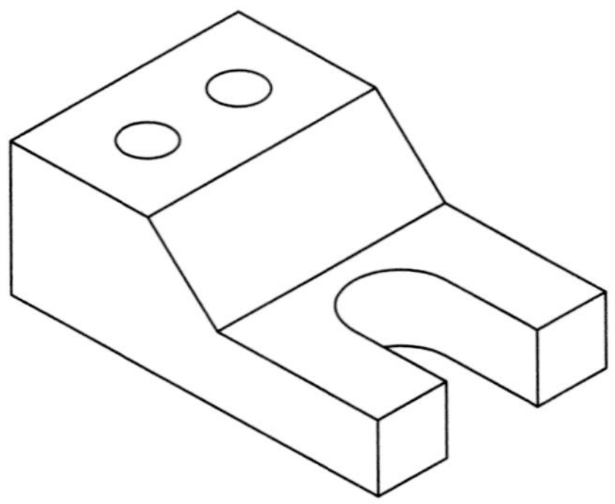

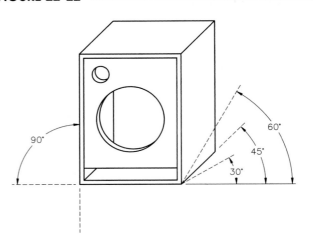

FIGURE 22-22

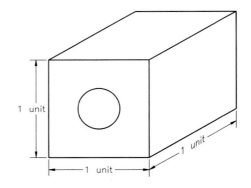

OBLIQUE DRAWING IN AutoCAD

Oblique drawings are characterized by having two axes at a 90 degree orientation. Typically, you should locate the <u>front face</u> of the object along these two axes. Since the object's characteristic shape is seen in the front view, an oblique drawing allows you to create all shapes parallel to the front face true size and shape as you would in a multi-view drawing. Circles on or parallel to the front face can be drawn as circles. The third axis, the receding axis, can be drawn at a choice of angles, 30, 45, or 60 degrees, depending on whether you want to show more of the top or the side of the object.

Figure 22-22 illustrates the axes orientation of an oblique drawing, including the choice of angles for the receding axis.

Another option allowed with oblique drawings is the measure-ment used for the receding axis. Using the full depth of the object along the receding axis is called <u>Cavalier</u> oblique drawing. This method depicts the object (a cube with a hole in this case) as having an elongated depth (Fig. 22-23).

FIGURE 22-23

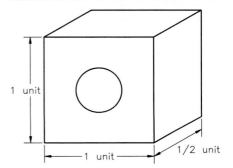

Using 1/2 of the true depth along the receding axis gives a more realistic pictorial representation of the object. This is called a <u>Cabinet</u> oblique (Fig. 22-24).

No functions or commands in AutoCAD are intended specifi-cally for oblique drawing. However, *Polar Snap* and *Polar Tracking* can simplify the process of drawing lines on the front face of the object and along the receding axis. The steps for cre-ating a typical oblique drawing are given next.

FIGURE 22-24

The object in Figure 22-25 is used for the example. From the dimensions given in the multiview, create a cabinet oblique with the receding axis at 45 degrees.

FIGURE 22-25

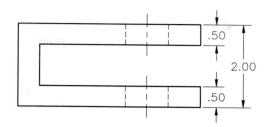

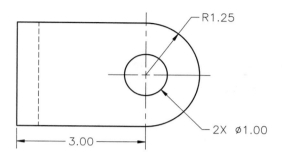

1. Create the characteristic shape of the front face of the object as shown in the front view (Fig. 22-26).

FIGURE 22-26

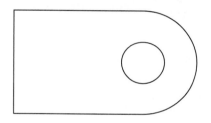

2. Use *Copy* with the multiple option to copy the front face back on the receding axis (Fig. 22-27). *Polar Snap* and *Polar Tracking* can be used to specify the "second point of displacement" as shown in Figure 22-27. Notice that a distance of .25 along the receding axis is used (1/2 of the actual depth) for this cabinet oblique.

FIGURE 22-27

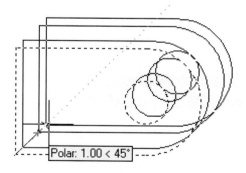

3. Draw the *Line* representing the edge on the upper-left of the object along the receding axis. Use *Endpoint OSNAP* to connect the *Lines*. Make a *Copy* of the *Line* or draw another *Line* .5 units to the right. Drop a vertical *Line* from the *Intersection* as shown (Fig. 22-28).

FIGURE 22-28

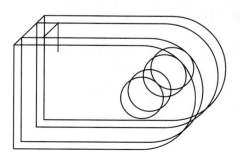

4. Use *Trim* and *Erase* to remove the unwanted parts of the *Lines* and *Circles* (those edges that are normally obscured) (Fig. 22-29).

FIGURE 22-29

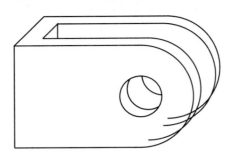

5. *Zoom* with a *window* to the lower-right corner of the drawing. Draw a *Line Tangent* to the edges of the arcs to define the limiting elements along the receding axis. *Trim* the unwanted segments of the arcs (Fig. 22-30).

FIGURE 22-30

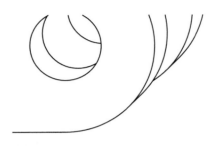

The resulting cabinet oblique drawing should appear like that in Figure 22-31.

FIGURE 22-31

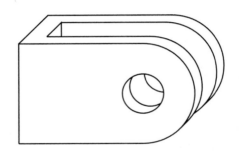

CHAPTER EXERCISES

Isometric Drawing

For Exercises 1, 2, and 3 create isometric drawings as instructed. To begin, use an appropriate template drawing and draw a *Pline* border and insert the **TBLOCK** drawing.

1. Create an isometric drawing of the cylinder shown in Figure 22-32. *Save* the drawing as **CYLINDER** and *Plot* so the drawing is *Scaled to Fit* on an A size sheet.

FIGURE 22-32

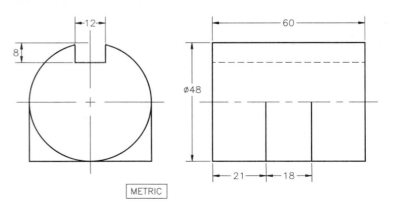

METRIC

2. Make an isometric drawing of the Corner
    Brace shown in Figure 22-33. *Save* the
    drawing as **CRNBRACE**. *Plot* at **1=1** scale
    on an A size sheet.

FIGURE 22-33

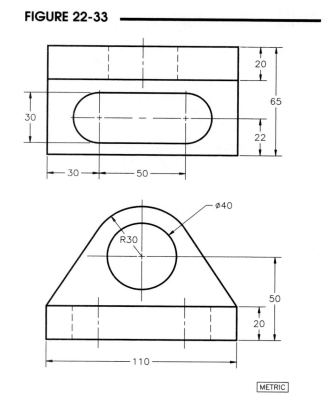

3. Draw the Support Bracket
    (Fig. 22-34) in isometric. The
    drawing can be *Plotted* at **1=1**
    scale on an A size sheet. *Save* the
    drawing and assign the name
    SBRACKET.

FIGURE 22-34

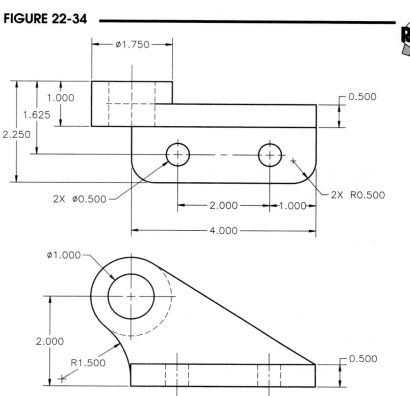

Oblique Drawing

For Exercises 4 and 5, create oblique drawings as instructed. To begin, use an appropriate template drawing and draw a *Pline* border and *Insert* the **TBLOCK** drawing.

4. Make an oblique cabinet projection of the Bearing shown in Figure 22-35. Construct all dimensions on the receding axis 1/2 of the actual length. Select the optimum angle for the receding axis to be able to view the 15 x 15 slot. *Plot* at **1=1** scale on an A size sheet. *Save* the drawing as **BEARING**.

FIGURE 22-35

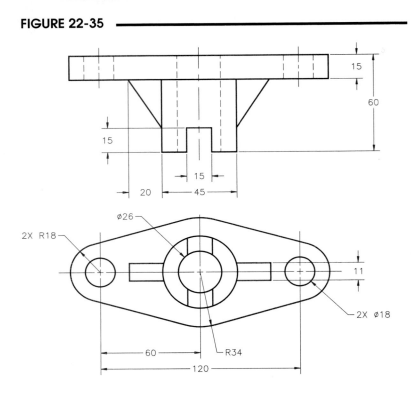

5. Construct a cavalier oblique drawing of the Pulley showing the circular view true size and shape. The illustration in Figure 22-36 gives only the side view. All vertical dimensions in the figure are diameters. *Save* the drawing as **PULLEY** and make a *Plot* on an A size sheet at **1=1**.

FIGURE 22-36

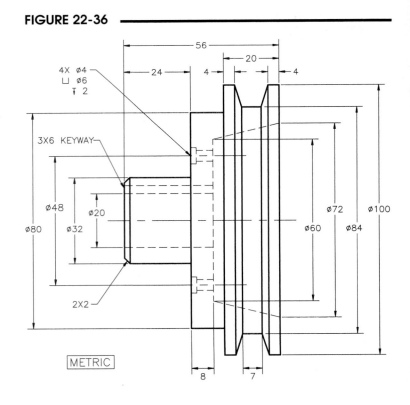

SECTION VIEWS AND HATCH PATTERNS

CHAPTER OBJECTIVES

After completing this chapter you should:

1. be able to use the *Hatch* command to apply associative hatch patterns;

2. know how to define a boundary for hatching using the *Pick Points* and *Select Objects* methods;

3. know how to create *Annotative* hatch patterns for use in multiple viewports;

4. be able to use *-Hatch* to create hatch lines and discard the boundary;

5. be able to drag and drop hatch patterns from a Tool Palette and be able to create your own Tool Palettes;

6. know how to use *Hatchedit* to modify parameters of existing hatch patterns in the drawing;

7. be able to edit hatched areas using *Trim*, Grips, selection options, and *Draworder*;

8. know how to draw cutting plane lines for section views.

CONCEPTS

A section view is a view of the interior of an object after it has been imaginarily cut open to reveal the object's inner details. A section view is typically one of two or more views of a multiview drawing describing the object. For example, a multiview drawing of a machine part may contain three views, one of which is a section view.

Hatch lines (also known as section lines) are drawn in the section view to indicate the solid material that has been cut through. Each combination of section lines is called a hatch <u>pattern</u>, and each pattern is used to represent a specific material. In full and half section views, hidden lines are omitted since the inside of the object is visible.

Hatch lines are used also for many other applications. For example, architectural elevation drawings often include wall sections and use hatch lines to indicate walls that have been "cut through" to reveal the structural components. In addition, floor plans often contain a solid hatch pattern to indicate wall sections, as if the walls were cut through to reveal the "floor."

For mechanical drawings, a cutting plane line is drawn in an adjacent view to the section view to indicate the plane that imaginarily cuts through the object. Arrows on each end of the cutting plane line indicate the line of sight for the section view. A thick dashed or phantom line should be used for a cutting plane line.

This chapter discusses the AutoCAD methods used to draw hatch lines for section views and related cutting plane lines. The *Hatch* command allows you to select an enclosed area and select the hatch pattern and the parameters for the appearance of the hatch pattern; then AutoCAD automatically draws the hatch (section) lines. Existing hatch lines in the drawing can be modified using *Hatchedit*.

DEFINING HATCH PATTERNS AND HATCH BOUNDARIES

A hatch pattern is composed of many lines that have a particular linetype, spacing, and angle. Many standard hatch patterns are provided by AutoCAD for your selection. Rather than having to draw each section line individually, you are required only to specify the area to be hatched and AutoCAD fills the designated area with the selected hatch pattern. An AutoCAD hatch pattern is inserted as <u>one object</u>. For example, you can *Erase* the inserted hatch pattern by selecting only one line in the pattern, and the entire pattern in the area is *Erased*.

FIGURE 23-1

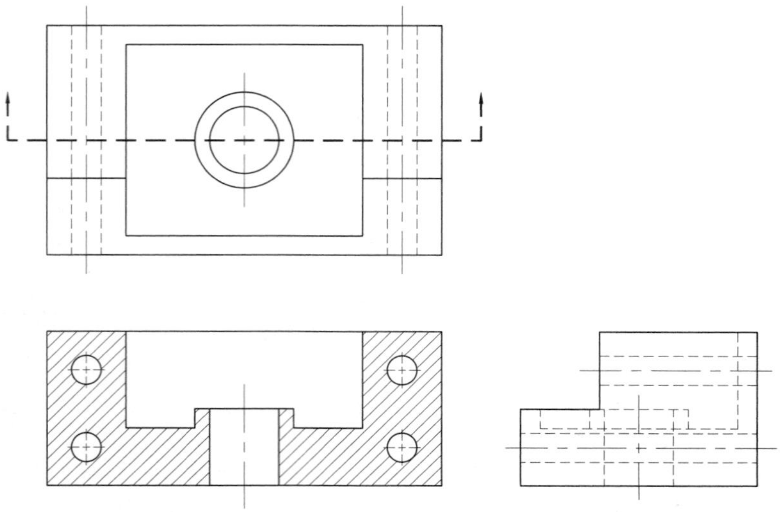

In a typical section view (Fig. 23-1), the hatch pattern completely fills the area representing the material that has been cut through. With the *Hatch* command you can define the boundary of an area to be hatched simply by pointing inside of an enclosed area.

The *Hatch* command is used to create hatch lines in the area that you specify. *Hatch* operates in a dialog box mode. You can select the desired hatch pattern, then pick inside the area you want to fill, and *Hatch* automatically finds the boundary and fills it with the pattern. You can also drag and drop hatch patterns into a closed area using the Tool Palettes.

Hatch patterns created with *Hatch* are <u>associative</u>. Associative hatch patterns are associated to the boundary geometry such that when the shape of the boundary changes (by *Stretch, Scale, Rotate, Move, Properties, Grips,* etc.), the hatch pattern automatically reforms itself to conform to the new shape (Fig. 23-2). For example, if a design change required a larger diameter for a hole, *Properties* could be used to change the diameter of the hole, and the surrounding section lines would automatically adapt to the new diameter.

FIGURE 23-2

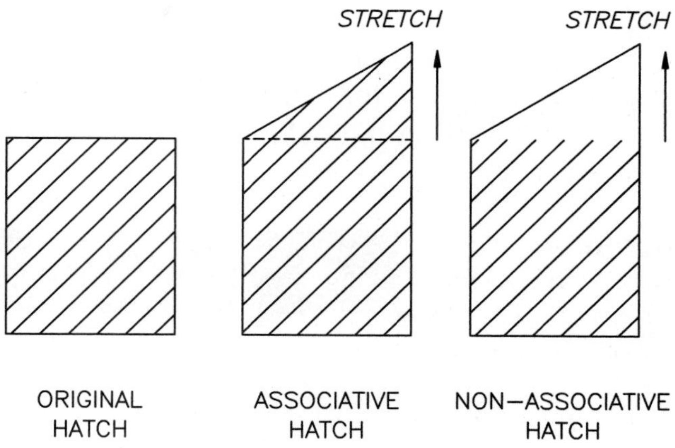

ORIGINAL HATCH   ASSOCIATIVE HATCH   NON–ASSOCIATIVE HATCH

Once the hatch patterns have been drawn, any feature of the existing hatch pattern (created with *Hatch*) can be changed retroactively using *Hatchedit*. The *Hatchedit* dialog box gives access to the same options that were used to create the hatch (in the *Hatch and Gradient* dialog box). Changing the scale, angle, or pattern of any existing section view in the drawing is a simple process.

Hatch patterns can also be assigned an *Annotative* property. Creating annotative hatch patterns is useful if you want to display the same hatched areas at different scales in multiple viewports.

2008

Steps for Creating a Section View Using the *Hatch* Command

1. Create the view that contains the area to be hatched using typical draw commands such as *Line, Arc, Circle,* or *Pline*. If you intend to have text or dimensions inside the area to be hatched, add them before hatching.

2. Invoke the *Hatch* command. The *Hatch and Gradient* dialog box appears (see Fig. 23-3).

3. Specify the *Type* to use. Select the desired pattern from the *Pattern* drop-down list or select the *Swatch* tile to allow you to select from the *Hatch Pattern Palette* image tiles.

4. Specify the *Scale* and *Angle* in the dialog box.

5. Define the area to be hatched by PICKing an internal point (*Pick points* button) or by individually selecting the objects (*Select objects* button).

6. If needed, specify any other parameters, such as *Hatch origin, Gap tolerance, Island detection style,* etc.

7. *Preview* the hatch to make sure everything is as expected. Adjust hatching parameters as necessary and *Preview* again.

8. Apply the hatch by selecting *OK*. The hatch pattern is automatically drawn and becomes an associated object in the drawing.

9. If other areas are to be hatched, additional internal points or objects can be selected to define the new area for hatching. The parameters used previously appear again in the *Hatch and Gradient* dialog box by default. You can also *Inherit Properties* from a previously applied hatch.

10. For mechanical drawings, draw a cutting plane line in a view adjacent to the section view. A *Lineweight* is assigned or a *Pline* with a *Dashed* or *Phantom* linetype is used. Arrows at the ends of the cutting plane line indicate the line of sight for the section view.

11. If any aspect of the hatch lines needs to be edited at a later time, *Hatchedit* can be used to change those properties. If the hatch boundary is changed by *Stretch, Rotate, Scale, Move, Properties*, etc., the hatched area will conform to the new boundary.

Hatch

Pull-down Menu	Command (Type)	Alias (Type)	Short-cut	Screen (side) Menu	Tablet Menu
Draw *Hatch...*	*Hatch* or *-Hatch*	*BH* or *H*	...	DRAW 2 *Hatch*	*P,9*

Hatch allows you to create hatch lines for a section view (or for other purposes) by simply PICKing inside a closed boundary. A closed boundary refers to an area completely enclosed by objects. *Hatch* locates the closed boundary automatically by creating a <u>temporary *Pline*</u> that follows the outline of the hatch area, fills the area with hatch lines, and then deletes the boundary (default option) after hatching is completed. *Hatch* ignores all objects or parts of objects that are not part of the boundary.

Any method of invoking *Hatch* yields the *Hatch and Gradient* dialog box (Fig. 23-3). Typically, the first step in this dialog box is the selection of a hatch pattern.

FIGURE 23-3

After you select the *Pattern* and other parameters for the hatch in the *Hatch and Gradient* dialog box, you select *Pick points* or *Select objects* to return to the drawing and indicate the area to fill with the pattern. While in the drawing, you can right-click to produce a shortcut menu, providing access to other options without having first to return to the dialog box (Fig. 23-4).

FIGURE 23-4

Hatch and Gradient Dialog Box—Hatch Tab

Type

This option allows you to specify the type of the hatch pattern: *Predefined, User-defined,* or *Custom.* Use *Predefined* for standard hatch pattern styles that AutoCAD provides (in the ACAD.PAT file).

Predefined

There are two ways to select from *Predefined* patterns: (1) select the *Swatch* tile to produce the *Hatch Pattern Palette* dialog box displaying hatch pattern names and image tiles (Fig. 23-5), or (2) select the *Pattern:* drop-down list to PICK the pattern name.

The *Hatch Pattern Palette* dialog box allows you to select a predefined pattern by its image tile or by its name. There are four tabs of image tiles: *ANSI, ISO, Other Predefined,* and *Custom.*

FIGURE 23-5

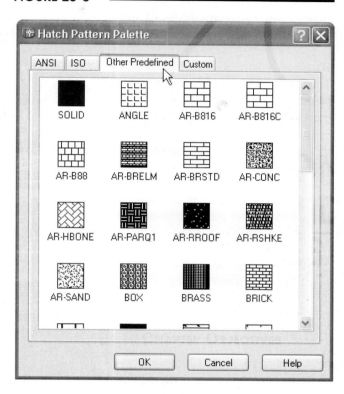

User-defined

To define a simple hatch pattern "on the fly," select the *User-defined* tile. This causes the *Pattern* and *Scale* options to be disabled and the *Angle, Spacing,* and *Double* options to be enabled. Creating a *User-defined* pattern is easy. Specify the *Angle* of the lines, the *Spacing* between lines, and optionally create *Double* (perpendicular) lines. All *User-defined* patterns have continuous lines.

Custom

Custom patterns are previously created user-defined patterns stored in other than the ACAD.PAT file. Custom patterns can contain continuous, dashed, and dotted line combinations. See the AutoCAD Customization Guide for information on creating and saving custom hatch patterns.

Pattern

Selecting the *Pattern* drop-down list (see Fig. 23-3) displays the name of each predefined pattern. Making a selection dictates the current pattern and causes the pattern to display in the *Swatch* window. The small button with ellipsis (…) just to the right of the *Pattern:* name drop-down list produces the *Hatch Pattern Palette.*

Swatch

Click in the *Swatch* tile to produce the *Hatch Pattern Palette* dialog box (see Fig. 23-5). You can select from *ANSI, ISO, Other Predefined,* and *Custom* (if available) hatch patterns.

TIP ✔ Hatch patterns are created using the current linetype. Therefore, the <u>*Continuous* linetype should be set current</u> when hatching to ensure the selected area is filled with the pattern as it appears in the image tile. After selecting a pattern, specify the desired *Scale* and *Angle* of the pattern or ISO pen width.

Angle

The *Angle* specification determines the angle (slant) of the hatch pattern. The default angle of 0 represents whatever angle is displayed in the pattern's image tile. Any value entered deflects the existing pattern (as it appears in the swatch) by the specified value (in degrees). The value entered in this box is held in the *HPANG* system variable.

Scale

The value entered in this edit box is a scale factor that is applied to the existing selected pattern. Normally, this scale factor should be changed proportionally with changes in the drawing *Limits*. Like many other scale factors (*LTSCALE, DIMSCALE*), AutoCAD defaults are set to a value of 1, which is appropriate for the default *Limits* of 12 x 9. If you have calculated the drawing scale factor, enter that value in the *Scale* edit box (see Chapter 12 for information on the "Drawing Scale Factor"). The *Scale* value is stored in the *HPSCALE* system variable.

2008 The scale you specify in this edit box is always relative to the model space geometry. If you are creating *Annotative* hatch patterns to display in multiple viewports at different scales, the hatch pattern scale can be automatically adjusted for the viewport scale (see "Annotative Hatch Patterns").

ISO hatch patterns are intended for use with metric drawings; therefore, the scale (spacing between hatch lines) is much greater than for inch drawings. Because *Limits* values for metric sheet sizes are greater than for inch-based drawings (25.4 times greater than comparable inch drawings), the ISO hatch pattern scales are automatically compensated. If you want to use an ISO pattern with inch-based drawings, calculate a hatch pattern scale based on the drawing scale factor and multiply by .039 (1/25.4).

Double

Only for a *User-defined* pattern, check this box to have a second set of lines drawn at 90 degrees to the original set.

Relative to Paper Space

2008 This option is useful when you want to change the hatch pattern scale relative to paper space units for the current viewport. The viewport must be active, then use *Hatch* or *Hatchedit* to select this option. All other viewports display the hatch pattern scale relative to the one active viewport; therefore, this option is of limited use when you have more than one viewport in the drawing. If you plan to display the hatch pattern in multiple viewports at multiple scales, you may want to use the *Annotative* property (see "Annotative Hatch Patterns") or create multiple hatch patterns on different layers and display each layer only in the appropriate viewport.

Spacing

This option is enabled if *User-defined* pattern is specified. Enter a value for the distance between lines.

ISO Pen Width

You must select an ISO hatch pattern for this tile to be enabled. Selecting an *ISO Pen Width* from the drop-down list automatically sets the scale and enters the value in the *Scale* edit box. See "*Scale*."

Use Current Origin

By default, all hatch pattern origins correspond to the current coordinate system's 0,0,0 location. (0,0,0 is the default *HPORIGIN* system variable setting.) This option is acceptable for most hatch patterns.

Specified Origin

Some hatch patterns, such as brick patterns, need to be aligned with a point on the hatch boundary. Figure 23-6 illustrates a brick pattern using the default (*Current origin*) and a *Bottom left* origin. Click this option to make the following options available so you control the starting location of hatch pattern generation.

Click to Set New Origin

Allows you to select a point in the drawing to specify the new hatch origin point.

Default to Boundary Extents

This option automatically calculates a new origin based on the rectangular extents of the hatch boundary. You can specify one of the four corners of the extents or the boundary's center.

Store as Default Origin

If you use a *Specified Origin*, check this box to store the value of the new hatch origin in the *HPORIGIN* system variable for future hatches in the drawing.

FIGURE 23-6

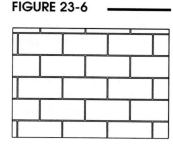

CURRENT ORIGIN

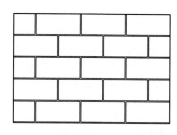

BOTTOM LEFT

Selecting the Hatch Area

Once the hatch pattern and options have been selected, you must indicate to AutoCAD what area(s) should be hatched. Either the *Add: Pick points* method, *Add: Select objects* method, or a combination of both can be used to accomplish this.

Add: Pick Points

This tile should be selected if you want AutoCAD to automatically determine the boundaries for hatching. You only need to select a point <u>inside</u> the area you want to hatch. The point selected must be inside a <u>closed shape</u>. When the *Add: Pick points* tile is selected, AutoCAD gives the following prompts:

```
Pick internal point or [Select objects/remove Boundaries]: PICK
Selecting everything...
Selecting everything visible...
Analyzing the selected data...
Analyzing the internal islands...
Pick internal point or [Select objects/remove Boundaries]:
```

When an internal point is PICKed, AutoCAD traces and highlights the boundary (Fig. 23-7). The interior area is then analyzed for islands to be included in the hatch boundary. Multiple boundaries can be designated by selecting multiple internal points.

FIGURE 23-7

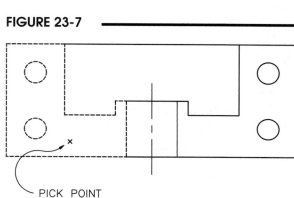

PICK POINT

The location of the point selected is usually not critical. However, the point must be PICKed inside the expected boundary. If there are any large gaps in the area, a complete boundary cannot be formed and a boundary error message appears.

Add: Select Objects

Alternately, you can designate the boundary with the *Select objects* method. Using the *Select Objects* method, <u>you</u> specify the boundary objects rather than let AutoCAD locate a boundary. With the *Select objects* method, no temporary *Pline* boundary is created as with the *Pick points* method. Therefore, the selected objects must form a closed shape with no large gaps or overlaps. If gaps exist or if the objects extend past the desired hatch area (Fig. 23-8), AutoCAD cannot interpret the intended hatch area correctly, and problems will occur.

FIGURE 23-8

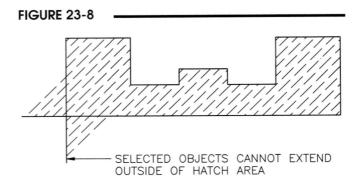

SELECTED OBJECTS CANNOT EXTEND OUTSIDE OF HATCH AREA

The *Select objects* option can be used after the *Pick points* method to select specific objects for *Hatch* to consider before drawing the hatch pattern lines. For example, if you had created text objects or dimensions within the boundary found by the *Pick points* method, you may then have to use the *Select objects* option to select the text and dimensions (if *Island detection* was off). Using this procedure, the hatch lines are automatically "trimmed" around the text and dimensions.

Remove Boundaries

PICKing this tile allows you to remove previously selected boundaries or to select specific islands (internal objects) to remove from those AutoCAD has found within the outer boundary. If hatch lines have been drawn, be careful to *Zoom* in close enough to select the desired boundary and not the hatch pattern.

Recreate Boundary

This option is always disabled in the *Hatch and Gradient* dialog box, but is available for editing hatches. See "*Hatchedit.*"

View Selections

Clicking the *View Selections* tile causes AutoCAD to highlight all selected boundaries. This can be used as a check to ensure the desired areas are selected.

Associative

This checkbox toggles on or off the associative property for hatch patterns. Associative hatch patterns automatically update by conforming to the new boundary shape if the boundary is changed (see Fig. 23-2). A non-associative hatch pattern is static even when the boundary changes.

Create Separate Hatches

By default, multiple hatches created with one use of the *Hatch* command are treated as one AutoCAD object, even when the hatches are within separate boundaries. In such a case, multiple hatch areas could be *Erased* by selecting only one of the hatched areas. Check *Create Separate Hatches* to ensure each boundary area forms a unique hatch object.

Preview

You should always use the *Preview* option after specifying the hatch parameters and selecting boundaries, but before you use *OK*. This option allows you to temporarily look at the hatch pattern in your drawing with the current settings applied and allows you to adjust the settings, if necessary, before using *OK*. After viewing the drawing, press the Esc key to redisplay the *Hatch and Gradient* dialog box, allowing you to make adjustments.

Expanded Dialog Box Options

Using the arrow button in the lower-right corner of the *Hatch and Gradient* dialog box you can expand the dialog box to reveal additional options as described below (Fig. 23-9).

Island Detection

Check this box to cause AutoCAD to automatically detect any internal areas (islands) inside the selected boundary, such as holes (*Circles*), text, or other closed shapes. *Island detection* is performed according to the following options.

FIGURE 23-9

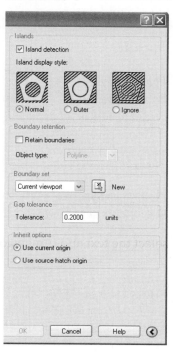

Normal

This should be used for most applications of *Hatch*. Text or closed shapes within the outer border are considered in such cases. Hatching will begin at the outer boundary and move inward, alternating between applying and not applying the pattern as interior shapes or text are encountered (Fig. 23-10).

Outer

This option causes AutoCAD to hatch only the outer closed shape. Hatching is turned off for all interior closed shapes (Fig. 23-10).

Ignore

Ignore draws the hatch pattern from the outer boundary inward, ignoring any interior shapes. The resulting hatch pattern is drawn through the interior shapes (Fig. 23-10).

FIGURE 23-10

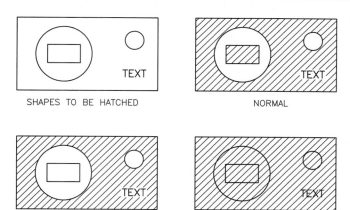

SHAPES TO BE HATCHED

NORMAL

OUTER

IGNORE

Retain Boundaries

When AutoCAD uses the *Pick points* method to locate a boundary for hatching, a temporary *Pline* or *Region* is created for hatching, then discarded after the hatching process. Checking this box forces AutoCAD to <u>keep</u> the boundary. When the box is checked, *Hatch* creates two objects—the hatch pattern and the boundary object. (You can specify whether you want to create a *Pline* or a *Region* boundary in the *Object Type* drop-down list.) Using this option and erasing the hatch pattern accomplishes the same results as using the *Boundary* command.

Object Type

Hatch creates *Polyline* or *Region* boundaries. This option is enabled only when the *Retain Boundaries* box is checked.

Boundary Set

By default, AutoCAD examines <u>all objects in the viewport</u> when determining the boundaries by the *Pick points* method. (*Current Viewport* is selected by default when you begin the *Hatch* command.) For complex drawings, examining all objects can take some time. In that case, you may want to specify a smaller boundary set for AutoCAD to consider. Clicking the *New* tile clears the dialog boxes and permits you to select objects or select a window to define the new set.

Gap Tolerance

FIGURE 23-11

Tolerance is set to 0 by default. In this case, all boundaries selected to hatch with the *Hatch* command must be <u>completely</u> closed—that is, no hatch boundaries can contain a "gap." However, if gaps exist in the selected boundary, you can set a *Tolerance* value to compensate for any boundary gaps (Fig. 23-11). You can set the *Gap tolerance* in the *Hatch and Gradient* dialog box or set the *HPGAPTOL* system variable. The *Gap tolerance* must be greater than the size of the gap in drawing units.

GAP TOLERANCE CAN BE SET
TO HATCH "OPEN" BOUNDARIES

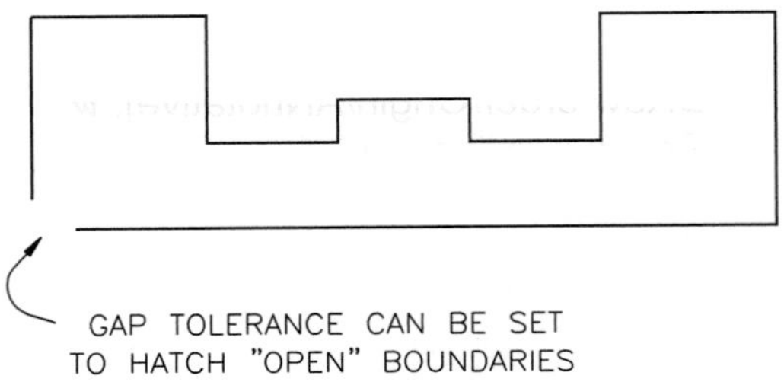

If you attempt to hatch an open area when no *Gap tolerance* is set or a gap exists that is greater than the setting for the *Gap tolerance, a Boundary Definition Error* warning box appears. However, if a gap exists that is within the *Gap tolerance*, the object will hatch correctly after answering *Yes* to the *Open Boundary Warning* box that appears.

-Hatch Command Line Options

-Hatch is the Command line alternative for the *Hatch and Gradient* dialog box. *-Hatch* must be typed at the keyboard since it is not available from the menus. Although you cannot create gradient hatches with *-Hatch*, most other options are available with this method.

```
Command: -hatch
Current hatch pattern:  ANSI31
Specify internal point or [Properties/Select objects/draW boundary/remove Boundaries/Advanced/
DRaw order/Origin/ANnotative]:
```

As you can see, most of the options in the dialog box version of this command are also available here. See "*Hatch*" for information on these options. The prompts for the *Properties* options and the *Advanced* options are shown next.

Properties

Use the *Properties* option to change the *Pattern, Scale,* and *Angle.*

Enter a pattern name or [?/Solid/User defined] <ANSI31>:
Specify a scale for the pattern <1.0000>:
Specify an angle for the pattern <0>:

Advanced

The *Advanced* option give access to these miscellaneous options:

Enter an option
[Boundary set/Retain boundary/Island detection/Style/Associativity/Gap tolerance/separate Hatches]:

Draw Boundary

The *Draw boundary* option of the *-Hatch* command is not available through the *Hatch and Gradient* dialog box. This option allows you to <u>pick points</u> to specify an area to hatch, and you can opt to discard the boundary after hatching. This procedure has the same results as creating a boundary using the *Line*, *Pline*, or other commands, then using *Hatch* to create the hatch pattern, then using *Erase* to remove the boundary. The command prompt sequence is as follows.

```
Command: -hatch
Current hatch pattern: ANSI31
Specify internal point or [Properties/Select objects/draW boundary/remove Boundaries/Advanced/
DRaw order/Origin/ANnotative]: w
Retain polyline boundary? [Yes/No] <N>: Enter
Specify start point: PICK
Specify next point or [Arc/Length/Undo]: PICK
Specify next point or [Arc/Close/Length/Undo]: PICK
Specify next point or [Arc/Close/Length/Undo]: PICK
Specify next point or [Arc/Close/Length/Undo]: close
Specify start point for new boundary or <Accept>:
```

With this method of *-Hatch*, you <u>specify points to define the boundary</u> rather than select objects (Fig. 23-12). The significance of this option is that you can create a hatch pattern without using existing objects as the boundary. In addition, you can specify whether you want to retain the boundary or not after the pattern is applied.

FIGURE 23-12

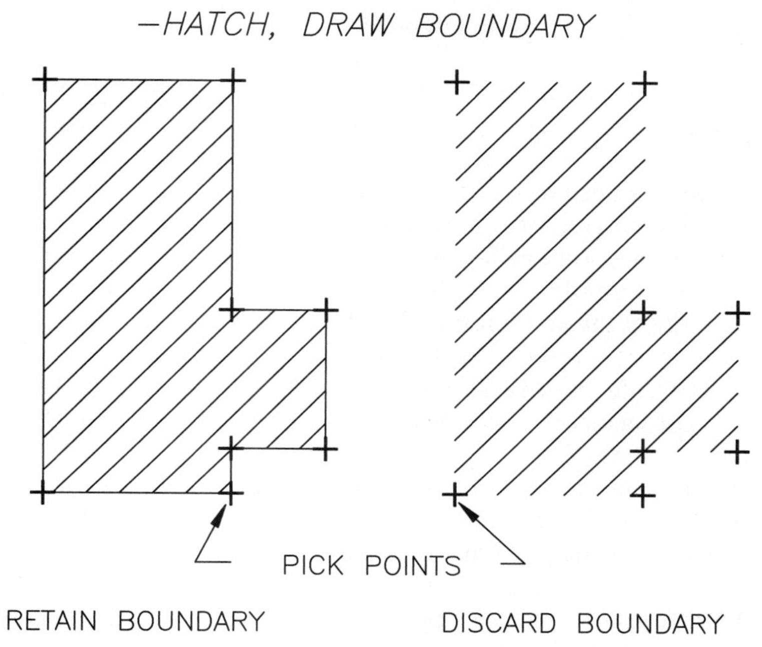

−HATCH, DRAW BOUNDARY

PICK POINTS

RETAIN BOUNDARY   DISCARD BOUNDARY

TIP   For special cases when you do not want a hatch area boundary to appear in the drawing (Fig. 23-13), you can use the *Draw boundary* option of *-Hatch* and opt to discard the boundary.

FIGURE 23-13

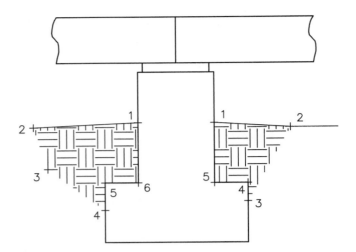

Annotative Hatch Patterns

For complex drawings that include multiple viewports that display the same geometry at different scales, it may be desirable to create annotative hatch patterns. Annotative hatch patterns automatically adjust for size when the *Annotative Scale* for each viewport is changed. Therefore, an annotative hatch pattern can appear the same scale when displayed in multiple viewports shown at different viewport scales. The following section is helpful if you plan to display the hatch patterns in multiple viewports at different scales. As an overview for creating and viewing annotative hatch patterns, follow these steps.

Steps for Creating and Setting the Scale for Annotative Hatch Patterns

1. Create the desired hatch pattern in model space using *Hatch*. Set the desired *Scale* as you would normally, using the "drawing scale factor" (DSF) as the hatch pattern *Scale* value.
2. In the *Hatch* or *Hatchedit* dialog box, select the *Annotative* checkbox to make the hatch pattern associative.
3. Create the desired viewports if not already created and display the desired geometry in each viewport.
4. In a layout tab when *PAPER* space is active (not in a viewport), enable the *Automatically add scales to annotative objects...* toggle in the lower-right corner of the Drawing Editor or set the *ANNOAUTOSCALE* variable to a positive value (1, 2, 3, or 4).
5. Set the desired scale for each viewport by double-clicking inside the viewport and then using the *VP Scale* pop-up list or *Viewports* toolbar. Alternately, select the viewport object (border) and use the *Properties* palette for the viewport. If the *Annotation Scale* is locked to the *VP Scale* (as it is by default), the annotation scale changes automatically to match the *VP Scale*. Setting the *VP Scale* automatically adjusts the annotation scale for all annotative objects for the viewport.

For example, assume you are drawing a section of a mechanical assembly and plan to print on an ANSI "A" size sheet (11" x 8.5"). Two viewports will be created to display the sectioned assembly in two different scales.

In order to draw the section in the *Model* tab full size, you change the *Limits* to match the sheet size (11,8.5), resulting in a drawing scale factor = 1. After creating the assembly views, you are ready to create the hatch patterns. Following steps 1 and 2, use *Hatch* to select the desired hatch patterns. Set the hatch *Scale* to 1 (scale = DSF) and select the *Annotative* checkbox. Apply the desired hatch patterns to the sectioned parts.

Next, following steps 4 through 6, switch to a *Layout* tab, set it up for the desired plot or print device using the *Page Setup Manager,* and create the two viewports. Ensure the *Automatically add scales to annotative objects…* is toggled on. Set the viewport scale for each viewport using the *VP Scale* pop-up list as shown in Figure 23-14. The *Annotation Scale* is locked to the *VP Scale* (by default) so its setting changes accordingly as the *VP Scale* changes. The hatch patterns should then appear in each viewport at the appropriate scale since the hatch scale in each viewport is automatically adjusted for the viewport scale. Note that the hatch patterns for the two viewports appear the same size in Figure 23-14.

FIGURE 23-14

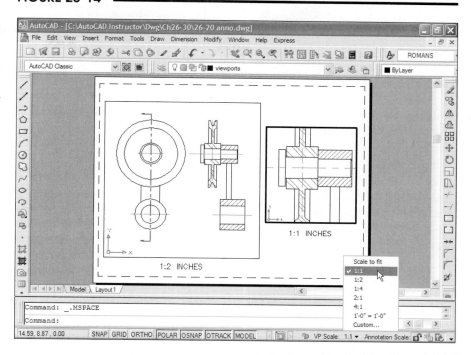

When using annotative hatches, it is simplest to use the drawing scale factor directly as the hatch *Scale.* Since the "scale" for hatch patterns is not based on inches or mm, the annotative feature is valid only for use with multiple viewports. On the other hand, when using annotative text, you can set the text height equal to the paper text height, and then create the text in model space at that scale. Next, set the annotative scale for the *Model* tab using the *Annotation Scale* pop-up list. This procedure changes the text to be properly sized for the model space.

Also different than annotative text and dimension objects, new annotative hatch objects are not created when different *VP Scales* are selected. The existing hatch is simply adjusted for the *VP Scale.* For annotative text and dimension objects, when the *Automatically add scales to annotative objects…* toggle is on and a new scale is selected from the *VP Scale* list, new annotative objects are created as a copies of the originals, but in the new viewport scale.

An alternate method for creating hatch patterns to appear in multiple viewports at different scales is to create hatches of different scales on different layers, then control the layer visibility for the appropriate viewports to display only the correctly scaled hatch pattern.

Assigning the *Annotative* Property Retroactively

You can change non-annotative hatches in the drawing to annotative by two methods. First, you can double-click on the hatch pattern to produce the *Hatchedit* dialog box and check the *Annotative* box as described previously. You can also invoke the *Properties* palette for a hatched area in a drawing. The *Properties* palette displays an entry for the *Annotative* property for selected hatches. You can use this section of the *Properties* palette to convert non-annotative hatches to annotative by changing the *No* value to *Yes.* In this case, a new entry appears for *Annotative Scale.*

Toolpalettes

Pull-down Menu	Command (Type)	Alias (Type)	Short-cut	Screen (side) Menu	Tablet Menu
Tools Tool Palettes Window	Toolpalettes	TP	Ctrl+3	...	...

 TIP You can use Tool Palettes to drag-and-drop hatch patterns into your drawing. The process is simple. Open Tool Palettes by any method and locate the palette that contains the hatch pattern you want to use. Next, drag-and-drop the hatch from the palette directly into the closed area you want to fill with the pattern (Fig. 23-15).

FIGURE 23-15

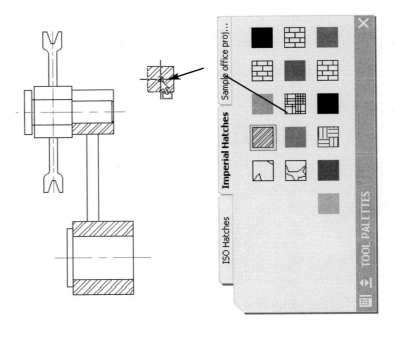

If you right-click on any tool in the palette, its *Tool Properties* dialog box appears (Fig. 23-16). This dialog box is specific to the particular tool you right-clicked, and allows you to specify settings for the hatch such as *Angle* and *Scale*.

FIGURE 23-16

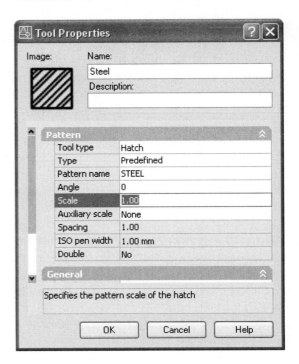

EDITING HATCH PATTERNS AND BOUNDARIES

Hatchedit

Pull-down Menu	Command (Type)	Alias (Type)	Short-cut	Screen (side) Menu	Tablet Menu
Modify Object > Hatch...	*Hatchedit* or -*Hatchedit*	HE	...	*MODIFY1 Hatchedt*	Y,16

Hatchedit allows you to modify an <u>existing</u> hatch pattern in the drawing. This feature of AutoCAD makes the hatching process more flexible because you can hatch several areas to quickly create a "rough" drawing, then retroactively fine-tune the hatching parameters when the drawing nears completion with *Hatchedit*.

You can produce the *Hatchedit* dialog box (Fig. 23-17) by any method shown in the Command table as well as double-clicking on any hatched area in the drawing (assuming *Dblclkedit* is set to *On* and *PICKSTYLE* is set to 1). The dialog box provides options for changing the *Pattern, Scale, Angle,* and *Type* properties of the existing hatch.

Apparent in Figure 23-17, the *Hatch Edit* dialog box is essentially the same as the *Hatch and Gradient* dialog box. Therefore, when you edit a hatch pattern, all the options are available that were originally used to select the boundary and create the hatch pattern. Note that you can use *Hatchedit* to retroactively assign the *Annotative* property to hatch patterns. One additional option is available only when editing a hatch pattern—*Recreate boundary*.

FIGURE 23-17

Recreate Boundary

This option allows you to create a duplicate boundary of the selected hatch area. The new boundary is created "on top of" the original boundary, so you have to use *Move, Last,* to see the recreated boundary (Fig. 23-18). When you select the hatch area to edit, then pick the *Recreate boundary* option, the following prompt appears:

Enter type of boundary object [Region/Polyline] <Polyline>: **Enter** or (**option**)
Reassociate hatch with new boundary? [Yes/No] <N>: **Enter** or (**option**)

FIGURE 23-18

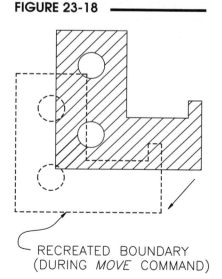

RECREATED BOUNDARY (DURING *MOVE* COMMAND)

You have the choice to create a new shape as a *Pline* or a *Region* object. If you choose not to reassociate the hatch with the new boundary, only a duplicate boundary is created. Selecting "Yes" to reassociate the hatch with the new boundary causes the original hatch pattern to be "connected" to the new boundary rather than the old boundary.

Using Grips with Hatch Patterns

Grips can be used effectively to edit hatch pattern boundaries of associative hatches. Do this by selecting the boundary with the pickbox or automatic window/crossing window (as you would normally activate an object's grips). Associative hatches retain the association with the boundary after the boundary has been changed using grips (see Fig. 23-2, earlier in this chapter).

FIGURE 23-19

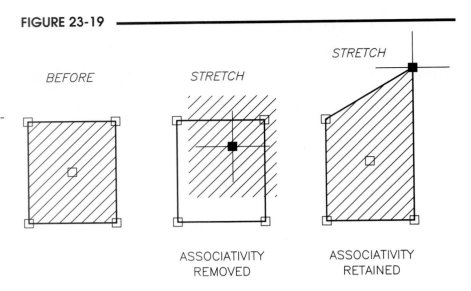

BEFORE

STRETCH

STRETCH

ASSOCIATIVITY REMOVED

ASSOCIATIVITY RETAINED

TIP

Beware, when selecting the hatch boundary, ensure you edit the boundary and not the hatch object. If you edit the hatch object only, the associative feature is lost. This can happen if you select the hatch object grip (Fig. 23-19).

If you activate grips on both the hatch object and the boundary and make a boundary grip hot, the boundary and related hatch retain associativity (see Fig. 23-19).

Using *Trim* with Hatch Patterns

You can use the *Trim* command to trim a hatch object created with *Hatch*. Simply create or use an existing object as a "cutting edge," then invoke *Trim* and trim the hatch object as you would any other *Line, Circle, Arc,* or other object that you would normally trim (Fig. 23-20). Beware, however, because trimming a hatch pattern converts the hatch to a non-associative hatch object. If you then *Move, Scale, Stretch,* or change the boundary in some other way, the hatch pattern will not reform to match the boundary.

FIGURE 23-20

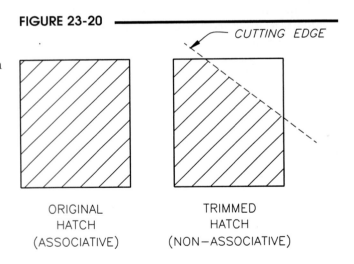

CUTTING EDGE

ORIGINAL HATCH (ASSOCIATIVE)

TRIMMED HATCH (NON—ASSOCIATIVE)

FILLMODE for Solid *Hatch* Fills

A *Solid* hatch pattern is available. This feature is welcomed by professionals who create hatch objects such as walls for architectural drawings and require the solid filled areas. There are many applications for this feature.

Solid-filled hatch areas have versatility because you can control the *Color* of the solid fill. By default, the color of the *Solid* hatch pattern (just like any other hatch pattern) assumes the current *Color* setting. In most cases, the current *Color* setting is *ByLayer*; therefore, the *Solid* hatch pattern assumes the current layer color. You can also assign the *Solid* hatches object-specific *Color,* in which case it is possible to have multiple *Solid* filled areas in different colors all on the same layer.

You can use the *Fill* command (or *FILLMODE* system variable) to control the visibility of the solid filled areas. *Fill* also controls the display of solid filled TrueType fonts. If you want to display solid filled *Hatch* objects as solid, set *Fill* to *On* (Fig. 23-21). Setting *Fill* to *Off* causes AutoCAD <u>not</u> to display the solid filled areas. *Regen* must be used after *Fill* to display the new visibility state. This solid fill control can be helpful for saving toner or ink during test prints or speeding up plots when much solid fill is used in a drawing.

FIGURE 23-21

FILL ON   FILL OFF

Draworder

Pull-down Menu	Command (Type)	Alias (Type)	Short-cut	Screen (side) Menu	Tablet Menu
Tools Draw Order >	Draworder	DR	...	TOOLS 1 Drawordr	T,9

It is possible to use solid hatch patterns and raster images in combination with filled text, other hatch patterns and solid images, etc. With these solid areas and images, some control of which objects are in "front" and "back or "above" and "under" must be provided. This control is provided by the *Draworder* command.

For example, if a company logo were created using filled *Text*, a *Sand* hatch pattern, and a *Solid* hatched circle, the object that was created last would appear "in front" (Fig. 23-22, top logo). The *Draworder* command is used to control which objects appear in front and back or above and under. This capability is necessary for creating prints and plots in black and white and in color.

FIGURE 23-22

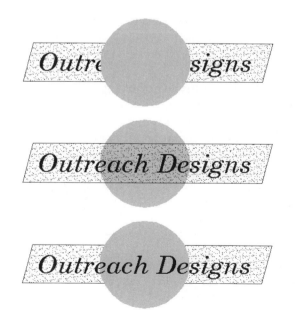

```
Command: draworder
Select objects: PICK
Select objects: Enter
Enter object ordering option [Above objects/Under
objects/Front/Back] <Back>: (option)
Regenerating drawing.
Command:
```

DRAWING CUTTING PLANE LINES

In mechanical drawings, most section views (full, half, and offset sections) require a cutting plane line to indicate the plane on which the object is cut. The cutting plane line is drawn in a view <u>adjacent</u> to the section view because the line indicates the plane of the cut from its edge view. (In the section view, the cutting plane is perpendicular to the line of sight, therefore, not visible as a line.)

Standards provide two optional line types for cutting plane lines. In AutoCAD, the two linetypes are *Dashed* and *Phantom,* as shown in Figure 23-23. Arrows at the ends of the cutting plane line indicate the <u>line-of-sight</u> for the section view.

FIGURE 23-23

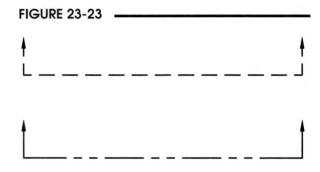

Cutting plane lines should be drawn or plotted in a heavy lineweight. Two possible methods can be used to accomplish this: use a *Pline* with *Width* or assign a *Lineweight* to the line or layer. If you prefer drawing the cutting plane line using a *Pline,* use a *Width* of .02 or .03 times the drawing scale factor. Assigning a *Lineweight* to the line or layer is a simpler method. A *Lineweight* of .8mm or .031" is appropriate for cutting plane lines.

CHAPTER EXERCISES

For the following exercises, create the section views as instructed. Use an appropriate template drawing for each unless instructed otherwise. Include a border and title block in the layout for each drawing.

1. ***Open*** the **SADDLE** drawing that you created in Chapter 21. Convert the front view to a full section. ***Save*** the drawing as **SADL-SEC.** *Plot* the drawing at **1=1** scale.

FIGURE 23-24

2. Make a multiview drawing of the Bearing shown in Figure 23-24. Convert the front view to a full section view. Add the necessary cutting plane line in the top view. ***Save*** the drawing as **BEAR-SEC.** Make a *Plot* at full size.

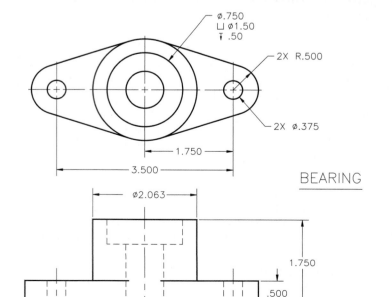

3. Create a multiview drawing of the Clip shown in Figure 23-25. Include a side view as a full section view. You can use the **CLIP** drawing you created in Chapter 21 and convert the side view to a section view. Add the necessary cutting plane line in the front view. *Plot* the finished drawing at **1=1** scale and *SaveAs* **CLIP-SEC**.

FIGURE 23-25

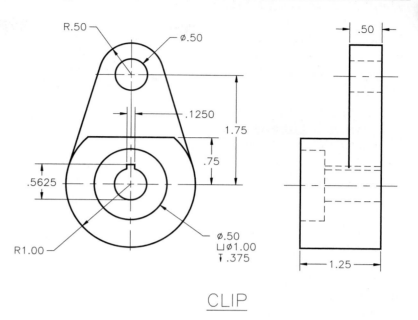

CLIP

4. Create a multiview drawing, including two full sections of the Stop Block as shown in Figure 23-26. Section B–B' should replace the side view shown in the figure. *Save* the drawing as **SPBK-SEC**. *Plot* the drawing at **1=2** scale.

FIGURE 23-26

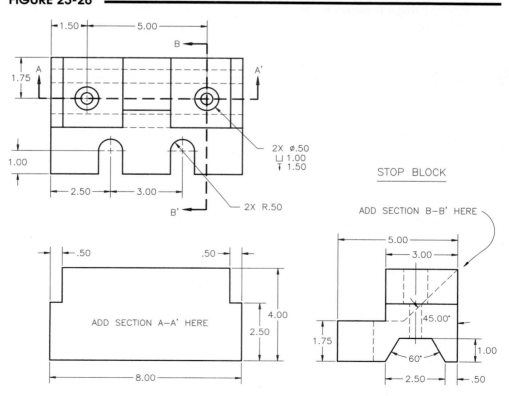

STOP BLOCK

5. Make a multiview drawing, including a half section of the Pulley (Fig. 23-27). All vertical dimensions are diameters. Two views (including the half section) are sufficient to describe the part. Add the necessary cutting plane line. *Save* the drawing as **PUL-SEC** and make a *Plot* at **1:1** scale.

FIGURE 23-27

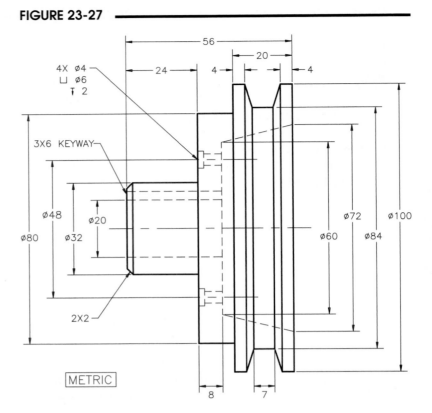

6. Draw the Grade Beam foundation detail in Figure 23-28. Do not include the dimensions in your drawing. Use the *Hatch* command to hatch the concrete slab with **AR-CONC** hatch pattern. Use *Sketch* to draw the grade line and *Hatch* with the **EARTH** hatch pattern. *Save* the drawing as **GRADBEAM**.

FIGURE 23-28

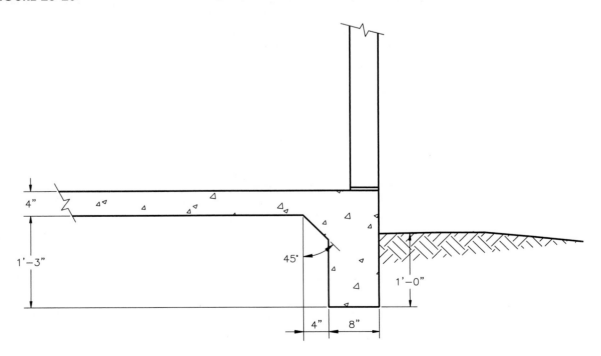

7. **Tool Palettes, Solid Fill**

A. *Open* the **OFFICE-REV** drawing that you last worked on in Chapter 20. Next, open the **Tool Palettes**.

B. Make a new *Layer* and name it **CARPET**. Accept the default color, linetype, etc. Make **CARPET** the *Current* layer.

C. Draw a *Line* across the doorway threshold of one of the two offices. Next, drag-and-drop a *Solid* pattern from the standard *Hatches* palette into the office. If problems occur, check to ensure that there are no gaps in the office walls.

D. When the solid pattern appears in the office, notice that the pattern may cover the outlines of the walls and furniture items. Use the *Draworder* command to move the new hatch pattern to the *Back*. Your drawing should look similar to that in Figure 23-29.

FIGURE 23-29

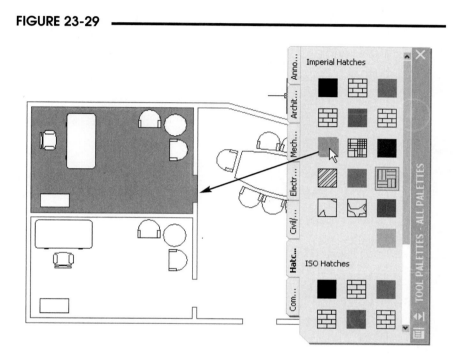

E. Use *SaveAs* to save and rename the drawing to **OFFICE-REV-2**.

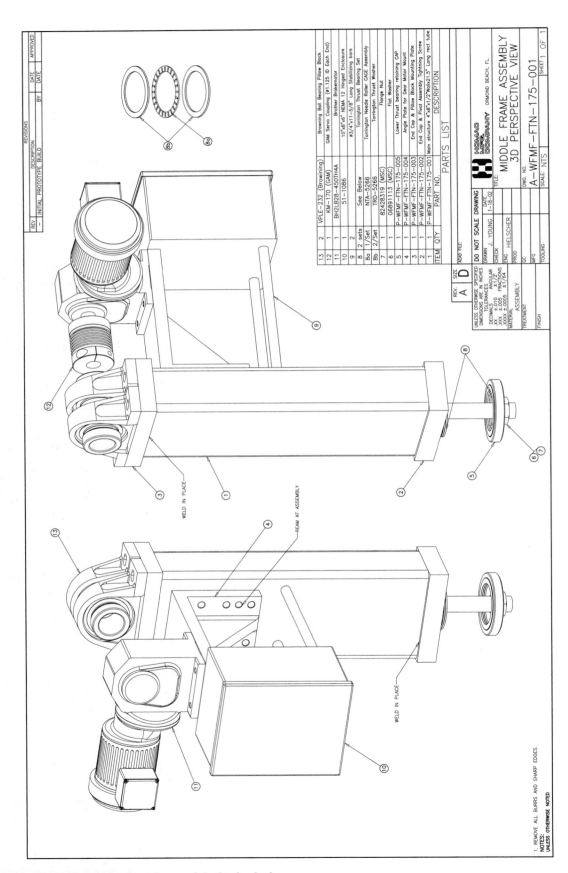

Drawing content

REVISIONS

REV	DESCRIPTION	DATE	APPROVED
		BY	DATE
–	INITIAL PROTOTYPE BUILD		

PARTS LIST

ITEM	QTY	PART NO.	DESCRIPTION
13	2	VPLE-232 (Browning)	Browning Ball Bearing Pillow Block
12	1	KM-170 (GAM)	GAM Servo Coupling (Ø1.125 ID Each End)
11	1	BH2LB28-450TH4A	Brother Brakemotor
10	1	51-1086	10"x8"x8" NEMA 12 Hinged Enclosure
9	2	See Below	⌀3/4"x11-5/8" Long Stabilizing bars
8	2 sets	NTA-5266	Torrington Thrust Bearing Set
8a	1/Set		Torrington Needle Roller CAGE Assembly
8b	2/Set	TRD-5266	Torrington Thrust Washer
7	1	82428319 (MSC)	Flange Nut
6	1	0689113 (MSC)	Flat Washer
5	1	P-WFMF-FTN-175-005	Lower Thrust bearing retaining CAP
4	1	P-WFMF-FTN-175-004	Angle Plate for Gear Motor Mount
3	1	P-WFMF-FTN-175-003	End Cap & Pillow Block Mounting Plate
2	1	P-WFMF-FTN-175-002	End Cap & Pivot Assembly Tightning Screw
1	1	P-WFMF-FTN-175-001	Main structure 4"x8"x1/2"Walls x21.5" Long rect tube

HEDMAS MFG COMPANY ORMOND BEACH, FL.

TITLE: **MIDDLE FRAME ASSEMBLY 3D PERSPECTIVE VIEW**

DWG. NO. **A-WFMF-FTN-175-001**

SCALE: NTS   SHEET 1 OF 1

DO NOT SCALE DRAWING		ACAD FILE:
DRAWN	J. YOUNG	DATE 1-18-02
CHECK	HIELSCHER	
ENG		
PROD		
QC		
MFG		
TOOLING		

UNLESS OTHERWISE SPECIFIED
DIMENSIONS ARE IN INCHES
TOLERANCES
DECIMALS   ANGULAR
.XX ±.010   ±1/2
.XXX ±.005   FRACTIONS
.XXXX ±.0005 ±1/64

REV **A**   SIZE **D**

MATERIAL — ASSEMBLY
TREATMENT
FINISH

WELD IN PLACE
REAM AT ASSEMBLY
WELD IN PLACE

NOTES:
1. REMOVE ALL BURRS AND SHARP EDGES.
UNLESS OTHERWISE NOTED

24

AUXILIARY
VIEWS

CHAPTER OBJECTIVES

After completing this chapter you should:

1. be able to use the *Rotate* option of *Snap* to change the angle of the *SNAP, GRID*, and *ORTHO*;

2. be able to set the *Increment angle* and *Additional angles* for creating auxiliary views using *Polar Tracking*;

3. know how to use the *Offset* command to create parallel line copies;

4. be able to use *Xline* and *Ray* to create construction lines for auxiliary views.

CONCEPTS

AutoCAD provides no commands explicitly for the creation of auxiliary views in a mechanical drawing or for angular sections of architectural drawings. However, four particular features that have been discussed previously can assist you in the construction of these drawings. Those features are *SNAP* rotation, *Polar Tracking*, the *Offset* command, and the *Xline* and *Ray* commands.

In a mechanical drawing, an auxiliary view may be needed in addition to the typical views (top, front, side). An auxiliary view is one that is normal (the line-of-sight is perpendicular) to an inclined surface of the object. Therefore, the entire view is constructed at an angle by projecting in a 90-degree direction from the edge view of an inclined surface in order to show the true size and shape of the inclined surface. Architectural drawings often include a portion of a plan drawn at some angle, such as a ranch house with one wing at a 45-degree angle to the main body of the house. In either case and depending on the object, the view or rooms could be drawn at any angle, so lines are typically drawn parallel and perpendicular relative to that section of the drawing. Therefore, the *SNAP* rotation feature, *Polar Tracking*, the *Offset* command, and the *Xline* and *Ray* commands can provide assistance in this task.

FIGURE 24-1

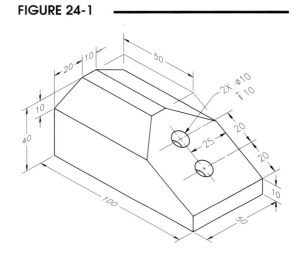

An example mechanical part used for the application of these AutoCAD features related to auxiliary view construction is shown in Figure 24-1. As you can see, there is an inclined surface that contains two drilled holes. To describe this object adequately, an auxiliary view should be created to show the true size and shape of the inclined surface. Although this chapter uses only this mechanical example, these same techniques would be used for architectural drawings or any other drawings that include views or entire sections of a drawing constructed at some angle.

CONSTRUCTING AN AUXILIARY VIEW

Setting Up the Principal Views

To begin this drawing, the typical steps are followed for drawing setup (Chapter 12). Because the dimensions are in millimeters, *Limits* should be set accordingly. For example, to provide enough space to draw the views full size and to plot full size on an A sheet, *Limits* of 279 x 216 are specified.

FIGURE 24-2

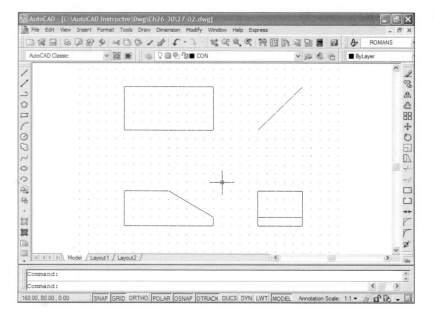

In preparation for the auxiliary view, the principal views are "blocked in," as shown in Figure 24-2. The purpose of this step is to ensure that the desired views fit and are optimally spaced within the allocated *Limits*. If there is too little or too much room, adjustments can be made to the *Limits*. Notice that space has been allotted between the views for a partial auxiliary view to be projected from the front view.

Before additional construction on the principal views is undertaken, initial steps in the construction of the partial auxiliary view should be performed. The projection of the auxiliary view requires drawing lines perpendicular and parallel to the inclined surface. One or more of the three alternatives (explained next) can be used.

Using *Snap Rotate* and *ORTHO*

One possibility to construct an auxiliary view is to use the *Snap* command with the *Rotate* option. This action permits you to rotate the *SNAP* to any angle about a specified base point. The *GRID* automatically follows the *SNAP*. Turning *ORTHO ON* forces *Lines* to be drawn orthogonally with respect to the rotated *SNAP* and *GRID*.

Figure 24-3 displays the *SNAP* and *GRID* after rotation. The command syntax is given below.

In this figure, the cursor size is changed from the default 5% of screen size to 100% of screen size to help illustrate the orientation of the *SNAP, GRID,* and cursor when *SNAP* is *Rotated*. You can change the cursor size in the *Display* tab of the *Options* dialog box.

NOTE: Since AutoCAD 2007 the *Rotate* option of *Snap* does not appear as an option in the *Drafting Settings* dialog box or on the Command line. However, the *Rotate* option still functions by typing the letter *R* to activate it.

FIGURE 24-3

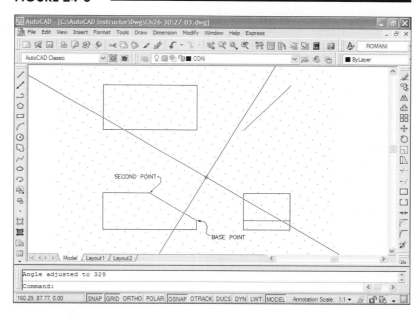

Command: **snap**
Specify snap spacing or [ON/OFF/Aspect/Style/Type] <0.5000>: **r**
Specify base point <0.0000,0.0000>: **PICK** or (**coordinates**) (PICK starts a rubberband line.)
Specify rotation angle <0>: **PICK** or (**value**) (PICK to specify second point to define the angle. See Fig. 24-3.)
Command:

PICK (or specify coordinates for) the endpoint of the *Line* representing the inclined surface as the "base point." At the "rotation angle:" prompt, a value can be entered or another point (the other end of the inclined *Line*) can be PICKed. Use *OSNAP* when PICKing the *Endpoints*. If you want to enter a value but don't know what angle to rotate to, use *List* to display the angle of the inclined *Line*. The *GRID, SNAP,* and crosshairs should align with the inclined plane as shown in Figure 24-3.

(An option for simplifying construction of the auxiliary view is to create a new *UCS* [User Coordinate System] with the origin at the new base point. Use the *3Point* option and turn on *ORTHO* to select the three points.

After rotating the *SNAP* and *GRID*, the partial auxiliary view can be "blocked in," as displayed in Figure 24-4. Begin by projecting *Lines* up from and perpendicular to the inclined surface. (Make sure *ORTHO* is *ON*.) Next, two *Lines* representing the depth of the view should be constructed parallel to the inclined surface and perpendicular to the previous two projection lines. The depth dimension of the object in the auxiliary view is equal to the depth dimension in the top or right view. *Trim* as necessary.

Locate the centers of the holes in the auxiliary view and construct two *Circles*. It is generally preferred to construct circular shapes in the view in which they appear as circles, then project to the other views. That is particularly true for this type of auxiliary since the other views contain ellipses. The centers can be located by projection from the front view or by *Offsetting Lines* from the view outline.

Next, project lines from the *Circles* and their centers back to the inclined surface (Fig. 24-5). Use of a hidden line layer can be helpful here. While the *SNAP* and *GRID* are rotated, construct the *Lines* representing the bottom of the holes in the front view. (Alternately, *Offset* could be used to copy the inclined edge down to the hole bottoms; then *Trim* the unwanted portions of the *Lines*.)

FIGURE 24-4

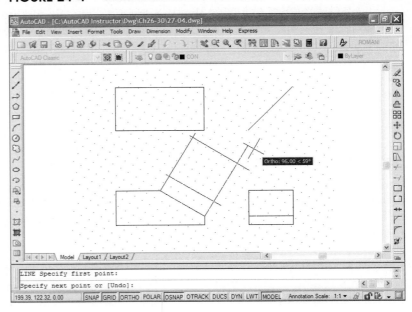

FIGURE 24-5

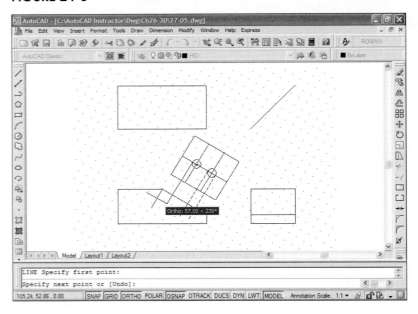

Rotating *SNAP* Back to the Original Position

Before details can be added to the other views, the *SNAP* and *GRID* should be rotated back to the original position. It is very important to rotate back using the <u>same base point</u>. Fortunately, AutoCAD remembers the original base point so you can accept the default for the prompt.

Next, enter a value of **0** when rotating back to the original position. (When using the *Snap Rotate* option, the value entered for the angle of rotation is absolute, not relative to the current position. For example, if the *Snap* was rotated to 45 degrees, rotate back to 0 degrees, not -45.)

> Command: **snap**
> Specify snap spacing or [ON/OFF/Aspect/Style/Type] <2.00>: **r**
> Specify base point <150.00,40.00>: **Enter** (AutoCAD remembers the previous base point.)
> Specify rotation angle <329>: **0**
> Command:

Construction of multiview drawings with auxiliaries typically involves repeated rotation of the *SNAP* and *GRID* to the angle of the inclined surface and back again as needed.

With the *SNAP* and *GRID* in the original position, details can be added to the other views as shown in Figure 24-6. Since the two circles appear as ellipses in the top and right side views, project lines from the circles' centers and limiting elements on the inclined surface. Locate centers for the two *Ellipses* to be drawn in the top and right side views.

FIGURE 24-6

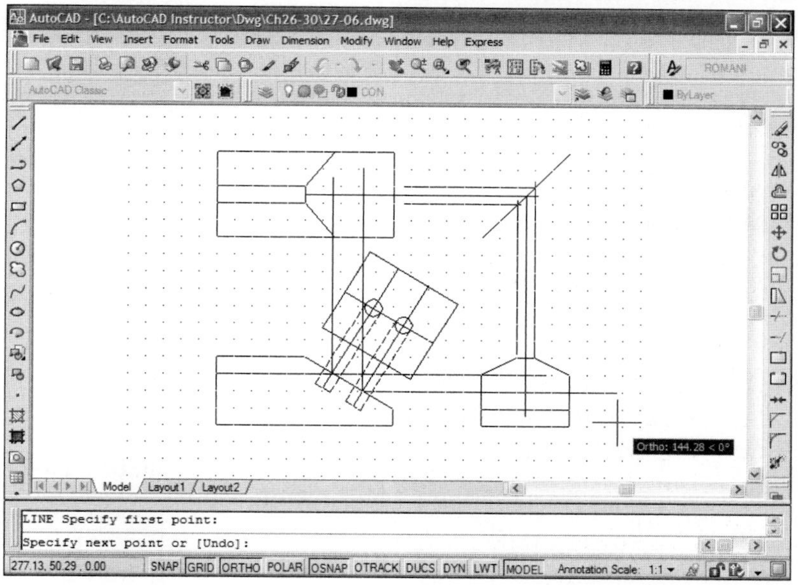

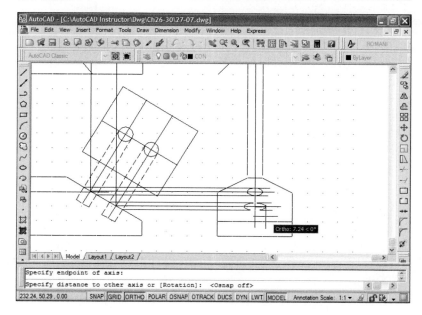

Use the *Ellipse* command to construct the ellipses in the top and right side views. Using the *Center* option of *Ellipse*, specify the center by PICKing with the *Intersection OSNAP* mode. *OSNAP* to the appropriate construction line *Intersection* for the first axis endpoint. For the second axis endpoint (Fig. 24-7), use the actual circle diameter, since that dimension is not foreshortened.

FIGURE 24-7

The remaining steps for completing the drawing involve finalizing the perimeter shape of the partial auxiliary view and *Copying* the *Ellipses* to the bottom of the hole positions. The *SNAP* and *GRID* should be rotated back to project the new edges found in the front view (Fig. 24-8).

At this point, the multiview drawing with auxiliary view is ready for centerlines, dimensioning, setting up a layout, and construction or insertion of a border and title block.

FIGURE 24-8

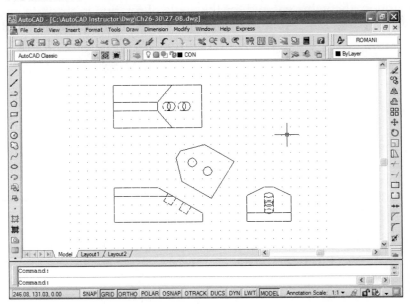

Using *Polar Tracking*

Polar Tracking can be used to create auxiliary views by facilitating construction of *Lines* at specific angles. To use *Polar Tracking* for auxiliary view construction, first use the *List* command to determine the angle of the inclined surface you want to project from. (Keep in mind that, by default, AutoCAD reports angles in whole numbers [no decimals or fractions], so use the *Units* command to increase the *Precision* of *Angular* units before using *List*.)

Once the desired angles are determined, specify the *Polar Angle Settings* in the *Drafting Settings* dialog box. Access the dialog box by right-clicking on the word *POLAR* at the Status Bar, typing *Dsettings*, or selecting *Drafting Settings* from the *Tools* pull-down menu. In the *Polar Tracking* tab of the dialog box, select the desired angles if appropriate from the *Increment angle* drop-down list. If the inclined plane is not at a regular angle offered from the drop-down list, specify the desired angle in the *Additional angles* edit box by selecting *New* and inputting the angles. Enter four angles in 90-degree increments (Fig. 24-9). Once the angles are set, ensure *POLAR* is on.

FIGURE 24-9

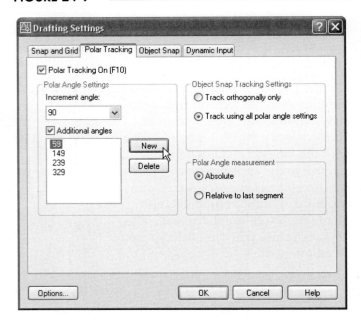

You may also want to set a *Polar Increment* in the *Snap and Grid* tab of the dialog box. This action makes the line lengths snap to regular intervals. (See Chapter 3 for more information on *Polar Snap* and *Polar Tracking*.)

Using *OSNAP*, begin constructing *Lines* perpendicular to the inclined plane. *Polar Tracking* should facilitate the line construction at the appropriate angles (Fig. 24-10).

FIGURE 24-10

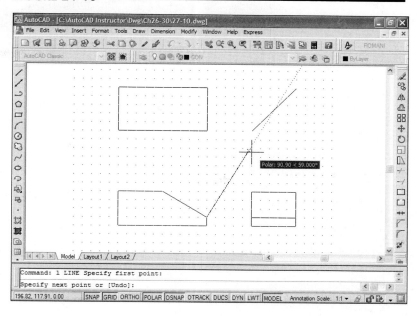

Perpendicular *Line* construction is also assisted by *Polar Tracking* (Fig. 24-11). Continue with this process, constructing necessary lines for the auxiliary view. The remainder of the auxiliary drawing, as described on previous pages (see Fig. 24-5 through Fig. 24-8), could be constructed using *Polar Tracking*. The advantage of this method is that drawing horizontal and vertical *Lines* is also possible while *Polar Tracking* is on.

Object Snap Tracking can also be employed to construct objects aligned (at the specified polar angles) with object snap locations (*Endpoint, Midpoint, Intersection, Extension*, etc.). See Chapter 7 for more information on *Object Snap Tracking*.

FIGURE 24-11

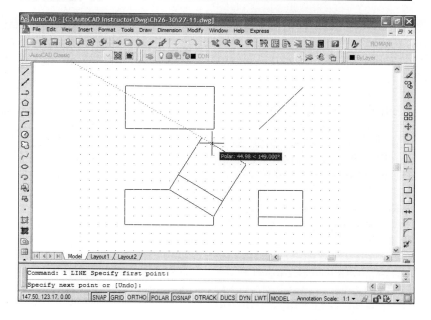

Using the *Offset* Command

Another possibility, and an alternative to the *SNAP* rotation, is to use *Offset* to make parallel *Lines*. This command can be particularly useful for construction of the "blocked in" partial auxiliary view because it is not necessary to rotate the *SNAP* and *GRID*.

Offset

Pull-down Menu	Command (Type)	Alias (Type)	Short-cut	Screen (side) Menu	Tablet Menu
Modify *Offset*	*Offset*	*O*	...	*MODIFY1* *Offset*	*V,17*

Invoke the *Offset* command and specify a distance. The first distance is arbitrary. Specify an appropriate value between the front view inclined plane and the nearest edge of the auxiliary view (20 for the example). *Offset* the new *Line* at a distance of 50 (for the example) or PICK two points (equal to the depth of the view).

 Note that the *Offset* lines have lengths equal to the original and therefore require no additional editing (Fig. 24-12).

FIGURE 24-12

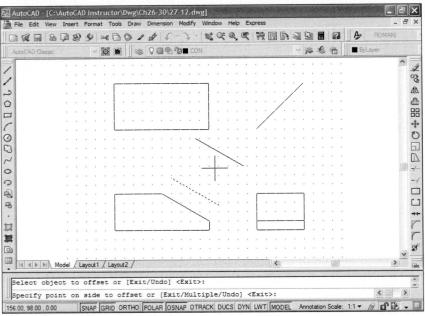

Next, two *Lines* would be drawn between *Endpoints* of the existing offset lines to complete the rectangle. *Offset* could be used again to construct additional lines to facilitate the construction of the two circles in the partial auxiliary view (Fig 24-13).

From this point forward, the construction process would be similar to the example given previously (Figs. 24-5 through 24-8). Even though *Offset* does not require that the *SNAP* be *Rotated*, the complete construction of the auxiliary view could be simplified by using the rotated *SNAP* and *GRID* in conjunction with *Offset*.

FIGURE 24-13

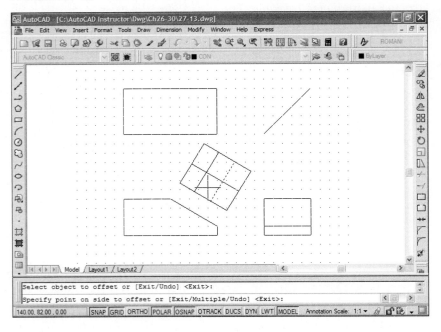

Using the *Xline* and *Ray* Commands

As a fourth alternative for construction of auxiliary views, the *Xline* and *Ray* commands could be used to create construction lines.

Xline

Pull-down Menu	Command (Type)	Alias (Type)	Short-cut	Screen (side) Menu	Tablet Menu
Draw Construction Line	*Xline*	*XL*	...	*DRAW 1 Xline*	*L,10*

Ray

Pull-down Menu	Command (Type)	Alias (Type)	Short-cut	Screen (side) Menu	Tablet Menu
Draw Ray	*Ray*	...	...	*DRAW 1 Ray*	*K,10*

The *Xline* command offers several options shown below:

Command: **xline**
Specify a point or [Hor/Ver/Ang/Bisect/Offset]:

The *Ang* option can be used to create a construction line at a specified angle. In this case, the angle specified would be that of the inclined plane or perpendicular to the inclined plane. The *Offset* option works well for drawing construction lines parallel to the inclined plane, especially in the case where the angle of the plane is not known.

Figure 24-14 illustrates the use of *Xline Offset* to create construction lines for the partial auxiliary view. The *Offset* option operates similarly to the *Offset* command described previously. Remember that an *Xline* extends to infinity but can be *Trimmed*, in which case it is converted to *Ray* (*Trim* once) or to a *Line* (*Trim* twice). See Chapter 15 for more information on the *Xline* command.

The *Ray* command also creates construction lines; however, the *Ray* has one anchored point and the other end extends to infinity.

FIGURE 24-14

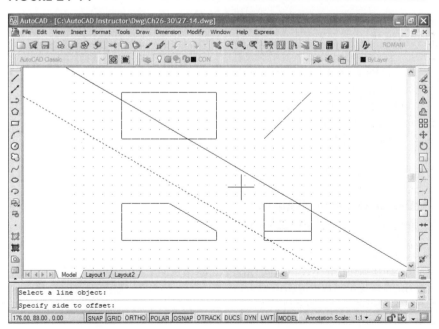

Command: **ray**
Specify start point: **PICK** or (**coordinates**)
Specify through point: **PICK** or (**coordinates**)

Rays are helpful for auxiliary view construction when you want to create projection lines perpendicular to the inclined plane. In Figure 24-15, two *Rays* are constructed from the *Endpoints* of the inclined plane and *Perpendicular* to the existing *Xlines*. Using *Xlines* and *Rays* in conjunction is an <u>excellent method</u> for "blocking in" the view.

There are two strategies for creating drawings using *Xlines* and *Rays*. First, these construction lines can be created on a separate layer and set up as a framework for the object lines. The object lines would then be drawn on top of the construction lines using *Osnaps*, but would be

FIGURE 24-15

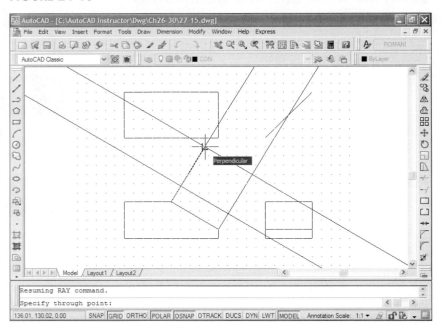

drawn on the object layer. The construction layer would be *Frozen* for plotting. The other strategy is to create the construction lines on the object layer. Through a series of *Trims* and other modifications, the *Xlines* and *Rays* are transformed to the finished object lines.

Constructing Full Auxiliary Views

The construction of a full auxiliary view begins with the partial view. After initial construction of the partial view, begin the construction of the full auxiliary by projecting the other edges and features of the object (other than the inclined plane) to the existing auxiliary view.

The procedure for constructing full auxiliary views in AutoCAD is essentially the same as that for partial auxiliary views. Use of the *Offset, Xline,* and *Ray* commands, *SNAP* and *GRID* rotation, and *Polar Tracking* should be used as illustrated for the partial auxiliary view example. Because a full auxiliary view is projected at the same angle as a partial, the same rotation angle and basepoint would be used for the *SNAP,* or the same *Increment angle* or *Additional angles* should be used for *Polar Tracking.*

CHAPTER EXERCISES

For the following exercises, create the multiview drawing, including the partial or full auxiliary view as indicated. Use the appropriate template drawing based on the given dimensions and indicated plot scale.

1.   Make a multiview drawing with a partial auxiliary view of the example used in this chapter. Refer to Figure 24-1 for dimensions. *Save* the drawing as **CH24EX1**. *Plot* on an A size sheet at **1=1** scale.

2. Recreate the views given in Figure 24-16 and add a partial auxiliary view. *Save* the drawing as **CH24EX2** and *Plot* on an A size sheet at **2=1** scale.

FIGURE 24-16

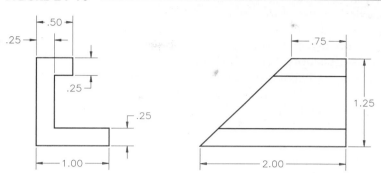

3. Recreate the views shown in Figure 24-17 and add a partial auxiliary view. *Save* the drawing as **CH27EX3.** Make a *Plot* on an A size sheet at **1=1** scale.

FIGURE 24-17

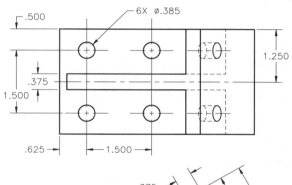

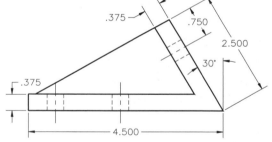

4. Make a multiview drawing of the given views in Figure 24-18. Add a full auxiliary view. *Save* the drawing as **CH24EX4**. *Plot* the drawing at an appropriate scale on an A size sheet.

FIGURE 24-18

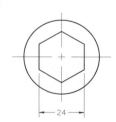

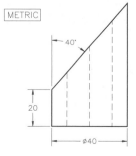

5. Draw the front, top, and a partial auxiliary view of the holder. During construction, note that the back corner indicated with hidden lines in Figure 24-19 is not a "clean" intersection, but is actually composed of two separate vertical edges. Make a *Plot* full size. Save the drawing as **HOLDER**.

FIGURE 24-19

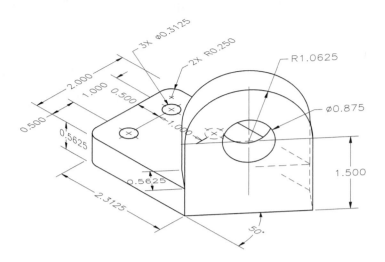

6. Draw three principal views and a full auxiliary view of the V-block shown in Figure 24-20. *Save* the drawing as **VBLOCK**. *Plot* to an accepted scale.

FIGURE 24-20

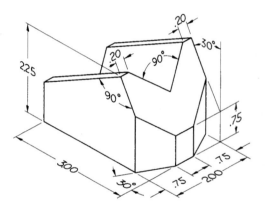

7. Draw two principal views and a partial auxiliary of the angle brace (Fig. 24-21).

    Save as **ANGLBRAC.** Make a plot to an accepted scale on an A or B size sheet. Convert the fractions to the current ANSI standard, decimal inches.

FIGURE 24-21

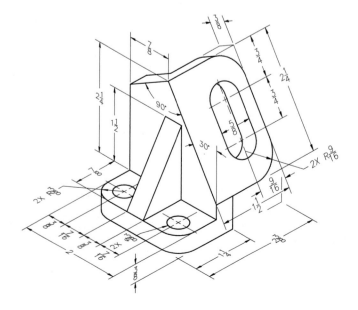

DIMENSIONING

CHAPTER OBJECTIVES

After completing this chapter you should:

1. be able to create linear dimensions with *Dimlinear*;

2. be able to append *Dimcontinue* and *Dimbaseline* dimensions to existing dimensions;

3. be able to create *Angular*, *Diameter*, and *Radius* dimensions;

4. know how to affix notes to drawings with *Leaders*;

5. know that *Dimordinate* can be used to specify Xdatum and Ydatum dimensions;

6. be able to use the *Qdim* command to quickly create a variety of associative dimensions;

7. know the possible methods for editing dimensions and dimensioning text.

CONCEPTS

As you know, drawings created with CAD systems should be constructed with the same size and units as the real-world objects they represent. In this way, the features of the object that you apply dimensions to (lengths, diameters, angles, etc.) are automatically measured by AutoCAD and the correct values are displayed in the dimension text. So if the object is drawn accurately, the dimension values will be created correctly and automatically. Generally, dimensions should be created on a separate layer named DIMENSIONS, or similar, and dimensions should be created in model space.

The main components of a dimension are:

1. Dimension line
2. Extension lines
3. Dimension text (usually a numeric value)
4. Arrowheads or tick marks

AutoCAD dimensioning is <u>semi-automatic</u>. When you invoke a command to create a linear dimension, AutoCAD only requires that you PICK an object or specify the extension line origins (where you want the extension lines to begin) and PICK the location of the dimension line (distance from the object). AutoCAD then measures the feature and draws the extension lines, dimension line, arrowheads, and dimension text (Fig. 25-1).

For linear dimensioning commands, there are <u>two ways</u> to specify placement for a dimension in AutoCAD: you can PICK the <u>object</u> to be dimensioned or you can PICK the two <u>extension line origins</u>. The simplest method is to select the object because it requires only one PICK (Fig. 25-2):

> Command: **dimlinear**
> Specify first extension line origin or <select object>: **Enter**
> Select object to dimension: **PICK**

The other method is to PICK the extension line origins (Fig. 25-3). <u>Osnaps should be used to PICK the object (endpoints in this case) so that the dimension is associated with the object</u>.

> Command: **dimlinear**
> Specify first extension line origin or <select object>: **PICK**
> Specify second extension line origin: **PICK**

Once the dimension is attached to the object, you specify how far you want the dimension to be placed from the object (called the "dimension line location").

FIGURE 25-1

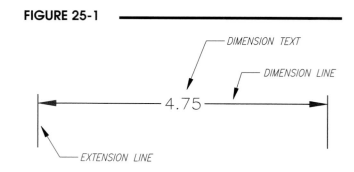

FIGURE 25-2

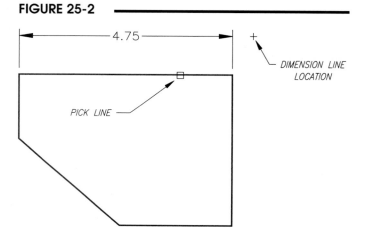

FIGURE 25-3

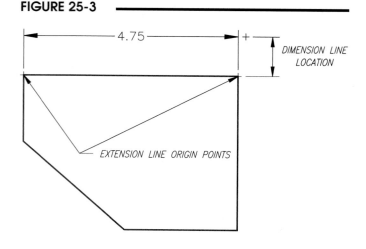

Dimensioning in AutoCAD is <u>associative</u> (by default). Because the extension line origins are "associated" with the geometry, the dimension text automatically updates if the geometry is *Moved*, *Stretched*, *Rotated*, *Scaled*, or edited using grips.

Because dimensioning is semi-automatic, <u>dimensioning variables</u> are used to control the way dimensions are created. Dimensioning variables can be used to control features such as text or arrow size, direction of the leader arrow for radial or diametrical dimensions, format of the text, and many other possible options. Groups of variable settings can be named and saved as <u>Dimension Styles</u>. Dimensioning variables and dimension styles are discussed in Chapter 26.

Dimensioning commands can be invoked by the typical methods. The *Dimension* pull-down menu contains the dimension creation and editing commands (Fig. 25-4). A *Dimension* toolbar (Fig. 25-5) can also be activated by using the *Toolbars* list. If you use the *2D Drafting & Annotation* workspace, a *Dimensions* panel is available on the dashboard.

FIGURE 25-4

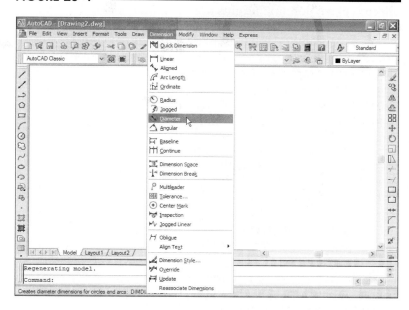

FIGURE 25-5

DIMENSION DRAWING COMMANDS

Dimlinear

Pull-down Menu	Command (Type)	Alias (Type)	Short-cut	Screen (side) Menu	Tablet Menu
Dimension Linear	Dimlinear	DIMLIN or DLI	...	DIMNSION Linear	W,5

Dimlinear creates a <u>horizontal, vertical, or rotated</u> dimension. If the object selected is a horizontal line (or the extension line origins are horizontally oriented), the resulting dimension is a horizontal dimension. This situation is displayed in the previous illustrations (see Fig. 25-2 and Fig. 25-3), or if the selected object or extension line origins are vertically oriented, the resulting dimension is vertical (Fig. 25-6).

FIGURE 25-6

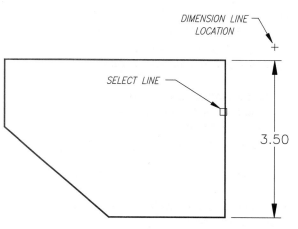

When you dimension an inclined object (or if the selected extension line origins are diagonally oriented), a vertical <u>or</u> horizontal dimension can be made, depending on where you drag the dimension line in relation to the object. If the dimension line location is more to the side, a vertical dimension is created (Fig. 25-7), or if you drag farther up or down, a horizontal dimension results (Fig. 25-8).

FIGURE 25-7

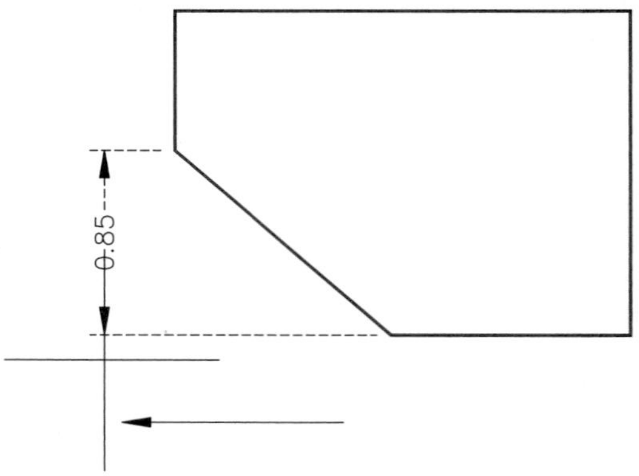

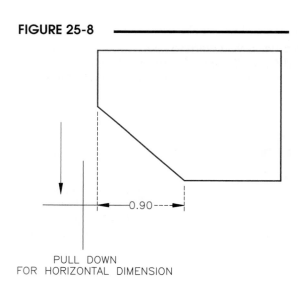

PULL TO SIDE
FOR VERTICAL DIMENSION

 If you select the extension line origins, it is very important to PICK the <u>object's endpoints</u> if the dimensions are to be truly associative (associated with the geometry). *OSNAP* should be used to find the object's *Endpoint, Intersection,* etc.

> Command: ***dimlinear***
> Specify first extension line origin or <select object>: **PICK** (use *Osnaps*)
> Specify second extension line origin: **PICK** (use *Osnaps*)
> Specify dimension line location or [Mtext/Text/Angle/Horizontal/Vertical/Rotated]: **PICK** (where you want the dimension line to be placed)
> Dimension text = *n.nnnn*
> Command:

When you pick the location for the dimension line, AutoCAD automatically measures the object and inserts the correct numerical value. The other options are explained next.

FIGURE 25-8

PULL DOWN
FOR HORIZONTAL DIMENSION

Rotated

If you want the dimension line to be drawn at an angle instead of vertical or horizontal, use this option. Selecting an inclined line, as in the previous two illustrations, would normally create a horizontal or vertical dimension. The *Rotated* option allows you to enter an <u>angular value</u> for the dimension line to be drawn. For example, selecting the diagonal line and specifying the appropriate angle would create the dimension shown in Figure 25-9. This object, however, could be more easily dimensioned with the *Dimaligned* command (see "*Dimaligned*").

FIGURE 25-9

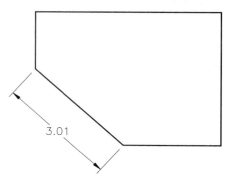

A *Rotated* dimension should be used when the geometry has "steps" or any time the desired dimension line angle is different than the dimensioned feature (when you need extension lines of different lengths). Figure 25-10 illustrates the result of using a *Rotated* dimension to give the correct dimension line angle and extension line origins for the given object. In this case, the extension line origins were explicitly PICKed. The feature of *Rotated* that makes it unique is that you specify the <u>angle</u> that the dimension line will be drawn.

FIGURE 25-10

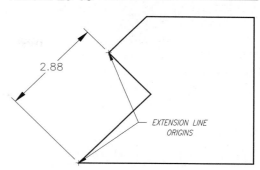

Text

Using the *Text* option allows you to enter any value or characters in place of the AutoCAD-measured text. The measured value is given as a reference at the command prompt.

> Command: ***dimlinear***
> Specify first extension line origin or <select object>: **PICK**
> Specify second extension line origin: **PICK**
> Specify dimension line location or
> [Mtext/Text/Angle/Horizontal/Vertical/Rotated]: ***T***
> Enter dimension text <*n.nnnn*>: (**text**) or (**value**)

FIGURE 25-11

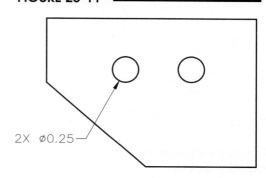

Entering a value or text at the prompt (above) causes AutoCAD to display that value or text instead of the AutoCAD-measured value. If you want to keep the AutoCAD-measured value but add a prefix or suffix, use the less-than, greater-than symbols (< >) to represent the actual (AutoCAD) value:

> Enter dimension text <0.25>: **2X <>**
> Specify dimension line location or [Mtext/Text/Angle/Horizontal/Vertical/Rotated]: **PICK**
> Dimension text = 0.25
> Command:

The "2X <>" response produces the dimension text shown in Figure 25-11.

NOTE: Changing the AutoCAD-measured value should be discouraged. If the geometry is drawn accurately, the dimensional value is correct. If you specify other dimension text, the text value is <u>not</u> updated in the event of *Stretching, Rotating,* or otherwise editing the associative dimension.

Mtext

This option allows you to enter text using the *Text Formatting* Editor. With the *Mtext* option, the actual AutoCAD-supplied text appears in the editor instead of being represented with less-than and greater-than symbols (<>). It is not recommended to change the AutoCAD-supplied text; however, you can enter any text before or after. Note that the AutoCAD-supplied text appears in a dark background, whereas any text you enter appears with a lighter background (Fig. 25-12). Keep in mind that you have the power to change fonts, text height, bold, italic, underline, and stacked text or fractions using this editor.

FIGURE 25-12

Angle

This creates text drawn at the angle you specify. Use this for special cases when the text must be drawn to a specific angle other than horizontal. (It is also possible to make the text automatically align with the angle of the dimension line using the *Dimension Style Manager*. See Chapter 26.)

Horizontal

Use the *Horizontal* option when you want to force a horizontal dimension for an inclined line and the desired placement of the dimension line would otherwise cause a vertical dimension.

Vertical

This option forces a *Vertical* dimension for any case.

Dimaligned

Pull-down Menu	Command (Type)	Alias (Type)	Short-cut	Screen (side) Menu	Tablet Menu
Dimension Aligned	Dimaligned	DIMALI or DAL	...	DIMNSION Aligned	W,4

An *Aligned* dimension is aligned with (at the same angle as) the selected object or the extension line origins. For example, when *Aligned* is used to dimension the angled object shown in Figure 25-13, the resulting dimension aligns with the *Line*. This holds true for either option—PICKing the object or the extension line origins. If a *Circle* is PICKed, the dimension line is aligned with the selected point on the *Circle* and its center. The command syntax for the *Dimaligned* command accepting the defaults is:

FIGURE 25-13

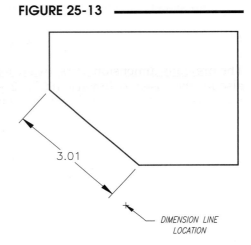

3.01

DIMENSION LINE LOCATION

```
Command: dimaligned
Specify first extension line origin or <select object>: PICK
Specify second extension line origin: PICK
Specify dimension line location or [Mtext/Text/Angle]: PICK
Dimension text = n.nnnn
Command:
```

The three options (*Mtext/Text/Angle*) operate similar to those for *Dimlinear*.

Mtext

The *Mtext* option calls the *Text Formatting* Editor. You can alter the AutoCAD-supplied text value or modify other visible features of the dimension text such as fonts, text height, bold, italic, underline, stacked text or fractions, and text style (see "*Dimlinear*," *Mtext*).

Text

You can change the AutoCAD-supplied numerical value or add other annotation to the value in command line format (see "*Dimlinear*," *Text*).

Angle

Enter a value for the angle that the text will be drawn.

TIP The typical application for *Dimaligned* is for dimensioning an angled but <u>straight</u> feature of an object, as shown in Figure 25-13. *Dimaligned* should not be used to dimension an object feature that contains "steps," as shown in Figure 25-10. *Dimaligned* always draws <u>extension lines of equal length</u>.

Dimbaseline

Pull-down Menu	Command (Type)	Alias (Type)	Short-cut	Screen (side) Menu	Tablet Menu
Dimension Baseline	*Dimbaseline*	*DIMBASE* or *DBA*	...	*DIMNSION Baseline*	...

Dimbaseline allows you to create a dimension that uses an extension line origin from a previously created dimension. Successive *Dimbaseline* dimensions can be used to create the style of dimensioning shown in Figure 25-14.

A baseline dimension must be connected to an existing dimension. If *Dimbaseline* is invoked immediately after another dimensioning command, you are required only to specify the second extension line origin since AutoCAD knows to use the <u>previous</u> dimension's <u>first</u> extension line origin:

    Command: **dimbaseline**
    Specify a second extension line origin or [Undo/Select]
    <Select>: **PICK**
    Dimension text = *n.nnnn*

FIGURE 25-14
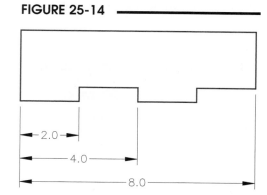

The <u>previous dimension's first extension line</u> is used also for the baseline dimension (Fig. 25-15). Therefore, you specify only the second extension line origin. Note that you are not required to specify the dimension line location. AutoCAD spaces the new dimension line automatically, based on the setting of the dimension line increment variable (Chapter 26).

FIGURE 25-15
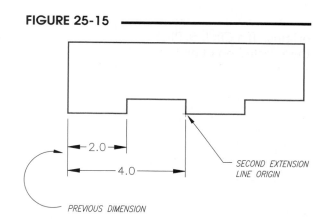

If you wish to create a *Dimbaseline* dimension using a dimension other than the one just created, use the "Select" option (Fig. 25-16):

    Command: **dimbaseline**
    Specify a second extension line origin or
    [Undo/Select] <Select>: **S** or **Enter**
    Select base dimension: **PICK**
    Specify a second extension line origin or
    [Undo/Select] <Select>: **Enter**
    Dimension text = *n.nnnn*

FIGURE 25-16
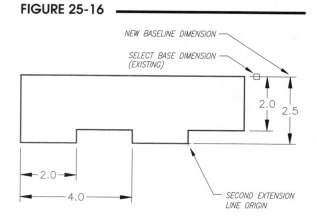

The extension line selected as the base dimension becomes the first extension line for the new *Dimbaseline* dimension.

The *Undo* option can be used to undo the last baseline dimension created in the current command sequence.

Dimbaseline can be used with rotated, aligned, angular, and ordinate dimensions.

Dimcontinue

	Pull-down Menu	Command (Type)	Alias (Type)	Short-cut	Screen (side) Menu	Tablet Menu
	Dimension Continue	Dimcontinue	DIMCONT or DCO	...	DIMNSION Continue	...

Dimcontinue dimensions continue in a line from a previously created dimension. *Dimcontinue* dimension lines are attached to, and drawn the same distance from, the object as an existing dimension.

Dimcontinue is similar to *Dimbaseline* except that an existing dimension's <u>second</u> extension line is used to begin the new dimension. In other words, the new dimension is connected to the <u>second</u> extension line, rather than to the <u>first</u>, as with a *Dimbaseline* dimension (Fig. 25-17).

The command syntax is as follows:

Command: **dimcontinue**
Specify a second extension line origin or
[Undo/Select] <Select>: **PICK**
Dimension text = *n.nnnn*

FIGURE 25-17 ───────────

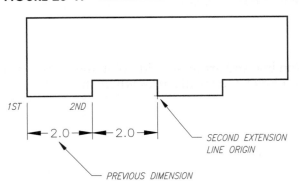

Assuming a dimension was just drawn, *Dimcontinue* could be used to place the next dimension, as shown in Figure 25-17.

 If you want to create a continued dimension and attach it to an extension line <u>other</u> than the previous dimension's second extension line, you can use the *Select* option to pick an extension line of any other dimension. Then, use *Dimcontinue* to create a continued dimension from the selected extension line (Fig. 25-18).

The *Undo* option can be used to undo the last continued dimension created in the current command sequence. *Dimcontinue* can be used with rotated, aligned, angular, and ordinate dimensions.

FIGURE 25-18 ───────────

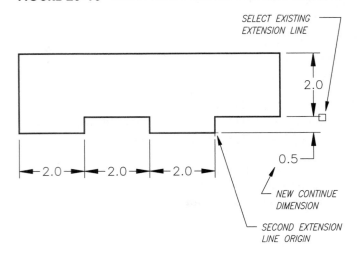

Dimdiameter

Pull-down Menu	Command (Type)	Alias (Type)	Short-cut	Screen (side) Menu	Tablet Menu
Dimension Diameter	Dimdiameter	DIMDIA or DDI	...	DIMNSION Diameter	X,4

The *Dimdiameter* command creates a diametrical dimension by selecting any *Circle*. Diametrical dimensions should be used for full 360 degree *Circles* and can be used for *Arcs* of more than 180 degrees.

```
Command: dimdiameter
Select arc or circle: PICK
Dimension text = n.nnnn
Specify dimension line location or [Mtext/Text/Angle]: PICK
Command:
```

You can PICK the circle at any location. AutoCAD allows you to adjust the position of the dimension line to any angle or length (Fig. 25-19). Dimension lines for diametrical or radial dimensions should be drawn to a regular angle, such as 30, 45, or 60 degrees, never vertical or horizontal.

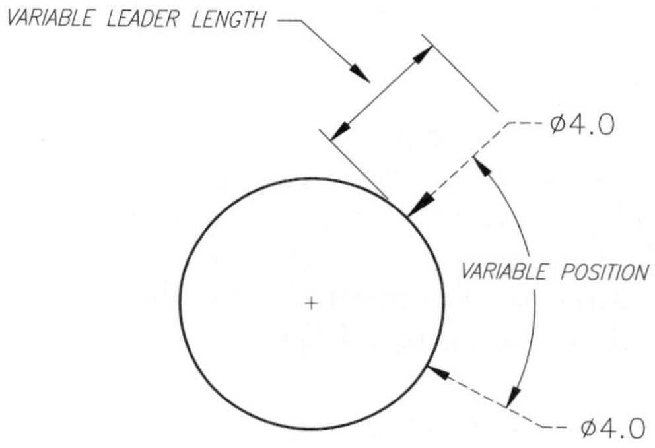

A typical diametrical dimension appears as the example in Figure 25-19. According to ANSI standards, a diameter dimension line and arrow should point inward (toward the center) for holes and small circles where the dimension line and text do not fit within the circle. Use the default variable settings for *Dimdiameter* dimensions such as this.

FIGURE 25-19

VARIABLE LEADER LENGTH
Ø4.0
VARIABLE POSITION
Ø4.0

For dimensioning large circles, ANSI standards suggest an alternate method for diameter dimensions where sufficient room exists for text and arrows inside the circle (Fig. 25-20). To create this style of dimensioning, set the variables to *Text* and *Place text manually when dimensioning* in the *Fit* tab of the *Dimension Style Manager* (see Chapter 26).

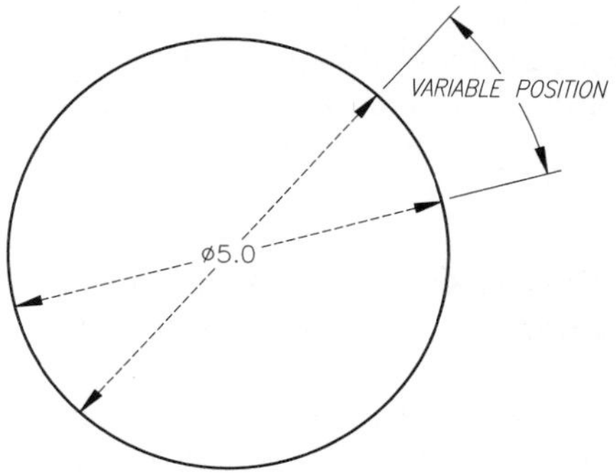

Notice that the *Diameter* command creates center marks at the *Circle's* center. Center marks can also be drawn by the *Dimcenter* command (discussed later). AutoCAD uses the center and the point selected on the *Circle* to maintain its associativity.

FIGURE 25-20

VARIABLE POSITION
Ø5.0

Mtext/Text

The *Mtext* or *Text* options can be used to modify or add annotation to the default value. The *Mtext* option summons the *Text Formatting* Editor and the *Text* option uses Command line format. Both options operate similar to the other dimensioning commands (see *Dimlinear, Mtext,* and *Text*).

Notice that with diameter dimensions AutoCAD automatically creates the Ø (phi) symbol before the dimensional value. This is the latest ANSI standard for representing diameters. If you prefer to use a prefix before or suffix after the dimension value, it can be accomplished with the *Mtext* or *Text* option.

TIP Remember that the < > symbols represent the AutoCAD-measured value so the additional text should be inserted on <u>either side</u> of the symbols. Inserting a prefix by this method <u>does not override</u> the Ø (phi) symbol (Fig. 25-21).

A prefix or suffix can alternately be added to the measured value by using the *Dimension Style Manager* and entering text or values in the *Prefix* or *Suffix* edit boxes. Using this method, however, <u>overrides</u> the Ø symbol (see Chapter 26).

Angle

With this option, you can specify an angle (other than the default) for the text to be drawn by entering a value.

FIGURE 25-21

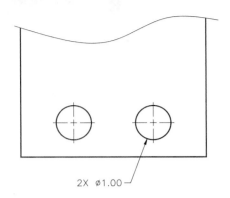

2X Ø1.00

Dimradius

Pull-down Menu	Command (Type)	Alias (Type)	Short-cut	Screen (side) Menu	Tablet Menu
Dimension Radius	Dimradius	DIMRAD or DRA	...	DIMNSION Radius	X,5

Dimradius is used to create a dimension for an arc of anything less than half of a circle. ANSI standards dictate that a *Radius* dimension line should point outward (from the arc's center), unless there is insufficient room, in which case the line can be drawn on the outside pointing inward, as with a leader. The text can be located inside an arc (if sufficient room exists) or is forced outside of small *Arcs* on a leader.

```
Command: dimradius
Select arc or circle: PICK
Dimension text = n.nnnn
Specify dimension line location or [Mtext/Text/Angle]: PICK
Command:
```

Assuming the defaults, a *Dimradius* dimension can appear on either side of an arc, as shown in Figure 25-22. Placement of the dimension line is variable. Dimension lines for arcs and circles should be positioned at a regular angle such as 30, 45, or 60 degrees, never vertical or horizontal.

When the radius dimension is dragged outside of the arc, a center mark is automatically created (Fig. 25-22). When the radius dimension is dragged inside the arc, no center mark is created. (Center marks can be created using the *Center* command discussed next.)

According to ANSI standards, the dimension line and arrow should point <u>outward</u> from the center for radial dimensions (Fig. 25-23). The text can be placed inside or outside of the *Arc*, depending on how much room exists. To create a *Dimradius* dimension to comply with this standard, dimension variables must be changed from the defaults. To achieve *Dimradius* dimensions as shown in Figure 25-23, set the variables to *Text* and *Place text manually when dimensioning* in the *Fit* tab of the *Dimension Style Manager* (see Chapter 26).

FIGURE 25-22

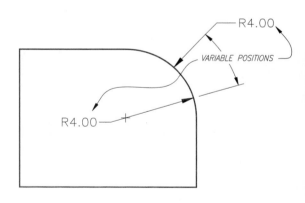

R4.00

VARIABLE POSITIONS

R4.00

FIGURE 25-23

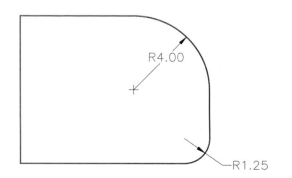

R4.00

R1.25

 TIP

For very small radii, such as that shown in Figure 25-24, there is insufficient room for the text and arrow to fit inside the *Arc*. In this case, AutoCAD automatically forces the text outside with the leader pointing inward toward the center. No changes have to be made to the default settings for this to occur.

FIGURE 25-24

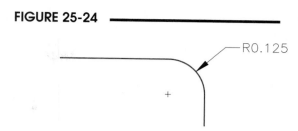

Mtext/Text

These options can be used to modify or add annotation to the default value. AutoCAD automatically inserts the letter "R" before the numerical text whenever a *Dimradius* dimension is created. This is the correct notation for radius dimensions. The *Mtext* option calls the *Text Formatting* Editor, and the *Text* option uses the Command line format for entering text. With the *Text* option AutoCAD uses the < > symbols to represent the AutoCAD-supplied value. Entering text inside the < > symbols overrides the measured value. Entering text before the symbols adds a prefix without overriding the "R" designation. Alternately, text can be added by using the *Prefix* and *Suffix* options of the *Dimension Style Manager* series; however, a *Prefix* entered in the edit box replaces the letter "R." (See Chapter 26.)

Angle

With this option, you can specify an angle (other than the default) for the text to be drawn by entering a value.

Dimcenter

Pull-down Menu	Command (Type)	Alias (Type)	Short-cut	Screen (side) Menu	Tablet Menu
Dimension Center Mark	Dimcenter	DCE	...	DIMNSION Center	X,2

The *Dimcenter* command draws a center mark on any selected *Arc* or *Circle*. As shown earlier, the *Dimdiameter* command and the *Dimradius* command sometimes create the center marks automatically.

The command requires you only to select the desired *Circle* or *Arc* to acquire the center marks:

    Command: **dimcenter**
    Select arc or circle: **PICK**
    Command:

No matter if the center mark is created by the *Center* command or by the *Diameter* or *Radius* commands, the center mark can be either a small cross or complete center lines extending past the *Circle* or *Arc* (Fig. 25-25). The type of center mark drawn is controlled by the *Mark* or *Line* setting in the *Lines and Arrows* tab of the *Dimension Style Manager*. It is suggested that a *Dimension Style* be adjusted with these settings just for drawing center marks. (See Chapter 26.)

When you are dimensioning, short center marks should be used for *Arcs* of less than 180 degrees, and full center lines should be drawn for *Circles* and for *Arcs* of 180 degrees or more (Fig. 25-26).

FIGURE 25-25

FIGURE 25-26

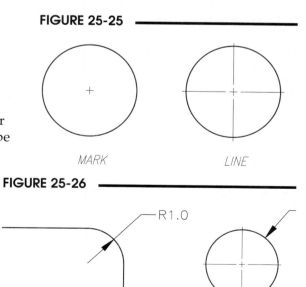

TIP

NOTE: Since the center marks created with the *Dimcenter* command are <u>not</u> associative, they may be *Trimmed*, *Erased*, or otherwise edited. The center marks created with the *Dimradius* or *Dimdiameter* commands <u>are</u> associative and cannot be edited.

Dimangular

Pull-down Menu	Command (Type)	Alias (Type)	Short-cut	Screen (side) Menu	Tablet Menu
Dimension Angular	Dimangular	DIMANG or DAN	...	DIMNSION Angular	X,3

The *Dimangular* command provides many possible methods of creating an angular dimension.

A typical angular dimension is created between two *Lines* that form an angle (of other than 90 degrees). The dimension line for an angular dimension is radiused with its center at the vertex of the angle (Fig. 25-27). A *Dimangular* dimension automatically adds the degree symbol (°) to the dimension text. The dimension text format is controlled by the current settings for *Units* in the *Dimension Style* dialog box.

FIGURE 25-27

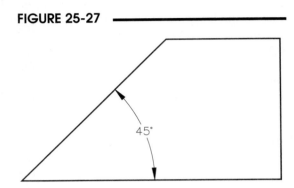

AutoCAD automates the process of creating this type of dimension by offering options within the command syntax. The default options create a dimension, as shown here:

    Command: **dimangular**
    Select arc, circle, line, or <specify vertex>: **PICK**
    Select second line: **PICK**
    Specify dimension arc line location or [Mtext/Text/Angle/Quadrant]: **PICK**
    Dimension text = *nn*
    Command:

Dimangular dimensioning offers some very useful and easy-to-use options for placing the desired dimension line and text location.

At the "Specify dimension arc line location or [Mtext/Text/Angle/Quadrant]:" prompt, you can move the cursor around the vertex to dynamically display possible placements available for the dimension. The dimension can be placed in any of four positions as well as any distance from the vertex (Fig. 25-28). Extension lines are automatically created as needed.

FIGURE 25-28

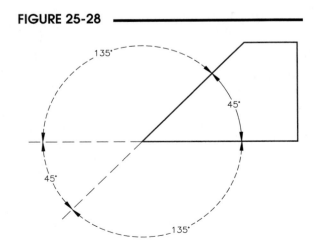

On the other hand, you can use the *Quadrant* option to lock the dimension lines to one of the four "quadrants" while you drag only the text outside the quadrant (see the four "quadrants" in Fig. 25-28). An arc extension line is then added from the quadrant to the text.

The *Dimangular* command offers other options, including dimensioning angles for *Arcs*, *Circles*, or allowing selection of any three points. If you select an *Arc* in response to the "Select arc, circle, line, or <specify vertex>:" prompt, AutoCAD uses the *Arc's* center as the vertex and the *Arc's* endpoints to generate the extension lines. You can select either angle of the *Arc* to dimension (Fig. 25-29).

FIGURE 25-29

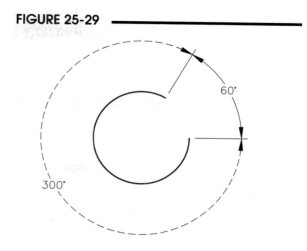

If you select a *Circle*, AutoCAD uses the PICK point as the first extension line origin. The second extension line origin does not have to be on the *Circle*, as shown in Figure 25-30.

FIGURE 25-30

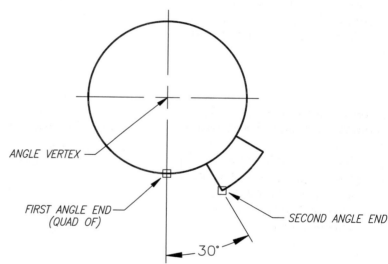

If you press Enter in response to the "Select arc, circle, line, or <specify vertex>:" prompt, AutoCAD responds with the following:

Specify angle vertex: **PICK**
Specify first angle endpoint: **PICK**
Specify second angle endpoint: **PICK**

This option allows you to apply an *Angular* dimension to a variety of shapes.

Dimbreak

Pull-down Menu	Command (Type)	Alias (Type)	Short-cut	Screen (side) Menu	Tablet Menu
Dimension > Dimension Break	Dimbreak	...	...	...	...

According to industry standards for architecture and engineering, you should not cross a dimension line with another dimension line or extension line (although extension lines can cross each other and can cross object lines). However, there are some cases where you cannot avoid crossing a dimension line with another dimension or extension line. For example in Figure 25-31, the "0.40" dimension is crossed by the extension line of the other two dimensions, but even with other arrangements of these dimensions there is no way to avoid crossing a dimension line.

FIGURE 25-31

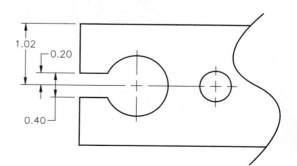

2008

In cases such as that shown, you can use the *Dimbreak* command to create a space in the extension or dimension line that causes the infraction. The Command line reports the following:

> Command: **dimbreak**
> Select a dimension or [Multiple]: **PICK** (select the line on you want to break)
> Select object to break dimension or [Auto/Restore/Manual] <Auto>: **PICK** or **Enter**
> Command:

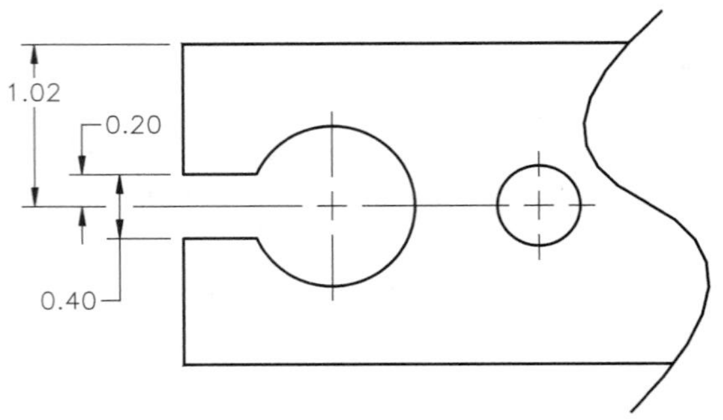

When dealing with two dimensions, at the "Select a dimension or [Multiple]:" prompt, <u>select the line to be broken</u>. For the example in Figure 25-32, select the horizontal center/extension line that crosses the "0.40" dimension. At the "Select object to break dimension" prompt, select the dimension line that is to remain unbroken. Alternately, pressing Enter at the next prompt automatically creates the break.

NOTE: Although you can also break a dimension line using an object line with *Dimbreak*, this practice should be avoided since it violates industry standards. The options are described next.

FIGURE 25-32

Multiple
Use this option to specify multiple dimensions to add breaks to or remove breaks from.

Auto
The *Auto* option automatically places breaks on the selected dimension or extension line where any other dimensions intersect. Any dimension break created using this option is automatically updated when either the dimension or an intersecting dimension is modified. However, when a new dimension is drawn that crosses a dimension that has previously created dimension breaks, no new breaks are applied at the intersecting points along the dimension object. To add new dimension breaks, you must use *Dimbreak* again.

Restore
Use this option to remove previously created breaks from the selected dimensions.

Manual
This option works similarly to the *Break* command where you can manually select the two points for the break. Dimension breaks can be added to dimensions even when objects do not intersect the dimension or extension lines. You can create only one manual dimension break at a time.

Leader

Pull-down Menu	Command (Type)	Alias (Type)	Short-cut	Screen (side) Menu	Tablet Menu
...	*Leader*	*LEAD*	...	...	...

The *Leader* command (not *Qleader*) allows you to create an associative leader similar to that created with the *Diameter* command. The *Leader* command is intended to give dimensional notes such as the manufacturing or construction specifications shown next.

2008

Command: *leader*
Specify leader start point: **PICK**
Specify next point: **PICK**
Specify next point or [Annotation/Format/Undo] <Annotation>: **Enter**
Enter first line of annotation text or <options>: **CASE HARDEN**
Enter next line of annotation text: **Enter**
Command:

At the "start point:" prompt, select the desired location for the arrow. You should use an *Osnap* option (such as *Nearest*, in this case) to ensure the arrow touches the desired object.

A short horizontal line segment called the "hook line" is automatically added to the last line segment drawn if the leader line is 15 degrees or more from horizontal. Note that command syntax for the *Leader* in Figure 25-33 indicates only one line segment was PICKed. A *Leader* can have as many segments as you desire.

FIGURE 25-33

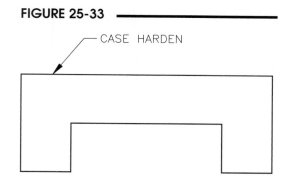

If you do not enter text at the "Annotation:" prompt, another series of options are available:

Command: *leader*
Specify leader start point: **PICK**
Specify next point: **PICK**
Specify next point or [Annotation/Format/Undo] <Annotation>: **Enter**
Enter first line of annotation text or <options>: **Enter**
Enter an annotation option [Tolerance/Copy/Block/None/Mtext] <Mtext>:

Format

This option produces another list of choices:

Enter leader format option [Spline/STraight/Arrow/None] <Exit>:

Spline/STraight

You can draw either a *Spline* or straight version of the leader line with these options. The resulting *Spline* leader line has the characteristics of a normal *Spline*. An example of a *Splined* leader is shown in Figure 25-34.

FIGURE 25-34

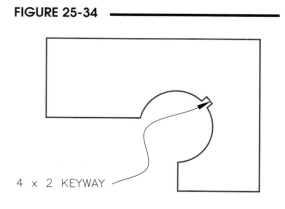

Arrow/None

This option draws the leader line with or without an arrowhead at the start point.

Annotation

This option prompts for text to insert at the end of the *Leader* line.

Mtext

The *Text Formatting* Editor appears with this option. Text can be entered into paragraph form and you can use the *Mtext* options (see "*Mtext*," Chapter 18).

Tolerance

This option produces a feature control frame using the *Geometric Tolerances* dialog boxes.

Copy

You can copy existing *Text* or *Mtext* objects from the drawing to be placed at the end of the *Leader* line. The copied object is associated with the *Leader* line.

Block

An existing *Block* of your selection can be placed at the end of the *Leader* line. The same prompts as the *Insert* command are used.

None

Using this option draws the *Leader* line with no annotation.

Undo

This option undoes the last vertex point of the *Leader* line.

Mleader

Pull-down Menu	Command (Type)	Alias (Type)	Short-cut	Screen (side) Menu	Tablet Menu
Dimension > Multileader	*Mleader*	*MLD*	...	...	...

The *Mleader* command creates a "multileader" object. A multileader is similar to a *Qleader*; however, you have more options and flexibility for annotating drawings using *Mleader*. For example, with a multi-leader, you can place the arrowhead first, the tail first, or the text content first.

A *Multileader* toolbar is available (Fig. 25-35). This toolbar contains buttons to access all related multileader commands. Included are the *Mleaderstyle* command which allows you to specify the parameters of the leader and text content, and the style drop-down list that allows you to select from existing styles.

FIGURE 25-35 ────────

The *Mleader* command creates multileaders. The default prompt is below:

        Command: **mleader**
        Specify leader arrowhead location or [leader Landing first/Content first/Options] <Options>: **PICK**
        Specify leader landing location: **PICK**
        Command:

Using the default prompts, similar to *Leader* and *Qleader*, you specify the location for the arrowhead first, then the other end of the leader line called the "landing." When the two end points of the line are selected, the *Text Formatting* editor appears. Key in the desired text content and select *OK* in the editor to end the command sequence.

Leader Landing First

Use this option to start drawing the multileader by selecting the location of the "landing" for the leader (Fig. 25-36).

Content First

Content First allows you to place the location of the text content first, then pick the location for the arrowhead (Fig. 25-36).

FIGURE 25-36 ────────

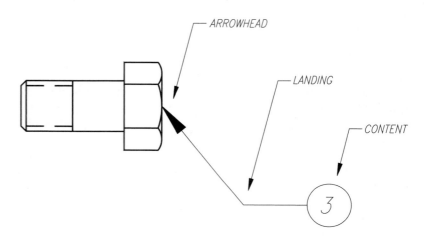

Options

The format of the leader and content is controlled by the *Multileader Style* and/or *Options* section of the *Mleader* command. Entering "O" at the "Specify leader arrowhead location or [leader Landing first/Content first/Options] <Options>:" prompt provides the following prompt:

Enter an option [Leader type/leader lAnding/Content type/Maxpoints/First angle/Second angle/eXit options] <eXit options>:

Mleaderedit

Pull-down Menu	Command (Type)	Alias (Type)	Short-cut	Screen (side) Menu	Tablet Menu
Modify > *Object >* *Multileader >* *Add, Remove*	*Mleaderedit*	MLE	...	...	...

Use *Mleaderedit* to add leaders to or remove leaders from an existing multileader.

If you add leaders, the new leaders automatically connect to an end of the existing landing line or text content depending on the direction of the new leader (Fig. 25-37). For example, to add a new leader line, use the following prompts:

FIGURE 25-37

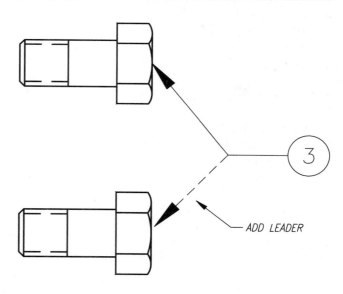

    ADD LEADER

Command: **mleaderedit**
Select a multileader: **PICK**
Select an option [Add leader/Remove leader] <Add leader>: **A**
Specify leader arrowhead location: **PICK**
Specify leader arrowhead location: **Enter**
Command:

Add Leader

Use this option to add a leader line to the selected multileader object. The new leader line is added to the left or right of the selected multileader, depending on the location of the cursor when you pick (Fig. 25-37). In the case that more than two leader points are defined in the current multileader style, you are prompted to specify additional points.

Remove Leader

Simply select a leader line from the selected multileader object to remove.

Mleaderalign

	Pull-down Menu	Command (Type)	Alias (Type)	Short-cut	Screen (side) Menu	Tablet Menu
	Modify > *Object >* *Multileader >* *Align*	*Mleaderalign*	*MLA*	...	...	...

Mleaderalign is used to align the text content for a group of multileaders. For example, assume your drawing contained two multi-leaders as shown in Figure 25-38. In this case, the text content does not align vertically.

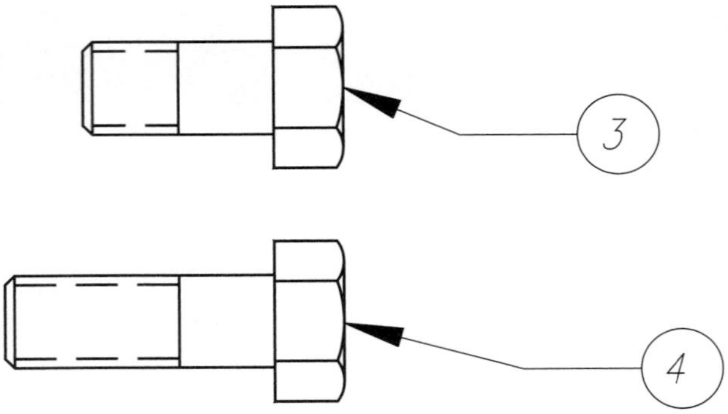

To remedy this inconsistency, use *Mleaderalign*. Follow the prompts shown next.

> Command: **mleaderalign**
> Select multileaders: **PICK**
> Select multileaders: **PICK**
> Select multileaders: **Enter**
> Current mode: Use current spacing
> Select multileader to align to or [Options]: **PICK**
> Specify direction: **PICK**
> Command:

After selecting the two leaders, select one of the leaders to use as the reference at the "Select multileader to align to" prompt. At the "Specify direction" prompt, an alignment vector appears (attached to the cursor) for you to use as a guide. Swing this guide to the desired orientation, click, and the text content automatically aligns (Fig. 25-39). The following options are available:

> Select multileader to align to or [Options]: **o**
> Enter an option [Distribute/make leader segments Parallel/specify Spacing/Use current spacing] <Use current spacing>:

FIGURE 25-38

FIGURE 25-39

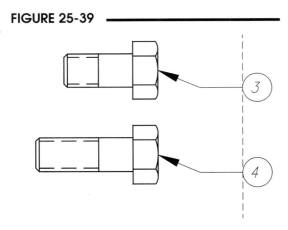

Distribute
This option causes the content to be spaced evenly between two selected points that you specify.

Make Leader Segments Parallel
This option makes the selected multileader lines parallel. This option does not align the text content. You may want to first use the *Use current spacing* option to align the text, then use *Make leader lines parallel*.

Specify Spacing
Enter a value to specify the spacing between the landing lines of the selected multileaders. You then indicate the desired direction between the landing lines in the "Specify direction" step.

Use Current

This option aligns the text content (vertically or horizontally) based on the multileader you select at the "Select multileader to align to" prompt.

Mleadercollect

Pull-down Menu	Command (Type)	Alias (Type)	Short-cut	Screen (side) Menu	Tablet Menu
Modify > *Object >* *Multileader >* *Collect*	*Mleaderalign*	*MLC*	...	...	...

Mleadercollect simply combines multiple leader lines into one, then stacks the text content based on your specification. For example, assume you have an assembly with two multileaders indicating the components with part number bubbles (Fig. 25-40). You can use *Mleadercollect* to combine the leaders into one leader containing both text bubbles.

Command: **mleadercollect**
Select multileaders: **PICK**
Select multileaders: **PICK**
Select multileaders: **Enter**
Specify collected multileader location or
[Vertical/Horizontal/Wrap] <Horizontal>: **PICK**
Command:

FIGURE 25-40

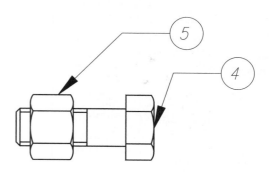

The resulting drawing (Fig. 25-41) would display only one leader containing both the original bubbles of text. The possible methods for collecting the text content are given next.

FIGURE 25-41

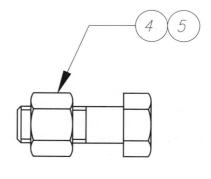

Vertical

With this option the multiple text objects are placed in a vertical orientation.

Horizontal

Similarly, this option places the multileader collection in a horizontal orientation.

Wrap

When you combine many leaders, the collection of text contents can create a long string. In such a case, use *Wrap* to specify a width for a wrapped multileader collection. Specify either a *Width* or a maximum *Number* of blocks per row in the multileader collection.

Dimordinate

Pull-down Menu	Command (Type)	Alias (Type)	Short-cut	Screen (side) Menu	Tablet Menu
Dimension *Ordinate*	*Dimordinate*	*DIMORD* *or DOR*	...	*DIMNSION* *Ordinate*	*W,3*

Ordinate dimensioning is a specialized method of dimensioning used in the manufacturing of flat components such as those in the sheet metal industry. Because the thickness (depth) of the parts is uniform, only the width and height dimensions are specified as Xdatum and Ydatum dimensions.

Dimordinate dimensions give an Xdatum or a Ydatum distance between object "features" and a reference point on the geometry treated as the origin, usually the lower-left corner of the part. This method of dimensioning is relatively simple to create and easy to understand. Each dimension is composed only of one leader line and the aligned numerical value (Fig. 25-42).

FIGURE 25-42

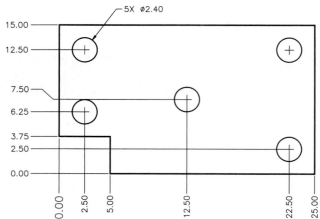

TIP

To create ordinate dimensions in AutoCAD, the *UCS* command should be used first to establish a new 0,0 point. *UCS*, which stands for user coordinate system, allows you to establish a new coordinate system with the origin and the orientation of the axes anywhere in 3D space . In this case, we need to change only the location of the origin and leave the orientation of the axes as is. Type *UCS* and use the *Origin* option to PICK a new origin as shown (Fig. 25-43).

FIGURE 25-43

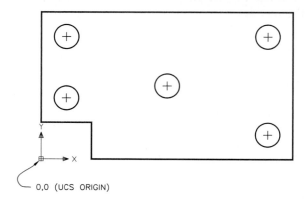

When you create a *Dimordinate* dimension, AutoCAD only requires you to (1) PICK the object "Feature" and then (2) specify the "Leader endpoint:". The dimension text is automatically aligned with the leader line.

It is not necessary in most cases to indicate whether you are creating an Xdatum or a Ydatum. Using the default option of *Dimordinate*, AutoCAD makes the determination based on the direction of the leader you specify (step 2). If the leader is <u>perpendicular</u> (or almost perpendicular) to the X axis, an Xdatum is created. If the leader is (almost) <u>perpendicular</u> to the Y axis, a Ydatum is created.

The command syntax for a *Dimordinate* dimension is this:

> Command: **dimordinate**
> Specify feature location: **PICK**
> Specify leader endpoint or
> [Xdatum/Ydatum/Mtext/Text/Angle]: **PICK**
> Dimension text = *n.nnnn*
> Command:

FIGURE 25-44

A *Dimordinate* dimension is created in Figure 25-44 by PICKing the object feature and the leader endpoint. That's all there is to it. The dimension is a Ydatum; yet AutoCAD automatically makes that determination, since the leader is perpendicular to the Y axis. It is a good practice to turn *ORTHO* or *POLAR ON* in order to ensure the leader lines are drawn horizontally or vertically.

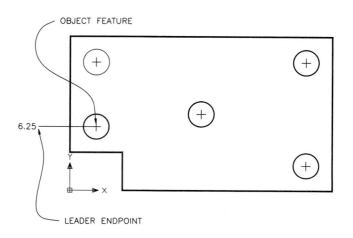

Qdim

Pull-down Menu	Command (Type)	Alias (Type)	Short-cut	Screen (side) Menu	Tablet Menu
Dimension *QDIM*	*Qdim*	...	...	...	W,1

Qdim (Quick Dimension) creates associative dimensions. *Qdim* simplifies the task of dimensioning by creating <u>multiple dimensions with one command</u>. *Qdim* can create *Continuous, Baseline, Radius, Diameter,* and *Ordinate* dimensions.

```
Command: qdim
Associative dimension priority = Endpoint
Select geometry to dimension: PICK
Select geometry to dimension: Enter
Specify dimension line position, or [Continuous/Staggered/Baseline/Ordinate/Radius/Diameter/
datumPoint/Edit/seTtings] <Continuous>: PICK
```

When *Qdim* prompts to "Select geometry to dimension," you can specify the geometry using any selection method such as a window, crossing window, or pickbox. For example, *Qdim* is used to dimension the geometry shown in Figure 25-45. Here, a crossing window is used to select several features on one side of the shape. The only other step is to select where (how far from the geometry) you want the dimensions to appear.

FIGURE 25-45

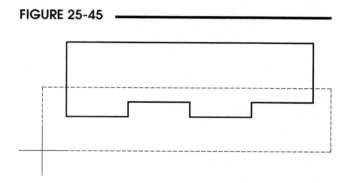

The resulting dimensions differ based on the *Qdim* options used and the current *Dimension Style* settings (see Chapter 26 for information on Dimension Styles). Each *Qdim* option and the resulting dimensions are described next.

SeTtings

This option specifies the priority object snap *Qdim* uses to automatically draw extension line origins. *Endpoint* is the default; however, *Intersection* is useful if you want to dimension to lines (such as center-lines) that cross your objects.

Continuous

The *Continuous* option creates a string of dimensions in one row, similar to using *Dimlinear* followed by *Dimcontinue*. The resulting dimensions for our example are shown in Figure 25-46.

FIGURE 25-46

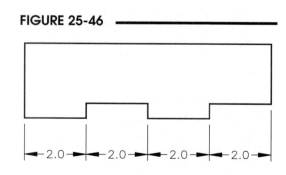

Baseline

Use the *Baseline* option of *Qdim* to create a stack of dimensions all generated from one baseline, similar to using *Dimlinear* followed by *Dimbaseline* (Fig. 25-47). The baseline end of the stack of dimensions is nearest the UCS (User Coordinate System) origin (0,0) by default. To change the baseline end of the stack of dimensions, use the *datumPoint* option of *Qdim* to specify a new origin point for the desired baseline end of the geometry (see the *datumPoint* option later in this section).

FIGURE 25-47

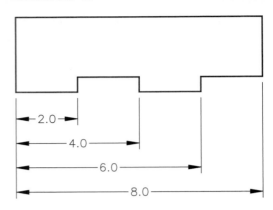

Staggered

The *Staggered* option creates a stack of dimensions alternating from one end or alternating outward from a central feature (Fig. 25-48). This type of dimension series is generally used only for parts that are symmetrical (denoted by a centerline), otherwise the dimension set would be incomplete. Note that the extension line sets (two extension lines for each dimension) are generated from the selected geometric features; therefore, dimensioning geometry with an odd number of features results in one feature not being dimensioned.

FIGURE 25-48

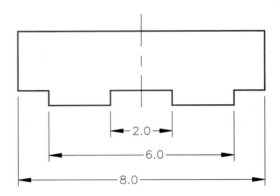

Ordinate

Using *Qdim* with the *Ordinate* option creates a string of ordinate dimensions, similar to using *Dimordinate* (Fig. 25-49). This style of dimensioning is used in manufacturing primarily for stamping or cutting parts of uniform thickness, such as in the sheet metal industry.

In ordinate dimensioning, each dimensional value specifies one "ordinate," or distance from a reference corner or datum point. By default, AutoCAD uses the current UCS or WCS origin as the datum point (note the values in Fig. 25-49). You can use the *datumPoint* option to designate one corner of the part to use as the reference origin (see *datumPoint* next).

FIGURE 25-49

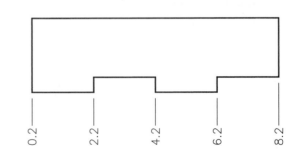

datumPoint

Typically, use the *datumPoint* option to specify a reference corner, or origin, of the geometry before creating ordinate dimensions. At the "Select new datum point:" prompt, use an *Endpoint* or other *Osnap* mode to select the desired datum point on the part. In Figure 25-50, the lower-left corner of the geometry was specified as the *datumPoint* before creating the ordinate dimensions.

FIGURE 25-50

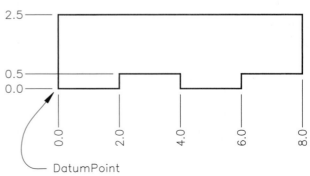

Radius

The *Continuous* option creates a series of radial dimensions similar to using *Dimradius* (Fig. 25-51). You can select multiple arcs to dimension with a window rather than picking the arcs individually because any linear geometry is automatically filtered out and not dimensioned. The resulting dimensions all have similar dimension line features, such as leader line length and other variables.

FIGURE 25-51

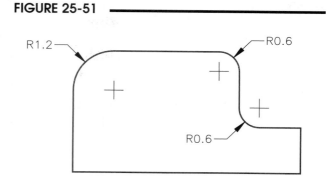

Diameter

Use the *Diameter* option to create a series of diametrical dimensions for circles or holes, similar to using *Dimdiameter*. Since any linear geometry is automatically filtered out when you select geometry to dimension, you can use a window rather than picking the circles individually. For example, selecting all geometry displayed in Figure 25-52 results in only two diametrical dimensions.

FIGURE 25-52

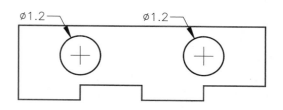

Edit

Once multiple dimensions are created, such as in Figure 25-46, you can use the *Edit* option to *Add* or *Remove* dimensions. Enter *E* at the *Qdim* prompt to select the options.

```
Command: qdim
Select geometry to dimension: PICK
Select geometry to dimension: Enter
Specify dimension line position, or [Continuous/Staggered/Baseline/Ordinate/Radius/
Diameter/datumPoint/Edit] <Continuous>: e
Indicate dimension point to remove, or [Add/eXit] <eXit>:
```

When the *Edit* option is invoked, small markers appear at the extension line origins called dimension points (Fig. 25-53). You can *Add* or *Remove* dimensions by adding or removing dimension points. For example, using the *Remove* option, three dimensions are reduced to one by removing the indicated dimension points.

FIGURE 25-53

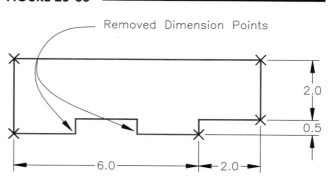

EDITING DIMENSIONS

Grip Editing Dimensions

Grips can be used effectively for editing dimensions. Any of the grip options (STRETCH, MOVE, ROTATE, SCALE, and MIRROR) is applicable. Depending on the type of dimension (linear, radial, angular, etc.), grips appear at several locations on the dimension when you activate the grips by selecting the dimension at the Command: prompt. Associative dimensions offer the most powerful editing possibilities, although nonassociative dimension components can also be edited with grips. There are many ways in which grips can be used to alter the measured value and configuration of dimensions.

Figure 25-54 shows the grips for each type of associative dimension. Linear dimensions (horizontal, vertical, and rotated) and aligned and angular dimensions have grips at each extension line origin, the dimension line position, and a grip at the text. A diameter dimension has grips defining two points on the diameter as well as one defining the leader length. The radius dimension has center and radius grips as well as a leader grip.

FIGURE 25-54

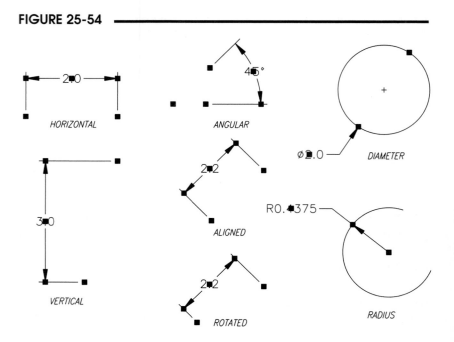

With dimension grips, a wide variety of editing options are possible. Any of the grips can be PICKed to make them **hot** grips. All grip options are valid methods for editing dimensions.

For example (Fig. 25-55), a horizontal dimension value can be increased by stretching an extension line origin grip in a horizontal direction. A vertical direction movement changes the length of the extension line. The dimension line placement is changed by stretching its grips. The dimension text can be stretched to any position by manipulating its grip.

FIGURE 25-55

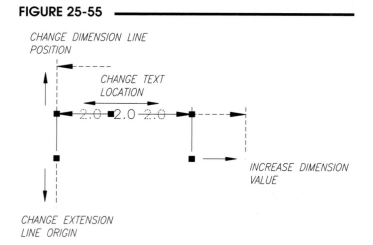

An angular dimension can be increased by stretching the extension line origin grip. The numerical value automatically updates. Stretching a rotated dimension's extension line origin allows changing the length of the dimension as well as the length of the extension line. An aligned dimension's extension line origin grip also allows you to change the aligned <u>angle</u> of the dimension (Fig. 25-56).

FIGURE 25-56

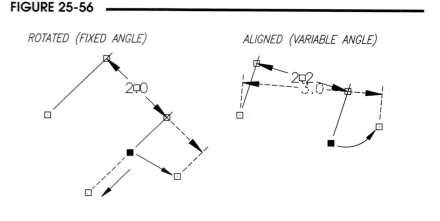

Rotating the <u>center</u> grip of a radius dimension (with the ROTATE option) allows you to reposition the location of the dimension around the *Arc*. Note that the text remains in its original horizontal orientation (Fig. 25-57).

FIGURE 25-57

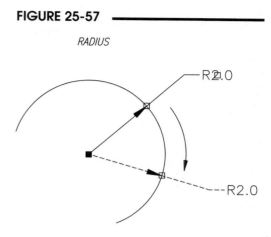

RADIUS

You can move the dimension <u>text</u> independent of the dimension line with grips for dimensions that have a *DIMTMOVE* variable setting of 2. Use an existing dimension, and change the *DIMTMOVE* setting to 2 (or change the *Fit* setting in the *Dimension Style Manager* to *Over the dimension line without a leader*). The dimension text can be moved to any location without losing its associativity. See *"Fit (DIMTMOVE),"* Chapter 26.

TIP

Many other possibilities exist for editing dimensions using grips. Experiment on your own to discover some of the possibilities that are not shown here.

Dimension Right-Click Menu

If you select any dimension, you can right-click to produce a shortcut menu (Fig. 25-58). This menu contains options that allow you to change the text position, the text precision, the *Dimension Style* of the dimension, the annotative scale, and to flip the arrows.

FIGURE 25-58

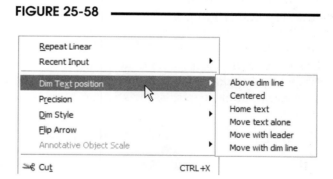

Dim Text Position

Several options are available on this cascading menu. The *Centered* and *Home Text* options are duplicates of the *Dimtedit* command options of the same names (see *"Dimtedit"*). The *Move text alone, Move with leader,* and *Move with dim line* options change the settings of the *DIMTMOVE* system variable for the dimension. The *Move* options operate best with grips on. For example, you can select the *Move text alone* option, make the text grip hot, then drag the text to any position in the drawing while still retaining the associative quality of the dimension (Fig. 25-59). (For more information on these options, see *"DIMTMOVE"* in Chapter 26.)

FIGURE 25-59

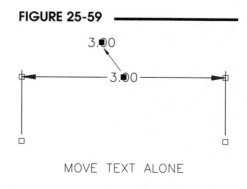

MOVE TEXT ALONE

Precision

Use this cascading menu to select from a list of decimal or fractional precision values. For example, you could change a dimension text value of "5.0000" to "5.00" with this menu. Your choice changes the *DIMDEC* system variable for the selected dimension. (For more information, see "*DIMDEC*" in Chapter 26.)

Dim Style

Use this option to change the selected dimension from one *Dimension Style* to another. Changing a dimension's style affects the appearance characteristics of the dimension. (For more information on *Dimension Styles*, see Chapter 26.)

Flip Arrow

Use this option to flip the arrow nearest to the point you right-click to produce the menu. In other words, if both arrows pointed from the center of the dimension out to the extension lines (default position), flipping both arrows would cause the arrows to be on the outside of the extension lines pointing inward (Fig. 25-60).

FIGURE 25-60

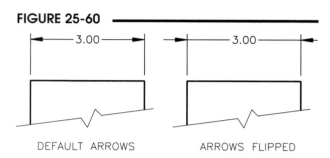

DEFAULT ARROWS   ARROWS FLIPPED

Although the *Centered* and *Home Text* options are duplicates of the *Dimtedit* command, the other options in this shortcut menu actually change system variables for the individual dimensions. These options do not change the variable settings for the *Dimension Style* of the selected dimension, but create dimension style <u>overrides</u> for the individual dimension. (For more information on *Dimension Styles* and dimension style overrides, see Chapter 26.)

Dimension Editing Commands

Several commands are provided to facilitate easy editing of existing dimensions in a drawing. Most of these commands are intended to allow variations in the appearance of dimension <u>text</u>. These editing commands operate with associative and nonassociative dimensions, but not with exploded dimensions.

Dimtedit

Pull-down Menu	Command (Type)	Alias (Type)	Short-cut	Screen (side) Menu	Tablet Menu
Dimension *Align Text >*	*Dimtedit*	*DIMTED*	...	*DIMNSION* *Dimtedit*	*Y,2*

Dimtedit (text edit) allows you to change the position or orientation of the text for a single associative dimension. To move the position of text, this command syntax is used:

```
Command: Dimtedit
Select dimension: PICK
Specify new location for dimension text or
[Left/Right/Center/Home/Angle]: PICK
```

At the "Specify new location for dimension text" prompt, drag the text to the desired location. The selected text and dimension line can be changed to any position, while the text and dimension line retain their associativity (Fig. 25-61).

FIGURE 25-61

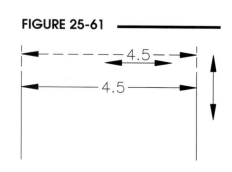

Angle

The *Angle* option works with any *Horizontal, Vertical, Aligned, Rotated, Radius,* or *Diameter* dimensions. You are prompted for the new <u>text</u> angle (Fig. 25-62):

FIGURE 25-62

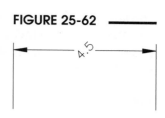

    Command: **dimtedit**
    Select dimension: **PICK**
    Specify new location for dimension text or [Left/Right/Center/Home/Angle]: **a**
    Specify angle for dimension text: **45**
    Command:

Home

The text can be restored to its original (default) rotation angle with the *Home* option. The text retains its right/left position.

Right/Left

FIGURE 25-63

The *Right* and *Left* options automatically justify the dimension text at the extreme right and left ends, respectively, of the dimension line. The arrow and a short section of the dimension line, however, remain between the text and closest extension line (Fig. 25-63).

Center

The *Center* option brings the text to the center of the dimension line and sets the rotation angle to 0 (if previously assigned; Fig. 25-63).

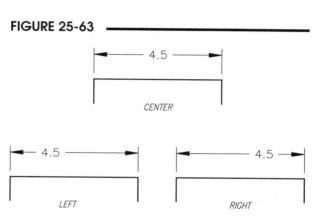

Dimedit

Pull-down Menu	Command (Type)	Alias (Type)	Short-cut	Screen (side) Menu	Tablet Menu
Dimension Oblique	Dimedit	DIMED or DED	...	DIMNSION Dimedit	Y,1

The *Dimedit* command allows you to change the angle of the extension lines to an obliquing angle and provides several ways to edit dimension text. Two of the text editing options (*Home, Rotate*) duplicate those of the *Dimtedit* command. Another feature (*New*) allows you to change the text value and annotation.

    Command: **dimedit**
    Enter type of dimension editing [Home/New/Rotate/Oblique] <Home>:

Home

This option moves the dimension text back to its original angular orientation angle without changing the left/right position. *Home* is a duplicate of the *Dimtedit* option of the same name.

New

You can change the text value of any existing dimension with this option. When you invoke this option, the *Text Formatting* editor appears with "0.0000" as the string of text in the dark background. Enter the desired text, press *OK*, then you are prompted to select a dimension. The text you entered is applied to the selected dimension.

TIP The key feature of the New option is that the original <u>AutoCAD-measured value can be restored</u> in the case that it was "manually" changed for some reason. Simply invoke *Dimedit* and the *New* option, then <u>do not enter any new text</u> in the *Text Formatting* Editor, and press the *OK* button. As prompted next, select the desired dimension, and the original AutoCAD-measured value is automatically restored.

Rotate

AutoCAD prompts for an angle to rotate the text. Enter an absolute angle (relative to angle 0). This option is identical to the *Angle* option of *Dimtedit*.

Oblique

This option is unique to *Dimedit*. Entering an angle at the prompt affects the extension lines. Enter an absolute angle:

Enter obliquing angle (press ENTER for none):

Normally, the extension lines are perpendicular to the dimension lines. In some cases, it is desirable to set an obliquing angle for the extension lines, such as when dimensions are crowded and hard to read (Fig. 25-64) or for dimensioning isometric drawings.

FIGURE 25-64 ────

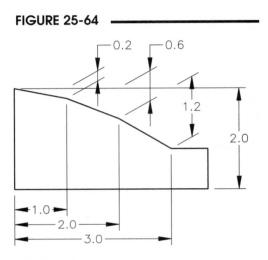

Properties

	Pull-down Menu	Command (Type)	Alias (Type)	Short-cut	Screen (side) Menu	Tablet Menu
	Modify Properties...	*Properties*	*MO or PR*	(Edit Mode) *Properties*	*MODIFY1 Modify*	*Y,14*

The *Properties* command (discussed in Chapter 16) can be used effectively for a wide range of editing purposes. Double-clicking on any dimension produces the *Properties* palette. The palette gives a list of all properties of the dimension including dimension variable settings.

FIGURE 25-65 ────

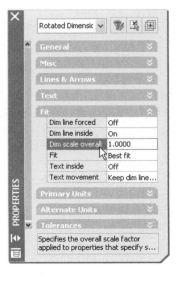

The dimension variable settings for the selected dimension are listed in the *Lines & Arrows*, *Text*, *Fit*, *Primary Units*, *Alternate Units*, and *Tolerances* sections. Each section is expandable, so you can change any property (dimension variable setting) in each section for the dimension (Fig. 25-65). Each entry in the left column represents a dimension variable; changing the entries in the right column changes the variable's setting.

Each of the six categories (*Lines & Arrows*, *Text*, *Fit*, *Primary Units*, *Alternate Units*, and *Tolerances*) corresponds to the tabs in the *Dimension Style Manager*. Typically, you create dimension styles in the *Dimension Style Manager* by selecting settings for the <u>dimensioning variables</u>, then draw the dimensions using one of the styles. Each style has different settings for dimension variables, so the resulting dimensions for each style appear differently.

Using the *Properties* palette, you can change the dimension variable settings for any one or more dimensions retroactively. Making a change to a dimension variable with this method does not change the previously created dimension style, but it does create an <u>override for that particular dimension</u>. For more information on dimension variables, dimension styles, and dimension style overrides, see Chapter 26, Dimension Styles and Variables.

Associative Dimensions

AutoCAD creates associative dimensions by default. Associative dimensions contain "definition points" that are associated to the related geometry. For example, when you use the *Dimlinear* command and then <u>Osnap</u> to two points on the drawing geometry in response to the "Specify first extension line origin" and "Specify second extension line origin," two definition points are created at the ends of the extension lines and are attached, or "associated," to the geometry. You may have noticed these small dots at the ends of the extension lines. The definition points are generally located where you PICK, such as at the extension line origins for most dimensioning commands, or the arrowhead tips for the leader commands and the arrowhead tips and centers for diameter and radius commands. All of the dimensioning commands are associative by default except *Qleader*, *Leader*, *Center Mark*, and *Tolerance*.

When you create the first associative dimension, AutoCAD automatically creates a new layer called DEFPOINTS. The layer is set to not plot and that property cannot be changed. AutoCAD manages this layer automatically, so do not attempt to alter this layer.

There are two advantages to associative dimensions. First, if the associated geometry is modified by typical editing methods, the associated definition points automatically change; therefore, the dimension line components (dimension lines and extension lines) automatically change and the numerical value automatically updates to reflect the geometry's new length, angle, radius, or diameter. Second, existing associative dimensions in a drawing automatically update when their dimension style is changed (see Chapter 26 for information on dimension styles).

Examples of typical editing methods that can affect associative dimensions are: *Chamfer*, *Extend*, *Fillet*, *Mirror*, *Move*, *Rotate*, *Scale*, *Stretch*, *Trim* (linear dimensions only), *Array* (if rotated in a *Polar* array), and grip editing options.

Dimensions can have three possible associativity settings controlled by the *DIMASSOC* system variable.

Associative dimensions	(*DIMASSOC* = 2) Use *Osnap* to create associative dimensions. Associative dimensions automatically adjust their locations, orientations, and measurement values when the geometric objects associated with them are modified. Associative dimensions in a drawing will update when the dimension style is modified. These fully associative dimensions are used in AutoCAD 2002 through the current release.
Nonassociative dimensions	(*DIMASSOC* = 1) Nonassociative dimensions do not change automatically when the geometric objects they measure are modified. These dimensions have definition points, but the definition points must be included (selected) if you want the dimensions to change when the geometry they measure is modified. The existing dimensions in a drawing will update when the dimension style is modified. In AutoCAD 2000 and previous releases, these dimensions were called associative.

If you want to determine whether a dimension is associative or nonassociative, you can use the *Properties* palette or the *List* command. You can also use the *Quick Select* dialog box to filter the selection of associative or nonassociative dimensions. A dimension is considered associative even if only one end of the dimension is associated with a geometric object. Associativity is *not* maintained between a dimension and a block reference if the block is redefined. Associativity is not possible when dimensioning *Multiline* objects.

Remember that the associativity status for newly created dimensions (associative, nonassociative, or exploded) is controlled by the *DIMASSOC* system variable (2, 1, 0, respectively). You can set the *DIMASSOC* system variable at the Command line or use the *Options* dialog box (Fig. 25-66). In the *User Preferences* tab, a check in the *Make new dimensions associative* box (right center) sets *DIMASSOC* to 2, while no check in this box sets *DIMASSOC* to 1. The *DIMASSOC* setting does not affect existing dimensions, only subsequently created ones. The *DIMASSOC* variable is not saved with a dimension style.

FIGURE 25-66

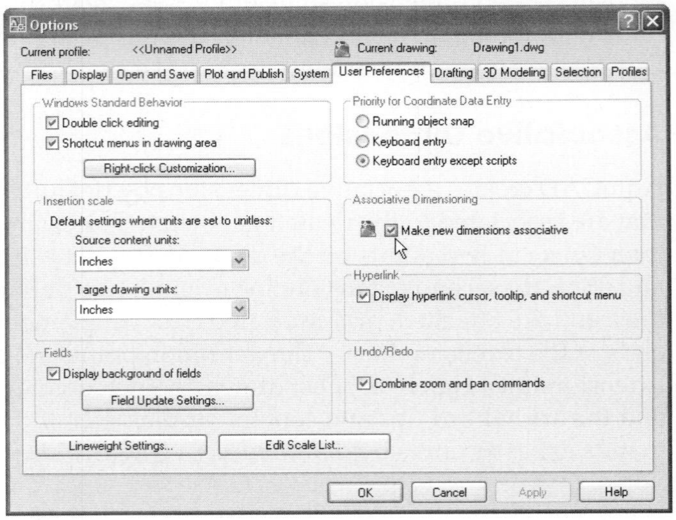

Dimreassociate

Pull-down Menu	Command (Type)	Alias (Type)	Short-cut	Screen (side) Menu	Tablet Menu
Dimension Reassociate Dimensions	*Dimreassociate*	...	...	...	...

With *Dimreassociate*, you can convert nonassociative dimensions to associative dimensions. In addition, you can change the feature of the drawing geometry (line endpoint, etc.) that an existing definition point is associated with. This command is especially helpful in several situations, such as:

if you are updating a drawing completed in AutoCAD 2000 or a previous release and want to convert the dimensions to fully associative AutoCAD 2008 dimensions;

if you created nonassociative dimensions in AutoCAD 2008 but want to convert them to associative dimensions;

if you altered the position of an associative dimension's definition points accidentally with grips or other method and want to reattach them;

if you want to change the attachment point for an existing associative dimension.

The *Dimreassociate* command prompts for the geometric features (line endpoints, etc.) that you want the dimension to be associated with. Depending on the type of dimension (linear, radial, angular, diametrical, etc.), you are prompted for the appropriate geometric feature as the attachment point.
For example, if a linear dimension is selected to reassociate, AutoCAD prompts you to specify (select attachment points for) the extension line origins.

Command: **dimreassociate**
Select dimensions to reassociate ...
Select objects: **PICK** (linear dimension selected)
Select objects: **Enter**
Specify first extension line origin or [Select object] <next>: **PICK**
Specify second extension line origin <next>: **PICK**
Command:

When you select the "dimensions to reassociate," the associative and nonassociative elements of the selected dimension(s) are displayed one at a time. A marker is displayed for each extension line origin (or other definition point) for each dimension selected (Fig. 25-67). If the definition point is not associated, an "X" marker appears, and if the definition point is associated, an "X" within a box appears. To reassociate a definition point, simply pick the new object feature you want the dimension to be attached to. If an extension line is already associated and you do not want to make a change, press Enter when its definition point marker is displayed. Figure 25-67 displays a partly associated linear dimension with its extension line origin markers during the *Dimreassociate* command.

FIGURE 25-67

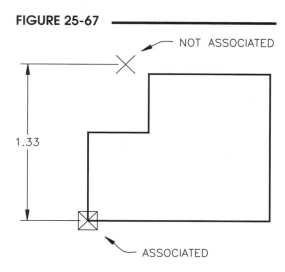

Dimdisassociate

Pull-down Menu	Command (Type)	Alias (Type)	Short-cut	Screen (side) Menu	Tablet Menu
...	*Dimdisassociate*	...	...	...	...

The *Dimdisassociate* command automatically converts selected associative dimensions to nonassociative. The command prompts for you to select only the dimensions you want to disassociate, then AutoCAD makes the conversion and reports the number of dimensions disassociated.

Command: **dimdisassociate**
Select dimensions to disassociate ...
Select objects: **PICK**
Select objects: **Enter**
nn disassociated.
Command:

You can use any selection method including *Qselect* to "Select objects." *Dimdisassociate* filters the selection set to include only associative dimensions that are in the current space (model space or paper space layout) and not on locked layers.

DIMENSIONING VARIABLES INTRODUCTION

Now that you know the commands used to create dimensions, you need to know how to control the appearance of dimensions. For example, after you create dimensions in a drawing, you may need to adjust the size of the dimension text and arrowheads, or change the text style, or control the number of decimal places appearing on the dimension text.

The appearance of dimensions is controlled by setting the dimensioning variables. You can ensure an entire group of dimensions appear the same by using one dimension style for the dimensions. A dimension style is the set of variables that all the dimension objects in the group reference.

TIP

Although you can change many settings that control the appearance of dimensions, there is one dimensioning setting that is the most universal, that is, the *Scale for dimension features*. The *Scale for dimension features* controls all of the size-related features of dimension objects, such as the arrowhead size, the text size, the gap between extension lines and the object, and so on. Even though each of those features can be changed individually, changing the *Scale for dimension features* controls <u>all</u> of the size-related features of dimensions together.

There are three methods for controlling the scale for dimensioning features:

1.   set an *Overall Scale* value;
2.   make the dimensions *Annotative*;
3.   set the dimension scale based on a (viewport) layout.

Using the *Overall Scale* method is the simplest and most widely used method. With this method, dimensions are displayed at a static scale—the same scale (in relation to the geometry) in all viewports and layouts. The *Overall Scale* value should be set based on the drawing scale factor. The *Overall Scale* value is stored in the *DIMSCALE* variable.

Annotative dimensions can be made to automatically change size when the *Annotation Scale* or *VP scale* is changed. This method is recommended when the drawing will be displayed in multiple viewports at different scales. This setting changes the *DIMANNO* variable (0 = off, 1 = on). This option works similarly to annotative text (Chapter 18).

The *Scale dimensions to a layout* (more specifically, scale to a viewport) method is useful when you want to set the size of dimension features based on one viewport at a time. This is an older method of automatically adjusting the dimension scale based on the viewport scale, but it is limited in that only one scale can be displayed at a time. This setting changes the *DIMSCALE* variable to 0.

<u>It is highly recommended that you use the *Overall Scale* method while you learn the dimensioning basics.</u> Since you will at least initially be creating simpler drawings to be displayed in one scale, setting the *Overall Scale* will avoid having to deal with complexities that are unnecessary as you learn the fundamentals.

To change the *Overall Scale*, open the *Dimension Style Manager* (by selecting it from the *Dimension* pull-down menu, *Dimension* toolbar, or by typing *D*). Select the *Modify* button, access the *Fit* tab, and enter the desired value in the *Use overall scale of* edit box (Fig. 25-68). This action changes the overall size for all dimensions created using that particular dimension style.

FIGURE 25-68

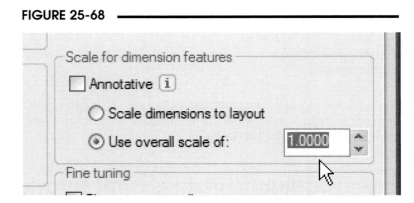

For example, assume you created a drawing and applied the first few dimensions, but noticed that the dimension text was so small that it was barely readable. Instead of changing the dimension variable that controls the size of just the dimension text, change the *Overall Scale*.

This practice ensures that the text, arrowheads, gaps, extension, and all size-related features of the dimension are changed together and are therefore correctly proportioned. Using the *Dimension Style Manager* to change this variable also ensures that all the dimensions created in that dimension style will have the same size.

Chapter 26 explains all the dimensioning variables, including *Overall Scale* and annotative dimensions, as well as dimension styles. While working on the exercises in this chapter, you will be able to create dimensions without having to change any dimension variables or dimension styles, although you may want to experiment with the *Overall Scale*. Knowledge and experience with <u>both</u> the dimensioning commands and dimensioning variables is essential for AutoCAD users in a real-world situation. Make sure you read Chapter 26 before attempting to dimension drawings other than those in these exercises.

CHAPTER EXERCISES

Only four exercises are offered in this chapter to give you a start with the dimensioning commands. Many other dimensioning exercises are given at the end of Chapter 26, Dimension Styles and Variables. Since most dimensioning practices in AutoCAD require the use of dimensioning variables and dimension styles, that information should be discussed before you can begin dimensioning effectively.

The units that AutoCAD uses for the dimensioning values are based on the *Unit format* and the *Precision* settings in the *Primary Units* tab of the *Dimension Style Manager*. You can dimension these drawings using the default settings, or you can change the *Units* and *Precision* settings for each drawing to match the dimensions shown in the figures. (Type *D*, select *Modify*, then the *Primary Units* tab.)

Other dimensioning features may appear different than those in the figures because of the variables set in the AutoCAD default STANDARD dimension style. For example, the default settings for diameter and radius dimensions may draw the text and dimension lines differently than you desire. After reading Chapter 26, those features in your exercises can be changed retroactively by changing the dimension style.

1. *Open* the **PLATES** drawing that you created in Chapter 9 Exercises. *Erase* the plate on the right. Use *Move* to move the remaining two plates apart, allowing 5 units between. Create a *New* layer called **DIM** and make it *Current*. Dimension the two plates, as shown in Figure 25-69. *Save* the drawing as **PLATES-D**.

FIGURE 25-69

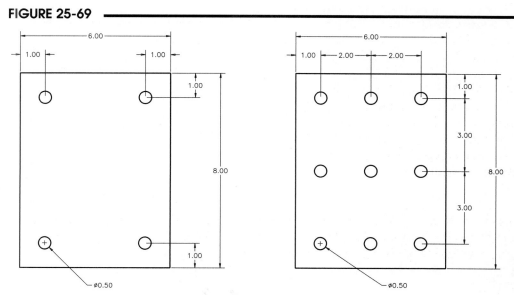

2. *Open* the **PEDIT1** drawing. Create a *New* layer called **DIM** and make it *Current*. Dimension the part as shown in Figure 25-70. *Save* the drawing as **PEDIT1-DIM**. Draw a *Pline* border of *.02 width* and *Insert* **TBLOCK**. *Plot* on an A size sheet using *Scale to Fit*.

FIGURE 25-70

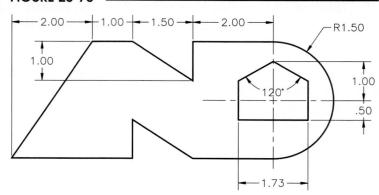

3. *Open* the **GASKETA** drawing that you created in Chapter 9 Exercises. Create a *New* layer called **DIM** and make it *Current*. Dimension the part as shown in Figure 25-71. *Save* the drawing as **GASKETD.**

   Draw a *Pline* border with *.02 width* and *Insert* **TBLOCK** with an **8/11** scale factor. *Plot* the drawing *Scaled to Fit* on an A size sheet.

FIGURE 25-71

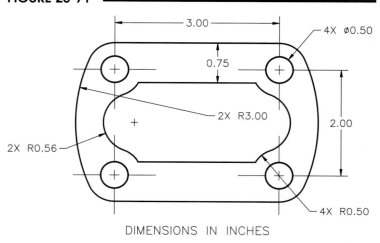

DIMENSIONS IN INCHES

4. *Open* the **BARGUIDE** multiview drawing that you created in Chapter 21 Exercises. Create the dimensions on the **DIM** layer. Keep in mind that you have more possibilities for placement of dimensions than are shown in Figure 25-72.

FIGURE 25-72

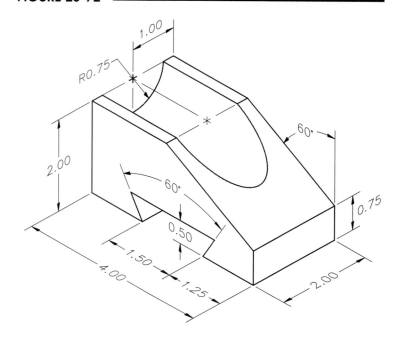

26

DIMENSION STYLES AND VARIABLES

CHAPTER OBJECTIVES

After completing this chapter you should:

1. be able to control dimension variable settings using the *Dimension Style Manager*;

2. be able to save and restore dimension styles with the *Dimension Style Manager*;

3. know how to create dimension style families and specify variables for each child;

4. know how to create and apply dimension style overrides;

5. be able to modify dimensions using *Update, Dimoverride, Properties,* and *Matchprop*;

6. know the guidelines for dimensioning;

7. be able to create annotative dimensions.

CONCEPTS

Dimension Variables

Since a large part of AutoCAD's dimensioning capabilities are automatic, some method must be provided for you to control the way dimensions are drawn. A set of 80 <u>dimension variables</u> allows you to affect the way dimensions are drawn by controlling sizes, distances, appearance of extension and dimension lines, and dimension text formats.

An example of a dimension variable is *DIMSCALE* (if typed) or *Overall Scale* (if selected from a dialog box, Fig. 26-1). This variable controls the overall size of the dimension features (such as text, arrowheads, and gaps). Changing the value from 1.00 to 1.50, for example, makes all of the size-related features of the drawn dimension 1.5 times as large as the default size of 1.00 (Fig. 26-2). Other examples of features controlled by dimension variables are arrowhead type, orientation of the text, text style, units and precision, suppression of extension lines, fit of text and arrows (inside or outside of extension lines), and direction of leaders for radii and diameters (pointing in or out). The dimension variable changes that you make affect the <u>appearance</u> of the dimensions that you create.

FIGURE 26-1

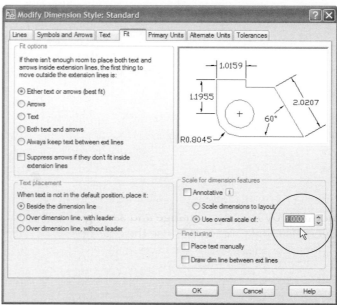

There are two basic ways to control dimensioning variables:

1. Use the *Dimension Style Manager* (Fig. 26-3).

2. Type the dimension variable name in Command line format.

The dialog boxes employ "user-friendly" terminology and selection, while the Command line format uses the formal dimension variable names but cannot be used for all settings.

FIGURE 26-2

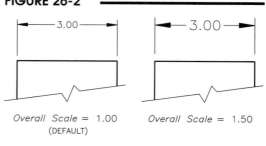

Overall Scale = 1.00
(DEFAULT)

Overall Scale = 1.50

Changes to dimension variables are usually made <u>before</u> you create the affected dimensions. Dimension variable changes are <u>not always retroactive</u>, in contrast to *LTSCALE*, for example, which can be continually modified to adjust the spacing of existing non-continuous lines. Changes to dimensioning variables affect <u>existing</u> dimensions only when those changes are *Saved* to an existing *Dimension Style* that was in effect when previous dimensions were created. Generally, dimensioning variables should be set <u>before</u> creating the desired dimensions although it is possible to modify dimensions and *Dimension Styles* retroactively.

FIGURE 26-3

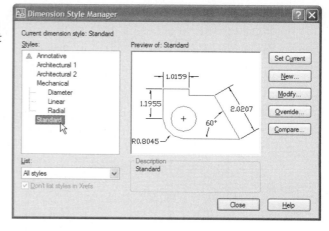

Dimension Styles

All dimensions are part of a <u>dimension style</u>. The default dimension style is called *Standard* (the *Annotative* style is also supplied). The *Standard* dimension style has the all the default dimension variable settings for creating dimensions with a typical size and appearance. Similar to layers, you can create, name, and specify settings for any number of dimension styles. Each dimension style contains the dimension variable settings that you select. <u>A dimension style is a group of dimension variable settings that has been saved under a name you assign</u>. When you set a dimension *Current*, AutoCAD remembers and resets that <u>particular combination of dimension variable settings</u>.

You can create a dimension style for each particular "style" of dimension. Each time you want to draw a particular style of dimension, select the style name from the list, make it the "current" style, then begin drawing dimensions.

To create a new dimension style, select the *New* button in the *Dimension Style Manager* (see Fig. 26-3), then use the seven tabs that appear in the dialog box to specify the appearance of the dimensions. Selecting options in these tabs (see Fig. 26-13) actually sets values for related dimension variables that are saved in the drawing. When a dimension style is *Set Current*, the dimensions you create with that style appear with the dimension variable settings you specified. Creating dimensions with the current *Dimension Style* is similar to creating text with the current *Text Style*—that is, the objects created take on the current settings assigned to the style.

Another advantage of using dimension styles is that <u>existing</u> dimensions in the drawing can be globally modified by making a change and saving it to the dimension style(s). To do this, select *Modify* in the *Dimension Style Manager* or use the *Save* option of *-Dimstyle* after changing a dimension variable in Command line format. This action saves the changes for newly created dimensions as well as <u>automatically updating existing dimensions in the style</u>.

Dimension Style Families

In the previous chapter, you learned about the various <u>types</u> of dimensions, such as linear, angular, radial, diameter, ordinate, and leader. You may want the appearance of each of these types of dimensions to vary, for example, all radius dimensions to appear with the dimension line arrow pointing out and all diameter dimensions to appear with the dimension line arrow inside pointing in. It is logical to assume that a new dimension style would have to be created for each variation. However, a new dimension style name is not necessary for each type of dimension. AutoCAD provides <u>dimension style families</u> for the purpose of providing variations within each named dimension style.

The dimension style <u>names</u> that you create are assigned to a dimension style <u>family</u>. Each dimension style family can have six <u>children</u>. The children are <u>linear, angular, diameter, radial, ordinate, and leader</u>. Although the children take on the dimension variable settings assigned to the family, you can assign special settings for one or more children. For example, you can make a radius dimension appear slightly different from a linear dimension in that family. Do this by selecting the *New* button in the *Dimension Style Manager*, then selecting the type of dimension (*linear, radius, diameter,* etc.) you want to change from the *Create New Dimension Style* dialog box that appears (Fig. 26-4).

FIGURE 26-4

Select *Continue* to specify settings for the child. When a dimension is drawn, AutoCAD knows what type of dimension is created and applies the selected settings to that child.

Dimension styles that have children (special variable settings set for the *linear, radial, diameter*, etc. dimensions) appear in the list of styles in the *Dimension Style Manager* as sub-styles branching from the parent name. For example, notice the "Mechanical" style in Figure 26-3 has special settings for its *Diameter, Linear*, and *Radial* children.

For example, you may want to create the "Mechanical" dimension style to draw the arrows and dimension lines inside the arc but drag the dimension text outside of the arc for *Radial* dimensions (as shown in Fig. 26-5). To do this, create a new child for the "Mechanical" family as previously described, and set "Mechanical" as the current style. When a *Radius* dimension is drawn, the dimension line, text, and arrow should appear as shown in Figure 26-5.

FIGURE 26-5

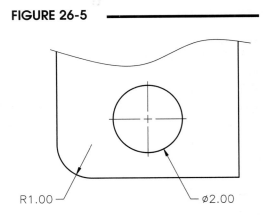

R1.00   Ø2.00

In summary, a dimension style family is simply a set of dimension variables related by name. Variations within the family are allowed, in that each child (type of dimension) can possess its own subset of variables. Therefore, each child inherits all the variables of the family in addition to any others that may be assigned individually.

AutoCAD refers to these children as $0, $2, $3, $4, $6, and $7 as suffixes appended to the family name such as "Standard$0." AutoCAD automatically applies the appropriate suffix code to the dimension style name when a child is created. Although the codes do not appear in the *Dimension Style Manager*, you can display the code by using the *List* command and selecting an existing child dimension.

Child Type	Suffix Code
linear	$0
angular	$2
diameter	$3
radial	$4
ordinate	$6
leader	$7

Dimension Style Overrides

If you want only a few dimensions in a dimension style to have a special appearance, you can create a dimension style override. An override does not normally affect the parent style (unless you save the override setting to the style). Therefore, existing dimensions in the drawing are not affected by a dimension style override. You can create a dimension style override two ways.

The more practical method of creating a dimension style override is to select an existing dimension, then use the *Properties* palette to change the desired variable setting. This method (explained in "Modifying Existing Dimensions") creates a dimension style override <u>for the selected dimension only</u> and does not affect any other dimensions in the style.

Alternately, you can create a dimension style override in the *Dimension Style Manager*, then create the dimensions. First, make the desired style *Current*, then select the *Override* button. Proceed to specify the desired variable settings from the seven tabs that appear. A new branch under the family name appears in the *Dimension Style Manager* styles list named "<style overrides>" (see Fig. 26-7 under "Architectural 2"). Finally, create the new dimensions. To clear the overrides, simply make another style current. (Alternately, type any dimension variable name at the Command line and change the value. The override is applied only to the newly created dimensions in the current style and clears when another style is made current.)

DIMENSION STYLES

Dimstyle and Ddim

Pull-down Menu	Command (Type)	Alias (Type)	Short-cut	Screen (side) Menu	Tablet Menu
Dimension Style...	*Dimstyle* or *Ddim*	DIMSTY, DST, or D	...	*DIMNSION Dimstyle*	Y,3

The *Dimstyle* or *Ddim* command produces the *Dimension Style Manager* (Fig. 26-6). This is the primary interface for creating new dimension styles and making existing dimension styles current. This dialog box also gives access to seven tabs that allow you to change dimension variables. Features of the *Dimension Style Manager* that appear on the opening dialog box (shown in Fig. 26-6) are described in this section. The tabs that allow you to change dimension variables are discussed in detail later in this chapter in the section "Dimension Variables."

FIGURE 26-6

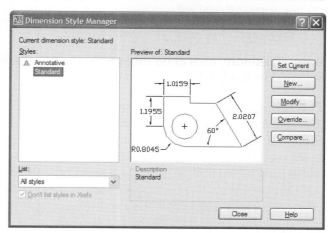

The *Dimension Style Manager* makes the process of creating and using dimension styles easier and more visual than using Command line format. Using the *Dimension Style Manager*, you can *Set Current*, create *New* dimension styles, *Modify* existing dimension styles, and create an *Override*. You also have the ability to view a list of existing styles including children and overrides; see a preview of the selected style, child, or override; examine a description of a dimension style as it relates to any other style; and make an in-depth comparison between styles. You can also control the entries in the list to include or not include Xreferenced dimension styles. The options are explained below.

Styles

This list displays all existing dimension styles in the drawing, including *Xrefs*, depending on your selection in the *List* section below. The list includes the parent dimension styles, the children that are shown on a branch below the parent style, and any overrides also branching from family styles.

The buttons on the right side of the *Dimension Style Manager* affect the highlighted style in the list. For example, to modify an existing style, select the style name from the list and PICK the *Modify* button. If you want to create a *New* style based on an existing style, select the style name you want to use as the "template," then select *New*.

Select (highlight) any dimension style from the *Styles* list to display the appearance of dimensions for that style in the *Preview* tile. Selecting a style from the list also makes the *Description* area list the dimension variables settings compared to the current dimension style.

You can right-click to use a shortcut menu for the selected dimension style (Fig. 26-7). The shortcut menu allows you to *Set Current*, *Rename*, or *Delete* the selected dimension style. Only dimension styles that are unreferenced (no dimensions have been created using the style) can be deleted.

If you have created a style override (by pressing the *Override* button, then setting variables), you can save the overrides to the dimension style using the right-click shortcut menu. Normally, overrides are automatically discarded when another dimension style is made current. This shortcut menu is the only method available in the *Dimension Style Manager* to save overrides. (You can also use the *-Dimstyle* command to save overrides.)

FIGURE 26-7

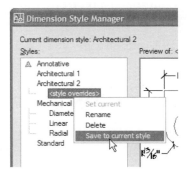

List

Select *All Styles* to display the list of all the previously created or imported dimension styles in the drawing. Selecting *Styles In Use* displays only styles that have been used to create dimensions in the drawing—styles that are saved in the drawing but are not referenced (no dimensions created in the style) do not appear in the list.

Don't List Styles in Xrefs

If the current drawing references another drawing (has an *Xref* attached), the *Xref's* dimension styles can be listed or not listed based on your selection here.

Preview

The *Preview* tile displays the appearance of the highlighted dimension style. When a dimension style name is selected, the *Preview* tile displays an example of all dimension types (linear, radial, etc.) and how the current variable settings affect the appearance of those dimensions. If a child is selected, the *Preview* tile displays only that dimension type and its related appearance. For example, Figure 26-8 displays the appearance of the selected child, *Mechanical*: *Radial*.

FIGURE 26-8

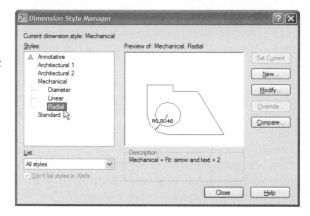

Description

The description area is a valuable tool for determining the variable settings for a dimension style. When a dimension style from the list is selected, the *Description* area lists the differences in variable settings from the current style. For example, assume the "Architectural 2" dimension style was created using "Standard" as the template. To display the dimension variable settings that have been changed (the differences between Standard and Architectural 2), make Standard the current style, then select Architectural 2 and examine the description area, as shown in Figure 26-9.

FIGURE 26-9

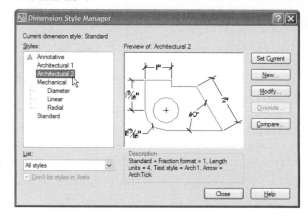

You can use the same procedure to display the variable settings assigned to a child (the differences between the family style and the child), as shown in the *Description* area in Figure 26-8.

Set Current

Select the desired existing style from the list on the left, then select *Set Current*. The current style is listed at the top of the *Dimension Style Manager*. When you *Close* the dialog box, the current style is used for drawing new dimensions. If you want to set an *Override* to an existing style, you must first make it the current style.

New

Use the *New* button to create a new dimension style. The *New* button invokes the *Create New Dimension Style* dialog box (Fig. 26-10). The options are discussed next.

FIGURE 26-10

New Style Name

Enter the desired new name in the *New Style Name* edit box. Initially the name that appears is "Copy of (current style)."

Start With

After assigning a new name, select any existing dimension style to use as a "template" from the *Start With* drop-down list. Initially (until dimension variable changes are made for the new style), the new dimension style is actually a copy of the *Start With* style (has identical variable settings). By default, the current style appears in the *Start With* edit box.

Use For

If you want to create a dimension style family, select *All Dimensions*. In this way, all dimension types (*linear, radial, diameter,* etc.) take on the new variable settings. If you want to specify special dimension variable settings for a child, select the desired dimension type from the drop-down list (see Fig. 26-4).

Continue

When the new name and other options have been selected, press the *Continue* button to invoke the *New Dimension Style* dialog box. Here you select from the seven tabs to specify settings for any dimension variable to apply to the new style. See the "Dimension Variables" section for information on setting dimension variables.

NOTE: Generally you should not change the Standard dimension style. Rather than change the Standard style, create *New* dimension styles using Standard as the base style. In this way, it is easy to restore and compare to the default settings by making the Standard style current. If you make changes to the Standard style, it can be difficult to restore the original settings.

Modify

Use this button to change dimension variable settings for an existing dimension style. First, select the desired style name from the list, then press *Modify* to produce the *Modify Dimension Style* dialog box. Using *Modify* to change dimension variables automatically saves the changes to the style and <u>updates existing dimensions</u> in the style, as opposed to using *Override*, which does not change the style or update existing dimensions. See the "Dimension Variables" section for information on setting dimension variables.

Override

A dimension style override is a temporary variable setting that affects only new dimensions created with the override. An override does not affect existing dimensions previously created with the style. You can set an override only for the current style, and the overrides are automatically cleared when another dimension style is made current, unless you use the right-click shortcut menu and select *Save to current style* (see Fig. 26-7).

To create a dimension style override, you must first select an existing style from the list and make it the current style. Next, select *Override* to produce the *Override Current Style* dialog box. See the "Dimension Variables" section for information on setting dimension variables to act as overrides.

Compare

This feature is very useful for examining variable settings for any dimension style. Press the *Compare* button to produce the *Compare Dimension Styles* dialog box (Fig. 26-11). This dialog box lists only dimension variable settings; changes to variable settings cannot be made from this interface.

FIGURE 26-11

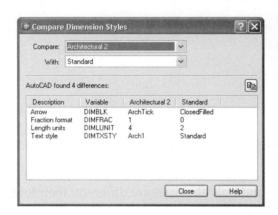

The *Compare Dimension Styles* dialog box lists the differences between the style in the *Compare* box and the style in the *With* box. For example, assume you created a new style named "Architectural 2" from the "Standard" style, but wanted to know what variable changes you had made to the new style. Select Architectural 2 in the *Compare* box and Standard in the *With* box. The variable changes made to Architectural 2 are displayed in the central area of the box. Also note that the settings for each variable are listed for each of the two styles.

A useful feature of this dialog box is that the formal variable names are listed in the *Variable* column. Many experienced users of AutoCAD prefer to use the formal variable names since they do not change from release to release as the dialog box descriptions do.

You can also use the *Compare Dimension Styles* dialog box to give the entire list of variables and related settings for one dimension style. Do this by selecting the desired dimension style name from the *Compare* drop-down list and select *<none>* from the *With* list (Fig. 26-12).

FIGURE 26-12

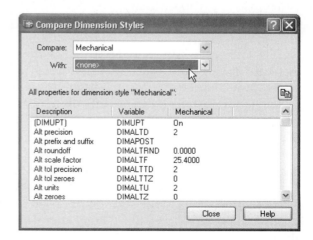

DIMENSION VARIABLES

Now that you understand how to create and use dimension styles and dimension style families, let's explore the dimension variables. <u>Two primary methods</u> can be used to set dimension variables: the *Dimension Styles Manager* and Command line format. First, we will examine the dialog box (accessible through the *Dimension Styles Manager*) that allows you to set dimension variables. Dimension variables can alternately be set by typing the formal name of the variable (such as *DIMSCALE*) and changing the desired value (discussed after this section on dialog boxes). The dialog box offers a more "user-friendly" terminology, edit boxes, checkboxes, drop-down lists, and a preview image tile that automatically reflects the variables changes you make.

Changing Dimension Variables Using the Dialog Box Method

In this section, the dialog box that contains the seven tabs for changing variables is explained (see Fig. 26-13). The seven tabs indicate different groups of dimension variables. The tab names are:

> *Lines, Symbols and Arrows, Text, Fit, Primary Units, Alternate Units, Tolerances*

Access to this dialog box from the *Dimension Style Manager* is accomplished by selecting *New, Modify,* or *Override.* Practically, there is only one dialog box that allows you to change dimension variables; however, you might say there are three dialog boxes since the title of the box changes based on your selection of *New, Modify,* or *Override.* Depending on your selection, you can invoke the *Create Dimension Style, Modify Dimension Style,* or *Override Current Style* dialog box. The options in the boxes are identical and only the titles are different; therefore, we will examine only one dialog box.

In the following pages, the heading for the paragraphs below include the dialog box option and the related formal dimension variable name in parentheses. The following figures typically display the AutoCAD default setting for a particular variable and an example of changing the setting to another value. These AutoCAD default settings (in the "Standard" dimension style) are for the ACAD.DWT template drawing and for *Start from Scratch, Imperial Default Settings.* If you use the *Metric Default Settings* or other template drawings, dimension variable settings may be different based on dimension styles that exist in the drawing other than Standard.

Lines Tab

The options in this tab change the appearance of dimension lines and extension lines (Fig. 26-13). The preview image automatically changes based on the settings you select in the *Dimension Lines* and *Extension Lines* sections.

FIGURE 26-13 ——————

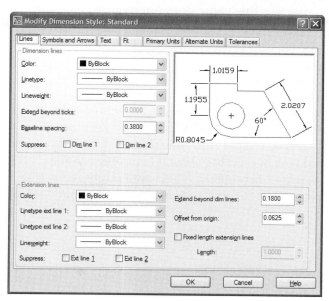

Dimension Lines Section (Lines Tab)

Color (DIMCLRD)

The *Color* drop-down list allows you to choose the color for the dimension line. Assigning a specific color to the dimension lines, extension lines, and dimension text gives you more control when color-dependent plot styles are used because you can print or plot these features with different line widths, colors, etc. This feature corresponds to the *DIMCLRD* variable (dim color dimension line).

Linetype (DIMLTYPE)

Generally, linetypes for dimensions are *Continuous*; however, for special applications you may want to change this setting (Fig. 26-14). Select from the drop-down list. This setting is stored in the *DIMLTYPE* variable.

FIGURE 26-14 ────────

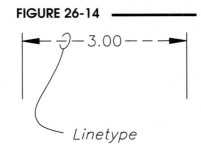

Lineweight (DIMLWD)

Use this option to assign lineweight to dimension lines (Fig. 26-15). Select any lineweight from the drop-down list or enter values in the *DIMLWD* variable (dim lineweight dimension line). Values entered in the variable can be -1 (*ByLayer*), -2 (*ByBlock*), 25 (*Default*), or any integer representing 100th of mm, such as 9 for .09mm.

FIGURE 26-15 ────────

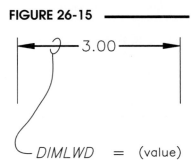

Extend beyond ticks (DIMDLE)

The *Extend beyond ticks* edit box is disabled unless the *Oblique, Integral,* or *Architectural Tick* arrowhead type is selected (in the *Arrowheads* section). The *Extend* value controls the length of dimension line that extends past the dimension line (Fig. 26-16). The value is stored in the *DIMDLE* variable (dimension line extension). Generally, this value does not require changing since it is automatically multiplied by the *Overall Scale* (DIMSCALE).

FIGURE 26-16 ────────

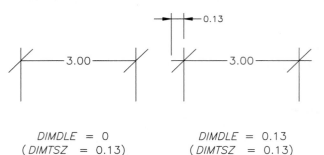

Baseline spacing (DIMDLI)

The *Baseline spacing* edit box reflects the value that AutoCAD uses in baseline dimensioning to "stack" the dimension line above or below the previous one (Fig. 26-17). This value is held in the *DIMDLI* variable (dimension line increment). This value rarely requires input since it is affected by *Overall Scale* (DIMSCALE).

FIGURE 26-17 ────────

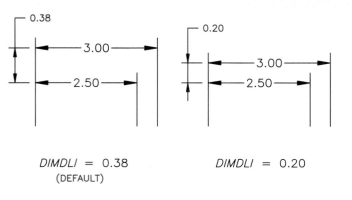

Suppress Dim Line 1, Dim Line 2 (DIMSD1, DIMSD2)

This area allows you to suppress (not draw) the *1st* or *2nd* dimension line or both (Fig. 26-18). The <u>first</u> dimension line would be on the "First extension line origin" side or nearest the end of object PICKed in response to "Select object to dimension." These toggles change the *DIMSD1* and *DIMSD2* dimension variables (suppress dimension line 1, 2).

FIGURE 26-18

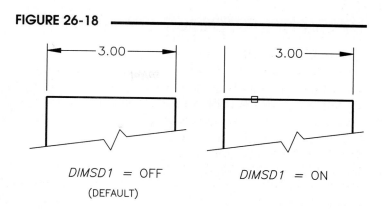

DIMSD1 = OFF
(DEFAULT)

DIMSD1 = ON

Extension Line Section (Lines Tab)

Color (DIMCLRE)

The *Color* drop-down list allows you to choose the color for the extension lines. You can also activate the standard *Select Color* dialog box. The color assignment for extension lines is stored in the *DIMCLRE* variable (dim color extension line).

Linetype Ext Line 1, Ext Line 2 (DIMLTEX1, DIMLTEX2)

These drop-down lists allow you to set the linetype for the first and second extension lines (Fig. 26-19). The linetype settings for extension line 1 and extension line 2 are stored in the *DIMLTEX1* and *DIMLTEX2* variables, respectively.

FIGURE 26-19

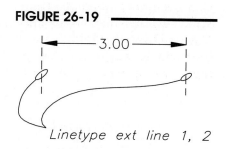

Linetype ext line 1, 2

Lineweight (DIMLWE)

This option is similar to that for dimension lines, only the lineweight is assigned to extension lines (Fig. 26-20). Select any lineweight from the drop-down list or enter values in the *DIMLWE* (dim lineweight extension line) variable (*ByLayer* = -1, *ByBlock* = -2, *Default* = 25, or any integer representing 100th of mm).

FIGURE 26-20

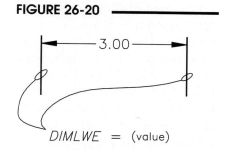

DIMLWE = (value)

Suppress Ext Line 1, Ext Line 2 (DIMSE1, DIMSE2)

This area is similar to the *Dimension Line* area that controls the creation of dimension lines but is applied to extension lines. This area allows you to suppress the *1st* or *2nd* extension line or both (Fig. 26-21). These options correspond to the *DIMSE1* and *DIMSE2* dimension variables (suppress extension line 1, 2).

FIGURE 26-21

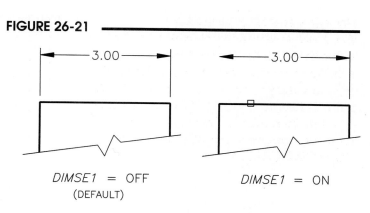

DIMSE1 = OFF
(DEFAULT)

DIMSE1 = ON

Extend beyond dim lines (DIMEXE)

The *Extend beyond dim lines* edit box reflects the value that AutoCAD uses to set the distance for the extension line to extend beyond the dimension line (Fig. 26-22). This value is held in the *DIMEXE* variable (extension line extension). Generally, this value does not require changing since it is automatically multiplied by the *Overall Scale (DIMSCALE)*.

FIGURE 26-22

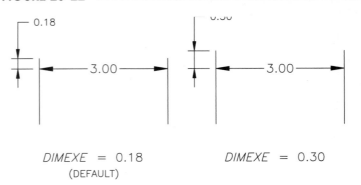

Offset from origin (DIMEXO)

The *Offset from origin* value specifies the distance between the origin points and the extension lines (Fig. 26-23). This offset distance allows you to PICK the object corners, yet the extension lines maintain the required gap from the object. This value rarely requires input since it is affected by *Overall Scale (DIMSCALE)*. The value is stored in the *DIMEXO* variable (extension line offset).

FIGURE 26-23

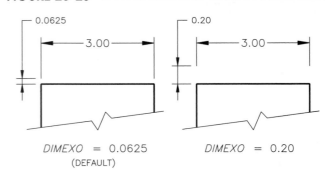

Fixed Length Extension Lines (DIMFXLON, DIMFXL)

This checkbox creates dimensions with extension lines all the same length (Fig. 26-24). The *Length* value determines the length of the extension lines measured from the dimension line toward the "extension line origin." The checkbox setting (on or off) is saved in the *DIMFXLON* variable, and the length value is stored in the *DIMFXL* variable.

FIGURE 26-24

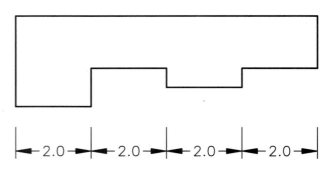

Symbols and Arrows Tab

Use this tab to set your preferences for the appearance of arrowheads, center marks, and special arc and radius symbols (Fig. 26-25). The image tile updates to display your choices for the *Arrowheads* and *Center Marks* sections only.

FIGURE 26-25

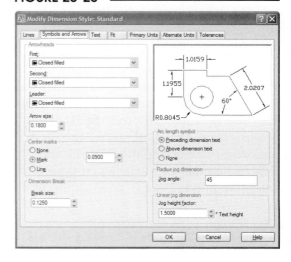

Arrowheads Section
(Symbols and Arrows Tab)

First, Second (DIMBLK, DIMSAH, DIMBLK1, DIMBLK2)

This area contains two drop-down lists of various arrowhead types, including dots and ticks. Each list corresponds to the *First* or *Second* arrowhead created in the drawn dimension (Fig. 26-26). The image tiles display each arrowhead type selected. Click in the first image tile to change both arrowheads, or click in each to change them individually. The variables affected are *DIMBLK, DIMSAH, DIMBLK1,* and *DIMBLK2*. *DIMBLK* specifies the *Block* to use for arrowheads if both are the same (Fig. 26-26). When *DIMSAH* is on (separate arrow heads), separate arrow heads are allowed for each end defined by *DIMBLK1* and *DIMBLK2*. You can enter any existing *Block* name in the *DIMBLK* variables.

FIGURE 26-26

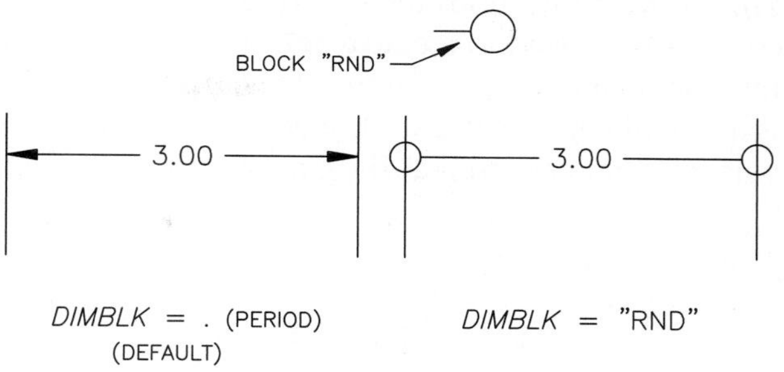

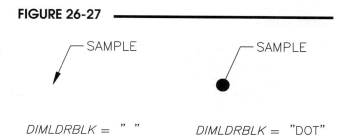

Leader (DIMLDRBLK)

This drop-down list specifies the arrow type for leaders (Fig. 26-27). This setting does not affect the arrowhead types for dimension lines. Select any option from the drop-down list or enter a value in the *DIMLDRBLK* variable (dim leader block).

FIGURE 26-27

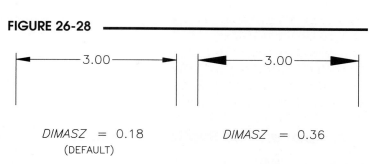

Arrow size (DIMASZ)

The size of the arrow can be specified in the *Arrow size* edit box. The *DIMASZ* variable (dim arrow size) holds the value (Fig. 26-28). Remember that this value is multiplied by *Overall Scale* (*DIMSCALE*).

FIGURE 26-28

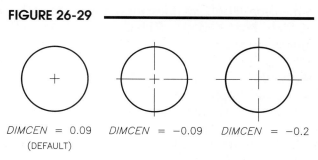

Other Sections (Symbols and Arrows Tab)

Center Marks (DIMCEN)

The list here determines how center marks are drawn when the dimension commands *Dimcenter, Dimdiameter,* or *Dimradius* are used. The image tile displays the *Mark, Line,* or *None* feature specified. This area actually controls the value of <u>one</u> dimension variable, *DIMCEN* by using a 0, positive, or negative value (Fig. 26-29). The *None* option enters a *DIMCEN* value of 0.

FIGURE 26-29

DIMCEN = 0.09   DIMCEN = −0.09   DIMCEN = −0.2
(DEFAULT)

Size (DIMCEN)

The *Size* edit box controls the size of the short dashes and extensions past the arc or circle. The value is stored in the *DIMCEN* variable (Fig. 26-29). Only positive values can be entered in this edit box.

Break Size

This value determines the size of the break when the *Dimbreak* command is used to create a space in a dimension line. The *Size* represents the length of line that is broken (Fig. 26-30). This setting is available only in the dialog box format. There is no dimension variable that can used at the Command line to access this value.

FIGURE 26-30

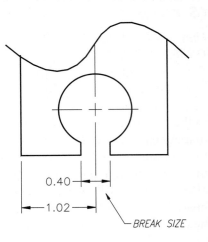

Text Tab

The options in this tab change the appearance, placement, and alignment of the dimension text (Fig. 26-31). The preview image automatically reflects changes you make in the *Text Appearance*, *Text Placement*, and *Text Alignment* sections.

FIGURE 26-31

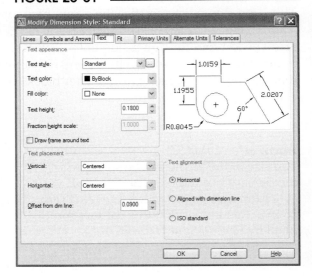

Text Appearance Section (Text Tab)

Text style (DIMTXSTY)

This feature of AutoCAD allows you to have different text styles for different dimension styles. The text styles are chosen from a drop-down list of <u>existing</u> styles in the drawing. You can also create a text style "on the fly" by picking the button just to the right of the *Text style* drop-down list, which produces the standard *Text Style* dialog box. The text style used for dimensions remains constant (as defined by the dimension style) and does not change when other text styles in the drawing are made current (as defined by *Style*, *Text*, or *Mtext*). The *DIMTXSTY* variable (dim text style) holds the text style name for the dimension style.

Text color (DIMCLRT)

Select this drop-down list to select a color or to activate the standard *Select Color* dialog box. The color choice is assigned to the dimension text only (Fig. 26-32). This is useful for controlling the text appearance for printing or plotting when color-dependent plot styles are used. The setting is stored in the *DIMCLRT* variable (dim color text).

FIGURE 26-32

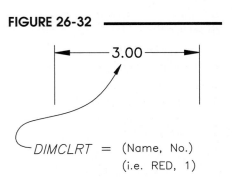

Fill Color (DIMTFILL, DIMTFILLCLR)

You can create a colored background for your dimensions using this option (Fig 26-33). Use the drop-down list to select a color or to activate the *Select Color* dialog box. You can also enter color name or number. The setting for type of background (none, drawing, or color) is saved in the *DIMTFILL* variable and the color value is saved in the *DIMTFILLCLR* variable.

FIGURE 26-33

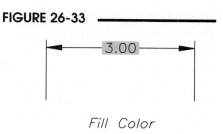

Fill Color

Text height (DIMTXT)

This value specifies the primary text height; however, this value is multiplied by the *Overall Scale* (*DIMSCALE*) to determine the actual drawn text height. Change *Text height* <u>only</u> if you want to increase or decrease the text height in relation to the other dimension components (Fig. 26-34). Normally, change the *Overall Scale* (*DIMSCALE*) to change all size-related features (arrows, gaps, text, etc.) proportionally. The text height value is stored in *DIMTXT*.

FIGURE 26-34

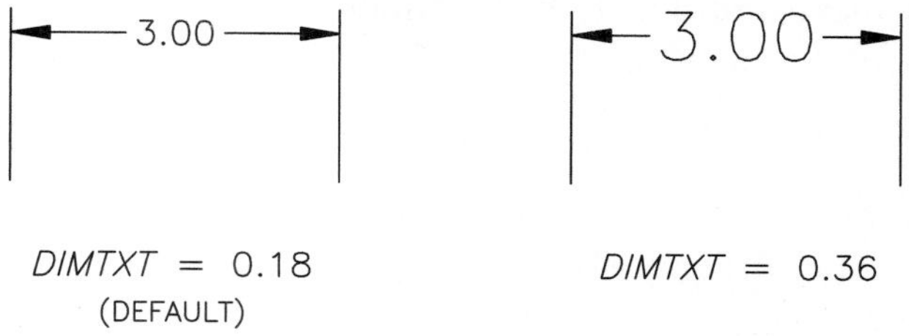

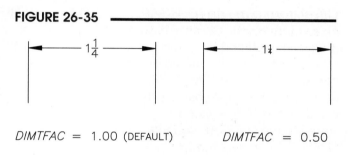

Fraction height scale (DIMTFAC)

When fractions or tolerances are used, the height of the fractional or tolerance text can be set to a proportion of the primary text height. For example, 1.0000 creates fractions and tolerances the same height as the primary text; .5000 represents fractions or tolerances at one-half the primary text height (Fig. 26-35). The value is stored in the *DIMTFAC* variable (dim tolerance factor).

FIGURE 26-35

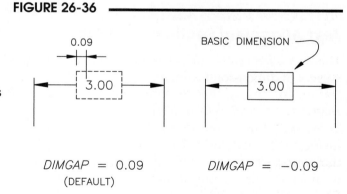

Draw frame around text (DIMGAP)

The "frame" or "gap" is actually an invisible box around the text that determines the offset from text to the dimension line. This option sets the *DIMGAP* variable to a negative value which makes the box visible (Fig. 26-36). This practice is standard for displaying a basic dimension. See also *Offset from dimension line* in the "Text Placement" section.

FIGURE 26-36

Text Placement Section (Text Tab)

Vertical (DIMTAD)

The *Vertical* option determines the vertical location of the text with respect to the dimension line. There are four possible settings that affect the *DIMTAD* variable (dim text above dimension line). *Centered*, the default option (*DIMTAD* = 0), centers the dimension text between the extension lines. The *Above* option places the dimension text above the dimension line except when the dimension line is not horizontal (*DIMTAD* = 1). *Outside* places the dimension text on the side of the dimension line farthest away from the extension line origin points—away from the dimensioned object (*DIMTAD* = 2). The *JIS* option places the dimension text to conform to Japanese Industrial Standards (*DIMTAD* = 3).

Horizontal (DIMJUST)

The *Horizontal* section determines the horizontal location of the text with respect to the dimension line (dimension justification). The default option (*DIMJUST* = 0) centers the text between the extension lines. The other four choices (*DIMJUST* = 1-3) place the text at either end of the dimension line in parallel and perpendicular positions (Fig. 26-37). You can use the *Horizontal* and *Vertical* settings together to achieve additional text positions.

FIGURE 26-37

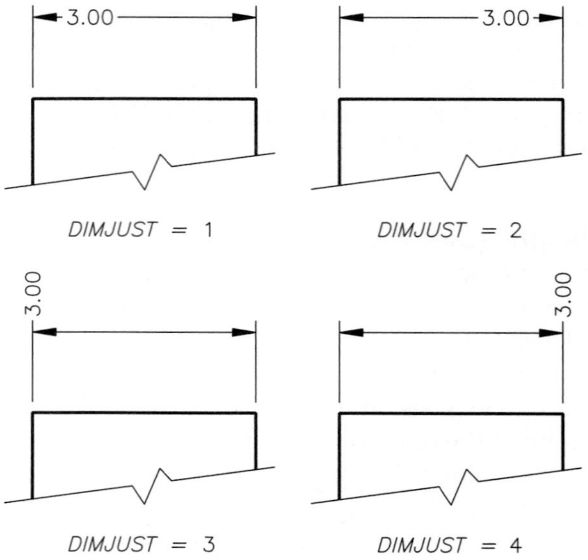

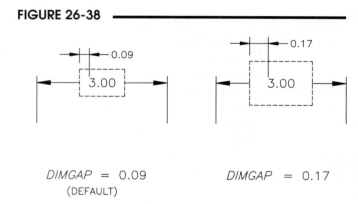

Offset from dim line (DIMGAP)

This value sets the distance between the dimension text and its dimension line. The offset is actually determined by an invisible box around the text (Fig. 26-38). Increasing or decreasing the value changes the size of the invisible box. The *Offset from dim line* value is stored in the *DIMGAP* variable. (Also see "*Draw frame around text.*")

FIGURE 26-38

DIMGAP = 0.09 (DEFAULT)   DIMGAP = 0.17

Text Alignment Section (Text Tab)

The *Text Alignment* settings can be used in conjunction with the *Text Placement* settings to achieve a wide variety of dimension text placement options.

Horizontal (DIMTIH, DIMTOH)

This radio button turns on the *DIMTIH* (dim text inside horizontal) and *DIMTOH* (dim text outside horizontal) variables. The text remains horizontal even for vertical or angled dimension lines (see Fig. 26-39 and Fig. 26-40, on the next page). This is the correct setting for mechanical drawings (other than ordinate dimensions) according to the ASME Y14.5M-1994 standard, section 1.7.5, Reading Direction.

FIGURE 26-39

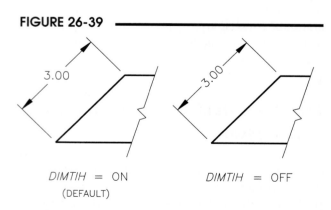

DIMTIH = ON (DEFAULT)   DIMTIH = OFF

Aligned with dimension line (DIMTIH, DIMTOH)

Pressing this radio button turns off the *DIMTIH* and *DIMTOH* variables so the text aligns with the angle of the dimension line (see Fig. 26-39 and Fig. 26-40).

ISO Standard (DIMTIH, DIMTOH)

This option forces the text inside the dimension line to align with the angle of the dimension line (*DIMTIH* = off), but text outside the dimension line is horizontal (*DIMTOH* = on).

FIGURE 26-40

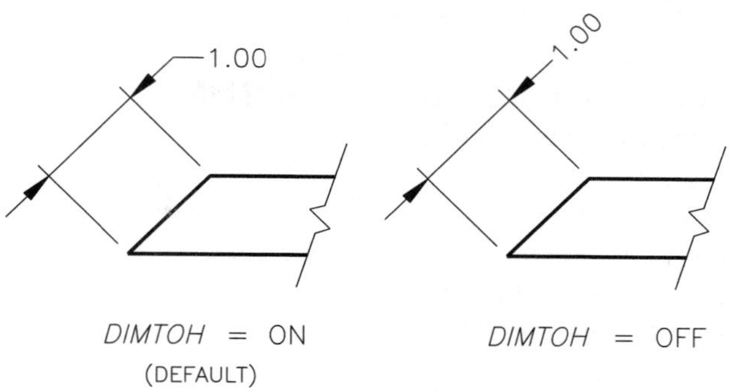

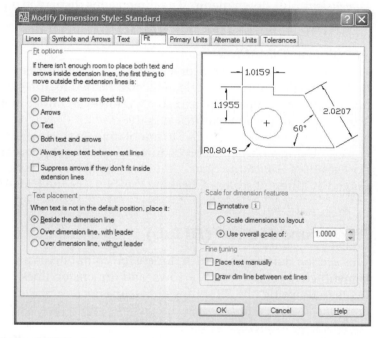

DIMTOH = ON
(DEFAULT)

DIMTOH = OFF

Fit Tab

The *Fit* tab allows you to determine the *Overall Scale* for dimensioning components, how the text, arrows, and dimension lines fit between extension lines, and how the text appears when it is moved (Fig. 26-41).

FIGURE 26-41

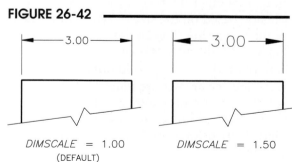

Scale for Dimension Features Section (*Fit* Tab)

Although this is not the first section in the dialog box, it is presented first because of its importance.

Overall Scale (DIMSCALE)

The *Overall Scale* value globally affects the scale of <u>all size-related features</u> of dimension objects, such as arrowheads, text height, extension line gaps (from the object), extensions (past dimension lines), etc. All other size-related (variable) values appearing in the dialog box series are <u>multiplied by the *Overall Scale*</u>. Notice how all dimensioning features (text, arrows, gaps, offsets) are all increased proportionally with the *Overall Scale* value (Fig. 26-42). Therefore, to keep all features proportional, <u>change this one setting rather than each of the others individually</u>.

FIGURE 26-42

DIMSCALE = 1.00
(DEFAULT)

DIMSCALE = 1.50

TIP

TIP

Although this area is located on the right side of the box, it is probably the most important option in the entire series of tabs. Because the *Overall Scale* should be set as a family-wide variable, setting this value is typically the <u>first step</u> in creating a dimension style (Fig. 26-43).

Changes in this variable should be based on the *Limits* and plot scale. You can use the drawing scale factor to determine this value. (See "Drawing Scale Factor," Chapter 12.) The *Overall Scale* value is stored in the *DIMSCALE* variable.

FIGURE 26-43

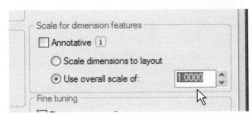

Scale dimensions to layout (DIMSCALE)

Checking this box allows you to retroactively change the size of the dimensions in a viewport based on the viewport scale by selecting all the dimensions (in the viewport), then using *Update* from the *Dimensions* pull-down menu. The dimensions display in a new size based on the viewport scale. Checking this box sets *DIMSCALE* to 0.

Annotative (DIMANNO)

Like setting the *Overall Scale* or selecting *Scale dimensions to layout*, the *Annotative* option is an alternate method of controlling the scale of all size-related features of dimension objects, such as arrowheads, text height, extension line gaps, and so on. However, rather than setting a static value as a multiplier for all size-related dimension features as is the case for *Overall Scale*, this option is dynamic since it creates annotative dimensions, similar to annotative text, annotative hatch patterns, and other annotative objects. Using a particular procedure, annotative objects can change scale based on the *VP Scale* (viewport scale) or *Annotative Scale*. Annotative objects are useful when you have multiple viewports displayed at different scales. Checking this box sets the *DIMANNO* variable to 1 (on).

2008

Fit Options Section (Fit Tab)

This section determines which dimension components are <u>forced outside the extension lines only if there is insufficient room</u> for text, arrows, and dimension lines. In most cases where space permits all components to fit inside, the *Fit* settings have no effect on placement. *Linear, Aligned, Angular, Baseline, Continue, Radius,* and *Diameter* dimensions apply. See "*Radius* and *Diameter* Variable Settings."

Either the text or the arrows (DIMATFIT)

This is the default setting for the STANDARD dimension style. AutoCAD makes the determination of whether the text or the arrows are forced outside based on the size of arrows and the length of the text string (Fig. 26-44). This option often behaves similarly to *Arrows*, except that if the text cannot fit, the text is placed outside the extension lines and the arrows are placed inside. However, if the arrows cannot fit either, both text and arrows are placed outside the extension lines. The *DIMATFIT* (dimension arrows/text fit) setting is 3.

FIGURE 26-44

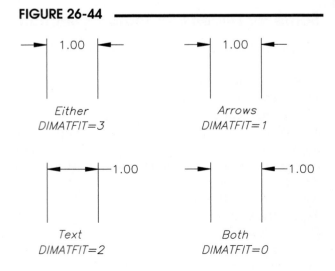

Arrows (DIMATFIT)

The *Arrows* option forces the arrows on the outside of the extension lines and keeps the text inside. If the text absolutely cannot fit, it is also placed outside the extension lines (see Fig. 26-44). This option sets *DIMATFIT* to 1.

Text (DIMATFIT)

The *Text* option places the text on the outside and keeps the arrows on the inside unless the arrows cannot fit, in which case they are placed on the outside as well (see Fig. 26-44). For this option, *DIMATFIT* = 2.

Both text and arrows (DIMATFIT)

The *Both text and arrows* option keeps the text and arrows together always. If space does not permit <u>both</u> features to fit between the extension lines, it places the text and arrows outside the extension lines (see Fig. 26-44). You can set *DIMATFIT* to 0 to achieve this placement.

Always keep text between ext lines (DIMTIX)

If you want the text to be forced between the extension line no matter how much room there is, use this option (Fig. 26-45). Pressing this radio button turns *DIMTIX* on (text inside extensions).

FIGURE 26-45

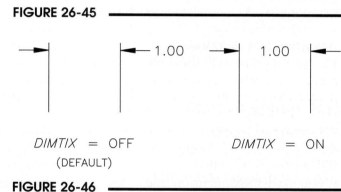

DIMTIX = OFF
(DEFAULT)

DIMTIX = ON

Suppress arrows if they don't fit (DIMSOXD)

When dimension components are forced outside the extension lines and there are many small dimensions aligned in a row (such as with *Continue* dimensions), the text, arrows, or dimension lines may overlap. In this case, you can prevent the arrows and the dimension lines from being drawn entirely with this option. This option suppresses the arrows and dimension lines <u>only</u> when they are forced outside (Fig. 26-46). The setting is stored as *DIMSOXD* = on (suppress dimension lines outside extensions).

FIGURE 26-46

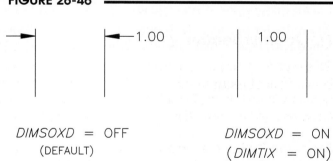

DIMSOXD = OFF
(DEFAULT)

DIMSOXD = ON
(DIMTIX = ON)

Text Placement Section (Fit Tab)

This section of the dialog box sets dimension text movement rules. When text is moved either by being automatically forced from between the dimension lines based on the *DIMATFIT* setting (the *Fit Options* above this section) or when you actually move the text with grips or by *Dimtedit*, these rules apply.

Beside the dimension line (DIMTMOVE)

This is the normal placement of the text—aligned with and beside the dimension line (Fig. 26-47). The text always moves when the dimension line is moved and vice versa. *DIMTMOVE* (dimension text move) = 0.

FIGURE 26-47

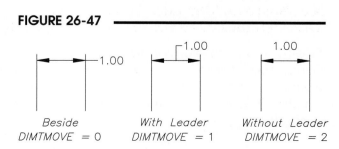

Beside
DIMTMOVE = 0

With Leader
DIMTMOVE = 1

Without Leader
DIMTMOVE = 2

Over the dimension line, with a leader (DIMTMOVE)

This option creates a leader between the text and the center of the dimension line whenever the text cannot fit between the extension lines or is moved using grips (see Fig. 26-47). *DIMTMOVE* = 1.

Over the dimension line, without a leader (DIMTMOVE)

Use this setting to have the text appear above the dimension line, similar to *DIMTMOVE* = 1, but without a leader. This occurs only when there is insufficient room for the text between the extension lines or when you move the text with grips. *DIMTMOVE* = 2.

FIGURE 26-48

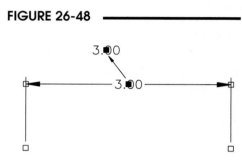

There is an important benefit to this setting (*DIMTMOVE* = 2). When there is sufficient room for the text and arrows between extension lines, this setting has no effect on the placement of the text. When there is insufficient room, the text moves above without a leader. In either case, if you prefer to move the text to another location with grips or using *Dimtedit*, the text moves as if were "detached" from the dimension line. The text can be moved independently to any location and the dimension retains its associatively. Figure 26-48 illustrates the use of grips to edit the dimension text.

TIP

Without Leader (DIMTMOVE=2)

Fine Tuning Section (Fit Tab)

Place text manually when dimensioning (DIMUPT)

When you press this radio button, you can create dimensions and move the text independently in relation to the dimension and extension lines as you place the dimension line in response to the "specify dimension line location" prompt (Fig. 26-49). Using this option is similar to using *Dimtedit* after placing the dimension. *DIMUPT* (dimension user-positioned text) is on when this box is checked.

FIGURE 26-49

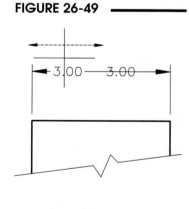

DIMUPT = ON

Always draw dim line between ext lines (DIMTOFL)

Occasionally, you may want the dimension line to be drawn inside the extension lines even when the text and arrows are forced outside. You can force a line inside with this option (Fig. 26-50). A check in this box turns *DIMTOFL* on (text outside, force line inside).

FIGURE 26-50

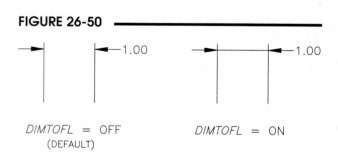

DIMTOFL = OFF
(DEFAULT)

DIMTOFL = ON

Radius and Diameter Variable Settings

For creating mechanical drawing dimensions according to ANSI standards, the default settings in AutoCAD are correct for creating *Diameter* dimensions but not for *Radius* dimensions. The following variable settings are recommended for creating *Radius* and *Diameter* dimensions for mechanical applications.

For *Diameter* dimensions, the default settings produce ANSI-compliant dimensions that suit most applications—that is, text and arrows are on the outside of the circle or arc pointing inward toward the center. For situations where large circles are dimensioned, *Fit Options* (*DIMATFIT*) and *Place text manually* (*DIMUPT*) can be changed to force the dimension inside the circle (Fig. 26-51).

For *Radius* dimensions the default settings produce incorrect dimensioning practices. Normally (when space permits) you want the dimension line and arrow to be inside the arc, while the text can be inside or outside. To produce ANSI-compliant *Radius* dimensions, set *Fit Options* (*DIMAT-FIT*) and *Place text manually* (*DIMUPT*) as shown in Figure 26-51. Radius dimensions can be outside the arc in cases where there is insufficient room inside.

FIGURE 26-51

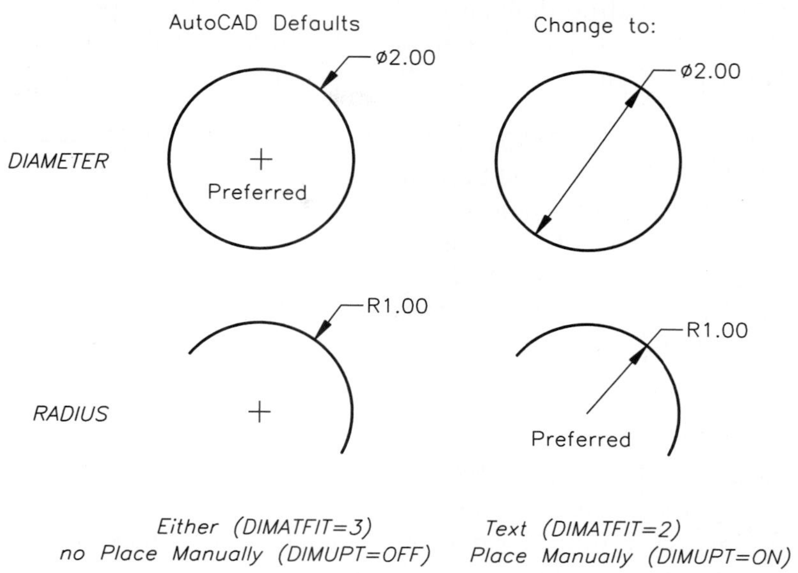

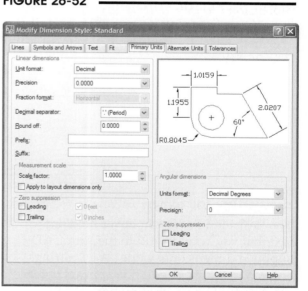

TIP

Primary Units **Tab**

FIGURE 26-52

The *Primary Units* tab controls the format of the AutoCAD-measured numerical value that appears with a dimension (Fig. 26-52). You can vary the numerical value in several ways such as specifying the units format, precision of decimal or fraction, prefix and/or suffix, zero suppression, and so on. These units are called primary units because you can also cause AutoCAD to draw additional or secondary units called *Alternate Units* (for inch <u>and</u> metric notation, for example).

Linear Dimensions **Section (*Primary Units* Tab)**

This section controls the format of all dimension types except *Angular* dimensions.

Unit format (DIMLUNIT)

The *Units format* section drop-down list specifies the type of units used for dimensioning. These are the same unit types available with the *Units* dialog box (*Decimal, Scientific, Engineering, Architectural,* and *Fractional*) with the addition of *Windows Desktop*. The *Windows Desktop* option displays AutoCAD units based on the settings made for units display in Windows Control Panel (settings for decimal separator and number grouping symbols). Remember that your selection affects the <u>units drawn in dimension objects, not the global drawing units</u>. The choice for *Units format* is stored in the *DIMLUNIT* variable (dimension linear unit). This drop-down list is disabled for an *Angular* family member.

Precision (DIMDEC)

The *Precision* drop-down list in the *Dimension* section specifies the number of places for decimal dimensions or denominator for fractional dimensions. This setting does not alter the drawing units precision. This value is stored in the *DIMDEC* variable (dim decimal).

Fraction format (DIMFRAC)

Use this drop-down list to set the fractional format. The choices are displayed in Figure 26-53. This option is enabled only when *DIMLUNIT* (*Unit format*) is set to 4 (*Architectural*) or 5 (*Fractional*).

FIGURE 26-53

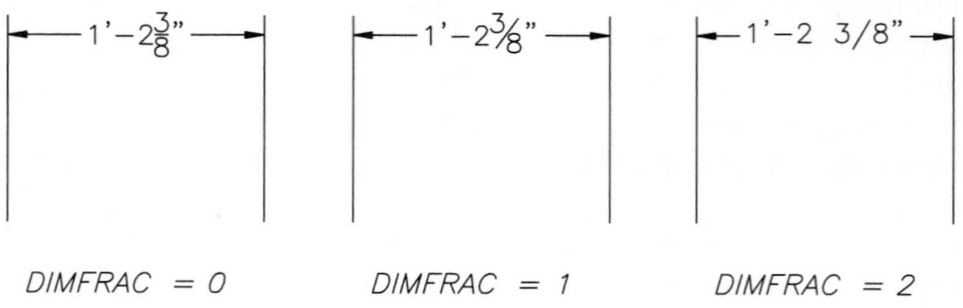

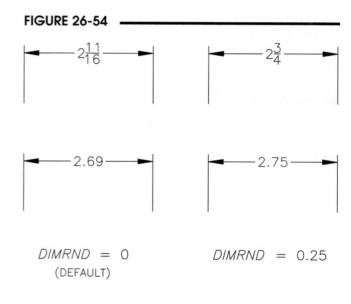

Decimal separator (DIMDSEP)

When you are creating dimensions whose unit format is decimal, you can specify a single-character decimal separator. Normally a decimal (period) is used; however, you can also use a comma (,) or a space. The character is stored in the *DIMDSEP* (dimension decimal separator) variable.

Round off (DIMRND)

Use this drop-down list to specify a precision for dimension values to be rounded. Normally, AutoCAD values are kept to 14 significant places but are rounded to the place dictated by the dimension *Precision* (*DIMDEC*). Use this feature to round up or down appropriately to the nearest specified decimal or fractional increment (Fig. 26-54).

FIGURE 26-54

Prefix/Suffix (DIMPOST)

The *Prefix* and *Suffix* edit boxes hold any text that you want to add to the AutoCAD-supplied dimensional value. A text string entered in the *Prefix* edit box appears before the AutoCAD-measured numerical value and a text string entered in the *Suffix* edit box appears after the AutoCAD-measured numerical value. For example, entering the string " mm" or a " TYP." in the *Suffix* edit box would produce text as shown in Figure 26-55. (In such a case, don't forget the space between the numerical value and the suffix.) The string is stored in the *DIMPOST* variable.

FIGURE 26-55

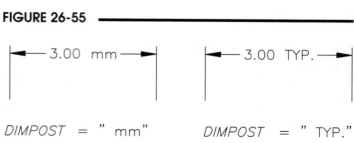

If you use the *Prefix* box to enter letters or values, any AutoCAD-supplied symbols (for radius and diameter dimensions) are overridden (not drawn). For example, if you want to specify that a specific hole appears twice, you should indicate by designating a "2X" before the diameter dimension. However, doing so by this method overrides the phi (Ø) symbol that AutoCAD inserts before the value. Instead, use the *Mtext/Dtext* options within the dimensioning command or use *Dimedit* or *Ddedit* to add a prefix to an existing dimension having an AutoCAD-supplied symbol. Remember that AutoCAD uses a dark background in the *Text Formatting* Editor to represent the AutoCAD-measured value, so place a prefix in front of the dark area.

Measurement Scale Section (*Primary Units* Tab)

Scale factor (DIMLFAC)

Any value placed in the *Scale factor* edit box is a <u>multiplier</u> for the AutoCAD-measured numerical value. The default is 1. Entering a 2 would cause AutoCAD to draw a value two times the actual measured value (Fig. 26-56). This feature might be used when a drawing is created in some scale other than the actual size, such as an enlarged detail view in the same drawing as the full view. Changing this setting is <u>unnecessary</u> when you create associative dimensions in paper space attached to objects in model space.

FIGURE 26-56

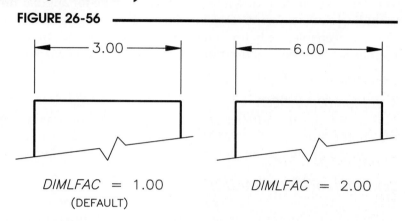

However, if you are using versions of AutoCAD previous to 2002, or if you are using nonassociative dimensions in paper space, change this setting to create dimensions that display other than the actual measured value. This setting is stored in the *DIMLFAC* variable (dimension length factor).

Apply to layout dimensions only (DIMLFAC)

Use this checkbox to apply the *Scale factor* to dimensions placed in a layout for <u>nonassociative</u> dimensions or for drawings in versions earlier than AutoCAD 2002. For example, assume you have a detail view displayed at 2:1 in a viewport and want to place nonassociative dimensions in paper space, set the *Scale factor* to .5 and check *Apply to layout dimensions only* so the measured values adjust for the viewport scale. A check in this box sets the *DIMLFAC* variable to the negative of the *Scale factor* value. Associative dimensions drawn in paper space automatically adjust for the viewport scale, so changing this setting is unnecessary.

Zero Suppression Section (*Primary Units* Tab)

The *Zero Suppression* section controls how zeros are drawn in a dimension when they occur. A check in one of these boxes means that zeros are <u>not drawn for that case</u>. This sets the value for *DIMZIN* (dimension zero indicator).

Leading/Trailing/0 Feet/0 Inches (DIMZIN)

Leading and *Trailing* are enabled for *Scientific, Decimal, Engineering* and *Fractional* units. The *0 Feet* and *0 Inches* checkboxes are enabled for *Architectural* and *Engineering* units.

For example, assume primary *Units format* was set to *Architectural* and *Zero Suppression* was checked for *0 Inches* only. Therefore, when a measurement displays feet and no inches, the 0 inch value is suppressed (Fig. 26-57, top dimension). On the other hand, since *0 Feet* is not checked, a measurement of less than 1 foot would report 0 feet.

FIGURE 26-57

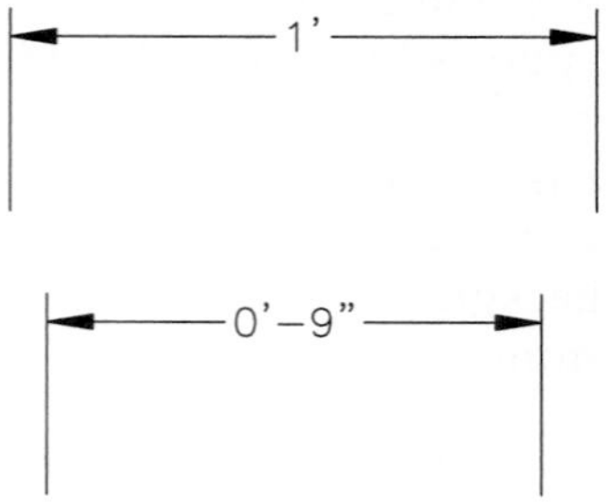

DIMZIN = 2

Angular Dimensions Section (*Primary Units* Tab)

Units format (DIMAUNIT)

The *Units format* drop-down list sets the unit type for angular dimensions, including *Decimal degrees*, *Degrees Minutes Seconds*, *Gradians*, and *Radians*. This list is enabled only for parent dimension style and *Angular* family members. The variable used for the angular units is *DIMAUNIT* (dimension angular units).

Precision (DIMADEC)

This option sets the number of places of precision (decimal places) for angular dimension text. The selection is stored in the *DIMADEC* variable (dimension angular decimals). This option is enabled only for the parent dimension style and *Angular* family member.

Zero Suppression (DIMAZIN)

Use these two checkboxes to set the desired display for angular dimensions when there are zeros appearing before or after the decimal. A check in one of these boxes means that zeros are <u>not drawn for that case</u>.

Changing Dimension Variables Using the Command Line Format

Dim...
(Variable
Name)

Pull-down Menu	Command (Type)	Alias (Type)	Short-cut	Screen (side) Menu	Tablet Menu
Tools *Inquiry >* *Set Variable*	*(Variable Name)*	...	...	*TOOLS 1* *Setvar*	*U,10*

As an alternative to setting dimension variables through the *Dimension Style Manager*, you can type the dimensioning variable name at the Command: prompt. There is a noticeable difference in the two methods—the dialog boxes use <u>different nomenclature</u> than the formal dimensioning variable names used in Command line format; that is, the dialog boxes use descriptive terms that, if selected, make the appropriate change to the dimensioning variable. The formal dimensioning variable names accessed by Command line format, however, all begin with the letters *DIM* and are accessible only by typing.

Another important but subtle difference in the two methods is the act of saving dimension variable settings to a dimension style. Remember that all drawn dimensions are part of a dimension style, whether it is STANDARD or some user-created style. When you change dimension variables by the Command line format, the changes become <u>overrides</u> until you use the *Save* option of the *-Dimstyle* command or the *Dimension Style Manager*. When a variable change is made, it becomes an <u>override that is applied to the current dimension style</u> and affects only the newly drawn dimensions. Variable changes must be *Saved* to become a permanent part of the style and to retroactively affect all dimensions created with that style.

All other dimensioning controls are accessible using the dimension variables. All these dimension variables except *DIMASSOC* are included in every dimension style.

In order to access and change a variable's setting by name, simply type the variable name at the Command: prompt. For dimension variables that require distances, you can enter the distance (in any format accepted by the current *Units* settings) or you can designate by (PICKing) two points.

For example, to change the value of the *DIMSCALE* to **.5**, this command syntax is used:

> Command: ***dimscale***
> Enter new value for dimscale <1.0000>: **.5**

For a complete list of dimension variables, see AutoCAD 2006, *Help, Command Reference, System Variables.*

MODIFYING EXISTING DIMENSIONS

Even with the best planning, a full understanding of dimension variables, and the correct use of *Dimension Styles*, it is probable that changes will have to be made to existing dimensions in the drawing due to design changes, new plot scales, or new industry/company standards. There are several ways that changes can be made to existing dimensions while retaining associativity and membership to a dimension style. The possible methods are discussed in this section.

Modifying a Dimension Style

Existing dimensions in a drawing can be modified by making one or more variable changes to the dimension style family or child, then *Saving* those changes to the dimension style. This process can be accomplished using either the *Dimension Style Manager* or Command line format. When you use the *Modify* option in the *Dimension Style Manager* or the *Save* option of the *-Dimstyle* command, the existing dimensions in the drawing <u>automatically update</u> to display the new variable settings.

Properties

Pull-down Menu	Command (Type)	Alias (Type)	Short-cut	Screen (side) Menu	Tablet Menu
Modify *Properties...*	*Properties*	*PR* *or CH*	(Edit Mode) *Properties* *or Crtl+1*	*MODIFY1* *Property*	*Y,12 to* *Y,13*

Remember that *Properties* can be used to edit existing dimensions. Using *Properties* and selecting one dimension displays the *Properties* palette with all of the selected dimension's properties and dimension variables.

Using *Properties* you can modify any aspect of one or more dimensions, including text, and access is given to the *Lines and Arrows, Text, Fit, Primary Units, Alternate Units,* and *Tolerances* categories. Any changes to dimension variables through *Properties* result in <u>overrides to the dimension object only</u> but do not affect the dimension style. *Properties* has essentially the same result as *Dimoverride* (see "Dimoverride"). *Properties* can also be used to modify the dimension text value. *Properties* of dimensions are also discussed in Chapter 25.

Creating Dimension Style Overrides and Using *Update*

You can modify existing dimensions in a drawing without making permanent changes to the dimension style by creating a dimension style override, then using *Update* to apply the new setting to an existing dimension. This method is preferred if you wish <u>to modify one or two dimensions</u> without modifying all dimensions referencing (created in) the style.

To do this, use either the Command line format or *Dimension Style Manager* to set the new variable. In the *Dimension Style Manager*, select *Override* and make the desired variable settings in the *Override Current Style* dialog box. This creates an override to the current style. In Command line format, simply enter the formal dimension variable name and make the change to create an override to the current style, then use *Update* to apply the current style settings plus the override settings to existing dimensions that you PICK.

The overrides remain in effect for the current style unless the variables are reset to the original values or until the overrides are cleared. You can <u>clear overrides</u> for a dimension style by making another dimension style current in the *Dimension Style Manager*.

Update

Pull-down Menu	Command (Type)	Alias (Type)	Short-cut	Screen (side) Menu	Tablet Menu
Dimension *Update*	*Dim* *Update*	*DIM* *UP*	...	*DIMNSION* *Update*	*Y,3*

Update can be used to update existing dimensions in the drawing to the current settings. The current settings are determined by the current dimension style and any dimension variable overrides that are in effect (see previous explanation, "Creating Dimension Style Overrides and Using *Update*"). This is an excellent method of modifying one or more existing dimensions without making permanent changes to the dimension style that the selected dimensions reference. *Update* has the same effect as using *-Dimstyle, Apply*.

Update is actually a Release 12 command. In Release 12, dimensioning commands could only be entered at the Dim: prompt. In the current release of AutoCAD, the command is given prominence by making available an *Update* button and an *Update* option in the *Dimension* pull-down menu. However, if you prefer to type, you must first type *Dim*, press Enter, and then enter *Update*. The command syntax is as follows:

```
Command: Dim
Dim: Update
Select objects: PICK (select a dimension object to update)
Select objects: Enter
Dim: press Esc or type Exit
Command:
```

For example, if you wanted to change the *DIMSCALE* of several existing dimensions to a value of 2, change the variable by typing *DIMSCALE*. Next use *Update* and select the desired dimension. That dimension is updated to the new setting. The command syntax is the following:

```
Command: Dimscale
New value for DIMSCALE <1.0000>: 2
Command: Dim
Dim: Update
Select objects: PICK  (select a dimension object to update)
Select objects: PICK  (select a dimension object to update)
Select objects: Enter
Dim: Exit
Command:
```

Beware, *Update* creates an override to the current dimension style. You should reset the variable to its original value unless you want to keep the override for creating other new dimensions. You can <u>clear overrides</u> for a dimension style by making another dimension style current in the *Dimension Style Manager*.

Dimoverride

Pull-down Menu	Command (Type)	Alias (Type)	Short-cut	Screen (side) Menu	Tablet Menu
Dimension Override	*Dimoverride*	*DIMOVER* or *DOV*	...	...	Y,4

Dimoverrride grants you a great deal of control to <u>edit existing dimensions</u>. The abilities enabled by *Dimoverride* are similar to the effect of using the *Properties* palette. *Dimoverride* enables you to <u>make variable changes</u> to dimension objects that exist in your drawing <u>without creating an override to the dimension style</u> that the dimension references (was created under). For example, using *Dimoverride*, you can make a variable change and select existing dimension objects to apply the change. The existing dimension does not lose its reference to the parent dimension style nor is the dimension style changed in any way. In effect, you can <u>override the dimension styles for selected dimension objects</u>. There are two steps: set the desired variable and select dimension objects to alter.

    Command: **dimoverride**
    Enter dimension variable name to override or [Clear overrides]: (**variable name**)
    Enter new value for dimension variable <current value>: (**value**)
    Enter dimension variable name to override: **Enter**
    Select objects: **PICK**
    Select objects: **PICK** or **Enter**
    Command:

The *Dimoverride* feature differs from creating dimension style overrides in two ways: (1) *Dimoverride* applies the changes to the selected dimension objects only, so the overrides are not appended to the parent dimension styles, only the objects; and (2) *Dimoverride* can be used once to change dimension objects referencing multiple dimension styles, whereas to make such changes to dimensions by creating dimension style overrides requires changing all the dimension styles, one at a time. The effect of using *Dimoverride* is essentially the same as using the *Properties* palette.

Dimoverride is useful as a "back door" approach to dimensioning. Once dimensions have been created, you may want to make a few modifications, but you do not want the changes to affect dimension styles (resulting in an update to all existing dimensions that reference the dimension styles). *Dimoverride* offers that capability. You can even make one variable change to affect all dimensions globally without having to change multiple dimension styles. For example, you may be required to make a test plot of the drawing in a different scale than originally intended, necessitating a new global *Dimscale*.

    Command: **dimoverride**
    Enter dimension variable name to override or [Clear overrides]: **dimscale**
    Enter new value for dimension variable <1.0000>: **.5**
    Enter dimension variable name to override: **Enter**
    Select objects: (window entire drawing) Other corner: 128 found
    Select objects: **Enter**
    Command:

This action results in having all the existing dimensions reflect the new *DIMSCALE*. No other dimension variables or any dimension styles are affected. Only the selected objects contain the overrides. After making the plot, *Dimoverride* can be used with the *Clear* option to change the dimensions (by object selection) back to their original appearance. The *Clear* option is used to clear overrides from <u>dimension objects, not from dimension styles</u>.

Clear
The *Clear* option removes the overrides from the <u>selected dimension objects</u>. It does not remove overrides from the current dimension style:

```
Command: dimoverride
Enter dimension variable name to override or [Clear overrides]: c
Select objects: PICK
Select objects: Enter
Command:
```

The dimension then displays the variable settings as specified by the dimension style it references without any overrides (as if the dimension were originally created without the overrides). Using *Clear* does not remove any overrides that are appended to the dimension style so that if another dimension is drawn, the dimension style overrides apply.

Matchprop

Pull-down Menu	Command (Type)	Alias (Type)	Short-cut	Screen (side) Menu	Tablet Menu
Modify *Match Properties*	*Matchprop*	*MA*	...	*MODIFY1* *Matchprp*	*Y,14 and* *Y,15*

Matchprop can be used to "convert" an existing dimension to the style (including overrides) of another dimension in the drawing. For example, if you have two linear dimensions, one has *Oblique* arrows and *Romans* text font (*Dimension Style* = "Oblique") and one is a typical linear dimension (*Dimension Style* = "Standard") as in Figure 26-58, "before."

FIGURE 26-58

You can convert the typical dimension to the "Oblique" style by using *Matchprop*, selecting the "Oblique" dimension as the "source object" (to match), then selecting the typical dimension as the "destination object" (to convert). The typical dimension then references the "Oblique" dimension style and changes appearance accordingly (Fig. 26-58, "after"). Note that *Matchprop* does not alter the dimension text value.

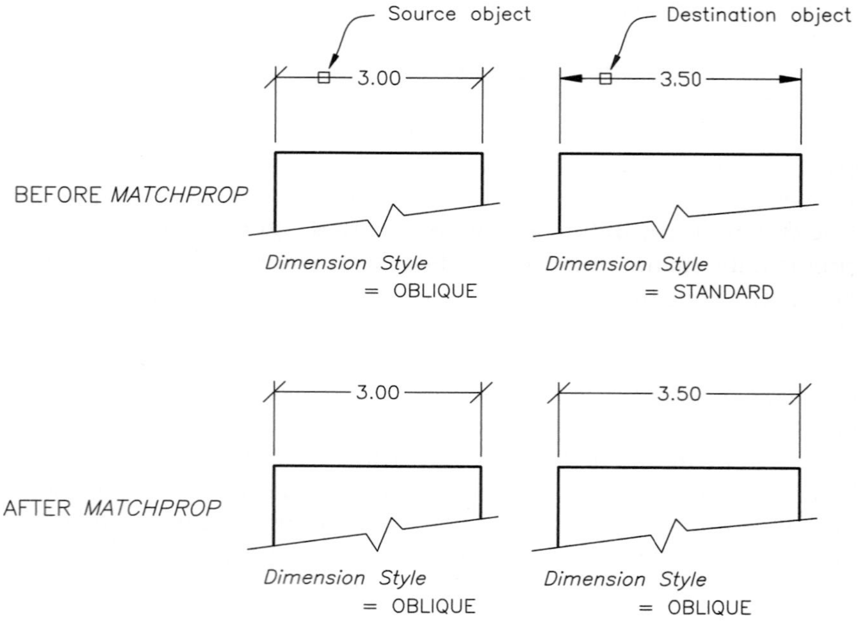

 Using the *Settings* option of the *Matchprop* command, you can display the *Property Settings* dialog box (Fig. 26-59). The *Dimension* box under *Special Properties* must be checked to "convert" existing dimensions as illustrated above.

Using *Matchprop* is a fast and easy method for modifying dimensions from one style to another. However, this method is applicable only if you have existing dimensions in the drawing with the desired appearance that you want others to match.

FIGURE 26-59

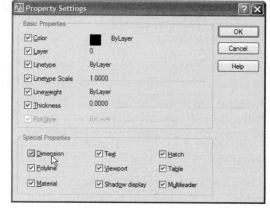

Dimension Right-Click Menu

FIGURE 26-60

If you select any dimension, you can right-click to produce a shortcut menu (Fig. 26-60). This menu contains options that allow you to change the text position, the text precision, and the *Dimension Style* of the dimension, and to flip the arrows. These options are also discussed in Chapter 25.

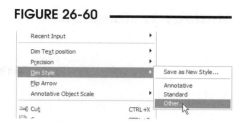

Dim Text Position

This cascading menu offers the *Centered* and *Home Text* options (duplicates of the *Dimtedit* command options), and the *Move text alone*, *Move with leader*, and *Move with dim line* options (*DIMTMOVE* variable settings). Using a *Move* option creates a *DIMTMOVE* <u>dimension style override</u> for the individual dimension.

Precision

Use this cascading menu to select from a list of decimal or fractional precision values. For example, you could change a dimension text value of "5.0000" to "5.00" with this menu. Your choice creates a <u>*DIMDEC* dimension style override</u> for the selected dimension.

Dim Style

Use this option to change the selected dimension from one dimension style to another. You can easily create a new dimension style or change the dimension to another style, similar to using *Match Properties*. Select *Other* to produce a list of all the dimension styles contained in the drawing.

Flip Arrow

Use this option to flip the arrow nearest to the point you right-click to produce the menu. In other words, if both arrows pointed from the center of the dimension out to the extension lines (default position), flipping both arrows would cause the arrows to be on the outside of the extension lines pointing inward (Fig. 26-61). The *DIMATFIT* variable can also be used to affect the position of the arrows.

FIGURE 26-61

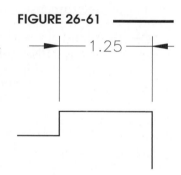

Annotative Object Scale

Several options are available from this menu item when an annotative dimension object is highlighted. You can add annotative scales to or delete scales from the highlighted object. When a highlighted annotative dimension has several scales, you can synchronize their positions.

2008

GUIDELINES FOR NON-ANNOTATIVE DIMENSIONING IN AutoCAD

Listed in this section are some guidelines to use for dimensioning a drawing using dimensioning variables and dimension styles. Although there are other strategies for dimensioning, two strategies are offered here as a framework so you can develop an organized approach to dimensioning.

In almost every case, dimensioning is one of the last steps in creating a drawing, since the geometry must exist in order to dimension it. You may need to review the steps for drawing setup, including the concept of drawing scale factor (Chapter 12).

Strategy 1. Dimensioning a Single Drawing

This method assumes that the fundamental steps have been taken to set up the drawing and create the geometry. Assume this has been accomplished:

Drawing setup is completed: *Units, Limits, Snap, Grid, Ltscale, Layers,* border, and titleblock.
The drawing geometry (objects comprising the subject of the drawing) has been created.

Now you are ready to dimension the drawing subject (of the multiview drawing, pictorial drawing, floor plan, or whatever type of drawing).

1.   Create a *Layer* (named DIM, or similar) for dimensioning if one has not already been created. Set *Continuous* linetype and appropriate color. Make it the *current* layer.

2.   Set the *Overall Scale (DIMSCALE)* based on drawing *Limits* and expected plotting size.

For plotted dimension text of 3/16":

Multiply *Overall Scale (DIMSCALE)* times the drawing scale factor. The default *Overall Scale* is set to 1, which creates dimensioning text of approximately 3/16" (default *Text Height:* or *DIMTXT* =.18) when plotted full size. All other size-related dimensioning variables' defaults are set appropriately.

For plotted dimension text of 1/8":

Multiply *Overall Scale* times the drawing scale factor, times .7. Since the *Overall Scale* times the scale factor produces dimensioning text of .18, then .18 x .7 = .126 or approximately 1/8". (See "Optional Method for Fixed Dimension Text Height.") (*Overall Scale* [*DIMSCALE*] is one variable that usually remains constant throughout the drawing and therefore should generally have the same value for every dimension style created.)

3.   Make the other dimension variable changes you expect you will need to create <u>most dimensions in the drawing</u> with the appearance you desire. When you have the basic dimension variables set as you want, *Save* this new dimension style family as TEMPLATE or COMPANY_STD (or other descriptive name) style. If you need to make special settings for types of dimensions (*Linear, Diameter,* etc.), create "children" at this stage and save the changes to the style. This style is the fundamental style to use for creating most dimensions and should be used as a template for creating other dimension styles. If you need to reset or list the original (default) settings, the STANDARD style can be restored.

4.   Create all the relatively simple dimensions first. These are dimensions that are easy and fast and require <u>no other dimension variable changes</u>. Begin with linear dimensions; then progress to the other types of dimensions.

5.   Create the special dimensions next. These are dimensions that require variable changes. Create appropriate dimension styles by changing the necessary variables, then save each set of variables (relating to a particular style of dimension) to an appropriate dimension style name. Specify dimension variables for the classification of dimension (children) in each dimension style when appropriate. The dimension styles can be created "on the fly" or as a group before dimensioning. Use TEMPLATE or COMPANY_STD as your base dimension style when appropriate.

For the few dimensions that require variable settings unique to that style, you can create dimension variable overrides. Change the desired variable(s), but do not save to the style so the other dimensions in the style are not affected.

6. When all of the dimensions are in place, make the final adjustments. Several methods can be used:

   A. If modifications need to be made <u>familywide</u>, change the appropriate variables and save the changes to the dimension style. This action automatically updates existing dimensions that reference that style.

   B. To modify the appearance of selected dimensions, use the *Properties* palette or the dimension right-click shortcut menu. Generally, these methods affect only the selected dimensions by creating dimension style overrides, so all dimensions in the style are not changed. You could also use *Dimoverride* for individual dimensions.

   C. Alternately, change the appearance of individual dimensions by creating dimension style overrides, then use *Update* or *-Dimstyle, Apply* to update selected dimensions. Clear the overrides by setting the original or another dimension style current.

   D. To change one dimension to adopt the appearance of another, use *Matchprop*. Select the dimension to match first, then the dimension(s) to convert. *-Dimstyle, Apply* can be used for this same purpose.

   E. Use *Dimoverride* to make changes to selected dimensions or to all dimensions <u>globally</u> by windowing the entire drawing. *Dimoverride* has the advantage of allowing you to assign the variables to change and selecting the objects to change all in one command. What is more, the changes are applied as overrides only to the selected dimensions but <u>do not alter the original dimension styles</u> in any way.

   F. If you want to change only the dimension text value, use *Properties* or *Dimedit, New*. *Dimedit, New* can also be used to reapply the AutoCAD-measured value if text was previously changed. The location of the text can be changed with *Dimtedit* or Grips.

   G. Grips can be used effectively to change the location of the dimension text or to move the dimension line closer or farther from the object. Other adjustments are possible, such as rotating a *Radius* dimension text around the arc.

Strategy 2. Creating Dimension Styles as Part of Template Drawings

1. Begin a *New* drawing or *Open* an existing *Template*. Assign a descriptive name.

2. Create a DIM *Layer* for dimensioning with *continuous* linetype and appropriate color and lineweight (if one has not already been created).

3. Set the *Overall Scale* accounting for the drawing *Limits* and expected plotting size. Use the guidelines given in Strategy 1, step 2. Make any other dimension variable changes needed for general dimensioning or required for industry or company standards.

4. Next, *Save* a dimension style named TEMPLATE or COMPANY_STD. This should be used as a template when you create most new dimension styles. The *Overall Scale* is already set appropriately for new dimension styles in the drawing.

5. Create the appropriate dimension styles for expected drawing geometry. Use TEMPLATE or COMPANY_STD dimension style as a base style when appropriate.

6. *Save* and *Exit* the newly created template drawing.

7. Use this template in the future for creating new drawings. Restore the desired dimension styles to create the appropriate dimensions.

Using a template drawing with prepared dimension styles is a preferred alternative to repeatedly creating the same dimension styles for each new drawing.

ANNOTATIVE DIMENSIONS

For complex drawings including multiple layouts and viewports, it may be desirable to create annotative objects. Annotative objects, such as annotative text, annotative dimensions, and annotative hatch patterns, automatically adjust for size when the *Annotative Scale* for each viewport is changed. That is, annotative objects can be automatically resized to appear "readable" in different viewports at different scales. The following section gives a quick overview of how to create annotative dimensions for use in multiple viewports at different scales.

Steps for Creating and Viewing Basic Annotative Dimensions

These steps assume the drawing geometry is created but dimensions have not been created.

1. Produce the *Dimension Style Manager* by any method.
2. In the *Dimension Style Manager*, create a *New* dimension style or *Modify* either the *Standard* or *Annotative* style. In the *Fit* tab, check the *Annotative* box if not already checked. Set the *Text Height* for dimensions to the actual paper space height you want to use. Set any other desired variable settings (*Text Style*, *Units Format* and *Precision*, etc.). Select the *OK* button.
3. In the *Model* tab, set the *Annotation Scale* for model space. You can use the *Annotation Scale* pop-up list or set the *CANNOSCALE* variable. Use the reciprocal of the "drawing scale factor" (DSF) as the annotation scale (AS) value, or AS = 1/DSF.
4. Create the dimensions in the *Model* tab. The dimensions should appear in the appropriate size. If you plan to create a viewport for a detailed view, you can wait to apply these dimensions later.
5. Create the desired viewports if not already created. Typically, one viewport is intended to show the entire model space geometry, and would therefore be displayed at about the same scale as that in model space. Other viewports may be created to display the geometry in other scales, such as for a detail view. Specify the desired plot device for the layouts using the *Page Setup Manager*.
6. In a layout tab when paper space is active (not in a viewport), enable the *Automatically add scales to annotative objects when the annotative scale changes* toggle in the lower-right corner of the Drawing Editor or set the *ANNOAUTOSCALE* variable to a positive value (1, 2, 3, or 4). This setting remains active for all layouts.
7. Next, activate the layout in which you intend to show the entire model space geometry. Set the desired scale for the viewport by double-clicking inside the viewport and using the *VP Scale* pop-up list or *Viewports* toolbar. (Alternately, select the viewport object [border] and use the *Properties* palette for the viewport.) Setting the *VP Scale* automatically adjusts the annotation scale for the viewport. The dimensions should appear in the appropriate size for the viewport. The dimensions appearing in this viewport do not change size if the *VP Scale* is the same scale that you selected for the *Annotation Scale* in model space.
8. Once the viewport scale is set, select the *Lock/Unlock Viewport* toggle in the lower-right corner of the Drawing Editor to lock the viewport. If you *Zoom* in an unlocked viewport, the scale can get out of synchronization with the selected scale.
9. Activate another viewport by double-clicking inside the viewport. Set the viewport scale for the viewport using the *VP Scale* pop-up list, *Viewports* toolbar, or *Properties* palette. Since the *Annotation Scale* is locked to the *VP Scale*, the annotation scale changes automatically to match the *VP Scale*.

10. Repeat step 9 for other viewports.
11. Create any other dimensions in detail views.
12. Make the necessary adjustments to the dimensions.

This procedure can become complex depending on the number of viewports and annotative scales.

Annotative Dimension Example

For example, assume you are drawing a stamped mechanical part. In order to draw the stamping in the *Model* tab full size, you change the default *Limits* (12, 9) by a factor of 4 to arrive at *Limits* of 48,36 (drawing scale factor = 4). After creating the part outline and mounting holes, you are ready to begin dimensioning.

Following steps 1 through 3, create a new dimension style and check the *Annotative* box in the *Fit* tab. Set any other desired variables, such as *Text Height*, *Text Style*, *Units Format* and *Precision*. Next, set the annotative scale for the *Model* tab using the *Annotation Scale* pop-up list in the lower-right corner of the Drawing Editor (Fig. 26-62). Using the drawing scale factor (DSF) of 4, set the *Annotation Scale* to 1:4 (or AS = 1/DSF). (Also see "Steps for Drawing Setup Using Annotative Objects" in Chapter 12.) Next (step 4), create the overall dimensions in the *Model* tab (Fig. 26-62).

FIGURE 26-62

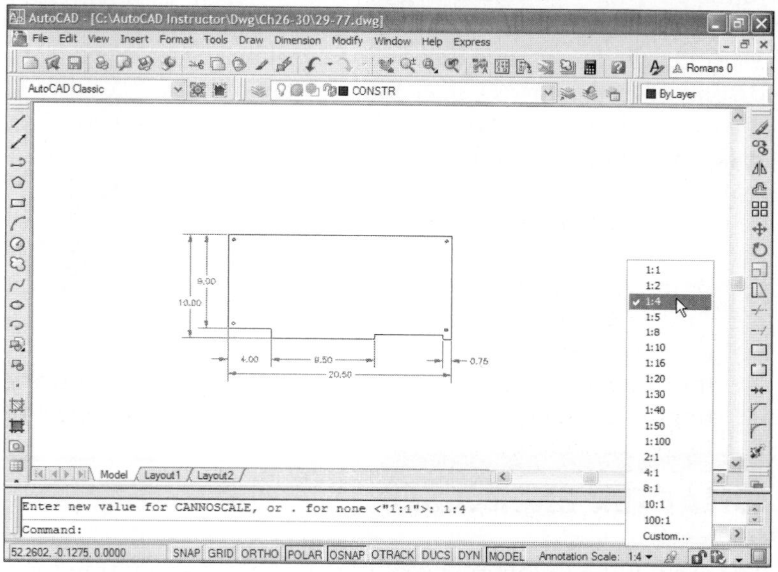

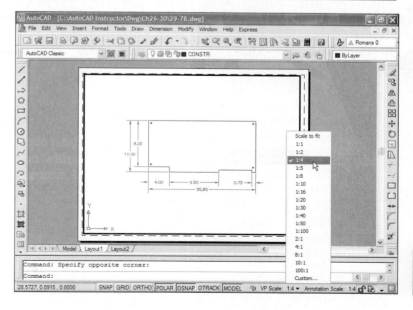

Next, following steps 5 and 6, switch to a *Layout* tab, set it up for the desired plot or print device using the *Page Setup Manager*, and create the a viewport to display the entire model space geometry. Set the viewport scale for the viewport using the *VP Scale* pop-up list as shown in Figure 26-63. Note that the *Annotation Scale* is locked to the *VP Scale* (by default) so its setting changes accordingly as the *VP Scale* changes. The dimensions may not appear to be changed in this viewport if the *VP Scale* is the same as that selected in model space for the *Annotation Scale*.

FIGURE 26-63

Note that at this point, using only one viewport, there is no real advantage to using annotative dimensions as opposed to non-annotative dimensions. The dimensions appear in the *Model* tab and *Layout* tab at the correct size, just as if non-annotative dimensions were used.

To illustrate setting the annotative scale for other viewports, create another layout with two viewports, one to display the entire stamping at 1:4 scale and the second viewport to display only a detail of one corner at 1:2 scale. Examining the display of the drawing in the two viewports at this point, the dimensions would be displayed at the same size relative to the geometry.

To change the display of the annotative text in the left viewport, make the viewport active and select 1:2 from the *VP Scale* pop-up list (Fig. 26-64). Once this is completed, the size of the geometry will be changed to exactly 1:2, and since the *Annotation Scale* is locked to the *VP Scale*, the annotative dimensions will be displayed at the correct size. Simply stated, the dimensions will appear at the same (paper space) size as those in the other viewport.

FIGURE 26-64

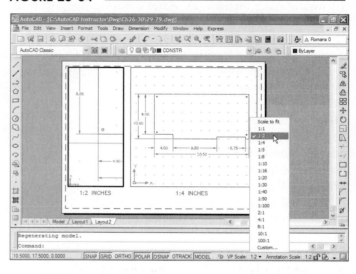

2008

CHAPTER EXERCISES

For each of the following exercises, use the existing drawings, as instructed. Create dimensions on the DIM (or other appropriate) layer. Follow the "Guidelines for Non-annotative Dimensioning in AutoCAD" given in the chapter, including setting an appropriate *Overall Scale* based on the drawing scale factor. Use dimension variables and create and use dimension styles when needed.

REUSE

1. **Dimension One View**

   Open the **HAMMER** drawing you created in Chapter 13 Exercises and use the *Saveas* command to rename it **HAMMER-DIM.** Add all necessary dimensions as shown in Chapter 13 Exercises, Figure 13-29. Create a *New* dimension style and set the variables to generate dimensions as they appear in the figure (set *Precision* to **0.00**, *Text Style* to *Romans* font, and *Overall Scale* to **1** or **1.3**). Create the

FIGURE 26-65

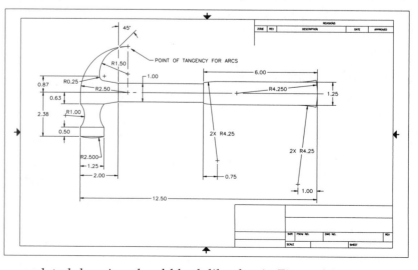

dimensions in model space. Your completed drawing should look like that in Figure 26-65.

2. Dimensioning a Multiview

Open the **SADDLE** drawing that you created in Chapter 21 Exercises. Set the appropriate dimensional *Units* and *Precision*. Add the dimensions as shown. Because the illustration in Figure 26-66 is in isometric, placement of the dimensions can be improved for your multiview. Use optimum placement for the dimensions. *Save* the drawing as **SADDL-DM** and make a *Plot* to scale.

FIGURE 26-66

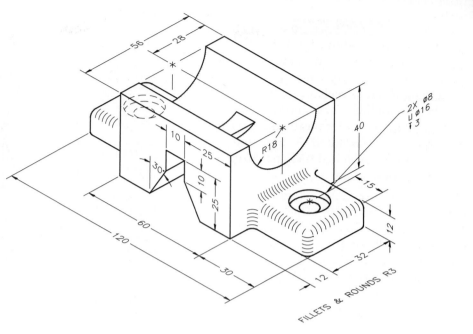

3. Architectural Dimensioning

Open the **OFFICE** drawing that you completed in Chapter 20. Dimension the floor plan as shown in Figure 26-67. Add *Text* to name the rooms. *Save* the drawing as **OFF-DIM** and make a *Plot* to an accepted scale and sheet size based on your plotter capabilities.

FIGURE 26-67

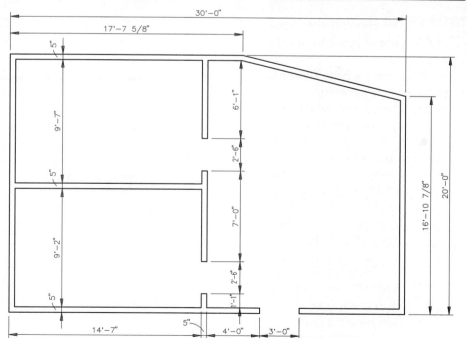

4. **Dimensioning an Auxiliary**

Open the **ANGLBRAC** drawing that you created in Chapter 24. Dimension as shown in Figure 26-68, but <u>convert the dimensions to *Decimal*</u> with **Precision** of **.000**. Use the "Guidelines for Dimensioning." Dimension the slot width as a *Limit* dimension—**.6248/.6255**. *Save* the drawing as **ANGL-DIM** and *Plot* to an accepted scale.

FIGURE 26-68

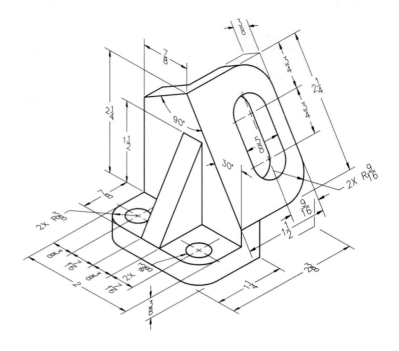

5. **Annotative Dimensioning**

A. *Open* the **EFF-APT2** drawing you worked with in Chapter 18. Make a *New* dimension style and set the variables to generate dimensions as they appear in Chapter 15 Exercises. In the *Fit* tab, check the *Annotative* box, then select *OK* to save the style. Create the dimensions in model space on a layer called **DIM**.

B. Create a *New Layout*. Use *Pagesetup* for the layout and select an appropriate *Plot Device* that can use a "**B**" size sheet, such as an HP 7475 plotter. Set the layout to a "**B**" size sheet. Next, use Design-Center to locate and insert the **ANSI-B title block** from the Template folder. If you do not have the ANSI-B title block on your computer, you can download it from **www.mhhe.com/leach, Student Resources, Download Exercise Drawings**. Make a new layer named **VPORTS** and create one viewport as shown in Figure 26-69. Set the viewport scale to display the apartment floor plan at **1/4"=1'** scale.

FIGURE 26-69

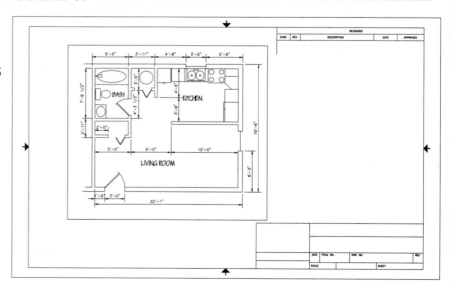

C. Create a smaller viewport on the right to display only one of the bathrooms. Turn off the toggle for *Automatically add scales to annotative objects when the annotation scale changes*. Set the *VP Scale* for the viewport to *1/2"=1'*. Create the dimensions for the plumbing wall (distance between centers of the fixtures) in paper space as shown in Figure 26-70. Label each viewport giving the scale. *Freeze* the **VPORTS** layer to achieve a drawing like that shown in Figure 26-70. Save the drawing as **APT-DIM**.

FIGURE 26-70

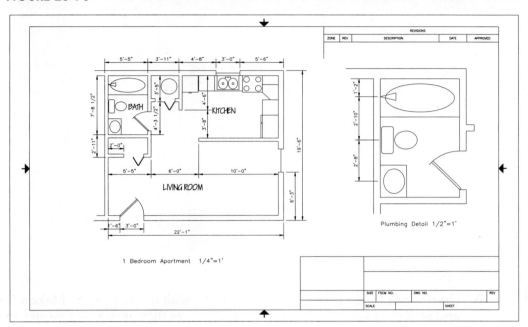

STADIUM NORTH ELEVATION.DWG, Courtesy of Autodesk, Inc.

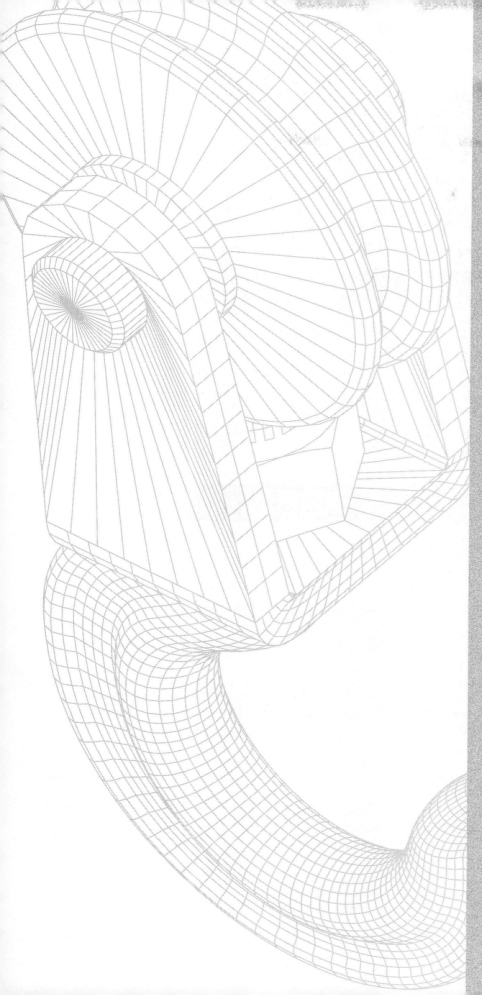

27

3D BASICS, NAVIGATION, DISPLAY, AND VIEWING

CHAPTER OBJECTIVES

After completing this chapter you should:

1. understand the AutoCAD 3D interface;

2. know the characteristics of wireframe, surface, and solid models;

3. know the formats for 3D coordinate entry;

4. understand the orientation of the World Coordinate system, the purpose of User Coordinate Systems, and the use of the *Ucsicon*;

5. be able to use 3D navigation and view commands to generate different views of 3D models;

6. be able to use the *Visual Styles* control panel to specify variations of wireframe, hidden, and shaded displays of 3D models.

THE AutoCAD 3D INTERFACE

The 3D interface in AutoCAD is different from the 2D interfaces that are available (*AutoCAD Classic* and *2D Drafting & Annotation*). In addition to the "look" of the interface, the functions for solid modeling construction, display, and rendering are quite different from the 2D commands. The full set of 3D functions in AutoCAD is discussed in the following chapters.

The 3D interface consists of the *Dashboard* on the right, preconfigured with 3D control panels, and a drawing template displaying a grid in perspective mode intended especially for creating 3D objects (Fig. 27-1).

FIGURE 27-1

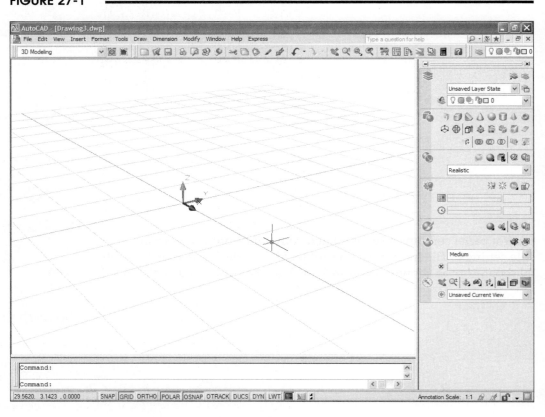

When you use AutoCAD, you have your choice of three workspaces: the *AutoCAD Classic* workspace (2D mode), the *2D Drafting and Annotation* workspace (2D mode), and the *3D Modeling* workspace (3D interface). By default, AutoCAD starts in 2D mode. It is likely that AutoCAD may have already been set up for you to begin working in one of the 2D modes in your particular office or laboratory. In that case, follow the instructions below.

To initiate the 3D interface and begin a 3D drawing:

FIGURE 27-2

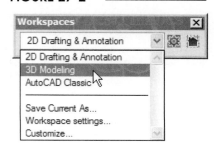

1. The *Workspaces* toolbar is normally visible near the upper-left corner of the Drawing Editor. If not visible, produce the *Workspaces* toolbar by any method. Select *3D Modeling* from the drop-down list (Fig. 27-2). This action removes some of the 2D draw and edit toolbars and produces the *Dashboard* set up with the 3D control panels. (The *Dashboard* and *Dashboardclose* commands can also be typed to open and close the *Dashboard*.)

2. Use the *New* command to begin a new drawing. When the *Select Template* dialog box appears, select the ACAD3D.DWT or ACADISO3D.DWT template (Fig. 27-3). If you intend to create 3D drawings continually, you might want to use the *Options* dialog box and change the *Default Template File Name for QNEW* to one of these 3D templates.

FIGURE 27-3

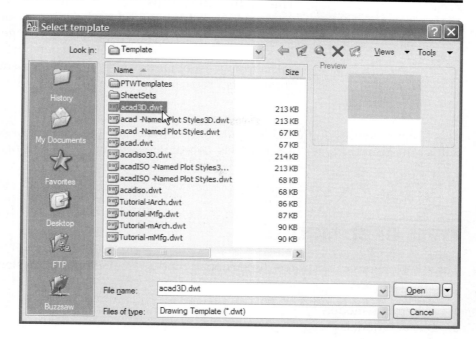

The *Dashboard*

The *Dashboard* consists of several individual control panels. Each control panel contains a set of related commands and/or system variable settings. The individual control panels are designated by horizontal separators as shown in Figure 27-4. Each control panel is displayed initially in its condensed form but can be expanded to display additional options by clicking on the down arrows (chevrons) that appear when you drag the pointer to the lower-left corner of the panel as shown in Figure 27-4, top panel. The seven control panels that appear by default (from top to bottom) on the *Dashboard* are listed here.

FIGURE 27-4

Layers	This is the same control panel used to control layers found in the *2D Drafting & Annotation* workspace.
3D Make	Use this control panel to create and edit solid model objects.
Visual Styles	Use the *Visual Styles* control panel to change the display of your 3D objects to wireframe, hidden, and various "shaded" displays.
Lights	This control panel is used to create light sources for the model.
Materials	You can select, edit, and apply material finishes, textures, and images to individual 3D objects.

Render   This control panel helps you set the desired specifications for a rendering, such as shadow, background, and resolution settings.

3D Navigate   The *3D Navigate* control panel is used to change your view of the 3D model. For example, you can switch between perspective and parallel projection, select from a list of views, or change your orientation with *3Dorbit* and other commands. You can also create simple "walk through" animations.

This chapter focuses on the *3D Navigate* control panel and the *Visual Styles* control panel. But first, you will learn about the types of 3D models, coordinate systems, and forms of coordinate entry.

TYPES OF 3D MODELS

Three basic types of 3D (three-dimensional) models created by CAD systems are used to represent actual objects. They are:

1. Wireframe models
2. Surface models
3. Solid models

These three types of 3D models range from a simple description to a very complete description of an actual object. The different types of models require different construction techniques, although many concepts of 3D modeling are the same for creating any type of model on any type of CAD system.

Wireframe Models

"Wireframe" is a good descriptor of this type of modeling. A wireframe model of a cube is like a model constructed of 12 coat-hanger wires. Each wire represents an <u>edge</u> of the actual object. The <u>surfaces</u> of the object are <u>not</u> defined; only the boundaries of surfaces are represented by edges. No wires exist where edges do not exist. The model is see-through since it has no surfaces to obscure the back edges. A wireframe model has complete dimensional information but contains no volume. Examples of wireframe models are shown in Figures 27-5 and 27-6.

FIGURE 27-5 ──────────────────

FIGURE 27-6 ──────────────────

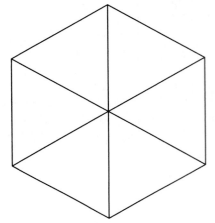

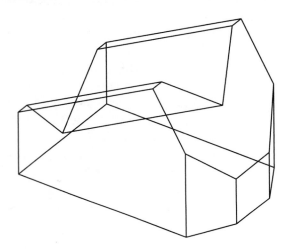

Wireframe models are relatively easy and quick to construct; however, they are not very useful for visualization purposes because of their "transparency." For example, does Figure 27-5 display the cube as if you are looking toward a top-front edge or looking toward a bottom-back edge? Wireframe models tend to have an optical illusion effect, allowing you to visualize the object from two opposite directions unless another visual clue, such as perspective, is given.

With AutoCAD a wireframe model is constructed by creating 2D objects in 3D space. The *Line, Circle, Arc,* and other 2D *Draw* and *Modify* commands are used to create the "wires," but 3D coordinates must be specified. The cube in Figure 27-5 was created with 12 *Line* segments. AutoCAD provides all the necessary tools to easily construct, edit, and view wireframe models.

A wireframe model offers many advantages over a 2D engineering drawing. Wireframe models are useful in industry for providing computerized replicas of actual objects. A wireframe model is dimensionally complete and accurate for all three dimensions. Visualization of a wireframe is generally better than a 2D drawing because the model can be viewed from any position or a perspective can be easily attained. The 3D database can be used to test and analyze the object <u>three dimensionally</u>. The sheet metal industry, for example, uses wireframe models to calculate flat patterns complete with bending allowances. A wireframe model can also be used as a foundation for construction of a surface model. Because a wireframe describes edges but not surfaces, wireframe modeling is appropriate to describe objects with planar or single-curved surfaces, but not compound curved surfaces.

Surface Models

Surface models provide a better description of an object than a wireframe, principally because the surfaces as well as the edges are defined. A surface model of a cube is like a cardboard box—all the surfaces and edges are defined, but there is nothing inside. Therefore, a surface model has volume but no mass. A surface model provides a better visual representation of an actual 3D object because the front surfaces obscure the back surfaces and edges from view. Figure 27-7 shows a surface model of a cube, and Figure 27-8 displays a surface model of a somewhat more complex shape. Notice that a surface model leaves no question as to which side of the object you are viewing.

FIGURE 27-7 ————————

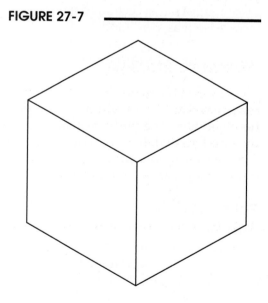

Surface models require a relatively tedious construction process. Each surface must be constructed individually. Each surface must be created in, or moved to, the correct orientation with respect to the other surfaces of the object. In AutoCAD, a surface can be constructed in a number of ways. Often, wireframe elements are used as a framework to build and attach surfaces. The complexity of the construction process of a surface is related to the number and shapes of its edges.

Because there are several generations of surfacing capabilities in AutoCAD, surface modeling in AutoCAD is somewhat ambiguous. Traditionally, surface modeling in AutoCAD was accomplished by creating polygon mesh surfaces such as a *Rulesurf*, *Tabsurf*, *Edgesurf*, or *Revsurf*. These older, "legacy," surfaces can be used to construct surface models as shown in Figure 27-8, but cannot be used in conjunction with solid models, and therefore these commands do not appear on the *Dashboard*.

FIGURE 27-8

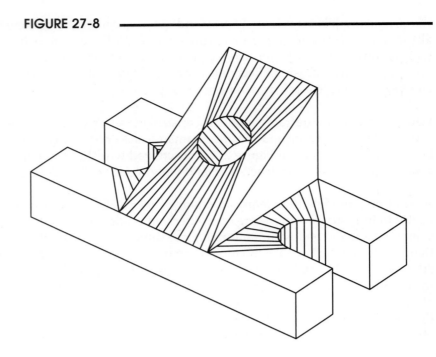

AutoCAD offers a number of newer surfaces much different than polygon meshes or regions. The *3D Make* control panel section of the *Dashboard* provides several commands that can be used to create surfaces as well as solid models. For example, *Extrude*, *Revolve*, *Sweep*, and *Loft* create a <u>surface</u> if applied to an "open" 2D object such as a *Line*, *Spline*, *Arc*, or *Pline*. However, *Extrude*, *Revolve*, *Sweep*, and *Loft* create a <u>solid</u> if applied to a "closed" 2D shape such as a *Circle*, *Ellipse*, *Region*, or closed *Pline*.

Surface models are capable of wireframe, hidden, or various shaded displays. A wireframe representation might be used during construction, but a hidden or shaded representation might be used to display the finished model.

Solid Models

Solid modeling is the most complete and descriptive type of 3D modeling. A solid model is a complete computerized replica of the actual object. A solid model contains the complete surface and edge definition, as well as description of the interior features of the object. If a solid model is cut in half (sectioned), the interior features become visible. Since a solid model is "solid," it can be assigned material characteristics and is considered to have mass. Because solid models have volume and mass, most solid modeling systems include capabilities to automatically calculate volumetric and mass properties.

Techniques for constructing solid models are generally much simpler and more unified than those for constructing surface models. AutoCAD uses a process known as Constructive Solid Geometry (CSG) modeling. CSG is characterized by its simple and straightforward construction techniques of combining "primitive" shapes (boxes, cylinders, wedges, etc.) utilizing Boolean operations (*Union*, *Subtract*, and *Intersect*). CSG models store the "history" (record of primitives and Boolean operations) used to construct the models.

The CSG construction typically begins by specifying dimensions for simple primitive shapes such as boxes or cylinders, then combining the primitives using Boolean operations to create a "composite" solid. Other primitives and/or composite solids can be combined by the same process. Several repetitions of this process can be continued until the desired solid model is finally achieved. Figure 27-9 displays a solid model constructed of simple primitive shapes combined by Boolean operations.

FIGURE 27-9

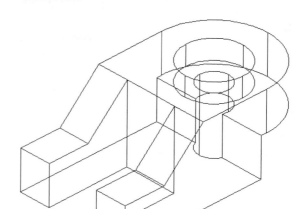

Solid models, like surface models, are capable of wireframe, hidden, or various shaded displays. For example, a wireframe display might be used during construction (see Fig. 27-9) and hidden or shaded representation used to display the finished model (Fig. 27-10).

FIGURE 27-10

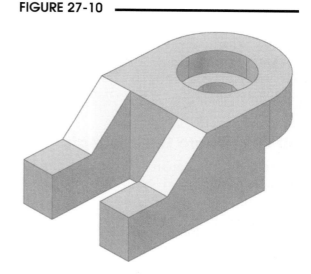

3D COORDINATE ENTRY

When creating a model in three-dimensional drawing space, the concept of the X and Y coordinate system, which is used for two-dimensional drawing, must be expanded to include the third dimension, Z, which is measured from the origin in a direction perpendicular to the plane defined by X and Y. Remember that two-dimensional CAD systems use X and Y coordinate values to define and store the location of drawing elements such as *Lines* and *Circles*.

FIGURE 27-11

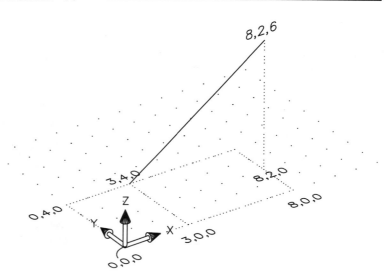

Likewise, a three-dimensional CAD system keeps a database of X, Y, and Z coordinate values to define locations and sizes of two- and three-dimensional elements. For example, a *Line* is a two-dimensional object, yet the location of its endpoints in three-dimensional space must be specified and stored in the database using X, Y, and Z coordinates (Fig. 27-11). The X, Y, and Z coordinates are always defined in that order, delineated by commas. The AutoCAD Coordinate Display (*Coords*) displays X, Y, and Z values.

The icon that appears (usually in the lower-left corner of the AutoCAD Drawing Editor) is the Coordinate System icon (sometimes called the UCS icon). When you draw in the 2D mode the icon indicates only the X and Y axis directions (Fig. 27-12).

FIGURE 27-12

When you change your display of an AutoCAD drawing to one of the 3D *Visual Styles* or begin a drawing using the ACAD3D.DWT or ACADISO3D.DWT drawing template, the icon changes to a colored, 3-pole icon (Fig. 27-13). This UCS icon displays the directions for the X (red), Y (green), and the Z (cyan) axes and is located by default at the origin, 0,0. Technically, the setting for *Visual Styles* controls which icon appears—the *2D wireframe* option displays the simple "wire" icon, whereas any 3D option displays the icon shown in Figure 27-13.

FIGURE 27-13

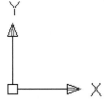

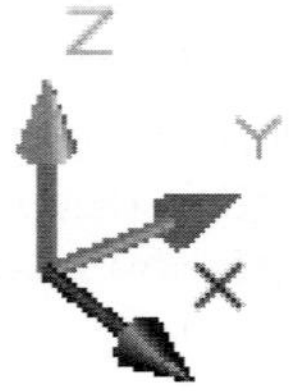

3D Coordinate Entry Formats

Because construction in three dimensions requires the definition of X, Y, <u>and</u> Z values, the methods of coordinate entry used for 2D construction must be expanded to include the Z value. The methods of command entry used for 2D construction are valid for 3D coordinates with the addition of a Z value specification. Relative polar coordinate specification (@dist<angle) is expanded to form two other coordinate entry methods available explicitly for 3D coordinate entry. The methods of coordinate entry for 3D construction follow:

1. **Mouse Input**   PICK   Use the cursor to select points on the screen. *OSNAP*, Object Snap Tracking, or point filters must be used to select a point in 3D space; otherwise, points selected are <u>on the XY plane</u>.

2. **Absolute Cartesian coordinates**   X,Y,Z   Enter explicit X, Y, and Z values relative to point 0,0.

3. **Relative Cartesian coordinates**   @X,Y,Z   Enter explicit X, Y, and Z values relative to the last point.

4. **Cylindrical coordinates (relative)**   @dist<angle,Z   Enter a distance value, an angle in the XY plane value, and a Z value, all relative to the last point.

5. **Spherical coordinates (relative)**   @dist<angle<angle   Enter a distance value, an angle <u>in</u> the XY plane value, and an angle <u>from</u> the XY plane value, all relative to the last point.

6. **Dynamic Input**   **dist,angle**   Enter a distance value in the first edit box, press
                              or **X,Y,Z**   Tab, enter an angular value in the second box.
                                                      You can also enter X,Y,Z coordinates separated
                                                      by commas to establish relative Cartesian coor-
                                                      dinates.

7. **Direct distance entry**   **dist,direction**   Enter (type) a value, and move the cursor in the
                                                      desired direction.

Cylindrical and spherical coordinates can be given <u>without</u> the @ symbol, in which case the location
specified is relative to point 0,0,0 (the origin). This method is useful if you are creating geometry centered around the origin. Otherwise, the @ symbol is used to establish points in space relative to the last
point.

Examples of several 3D coordinate entry methods are illustrated in the following section. In the illustrations, the orientation of the observer has been changed from the default plan view in order to enable the
visibility of the three dimensions.

Mouse Input

Figure 27-14 illustrates using the <u>inter-active</u> method to PICK a location in 3D space. *OSNAP* <u>must</u> be used to PICK in 3D space. Any point PICKed with the input device <u>without</u> *OSNAP* will result in a location <u>on the XY plane</u> (unless you are establishing a Dynamic UCS). In this example, the *Endpoint OSNAP* mode is used to establish the "Specify next point:" of a second *Line* by snapping to the end of an existing vertical *Line* at 8,2,6:

FIGURE 27-14

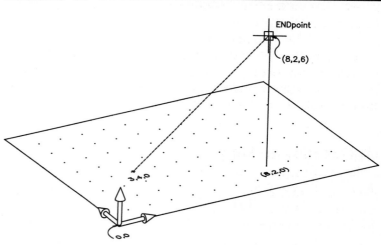

```
Command: line
Specify first point: 3,4,0
Specify next point or [Undo]: endpoint of PICK
```

Absolute Cartesian Coordinates

Figure 27-15 illustrates the <u>absolute</u> coordinate entry to draw the *Line*. The endpoints of the *Line* are given as explicit X,Y,Z coordinates:

FIGURE 27-15

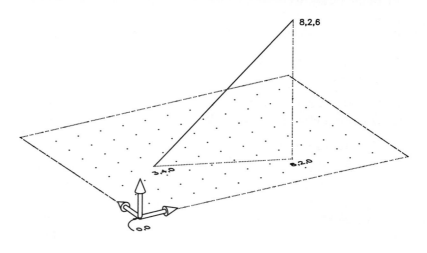

```
Command: line
Specify first point: 3,4,0
Specify next point or [Undo]: 8,2,6
```

Relative Cartesian Coordinates

Relative Cartesian coordinate entry is displayed in Figure 27-16. The "Specify first point:" of the *Line* is given in absolute coordinates, and the "Specify next point:" end of the *Line* is given as X,Y,Z values <u>relative</u> to the last point:

FIGURE 27-16

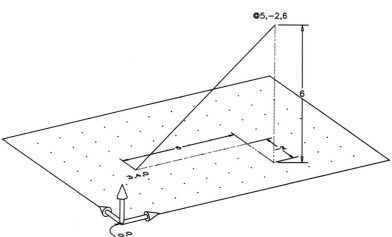

```
Command: line
Specify first point: 3,4,0
Specify next point or [Undo]: @5,
-2,6
```

Direct Distance Entry

With Direct Distance Entry, entering a distance value and indicating a direction with the pointer normally results in a point on the current XY plane. However, by turning on <u>*POLAR* or *ORTHO*</u>, you can draw in an X, Y, or Z direction from the "last point." For example, to draw a square in 3D space with sides of 3 units, begin by using *Line* and selecting the *Endpoint* of the existing diagonal *Line* as the "first point:". To draw the first 3-unit segment, turn on *POLAR* or *ORTHO*, move the cursor in the desired X, Y, or Z direction, and enter "3". Next, move the cursor in the desired (90 degree) direction and enter "3," and so on.

Object Snap Tracking

Object Snap Tracking functions in 3D similar to the way that *OSNAP* and Direct Distance entry operate—that is, when a point is "acquired" by one of the *OSNAP* modes, tracking occurs in the plane of the acquired point. To be more specific, when you acquire a point in 3D space (not on the XY plane), tracking vectors appear originating from the acquired object <u>in a plane parallel to the current XY plane</u>.

Dynamic Input

In the normal Polar format (distance and angle edit boxes), Dynamic Input operates only in the current XY plane. You can, however, create User Coordinate Systems in 3D space and use Dynamic Input on the XY plane of these coordinate systems. See "Coordinate Systems" next. With the default settings, if you instead enter an X, Y, and Z coordinate set separated by commas, Dynamic Input accepts the coordinates as relative Cartesian coordinates (from the last point).

COORDINATE SYSTEMS

In AutoCAD two kinds of coordinate systems can exist, the <u>World Coordinate System</u> (WCS) and one or more <u>User Coordinate Systems</u> (UCS). The World Coordinate System <u>always</u> exists in any drawing and cannot be deleted. The user can also create and save multiple User Coordinate Systems to make construction of a particular 3D geometry easier. <u>Only one coordinate system can be active</u> at any one time in any one view or viewport, either the WCS or one of the user-created UCSs. You can have more than one UCS active at any one time if you have several viewports active (model space or layout viewports).

The World Coordinate System (WCS) and WCS Icon

The World Coordinate System (WCS) is the default coordinate system in AutoCAD for defining the position of drawing objects in 2D or 3D space. The WCS is always available and cannot be erased or removed but is deactivated temporarily when utilizing another coordinate system created by the user (UCS). The icon that appears (by default) at the lower-left corner of the Drawing Editor (Fig. 27-13) indicates the orientation of the WCS. The icon, whose appearance (*ON, OFF*) is controlled by the *Ucsicon* command, appears for the WCS and for any UCS.

The orientation of the WCS with respect to Earth may be different among CAD systems. In AutoCAD the WCS has an architectural orientation such that the XY plane is a horizontal plane with respect to Earth, making Z the height dimension. A 2D drawing (X and Y coordinates only) is thought of as being viewed from above, sometimes called a <u>plan</u> view. This default 2D orientation is like viewing a floor <u>plan</u>—from above (Fig. 27-17). Therefore, in a 3D AutoCAD drawing, imagine this same orientation of objects in space; that is, X is the width dimension, Y is the depth dimension, and <u>Z is height</u>.

FIGURE 27-17

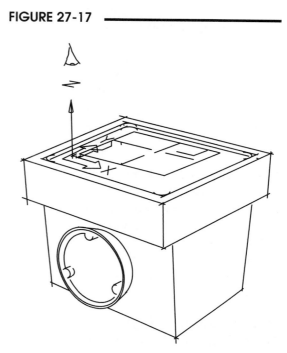

User Coordinate Systems (UCS) and Icons

There are no User Coordinate Systems that exist as part of the AutoCAD template drawings as they come "out of the box." UCSs are created to suit the 3D model when and where they are needed.

Creating geometry is relatively simple when having to deal only with X and Y coordinates, such as in creating a 2D drawing or when creating simple 3D geometry with uniform Z dimensions. However, 3D models containing complex shapes on planes not parallel with the XY plane are good candidates for UCSs. A User Coordinate System is thought of as a construction plane created to simplify creation of geometry on a specific plane or surface of the object. The user creates the UCS, aligning its XY plane with a surface of the object, such as along an inclined plane, with the UCS origin typically at a corner or center of the surface (Fig. 27-18).

FIGURE 27-18

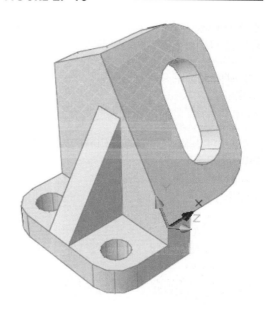

The user can then create geometry aligned with that plane by defining only X and Y coordinate values of the current UCS. The Coordinate Display (*Coords*) in the Status Bar always displays the X, Y and Z values of the <u>current</u> coordinate system, whether it is WCS or UCS. The *SNAP, GRID,* and *Polar Tracking* automatically align with the current coordinate system, providing *SNAP* points, tracking vectors, and enhanced visualization of the current XY "construction" plane (Fig. 27-19). Practically speaking, it is easier in most cases to specify only X and Y coordinates with respect to a specific plane on the object rather than calculating X, Y, and Z values with respect to the World Coordinate System. For example, a UCS would have been created on the inclined surface to construct the slotted hole shown in Figure 27-19.

User Coordinate Systems can be created by any of several options of the *UCS* command. Once a UCS has been created, it becomes the current coordinate system and remains current until another UCS is activated.

FIGURE 27-19

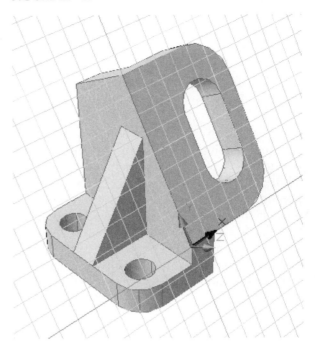

Dynamic UCSs

In addition to static UCSs created with the *UCS* command, you can use temporary coordinate systems called Dynamic User Coordinate Systems; however, <u>Dynamic User Coordinate Systems operate only with solid models</u>. A Dynamic User Coordinate System, or DUCS, appears automatically only during the creation of a primitive such as a *Box, Wedge*, or *Cylinder*. Use the *DUCS* toggle on the Status line to activate the Dynamic UCS feature.

When the *DUCS* toggle is on, a Dynamic UCS appears on the surface you select to establish the base of a primitive. You can select which surface you want by hovering your cursor over any surface of an existing solid during a primitive command. For example, Figure 27-20 displays a Dynamic UCS appearing on the inclined surface while constructing a *Box* primitive. The <u>Dynamic UCS disappears after the primitive has been created</u>.

FIGURE 27-20

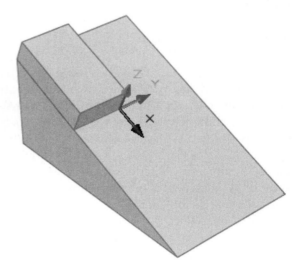

The Right-Hand Rule

AutoCAD complies with the right-hand rule for defining the orientation of the X, Y, and Z axes. The right-hand rule states that if your right hand is held partially open, the thumb, first, and middle fingers define positive X, Y, and Z directions, respectively.

More precisely, if the thumb and first two fingers are held out to be mutually perpendicular, the thumb points in the positive X direction, the first finger points in the positive Y direction, and the middle finger points in the positive Z direction (Fig. 27-21).

FIGURE 27-21

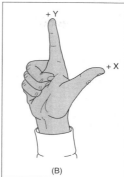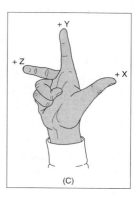

(A)   (B)   (C)

In this position, looking toward your hand from the tip of your middle finger is like the default AutoCAD viewing orientation in 2D mode—positive X is to the right, positive Y is up, and positive Z is toward you.

When you rotate geometry in 3D space or create a new User Coordinate System, it is possible that you want to rotate about the existing X, Y, or Z axis. In doing this you must indicate whether you want to rotate in a positive or negative direction. The right-hand rule is used to indicate the direction of positive rotation about any axis. Imagine holding your hand closed around any axis but with the thumb extended and pointing in the positive direction of the axis (Fig. 27-22). The direction your fingers naturally curl indicates the positive rotation direction about that axis. This principle applies to all three axes.

FIGURE 27-22

(Figures 27-21 and 27-22 are from *Technical Graphics Communication*, 2/E © 1997, Bertoline et al, McGraw-Hill. Reproduced with permission of The McGraw-Hill Companies.)

Ucsicon

Pull-down Menu	Command (Type)	Alias (Type)	Short-cut	Screen (side) Menu	Tablet Menu
View *Display >* *USC Icon*	*Ucsicon*	...	...	*VIEW 2* *UCSicon*	*L,2*

The *Ucsicon* command controls the appearance and positioning of the Coordinate System icon. In order to aid your visualization of the current UCS or the WCS, it is <u>highly</u> recommended that the Coordinate System icon be turned *ON* and positioned at the *ORigin*.

```
Command: ucsicon
Enter an option [ON/OFF/All/Noorigin/ORigin/Properties] <ON>: on
Command:
```

The *On* option causes the Coordinate System icon to appear (Fig. 27-23, lower left).

FIGURE 27-23

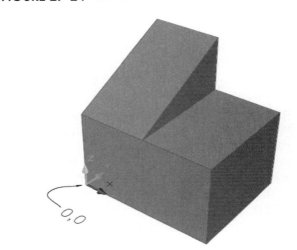

The *Origin* setting causes the icon to move to the origin of the current coordinate system (Fig. 27-24).

FIGURE 27-24

The icon may not always appear at the origin after using some viewing commands like *3Dorbit*. This is because the origin (0,0) may be located outside the border of the drawing screen (or viewport) area, which causes the icon to appear in the lower-left corner. In order to cause the icon to appear at the origin, try zooming out until the icon aligns itself with the origin.

Options of the *Ucsicon* command are:

ON	Turns the Coordinate System icon on.
OFF	Turns the Coordinate System icon off.
All	Causes the *Ucsicon* settings to be effective for all viewports.
Noorigin	Causes the icon to appear always in the lower-left corner of the screen, not at the origin.
ORigin	Forces the placement and orientation of the icon to align with the origin of the current coordinate system.

Properties

The *Properties* option produces the *UCS Icon* dialog box (Fig. 27-25). Ironically, the *UCS Icon Style* and the *UCS Icon Size* have no effect for the 3-pole colored icon that appears in 3D mode. Also, it is actually the *Visual Styles* option that determines which icon is displayed: the *2D wireframe* option displays the simple "wire" icon, whereas any 3D option displays the 3-pole colored icon. You can change the *UCS icon color* for model space and paper space. The default option for *Model space icon color* is either *Black* or *White*—the opposite of whatever you set for the *Model* tab background color in the *Options* dialog box.

FIGURE 27-25

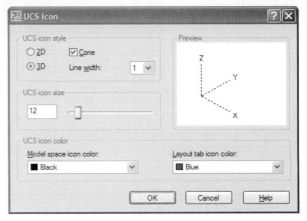

3D NAVIGATION AND VIEW COMMANDS

A "view" is a particular display of a 2D drawing or a 3D model. The term "navigate" is descriptive of the process of achieving a view of a 3D model—that is, you as the observer move around, move closer to, or move farther from, the 3D model to arrive at a particular view. The commands listed below allow you to change the direction from which you view a 3D model or otherwise affect the view of the 3D model. Most of these commands appear on the *3D Navigate* control panel; however, many are accessed by other menus, toolbars, or Command line entry.

Pan	This command is the same as the real-time *Pan* used for 2D drawings. This command enables you to drag 3D objects about the view interactively.
Zoom	Identical to the real-time *Zoom* used for 2D drawings, you can enhance your view of 3D objects by interactively zooming in and out. For a view in perspective projection, *Zoom* also changes the amount of perspective as the objects appear nearer or farther.
3Dorbit	*3Dorbit* controls interactive viewing of objects in 3D. You can dynamically view the object from any point, even while the object retains its visual style. The constrained orbit (default) mode locks the vertical edges to a vertical position.
3Dforbit	The free orbit mode of *3Dorbit* allows you to dynamically view the object from any point; however, vertical edges do not remain vertical. You can "spin" the view around in this mode.
3Dcorbit	The *3Dcorbit* command enables you to set the 3D objects into continuous motion. This mode is a continuous free orbit mode since the vertical axis is not locked and objects appear to spin in any direction.
PERSPECTIVE	Actually a system variable, this setting toggles the current view between parallel projection and perspective projection.
3Ddistance	Use this command to make objects appear closer or farther away. This command performs the same function as *Zoom*, in that it changes the amount of perspective as if you moved closer to or farther from the 3D objects.
View Presets	You can access preset views that change the view of the 3D model, such as a top, front, or isometric displays.
View	The *View* command produces the *View Manager*. Here you can create new views and manage the specifications for the *Current View*, *Model Views*, and *Layout Views*. You can also select from preset orthographic views and isometric viewpoints. This dialog box provides access to view settings such as camera and target locations and lens length.
Camera	You can create one or more cameras where each camera represents a view of the 3D model. You can specify a X,Y, and Z camera position, target position, and lens length for each camera. You can edit a camera's settings using grips or the *Properties* palette.
Vpoint	*Vpoint* is an old AutoCAD command that allows you to change your viewpoint of a 3D model. Three options are provided: *Vector, Tripod,* and *Rotate*.
Ddvpoint	Also a legacy command, this tool allows you to select a viewing angle in degrees.
Plan	The *Plan* command automatically gives the observer a plan (top) view of the object. The plan view can be with respect to the WCS (World Coordinate System) or an existing UCS (User Coordinate System).

| Grid | The *Grid* command offers several settings that affect the appearance of the grid in 3D mode and control how the grid lines react to zooming. |

| Vports | You can use *Vports* effectively during construction of 3D objects to divide the screen into several sections, each displaying a different standard 3D view. |

3D Navigate **Control Panel**

Many of the 3D navigation and view commands discussed in this section can be accessed through the *3D Navigate* control panel (Fig. 27-26). This control panel is the second panel from the top in the *Dashboard*, just below the *3D Make* control panel. In addition, a *3D Navigation* toolbar is available which contains several of the commands discussed in this section (Fig. 27-27). Other navigation and view commands can be accessed through the *View* pull-down menu.

FIGURE 27-26

FIGURE 27-27

When using the navigation commands such as *3Dpan*, *3Dzoom*, *3Dorbit*, *View*, *Vpoint*, *Plan*, etc., it is important to imagine that the <u>observer moves about the object</u> rather than imagining that the object rotates. The object and the coordinate system (WCS and icon) always remain <u>stationary</u> and always keep the same orientation with respect to Earth. Since the observer moves and not the geometry, the objects' coordinate values retain their integrity, whereas if the geometry rotated within the coordinate system, all coordinate values of the objects would change as the object rotated. The viewing commands change only the viewpoint of the <u>observer</u>.

Pan

Pull-down Menu	Command (Type)	Alias (Type)	Short-cut	Screen (side) Menu	Tablet Menu
View *Pan >* *Realtime*	*Pan*	*P*	VIEW *Pan*	*Pan*	*N,10-P,11*

The cursor changes to a hand cursor when this command is active. Identical to the real-time *Pan* command, you can click and drag the view of the 3D objects to any location in your drawing area. However, when any 3D visual style is displayed and *Perspective* is toggled on, *Pan* displays a true 3D effect. That is, panning slightly changes the direction from which the 3D objects are viewed. As an alternative to invoking this command, you can press and hold the mouse wheel to activate *Pan*. To end this command, press the Esc key or select *Exit* from the right-click shortcut menu.

Zoom

Pull-down Menu	Command (Type)	Alias (Type)	Short-cut	Screen (side) Menu	Tablet Menu
View *Zoom >* *Realtime*	*Zoom*	*Z*	VIEW *Zoom*	*Zoom*	*J,2-J,5*

This *Zoom* is identical to the real-time version of the *Zoom* command used in a 2D drawing. However, when viewing 3D objects in any 3D visual style and with *Perspective* toggled on, *Zoom* performs an actual 3D zoom; that is, zooming changes the distance the object appears from the observer. Technically, zooming changes the distance between the camera (observer) and the target (objects). Therefore, *Zoom* changes the size of the objects relative to screen size as well as the amount of perspective.

The *Zoom* cursor is a magnifying glass icon. Press and hold the left mouse button, then move up to zoom in (enlarge image) and down to zoom out (reduce image). As an alternative to invoking this command, you can turn the mouse wheel forward and backward to zoom in and out, respectively. To end this command, press the Esc key or select *Exit* from the right-click shortcut menu.

NOTE: In releases of AutoCAD previous to AutoCAD 2007, the *Zoom* command and the *3Dzoom* command changed only the size of the 3D objects relative to the screen but did not change the amount of perspective. In other words, the distance between the camera and target, and therefore the amount of perspective, did not change with *Zoom* or *3Dzoom*. The *3Ddistance* command was used to change the camera-target distance or the amount of perspective.

Since AutoCAD 2007, *Zoom* and *3Ddistance* perform the same action—change the size of the objects on the screen and change the camera-target distance. Also see NOTE in "*3Ddistance*."

3D Orbit Commands

AutoCAD offers a set of interactive 3D viewing commands, *3Dorbit*, *3Dforbit*, and *3Dcorbit*. When one of the three 3D orbit commands is activated, all commands and related options in the group are available through a right-click shortcut menu. These commands and options are discussed in the following sections. The 3D orbit commands all issue the same simple command prompt because they are interactive rather than Command-line driven.

Command: **3dorbit**
Press ESC or ENTER to exit, or right-click to display shortcut-menu.
Regenerating model.
Command:

With these commands you can manipulate the view of 3D models by dragging your cursor. You can control the view of the entire 3D model or of selected 3D objects. If you want to see only a few objects during the interactive manipulation, select those objects <u>before you invoke the command</u>, otherwise the entire 3D model is displayed during orbit.

3Dorbit

Pull-down Menu	Command (Type)	Alias (Type)	Short-cut	Screen (side) Menu	Tablet Menu
View *Orbit* > *Constrained Orbit*	*3Dorbit*	*3DO*	...	*VIEW* *3Dorbit*	*R,5*

3Dorbit controls viewing of 3D objects interactively. After activating the command, simply hold down the left mouse button and drag the pointer around the screen to change the view of the selected 3D objects. Drag up and down to view the objects from above or below, and drag left and right to view the objects from around the sides. All objects in the drawing are displayed in the interactive display unless individual objects are selected before invoking the command. As stated in the Command prompt, press Esc, Enter, or select *Exit* from the right-click shortcut menu to end the command.

Using the *3Dorbit* command forces the interactive movement to a "constrained orbit." That is, the Z axis is constrained from changing so it always remains in a vertical orientation. This limitation keeps all vertical edges of the 3D objects in a vertical position, just as real objects are oriented (Fig. 27-28). Use *3Dforbit* (free orbit) to change the orientation of the vertical axis.

NOTE: Remember that you are changing the <u>view</u>. When moving the cursor about the screen, it appears that the 3D objects rotate; however, in theory the <u>target</u> remains stationary while the observer, or <u>camera</u>, moves around the objects.

FIGURE 27-28

PERSPECTIVE

Pull-down Menu	Command (Type)	Alias (Type)	Short-cut	Screen (side) Menu	Tablet Menu
Perspective	...	...	...	...	...

PERSPECTIVE is actually a system variable, not a command. The *PERSPECTIVE* variable can be changed by command line or by selecting the *Perspective Projection* and *Parallel Projection* buttons on the *3D Navigate* control panel. A setting of 1 turns on the perspective projection and a setting of 0 displays parallel projection.

    Command: **perspective**
    Enter new value for PERSPECTIVE <1>:

Parallel projection is used for 2D drawing and it is automatically turned on when the *2D Wireframe* visual style is set. However, parallel projection displays 3D objects unrealistically because parallel edges of an object remain parallel in the display. A good example of this theory is shown in Figure 27-29 where the grid lines remain parallel to each other, no matter how far they extend away from the observer. Parallel projection does not take into account the distance the observer is from the object.

FIGURE 27-29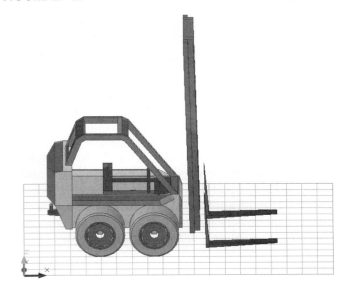

Perspective projection simulates the way we actually see 3D objects because it takes into account the distance the observer is from the object. In AutoCAD, this distance is determined by the distance from our eyes to the objects. Therefore, since objects that are farther away appear smaller, parallel edges on a 3D object appear to converge toward a point as they increase in distance. Figure 27-30 shows the same objects as the previous figure but in perspective projection. Perspective projection can be displayed (*PERSPECTIVE* is set to 1) for any 3D objects shown in any of the 3D visual styles.

FIGURE 27-30

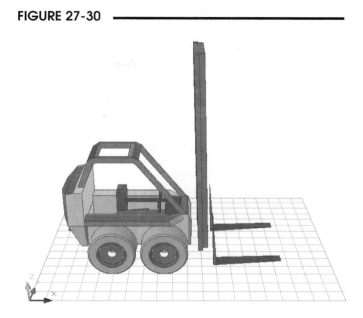

3Ddistance

Pull-down Menu	Command (Type)	Alias (Type)	Short-cut	Screen (side) Menu	Tablet Menu
View *Camera >* *Distance*	*3Ddistance*	...	...	...	...

The *3Ddistance* command changes the distance between the camera (observer) and the target (objects). When perspective projection is on, the amount of perspective changes as a function of camera-target distance. When you invoke *3Ddistance*, the icon changes to a double-sided arrow shown in perspective. Hold the left mouse button down and move the cursor upward to bring the objects closer (and to achieve more perspective), or move the cursor down to move the objects farther away (and achieve less perspective).

Since AutoCAD 2007, <u>the *Zoom* command serves the same function as *3Ddistance*</u>; therefore, you can use *Zoom* instead of *3Ddistance* to accomplish the same display. Normally, you think of using *Zoom* to change the size of objects on the screen; however, object size, camera-target distance, and amount of perspective are now interrelated in AutoCAD.

View Presets

The View Presets provide standard viewing directions for a 3D model, such as *Top*, *Front*, *Right*, and standard isometric viewing directions. The View Presets are not commands that can be entered at the Command line,

FIGURE 27-31

but are selections you can make from any of several methods described here. The View Presets are available from the *View* toolbar (Fig. 27-31) and from the *3D Navigate* control panel drop-down list (Fig. 27-32, on the next page). In addition, you can find the View Presets in the *3D Views* option from the *Views* pull-down menu (not shown) and in the *View Manager* (described in the next section).

Each of the 3D View options is described next. It is important to remember that for each view, imagine you are viewing the object from the indicated position in space. Imagine the observer changes position while the object remains stationary.

FIGURE 27-32

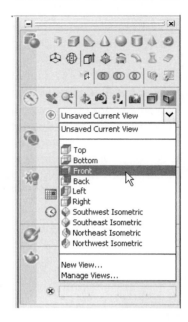

Top

The object is viewed from the top. Selecting this option shows the XY plane from above (the default orientation when you begin a *New* drawing). Notice the position of the WCS icon. This orientation should be used periodically during construction of a 3D model to check for proper alignment of parts.

Bottom

This option displays the object as if you are looking up at it from the bottom. The Coordinate System icon appears backward when viewed from the bottom.

Left

This option looks at the object from the left side.

Right

Imagine looking at the object from the right side. This view is used often as one of the primary views for mechanical part drawing.

Front

Selecting this option displays the object from the front view. This common view is used often during the construction process. The front view is usually the "profile" view for mechanical parts.

Back

Back displays the object as if the observer is behind the object. Remember that the object does not rotate; the observer moves.

SW Isometric

A *SW Isometric* is a view of the object as if the <u>viewer were looking from a southwest position</u>. Isometric views are used more than the "orthographic" views (front, top, right, etc.) for constructing a 3D model. Isometric viewpoints give more information about the model because you can see three dimensions instead of only two dimensions (as in the previously discussed views).

SE Isometric

The *Southeast Isometric* view is generally the first choice for displaying 3D geometry. This view shows the object as if the observer was located southeast in relation to the 3D objects. If the object is constructed with its base on the XY plane so that X is length, Y is depth, and Z represents height, this orientation shows the front, top, and right sides of the object. Try to use this viewpoint as your principal mode of viewing during construction. Note the orientation of the WCS icon.

NE Isometric

The *Northeast Isometric* view shows the right side, top, and back (if the object is oriented in the manner described earlier).

NW Isometric

This viewpoint allows the observer to look at the left side, top, and back of the 3D object.

View

Pull-down Menu	Command (Type)	Alias (Type)	Short-cut	Screen (side) Menu	Tablet Menu
View *Named Views*	*View*	*V*	...	*VIEW 1* *Ddview*	...

The *View* command produces the *View Manager* which provides several functions useful for viewing 3D models. You can assign a name, save, and restore viewed areas of a drawing. You can access information and controls to the current view, any camera, and any saved views. You can also use this manager to adjust the amount of perspective displayed for selected views.

For basic information on the *Set Current, New, Update Layers, Edit Boundaries,* and *Delete* functions of the *View Manger* see "*View*" in Chapter 10. The *View Manager* also provides access to the *View Presets* (described in the previous section) as shown in Figure 27-33.

FIGURE 27-33

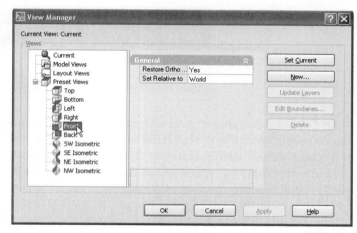

New

When you create a *New* view of a 3D model, AutoCAD actually creates a camera and inserts a camera icon into the drawing. The view settings and camera are saved under the view name. Therefore, you can access information about the view in the *View Manager* or by selecting the view's camera icon and using the *Properties* palette.

Preset Views

You can change your display to any of the *Preset Views* which are identical to those available from the *View* toolbar, the *3D Navigate* control panel drop-down list, or the *Views* pull-down menu, as explained earlier (see "View Presets"). To generate a 3D view using the *View Manager*, several steps are required. First, select the desired view preset from the list. Next, make the desired view current by selecting the *Set Current* button. Alternately, you can double-click on the desired view to make it current. Finally, select the *OK* button to produce the view.

Current and Model Views

The *View Manager* not only allows you to set views current, but provides access to information and controls for the current view and for any saved views. Note that both saved views and any cameras that have been created are listed under *Model Views*. You can control such settings as *Name*, *Category*, and *Visual Style*, as shown in Figure 27-34.

Note that the settings for *Camera* and *Target* are read-only in the *View Manager*. These values cannot be changed here since the *Camera* values are automatically generated using *3Dorbit* or similar navigation command, and the *Target* values can only be changed using the *3Dswivel* or *Camera* commands (and camera grips and *Properties* palette).

FIGURE 27-34

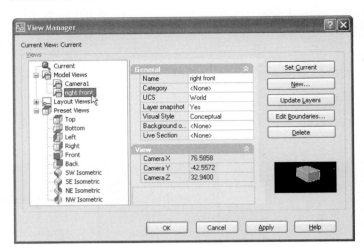

You can also change the settings for the *Lens length* and the *Field of View* for the *Current* view or any saved model view or camera. These settings affect the amount of perspective for the view.

Ddvpoint

Pull-down Menu	Command (Type)	Alias (Type)	Short-cut	Screen (side) Menu	Tablet Menu
View *3D Views >* *Viewpoint Presets...*	*Ddvpoint*	*VP*	...	*VIEW 1* *Ddvpoint*	*N,5*

The *Ddvpoint* command is an older command used mostly before the introduction of *3Dorbit*. *Ddvpoint* produces the *Viewpoint Presets* dialog box (Fig. 27-35). This tool is an interface for attaining viewpoints that could otherwise be attained with the *3Dorbit* or the *Vpoint* command.

Ddvpoint serves the same function as the *Rotate* option of *Vpoint*. You can specify angles *From: X Axis* and *XY Plane*. Angular values can be entered in the edit boxes, or you can PICK anywhere in the image tiles to specify the angles. PICKing in the enclosed boxes results in a regular angle (e.g., 45, 90, or 10, 30) while PICKing near the sundial-like "hands" results in irregular angles (Fig. 27-35). The *Set to Plan View* tile produces a plan view.

The *Relative to UCS* radio button calculates the viewing angles with respect to the current UCS rather than the WCS. Normally, the viewing angles should be absolute to WCS, but certain situations may require this alternative. Viewing angles relative to the current UCS can produce some surprising viewpoints if you are not completely secure with the model and observer orientation in 3D space.

FIGURE 27-35

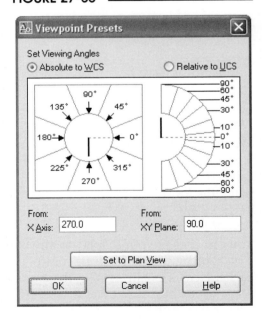

NOTE: When you use this tool, ensure that the *Absolute to WCS* radio button is checked unless you are sure the specified angles should be applied relative to the current UCS.

Plan

Pull-down Menu	Command (Type)	Alias (Type)	Short-cut	Screen (side) Menu	Tablet Menu
View *3D Views >* *Plan View >*	*Plan*	...	...	VIEW 1 *Plan*	N,3

This command is useful for quickly displaying a plan view of any UCS:

Command: **plan**
Enter an option [Current ucs/Ucs/World] <Current>: (**option**) or **Enter**
Regenerating model.
Command:

Responding by pressing Enter causes AutoCAD to display the plan (top) view of the current UCS. Typing *W* causes the display to show the plan view of the World Coordinate System. The *World* option does not cause the WCS to become the active coordinate system but only displays its plan view. Invoking the *UCS* option displays the following prompt:

Enter name of UCS or [?]:

The *?* displays a list of existing named UCSs. Entering the name of an existing UCS displays a plan view of that UCS.

Vports

Pull-down Menu	Command (Type)	Alias (Type)	Short-cut	Screen (side) Menu	Tablet Menu
View *Viewports >*	*Vports*	...	...	VIEW 1 *Vports*	M,3 and M,4

The *Vports* command creates viewports. Either model space viewports or paper space (layout) viewports are created, depending on which space is current when you activate *Vports*. (See Chapters 10 and 12 for basic information about *Vports* and using viewports for 2D drawings.)

This section discusses the use of viewports to enhance the construction of 3D geometry. Options of *Vports* are available with AutoCAD specifically for this application.

When viewports are created in the *Model* tab (model space), the screen is divided into sections like tiles, unlike paper space viewports. Tiled viewports (viewports in model space) affect only the screen display. The viewport configuration cannot be plotted. If the *Plot* command is used, only the current viewport is plotted.

To use viewports for construction of 3D geometry, use the *Vports* command to produce the *Viewports* dialog box (Fig. 27-36). Make sure you invoke *Vports* in the

FIGURE 27-36

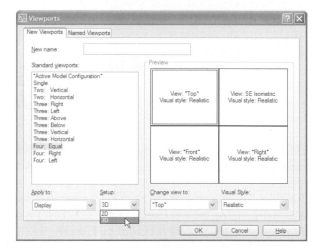

Model tab (model space). To generate a typical orthographic view arrangement in your drawing, first ensure that the current display is set to *Parallel projection*.

2007

Next, invoke the *Viewports* dialog box and select three or four viewports from the list in the *Standard viewports* list, then select the *3D* option in the *Setup:* drop-down list (see Fig. 27-36, bottom center). This action produces a typical top, front, southeast isometric, and side view, depending on whether you selected three or four viewports.

Assuming the options shown in Figure 27-36 were used with the FORKLIFT drawing, the configuration shown in Figure 27-37 would be generated automatically. Note, however, that the objects are not sized proportionally in the viewports. Since a *Zoom Extents* is automatically performed, each view is sized in relation to its viewport, not in relation to the other views.

FIGURE 27-37

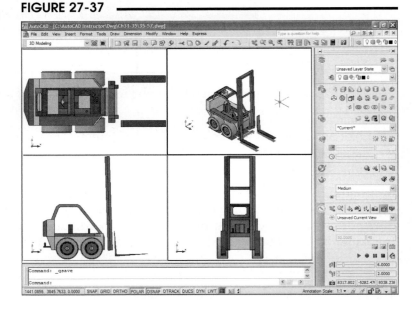

Next, if you desire all views to display the object proportionally, use *Zoom* with a constant scale factor (such as 1, not 1x) in each viewport to generate a display similar to that shown in Figure 27-38. You can then change other settings as desired for each viewport, such as *Visual style*, *Perspective*, etc.

FIGURE 27-38

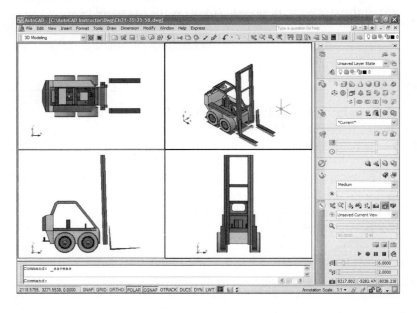

NOTE: If *Perspective Projection* is turned on before using the *3D* option in the *Setup:* drop-down list, the objects are sized equally in the respective viewports.

NOTE: The setting of *UCSORTHO* has an effect on the resulting UCSs when the *3D Setup* option of the *Viewport* dialog box is used. If *UCSORTHO* is set to 0, no new UCSs are created when *Vports* is used. However, if *UCSORTHO* is set to 1, a new UCS is generated for <u>each new orthographic view created</u> so that the XY plane of each new UCS is parallel with the view. In other words, each orthographic view (not the isometric) now shows a UCS with the XY plane parallel to the screen.

VISUAL STYLES

A "visual style" is a group of settings that specify the display of 3D objects. The display, or appearance, of 3D objects includes such characteristics as wireframe, hidden, or shaded displays, treatment of object edges, and application of shadows. Visual styles that display shaded surfaces on an object or that generate shadows generally increase the realistic appearance of 3D objects and can greatly enhance your visibility and understanding of the model, especially for complex geometry. Five preset "default" visual styles are available and can be applied to surface and solid models, whereas only wireframe displays are appropriate for wireframe models since those objects do not contain surfaces. You can specify a visual style for each viewport. When a visual style is applied, the model <u>retains</u> its appearance until you change it.

FIGURE 27-39

Applying visual styles replaces commands used in previous versions of AutoCAD such as *Shademode*, *Hide*, and *Hlsettings*. Instead of using many individual commands and typing system variables, you apply a visual style to the model and change the properties of the current visual style using the *Visual Styles* control panel or the *Vscurrent* command. The *Visual Styles* control panel is located near the center of the *Dashboard* (Fig. 27-39, shown in expanded mode). As soon as you apply a visual style or change its settings, you can see the effect in the viewport. Using the *Visual Styles* control panel allows you to select from five default visual styles and modify the current style to your needs. Any changes are applied to a visual style called "*Current*."

Changes you make to the *Current* style are saved with the drawing only if *Current* is the applied style when the drawing is saved. If you want to create new "named" visual styles, save them with the drawing, and apply them to any viewport or other drawing.

NOTE: Visual styles are saved in the drawing file. Whether you make changes to the *Current* visual style using the *Visual Styles* control panel and save them with the drawing or whether you make changes to the five default styles or any new "named" styles using the *Visual Styles Manager*, visual styles are saved with the drawing. Visual styles can be applied across multiple drawings only if named styles are exported to a tool palette or are copied and pasted using the *Visual Styles Manager*.

NOTE: The illustrations in this section may appear differently than those shown in your computer display depending on the type and configuration of your display driver.

Default Visual Styles

FIGURE 27-40

The drop-down list in the *Visual Styles* control panel provides access to the five default visual styles: *2D wireframe*, *3D Wireframe*, *3D Hidden*, *Conceptual*, and *Realistic*. Although each of these styles can be modified to your liking and other new styles created, a description and illustration for each of these styles is given next (Fig. 27-40).

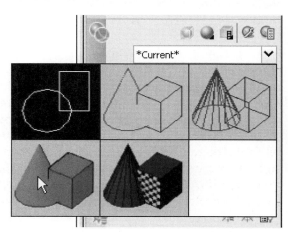

2D Wireframe

This option displays only lines and curves. The 3D geometry is defined only by the edges representing the surface boundaries. The model is displayed in parallel projection—perspective projection mode is disabled. This display uses the black model space background typically used for 2D drawing. The UCS icon is displayed as a "wireframe" icon. Also, linetypes and lineweights are visible. Any raster and OLE objects are also visible in the display. Generally used for 2D geometry, this option can be used for 3D geometry when lineweights and linetypes are important (Fig. 27-41).

FIGURE 27-41

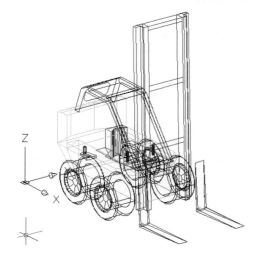

3D Wireframe

Generally used to display 3D geometry when a wireframe representation is needed, this mode displays the objects using lines and curves to represent the surface edges (Fig. 27-42). Since all edges are displayed, this option is useful for 3D construction and editing. The shaded 3D UCS icon appears in this mode. Linetypes and lineweights are not displayed and raster and OLE objects are not visible. The *Compass* is visible if turned on. The lines are displayed in the material colors if you have applied *Materials*, otherwise they appear in the defined line or layer color.

FIGURE 27-42

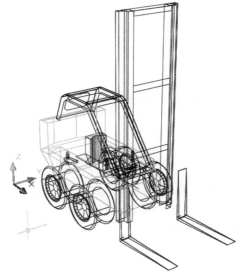

3D Hidden

Use this mode to display 3D objects similar to the *3D Wireframe* representation but with hidden lines suppressed. In other words, edges that would normally be obscured from view by opaque surfaces become hidden (Fig. 27-43).

NOTE: The *3D Hidden* visual style replaces the older *Hide* command. *Hide* can still be used with the *2D Wireframe* style to remove obscured (hidden) lines from the display; however, *Hide* in this mode is not persistent—using *Regen* or *3Dorbit* cancels the display generated by *Hide*. You can use *Hide* to display meshed surfaces for curved edges. *Hide* is also affected by the setting of the *DISPSILH* variable; that is, if *DISPSILH* is set to 1, no mesh lines appear on curved surfaces.

FIGURE 27-43

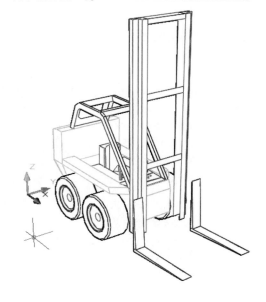

Conceptual

This option shades the objects and smoothes the edges between polygon faces. Shading is automatically generated by using default light sources to illuminate the objects and adjusting the illumination of surfaces based on the angle between the lights and surfaces. When 3D objects contain curved geometry, this option smoothes edges to display true curved geometry by eliminating the facet, or mesh, lines (Fig. 27-44). This option uses the *Gooch* face style which can show details better by softening the contrast between lighted areas and shadowed areas. The lighter areas display warm tones and darker areas display cool tones. The effect is less realistic, but it can make the details of the model easier to see by using color contrast instead of light/dark contrast. Shadows can be applied with this option.

FIGURE 27-44

Realistic

With this option the surfaces are also shaded like *Conceptual*, except here the *Realistic* face style is used. The *Realistic* face style is meant to produce the effect of realism by using the object or layer colors on the object surfaces and edges (Fig. 27-45). If *Materials* have been attached to the objects, the materials are displayed rather than the layer or object colors. Shadows can be applied also with the *Realistic* option.

In the two shaded visual styles (*Conceptual* and *Realistic*), faces are lighted by two distant light sources that move to follow the viewpoint of the observer as you move around the model. In this way, you never see a "dark side" of the model. The default lighting is available only when other lights, including the sun, are turned off.

FIGURE 27-45

Render

The *Render* command is used to create a realistic rendering of the 3D model (solid or surface) model. AutoCAD's rendering capabilities provide you with an amazing amount of control for generating photo-realistic renderings, including lighting and controls, backgrounds, shadows, reflections, texture mapping, and landscaping effects.

Printing 3D Models

You can print 3D models in any visual style. If you have a laserjet, inkjet, or electrostatic type printer configured as the plot device (not a pen plotter), you can select from *As Displayed, Wireframe, Hidden, 3D Hidden, 3D Wireframe, Conceptual*, or *Realistic* options from *Shadeplot* section of the (expanded) *Plot* dialog box. You could also select these options or the *Rendered* option from the *Properties* palette for the viewport.

Printing a Shaded Image From the *Model* Tab

If you want to print a shaded 3D drawing from the *Model* tab, use the *Shaded Viewport Options* section of the *Plot* dialog box, expanded section. Use the *Shade Plot* drop-down list to select from the *Wireframe, Hidden, 3D Hidden, 3D Wireframe, Conceptual,* or *Realistic* options (Fig. 27-46). Alternately, you can apply a visual style to the drawing display, then select *As Displayed* from this list to print the drawing as it appears on screen. You can also select from several resolution qualities for the print (see the "Shaded Viewport Options" section in Chapter 14).

FIGURE 27-46

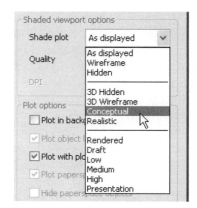

Printing a Shaded Image From a *Layout* Tab

If you are printing from a layout and want to specify a shaded image for one or more viewports, use the *Shadeplot* option of the viewport's *Properties* palette. Here you can select from the same options (Fig. 27-47) as in the *Plot* dialog box.

FIGURE 27-47

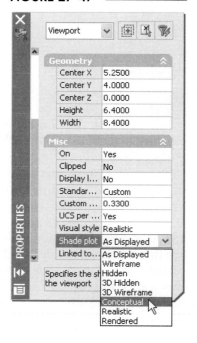

CHAPTER EXERCISES

1. What are the three types of 3D models?

2. What characterizes each of the three types of 3D models?

3. What kind of modeling techniques does AutoCAD's solid modeling system use?

4. What are the seven formats for 3D coordinate specification?

5. Examine the 3D geometry shown in Figure 27-48. Specify the designated coordinates in the specified formats below.

FIGURE 27-48

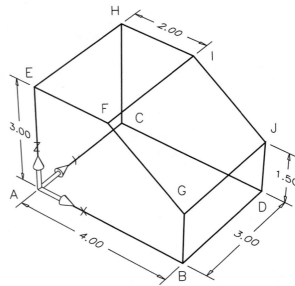

A. Give the coordinate of corner D in absolute Cartesian format.

B. Give the coordinate of corner F in absolute Cartesian format.

C. Give the coordinate of corner H in absolute Cartesian format.

D. Give the coordinate of corner J in absolute Cartesian format.

E. What are the coordinates of the line that define edge F-I?

F. What are the coordinates of the line that define edge J-I?

G. Assume point E is the "last point." Give the coordinates of point I in relative Cartesian format.

H. Point I is now the last point. Give the coordinates of point J in relative Cartesian format.

I. Point J is now the last point. Give the coordinates of point A in relative Cartesian format.

NOTE: For the following exercises you need to download the drawings from the McGraw-Hill/Leach Web site. Go to **www.mhhe.com/leach**, then find the page for **AutoCAD 2008 Instructor** and locate the exercise download drawings. Four of the drawings are used in these exercises: TRUCK MODEL, OPERA, WATCH, and R300-20. These drawings are previous release AutoCAD sample drawings and are provided courtesy of Autodesk.

6. In this exercise you will use a sample drawing and use the *Visual Styles* control panel to practice with the five default visual styles.

A. *Open* the **TRUCK MODEL** drawing. Select the *Model* tab if not already current.

B. In the *Visual Styles* control panel, select the **3D Hidden** visual style. Your display should look similar to that in Figure 27-49. Note that perspective projection should be on. If perspective projection is not on, select the **Perspective Projection** button.

FIGURE 27-49

C. In the *Visual Styles* control panel, select the **3D Wireframe** visual style. Do you see all the edges of the model?

D. Next, select the **2D Wireframe** visual style. Note that the display has been automatically changed to parallel projection mode for this visual style.

E. Type *Hide* at the command prompt. Do mesh lines occur on the truck?

F. Type *DISPSILH* and change the setting to **1**. Type *Hide* at the command prompt again. Did the mesh lines disappear? Change *DISPSILH* back to **0** and try the *Hide* command again.

G. Select the **3D Hidden** visual style again. Note that the truck is displayed in the parallel projection mode. Select the **Perspective Projection** button.

H. Next, select the *Conceptual* style. Finally, change the visual style to *Realistic*. Which of these two styles better display the contrast between surfaces?

I. *Exit* the Truck Model drawing and <u>do not save the changes</u>.

7. In this exercise, you open an AutoCAD sample drawing and use the *3D Navigate* control panel and the *View Manager* to practice with the preset views.

A. *Open* the **OPERA** drawing. Select the *Model* tab. *Freeze* layers **E-AGUAS, E-PREDI, E-TERRA, E-VIDRO,** and **O-PAREDE**. Use the *Visual Styles* control panel to generate a *Conceptual* display.

B. Select a *Top* view from the *3D Navigate* control panel. *Zoom* in if necessary to fully display the objects on the screen. Examine the view and notice the orientation of the coordinate system icon.

    NOTE: If *Zooming* is very slow, try changing the visual style to *3D Wireframe*.

C. Next use the *View* command to produce the *View Manager*. From the **Preset Views**, select a *SE Isometric* view, select *Set Current*, then *OK*. *Zoom* if necessary. Examine the icon and find north (Y axis). Consider how you (the observer) are positioned as if viewing from the southeast.

D. Use any method to produce a *Front* view. *Zoom* if needed. This is a view looking north. Notice the icon. The Y axis should be pointing away from you.

E. Generate a *Right* view. Produce a *Top* view again. *Zoom* whenever needed for a full-screen display.

F. Next, view the OPERA from a *SE Isometric* viewpoint once again. Generate a *3D Hidden* visual style. Your view should appear like that in Figure 27-50. Notice the orientation of the WCS icon on your screen.

FIGURE 27-50

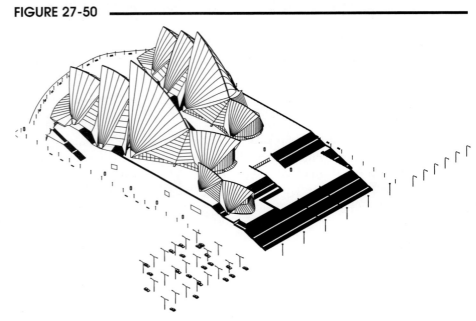

G. Finally, generate a *NW Isometric* and *NE Isometric*. In each case, examine the WCS icon to orient your viewing direction.

H. Experiment and practice more with *3D Navigate* control panel and the *View Manager*. Do not save the drawing.

8. In this exercise you will practice using the *3D Navigate* control panel and *Zoom*.

A. *Open* the **TRUCK MODEL** again that you used in Exercise 6. Do not change the current visual style throughout this exercise. Type "*UCS*," then use the *World* option. This action makes the World Coordinate System the current coordinate system. Use the *Ucsicon* command and ensure the icon is at the *Origin*. Notice the orientation of the truck with respect to the WCS. The X axis points outward from the right side of the truck, Y positive is to the rear of the truck, and Z positive is up.

B. Use the *3D Navigate* control panel to generate a *Top* view. Note the position of the X, Y, and Z axes. You should be looking down on the XY plane and Z points up toward you.

C. Generate a *SE Isometric* view. Note the position of the X, Y, and Z axes.

D. Use the *Parallel Projection* button to ensure perspective is turned off. Generate a *Right* view. Notice that AutoCAD automatically performs a *Zoom Extents* so all of the drawing appears on the screen. *Freeze* layer **BASE**. Now use *Zoom Extents* to display the truck at the maximum screen size.

E. Finally, use each of the following preset views in the order given and notice how the truck appears to change its position for each view. Remember that you should imagine that the viewer moves around the object for each view, rather than imagining that the object rotates.

SE Isometric, Top, SE Isometric, Bottom, SE Isometric, Right, SW Isometric, Left, SE Isometric, Front, SE Isometric

F. Close the drawing but do not save the changes.

9. Now you will get some practice with *3Dorbit* and the shortcut menu options.

A. *Open* the **WATCH** drawing. Activate the *Model* tab. Now generate a *SE Isometric* view to get a good look at the watch. Notice that the components are disassembled, similar to an exploded view assembly drawing.

B. Now invoke *3Dorbit*. Note that the *Grid* is on. Right-click to produce the shortcut menu, then under *Visual Aids* toggle off the *Grid*. In the shortcut menu, select *Visual Styles*, then *Realistic*.

C. Now drag your pointing device to change your viewpoint. Generate your choice for a view that shows the watch so all components are mostly visible. Right click and *Exit 3Dorbit*.

D. Activate *3Dorbit* again. Right click and select *Other Navigation Modes*, then *Free Orbit*. With this mode, try "rolling" the display by using your cursor outside the arc ball. Next, try placing your cursor on the small circle at the right or left quadrant of the arc ball. Hold down the left mouse button and drag left and right to spin the watch about an imaginary vertical axis. Try the same technique from the top and bottom quadrant to spin the watch about an imaginary horizontal axis.

E. Finally, switch back to *Constrained Orbit* and generate your choice of views that shows the watch so all components are visible. Right click and change to *Perspective* projection (you may then have to drag the cursor and change the display slightly to activate the perspective display).

View the watch from several directions and examine the perspective effect. Arrive at a view that displays all of the components clearly.

F.   Right click to produce the shortcut menu and select *Other Navigation Modes*, then *Continuous Orbit*. Hold down the left mouse button, drag the mouse, and release the button to put the watch in a continuous spin.

G.   Finally, *Exit* 3Dorbit. *Close* the drawing and <u>do not save changes</u>.

10. In this exercise, you will create a simple solid model to get an introduction to some of the solid modeling construction techniques discussed in Chapter 29. You will use this model again in these chapter exercises for practicing visual style edge and face settings. You will also use this model for creation of User Coordinate Systems in Chapter 28.

A.   Start AutoCAD in the 3D environment (ensure you are using the *3D Modeling* workspace and that the *Dashboard* is on). Begin a *New* drawing using the ACAD3D.DWT template drawing. Use *Save* and name the drawing **SOLID1.** Also make sure the *Ucsicon* is *On* and set to appear at the *ORigin*. Use the *Layer Properties Manager* to create a new layer named **MODEL** and set the layer's color (in the *Index Color* tab) to **253**. Make **MODEL** the *Current* layer.

B.   Type or select *Box* to create a solid box. When the "Specify first corner" prompt appears, enter **0,0,0**. Then move your cursor so the rectangle appears in positive X,Y space. At the "Specify other corner" prompt, enter **5,4**. Enter a value of **3** for the "height." Change your viewpoint to *SE Isometric*, then *Zoom* if necessary to view the box. Use *3Dorbit* to change the view <u>slightly</u>.

FIGURE 27-51 ─────────

C.   Change the visual style to *3D Wireframe*. Type or select the *Wedge*. When the prompts appear, use **0,0,0** as the "Corner of wedge." Then move your cursor again so the rectangle appears in positive X,Y space. At the "Specify other corner" prompt, enter **4, 2.5**. Specify **2** for the "height." Your solid model should appear similar to that shown in Figure 27-51.

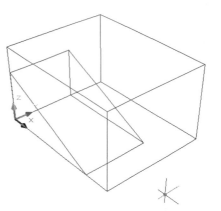

D.   Use the *Rotate* command and select <u>only the wedge</u>. Use **0,0** as the "base point" and enter **-90** as the "rotation angle." The model should appear as Figure 27-52.

FIGURE 27-52 ─────────

E.   Now use *Move* and select the wedge for moving. Use **0,0** as the "base point" and enter **0,4,3** as the "second point of displacement." *Union* the two Solids together and *Save* the drawing.

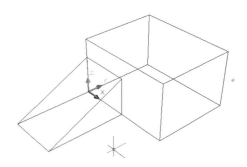

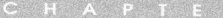

28

USER COORDINATE SYSTEMS

CHAPTER OBJECTIVES

After completing this chapter you should:

1. know how to create, *Save, Restore,* and *Delete* User Coordinate Systems using the *UCS* command;

2. know how to use the *UCS Manager* to create, *Save, Restore,* and *Delete* User Coordinate Systems;

3. be able to create UCSs by the *Origin, Zaxis, 3point, Object, Face, View, X, Y,* and *Z* methods;

4. be able to create *Orthographic* UCSs using the UCS Manager;

5. know the function of the *UCSVIEW, UCSORTHO,* and *UCSFOLLOW* system variables.

CONCEPTS

No User Coordinate Systems exist as part of any of the AutoCAD default template drawings as they come "out of the box." UCSs are created to simplify construction of the 3D model when and where they are needed. UCSs are applicable to all types of 3D models—wireframe models, surface models, and solid models.

When you create a multiview or other 2D drawing, creating geometry is relatively simple since you only have to deal with X and Y coordinates. However, when you create 3D models, you usually have to consider the Z coordinates, which makes the construction process more complex. Constructing some geometries in 3D can be very difficult, especially when the objects have shapes on planes not parallel with, or perpendicular to, the XY plane or not aligned with the WCS (World Coordinate System).

UCSs are created when needed to simplify the construction process of a 3D object. For example, imagine specifying the coordinates for the centers of the cylindrical shapes in Figure 28-1, using only world coordinates. Instead, if a UCS were created on the face of the object containing the cylinders (the inclined plane), the construction process would be much simpler. To draw on the new UCS, only the X and Y coordinates (of the UCS) would be needed to specify the centers, since anything drawn on that plane has a Z value of 0. The Coordinate Display (*Coords*) in the Status Bar always displays the X, Y, and Z values of the current coordinate system, whether it is the WCS or a UCS.

FIGURE 28-1

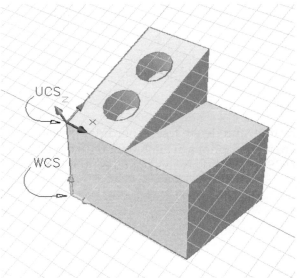

Generally, you would create a UCS aligning its XY plane with a surface of the object, such as along an inclined plane, with the UCS origin typically at a corner or center of the surface. Any coordinates that you specify (in any format) are assumed to be user coordinates (coordinates that lie in or align with the current UCS). Even when you PICK points interactively, they fall on the XY plane of the current UCS. The *SNAP, Polar Snap,* and *GRID* automatically align with the current coordinate system XY plane, enhancing your usefulness of the construction plane.

CREATING AND MANAGING USER COORDINATE SYSTEMS

You can create User Coordinate Systems several ways. The *Ucs* command is the traditional method for creating new coordinate systems and provides several options for orienting the new UCSs. The UCS Manager (*Ucsman* command) allows you to create orthographically positioned UCSs as well as name, save, and restore UCSs. You can also have UCSs created for you automatically when you create or restore an orthographic view.

You can create and save multiple UCSs in one drawing. Rather than recreating the same UCSs multiple times, you can assign a name and save the UCSs that you create so they can be restored at a later time. Although you can have multiple UCSs, only one UCS per viewport can be active at any one time.

Because you would normally want multiple UCSs for a 3D drawing, this chapter explains the features and assists you in creating and managing UCSs wisely. First the chapter discusses how to create UCSs, then naming, saving, and restoring UCSs are explained. At the end of the chapter, the system variables that control UCSs are discussed.

If you are constructing solid models, you can create Dynamic UCSs. Dynamic UCSs are temporary coordinate systems used for constructing a solid, but they cannot be saved or named. Dynamic UCSs are discussed in Chapter 38.

NOTE: When creating 3D models, it is recommended that you use the *Ucsicon* command with the *On* option to make the icon visible and the *Origin* option to force the icon to the origin of the current coordinate system.

TIP

Ucs

Pull-down Menu	Command (Type)	Alias (Type)	Short-cut	Screen (side) Menu	Tablet Menu
Tools Move UCS New UCS >	Ucs	...	...	TOOLS 2 UCS	W,7

The *UCS* command allows you to create, save, restore, and delete User Coordinate Systems. There are many options available for creating UCSs. The *UCS* command always operates only in Command line mode.

FIGURE 28-2

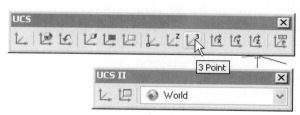

Command: **ucs**
Current ucs name: *WORLD*
Specify origin of UCS or [Face/NAmed/OBject/Previous/View/World/X/Y/Z/ZAxis] <World>:

Although the *UCS* command operates only in Command line mode, two *UCS* toolbars are available (Fig. 28-2). Choosing one of these icons activates the particular option of the *UCS* command.

You can also use the *Tools* pull-down menu to select an option of the *UCS* command. Suboptions are on a cascading menu (Fig. 28-3).

FIGURE 28-3

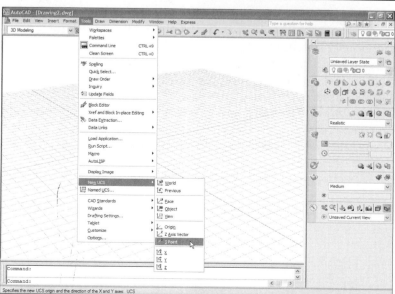

TIP When you create UCSs, AutoCAD prompts for points. These points can be entered as coordinate values at the keyboard or you can PICK points on existing 3D objects. *OSNAP* modes should be used to PICK points in 3D space (not on the current XY plane). In addition, understanding the right-hand rule is imperative when creating some UCSs (see Chapter 27).

2007 NOTE: While learning these options, it is suggested that you turn off Dynamic UCS (*DUCS* on the Status bar). Dynamic UCS affects the orientation of the UCS as you select points in 3D space. Dynamic UCS is explained in Chapter 38.

The options of the UCS command are explained next.

The default option of the *Ucs* command defines a new UCS using one, two, or three points.

```
Command: ucs
Current ucs name: *WORLD*
Specify origin of UCS or [Face/NAmed/OBject/Previous/View/World/X/Y/Z/ZAxis] <World>:
```

If you specify a <u>single point</u>, the origin of the current UCS shifts to the point you specify. The orientation of the new UCS (direction of the X, Y, and Z axes) remains the same as its previous position, only the origin location changes. This option is identical to the *Origin* option.

For example, if you respond to the "Specify origin of UCS" prompt by PICKing a new location using *Osnap*, the new UCS's 0,0,0 location is the point you select (Fig. 28-4). Note that the X,Y,Z orientation is the same as the previous UCS position. At the following prompt, press Enter to accept the first point.

```
Specify point on X-axis or <Accept>:
```

FIGURE 28-4

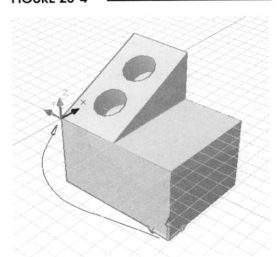

If you specify a <u>second point</u> (at the "Specify point on X-axis or <Accept>:" prompt), the UCS rotates around the previously specified origin point such that the positive X axis of the UCS passes through the new point you PICK.

For example, if you selected a second point as shown in Figure 28-5 as the "point on X axis," the coordinate system spins so the positive X direction aligns with the point. Press Enter to limit the input to two points.

```
Specify point on XY plane or <Accept>:
```

FIGURE 28-5

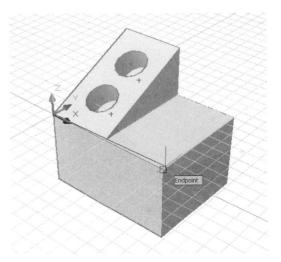

You can specify a <u>third point</u> at the prompt shown above. If you specify a third point, the UCS rotates around the X axis such that the positive Y half of the XY plane of the UCS contains the point. Remember that the first two points lock the origin and the direction for positive X, so the third point only specifies a location in the XY plane. This option (specifying three points at the default prompt) is identical to the separate *3Point* option.

For example, if the third point was specified on the inclined plane as shown in Figure 28-6, the XY plane of the resulting UCS would align itself with the inclined plane.

FIGURE 28-6

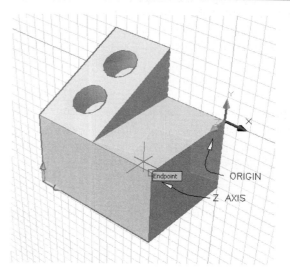

ORIGIN

The *Origin* option defines a new UCS by allowing you to specify a point for the new 0,0,0 location. The orientation of the X,Y, and Z axes remains the same as the previous UCS.

```
Command: ucs
Current ucs name: *WORLD*
Specify origin of UCS or [Face/NAmed/OBject/Previous/View/World/X/Y/Z/ZAxis] <World>: o
Specify new origin point <0,0,0>:
```

This option is identical to using the default option to PICK one point, then pressing Enter to accept that point as the new origin.

ZAxis

You can define a new UCS by specifying an <u>origin</u> and a direction for the <u>Z axis</u>. Only the two points are needed. The X <u>or</u> Y axis generally remains parallel with the current UCS XY plane, depending on how the Z axis is tilted (Fig. 28-7). AutoCAD prompts:

```
Specify new origin point or [Object] <0,0,0>: PICK
Specify point on positive portion of Z-axis
<0.0000,0.0000,0.0000>: PICK
```

FIGURE 28-7

3point

This new UCS is defined by (1) the <u>origin</u>, (2) a point on the X axis (positive direction), and (3) a point on the Y axis (positive direction) or <u>XY plane</u> (positive Y). This is the most universal of all the UCS options (works for most cases). It helps if you have geometry established that can be PICKed with *OSNAP* to establish the points (Fig. 28-8). The prompts are:

FIGURE 28-8

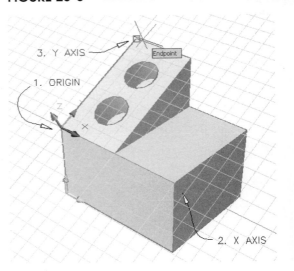

```
Command: ucs
Current ucs name:  *WORLD*
Specify origin of UCS or
    [Face/NAmed/OBject/Previous/View/
    World/X/Y/Z/ZAxis] <World>: 3
Specify new origin point <0,0,0>: PICK
Specify point on positive portion of X-axis
    <0.000,0.000,0.000>: PICK
Specify point on positive-Y portion of the UCS XY plane
    <0.000,0.000,0.000>: PICK
```

Object

This option creates a new UCS aligned with the selected object. You need to PICK only one point to designate the object. The origin of the new UCS is nearest the endpoint of the edge (or center of circular shape) you select. The orientation of the new UCS is based on the type of object selected and the XY plane <u>that was current when the object was created</u>.

Face

The *Face* option operates with <u>solids only</u>. A *Face* is a planar or curved surface on a solid object, as opposed to an edge. When selecting a *Face*, you can place the cursor <u>directly on the desired face</u> when you PICK, or select an edge of the desired face.

```
Specify origin of new UCS or [Face/NAmed/OBject/Previous/View/World/X/Y/Z/ZAxis] <World>: f
Select face of solid object: PICK
(Select desired face of 3D solid to attach new UCS.  Edges or faces can be selected.)
Enter an option [Next/Xflip/Yflip] <accept>:
```

FIGURE 28-9

Next
Since you cannot actually PICK a surface, or if you PICK an edge, AutoCAD highlights a surface near where you picked or adjacent to a selected edge (Fig. 28-9). Since there are always two surfaces joined by one edge, the *Next* option causes AutoCAD to highlight the other of the two possible faces. Press Enter to accept the highlighted face.

Xflip
Use this option to flip the UCS icon 180 degrees about the X axis. This action causes the Y axis to point in the opposite direction (see Fig. 28-9).

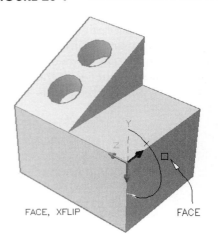

Yflip
This option causes the UCS icon to flip 180 degrees about the Y axis. The X axis then points in the opposite direction.

View

This *UCS* option creates a UCS parallel with the screen (perpendicular to the viewing angle) (Fig. 28-10). The UCS <u>origin</u> remains unchanged. This option is handy if you wish to use the current viewpoint and include a border, title, or other annotation. There are no options or prompts.

FIGURE 28-10

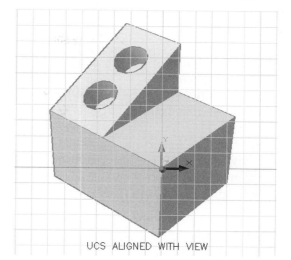

UCS ALIGNED WITH VIEW

X   Y   Z

Each of these options rotates the UCS about the indicated axis according to the right-hand rule (Fig. 28-11).

The Command prompt is:

Specify rotation angle about *n* axis <90>:

The angle can be entered by PICKing two points or by entering a value. This option can be repeated or combined with other options to achieve the desired location of the UCS. It is imperative that the right-hand rule is followed when rotating the UCS about an axis.

FIGURE 28-11

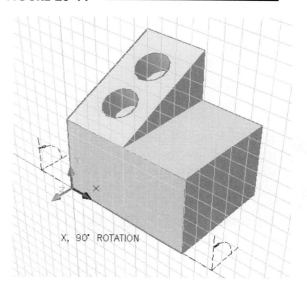

X, 90° ROTATION

Notice that in the prompt for the X, Y, or Z axis rotate options, the default rotation angle is 90 degrees ("<90>" as shown above). If you press Enter at the prompt, the axis rotates by the displayed amount (90 degrees, in this case). You can set the *UCSAXISANG* system variable to another value which will be displayed in brackets when you next use the X, Y, or Z, axis rotate options. Valid values are: 5, 10, 15, 18, 22.5, 30, 45, 90, 180.

Previous

Use this option to restore the previous UCS. AutoCAD remembers the ten previous UCSs used. *Previous* can be used repeatedly to step back through the UCSs.

World

Using this option makes the WCS (World Coordinate System) the current coordinate system.

Named

The *Named* option allows you to *Save* the current UCS, *Restore* previously named UCSs, and *Delete* named UCSs.

```
Command: ucs
Current ucs name:  *WORLD*
Specify origin of UCS or [Face/NAmed/OBject/Previous/View/World/X/Y/Z/ZAxis] <World>: na
Enter an option [Restore/Save/Delete/?]:
```

Save

Invoking this option prompts for the name you want to assign for saving the current UCS. Up to 256 characters can be used in the name. Entering a name causes AutoCAD to save the <u>current</u> UCS. The *?* option and the UCS Manager list all previously saved UCSs.

Restore

Any *Save*d UCS can be restored with this option. The *Restored* UCS becomes the <u>current</u> UCS. The *?* option and the UCS Manager list the previously saved UCSs.

Delete

You can remove a *Save*d UCS with this option. Entering the name of an existing UCS causes AutoCAD to delete it.

?

This option lists the named UCSs (UCSs that were previously *Saved*). Also listed are the origin coordinates and directions for the X, Y, and Z axes.

NOTE: It is important to remember that changing a UCS or creating a new UCS does not change the display (unless the *UCSFOLLOW* system variable is set to 1). Only one coordinate system can be active in a view. If you are using viewports, each viewport can have its own UCS if desired.

Dducsp

Pull-down Menu	Command (Type)	Alias (Type)	Short-cut	Screen (side) Menu	Tablet Menu
...	*Dducsp*	...	...	*TOOLS 2* *Ucsp*	*W,9*

The *Dducsp* command invokes the UCS Manager. It can be used to create new UCSs or to restore existing UCSs. The UCS Manager can also be used to set specific UCS-related system variables, described later.

The *Dducsp* command <u>activates specifically the Orthographic UCSs tab</u> of the UCS Manager (Fig. 28-12). Here you can create new UCSs. *Orthographic UCSs* <u>are not automatically saved</u> as named UCSs but are created as you select them; that is, you can create, save, and restore <u>additional but separate</u> UCSs with the names Front, Top, Right, etc.

FIGURE 28-12

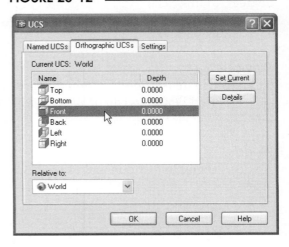

To create one of these *Orthographic UCSs*, you can <u>double-click</u> the desired UCS name, or highlight the desired UCS name (one click) and select the *Current* button. Press *OK* to return to the drawing to see the new UCS. Keep in mind that although the new UCSs are normal to the front, top, right, etc. orthographic views, they are <u>not necessarily attached to the front, top, or right faces of the objects</u>. The new UCSs typically have the same origin as the WCS, unless a new *Depth* is specified or a named UCS is selected in the *Relative to:* drop-down list.

Depth

Notice in the *Orthographic UCS* tab that each UCS can have a specified depth. *Depth* is a Z dimension or distance perpendicular to the XY plane to locate the new UCS origin. Select the *Depth* <u>value</u> of any UCS to produce the *Orthographic UCS Depth* dialog box (Fig. 28-13).

FIGURE 28-13

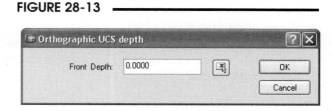

Details

Select the *Details* button to see the coordinates for the origin of the highlighted UCS. The direction vector coordinates for the X, Y, and Z axes are also listed. This has the same function as the *?* option of the *UCS* command.

Relative to:

The *Relative to:* drop-down list allows you to select any named UCS as the base UCS. When the *Front*, *Top*, *Right*, or other UCS option is selected, it is created in a orthographic orientation with respect to the origin and orientation of the selected named UCS. Typically, you want to <u>ensure *World* is selected as the base UCS</u> unless you have some other specific application. The setting in the *Relative to:* list is stored in the *UCSBASE* system variable.

UCS SYSTEM VARIABLES

AutoCAD offers a great amount of control and flexibility for UCSs, such as the ability to automatically create UCSs when creating views, saving UCSs with views, and having multiple UCSs current when using multiple viewports. Because of these options, it is easy to be overwhelmed if you are not aware of the system variables that control UCSs. This section is provided to help avoid confusion and to offer guidelines for UCS use. Not all UCS system variables are explained here, but only those that are typically accessed by a user.

UCSBASE

UCSBASE stores the name of the UCS that defines the origin and orientation when *Orthographic* UCSs are created using the *UCS, Orthographic* option or the *Orthographic UCSs* tab of the UCS Manager. Normally the World Coordinate System is used as the base UCS, so when a new orthographic UCS is created it is simply rotated about the WCS origin to be normal to the selected orthographic view (top, front, right, etc.). *UCSBASE* can also be set in the *Relative to:* drop-down list of the *Orthographic UCSs* tab of the UCS Manager (see Fig. 28-12).

For beginning users, it is suggested that this variable be left at its initial value, *World*. This variable is saved in the current drawing.

UCSVIEW

UCSVIEW determines whether the current UCS is saved with a <u>named</u> view. For example, suppose you create a new 3D viewpoint, such as a southeast isometric, to enhance your visualization while you construct one area of a model. You also create a UCS on the desired construction plane. With *UCSVIEW* set to 1, you then *Save* that view and assign a name such as RIGHT. Whenever the RIGHT view is later restored, the UCS is also restored, no matter what UCS is current before restoring the view. Without this control, the current UCS would remain active when the RIGHT view is restored. *UCSVIEW* has two possible settings.

0   The current UCS is not saved with a named view.
1   The current UCS is saved whenever a named view is
     saved (initial setting).

 UCSVIEW is easily remembered because it saves the *UCS* with the *VIEW*. Remember that <u>the view must be named to save its UCS</u>.

UCSORTHO

This variable determines whether an orthographic UCS is automatically created when an orthographic view is created or restored. For example, if *UCSORTHO* is set to 1 and you select or create a front view, a UCS is automatically set up to be parallel with the view. In this way, the XY plane of the UCS is always parallel with the screen when an orthographic view is restored. The settings are as follows.

0   Specifies that the UCS setting remains unchanged when an orthographic view is restored.
1   Specifies that the related orthographic UCS is automatically created or restored when an orthographic view is created or restored (initial setting).

 In order for *UCSORTHO* to operate as expected, you must select a preset orthographic view from the *View* pull-down menu, *Orthographic* option of the *View* command, or an orthographic view icon button. *UCSORTHO* does <u>not</u> operate with views created using *Vpoint*, *Plan*, the *3Dorbit* commands, or for any isometric preset views. *UCSORTHO* operates for views as well as viewports. *UCSORTHO* is saved in the current drawing.

A good exercise you can use to understand *UCSORTHO* is to use *Vports* to create four tiled viewports in model space and select the *3D Setup*. With *UCSORTHO* set to 1, the arrangement shown in Figure 28-14 would result. Here, each orthographic view has its own orthographically aligned UCS. Note the grid and UCS icon display in each viewport.

FIGURE 28-14

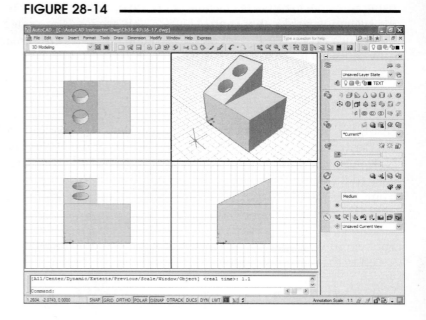

If *UCSORTHO* is set to 0 and the
Vports command is used to create a *3D
Setup*, the arrangement shown in
Figure 28-15 would be created. Notice
that the WCS is the only coordinate
system in all viewports. Note the
display of the grid and the UCS icons.

A point to help remember *UCSORTHO*
is when the view changes to an orthographic view, the UCS changes.

FIGURE 28-15 ——————————

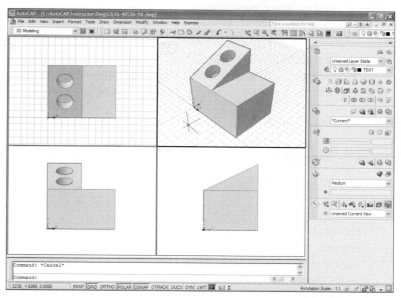

TIP

UCSFOLLOW

The *UCSFOLLOW* system variable, if set to 1 (*On*), causes the plan view of a UCS to be displayed automatically when a UCS is made current. *UCSFOLLOW* can be set separately for each viewport.

For example, consider the case shown
in Figure 28-16. Two tiled viewports
(*Vports*) are being used, and the *UCS-
FOLLOW* variable is turned *On* for the
left viewport only. Notice that a UCS
created on the inclined surface is
current. The left viewport shows the
plan view of this UCS automatically,
while the right viewport shows the
model in the same orientation no
matter what coordinate system is
current. If the WCS were made active,
the right viewport would keep the
same viewpoint, while the left would
automatically show a plan view of the
new UCS when the change was made.

FIGURE 28-16 ——————————

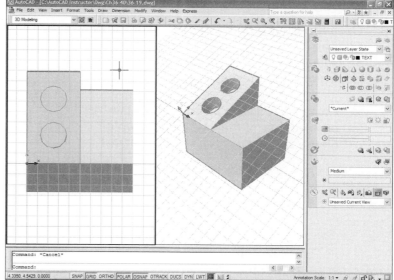

NOTE: When *UCSFOLLOW* is set to 1,
the display does not change until a
new UCS is created or restored.

The *UCSFOLLOW* settings are as follows.

0   Changing the UCS does not affect the view (initial setting).
1   Any UCS change causes a change to plan view of the new UCS in the viewport current when
     UCSFOLLOW is set.

The setting of *UCSFOLLOW* affects only model space. *UCSFOLLOW* is saved in the current drawing
and is viewport specific.

A point to help remember *UCSFOLLOW* is when the UCS changes, the view follows.

CHAPTER EXERCISES

1. Using the model in Figure 28-17, assume a UCS was created by rotating about the X axis 90 degrees from the existing orientation (World Coordinate System).

    A. What are the absolute coordinates of corner F?

    B. What are the absolute coordinates of corner H?

    C. What are the absolute coordinates of corner J?

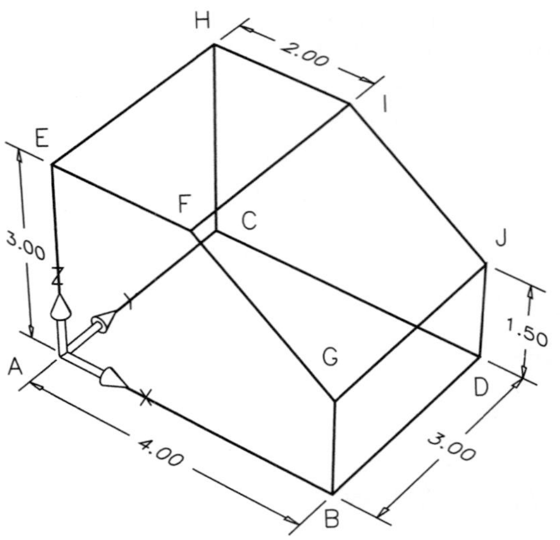

2. Using the model in Figure 28-17, assume a UCS was created by rotating about the X axis 90 degrees from the existing orientation (WCS) and then rotating 90 degrees about the (new) Y.

    A. What are the absolute coordinates of corner F?

    B. What are the absolute coordinates of corner H?

    C. What are the absolute coordinates of corner J?

FIGURE 28-17 ───────────────

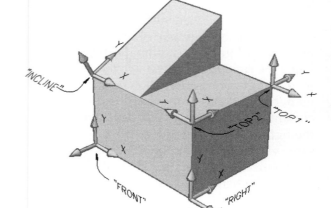

3. *Open* the **SOLID1** drawing that you created in Chapter 27, Exercise 10. View the object from a *SE Isometric* view. Make sure that the *UCSicon* is *On* and set to the *ORigin*.

    A. Create a *UCS* with a vertical XY plane on the front surface of the model as shown in Figure 28-18. *Save* the UCS as **FRONT**.

FIGURE 28-18 ───────────────

    B. Change the coordinate system back to the *World*. Now, create a *UCS* with the XY plane on the top horizontal surface and with the orientation as shown in Figure 28-18 as TOP1. *Save* the UCS as **TOP1**.

    C. Change the coordinate system back to the *World*. Next, create the *UCS* shown in the figure as TOP2. *Save* the UCS under the name **TOP2**.

    D. Activate the *WCS* again. Create and *Save* the **RIGHT** UCS.

    E. Use *SaveAs* and change the drawing name to **SOLIDUCS**.

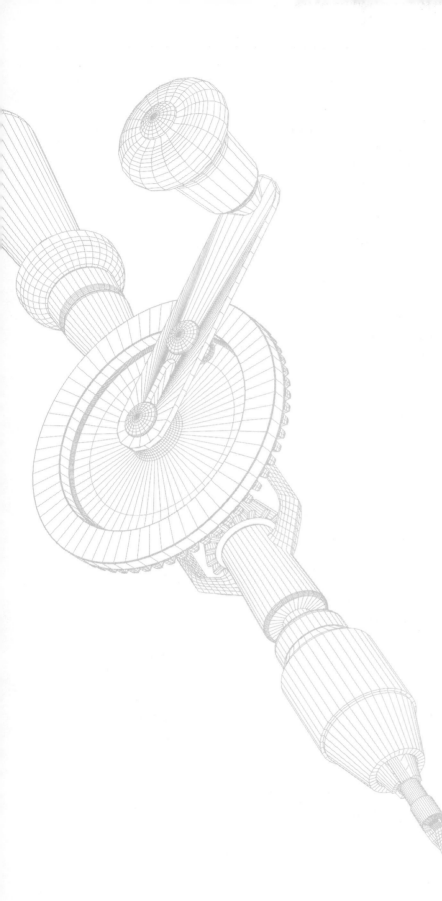

29

SOLID
MODELING
CONSTRUCTION

CHAPTER OBJECTIVES

After completing this chapter you should:

1. be able to create solid model primitives using the following commands: *Box, Wedge, Cone, Cylinder, Torus, Pyramid,* and *Sphere;*

2. be able to create swept solids from existing 2D shapes using the *Extrude, and Revolve* commands;

3. be able to move solids in 3D space using *3Dmove, 3Dalign,* and *3Drotate;*

4. be able to create new solids in specific relation to other solids using *Mirror3D* and *3DArray;*

5. be able to combine multiple primitives into one composite solid using *Union, Subtract,* and *Intersect;*

6. know how to create beveled edges and rounded corners using *Chamfer* and *Fillet.*

CONCEPTS

A solid model is a virtually complete representation of the 3D shape of a physical object. Solid modeling differs from wireframe or surface modeling in two fundamental ways: (1) the information is more complete in a solid model and (2) the method of construction of the model itself is relatively easy.

The basic solid modeling construction process is to use simple 3D solids, called "primitives," then combine them using "Boolean operations" to create a complex 3D model called a "composite solid." This modeling technique is known as Constructive Solid Geometry, or CSG, modeling. Although the process is relatively simple, you have tremendous flexibility during construction using tools such as UCSs and Dynamic UCSs, commands for moving and rotating, grip editing, and many other editing commands and techniques. AutoCAD uses a solid modeling engine, or "kernel," called ACIS.

Solid Model Construction Process

The process used to construct most composite solid models follow four general steps:

1. Construct simple 3D <u>primitive</u> solids such as *Box, Cylinder,* or *Wedge,* or create 2D shapes and convert them to 3D solids by *Extrude, Revolve, Sweep,* or *Loft.*

2. Create the primitives <u>in location</u> relative to the associated primitives using UCSs or <u>move, rotate, or mirror</u> the primitives into the desired location relative to the associated primitives.

3. Use <u>Boolean operations</u> (*Union, Subtract,* and *Intersect*) to combine the primitives to form a <u>composite solid</u>.

4. Make necessary design changes to features of a composite solid using a variety of editing tools such as grips, *Solidedit,* and using surfaces to *Slice* solids.

This chapter discusses the basic construction process organized into the following sections: 1) Solid Primitives Commands, 2) Dynamic User Coordinate Systems, 3) Commands to Move, Rotate, and Copy Solids, and 4) Boolean Operations. Each section is briefly described next. Chapter 30 discusses advanced editing techniques, such as grip editing, editing faces and edges, sectioning, and display variables.

Solid Primitives Commands

Solid primitives are the basic building blocks that make up more complex solid models. The commands that create 3D primitives "from scratch" are:

Box	Creates a solid box or cube
Cone	Creates a solid cone with a circular or elliptical base
Cylinder	Creates a solid cylinder with a circular or elliptical base
Polysolid	Creates a series of connected solid segments having a specified height and width
Pyramid	Creates a solid right pyramid with any number of sides
Sphere	Creates a solid sphere
Torus	Creates a solid torus
Wedge	Creates a solid wedge

Commands that create solid primitives from existing 2D shapes are:

Extrude	Converts a closed 2D object such as a *Pline, Circle, Region,* or *Planar Surface* to a solid by adding a Z dimension
Loft	Creates a solid using a series of 2D cross sections (curves or lines) that define the shape
Revolve	Converts a closed 2D object such as a *Pline, Circle,* or *Region* to a solid by revolving the shape about an axis
Sweep	Creates a solid by sweeping a closed planar profile along an open or closed 2D or 3D path

Commands to Move, Rotate, and Mirror Solids

It is not always efficient or practical to create every primitive solid initially in the desired location and/or orientation. In may cases primitives are created in another position, then they are moved, rotated, or otherwise aligned into the desired position before combining them into composite solids. Also, some primitives can be created from others by mirroring or arraying existing solids. This section of the chapter discusses the following commands:

Move	The command used for moving objects in 2D can also be used in 3D space
3Darray	Makes a polar array about any axis or a rectangular array with rows, columns, and levels
3Dalign	Connects two solids together by matching three alignment points on each solid
3Dmove	A temporary grip tool allows you to restrict the movement to an axis or a plane
3Drotate	A temporary grip tool allows you to specify an axis about which to rotate
Mirror3D	Creates a mirror copy of solids by specifying a mirror plane

Boolean Operations

Primitives are combined to create complex solids by using Boolean operations. The Boolean operators are listed below. An illustration and detailed description are given for each of the commands.

Union	Unions (joins) selected solids.
Subtract	Subtracts one set of solids from another.
Intersect	Creates a solid of intersection (common volume) from the selected solids.

Primitives are created at the desired location or are moved into the desired location before using a Boolean operator. In other words, two or more primitives can occupy the same space (or share some common space), yet are separate solids. When a Boolean operation is performed, the solids are combined or altered in some way to create one solid. AutoCAD takes care of deleting or adding the necessary geometry and displays the new composite solid complete with the correct configuration and lines of intersection.

FIGURE 29-1 _____

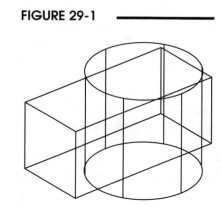

Consider the two solids shown in Figure 29-1. When solids are created, they can occupy the same physical space. A Boolean operation is used to combine the solids into one composite solid and it interprets the resulting utilization of space.

Union

Union creates a union of the two solids into one composite solid (Fig. 29-2). The lines of intersection between the two shapes are calculated and displayed by AutoCAD.

FIGURE 29-2 ─────────

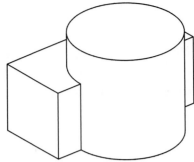

Subtract

Subtract removes one or more solids from another solid. ACIS calculates the resulting composite solid. The term "difference" is sometimes used rather than "subtract." In Figure 29-3, the cylinder has been subtracted from the box.

FIGURE 29-3 ─────────

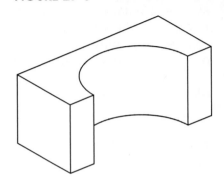

Intersect

Intersect calculates the intersection between two or more solids. When *Intersect* is used with *Regions* (2D surfaces), it determines the shared <u>area</u>. Used with solids, as in Figure 29-4, *Intersect* creates a solid composed of the shared <u>volume</u> of the cylinder and the box. In other words, the result of *Intersect* is a solid that has only the volume which is part of both (or all) of the selected solids.

FIGURE 29-4 ─────────

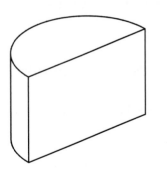

Chamfer and Fillet

The *Chamfer* command creates an angled edge between two surfaces, and a *Fillet* creates a rounded edge between two surfaces. Although the *Chamfer* command and the *Fillet* command are not Boolean operators, they are included in this section. With each of these commands, you specify dimensions to create a wedge primitive (with *Chamfer*) or a rounded primitive (with *Fillet*) and AutoCAD automatically performs the Boolean needed to add or subtract the primitive from the selected solids.

SOLID PRIMITIVES COMMANDS

This section explains the commands that allow you to create primitives used for construction of composite solid models. The commands allow you to specify the dimensions and the orientation of the solids. Once primitives are created, they are combined with other solids using Boolean operations to form composite solids.

NOTE: If you use a pointing device to PICK points, use *OSNAP* when possible to PICK points in 3D space. If you do not use *OSNAP*, the selected points are located on the current XY construction plane, so the true points may not be obvious. It is recommended that you use *OSNAP* or enter values.

The solid primitives commands can be accessed several ways. At the top of the *Dashboard* is the *3D Make* control panel that contains the solid primitives commands as well as other related editing commands. Figure 29-5 shows the control panel in its expanded configuration.

FIGURE 29-5

From the pull-down menus, select *Draw* then *Modeling* to produce the solid primitives commands on a cascading menu (Fig 29-6). Commands for moving solids and editing solids (including the Boolean commands) are located on menus branching from the *Modify* pull-down menu (not shown).

FIGURE 29-6

A toolbar is available that contains the solids primitives commands as well as the three Boolean commands. Note the name of the toolbar is *Modeling* (Fig. 29-7). A second toolbar is available (not shown) exclusively for editing solids.

FIGURE 29-7

In addition to these methods, each of the solid modeling commands are accessible by typing in the command name at the Command prompt, such as *Box* or *Wedge*.

Box

Pull-down Menu	Command (Type)	Alias (Type)	Short-cut	Screen (side) Menu	Tablet Menu
Draw Modeling > Box	*Box*	...	...	*DRAW 2 SOLIDS Box*	*J,7*

Box creates a solid box primitive to your dimensional specifications. The <u>base of the box is always oriented parallel to the current XY plane</u> (WCS, UCS, or Dynamic UCS). You can specify dimensions of the box by PICKing or by entering values. The box can be defined by (1) giving the corners of the base, then height, (2) by locating the center, then the corners and height, or (3) by giving each of the three dimensions.

    Command: **box**
    Specify first corner or [Center]: **PICK** or (**coordinates**) or **C**

Specify first corner

You can either PICK or supply coordinate values for the first corner. AutoCAD then responds with:

    Specify other corner or [Cube/Length]:

If you PICK the other corner interactively, the length and width of the base are parallel with the X and Y axes. AutoCAD then requests the height.

    Specify height or [2Point] <0.0000>:

Again, you can PICK or enter a value at this prompt. The *2Point* option allows you to PICK two points as the height distance.

Figure 29-8 shows a *Box* with the first corner at 0,0 and the other corner at 5,4. The height is 3. This box was created interactively by picking points. Note that the box is oriented with the edges aligned with the X,Y, and Z axes.

FIGURE 29-8 ——————————

Length

After supplying the first corner, you can use the *Length* option to specify three dimensions for the box: the length, width, and height. Note that the first two values supplied (*Length* and *Width*) are the dimensions of the base, but either can be longer or shorter (*Length* in this case is not necessarily the longer of the two dimensions, as the term implies). Next, specify the height.

    Specify other corner or [Cube/Length]: *l*
    Specify length: **PICK** or (**value**)
    Specify width: **PICK** or (**value**)
    Specify height or [2Point] <0.0000>: **PICK** or (**value**)

NOTE: Using the *Length* option, the <u>direction</u> of the length and width is <u>determined by the current loca-</u><u>tion of the crosshairs;</u> therefore, the edges of the box's base are not aligned with the X and Y axes unless *POLAR* or *ORTHO* is on. In addition, the box can be created in negative X,Y, and Z directions.

Cube

The *Cube* option requires only one dimension to create the *Box*. You can PICK a distance or enter a value. In either case, the <u>direction of the edges of the base is determined by the current location of the</u> <u>crosshairs.</u>

Center

With this option you first locate the center of the box, then supply the dimensions of the box by PICKing or entering values.

> Command: **box**
> Specify first corner or [Center]: **c**
> Specify center: **PICK** or (**coordinates**)
> Specify corner or [Cube/Length]:

Figure 29-9 displays a *Box* created using the *Center* method. Note that the center is the <u>volumetric</u> center, not the center of the base.

FIGURE 29-9 ───────────

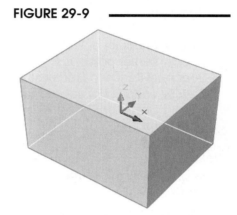

Cone

Pull-down Menu	Command (Type)	Alias (Type)	Short-cut	Screen (side) Menu	Tablet Menu
Draw *Modeling* > *Cone*	*Cone*	...	...	DRAW 2 SOLIDS *Cone*	*M,7*

FIGURE 29-10 ───

Cone creates a right circular or elliptical solid cone ("right" means the axis forms a right angle with the base). By the default method, you specify the center location, radius (or diameter), and height. With this option, the orientation of the cylinder is determined by the current UCS so that <u>the base lies on the XY plane and height is perpendicular</u> (in a Z direction). Alternately, a different orientation can be defined by using the *Axis endpoint* or *3Ppoint* option.

Center Point

Using the defaults (PICK or supply values for the center, radius, and height), the cone is generated in the orientation shown in Figure 29-10. The default prompts are shown as follows:

Command: **cone**
Specify center point of base or [3P/2P/Ttr/Elliptical]: **PICK** or (**coordinates**)
Specify base radius or [Diameter]: **PICK** or (**coordinates**)
Specify height or [2Point/Axis endpoint/Top radius]: **PICK** or (**value**)

3P/2P/Ttr

Note that you can define the location and diameter of the base using the *3Point*, *2Point*, or *Ttr* (tangent, tangent, radius) methods. These options work similar to creating a *Circle* by the same methods. Using the *2Point* and *Ttr* methods, the *Cone* base always lies in a plane parallel with the UCS, no matter which points in 3D space are selected. However, the *3Point* option allows a different orientation such that the base of the *Cone* lies in the plane defined by the three points. These options and principles operate similarly for creating a *Cylinder* or *Torus*.

Elliptical

FIGURE 29-11

This option draws a cone with an elliptical base. You specify two axis endpoints to define the elliptical base. The elliptical cone in Figure 29-11 was created using the following specifications :

    Command: **cone**
    Specify center point of base or [3P/2P/Ttr/Elliptical]: **e**
    Specify endpoint of first axis or [Center]: **c**
    Specify center point: **0,0**
    Specify distance to first axis: **2**
    Specify endpoint of second axis: **1**
    Specify height or [2Point/Axis endpoint/Top radius]: **3**
    Command:

2Point

When specifying the height, you can use this option to interactively specify two points to determine the height distance.

Axis endpoint

FIGURE 29-12

Invoking the *Axis endpoint* option (after the base has been established) displays the following prompt:

    Specify apex point: **PICK** or (**coordinates**)

Locating a point for the axis endpoint defines the height and orientation of the cone. The axis of the cone is aligned with the line between the specified center point of the base and the apex, and the height is equal to the distance between the two points. Figure 29-12 shows a *Cone* with the base at 0,0 and the *Axis endpoint* located on the X axis.

FIGURE 29-13

Top radius

The *Top radius* option allows you to create a frustum (a truncated right cone with the top plane parallel to the base). The following prompt is displayed:

    Specify height or [2Point/Axis endpoint/Top radius]: **t**
    Specify top radius <0.0000>: **PICK** or (**value**)
    Specify height or [2Point/Axis endpoint]: **PICK** or (**value**)

The top radius is the radius for the circle at the top of the frustum. Figure 29-13 shows a *Cone* with a base radius of 2 and a top radius of 1.

Cylinder

Pull-down Menu	Command (Type)	Alias (Type)	Short-cut	Screen (side) Menu	Tablet Menu
Draw Modeling > Cylinder	*Cylinder*	...	...	DRAW 2 SOLIDS *Cylinder*	L,7

Cylinder creates a cylinder with an elliptical or circular base with a center location, diameter, and height you specify. Default orientation of the cylinder is determined by the current UCS, such that the circular plane is coplanar with the XY plane and height is in a Z direction. However, the orientation can be defined otherwise by the *Axis endpoint* or *3Point* option.

FIGURE 29-14 ————

Center Point

The default options create a cylinder in the orientation shown in Figure 29-14:

```
Command: cylinder
Specify center point of base or [3P/2P/Ttr/Elliptical]: PICK or (coordi-
nates)
Specify base radius or [Diameter] <0.0000>: PICK or (value)
Specify height or [2Point/Axis endpoint] <0.0000>: PICK or (value)
Command:
```

3P/2P/Ttr

You can define the location and diameter of the base using the *3Point*, *2Point*, or *Ttr* (tangent, tangent, radius) methods. These options work similar to creating a *Circle* by the same methods. If you use the *2Point* and *Ttr* methods, you can select any points in 3D space to define the base of the *Cylinder*; however, the base always lies in a plane parallel with the UCS. On the other hand, if you use the *3Point* option, the base lies in the same plane as the three specified points. These options and principles are the same for creating a *Cone* base diameter or *Torus* diameter.

Elliptical

This option draws a cylinder with an elliptical base. The elliptical cone in Figure 29-15 was created using the following specifications.

FIGURE 29-15 ————

```
Command: cylinder
Specify center point of base or [3P/2P/Ttr/Elliptical]: e
Specify endpoint of first axis or [Center]: c
Specify center point: 0,0
Specify distance to first axis <0.0000>: 2
Specify endpoint of second axis: 1.5
Specify height or [2Point/Axis endpoint] <0.0000>: 3
```

2Point

When specifying the height, you can use this option to interactively specify two points to determine the height distance.

Axis endpoint

Invoking the *Axis endpoint* option (after the base has been established) allows you to determine both the height and orientation for
the cylinder.

FIGURE 29-16

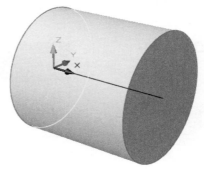

    Command: *cylinder*
    Specify center point of base or [3P/2P/Ttr/Elliptical]: **PICK**
    (Endpoint)
    Specify base radius or [Diameter] <0.0000>: **1.5**
    Specify height or [2Point/Axis endpoint] <0.0000>: *a*
    Specify axis endpoint: **PICK (Endpoint)**

The cylinder in Figure 29-16 was created using the *Axis endpoint* method. Here, the *Line* along the X axis
was previously established. The *Cylinder* was created by using *Endpoint* to snap to the two endpoints for
the "center point of base" and "axis endpoint" (see the prompts listed above).

Wedge

Pull-down Menu	Command (Type)	Alias (Type)	Short-cut	Screen (side) Menu	Tablet Menu
Draw Modeling > Wedge	*Wedge*	*WE*	...	*DRAW 2 SOLIDS Wedge*	*N,7*

Creating a *Wedge* solid primitive is similar to creating a *Box*
(see "*Box*"). Note that the prompts are the same, and therefore the methods for creating a *Wedge* are identical to those for
creating a *Box*. The difference between a *Wedge* and a *Box*
with the same dimensions is the *Wedge* contains half the
volume of the *Box*—a *Wedge* is like a *Box* cut in half diagonally.

FIGURE 29-17

    Command: *wedge*
    Specify first corner or [Center]: **PICK** or **(coordinates)**
    Specify other corner or [Cube/Length]: **PICK** or **(coor-
    dinates)**
    Specify height or [2Point] <0.0000>: **PICK** or **(value)**
    Command:

Similar to a *Box*, a *Wedge's* base is always parallel to the XY plane of the current coordinate system (WCS,
UCS, or Dynamic UCS). Additionally, <u>using the default option</u> ("Specify first corner"), the <u>*Wedge*
always slopes down toward second ("other") corner</u>. Notice the orientation of the *Wedge* in Figure 29-17,
which was created by PICKing the "first corner" at the origin and the "other corner" in an XY-positive
direction.

Length

As an alternative to specifying two corners and a height, you can use the *Length* option to specify three
dimensions for the *Wedge*: the length, width, and height. NOTE: Using the *Length* option, the <u>direction
of the length and width is <u>determined by the current location of the crosshairs</u>; therefore, the edges of
the *Wedge's* base are not aligned with the X and Y axes unless *POLAR* or *ORTHO* are on. In addition, the
<u>*Wedge* always slopes down toward the direction indicated for the *Length*</u>.

Cube

The *Cube* option requires only one dimension to create the *Wedge*. You can PICK a distance or enter a value. In either case, with this option the <u>*Wedge* always slopes down toward the direction indicated by the current location of the crosshairs.</u>

Center

With this option you first locate a "center" for the *Wedge*, then supply the dimensions by PICKing or entering values.

Command: **box**
Specify first corner or [Center]: *c*
Specify center: **0,0**
Specify corner or [Cube/Length]:

Figure 29-18 displays a *Wedge* created by specifying 0,0 as the *Center* method (see prompts above). Note that the <u>designated center is not the volumetric center, but the center of the sloping side of the *Wedge*.</u> If you imagine the *Wedge* as one half of a *Box*, the center would be the volumetric center for the imaginary *Box* (see "*Box*").

FIGURE 29-18 ———————

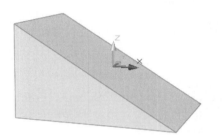

Sphere

Pull-down Menu	Command (Type)	Alias (Type)	Short-cut	Screen (side) Menu	Tablet Menu
Draw Modeling > Sphere	*Sphere*	...	...	DRAW 2 SOLIDS *Sphere*	K,7

Sphere allows you to create a sphere by defining its center point and radius or diameter.

Command: **sphere**
Specify center point or [3P/2P/Ttr]: **PICK** or (**coordinates**)
Specify radius or [Diameter] <0.0000>: **PICK** or (**value**)

Creating a *Sphere* with the default options and using 0,0 as the center would create the shape shown in Figure 29-19.

FIGURE 29-19 ———————

3P/2P/Ttr

As options, you can define the location and diameter of the *Sphere* using the *3Point*, *2Point*, or *Ttr* (tangent, tangent, radius) methods. For example, you could use the *3Point* option and *Osnap* to create a *Sphere* location and diameter based on three points on existing solids. These options are similar to the same options for creating a *Cone* base, *Cylinder* base, or *Torus*.

Torus

Pull-down Menu	Command (Type)	Alias (Type)	Short-cut	Screen (side) Menu	Tablet Menu
Draw Modeling > Torus	Torus	TOR	...	DRAW 2 SOLIDS Tours	O,7

Torus creates a torus (donut shaped) solid primitive. Three specifications are needed: (1) the center location, (2) the radius or diameter of the torus (from the center of the *Torus* to the centerline of the tube), and (3) the radius or diameter of the tube. AutoCAD prompts:

FIGURE 29-20

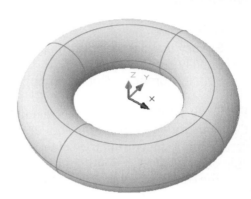

> Command: **torus**
> Specify center point or [3P/2P/Ttr]: **PICK** or (**coordinates**)
> Specify radius or [Diameter] <0.0000>: **PICK** or (**value**)
> Specify tube radius or [2Point/Diameter] <0.0000>: **PICK** or (**value**)
> Command:

Using the default ("Specify center point") option, the *Torus* is oriented such that the *Torus* diameter (tube centerline) lies in the XY plane of the current UCS and the rotational axis of the tube is parallel with the Z axis of the UCS. Figure 29-20 shows a *Torus* created using the default orientation and the center of the torus at 0,0,0. The *Torus* has a torus radius of 3 and a tube radius of 1. (To improve the visibility of this figure, the Visual style edge setting is *Isolines*.)

A self-intersecting torus is allowed with the *Torus* command. A self-intersecting torus is created by specifying a tube radius equal to or greater than the torus radius. Figure 29-21 illustrates a torus with a torus radius of 3 and a tube radius of 3.

FIGURE 29-21

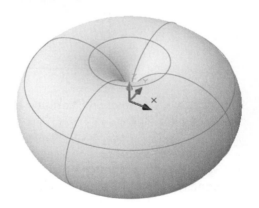

3P/2P/Ttr

Similar to the same options for drawing a *Circle*, you can define the location and diameter of the *Torus* using the *3Point*, *2Point*, or *Ttr* (tangent, tangent, radius) methods. For example, you could use the *3Point* option and *Osnap* to create a *Torus* location and diameter based on three points on existing solids. Using the *2Point* and *Ttr* methods, the *Torus* always lies in a plane parallel with the UCS, whereas the *3Point* option allows a different orientation. These options are similar to the same options for creating *Cone* base or *Cylinder* base.

Pyramid

Pull-down Menu	Command (Type)	Alias (Type)	Short-cut	Screen (side) Menu	Tablet Menu
Draw Modeling > Pyramid	Pyramid	PYR	...	DRAW 2 SOLIDS Pyramid	...

The *Pyramid* command creates a right pyramid with any number of sides.

With the default options, you specify the center of the base, the base radius, and the height. The resulting *Pyramid* has its base parallel to the XY plane of the current UCS, unless the *Axis endpoint* method is used.

FIGURE 29-22

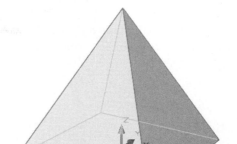

```
Command: pyramid
4 sides  Circumscribed
Specify center point of base or [Edge/Sides]: PICK or (coordinates)
Specify base radius or [Inscribed] <0.0000>: PICK or (value)
Specify height or [2Point/Axis endpoint/Top radius] <0.0000>: PICK or (value)
Command:
```

Figure 29-22 shows a four-sided pyramid with the base located at 0,0, a base radius (inscribed) of 2, and a height of 4.

Edge

Use the *Edge* option to PICK two points in space to define the length and orientation of one edge. The *Pyramid* base is generated on the left side of the defined line going from the first to the second point.

```
Specify center point of base or [Edge/Sides]: e
Specify first endpoint of edge: PICK or (coordinates)
Specify second endpoint of edge: PICK or (coordinates)
```

Sides

Use this option to specify the number of sides (not including the base) for the *Pyramid*. Any practical number of sides can be specified with 3 as the minimum.

Inscribed/Circumscribed

This feature specifies how the base radius is drawn—either inscribed within or circumscribed about an imaginary circle. This method is similar to defining the size of a *Polygon*.

2Point

This option is the same as for a *Cone* or *Cylinder*. You can PICK two points to specify the height of the *Pyramid* such that the resulting height is equal to the indicated distance; however, the *Pyramid* does not change its orientation from that set by the base radius.

Axis endpoint

This option sets the height as well as the orientation for *Pyramid*. The height is equal to the distance between the center of the base and the axis endpoint selected. For example, Figure 29-23 shows a *Pyramid* similar to that in the previous figure, but with the *Axis endpoint* selected at 4,0. This option is the same as for a *Cone* or *Cylinder*.

FIGURE 29-23

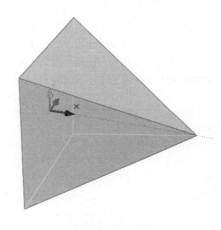

Top radius

Specifying a value for the *Top radius* creates a truncated pyramid such that the top surface is parallel with the base. Figure 29-24 displays a *Pyramid* with a base radius (inscribed) of 2, a *Top radius* of 1, and a height of 3.

FIGURE 29-24

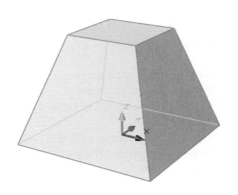

Extrude

Pull-down Menu	Command (Type)	Alias (Type)	Short-cut	Screen (side) Menu	Tablet Menu
Draw Modeling > Extrude	Extrude	EXT	...	DRAW 2 SOLIDS Extrude	P,7

Extrude means to add a third dimension (height) to an existing 2D shape. For example, if you *Extrude* a *Circle*, the resulting shape is a cylinder. If you *Extrude* a *Line*, the resulting shape is a planar surface. Therefore, <u>*Extrude* can create a surface or a solid</u>, depending on the 2D shape used. If you *Extrude* an <u>open</u> 2D shape (one having endpoints), such as a *Line*, *Arc*, or open *Pline*, a <u>surface</u> is created. (These ideas will be explained fully in Chapters 39 and 40.) However, if you *Extrude* an existing <u>closed</u> 2D shape such as a *Circle, Polygon, Rectangle, Ellipse, Pline, Spline, Region,* or planar surface, the resulting extrusion is a <u>solid</u>. For example, Figure 29-25 displays a closed *Pline* (before) to create a solid (after) using *Extrude*. Extruding solids is discussed in this section.

FIGURE 29-25

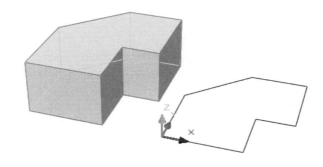

Extrude can simplify the creation of many solids that may otherwise take much more time and effort using typical primitives and Boolean operations. The versatility of this command lies in the fact that <u>any closed shape</u> that can be created by or converted to a *Pline, Spline, Region,* surface, etc., no matter how complex, can be transformed into a solid by *Extrude*. The closed 2D shape cannot be self-intersecting (crossing over itself). For example, the original shape used for the extrusion in Figure 29-26 (a closed *Pline*) contains straight and curved segments. Note that the original 2D shape was created on a vertical plane (the UCS XY plane), then extruded perpendicularly.

FIGURE 29-26

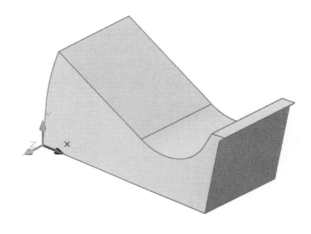

First, create the closed 2D shape, then invoke *Extrude*.

```
Command: extrude
Current wire frame density:  ISOLINES=4
Select objects to extrude: PICK
Select objects to extrude: Enter
Specify height of extrusion or [Direction/Path/Taper angle]: PICK or (value)
Command:
```

By default the 2D shape is consumed (deleted) by the *Extrude* command. The *DELOBJ* system variable controls if the 2D shape is deleted by *Extrude*.

Taper angle

A taper angle can be specified for the extrusion. Entering a <u>positive</u> taper angle causes the sides of the extrusion to slope <u>inward</u> (Fig. 29-27, right), while a negative taper angle causes the sides to slope outward (Fig. 29-27, left). This is helpful, for example, developing molded parts that require a slight draft angle to facilitate easy removal from the mold.

FIGURE 29-27

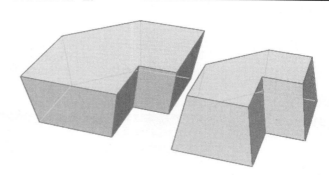

Direction

Normally, with the default method of *Extrude*, the selected 2D shape is extruded perpendicular to the plane of the shape regardless of the current UCS orientation, as shown in the previous three figures. However, you can use the *Direction* option to specify an <u>extrusion direction other than perpendicular</u>. Select two points to determine the direction.

```
Specify height of extrusion or [Direction/Path/Taper angle]: d
Specify start point of direction: PICK or (coordinates)
Specify end point of direction: PICK or (coordinates)
```

The two points of direction can be anywhere in 3D space and can be determined by selecting points on existing objects. The two points only indicate direction and do not change the orientation or position of the original 2D shape. For example, Figure 29-28 extrudes a *Circle* using the *Direction* option to *Osnap* to the *Endpoints* along the edge of a *Pyramid*.

FIGURE 29-28

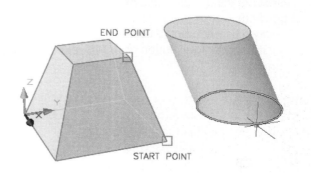

Path

The *Path* option allows you to sweep the 2D shape called a "profile" along an existing line or curve called a "path." The path can be composed of a *Line, Arc, Ellipse, Pline, Spline,* or *Helix* (or can be different shapes converted to a *Pline*). The path does not have to be in a plane (2D) but can be in 3D space, such as in the form of a helix.

Specify height of extrusion or [Direction/Path/Taper angle]: **p**
Select extrusion path or [Taper angle]: **PICK**

Figure 29-29 shows a *Circle* and a closed *Pline* (on the XY plane) extruded along a *Spline* path (that lies in the XZ plane). In this case, the same *Spline* path was selected for each extrusion. The path temporarily moves to the location of the profile for the extrusion. The 2D shapes that are extruded (shown highlighted) are consumed by default, but the path is not deleted. Notice how the original 2D shapes that were extruded change orientation as they extrude along the path.

FIGURE 29-29

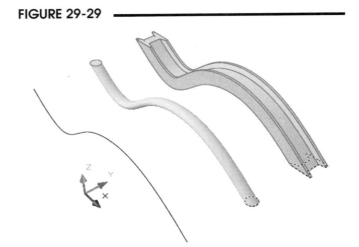

Revolve

Pull-down Menu	Command (Type)	Alias (Type)	Short-cut	Screen (side) Menu	Tablet Menu
Draw Modeling > Revolve	*Revolve*	*REV*	...	DRAW 2 SOLIDS *Revolve*	Q,7

Revolve allows you to revolve an open or closed 2D shape (profile) about a selected axis. Like *Extrude*, the resulting shape can be a surface or solid depending on the original shape used. *Revolve* creates a <u>solid</u> if the original profile used is "<u>closed</u>," such as a *Pline, Polygon, Circle, Ellipse, Spline*, or a *Region* object. The command syntax for *Revolve* (accepting the defaults) is:

Command: **revolve**
Current wire frame density: ISOLINES=4
Select objects to revolve: **PICK**
Select objects to revolve: **Enter**
Specify axis start point or define axis by [Object/X/Y/Z] <Object>: **PICK**
Specify axis endpoint: **PICK**
Specify angle of revolution or [STart angle] <360>: **Enter** or (**value**)
Command:

Using the default option of *Revolve*, you must specify two points to define the axis to revolve about. You can select two points on the 2D object as the axis. The profile in Figure 29-30 (right) was used to create the 3D object (left) by selecting two points on the vertical edge.

FIGURE 29-30

To define the axis of revolution using the default option or the *Object* option, you can select points on other objects or specify coordinates. <u>The profile is always revolved about an axis in the same plane as the profile</u>. However, the two points that define the axis for revolving do not have to be coplanar with the 2D shape as long as one of the points lies in the same plane and the axis is not perpendicular to the 2D shape. In this case, *Revolve* uses an axis <u>in the plane</u> of the revolved shape aligned with the direction of the selected points or object. The other options for selecting an axis of revolution are described next.

Object

A *Line* or single segment *Pline* can be selected for an axis. The positive axis direction is from the closest endpoint PICKed to the other end.

X/Y/Z

Uses the positive X, Y, or Z axis of the current UCS as the axis of rotation. However, keep in mind that an axis perpendicular to the plane of the profile cannot be used.

Start angle

After defining the axis of revolution, *Revolve* requests the number of degrees for the object to be revolved. Any angle can be entered. For example, one possibility for revolving a profile is shown in Figure 29-31 where the profile is generated through 270 degrees.

By default, the object is revolved from its current plane in a counter-clockwise direction. If you revolve the object less than 360 degrees, you can use the *Start angle* option to specify an angular value from the current 2D shape position to begin the *Revolve*.

FIGURE 29-31

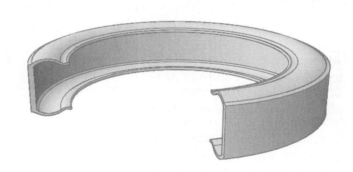

COMMANDS FOR MOVING SOLIDS

When you create the desired primitives, Boolean operations are used to construct the composite solids. However, the primitives must be in the correct position and orientation with respect to each other before Boolean operations can be performed. You can either create the primitives in the desired position during construction (by using UCSs) or move the primitives into position after their creation. Several methods that allow you to move solid primitives are explained in this section.

Move

Pull-down Menu	Command (Type)	Alias (Type)	Short-cut	Screen (side) Menu	Tablet Menu
Modify *Move*	*Move*	*M*	...	*MODIFY2* *Move*	*V,19*

The *Move* command that you use for moving 2D objects in 2D drawings can also be used to move 3D primitives. Generally, *Move* is used to change the position of an object in one plane (translation), which is typical of 2D drawings. *Move* can also be used to move an ACIS primitive in 3D space <u>if</u> OSNAPs or 3D coordinates are used.

Move operates in 3D just as you used it in 2D. Previously, you used *Move* only for repositioning objects in the XY plane, so it was only necessary to PICK or use X and Y coordinates. Using *Move* in 3D space requires entering X, Y, and Z values or using *OSNAPs*.

For example, to create the composite solid used in the figures in Chapter 28, a *Wedge* primitive was *Moved* into position on top of the *Box*. The *Wedge* was created at 0,0,0, then rotated. Figure 29-32 illustrates the movement using absolute coordinates described in the syntax below:

FIGURE 29-32

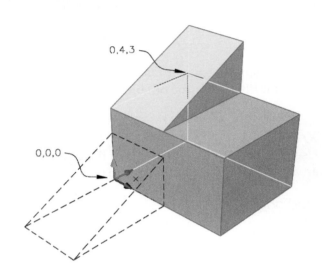

```
Command: move
Select objects: PICK
Select objects: Enter
Specify base point or displacement: 0,0
Specify second point of displacement or <use
first point as displacement>: 0,4,3
Command:
```

Alternately, you can use *OSNAPs* to select geometry in 3D space. Figure 29-33 illustrates the same *Move* operation using *Endpoint OSNAPs* instead of entering coordinates.

FIGURE 29-33

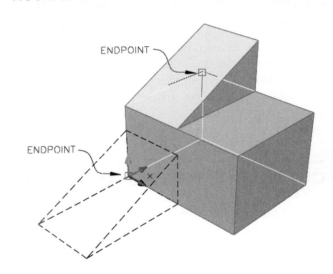

3Dalign

	Pull-down Menu	Command (Type)	Alias (Type)	Short-cut	Screen (side) Menu	Tablet Menu
	Modify *3D Operations >* *3D Align*	*3Dalign*	*3AL*	...	*MODIFY2* *3D Align*	...

3Dalign provides a means of aligning one 3D solid with another. *3Dalign* is very powerful because it automatically performs 3D <u>translation and rotation</u> if needed. All you have to do is select the points on two 3D objects that you want to align (connect).

Aligning one solid with another can be accomplished by connecting source points (on the solid to be moved) to destination points (on the stationary solid). Alternately, you can short cut this procedure by indicating a source plane or destination plane instead of selecting 3 points. You should always use *Osnap* modes to select the source and destination points to assure accurate alignment.

3Dalign performs a translation (like *Move*) and two rotations (like *Rotate*), each in separate planes to align the points as designated. The motion performed by *3Dalign* is actually done in three steps. For example, assume you wanted to move the *Wedge* on top of the *Box* and also reorient the *Wedge*. The command syntax for 3D alignment is as follows:

```
Command: 3dalign
Select objects: PICK
Select objects: Enter
Specify source plane and orientation ...
Specify base point or [Copy]: PICK  (S1)
Specify second point or [Continue] <C>: PICK (S2)
Specify third point or [Continue] <C>: PICK (S3)
Specify destination plane and orientation ...
Specify first destination point: PICK  (D1)
Specify second destination point or [eXit] <X>: PICK  (D2)
Specify third destination point or [eXit] <X>: PICK  (D3)
```

In this case, as shown in Figure 29-34, the "base point" (S1) is connected to the first destination point (D1). This first motion is a translation. <u>The base point and the first destination point always physically touch</u>.

FIGURE 29-34

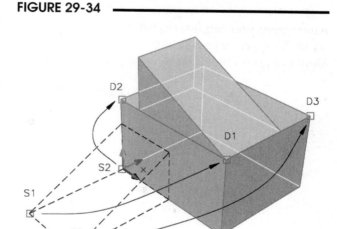

Next, the vector defined by the first and second source points (S1 and S2) is aligned with the vector defined by the first and second destination points (D1 and D2). The length of the segments between the first and second points on each object is not important because AutoCAD only considers the <u>vector direction</u>. This second motion is a rotation along one axis.

Finally, the third set of points are aligned similarly (S2-S3 with D2-D3). This third motion is a rotation along the other axis, completing the alignment.

In some cases, such as when cylindrical objects are aligned, only two sets of points have to be specified. For example, if aligning a shaft with a hole, the first set of points (source and destination) specify the attachment of the base of the shaft with the bottom of the hole. The second set of points specify the alignment of the axes of the two cylindrical shapes. A third set of points is not required because the radial alignment between the two objects is not important. When only two sets of points are specified, AutoCAD asks if you want to "continue" based on only two sets of alignment points.

You can also use the "source plane" method to align the *Wedge* with the *Box* as shown in Figure 29-35. This method uses the first source point (S1), or "base point," to indicate the XY plane of the source object. Once this base point is selected, press Enter to begin specifying the destination points, as shown in the following prompts.

FIGURE 29-35

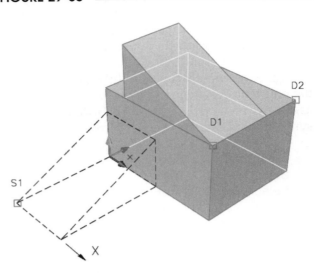

Command: *3dalign*
Select objects: **PICK**
Select objects: **Enter**
Specify source plane and orientation ...
Specify base point or [Copy]: **PICK** (S1)
Specify second point or [Continue] <C>: **Enter**
Specify destination plane and orientation ...
Specify first destination point: **PICK** (D1)
Specify second destination point or [eXit] <X>:
PICK (D2)
Specify third destination point or [eXit] <X>: **Enter**

Only the first two destination points need to be specified. *3Dalign* uses the positive X direction of the source plane to make the alignment between the first two destination points (D1 and D2). Since the planar orientation of the source and destination planes is the same, no other rotation is needed, so pressing Enter completes the alignment.

3Dmove

Pull-down Menu	Command (Type)	Alias (Type)	Short-cut	Screen (side) Menu	Tablet Menu
Modify *3D Operations >* *3D Move*	3DMove	3M	...	...	...

3Dmove can be used like to the *Move* command to move objects in 3D space. Even the prompts for *Move* and *3Dmove* are identical. However, using *3Dmove*, you can <u>constrain movements only to a particular axis or plane</u>. For example, you can use *3Dmove* to move a solid along the X axis only.

Using *3Dmove*, once you select the object(s) to move, the <u>move grip tool</u> is displayed. Next, select the desired base point, just as you would with the *Move* command. The move grip tool locks to the selected "base point." For example, in Figure 29-36, the move grip tool appears during the "Specify base point" prompt. In this case, one *Endpoint* of the *Wedge* is at the base point.

FIGURE 29-36

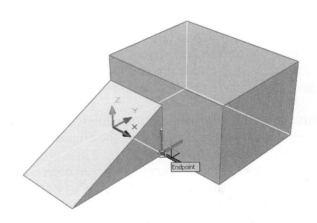

Command: *3dmove*
Select objects: **PICK**
Select objects: **Enter**
Specify base point or [Displacement] <Displacement>: **PICK**
Specify second point or <use first point as displacement>: **PICK**

When the move grip tool appears, select any <u>axis</u> to constrain the movement of the selected objects to the indicated grip tool axis. Once selected, the axis changes to a yellow-gold color. For example, selecting the X axis of the grip tool displays a vector along the axis (Fig. 29-37, left). Likewise, selecting the Y or Z axis highlights the appropriate move vector (Fig. 29-37, center and right).

FIGURE 29-37

You can also constrain the movement to a <u>plane</u>. Do this by selecting one of the small squares on the move grip tool. For example, if you want to move the selected object within the XZ plane, select the small square connecting the X and Z axes (Fig. 29-38).

FIGURE 29-38

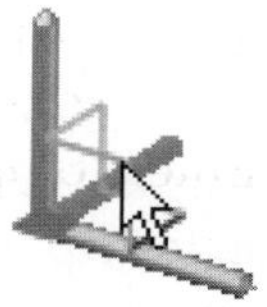

From the previous *Box* and *Wedge* example, selecting the Z axis on the move grip tool would constrain movement of the *Wedge* to a positive or negative Z direction, as shown in Figure 29-39. Next, move the *Wedge* to the desired location on the Z axis, then PICK or enter a value as the "second point."

FIGURE 29-39

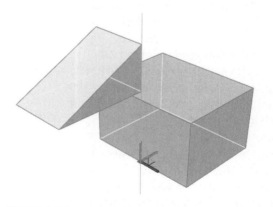

You can also use the Dynamic UCS feature with *3Dmove*. Either turn on the *DUCS* toggle on the Status bar or press Ctrl+D to turn on Dynamic UCS. This action causes the move grip tool to realign its orientation as you move the pointer over faces of solids (Fig. 29-40). The grip tool orients itself to the current plane depending on which edge of the face the pointer crosses. When the move grip tool displays the desired orientation, PICK to place the grip tool at the desired edge or corner (normally using *Osnap*).

FIGURE 29-40

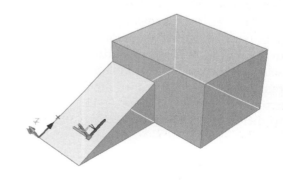

Once the grip tool is located on the desired plane, you can select any axis or plane indicator on the tool to constrain the direction of the move operation. For our example (Fig. 29-41), once the tool is oriented on the inclined plane of the *Wedge*, the movement of the *Wedge* is constrained to the grip tool's X axis. You can then PICK the desired "second point." If you enter coordinate values, they are relative to the selected plane.

FIGURE 29-41

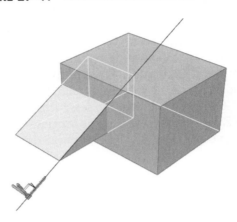

3Drotate

Pull-down Menu	Command (Type)	Alias (Type)	Short-cut	Screen (side) Menu	Tablet Menu
Modify *3D Operations >* *3D Rotate*	*3Drotate*	*3R*	...	...	*W,22*

3Drotate allows you to rotate solids about an axis. You specify the axis to rotate about by using the *3Drotate* grip tool. This grip tool displays only three possible rotational axes. However, since the grip tool can be located anywhere in 3D space and in any orientation using the *DUCS* feature, virtually any axis can be specified for the rotate action. The grip tool appears at the "Specify base point" prompt.

```
Command: 3drotate
Current positive angle in UCS: ANGDIR=counterclockwise  ANGBASE=0
Select objects: PICK
Select objects: Enter
Specify base point: PICK
Pick a rotation axis: PICK  (Select an axis handle to specify the rotation axis)
Specify angle start point: PICK
Specify angle end point: PICK
Command:
```

The rotate grip tool appears at the base point you specify. At the "Pick a rotation axis" prompt, select one of the circular "handles." Once a circular handle is selected, the related axis is displayed. For example, Figure 29-42 (left) displays selection of the X axis handle. Note the cursor location in the figure. When the handle is selected, the related axis vector remains on the screen until the rotate action is complete. Figure 29-42 (right) displays selection of the Z axis handle.

FIGURE 29-42

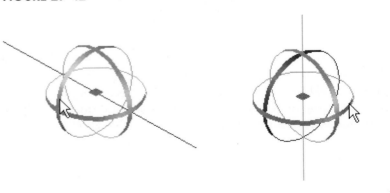

For example, suppose you wanted to rotate the wedge shown in Figure 29-43 about the Z axis at its nearest corner. Using *3Drotate*, select the wedge at the "Select objects" prompt, then select the nearest corner of the wedge at the "Specify base point" prompt to locate the *3Drotate* grip tool. At the "Pick a rotation axis" prompt, select the Z axis handle (note the cursor location).

FIGURE 29-43

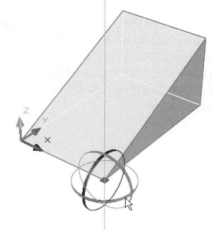

Next, you must specify an "angle start point." In our example, the *Endpoint* at the far right corner on the base is selected as the "angle start point." Finally, you can rotate the object about the selected point in real time at the "Specify angle endpoint" prompt, as shown in Figure 29-44.

FIGURE 29-44

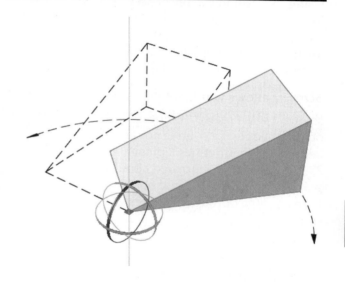

Mirror3D

Pull-down Menu	Command (Type)	Alias (Type)	Short-cut	Screen (side) Menu	Tablet Menu
Modify *3D Operations >* *3D Mirror*	*Mirror3D*	...	...	...	*W,21*

Mirror3D operates similar to the 2D version of the command *Mirror* in that mirrored replicas of selected objects are created. With *Mirror* (2D) the selected objects are mirrored about an axis. The axis is defined by a vector lying in the XY plane. With *Mirror3D*, selected objects are mirrored about a <u>plane</u>. *Mirror3D* provides multiple options for specifying the plane to mirror about:

```
Command: mirror3d
Select objects: PICK
Select objects: Enter
Specify first point of mirror plane (3 points) or [Object/Last/Zaxis/View/XY/YZ/ZX/3points] <3points>:
PICK or (option)
```

3points

The *3points* option mirrors selected objects about the plane you specify by selecting three points to define the plane. You can PICK points (with or without *OSNAP*) or give coordinates. *Midpoint OSNAP* is used to define the 3 points in Figure 29-45 A to achieve the result in B.

FIGURE 29-45

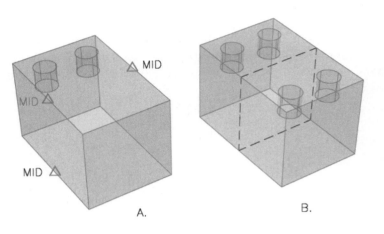

A.

B.

Object

Using this option establishes a mirroring plane with the plane of a 2D object. Selecting an *Arc* or *Circle* automatically mirrors selected objects using the plane in which the *Arc* or *Circle* lies. The plane defined by a *Pline* segment is the XY plane of the *Pline* when the *Pline* segment was created. Using a *Line* object or edge of a solid is not allowed because neither defines a plane. Figure 29-46 shows a box mirrored about the plane defined by the *Circle* object. Using *Subtract* produces the result shown in B.

FIGURE 29-46

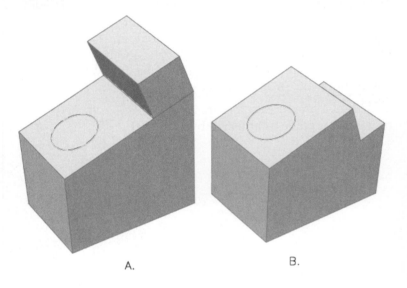

A.

B.

Last

Selecting this option uses the plane that was last used for mirroring.

Zaxis

With this option, the mirror plane is the XY plane perpendicular to a Z vector you specify. The first point you specify on the Z axis establishes the location of the XY plane origin (a point through which the plane passes). The second point establishes the Z axis and the orientation of the XY plane (perpendicular to the Z axis). Figure 29-47 illustrates this concept. Note that this option requires only two PICK points.

FIGURE 29-47

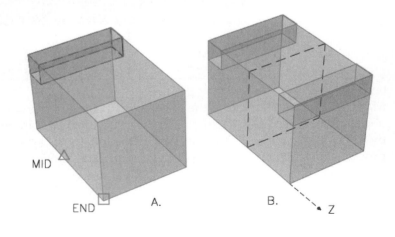

A.

B.

View

The *View* option of *Rotate3D* uses a mirroring plane <u>parallel</u> with the screen and perpendicular to your line of sight based on your current viewpoint. You are required to select a point on the plane. Accepting the default (0,0,0) establishes the mirroring plane passing through the current origin. Any other point can be selected. You must change your viewpoint to "see" the mirrored objects (Fig. 29-48).

FIGURE 29-48

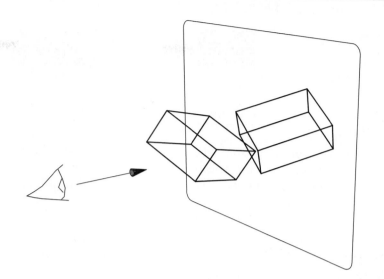

XY

This option situates a mirroring plane parallel with the current XY plane. You can specify a point through which the mirroring plane passes. Figure 29-49 represents a plane established by selecting the *Center* of an existing solid.

FIGURE 29-49

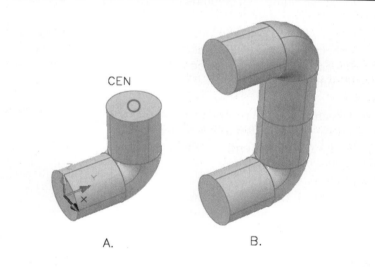

CEN

A.   B.

YZ

Using the *YZ* option constructs a plane to mirror about that is parallel with the current YZ plane. Any point can be selected through which the plane will pass (Fig. 29-50).

FIGURE 29-50

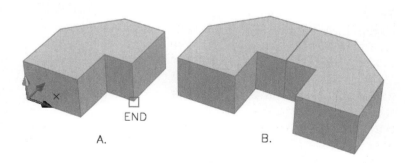

END

A.   B.

ZX

This option uses a plane parallel with the current ZX plane for mirroring.

3Darray

Pull-down Menu	Command (Type)	Alias (Type)	Short-cut	Screen (side) Menu	Tablet Menu
Modify *3D Operations >* *3D Array*	*3Darray*	*3A*	...	*MODIFY2* *3Darray*	*W,20*

Rectangular

With this option of *3Darray*, you create a 3D array specifying three dimensions—the number of and distance between rows (along the Y axis), the number/distance of columns (along the X axis), and the number/distance of levels (along the Z axis). Technically, the result is an array in a <u>prism</u> configuration <u>rather than a rectangle</u>.

```
Command: 3darray
Select objects: PICK
Select objects: Enter
Enter the type of array [Rectangular/Polar] <R>: r
Enter the number of rows (—-) <1>: (value)
Enter the number of columns (|||) <1>: (value)
Enter the number of levels (...) <1>: (value)
Specify the distance between rows (—-): PICK or (value)
Specify the distance between columns (|||): PICK or (value)
Specify the distance between levels (...): PICK or (value)
Command:
```

The selection set can be one or more objects. The entire set is treated as one object for arraying. All values entered must be positive.

Figures 29-51 and 29-52 illustrate creating a *Rectangular 3Darray* of a cylinder with 3 rows, 4 columns, and 2 levels.

FIGURE 29-51

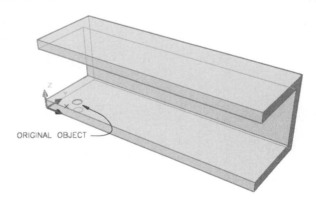

ORIGINAL OBJECT

The cylinders are *Subtracted* from the extrusion to form the finished part (Fig. 29-52).

FIGURE 29-52

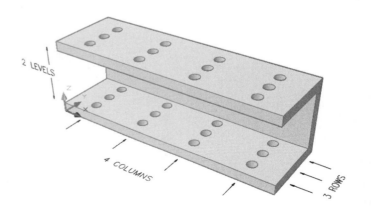

2 LEVELS

4 COLUMNS

3 ROWS

Polar

Similar to a *Polar Array* (2D), this option creates an array of selected objects in a <u>circular</u> fashion. The only difference in the 3D version is that an array is created about an <u>axis of rotation</u> (3D) rather than a point (2D). Specification of an axis of rotation requires two points in 3D space:

```
Command: 3darray
Select objects: PICK
Select objects: Enter
Enter the type of array [Rectangular/Polar] <R>: p
Enter the number of items in the array: (value)
Specify the angle to fill (+=ccw, -=cw) <360>: PICK or (value)
Rotate arrayed objects? [Yes/No] <Y>: y or n
Specify center point of array: PICK or (coordinates)
Specify second point on axis of rotation: PICK or (coordinates)
Command:
```

In Figures 29-53 and 29-54, a *3Darray* is created to form a series of holes from a cylinder. The axis of rotation is the center axis of the large cylinder specified by PICKing the *Center* of the top and bottom circles.

FIGURE 29-53

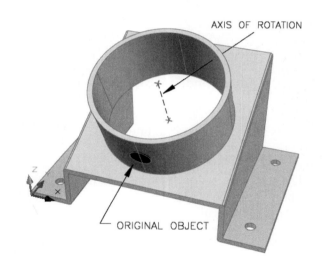

After the eight items are arrayed, the small cylinders are subtracted from the large cylinder to create the holes.

FIGURE 29-54

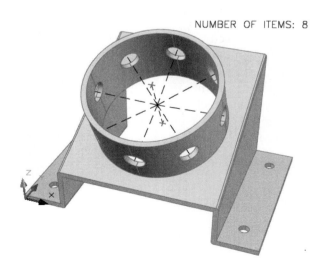

BOOLEAN OPERATION COMMANDS

Once the individual 3D primitives have been created and moved into place, you are ready to put together the parts. The primitives can be "assembled" or combined by Boolean operations to create composite solids. The Boolean operations found in AutoCAD are listed in this section: *Union*, *Subtract*, and *Intersect*.

Union

Pull-down Menu	Command (Type)	Alias (Type)	Short-cut	Screen (side) Menu	Tablet Menu
Modify Solid Editing > Union	Union	UNI	...	MODIFY2 Union	X,15

Union joins selected primitives or composite solids to form one composite solid. Usually, the selected solids occupy portions of the same space, yet are separate solids. *Union* creates one solid composed of the total encompassing volume of the selected solids. (You can union solids even if the solids do not overlap.) All lines of intersections (surface boundaries) are calculated and displayed by AutoCAD. Multiple solid objects can be unioned with one *Union* command:

        Command: **union**
        Select objects: **PICK** (Select two or more solids.)
        Select objects: **Enter** (Indicate completion of the selection process.)
        Command:

Two solid boxes are combined into one composite solid with *Union* (Fig. 29-55). The original two solids (A) share the same physical space. The resulting union (B) consists of the total contained volume. The new lines of intersection are automatically calculated and displayed. *Obscured edges* was used to enhance visualization.

FIGURE 29-55

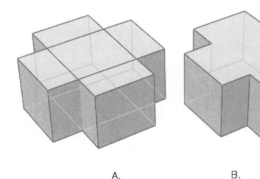

A.   B.

Because the volume occupied by any one of the primitives is included in the resulting composite solid, any redundant volumes are immaterial. The two primitives in Figure 29-56 A yield the same enclosed volume as the composite solid B. (*X-ray mode* has been used for this figure.)

FIGURE 29-56

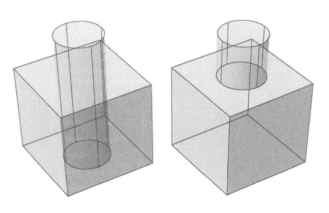

A.   B.

Multiple objects can be selected in response to the *Union* "Select objects:" prompt. It is not necessary, nor is it efficient, to use several successive Boolean operations if one or two can accomplish the same result. Two primitives that have coincident faces (touching sides) can be joined with *Union*. Several "blocks" can be put together to form a composite solid.

Figure 29-57 illustrates how several primitives having coincident faces (A) can be combined into a composite solid (B). <u>Only one *Union* is required</u> to yield the composite solid.

FIGURE 29-57 ————————————————

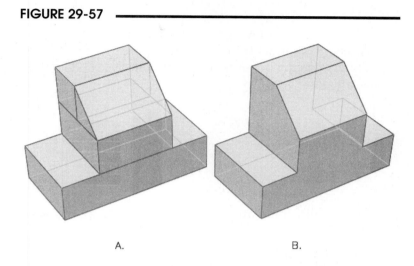

A.   B.

Subtract

Pull-down Menu	Command (Type)	Alias (Type)	Short-cut	Screen (side) Menu	Tablet Menu
Modify *Solid Editing >* *Subtract*	*Subtract*	*SU*	...	*MODIFY2* *Subtract*	*X,16*

Subtract takes the difference of one set of solids from another. *Subtract* operates with *Regions* as well as solids. When using solids, *Subtract* subtracts the <u>volume</u> of one set of solids from another set of solids. Either set can contain only one or several solids. *Subtract* requires that you first select the set of solids that will remain (the "source objects"), then select the set you want to subtract from the first:

```
Command: subtract
Select solids and regions to subtract from...
Select objects: PICK (select what you want to keep)
Select objects: Enter
Select solids and regions to subtract...
Select objects: PICK (select what you want to remove)
Select objects: Enter
Command:
```

The entire volume of the solid or set of solids that is subtracted is completely removed, leaving the remaining volume of the source set.

To create a box with a hole, a cylinder is located in the same 3D space as the box (see Fig. 29-58). *Subtract* is used to subtract the entire volume of the cylinder from the box. Note that the cylinder can have any height, as long as it is at least equal in height to the box.

Because you can select more than one object for the objects "to subtract from" and the objects "to subtract," many possible construction techniques are possible.

FIGURE 29-58

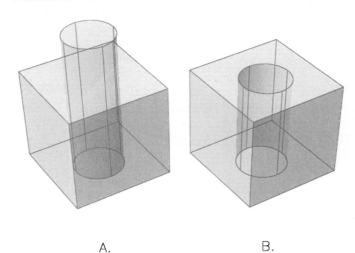

A.   B.

Figure 29-59 illustrates an *n*-way Boolean. The two boxes (A) are selected in response to the objects "to subtract from" prompt. The cylinder is selected as the objects "to subtract...". *Subtract* joins the source objects (identical to a *Union*) and subtracts the cylinder. The resulting composite solid is shown in B.

FIGURE 29-59

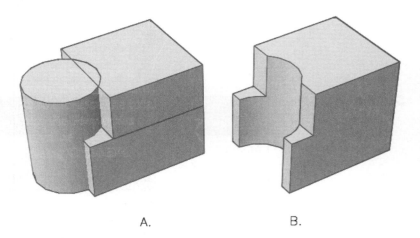

A.   B.

Intersect

Pull-down Menu	Command (Type)	Alias (Type)	Short-cut	Screen (side) Menu	Tablet Menu
Modify Solid Editing > Intersect	Intersect	IN	...	MODIFY2 Intrsect	X,17

Intersect creates composite solids by calculating the intersection of two or more solids. The intersection is the common volume <u>shared</u> by the selected objects. Only the 3D space that is <u>part of all</u> of the selected objects is included in the resulting composite solid. *Intersect* requires only that you select the solids from which the intersection is to be calculated.

    Command: **intersect**
    Select objects: **PICK** (Select all desired solids.)
    Select objects: **Enter** (Indicates completion of the selection process.)
    Command:

An example of *Intersect* is shown in Figure 29-60. The cylinder and the wedge share common 3D space (A). The result of the *Intersect* is a composite solid that represents that common space (B). (*X-ray mode* has been used for this figure.)

FIGURE 29-60

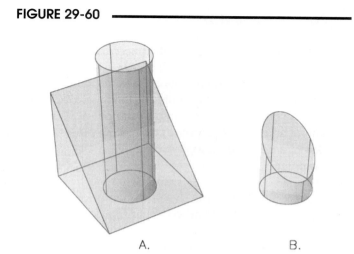

A.   B.

Intersect can be very effective when used in conjunction with *Extrude*. A technique known as <u>reverse drafting</u> can be used to create composite solids that may otherwise require several primitives and several Boolean operations. Consider the composite solid shown in Figure 29-57A. Using *Union*, the composite shape requires four primitives.

The two *Pline* "views" are extruded with *Extrude* to comprise the total volume of the desired solid (Fig. 29-61 A). Finally, *Intersect* is used to calculate the common volume and create the composite solid (B).

FIGURE 29-61

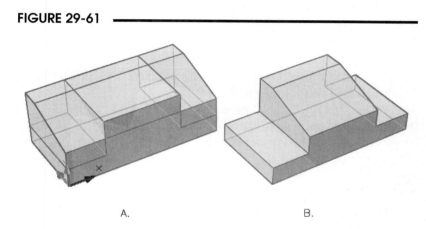

A.   B.

Chamfer

| | Command | Alias | Short- | Screen (side) | Tablet |
Pull-down Menu	(Type)	(Type)	cut	Menu	Menu
Modify				*MODIFY2*	
Chamfer	*Chamfer*	*CHA*	...	*Chamfer*	*W,18*

Chamfering is a machining operation that bevels a sharp corner. *Chamfer* chamfers selected edges of an AutoCAD solid as well as 2D objects. Technically, *Chamfer* (used with a solid) is a Boolean operation because it creates a wedge primitive and then adds to or subtracts from the selected solid.

When you select a solid, *Chamfer* recognizes the object as a solid and <u>switches to the solid version of prompts and options</u>. Therefore, all of the 2D options are not available for use with a solid, <u>only the "distances" method</u>. When using *Chamfer*, you must both select the "base surface" and indicate which edge(s) on that surface you wish to chamfer:

Command: **chamfer**
(TRIM mode) Current chamfer Dist1 = 0.5000, Dist2 = 0.5000
Select first line or [Polyline/Distance/Angle/Trim/Method/mUltiple]: **PICK** (Select solid at desired edge)
Base surface selection...
Enter surface selection option [Next/OK (current)] <OK>: **N** or **Enter**
Specify base surface chamfer distance <0.5000>: **(value)**
Specify other surface chamfer distance <0.5000>: **(value)**
Select an edge or [Loop]: **PICK** (Select edges to be chamfered)
Select an edge or [Loop]: **Enter**
Command:

FIGURE 29-62

When AutoCAD prompts to select the "base surface," only an edge can be selected since the solids are displayed in wireframe. When you select an edge, AutoCAD highlights one of the two surfaces connected to the selected edge. Therefore, you must use the "Next/<OK>:" option to indicate which of the two surfaces you want to chamfer (Fig. 29-62 A). The two distances are applied to the object, as shown in Figure 29-62 B.

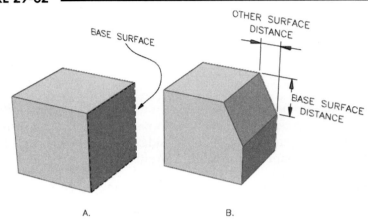

A.   B.

You can chamfer multiple edges of the selected "base surface" simply by PICKing them at the "Select an edge or [Loop]:" prompt (Fig. 29-63). If the base surface is adjacent to cylindrical edges, the bevel follows the curved shape.

FIGURE 29-63

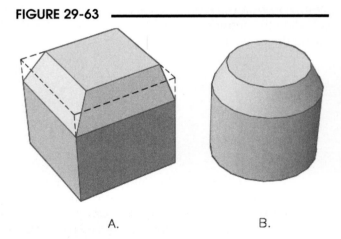

A.   B.

Loop

The *Loop* option chamfers the entire perimeter of the base surface. Simply PICK any edge on the base surface (Fig. 29-64).

    Select an edge or [Loop]: **1**
    Select an edge loop or [Edge]: **PICK** (Select edges to form loop)
    Select an edge or [Loop]: **Enter**
    Command:

FIGURE 29-64

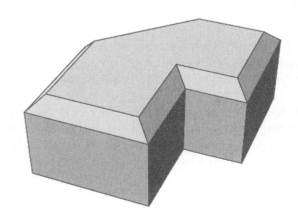

Edge

The *Edge* option switches back to the "Select edge" method.

Multiple

The *Multiple* option simply repeats the *Chamfer* command.

Fillet

Pull-down Menu	Command (Type)	Alias (Type)	Short-cut	Screen (side) Menu	Tablet Menu
Modify *Fillet*	*Fillet*	*F*	...	*MODIFY2* *Fillet*	*W,19*

Fillet creates fillets (concave corners) or rounds (convex corners) on selected solids, just as with 2D objects. Technically, *Fillet* creates a rounded primitive and automatically performs the Boolean needed to add or subtract it from the selected solids.

When using *Fillet* with a solid, the command <u>switches</u> to a special group of prompts and options for 3D filleting, and the <u>2D options become invalid</u>. After selecting the solid, you must specify the desired radius and then select the edges to fillet. When selecting edges to fillet, the edges must be PICKed individually. Figure 29-65 depicts concave and convex fillets created with *Fillet*. The selected edges are highlighted.

FIGURE 29-65

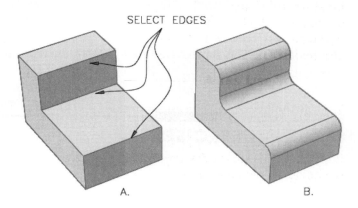

SELECT EDGES

A.   B.

```
Command: fillet
Current settings: Mode = TRIM, Radius = 0.5000
Select first object or [Undo/Polyline/Radius/Trim/Multiple]: PICK  (Select desired edge to fillet)
Enter fillet radius <0.5000>: (value)
Select an edge or [Chain/Radius]: Enter or PICK  (Select additional edges to fillet)
Command:
```

Curved surfaces can be treated with *Fillet*, as shown in Figure 29-66. If you want to fillet intersecting concave or convex edges, *Fillet* handles your request, provided you specify all edges in <u>one</u> use of the command. Figure 29-66 shows the selected edges (highlighted) and the resulting solid. Make sure you select <u>all</u> edges together (in one *Fillet* command).

FIGURE 29-66

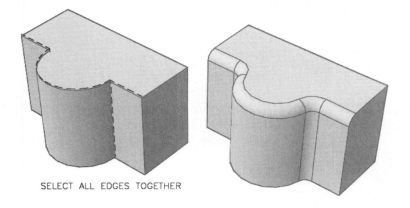

SELECT ALL EDGES TOGETHER

Chain

The *Chain* option allows you to fillet a series of connecting edges. Select the edges to form the chain (Fig. 29-67). If the chain is obvious (only one direct path), you can PICK only the ending edges, and AutoCAD will find the most direct path (series of connected edges):

FIGURE 29-67

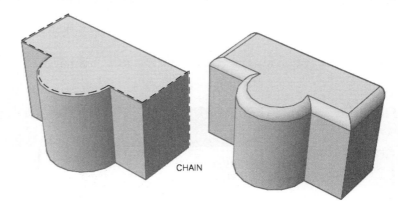

CHAIN

Select an edge or [Chain/Radius]: *c*
Select an edge chain or
[Edge/Radius]: **PICK**

Edge

This option cycles back to the "<Select edge>:" prompt.

Radius

This method returns to the "Enter radius:" prompt.

Multiple

The *Multiple* option causes the *Fillet* command to repeat.

CHAPTER EXERCISES

1. What are the typical four steps for creating composite solids?

2. Consider the two solids in Figure 29-68. They are two extruded hexagons that overlap (occupy the same 3D space).

FIGURE 29-68

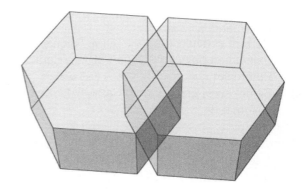

A. Sketch the resulting composite solid if you performed a *Union* on the two solids.

B. Sketch the resulting composite solid if you performed an *Intersect* on the two solids.

C. Sketch the resulting composite solid if you performed a *Subtract* on the two solids.

3. Begin a drawing and assign the name **CH29EX3.**

FIGURE 29-69 ⸺⸺⸺⸺⸺⸺

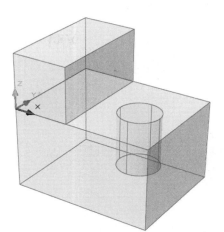

A. Create a **box** with the lower-left corner at **0,0,0**. The **Lengths** are **5**, **4**, and **3**.

B. Create a second **box** using a Dynamic UCS as shown in Figure 29-69 (use the **ORigin** option). The *box* dimensions are **2 x 4 x 2**.

C. Create a **Cylinder**. Use a Dynamic UCS as in the previous step. The *cylinder Center* is at **3.5,2** (of the *UCS*), the **Diameter** is **1.5**, and the **Height** is **-2**.

D. **Save** the drawing.

E. Perform a **Union** to combine the two boxes. Next, use **Subtract** to subtract the cylinder to create a hole. The resulting composite solid should look like that in Figure 29-70. **Save** the drawing.

FIGURE 29-70 ⸺⸺⸺⸺⸺⸺

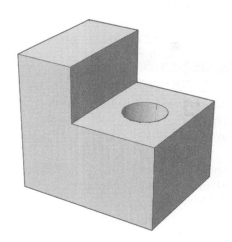

4. Begin a drawing and assign the name **CH29EX4.**

A. Create a **Wedge** at point **0,0,0** with the **Lengths** of **5**, **4**, **3**.

B. Create a **3point UCS** option with an orientation indicated in Figure 29-71. Create a **Cone** with the **Center** at **2,3** (of the *UCS*) and a **Diameter** of **2** and a **Height** of **-4**.

FIGURE 29-71 ⸺⸺⸺⸺⸺⸺

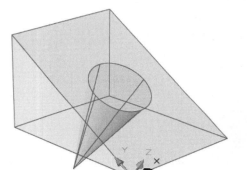

C. *Subtract* the cone from the wedge. The resulting
    composite solid should resemble Figure 29-72.
    Save the drawing.

FIGURE 29-72

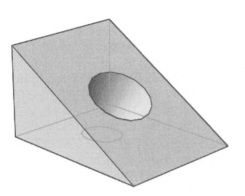

5. Begin a drawing and assign the name **CH29EX5.** Display a *Plan* view.

A. Create 2 *Circles* as shown in Figure 29-73,
    with dimensions and locations as specified.
    Use *Pline* to construct the rectangular shape.
    Combine the 3 shapes into a *Region* by using
    the *Region* and *Union* commands <u>or</u> convert-
    ing the <u>outside</u> shape into a *Pline* using *Trim*
    and *Pedit*.

FIGURE 29-73

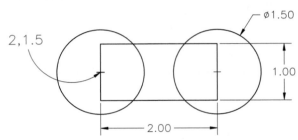

B. Change the display to an isometric-type view.
    Extrude the *Region* or *Pline* with a **Height** of **3**
    (no *taper angle*).

C. Create a *Box* with the lower-left corner at **0,0**.
    The *Lengths* of the box are **6, 3, 3**.

D. *Subtract* the extruded shape from the box. Your composite solid should look like that in
    Figure 29-74. *Save* the drawing.

FIGURE 29-74

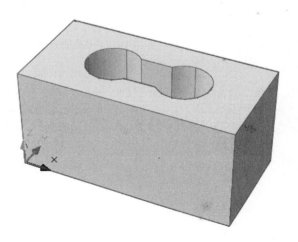

6. Begin a drawing and assign the name **CH29EX6.** Display a *Plan* view.

   A. Create a closed *Pline* shape symmetrical about the X axis with the locational and dimensional specifications given in Figure 29-75.

   B. Change to an isometric-type view. Use *Revolve* to generate a complete circular shape from the closed *Pline*. Revolve about the **Y** axis.

   C. Create a *Torus* with the *Center* at **0,0**. The *Radius of torus* is **3** and the *Radius of tube* is **.5**. The two shapes should intersect.

   D. Generate a display like Figure 29-76.

   E. Create a *Cylinder* with the *Center* at **0,0,0**, a *Radius* of **3**, and a *Height* of **8**.

   F. Use *3DRotate* to rotate the revolved *Pline* shape **90** degrees about the X axis (the *Pline* shape that was previously converted to a solid—not the torus). Next, move the shape up (positive Z) **6** units with *3DMove*.

   G. Move the torus up **4** units with *3DMove*.

   H. The solid primitives should appear as those in Figure 29-77.

FIGURE 29-75

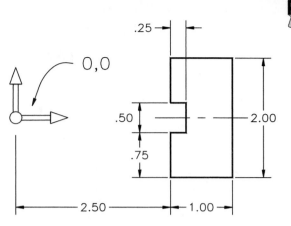

FIGURE 29-76

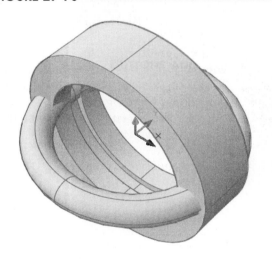

FIGURE 29-77

I. Use *Subtract* to subtract both revolved shapes from the cylinder.   The solid should resemble that in Figure 29-78. *Save* the drawing.

FIGURE 29-78

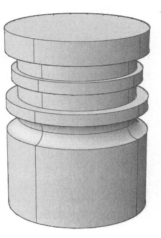

7. Begin a *New* drawing or use a *Template.*
    Assign the name **FAUCET1.**

A. Draw 3 closed *Pline* shapes, as shown in Figure 29-79. Assume symmetry about the longitudinal axis. Use the WCS and create 2 new *UCS*s for the geometry. Use *3point Arcs* for the "front" profile. *Save* the drawing as **FAUCET2**.

FIGURE 29-79

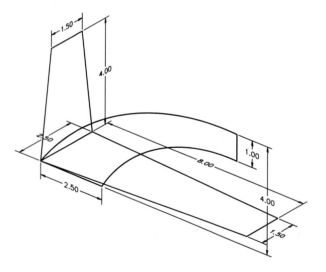

B. *Extrude* each of the 3 profiles into the same space. Make sure you specify the correct positive or negative *Height* value.

FIGURE 29-80

C. Finally, use *Intersect* to create the composite solid of the Faucet (Fig. 29-80). *Save* the drawing. Use *Saveas* and assign the name **FAUCET1**.

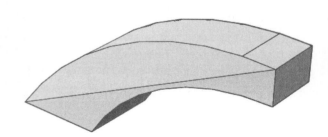

D. (Optional) Create a nozzle extending down from the small end. Then create a channel for the water to flow (through the inside) and subtract it from the faucet.

9. Construct a solid model of the bar guide in Figure 29-81. Strive for the most efficient design. It is possible to construct this object with one *Extrude* and one *Subtract*. Save the model as **BGUID-SL.**

FIGURE 29-81

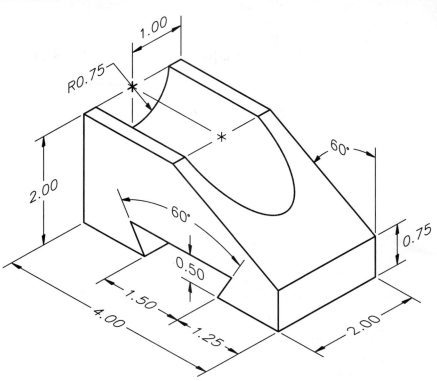

10. Create a solid model of the pulley (Fig. 29-82). All vertical dimensions are diameters. Orientation of primitives is critical in the construction of this model. Try creating the circular shapes on the XY plane (circular axis aligns with Z axis of the WCS). After the construction, use *3DRotate* to align the circular axis of the composite solid with the Y axis of the WCS. *Save* the drawing as **PULLY-SL.**

FIGURE 29-82

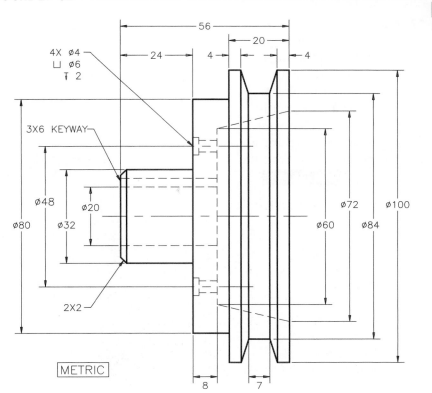

11. Construct a solid model of the saddle shown in Figure 29-83. An efficient design can be utilized by creating **Pline** profiles of the top "view" and the front "view,". Use **Extrude** and **Intersect** to produce a composite solid. Additional Boolean operations are required to complete the part. Save the drawing as SADL-SL.

FIGURE 29-83

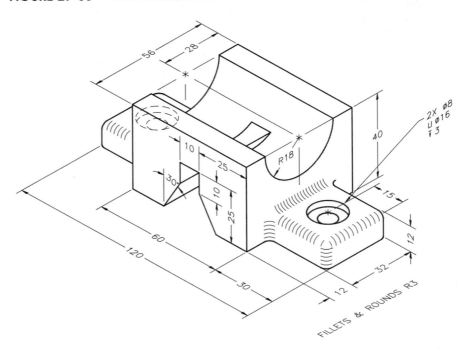

12. Make a solid model of the V-block shown in Figure 29-84. Several strategies could be used for construction of this object. Strive for the most efficient design. Plan your approach by sketching a few possibilities. Save the model as VBLOK-SL.

FIGURE 29-84

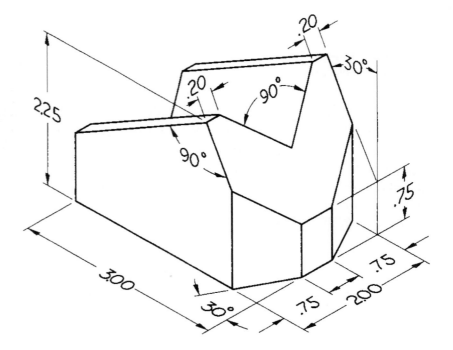

SOLID MODEL EDITING

CHAPTER OBJECTIVES

After completing this chapter you should:

1. be able to edit the properties of solid models using the *Properties* palette;

2. be able to edit primitive solids and composite solids using grips and the move and rotate grip tools;

3. be able to use subobject selection to edit faces, edges, and vertices of solids;

4. know how to use *Solidedit* to accomplish a variety of solid editing tasks;

5. know how to pass a live slicing plane through solids using *Sectionplane*;

6. be able to create surfaces using *Planesurf*, *Extrude*, *Revolve*, *Sweep*, *Loft*, and use the surfaces to *Slice* solids;

7. be able to calculate and export solid geometry data.

CONCEPTS

As you learned in Chapter 29, the basic solid modeling construction process is to create simple 3D solids, called "primitives," then combine them using "Boolean operations" to create a complex 3D model called a "composite solid." A number of primitive types and construction techniques were discussed as well as strategies to augment the construction process, including using Dynamic User Coordinate Systems.

The topics in this chapter are arranged as follows:

Editing Primitives using the *Properties* Palette

Editing Solids Using Grips

Solids Editing Commands: *Solidedit*, *Slice*, and *Sectionplane*

Using the information from Chapter 29 and the solid editing techniques discussed in this chapter, you should be able to create practically any shape or configuration of solid model.

EDITING PRIMITIVES USING THE *PROPERTIES* PALETTE

You can use the *Properties* palette to edit basic dimensional characteristics of solid primitives. Even if a primitive has been combined into a composite solid using a Boolean operation, it can be edited by holding down the Ctrl key and selecting the desired primitive as a "subobject." Editing primitives using the *Properties* palette is fairly simple and straight-forward. This section explains the following procedures:

Editing *Properties* of individual primitives

Composite solid subobject selection

Editing *Properties* of composite solid subobjects (primitives)

The illustrations in this section display grips on the selected primitives only to indicate which primitives are selected for editing. Grips are not discussed, however, until the next section of this chapter.

Editing *Properties* of Individual Primitives

The basic dimensional and positional properties of individual solids are available for editing using the *Properties* palette. The procedure is simple: with the *Properties* palette displayed on the screen, select the desired primitive, then change the desired values in the *Geometry* section of the *Properties* palette.

For example, assume you want to change the height of a *Box* primitive. Selecting the box displays the primitive's properties in the *Properties* palette (Fig. 30-1). (Notice that the grips appear on the primitive when selected, but the grips are not needed to access the *Properties* palette.) Find the *Height* edit box in the *Geometry* section of the palette and enter the desired value.

FIGURE 30-1

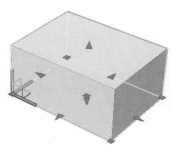

Notice that changing a dimensional or positional value in the *Properties* palette immediately changes the primitive. In this case, the *Height* of the box was changed from 2.0000 to 4.0000.

FIGURE 30-2

The positional values of primitives can be changed by the same method. Values that appear in the *Position X*, *Position Y*, and *Position Z* edit boxes represent the center of the base. This location is also indicated by a square grip at the base's center. Note that in our example the move grip tool appears at 0,0,0. In Figure 30-3 the *Position Y* value was changed from 1.5000 to 4.0000.

FIGURE 30-3

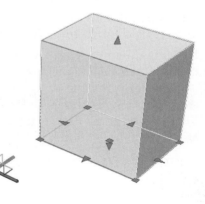

2007

Although all primitives have *Position X*, *Position Y*, and *Position Z* properties, each primitive type has distinct dimensional values. For example, a *Cylinder* has a field for *Radius*, *Height*, and *Elliptical* (Fig. 30-4). If *Elliptical* is set to *Yes*, additional fields appear for *Major radius* and *Minor radius*.

A *Cone* primitive has similar dimensional properties to a *Cylinder* with the addition of a dimensional value for *Top radius* (see "Grip Editing Primitives"). A *Pyramid*, as another example, has property fields for *Base radius*, *Top radius*, (number of) *Sides*, *Height*, and *Rotation*. Therefore, each primitive contains properties based on the dimensional features of the shape and the specifications given to create the solid.

FIGURE 30-4

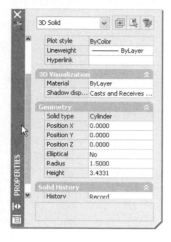

Composite Solid Subobject Selection

A <u>subobject</u> is a component of a solid. Therefore, an individual primitive that is part of a composite solid is a subobject. You can select any subobject by <u>holding down the Ctrl key</u>, then hovering the cursor over a subobject until it becomes highlighted, then using the left mouse button to PICK it. For example, assume a composite solid is constructed from two *Box* primitives as shown in Figure 30-5. Note that the two *Boxes* are combined by a *Union* since there are no edges displayed between the two primitives.

FIGURE 30-5

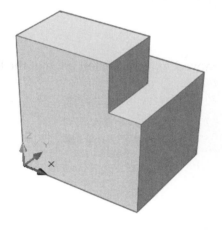

Holding down the Ctrl key forces the cursor to change from the crosshairs to a small pickbox. Hovering the small pickbox over the smaller *Box* primitive on top displays a "highlighted" effect for the small *Box* only, as shown in Figure 30-6. While it is highlighted, PICKing the small *Box* primitive displays its grips (assuming grips are turned on) as shown in Figure 30-7.

FIGURE 30-6

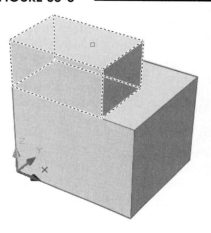

Editing *Properties* of Composite Solids

To change the properties of a composite solid subobject using the *Properties* palette, hold down the Ctrl key, hover over the desired primitive until it becomes highlighted, then PICK. The *Properties* palette should display the locational and dimensional properties of the primitive solid only, not the properties of the composite solid (Fig. 30-7). Although the primitive's grips are displayed when it is selected, grips do not have to be turned on to edit the primitive using *Properties*.

FIGURE 30-7

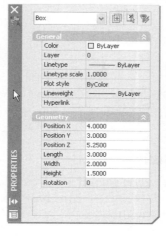
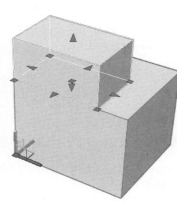

For example, assume you want to change both the location and dimensions of the subobject (*Box* primitive). Once the subobject is selected, change the desired values in the *Geometry* section of the *Properties* palette (Fig. 30-8). Note that in our example, the *Position Y* was changed from 3.0000 to 3.5000, so the *Box* moved in from the left edge of the base feature. In addition, the *Length* value was changed from 3.0000 to 2.0000. Note that the *Length*, *Width*, and *Height* are relative to the center of the base of the primitive; therefore, the length was reduced by 0.5000 on each end of the subobject.

FIGURE 30-8

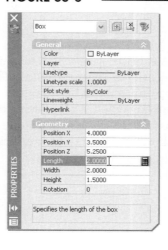
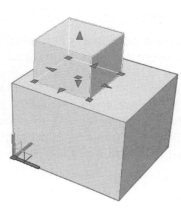

The important concept to note is that when you edit individual primitives (subobjects) of a composite solid using *Properties* or grips, the composite solid remains one unit. In other words, although the small box in our example changed dimensions and location, it is still *Unioned* to the large box.

EDITING SOLIDS USING GRIPS

The most intuitive method for editing solids is using the grips available on solids. When you select a solid, the grips appear. The grips allow you to change the solids in different ways, depending on the type of primitive and the type (shape) of grip.

You can also edit individual subobjects within solids. "Subobject" is the term AutoCAD uses to describe a component of a solid. You can select composite solids to edit, individual primitives contained in composite solids, and even individual faces, edges, and vertices.

There are numerous possibilities for editing solids using grips since this editing method is so dynamic and flexible. Every situation cannot be described in this section; therefore, only the basic principles and most common possibilities are discussed. This section will explain the following concepts:

Grip editing primitive solids
Move and rotate grip tools
Grip editing composite solids
Subobjects and selection
Grip editing faces, edges, and vertices
Selecting overlapping primitives
History settings

NOTE: To use grip editing features for solids, make sure grips are turned on. Do this by setting the *GRIPS* variable to 1 or checking *Enable Grips* in the *Selection* tab of the *Options* dialog box.

Grip Editing Primitive Solids

Assuming grips are turned on, when a solid is selected, one or more grips appear. Each type of solid primitive displays its own special types of grips based on the geometric characteristics of the primitive. For example, a *Box* is defined by width, height, and depth, so appropriate grips appear allowing you to change those characteristics. A *Cylinder*, however, is defined by a base diameter and height, so grips especially for those features appear. Several examples of solids and the related grips are described in this section.

Box

When you select a *Box* primitive, grips appear that allow you change the dimensional and locational characteristics. The shape of the grip informs you how the primitive can be changed. For example, note the grips that appear when a *Box* is selected (Fig. 30-9). The arrow-shaped grips indicate the shape can be changed by stretching the grip forward or backward in the specified direction. Square grips indicate motion in any direction.

FIGURE 30-9

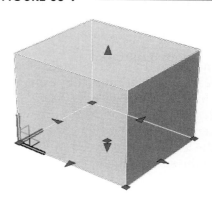

Change the dimensions of the *Box* using any of the arrow-shaped grips. For example, select a grip to make it hot, then stretch the face of the solid by moving the grip inward or outward as illustrated in Figure 30-10. The arrow grips stretch <u>only</u> in the indicated direction. Polar Tracking does not have to be turned on, but may help indicate the current length and can be used in conjunction with Polar Snap. You can also enter a value at the command prompt to specify a distance to stretch the solid.

FIGURE 30-10

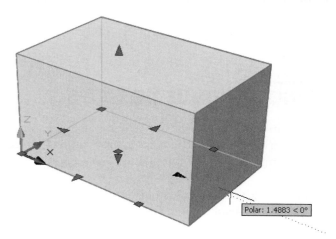

Specify point location or [Base point/
Undo/eXit]: **PICK** or (**value**)

Stretch any other face inward or outward using the appropriate arrow grip. For example, change the height of the *Box* by stretching the arrow grip on top or bottom (Fig. 30-11).

FIGURE 30-11

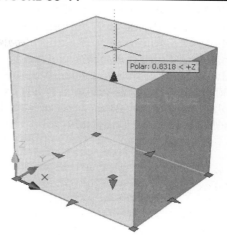

The square grips indicate that you can stretch the corner of the solid in any direction. For example, you can select a corner (square) grip to make it hot, then shorten one dimension while lengthening another (Fig. 30-12). The same prompt appears at the Command line where you can enter coordinates for the new location of the corner.

FIGURE 30-12

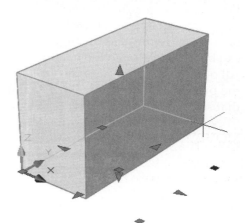

Specify point location or [Base point/Undo/eXit]: **PICK** or (**coordinates**)

If you select the square grip at the center of the base, the location of the entire primitive can be changed as shown in Figure 30-13. Polar Tracking can be used to ensure movement along a specified angle.

When the solid's "base" grip is selected (the square grip at the center of the base), the Command prompt changes to the standard grip options.

FIGURE 30-13

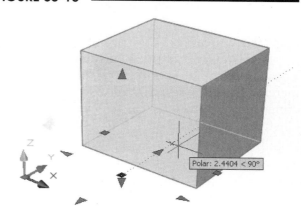

** STRETCH **
Specify stretch point or [Base point/
Copy/Undo/eXit]:

Cycle through the five options (STRETCH, MOVE, ROTATE, MIRROR, and SCALE) to change the entire primitive using the desired option. For example, you can rotate the entire primitive by making the base grip hot, then pressing Enter or spacebar to cycle to ROTATE, then dynamically rotating the solid in the current XY plane (Fig. 30-14). Keep in mind that STRETCH and MOVE accomplish the same results. Also, MIRROR has an effect on the solid only if the Copy suboption is used.

FIGURE 30-14

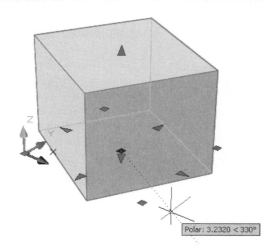

Wedge

Editing a *Wedge* primitive is similar to editing a *Box* in that the grips that appear on each are similar. The height grip on a *Wedge* appears at the highest point of the solid and can be used to stretch the height as shown in Figure 30-15. All other grip options are the same as for a *Box*, including the prompt that appears on the Command line. Remember that selecting the grip at the center of the base allows you to cycle through the five normal grip options (STRETCH, MOVE, ROTATE, MIRROR, and SCALE).

FIGURE 30-15

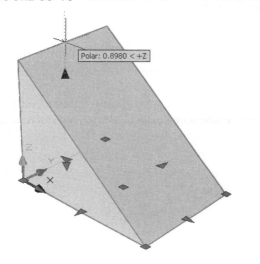

Cone

Grips that appear on a *Cone* primitive when selected are displayed in Figure 30-16. Remember that a *Cone* is dimensionally defined by specifying the base radius and height; therefore, grips appear that allow you to change those dimensional characteristics. For example, Figure 30-16 illustrates dragging the top grip downward to change the height of the solid.

FIGURE 30-16

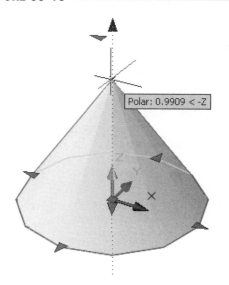

Also remember that a *Cone* can have a top radius. In such a case, the cone is a frustum (truncated cone). Even if a top radius was not defined initially and a normal cone was created, a top radius grip appears (see previous Fig. 30-16). Figure 30-17 displays the process of dragging the top radius grip to create a frustum.

FIGURE 30-17

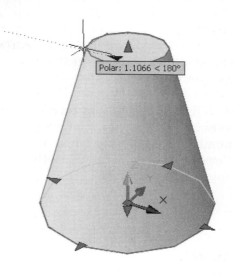

Polar: 1.1066 < 180°

Sphere

In the case of a *Sphere,* four arrow-shaped grips appear (not shown). Therefore, you can change only the diameter dimension. In addition, a center grip appears for changing location.

Cylinder

A *Cylinder* is defined by the base radius and height, so the related grips appear when the primitive is selected as shown in Figure 30-18. You can change the height at the top or bottom grip or change the radius using one of the four arrow grips. Move the solid or access the five normal grip options by selecting the grip at the center of the base.

FIGURE 30-18

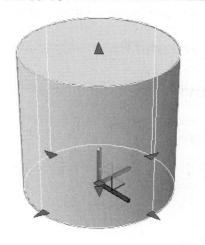

Pyramid

A *Pyramid* is also defined by the base radius and height. As you can see in Figure 30-19, selecting a *Pyramid* primitive reveals grips at each corner and along each side of the base; however, you can changes only the base diameter with these eight grips. Selecting any one of the corner grips or side grips changes all corners and sides uniformly. One side or corner cannot be changed individually because a *Pyramid* primitive is a right pyramid (the apex is always directly above the center of the base).

FIGURE 30-19

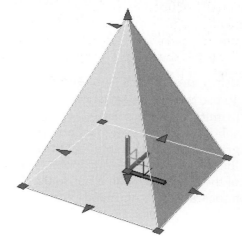

Note that a *Pyramid* can have a top radius, similar to a *Cone*. Therefore, a related top radius grip can be used to create a top radius for a *Pyramid* originally defined with no top radius, as shown in Figure 30-20. Likewise, you can change a previously defined top radius to a *Pyramid* with an apex (top radius of 0).

NOTE: Although a *Pyramid* can be defined initially with as many sides as you want, you cannot change the original number of sides using grips. You can, however, use the *Properties* palette to reset the original number of sides to any practical number.

FIGURE 30-20

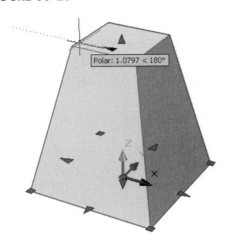

Torus

When a *Torus* primitive is selected for grip editing (not shown), four arrow grips appear allowing you to change the tube radius. In addition, one arrow grip allows you to change the torus radius. You can also use the grip at the center to change the location of the primitive or to access the five normal grip editing options (STRETCH, MOVE, ROTATE, MIRROR, and SCALE).

Move and Rotate Grip Tools

Once the desired solids (primitives or composite solids) have been selected for editing, two things happen—the cold grip(s) appear and the move grip tool appears at the current UCS location. At this point, you have a choice to use the normal grip options (STRETCH, MOVE, ROTATE, SCALE, MIRROR) or use the move grip tool and rotate grip tool. *DUCS* has no effect on these options. Follow one of these procedures:

A. To use the normal grip options, simply select the base grip for the solid, face, edge, or vertex to make the grip hot. The STRETCH option appears at the command prompt. You can cycle through all <u>five</u> options by pressing Enter.

B. To use the move and rotate grip tools, hover the cursor over the solid's base grip (the square grip generally at the center of the base) until the move tool automatically relocates to the grip location. Select an axis or plane on the tool. The STRETCH option appears at the command prompt; however, only the STRETCH, MOVE, and ROTATE options can be used. If an axis of the grip tool is selected, the selected solid stretches, moves, or rotates along the axis. If a plane from the tool is selected (the small square area connecting two axes), the selected solid stretches, moves, or rotates within the plane.

Using the Normal Five Grip Options

To use the normal grip options, for example, assume a *Box* primitive solid is selected for editing. Once selected, the solid becomes highlighted and the grips appear as cold grips (Fig. 30-21). Note that the move grip tool appears at the UCS location. At this point, select the grip at the center of the base and make it hot so the STRETCH option appears at the Command prompt.

FIGURE 30-21

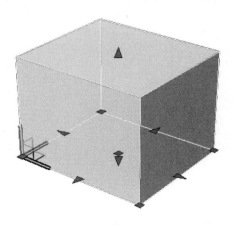

Once the base grip is hot, the move grip tool disappears and the UCS reappears. You can cycle through all five normal grip options: STRETCH, MOVE, ROTATE, SCALE, and MIRROR. Figure 30-22 displays a MOVE in progress. Using the normal grip options, when you cycle to the ROTATE grip option the selected solid rotates about the hot grip in the XY plane of the grip as shown previously in Figure 30-14.

FIGURE 30-22

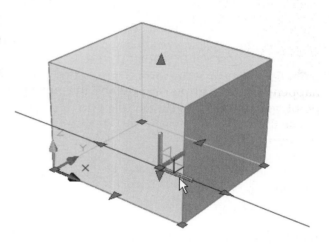

Using the Move and Rotate Grip Tools

On the other hand, to use the move and rotate grip tools with our example, assume the *Box* primitive is selected for editing as displayed previously in Figure 30-21. Rather than selecting the base grip to make it hot, <u>hover over the grip</u> until the move tool automatically relocates to the location of the base grip. At this point, select the desired axis or plane on the grip tool. Figure 30-23 displays this process during selection of the X axis of the grip tool. The grip is now considered hot and the STRETCH option appears at the Command prompt. Either the STRETCH or MOVE option could be used to move the solid along the grip tool's X axis. Pressing the Esc key allows you to select a different axis or plane on the move grip tool.

FIGURE 30-23

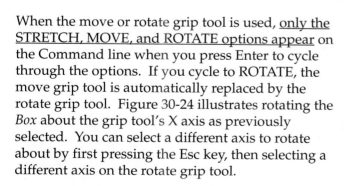

When the move or rotate grip tool is used, <u>only the STRETCH, MOVE, and ROTATE options appear</u> on the Command line when you press Enter to cycle through the options. If you cycle to ROTATE, the move grip tool is automatically replaced by the rotate grip tool. Figure 30-24 illustrates rotating the *Box* about the grip tool's X axis as previously selected. You can select a different axis to rotate about by first pressing the Esc key, then selecting a different axis on the rotate grip tool.

FIGURE 30-24

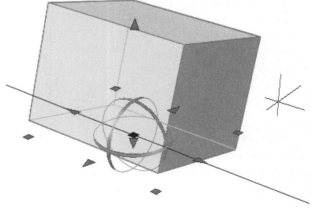

Grip Editing Composite Solids

When you select a composite solid for editing with grips, only the one grip appears at the center of the base as shown in Figure 30-25. The other grips that appear on individual primitives do not appear by default, but can be selected by holding down the Ctrl key and hovering the cursor over a primitive, then selecting it (see "Subobjects and Selection" next).

FIGURE 30-25

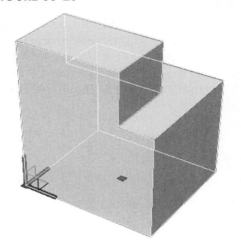

The single base grip that appears when a composite solid is selected, however, allows you to edit the composite solid in the same way as described in previous section "Move and Rotate Grip Tools." That is, if you select the base grip so it becomes hot, you can cycle through all five normal grip options: STRETCH, MOVE, ROTATE, SCALE, and MIRROR. For example, using the normal grip options, once the base grip is hot, you cycle to the ROTATE option to rotate the composite solid in the XY plane of the grip (Fig. 30-26.)

FIGURE 30-26

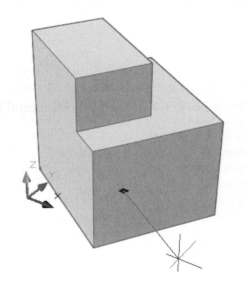

Similar to editing primitives with grips, to use the move and rotate grip tools, hover over the grip until the move tool automatically relocates to the location of the base grip. At this point, select the desired axis or plane on the grip tool (Fig. 30-27). You can press the Esc key to select a different axis or plane on the move grip tool. Remember that only the STRETCH, MOVE, and ROTATE options appear on the Command line when you use the move and rotate grip tools.

FIGURE 30-27

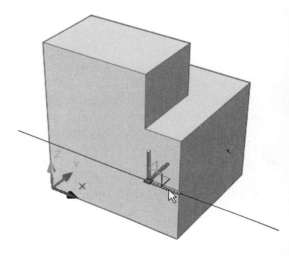

For example, if you cycle to ROTATE, the move grip tool is automatically replaced by the rotate grip tool. By default the composite solid rotates about the selected axis (Fig. 30-28). You can select a different axis to rotate about by first pressing the Esc key, then selecting a different axis on the rotate grip tool.

FIGURE 30-28

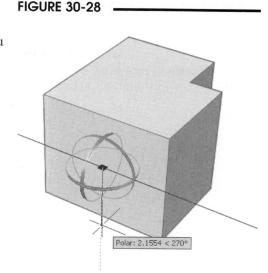

Polar: 2.1554 < 270°

Subobjects and Selection

A "subobject" is a component of a solid. A primitive within a composite solid is considered a subobject. Also, a face, edge, or vertex is a subobject of a primitive solid. The grips that appear on subobjects are exclusive to that type of subobject depending on the type of subobject: edge, vertex, face, or type of primitive.

As you already know, to edit a solid (not a subobject) using grips, you simply select the solid with the cursor for grips to appear. However, to select a subobject, you must hold down the Ctrl key, hover the cursor over the desired subobject until it becomes highlighted, then use the left mouse button to PICK the subobject.

With the default settings, there can be <u>two levels of subobjects</u>. Selecting the desired subobjects depends on the "level" of the subobject. For a first level subobject, hold down the Ctrl key, then PICK it. To select a second level subobject, repeat this process. The two levels are outlined here:

> <u>First level subobject</u>, select once using Ctrl:
> Primitive within a composite solid
> Edge, face or vertex of a stand-alone primitive solid
>
> <u>Second level subobject</u>, select twice using Ctrl:
> Edge, face, or vertex of a primitive within a composite solid

NOTE: A second level subobject is available only when the solid's *History* property is set to *Record*. See "*History* Settings."

FIGURE 30-29

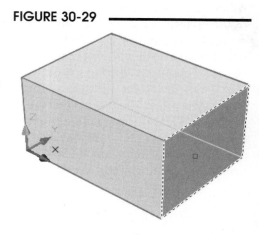

Selecting a first-level subobject requires a single selection. For example, to select a primitive within a composite solid, hold the Ctrl key to highlight, then select the primitive. This sequence is displayed previously in Figures 30-5, 30-6, and 30-7.

To select an edge, face, or vertex within a stand-alone primitive (a first level subobject), hold the Ctrl key to highlight, then select the desired edge, face or vertex. Figure 30-29 displays using Ctrl to highlight a face of a *Box* primitive. Once selected, the grip related to the feature appears. A face, edge, and vertex each have only one grip.

Selecting a second-level subobject (a face, edge, or vertex within a composite solid) requires a double selection. For example, assume a composite solid is constructed of two *Box* primitives and you want to edit one edge of the composite solid. First press and hold the Ctrl key, hover over and select the desired primitive as shown in Figure 30-30.

FIGURE 30-30

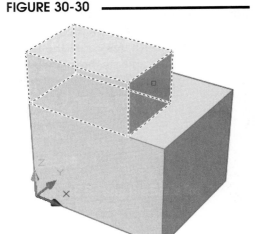

Once the grips appear on the primitive, select the desired edge, face, or vertex using the same method. That is, press and hold the Ctrl key to hover over and select the desired edge, face, or vertex. In our example, once grips appear on the previously selected primitive, press Ctrl to hover and select the desired edge of the highlighted primitive. The edge grip appears as shown in Figure 30-31.

FIGURE 30-31

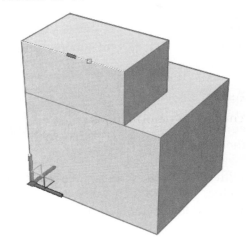

Grip Editing Faces, Edges, and Vertices

An individual face, edge, and vertex on a solid has a distinct grip shape. Each type of grip can be used to edit the object feature in predefined ways. Similar to using grips to edit primitive solids and composite solids, you can chose to edit with or without the move and rotate grip tools. The methods for activating the move and rotate tools for faces, edges, and vertices are the same as previously described for primitives and composite solids. When not using the move and rotate grip tools, all five normal grip options are available (STRETCH, MOVE, ROTATE, SCALE, and MIRROR). Using the move and rotate grip tools, only the STRETCH, MOVE, and ROTATE options can be used.

Edge Grips

Several examples for using edge grips are described and illustrated in this section; however, many more possibilities exist. Using our composite solid example, assume the top right edge grip was selected for editing. Without using the move and rotate grip tools, you can cycle to all five grip options. Using the STRETCH option would allow you to stretch the selected edge in any direction as shown in Figure 30-32. Note that other edges and faces are also affected; however, only the primitive containing the selected edge is changed.

FIGURE 30-32

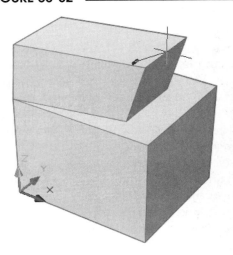

Cycling to the ROTATE option allows you to rotate the edge. Note in Figure 30-33 that the face containing the selected edge is rotated but the length of the edge itself does not change. Other primitives in a composite solid are not affected.

FIGURE 30-33

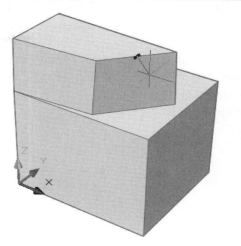

The SCALE option allows you to change the length of the selected edge (Fig. 30-34). Other edges and faces are affected but only those on the primitive containing the edge.

When the MOVE (with or without the move and rotate grip tools) and MIRROR grip options are used with an edge grip, the entire primitive is moved or mirrored without any other changes or deformations.

FIGURE 30-34

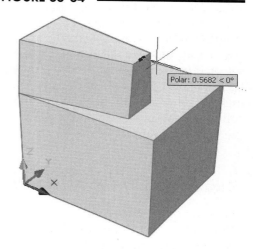

If you want to use the move and rotate grip tools, press Ctrl and hover over the grip until the move tool relocates to the grip position. Select the desired axis or plane on the grip tool. For example, you could select the grip tool's Z axis to ensure the selected edge grip moves only in the Z direction, as indicated in Figure 30-35. Press the Esc key to select another axis or plane.

FIGURE 30-35

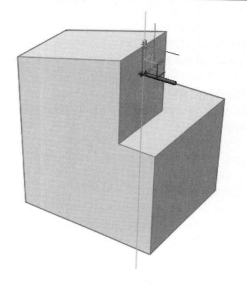

When you cycle to the ROTATE option, the rotate grip tool appears. By default, rotation occurs about the previously selected axis. Press Esc to select another axis on the rotate tool. In Figure 30-36, the rotate tool's X axis is used. Beware since the selected edge does not change length using any option other that SCALE; therefore, the primitive may become deformed depending on the axis of rotation as shown in Figure 30-36.

FIGURE 30-36

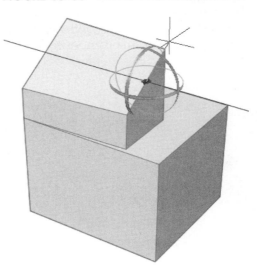

Face Grips

A face grip of a primitive solid or of a composite solid can be selected using the same procedure as explained earlier (using the Ctrl key). Faces of solids are usually planar but can include some curved faces. Grips that appear on faces of solids appear as a circle on the plane of the face as shown on the vertical front face of the *Wedge* primitive in Figure 30-37.

FIGURE 30-37

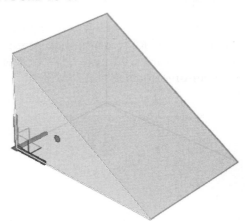

One of the most useful applications of using face grips is to STRETCH a face perpendicular to the plane of the face, similar to the way a shape would be *Extruded*. For example, using the move grip tool, the vertical face could be STRETCHed to change the depth of the *Wedge* as shown in Figure 30-38.

FIGURE 30-38

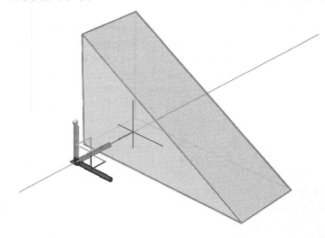

Another application is shown in Figure 30-39. In this case, the right face of a Pyramid (with a top radius) was selected. The illustration shows the face being STRETCHed using the move grip tool.

FIGURE 30-39

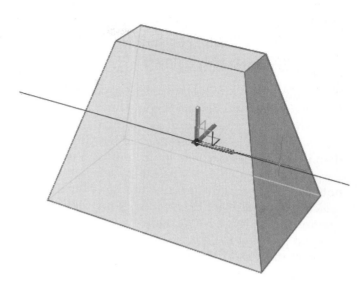

Vertex Grips

A vertex grip can be selected by using Ctrl to select any vertex of a solid. For example the upper-right corner vertex has been selected on the *Box* primitive shown in Figure 30-40.

FIGURE 30-40

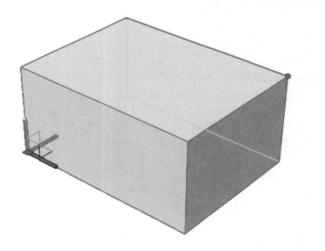

A vertex grip can be STRETCHed as shown in Figure 30-41. Only the selected vertex is affected along with any connected faces and edges. Interestingly, only the STRETCH can be used effectively with vertex grips. Since a vertex represents one point (with no length or dimension), it cannot be scaled or rotated. If the MOVE or MIRROR option is used, the entire solid is affected.

NOTE: Keep in mind that multiple grips can be selected. For example, you may want to ROTATE two faces together. Accomplish this by holding the Ctrl key down while selecting each face. Additionally, any combination of grip types can be selected. For example, you could select one face grip, one edge grip, and one vertex grip.

FIGURE 30-41

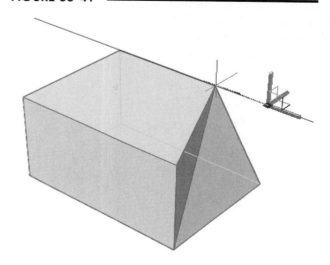

Selecting Overlapping Subobjects

In some cases the subobject you want to select may be behind, or overlapping, another subobject, and therefore, difficult to PICK. When you "hover" over composite solids while holding Ctrl, the solid in the foreground is detected first. For example, the location of the pickbox shown previously in Figure 30-33 would select the top *Box* primitive by default.

FIGURE 30-42 ─────────

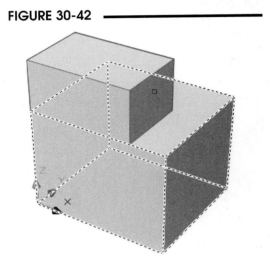

When subobjects are overlapping and selection preview is turned on (hovered objects become highlighted), you can cycle through the subobjects by rolling over the subobject on top to highlight it, and pressing and holding the Ctrl key and then pressing the spacebar continuously. When the required subobject is highlighted, left-click to select it. For example, note the location of the pickbox in Figure 30-42. Making this selection would result in selection of the top *Box*, as shown previously in Figure 30-30. However, holding down the Ctrl key and pressing the spacebar would cycle to select the large *Box* primitive under the small *Box* as shown in Figure 30-42. If selection preview is turned off, and more than one subobject is under the pickbox, you can cycle through the subobjects until the correct one is selected by pressing and holding Ctrl+spacebar and then left-clicking.

History Settings

Solid History

By default, a composite solid's *History* is set to *Record*. This default setting allows you to select individual primitives that make up the composite solid, as described in the previous sections. Using this scheme, selecting a second level subobject (a face, edge, or vertex within a primitive) requires a second Ctrl and PICK.

However, if you always want to select faces, edges, and vertices within composite solids without first selecting the individual primitive, you can change the solid's *History* setting to *None* using the *Properties* palette (Fig. 30-43). With this setting, you can select any face, edge, or vertex without having to select the primitive first. For example, you could directly select the top face on the large *Box* primitive without having to first select the *Box* itself. Remember that using the *Properties* palette allows you to change properties for the selected object(s) only.

FIGURE 30-43 ─────────

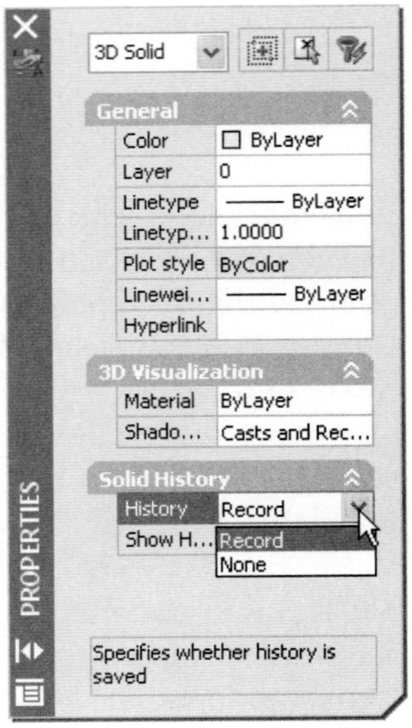

For example, assume the *History* setting for the composite solid shown in Figure 30-44 was changed to *None*. Selection of the L-shaped face would be possible only with this setting. Otherwise, with a *History* setting of *Record*, the entire faces of the *Box* primitives would be selected and highlighted.

NOTE: Once *History* is set to *None*, the history of the composite solid's construction is deleted and cannot be retrieved. Setting the *History* back to *Record* begins a new record.

SOLIDHIST

The *SOLIDHIST* system variable sets the history for <u>all</u> solids, both new and existing. This setting overrides the *History* setting in an individual solid's *Properties* palette. When *SOLIDHIST* is set to the default of 1, composite solids retain the history of the original objects contained in the composite.

0   Sets the *History* property to *None* for all solids. No history is retained.
1   Sets the *History* property to *Record* for all solids. All solids retain a history of their original objects.

Show History

A composite solid is created by using multiple primitives combined by Boolean operations. For example, assume a composite solid was created by using *Subtract* to remove a small *Box* primitive from a large *Box* primitive as shown in Figure 30-45. Normally, all of the individual primitives are not visible unless you select them as subobjects. In our example, the small *Box* primitive is not visible.

If a composite solid's *Show History* property is set to *Yes* in the *Properties* palette, the individual primitives that make up the composite solid can be viewed. To do this, select the composite solid and change its *Show History* setting to *Yes* as shown in Figure 30-46.

FIGURE 30-44

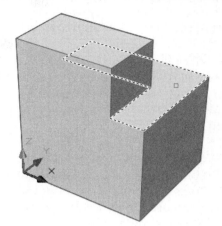

FIGURE 30-45

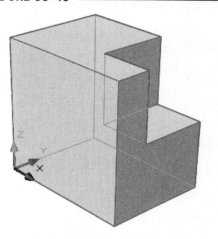

FIGURE 30-46

When a composite solid's *Show History* setting is set to *Yes*, the individual primitives that were used to create the solid are visible at all times as shown in Figure 30-47.

FIGURE 30-47

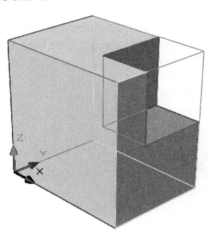

SHOWHIST

The *SHOWHIST* system variable controls the *Show History* property for <u>all</u> solids in a drawing. Depending on the setting, an individual solid's *Show History* setting in the *Properties* palette can be overridden.

0   Sets the *Show History* property to *No* (read-only) for all solids. Overrides the individual *Show History* property settings for solids. You cannot view the original objects that were used to create the solid.

1   Does not override the individual *Show History* property settings for solids.

2   Displays the history of all solids by overriding the individual *Show History* property settings for solids. You can view the original objects that were used to create the solid.

SOLIDS EDITING COMMANDS

In addition to using grips to edit composite solids, AutoCAD offers several commands that allow you to change the geometry.

Solidedit   This command has options for extruding, moving, rotating, offsetting, tapering, copying, coloring, separating, shelling, cleaning, checking, and deleting faces and edges of 3D solids.

Slice   Use this command to cut through a solid using a plane or a surface. You can keep both parts of the solid or delete part.

Presspull   This unique command combines creating an extrusion and performing a Boolean all in one command. The extrusion is created on a solid surface and then pushed into the solid's face (*Subtract*) or pulled out from the face (*Union*).

Sectionplane   Use this command to create an imaginary variable cutting plane, similar to a clipping plane, to view the inside of a solid.

Interfere   *Interfere* checks two or more solids to see if they overlap in 3D space. You can create a new solid from an overlapping volume.

Imprint   Imprint creates an "edge" element on a solid that can be used for editing with grips.

Although most of the solid modeling commands are available using the *3D Make control panel* in the upper-right corner of the Drawing Editor, some of the editing commands are not and must be invoked through another menu or by typing.

For example, all options of the *Solidedit* command are accessible only from the *Solids Editing* cascading menu from the *Modify* pull-down menu (Fig. 30-48).

FIGURE 30-48

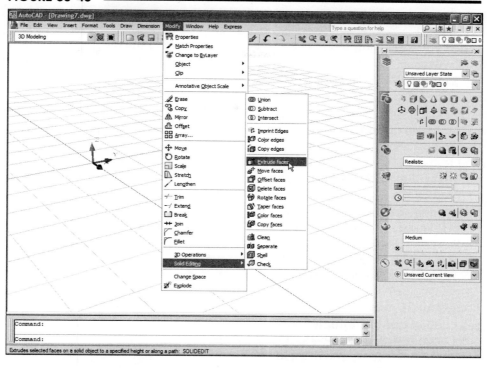

Solidedit

Pull-down Menu	Command (Type)	Alias (Type)	Short-cut	Screen (side) Menu	Tablet Menu
Modify *Solids Editing >*	*Solidedit*	...	...	...	...

The *Solidedit* command has several "levels" of options. The first level prompts you to specify the type of geometry you want to edit—*Face*, *Edge*, or *Body*. The second level of options depends on your first level response as shown below.

```
Command: solidedit
Solids editing automatic checking:  SOLIDCHECK=1
Enter a solids editing option [Face/Edge/Body/Undo/eXit] <eXit>: f
Enter a face editing option
[Extrude/Move/Rotate/Offset/Taper/Delete/Copy/coLor/Undo/eXit] <eXit>:
```

Solidedit operates on three types of geometry: *Edges, Faces* and *Bodies*. An *Edge* is defined as the common edge between two surfaces that has the appearance of a line, arc, circle, or spline (Fig. 30-49 A). *Faces* are planar or curved surfaces of a 3D object (Fig. 30-49 B). A *Body* is defined as an existing 3D solid or a non-solid shape created with a *Solidedit* option.

<u>Object selection is an integral part of using</u> <u>*Solidedit*</u>. For example, once you have specified the type of geometry and the particular editing option, the *Solidedit* command prompts to select *Edges, Faces,* or a *Body*.

FIGURE 30-49

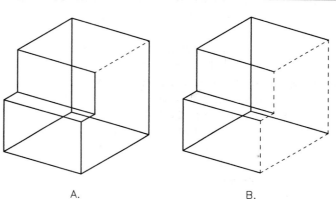

A.   B.

Select faces or [Undo/Remove]: Select a face or enter an option
Select faces or [Undo/Remove/ALL]: Select a face or enter an option

When you are prompted to select *Faces* or *Edges*, select the desired face or edge directly with the pickbox. You will most likely go through a series of adding and removing geometry until the desired set of lines is highlighted. Remember that you can hold down the Shift key and select objects to *Remove* them from the selection set.

If you use this command often, it may help to change the composite solid's *History* setting to *None* using the *Properties* palette (see previous discussion, "History Settings"). Setting the *History* to *None* allows you to use the Ctrl key while hovering over solids to select individual faces and edges rather than the entire primitives that make up the solid.

 TIP Ensure that you highlight exactly the intended geometry before you proceed with editing. If an incorrect set is selected, you can get unexpected results or an error message can appear such as that below.

Modeling Operation Error:
    No solution for an edge.

For example, many of the *Face* options allow selection of one or multiple faces. In Figure 30-50 both selection sets are valid faces or face combinations, but each will yield different results. Figure 38-50 A indicates selection of one face and B indicates selection of several faces defining an entire primitive. Add and remove faces or edges until you achieve the desired geometry.

FIGURE 30-50

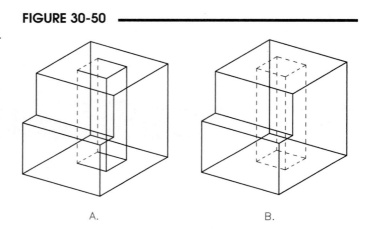

A.   B.

Face Options

The *Face* options edit existing surfaces and create new surfaces. A *Face* is a planar or curved surface. *Faces* that are part of existing 3D solids can be altered to change the configuration of the 3D composite solid. Individual *Faces* can be edited and multiple *Faces* comprising a primitive can be edited. New surfaces can be created from existing *Faces*, but entirely new independent solids cannot be created using these editing tools.

Extrude

 The *Extrude* option of *Solidedit* allows you to extrude any *Face* of a 3D object in a similar manner to using the *Extrude* command to create a 3D solid from a *Pline*. This capability is extremely helpful if you need to make a surface on a 3D solid taller, shorter, or longer. You can select one or more faces to extrude at one time.

Command: *solidedit*
Solids editing automatic checking: SOLIDCHECK=1
Enter a solids editing option [Face/Edge/Body/Undo/eXit] <eXit>: *f*
Enter a face editing option
[Extrude/Move/Rotate/Offset/Taper/Delete/Copy/coLor/Undo/eXit] <eXit>: *e*
Select faces or [Undo/Remove]: **PICK**

Select faces or [Undo/Remove/ALL]: **PICK** or remove
Select faces or [Undo/Remove/ALL]: **Enter**
Specify height of extrusion or [Path]: **(value)**
Specify angle of taper for extrusion <0>: **Enter** or **(value)**
Solid validation started.
Solid validation completed.

For example, Figure 30-51 illustrates the extrusion of a face to make the 3D object taller, where A indicates the selected (highlighted) face and B shows the result. This extrusion has a 0 degree taper angle.

FIGURE 30-51

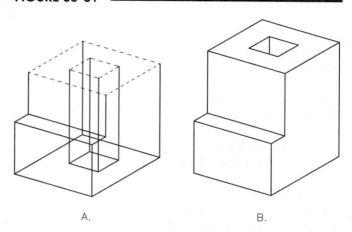

A.   B.

Specifying a positive angle tapers the face inward (Fig. 30-52 A) and specifying a negative angle tapers the face outward (B) *Height* is always perpendicular to the selected *Face*, not necessarily vertical.

FIGURE 30-52

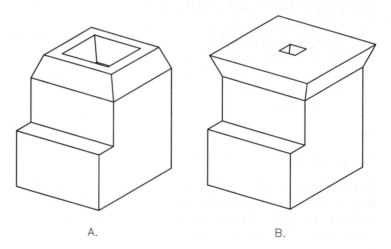

A.   B.

An internal face can be selected for extrusion. In Figure 30-53, a single face is selected (A) and the *Height* value specified is greater than the distance to the outer face of the solid, creating an open side on the object (B). *Extrude* allows an internal face to pass through an external face as shown in Figure 30-53 B. Several other options, such as *Move, Rotate,* and *Offset* also allow the interior primitives to pass through the bounding box.

FIGURE 30-53

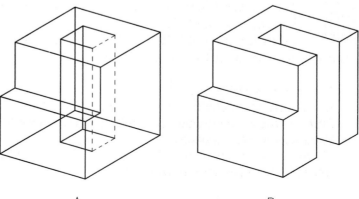

A.   B.

Faces can be extruded along a *Path*. The *Path* object can be a *Line*, *Circle*, *Arc*, *Ellipse*, *Pline*, or *Spline*. For example, Figure 30-54 illustrates extrusion of a face along a *Line* path (A) to yield the result shown on the right (B).

FIGURE 30-54

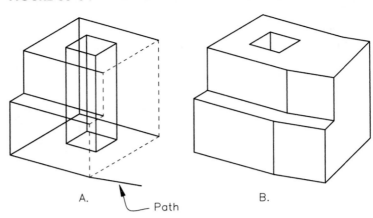

A.   Path   B.

 NOTE: Entering a <u>positive</u> value at the "Specify height of extrusion or [Path]:" prompt <u>increases</u> the volume of the solid and entering a <u>negative</u> value <u>decreases</u> the volume of the solid. For example, entering a positive value for extruding the face in Figure 30-53 A would result in a smaller hole.

Move

 With the *Move* option, you can move a single face of a 3D solid or move an entire primitive within a 3D solid. This capability is particularly helpful during geometry construction or when there is a design change and it is required to alter the location of a hole or other feature within the confines of the composite 3D model.

```
[Extrude/Move/Rotate/Offset/Taper/Delete/Copy/coLor/Undo/eXit] <eXit>: m
Select faces or [Undo/Remove]: PICK
Select faces or [Undo/Remove/ALL]: PICK  or remove
Select faces or [Undo/Remove/ALL]: Enter
Specify a base point or displacement: PICK
Specify a second point of displacement: PICK
```

For example, Figure 30-55 illustrates using the *Move* option to relocate a hole primitive (four faces) within a composite solid. In this case you must ensure all faces, and only the faces, comprising the primitive are selected.

FIGURE 30-55

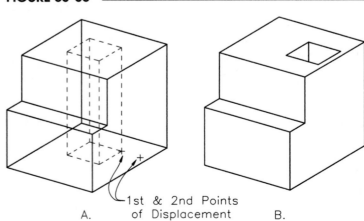

A.   1st & 2nd Points of Displacement   B.

With the *Move* option the internal features <u>can</u> be moved to extend into or past the "bounding box" of the composite solid, as they can with *Extrude* (see Fig. 30-53).

Another possibility for *Move* is to move only one or selected faces to alter the configuration of an internal feature of a composite solid. An example is shown in Figure 30-56, where only one face is selected to move. Compare the results in Figures 30-55 B and 30-56 B.

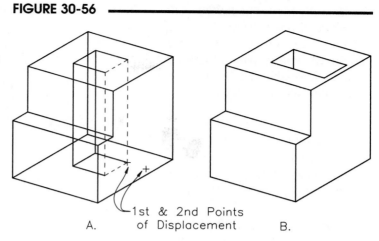

FIGURE 30-56

1st & 2nd Points of Displacement

A.   B.

Keep in mind that *Move* can be used to move a singular external face, achieving the same result as *Extrude* (see previous Fig. 30-51).

Offset

The 2D *Offset* command makes a parallel copy of a *Line, Arc, Circle, Pline,* etc. The *Offset* option of *Solidedit* makes a 3D offset. A simple application is making a hole larger or smaller.

```
[Extrude/Move/Rotate/Offset/Taper/Delete/Copy/coLor/Undo/eXit] <eXit>: O
Select faces or [Undo/Remove]: PICK
Select faces or [Undo/Remove/ALL]: PICK  or remove
Select faces or [Undo/Remove/ALL]: Enter
Specify the offset distance: PICK or (value)
```

You can offset through a point or specify an offset distance. Specify a positive value to increase the volume of the solid or a negative value to decrease the volume of the solid.

Figure 30-57 demonstrates how an internal feature, such as a hole, can be *Offset* to effectively change the volume of the hole. Since positive values increase the size of the solid, a negative value was used here. In other words, holes inside a solid become smaller when the solid is *Offset* larger.

FIGURE 30-57

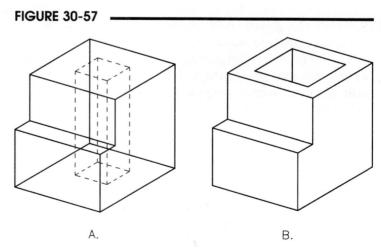

A.   B.

Delete

This option of *Solidedit* deletes faces from composite solids. Although you cannot delete one planar face from a solid, you can delete one curved face, such as a cylinder, that comprises an entire primitive. You can also delete multiple faces that comprise a primitive or entire feature.

[Extrude/Move/Rotate/Offset/Taper/Delete/Copy/coLor/Undo/eXit] <eXit>: **d**
Select faces or [Undo/Remove]: **PICK**
Select faces or [Undo/Remove/ALL]: **PICK** or remove
Select faces or [Undo/Remove/ALL]: **Enter**
Solid validation started.
Solid validation completed.

As an example, Figure 30-58 A displays two selected faces to be deleted from a composite solid. The result is displayed in B. In this case, <u>both faces must be highlighted</u> for the deletion to operate.

For the same composite solid, if all <u>four</u> internal faces comprising the hole are selected, as shown in previous Figure 30-57 A, the hole could be deleted.

FIGURE 36-58

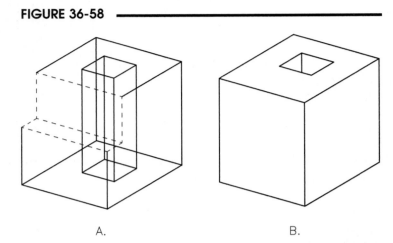

A.   B.

Rotate

Rotate is helpful during geometry construction and for design changes when components within a composite solid must be rotated in some way.

[Extrude/Move/Rotate/Offset/Taper/Delete/Copy/coLor/Undo/eXit] <eXit>: **r**
Select faces or [Undo/Remove]: **PICK**
Select faces or [Undo/Remove/ALL]: **PICK** or remove
Select faces or [Undo/Remove/ALL]: **Enter**
Specify an axis point or [Axis by object/View/Xaxis/Yaxis/Zaxis] <2points>: **PICK**
Specify the second point on the rotation axis: **PICK**
Specify a rotation angle or [Reference]: **PICK** or (**value**)

Several methods of rotation are possible based on the selected axis. These options are essentially the same as those available with the *3DRotate* command (see "*3DRotate*" discussed previously).

Taper

Taper angles a face. You can use *Taper* to change the angle of planar or curved faces (Fig. 30-59, on the next page).

[Extrude/Move/Rotate/Offset/Taper/Delete/Copy/coLor/Undo/eXit] <eXit>: **t**
Select faces or [Undo/Remove]: **PICK**
Select faces or [Undo/Remove/ALL]: **PICK** or remove
Select faces or [Undo/Remove/ALL]: **Enter**
Specify the base point: **PICK**
Specify another point along the axis of tapering: **PICK**
Specify the taper angle: **Enter** or (**value**)

The rotation of the taper angle is determined by the order of selection of the base point and second point. Although AutoCAD prompts for the "axis of tapering," the two points actually determine a reference line from which the taper angle is applied. The taper starts at the first base point and tapers away from the second base point. The rotational axis passes through the first base point and is <u>perpendicular</u> to the line between the base points. <u>The line between the base points does not specify the rotational axis, as implied</u>.

FIGURE 30-59

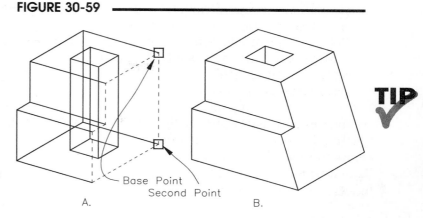

Base Point
Second Point

A.   B.

To explain this idea, Figure 30-59 A illustrates the selection of a planar face, the base point, and the second point. An angle of –20 degrees (negative value) is entered to achieve the results shown in B. Positive angles taper the face inward (toward the solid) and negative values taper the face outward (away from the solid).

For some cases, either *Taper* or *Rotate* could be used to achieve the same results. For example, both Figure 30-59 B could be attained using *Taper* or *Rotate* (although different points must be selected depending on the option used).

Copy

As you would expect, the *Copy* option copies *Faces*. However, the resulting objects are *Regions* or *Bodies*.

```
[Extrude/Move/Rotate/Offset/Taper/Delete/Copy/coLor/Undo/eXit] <eXit>: c
Select faces or [Undo/Remove]: PICK
Select faces or [Undo/Remove/ALL]: PICK  or remove
Select faces or [Undo/Remove/ALL]: Enter
Specify the base point or displacement: PICK  or (value)
Specify a second point of displacement: PICK
```

Although any *Face* can be selected (planar or curved), Figure 30-60 shows how *Copy* can be used to create a *Region* from a planar face. Typically, the *Copy* option would be used to create a new *Region* from existing 3D solid geometry. The *Extrude* command could then be used on the *Region* to form a new 3D solid from the original face.

FIGURE 30-60

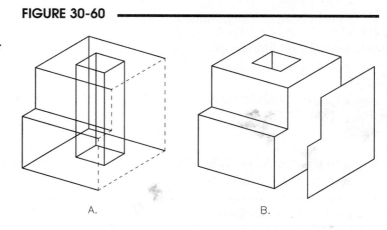

A.   B.

If a curved *Face* is selected to *Copy*, the resulting object is a *Body*. This type of body is a curved surface, not a solid.

Using the same composite model as in previous figures, assume you were to *Copy* only the five vertical faces indicated in Figure 30-61 A. After selecting the second point of displacement (B), notice only those five vertical surfaces result. Not included are the faces defining the hole or any horizontal faces. Therefore, you can select *All* faces comprising a composite solid to *Copy*, and the resulting model is a surface model composed of *Regions*.

FIGURE 30-61

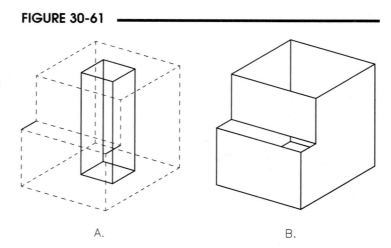

A.

B.

Edge Options

Edges are lines or curves that define the boundary between *Faces*. Copying an *Edge* would result in a 2D object. The selection process for *Edges* is critical, but not as involved as selecting *Faces*. You can use multiple selection methods (the same options as with *Edges*), and selection may involve a process of adding and removing objects. As with *Face* selection, make sure you have the exact desired set highlighted before continuing with the procedure. The command syntax is as follows.

```
Command: solidedit
Solids editing automatic checking:  SOLIDCHECK=1
Enter a solids editing option [Face/Edge/Body/Undo/eXit] <eXit>: edge
Enter an edge editing option [Copy/coLor/Undo/eXit] <eXit>:
```

Copy

The *Copy* option of *Edge* creates wireframe elements, not surfaces or solids. Copied edges become 2D objects such as a *Line, Arc, Circle, Ellipse,* or *Spline*. These elements could be used to create other 3D models such as wireframes, surfaces, or solids.

```
Enter an edge editing option [Copy/coLor/Undo/eXit] <eXit>: c
Select edges or [Undo/Remove]: PICK
Select edges or [Undo/Remove]: PICK  or remove
Select edges or [Undo/Remove]: Enter
Specify a base point or displacement: PICK
Specify a second point of displacement: PICK
```

One or multiple *Edges* can be selected. After indicating the second point of displacement, the selected edges are extracted from the solid and copied to the new location (Fig. 30-62).

Keep in mind that the *Copied Edges* are wireframe elements. These new elements can be used for construction of other 3D geometry, such as with the *Body, Imprint* option (see "*Body* Options" later in this section).

FIGURE 30-62

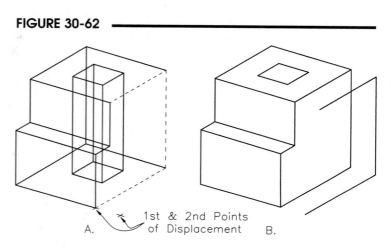

A.

1st & 2nd Points of Displacement

B.

Body Options

A *Body* is typically a 3D solid. Any primitive or composite solid can be selected as a *Body*. However, some *Solidedit* options can create a *Body* that is a non-solid. For example, selecting a curved *Face* and using *Copy* creates a curved surface that AutoCAD lists as a *Body*. Several options are offered here to alter the configuration of a *Body* or to create or edit 2D geometry used to interact with a *Body*.

Imprint

Imprint is used to attach a 2D object on an existing face of a solid (*Body*). The new *Imprint* can then be used with *Extrusion* to create a new 3D solid.

Objects that can be used to make the *Imprint* can be an *Arc, Circle, Line, Pline, Ellipse, Spline, Region,* or *Solid*. The selected object to *Imprint* must touch or intersect the solid in some way, such as when a *Circle* lies partially on a *Face,* or when two solid volumes overlap. When *Imprint* is used, one or more 2D components becomes attached, or imprinted, to an existing 3D solid face. The resulting *Imprint* has little usefulness of itself, but is an intermediate step to creating a new 3D solid shape on the existing solid.

```
[Imprint/sePaparate solids/Shell/cLean/Check/Undo/eXit] <eXit>: i
Select a 3D solid: PICK
Select an object to imprint: PICK
Delete the source object <N>: y
Select an object to imprint: Enter
```

For example, Figure 30-63 A displays a 3D solid with a *Circle* that is on the same plane as the vertical right face of the solid. The *Imprint* option is used, the solid is selected, and the *Circle* is selected as the "object to imprint." Answering "Y" to "Delete the source object," the resulting *Imprint* is shown in B.

FIGURE 30-63

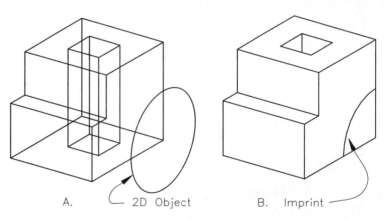

A.   2D Object   B.   Imprint

The *Imprint* creates a separate *Face* on the object—in this case, a total of two coplanar *Faces* are on the same vertical surface of the object. The new *Face* can be treated independently for use with other *Solidedit* options. For example, *Extrude* could be used to create a new solid feature on the composite solid (Fig. 30-64).

FIGURE 30-64

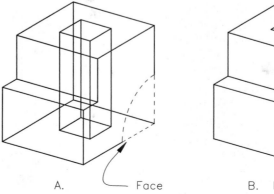
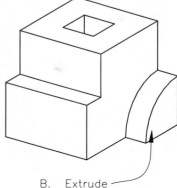

A.   Face   B.   Extrude

Clean

 The *Clean* option deletes any 2D geometry on the solid. For example, you may use *Imprint* to "attach" 2D geometry to a *Body* (solid) for the intention of using *Extrude*. *Imprinted* objects become permanently attached to the *Body* and cannot be removed by *Erase*. If you need to remove an *Imprint*, use the *Clean* option to do so. One use of *Clean* removes all *Imprints*.

    [Imprint/seParate solids/Shell/cLean/Check/Undo/eXit] <eXit>: **l**
    Select a 3D solid: **PICK**

Separate

 It is possible in AutoCAD to *Union* two solids that do not occupy the same physical space and have two distinct volumes. In that case, the two shapes appear to be separate, but are treated by AutoCAD as one object (if you select one, both become highlighted).

 Occasionally you may intentionally or inadvertently use *Union* to combine two or more separate (not physically touching) objects. *Shell* and *Offset*, when used with solids containing holes, can also create two or more volumes from one solid. *Separate* can then be used to disconnect these into discrete independent solids. *Separate* <u>cannot</u> be used to disconnect or "break down" *Unioned*, *Subtracted*, or *Intersected* solids that form one volume.

    [Imprint/seParate solids/Shell/cLean/Check/Undo/eXit] <eXit>: **p**
    Select a 3D solid: **PICK**

Shell

 Shell converts a solid into a thin-walled "shell." You first select a solid to shell, then specify faces to remove, and enter an offset distance. An example would be to convert a solid cube into a hollow box—the thickness of the walls equal the "shell offset distance." If no faces are removed from the selection set, the box would have no openings. Faces that are removed become open sides of the box.

    [Imprint/seParate solids/Shell/cLean/Check/Undo/eXit] <eXit>: **s**
    Select a 3D solid: **PICK**
    Remove faces or [Undo/Add/ALL]: **PICK**
    Remove faces or [Undo/Add/ALL]: **PICK**
    Remove faces or [Undo/Add/ALL]: **Enter**
    Enter the shell offset distance: (**value**)
    Solid validation started.
    Solid validation completed.

 A positive value entered at the "shell offset distance" prompt creates the wall thickness outside of the original solid, whereas a negative value creates the wall thickness inside the existing boundary.

An example is shown in Figure 30-65. The left object (A) indicates the solid with all the selected faces highlighted and the faces removed from the selection set not highlighted. A negative offset distance is used to yield the shape shown on the right (B).

FIGURE 30-65

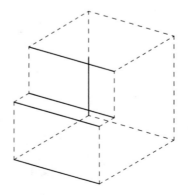

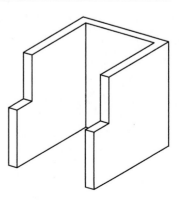

A.

B.

Check

Use this option to validate the 3D solid object as a valid ACIS solid, independent of the *SOLID-CHECK* system variable setting.

```
[Imprint/seParate solids/Shell/cLean/Check/Undo/eXit] <eXit>: c
Select a 3D solid: PICK
This object is a valid ACIS solid.
```

The *SOLIDCHECK* variable turns the automatic solid validation on and off for the current AutoCAD session. By default, *SOLIDCHECK* is set to 1, or on, to validate the solid. The *Solidedit* command displays the current status of the variable as shown in the command syntax below.

```
Command: solidedit
Solids editing automatic checking:  SOLIDCHECK=0
Enter a solids editing option [Face/Edge/Body/Undo/eXit] <eXit>
```

Slice

Pull-down Menu	Command (Type)	Alias (Type)	Short-cut	Screen (side) Menu	Tablet Menu
Modify 3D Operations > Slice	Slice	SL	...	DRAW 2 SOLIDS Slice	...

The *Slice* command is a separate command (not part of the *Solidedit* options). However, *Slice* is a useful command for solids construction and editing for special cases and is therefore discussed in this chapter for those purposes.

FIGURE 30-66

A solid that contains oblique surfaces is a likely candidate for using *Slice*. For example, the simplest method of construction for the solid shown in Figure 30-66 is to create the rectilinear shapes (using *Box*), then use *Slice* to create the oblique surfaces.

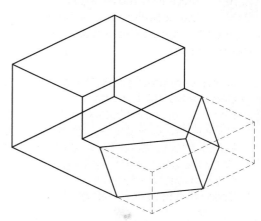

With the *Slice* command you can cut a solid into two "halves" by defining a plane in any orientation. You can keep both sides of the resulting solid(s) or PICK one side to keep. (If in doubt as to which side to keep or how to select it, keep both sides, then use *Erase* to remove the unwanted side.)

```
Command: slice
Select objects to slice: PICK
Select objects to slice: Enter
Specify start point of slicing plane or [planar Object/Surface/Zaxis/View/XY/YZ/ZX/3points] <3points>:
(option specific prompts)
Specify a point on desired side or [keep Both sides] <Both>: PICK or Enter
Command:
```

One oblique surface for the solid (Fig. 30-67) is created by first drawing a 2D *Line* on the solid, then defining a *3point* slicing plane using *OSNAPs* (*Endpoint*, *Endpoint* of *Line*, then *Midpoint* of solid edge). This figure shows the resulting solids after using *Slice* and keeping both sides. Duplicating this procedure for the other oblique surface creates the solid shown previously in Figure 30-66.

FIGURE 30-67

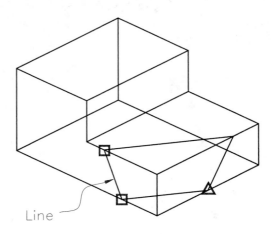

Line

TIP

Slice can be used for many applications, such as an alternative to using *Wedge*, *Chamfer*, or *Solidedit*, *Taper*. Additionally, if you wanted to cut a section of a solid away, such as creating a notch or a half section (cutting away one quarter of an object), consider making multiple slicing planes with *Slice*, keeping all sides, *Erase* the unwanted section, then using *Union* to join the sections back into one solid.

Slice can also be used to create a true solid section. *Slice* cuts a solid or set of solids on a specified cutting plane. The original solid is converted to two solids. You have the option to keep both halves or only the half that you specify. Examine the object in Figure 30-68.

FIGURE 30-68

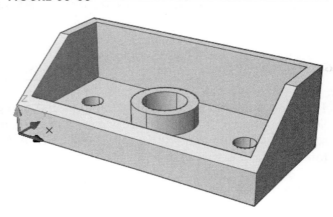

To create a true section of the example, use *Slice*. The *ZX* method is used to define the cutting plane midway through this solid. Next, the new solid to retain was specified by PICKing a point on that geometry. The resulting sectioned solid is shown in Figure 30-69. Note that the solid on the near side of the cutting plane was not retained.

FIGURE 30-69

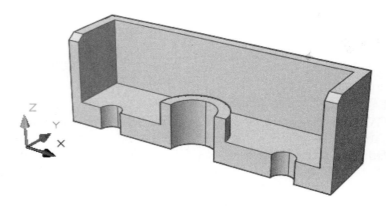

Entering **B** at the "keep both sides" prompt retains the solids on both sides of the cutting plane. Otherwise, you can pick a point on either side of the plane to specify which half to <u>keep</u>.

Sectionplane

Pull-down Menu	Command (Type)	Alias (Type)	Short-cut	Screen (side) Menu	Tablet Menu
Draw Modeling > Section Plane	*Sectionplane*	*SPANE*	...	...	...

Sectionplane creates a translucent clipping plane that can pass through a solid to reveal the inside features of the solid. The section plane is actually a section <u>object</u> that can be moved or revolved to dynamically display the solid's interior features. The section plane can contain "jogs" to create an offset cutting plane. The section object is manipulated using grips that appear on the section plane object.

Command: **sectionplane**
Select a face or any point to locate section line or [Draw section/Orthographic]: **PICK**
Specify through point: **PICK**

You can select a face to create a section plane or specify two points to create the ends of the section plane. Selecting a face on a solid aligns the section object parallel to that face. For example, selecting the right face at the first prompt would create the section plane shown in Figure 30-70.

FIGURE 30-70

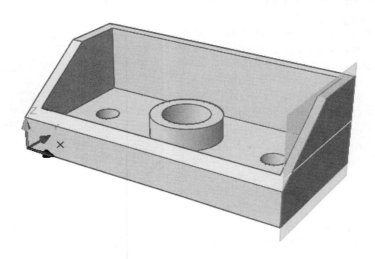

Selecting any point that is not on a face creates the first point of a section object independent of the solid. The second point creates the section object. For example, selecting two points just to the right and left of the composite solid shown in Figure 30-71 would produce a section plane crossing through the center of the solid. Note that the section plane is created perpendicular to the XY plane.

FIGURE 30-71

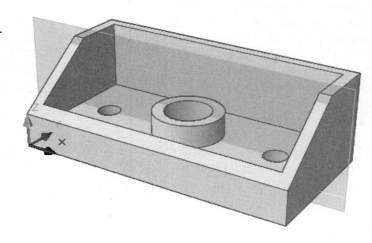

Section Plane Grips

Note that selecting the section plane object makes its grips appear (Fig. 30-72). Grips can be used to rotate the section plane to a different orientation, change the length of the section plane, flip the section plane, and drag the section plane into a new position. For example, Figure 30-72 displays the direction grip being dragged through the solid. PICKing the point at the center of the hole would move the section plane to that location, and the object would appear cut in half as shown in the figure.

FIGURE 30-72

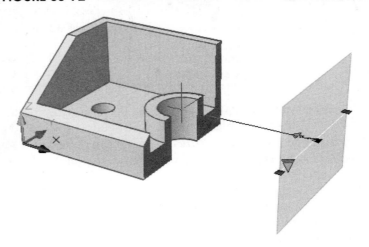

Once one of the square grips at the end of the section plane object is made hot, the normal five grip options appear (STRETCH, MOVE, ROTATE, SCALE, and MIRROR). For example, you can rotate the section plane by making a square grip hot, then cycling to the ROTATE option as shown in Figure 30-73.

FIGURE 30-73

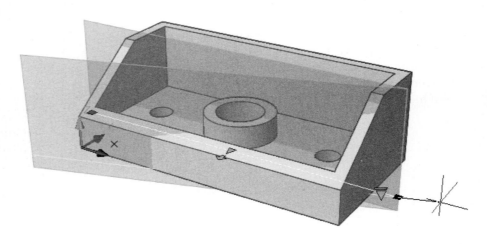

Live Sectioning

To make solid geometry on one side of the section plane appear invisible, the *Live Section* property for the section plane object must be set to *Yes*. When a section plane is created using the "face" option, *Live Section* property is set to *Yes* by default as shown previously in Figure 30-72. However, when you create a section plane by selecting points, *Live Section* is set to *No*. For example, note that in Figure 30-71 the section plane passes through the center of the solid, yet both sides of the plane are visible.

FIGURE 30-74

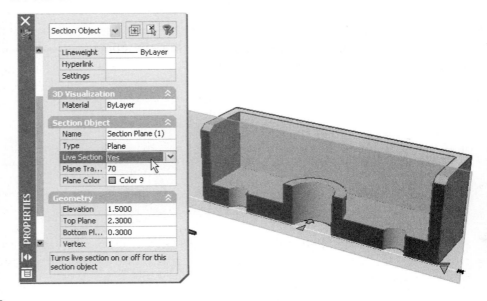

Using the *Properties* palette to set the *Live Section* to *Yes* makes the solid geometry on one side of the plane invisible as shown in Figure 30-73. The "flip" arrow grip determines which side of the section plane is invisible.

Draw Section

The *Draw section* option allows you to create an offset section plane. Offset section planes are used when "jogs" in the section line are needed to pass through multiple features of the solid as shown in Figure 30-75.

FIGURE 30-75

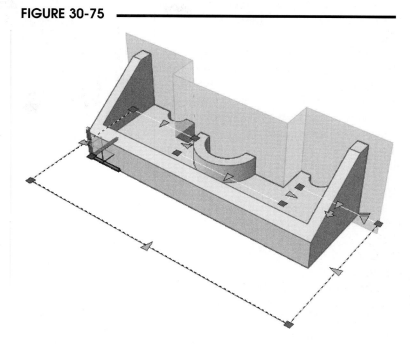

> Command: **sectionplane**
> Select face or any point to locate section line or [Draw section/Orthographic]: **d**
> Specify start point: **PICK**
> Specify next point: **PICK**
> Specify next point or ENTER to complete: **Enter**
> Specify next point in direction of section view: **PICK**

This option creates a section object with live sectioning turned off. Note that the section boundary is turned on by default (see "Section Object States"). If you select the section plane object so the grips appear, you can change several features of the section plane object. Stretch grips for each section of the section plane object allow you to change the location for the jogged sections (Fig. 30-75).

Orthographic

This option is very useful for creating typical sections. *Orthographic* aligns the section object to a specific orthographic orientation relative to the UCS. The resulting section contains all 3D objects in the model. If a single composite solid is located correctly with respect to the USC (lower-left corner at 0,0,0 and front side along the X axis), AutoCAD uses the center of volume of the solid (or multiple solids) to create the new section plane object.

FIGURE 30-76

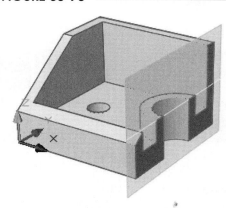

> Command: **sectionplane**
> Select face or any point to locate section line or [Draw section/Orthographic]: **o**
> Align section to: [Front/Back/Top/Bottom/Left/Right]:

For example, using the *Right* option creates the section plane shown in Figure 30-76. Using the *Front* option produces a section plane as shown in Figure 30-74. Note that this option creates a section object with *Live Sectioning* turned on.

Section Object States

Section objects have three states: *Section Plane* (a 2D plane), *Section Boundary* (a 2D box), and *Section Volume* (a 3D box). Grips that appear for each state allow you to make adjustments to the length, width, and height of the cutting area. Solids inside the cutting area are affected by the section plane object, whereas solids outside the area are not affected. Change the section state by selecting the state grip as shown in Figure 30-77.

FIGURE 30-77

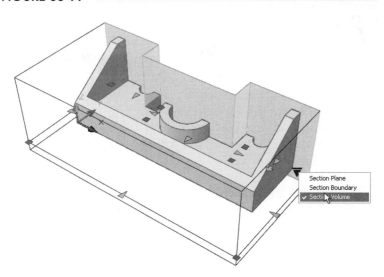

Section Plane	The cutting plane extends infinitely in all directions. Only the section line and transparent section plane indicator are displayed (Fig. 30-76).
Section Boundary	The cutting plane along the Z axis extends infinitely. A 2D box shows the XY extents of the cutting plane (Fig. 30-75).
Section Volume	A 3D box shows the extents of the cutting plane (Fig. 30-77).

USING SOLIDS DATA

AutoCAD offers a few commands that allow you to extract, import, and export data from solid model geometry. The *Massprop* command calculates a variety of properties for selected solids. Commands available for exporting and importing solid model geometric data for use with other programs are *Stlout*, *Acisin*, and *Acisout*.

Massprop

Pull-down Menu	Command (Type)	Alias (Type)	Short-cut	Screen (side) Menu	Tablet Menu
Tools *Inquiry >* *Region, Mass Properties*	*Massprop*	...	...	*TOOLS 1* *Massprop*	*U,7*

Since solid models define a complete description of the geometry, they are ideal for mass properties analysis. The *Massprop* command automatically computes a variety of mass properties.

Mass properties are useful for a variety of applications. The data generated by the *Massprop* command can be saved to an .MPR file for future exportation in order to develop bills of material, stress analysis, kinematics studies, and dynamics analysis (Fig. 30-78).

FIGURE 30-78

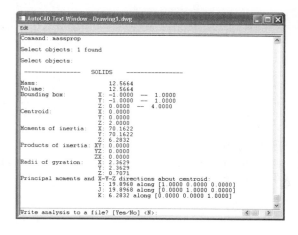

CHAPTER EXERCISES

1. **Editing Primitives using** *Properties*

FIGURE 30-79

Open the **CH29EX3** drawing that you created in Chapter 29 Exercises. Invoke the *Properties* palette. Hold down the **Ctrl** key while hovering and selecting the hole (*Cylinder* primitive) so properties for the primitive appear in the palette (Fig. 30-79). Edit the hole to have a new *Radius* of **1.0000** and a *Height* of **1.0000**.

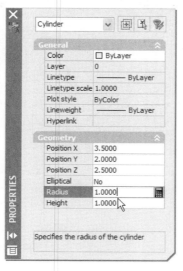
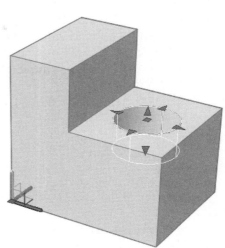
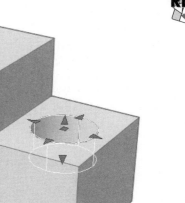

(Note that the "height" is the dimension in the negative Z direction in this case.) Use *SaveAs* and name the drawing **CH30EX1**.

2. **Editing Primitives using** *Properties*

FIGURE 30-80

If not already open, *Open* the **CH29EX1** drawing that you just created in the previous exercise. Invoke the *Properties* palette again. This time, hover and select the to *Box* primitive so its properties appear in the palette (Fig. 30-80). Edit the *Width* dimension of the *Box* to a value of **3.0000**. Use *SaveAs* and name the drawing **CH30EX2**.

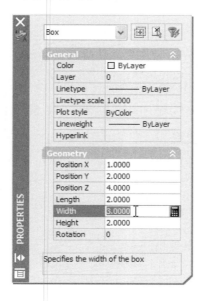
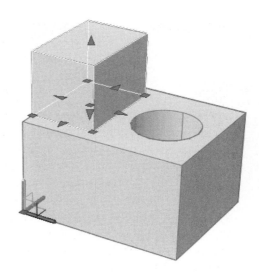

3.   **Editing Primitives using Grips**

Open the **CH29EX5** drawing that you created in Chapter 29 Exercises. Turn on *POLAR*. Hold down the **Ctrl** key while hovering and selecting the *Box* primitive. Select the grip in the center of the base on the right side to make it hot, then stretch the grip 3 units toward the center of the solid (Fig. 30-81). This action effectively "cuts" the solid in half. Use *SaveAs* and name the drawing **CH30EX3**.

FIGURE 30-81

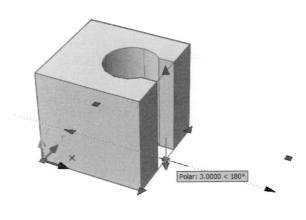

4.   **Editing Primitives using Grips**

Open the **CH29EX6** drawing that you created in Chapter 29 Exercises. Ensure *POLAR* is on. Hold down the **Ctrl** key while hovering and selecting the *Torus* primitive (Fig. 30-82). Select the appropriate grip to stretch the <u>tube</u> radius, not the torus radius. Enter a value of **.75** to increase the total tube radius (originally .50) to 1.25. Invoke the *Properties* palette to ensure the new *Tube Radius* is a value of **1.2500**. Use *SaveAs* and name the drawing **CH30EX4**.

FIGURE 30-82

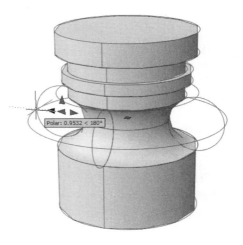

5.   **Editing Subobjects using the Move Grip Tool**

For this exercise, assume you are working as a designer of kitchen and bathroom fixtures. You have been contracted to design a kitchen faucet based on the style of a recently designed bathroom faucet. You can easily alter the existing bathroom design using the move grip tool to extend the faucet for use with a kitchen sink.

FIGURE 30-83

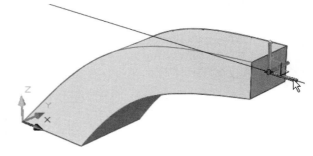

A.   *Open* the **FAUCET1** drawing that you created in Chapter 29 Exercises. Use *SaveAs* and assign the name **FAUCETK**.

B. Turn on *POLAR*. Hold down the **Ctrl** key while hovering and selecting the vertical <u>face</u> at the end of the faucet (Fig. 30-83). A small circular grip should appear on the face. Hover the cursor over the grip until the move grip tool relocates to the grip (Fig. 30-83). Select the X axis of the move grip tool.

C. Stretch the end face of the faucet in a positive X direction using Polar Tracking. Enter a value of **3** to make the faucet 3 units longer.

FIGURE 30-84 ———————————

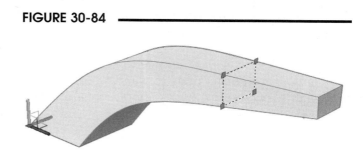

D. Depending on your setting of the *DELOBJ* system variable, some of the original geometry used to create the faucet may still remain. If the original *Rectangle* is visible as shown in Figure 30-84, select the object so it becomes highlighted, then use *Erase* to delete it.

6. **Editing using *Solidedit***

A design change is needed to change the angle of the inclined surface on the Bar Guide by 10 degrees. Make this change using the *Rotate Faces* option of *Solidedit*.

A. *Open* the **BGUID-SL** drawing from the Chapter 29 Exercises. Use *SaveAs* and assign the name **BGUID-SL-2**.

FIGURE 30-85 ———————————

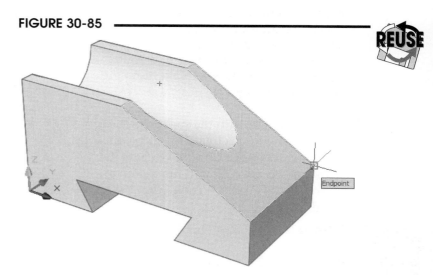

B. Invoke the *Solidedit* command and the *Rotate Faces* option. Select the inclined plane as the face to rotate. Use *Endpoint Osnap* to specify 2 points to define the axis of rotation. Select the endpoints of the bottom edge of the surface as shown in Figure 30-85. Rotate the surface **10** degrees. *Save* the drawing.

FIGURE 30-86 ———————————

7. **Using *Sectionplane***

A. *Open* the **PULLY-SL** drawing. Use *SaveAs* and assign the name **PULLY-SC**. Ensure your UCS is identical to that shown in Figure 30-86 such that the origin is at the front center, the X axis points to the right, the Y axis goes through the axis of the hole, and Z points upward. If your UCS is not in this orientation, use the *UCS* command to make a new UCS in this orientation.

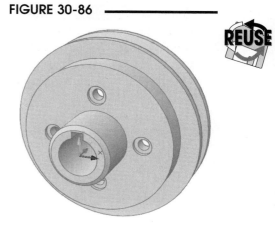

B. Invoke *Sectionplane*. Use the *Orthographic* option and align the section to the *Right* side. A new section plane should appear as shown in Figure 30-87.

FIGURE 30-87

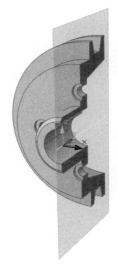

C. Experiment with the section plane by selecting it to activate the grips, then using the move grip tool to pass the section plane through the pulley to reveal its inner features.

D. Invoke the *Properties* palette, select the section plane object, and set the *Live Section* property to *No*. The section plane should appear in the drawing, but the pulley should not appear cut. *Save* the drawing.

8. **Using** *Sectionplane*

A. *Open* the **SADL-SL** drawing from the Chapter 28 Exercises. Use *SaveAs* and assign the name **SADL-SC**.

FIGURE 30-88

B. Invoke *Sectionplane* to create a new section plane like that shown in Figure 30-88.

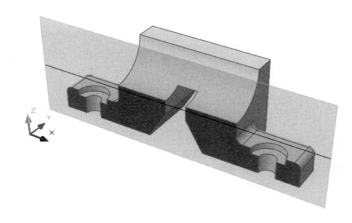

C. If you have problems, ensure your UCS is identical to that shown in Figure 30-88. Also check to make sure *Live Sectioning* is on.

D. Experiment with the section plane by selecting it to activate the grips, then using the move grip tool to pass the section plane through the pulley to reveal its inner features. *Save* the drawing.

9. **Calculating** *Mass Properties*

Open the **SADL-SL** drawing that you created in Chapter 28 Exercises. Calculate *Mass Properties* for the saddle. Write the report out to a file named **SADL-SL.MPR**. Use a text editor to examine the file.

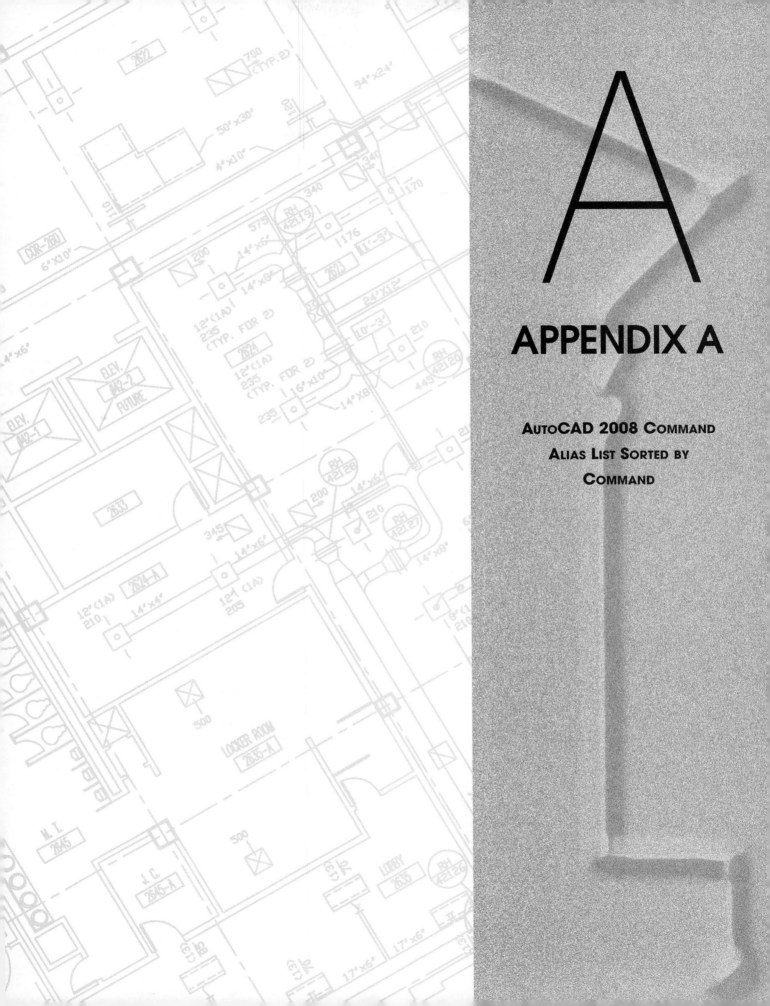

A

APPENDIX A

AUTOCAD 2008 COMMAND
ALIAS LIST SORTED BY
COMMAND

APPENDIX A

AutoCAD 2008 Command Alias List Sorted by Command

Command	Alias	Command	Alias
3DALIGN	3AL	DDEDIT	ED
3DARRAY	3A	DDGRIPS	GR
3DFACE	3F	DDVPOINT	VP
3DMOVE	3M	DIMALIGNED	DAL
3DORBIT	3DO	DIMANGULAR	DAN
3DORBIT	ORBIT	DIMARC	DAR
3DPOLY	3P	DIMBASELINE	DBA
3DROTATE	3R	DIMCENTER	DCE
3DWALK	3DNAVIGATE	DIMCONTINUE	DCO
3DWALK	3DW	DIMDIAMETER	DDI
ADCENTER	ADC	DIMDISASSOCIATE	DDA
ADCENTER	DC	DIMEDIT	DED
ADCENTER	DCENTER	DIMJOGGED	DJO
ALIGN	AL	DIMJOGGED	JOG
APPLOAD	AP	DIMJOGLINE	DJL
ARC	A	DIMLINEAR	DLI
AREA	AA	DIMORDINATE	DOR
ARRAY	AR	DIMOVERRIDE	DOV
-ARRAY	-AR	DIMRADIUS	DRA
ATTDEF	ATT	DIMREASSOCIATE	DRE
-ATTDEF	-ATT	DIMSTYLE	D
ATTEDIT	ATE	DIMSTYLE	DST
-ATTEDIT	-ATE	DIST	DI
-ATTEDIT	ATTE	DIVIDE	DIV
ATTIPEDIT	ATI	DONUT	DO
BACTION	AC	DRAWINGRECOVERY	DRM
BCLOSE	BC	DRAWORDER	DR
BEDIT	BE	DSETTINGS	DS
BLOCK	B	DSETTINGS	SE
-BLOCK	-B	DVIEW	DV
BOUNDARY	BO	ELLIPSE	EL
-BOUNDARY	-BO	ERASE	E
BPARAMETER	PARAM	EXPLODE	X
BREAK	BR	EXPORT	EXP
BSAVE	BS	-EXPORTTOAUTOCAD	AECTOACAD
BVSTATE	BVS	EXTEND	EX
CAMERA	CAM	EXTERNALREFERENCES	ER
CHAMFER	CHA	EXTRUDE	EXT
CHANGE	-CH	FILLET	F
CHECKSTANDARDS	CHK	FILTER	FI
CIRCLE	C	FLATSHOT	FSHOT
COLOR	COL	GEOGRAPHICLOCATION	GEO
COLOR	COLOUR	GEOGRAPHICLOCATION	NORTH
COMMANDLINE	CLI	GEOGRAPHICLOCATION	NORTHDIR
COPY	CO	GRADIENT	GD
COPY	CP	GROUP	G
CTABLESTYLE	CT	-GROUP	-G
CYLINDER	CYL	HATCH	BH
DATAEXTRACTION	DX	HATCH	H
DATALINK	DL	-HATCH	-H
DATALINKUPDATE	DLU	HATCHEDIT	HE
DBCONNECT	DBC	HIDE	HI

Command	Alias	Command	Alias
IMAGE	IM	POINT	PO
-IMAGE	-IM	POINTLIGHT	FREEPOINT
IMAGEADJUST	IAD	POLYGON	POL
IMAGEATTACH	IAT	POLYSOLID	PSOLID
IMAGECLIP	ICL	PREVIEW	PRE
IMPORT	IMP	PROPERTIES	CH
INSERT	I	PROPERTIES	MO
-INSERT	-I	PROPERTIES	PR
INSERTOBJ	IO	PROPERTIES	PROPS
INTERFERE	INF	PROPERTIESCLOSE	PRCLOSE
INTERSECT	IN	PSPACE	PS
JOIN	J	PUBLISHTOWEB	PTW
LAYER	LA	PURGE	PU
-LAYER	-LA	-PURGE	-PU
LAYERSTATE	LAS	PYRAMID	PYR
LAYERSTATE	LMAN	QLEADER	LE
-LAYOUT	LO	QUICKCALC	QC
LENGTHEN	LEN	QUICKCUI	QCUI
LINE	L	QUIT	EXIT
LINETYPE	LT	RECTANG	REC
-LINETYPE	-LT	REDRAW	R
LINETYPE	LTYPE	REDRAWALL	RA
-LINETYPE	-LTYPE	REGEN	RE
LIST	LI	REGENALL	REA
LIST	LS	REGION	REG
LTSCALE	LTS	RENAME	REN
LWEIGHT	LINEWEIGHT	-RENAME	-REN
LWEIGHT	LW	RENDER	RR
MARKUP	MSM	RENDERCROP	RC
MATCHPROP	MA	RENDERPRESETS	RP
MATERIALS	MAT	RENDERWIN	RW
MEASURE	ME	REVOLVE	REV
MIRROR	MI	ROTATE	RO
MIRROR3D	3DMIRROR	RPREF	RPR
MLEADER	MLD	SCALE	SC
MLEADERALIGN	MLA	SCRIPT	SCR
MLEADERCOLLECT	MLC	SECTION	SEC
MLEADEREDIT	MLE	SECTIONPLANE	SPLANE
MLEADERSTYLE	MLS	SETVAR	SET
MLINE	ML	SHADEMODE	SHA
MOVE	M	SHEETSET	SSM
MSPACE	MS	SLICE	SL
MTEXT	MT	SNAP	SN
MTEXT	T	SOLID	SO
-MTEXT	-T	SPELL	SP
MVIEW	MV	SPLINE	SPL
OFFSET	O	SPLINEDIT	SPE
OPTIONS	OP	STANDARDS	STA
OSNAP	OS	STRETCH	S
-OSNAP	-OS	STYLE	ST
PAN	P	SUBTRACT	SU
-PAN	-P	TABLE	TB
-PARTIALOPEN	PARTIALOPEN	TABLESTYLE	TS
PASTESPEC	PA	TABLET	TA
PEDIT	PE	TEXT	DT
PLINE	PL	THICKNESS	TH
PLOT	PRINT	TILEMODE	TI

Command	Alias	Command	Alias
TOLERANCE	*TOL*	*VPOINT*	*-VP*
TOOLBAR	*TO*	*VSCURRENT*	*VS*
TOOLPALETTES	*TP*	*WBLOCK*	*W*
TORUS	*TOR*	*-WBLOCK*	*-W*
TRIM	*TR*	*WEDGE*	*WE*
UCSMAN	*UC*	*XATTACH*	*XA*
UNION	*UNI*	*XBIND*	*XB*
UNITS	*UN*	*-XBIND*	*-XB*
-UNITS	*-UN*	*XCLIP*	*XC*
VIEW	*V*	*XLINE*	*XL*
-VIEW	*-V*	*XREF*	*XR*
VISUALSTYLES	*VSM*	*-XREF*	*-XR*
-VISUALSTYLES	*-VSM*	*ZOOM*	*Z*

AutoCAD 2008 Commands Called by Discontinued Commands or Aliases

2008 Command	Old Alias or Command	2008 Command	Old Alias or Command
ADCENTER	*CONTENT*	*DSVIEWER*	*AV*
ATTDEF	*DDATTDEF*	*INSERT*	*DDINSERT*
ATTEDIT	*DDATTE*	*INSERT*	*INSERTURL*
ATTEXT	*DDATTEXT*	*LAYER*	*DDLMODES*
BLOCK	*ACADBLOCKDIA-*	*LEADER*	*LEAD*
LOG		*LINETYPE*	*DDLTYPE*
BLOCK	*BMAKE*	*LIST*	*SHOWMAT*
BLOCK	*BMOD*	*MATCHPROP*	*PAINTER*
BOUNDARY	*BPOLY*	*MATERIALMAP*	*SETUV*
COLOR	*DDCOLOR*	*MATERIALS*	*FINISH*
COPY	*CP*	*MATERIALS*	*RMAT*
DBCONNECT	*AEX*	*OPEN*	*DXFIN*
DBCONNECT	*ALI*	*OPEN*	*OPENURL*
DBCONNECT	*ARO*	*OPTIONS*	*PREFERENCES*
DBCONNECT	*ASQ*	*OSNAP*	*DDOSNAP*
DBCONNECT	*AAD*	*PLOT*	*DWFOUT*
DBCONNECT	*ASE*	*PLOTSTAMP*	*DDPLOTSTAMP*
DIMALIGNED	*DIMALI*	*PROPERTIES*	*DDCHPROP*
DIMANGULAR	*DIMANG*	*PROPERTIES*	*DDMODIFY*
DIMBASELINE	*DIMBASE*	*RECTANG*	*RECTANGLE*
DIMCONTINUE	*DIMCONT*	*RENDERENVIRONMENT*	*FOG*
DIMDIAMETER	*DIMDIA*	*RENDERPRESETS*	*RFILEOPT*
DIMEDIT	*DIMED*	*RENDERWIN*	*RENDSCR*
DIMLINEAR	*DIMHORIZONTAL*	*SAVE*	*SAVEURL*
DIMLINEAR	*DIMLIN*	*SAVEAS*	*DXFOUT*
DIMLINEAR	*DIMROTATED*	*SHADEMODE*	*SHADE*
DIMLINEAR	*DIMVERTICAL*	*STYLE*	*DDSTYLE*
DIMORDINATE	*DIMORD*	*TEXT*	*DTEXT*
DIMOVERRIDE	*DIMOVER*	*TILEMODE*	*TM*
DIMRADIUS	*DIMRAD*	*UCS*	*DDUCS*
DIMSTYLE	*DDIM*	*UCSMAN*	*DDUCS*
DIMSTYLE	*DIMSTY*	*UCSMAN*	*DDUCSP*
DIMTEDIT	*DIMTED*	*UNITS*	*DDUNITS*
DONUT	*DOUGHNUT*	*VIEW*	*DDVIEW*
DSETTINGS	*DDRMODES*	*VPORTS*	*VIEWPORTS*
		WBLOCK	*ACADWBLOCK-DIALOG*

B

APPENDIX B

BUTTONS AND SPECIAL KEYS

APPENDIX B

BUTTONS AND SPECIAL KEYS

Mouse and Digitizing Puck Buttons

Depending on the type of mouse or digitizing puck used for cursor control, a different number of buttons are available. The *User Preferences* tab of the *Options* dialog box can be used to control the appearance of shortcut menus and to customize the actions of a right-click. In any case, the buttons have the following default settings.

#1 (left mouse)	**PICK**	Used to select commands or point to locations on screen.
#2 (right mouse)	**Shortcut menu or Enter**	Generally, activates a shortcut menu (see Chapter 1, "Shortcut Menus" and Chapter 2, "Windows Right-Click Shortcut Menus"). Otherwise, performs the same action as the Enter key on the keyboard.
press (center wheel)	*Pan*	Activates the realtime *Pan* command.
turn (center wheel)	*Zoom*	Activates the realtime *Zoom* command.

Function (F) Keys

Function keys in AutoCAD offer a quick method of turning on or off (toggling) drawing aids.

F1	*Help*	Opens a help window providing written explanations on commands and variables (see Chapter 5).
F2	*Flipscreen*	Activates a text window showing the previous command line activity (see Chapter 1).
F3	*Osnap Toggle*	If Running Osnaps are set, toggling this key temporarily turns the Running Osnaps off so that a point can be picked without using Osnaps. If no Running Osnaps are set, F3 produces the *Object Snap* tab of the *Drafting Settings* dialog box (see Chapter 7).
F4	*Tablet*	Turns the *TABMODE* variable on or off. If *TABMODE* is on, the digitizing tablet can be used to digitize an existing paper drawing into AutoCAD.
F5	*Isoplane*	When using an *Isometric* style *SNAP* and *GRID* setting, toggles the crosshairs (with *ORTHO* on) to draw on one of three isometric planes (see Chapter 22).
F6	*DUCS*	Toggles *Dynamic UCS* on and off (see Chapter 38).
F7	*GRID*	Turns the *GRID* on or off (see Chapter 1, "Drawing Aids").
F8	*ORTHO*	Turns *ORTHO* on or off (see Chapter 1, "Drawing Aids").
F9	*SNAP*	Turns *SNAP* (Grid Snap or Polar Snap) on or off (see Chapter 1, "Drawing Aids" and Chapter 3, "Polar Tracking and Polar Snap").
F10	*POLAR*	Turns Polar Tracking on or off (see Chapter 1, "Drawing Aids" and Chapter 3, "Polar Tracking and Polar Snap").
F11	*OTRACK*	Turns Object Snap Tracking on or off (see Chapter 7).
F12	*DYN*	Toggles Dynamic Input (see Chapter 1, "Drawing Aids" and Chapter 3, "Dynamic Input").

Control Key Sequences (Accelerator Keys)

Accelerator keys (holding down the Ctrl key and pressing another key simultaneously) invoke regular AutoCAD commands or produce special functions. Several have the same duties as F3 through F11.

Ctrl+A	*Select*	Selects all objects.
Crtl+B (F9)	*SNAP*	Turns *SNAP* on or off.
Ctrl+C	*Copyclip*	Copies the highlighted objects to the Windows clipboard.
Ctrl+D (F6)	*DUCS*	Toggles *Dynamic UCS* on and off (see Chapter 38).
Ctrl+E (F5)	*Isoplane*	When using an *Isometric* style *SNAP* and *GRID* setting, toggles the crosshairs (with *ORTHO* on) to draw on one of three isometric planes.
Ctrl+F (F3)	*Osnap Toggle*	If Running Osnaps are set, pressing Ctrl+F temporarily turns off the Running Osnaps so that a point can be picked without using Osnaps. If there are no Running Object Snaps set, Ctrl+F produces the *Osnap Settings* dialog box. This dialog box is used to turn on and off Running Object Snaps (see Chapter 7).
Ctrl+G (F7)	*GRID*	Turns the *GRID* on or off.
Ctrl+H	*PICKSTYLE*	Toggles *PICKSTYLE* On (1) and Off (0) and toggles selectable *Groups* on or off.
Ctrl+J	*Enter*	Executes the last command.
Ctrl+K	*Hyperlink*	Activates the *Hyperlink* command.
Ctrl+L (F8)	*ORTHO*	Turns *ORTHO* on or off.
Ctrl+N	*New*	Invokes the *New* command to start a new drawing.
Ctrl+O	*Open*	Invokes the *Open* command to open an existing drawing.
Ctrl+P	*Plot*	Produces the *Plot* dialog box for creating and controlling prints and plots.
Ctrl+Q	*Exit*	Exits AutoCAD.
Ctrl+S	*Qsave*	Performs a quick save or produces the *Saveas* dialog box if the file is not yet named.
Ctrl+T (F4)	*Tablet*	Turns the *TABMODE* variable on or off. If *TABMODE* is on, the digitizing tablet can be used to digitize an existing paper drawing into AutoCAD.
Ctrl+U (F10)	*POLAR*	Turns *Polar Tracking* on or off.
Ctrl+V	*Pasteclip*	Pastes the clipboard contents into the current AutoCAD drawing.
Ctrl+W (F11)	*OTRACK*	Turns *Object Snap Tracking* on or off.
Ctrl+X	*Cutclip*	Cuts the highlighted objects from the drawing and copies them to the Windows clipboard.
Ctrl+Y	*Redo*	Invokes the *Redo* command.
Ctrl+Z	*Undo*	Undoes the last command.
Ctrl+0	*Clean Screen*	Toggles *Clean Screen* on and off.
Ctrl+1	*Properties*	Toggles the *Properties* palette.
Ctrl+2	*DesignCenter*	Toggles DesignCenter.
Ctrl+3	*Toolpalettes*	Toggles the *Tool Palettes* window.
Ctrl+4	*Sheet Set Manager*	Toggles the *Sheet Set Manager*.
Ctrl+6	*dbConnect*	Toggles the *DBConnect Manager*.
Ctrl+7	*Markup Set Manager*	Toggles the *Markup Set Manager*.
Ctrl+8	*QuickCalc*	Opens the *Sheet Set Manager*.
Ctrl+9	*Command Line*	Opens the *Command Line Window* dialog box.
Alt+F8	*Macros*	Opens the *Macros* dialog box.
Alt+F11	*Visual Basic Editor*	Opens the Visual Basic Editor.

Temporary Override Keys

Temporary override keys temporarily turn on or off one of the drawing aids that can otherwise be toggled on the Status Bar or in the *Drafting Settings* dialog box. Each function has a key combination for each (left and right) side of the keyboard.

Keyboard Left	Keyboard Right	Drawing Aid
Shift	**Shift**	Toggles *ORTHO*
Shift+A	**Shift+'**	Toggles *OSNAP*
Shift+X	**Shift+.**	Toggles *POLAR*
Shift+Q	**Shift+]**	Toggles *OTRACK*
Shift+E	**Shift+P**	*Endpoint* Osnap override
Shift+V	**Shift+M**	*Midpoint* Osnap override
Shift+C	**Shift+,**	*Center* Osnap override
Shift+D	**Shift+L**	Disable all snapping and tracking
Shift+S	**Shift+;**	Enables Osnap enforcement
Shift+Z	**Shift+/**	Toggles *DUCS*

Special Key Functions

Esc   The Escape key cancels a command, menu, or dialog box or interrupts processing of plotting or hatching.

Space bar   In AutoCAD, the space bar performs the same action as the Enter key or #2 button. Only when you are entering text into a drawing does the space bar create a space.

Enter   If Enter or Space bar is pressed when no command is in use (the open Command: prompt is visible), the last command used is invoked again.

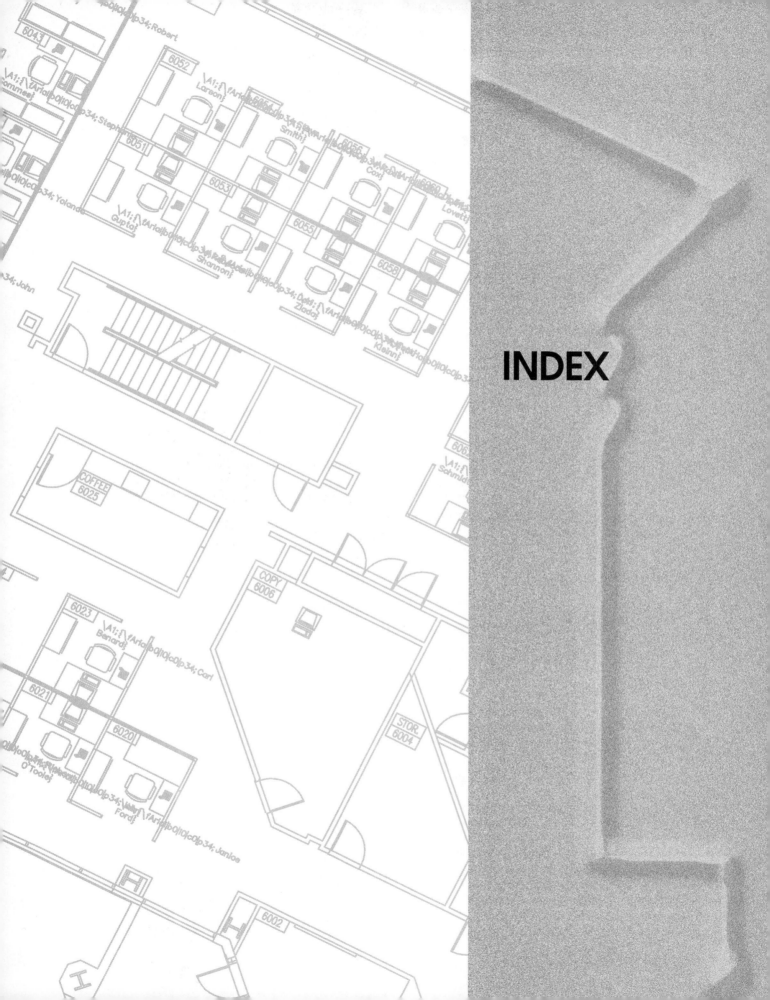

INDEX